Applied Probability and Statistics (Continued)

CHAKRAVARTI, LAHA and ROY · Handbook of Methods of Applied Statistics, Vol. I
CHAKRAVARTI, LAHA and ROY · Handbook of Methods of Applied Statistics, Vol. II
CHERNOFF and MOSES · Elementary Decision Theory
CHIANG · Introduction to Stochastic Processes in Biostatistics
CLELLAND, deCANI, BROWN, BURSK, and MURRAY · Basic Statistics with Business Applications
COCHRAN · Sampling Techniques, *Second Edition*
COCHRAN and COX · Experimental Designs, *Second Edition*
COX · Planning of Experiments
COX and MILLER · The Theory of Stochastic Processes
DAVID · Order Statistics
DEMING · Sample Design in Business Research
DODGE and ROMIG · Sampling Inspection Tables, *Second Edition*
DRAPER and SMITH · Applied Regression Analysis
GOLDBERGER · Econometric Theory
GUTTMAN and WILKS · Introductory Engineering Statistics
HALD · Statistical Tables and Formulas
HALD · Statistical Theory with Engineering Applications
HOEL · Elementary Statistics, *Second Edition*
HUANG · Regression and Econometric Methods
JOHNSON and LEONE · Statistics and Experimental Design: In Engineering and the Physical Sciences, Volumes I and II
LANCASTER · The Chi Squared Distribution
MILTON · Rank Order Probabilities: Two-Sample Normal Shift Alternatives
PRABHU · Queues and Inventories: A Study of Their Basic Stochastic Processes
SARHAN and GREENBERG · Contributions to Order Statistics
SEAL · Stochastic Theory of a Risk Business
WILLIAMS · Regression Analysis
WOLD and JURÉEN · Demand Analysis
WONNACOTT and WONNACOTT · Econometrics
YOUDEN · Statistical Methods for Chemists

Tracts on Probability and Statistics

BILLINGSLEY · Ergodic Theory and Information
BILLINGSLEY · Convergence of Probability Measures
CRAMÉR and LEADBETTER · Stationary and Related Stochastic Processes
RIORDAN · Combinatorial Identities
TAKÁCS · Combinatorial Methods in the Theory of Stochastic Processes

Multiple Time Series

A WILEY PUBLICATION IN APPLIED STATISTICS

Multiple Time Series

E. J. HANNAN
The Australian National University
Canberra

John Wiley and Sons, Inc.
New York · London · Sydney · Toronto

10 9 8 7 6 5 4 3 2 1

Library of Congress Catalogue Card Number: 77-112847

ISBN 0 471 34805 8

Printed in the United States of America

Preface

The subject of time series analysis has intimate connections with a wide range of topics, among which may be named statistical communication theory, the theory of prediction and control, and the statistical analysis of time series data. The last of these is to some extent subsidiary to the other two, since its purpose, in part at least, must be to provide the information essential to the application of those theories. However, it also has an existence of its own because of its need in fields (e.g., economics) in which at present well-developed, exact theories of control are not possible. It is with the third of these topics that this book is concerned. It extends beyond that in two ways. The first and most important is by the inclusion in the first half of the book of a fairly complete treatment of the underlying probability theory for second-order stationary process. Although this theory is for the most part classical and available elsewhere in book form, its understanding is an essential preliminary to the study of time series analysis and its inclusion is inevitable. I have, however, included a certain amount of material over and above the minimum necessary, in relation, for example, to nonlinear filters and random processes in space as well as time. The statistical development of this last subject is now fragmentary but may soon become important.

The second additional topic is the theory of prediction, interpolation, signal extraction, and smoothing of time series. The inclusion of this material seems justified for two reasons. The first arises from the understanding that the classical "Wiener-Kolmogoroff" theories give of the structure of time series. This understanding is needed, in part at least, for statistical developments (e.g., identification problems and problems associated with the relation between the eigenvalues of the covariance matrix and the spectrum). The second reason is that these developments are becoming important to people who are statisticians concerned with time series (e.g., in missile trajectory estimation and economics). Of course, some of the more practically valuable work here is recent and would require a separate volume for an adequate treatment, but some indications concerning it are needed.

There is one other characteristic that a modern book on time series must have and that is the development of the theory and methods for the case in which multiple measurements are made at each point, for this is usually the case.

Having decided on the scope of the book, one must consider the manner in which the material will be presented and the level of the presentation. This book sets out to give the theory of the methods that appear to be important in time series analysis in a manner that it is hoped will lead finally to an understanding of the methods as they are to be used. On the whole it presents final formulas but often does not discuss computational details and it does not give computer programs. (For the most part the methods discussed are already programmed and these programs are available.) With minor exceptions numerical examples are not given. It is not a book on "practical time series analysis" but on the theory of that subject. There is a need for books of the first kind, of course, but also of this second kind, as any time series analyst knows from requests for references to a definitive discussion of the theory of this or that topic. The level of presentation causes problems, for the theory is both deep and mathematically unfamiliar to statisticians. It would probably be possible to cover the underlying probability theory more simply than has been done by making more special assumptions (or by making the treatment less precise). To make the book more accessible a different device has been used and that is by placing the more difficult or technical proofs in chapter appendices and starring a few sections that can be omitted. It is assumed that the reader knows probability and statistics up to a level that can be described as familiarity with the classic treatise *Mathematical Methods of Statistics* by Harald Cramér. A mathematical appendix which surveys some needed elementary functional analysis and Fourier methods has been added.

Some topics have not been fully discussed, partly because of the range of my interests and partly because of the need to keep the length of the book within reasonable bounds. I have said only a small amount about the spectra of higher moments. This is mainly because the usefulness of this spectral theory has not yet been demonstrated. (See the discussion in Chapter II, Section 8.) Little also has been said about nonstationary processes, and particularly about their statistical treatment. This part of the subject is fragmented at the moment. Perhaps, of necessity, it always will be. A third omission is of anything other than a small discussion of "digitized" data (e.g., "clipped signals" in which all that is recorded is whether the phenomenon surpassed a certain intensity). There is virtually no discussion of the sample path behavior of Gaussian processes, for this subject has recently been expertly surveyed by Cramér and Leadbetter (1967) and its inclusion here is not called for. I have also not discussed those inference procedures for point processes based on the times of occurrence of the events in the process (as distinct from the intervals between these times). This has recently been surveyed by Cox and Lewis (1966). Finally, the second half of the book (on inference problems) discusses only the discrete time case. This is justified by the dominance of digital computer techniques.

I have not attempted to give anything approaching a complete bibliography of writing on time series. For the period to 1959 a very complete listing is available in Wold (1965). It is hoped that the references provided herein will allow the main lines of development of the subject to the present time to be followed by the reader.

I have many people to thank for help. The book developed from a course given at The Johns Hopkins University, Baltimore, Maryland, and an appreciable part of the work on it was supported by funds from the United States Air Force. The book's existence is due in part to encouragement from Dr. G. S. Watson. Dr. C. Rhode at Johns Hopkins and R. D. Terrell, P. Thomson, and D. Nicholls at the Australian National University have all read parts of the work and have corrected a number of errors in its preliminary stages. The typing was entirely done, many times over, by Mrs. J. Radley, to whom I am greatly indebted.

E. J. HANNAN

Canberra, Australia
April, 1970

Contents

† These topics are special and may be omitted.

PART ONE

Basic Theory

CHAPTER I

Introductory Theory

1. INTRODUCTION

In a wide range of statistical applications the data to hand consist of a time series, that is, a series of observations, $x_j(n); j = 1, \ldots, p; n = 1, \ldots, N$, made sequentially through time. Here j indexes the different measurements made at each time point n. Sets of observations of this kind dominate in some of the earth sciences (e.g., meteorology, seismology, oceanography, and geomorphology) and the social sciences (particularly economics). They are of great importance in other fields also; for example, in medical research (electrocardiograms and electroencephalograms) and in connection with problems of estimating missile trajectories. Although we have so far thought of n as measuring the passage of time, there are situations in which it might be a space variable. The $x_j(n)$ might be measurements made at equidistant points along a strip of material (e.g., a strip of newsprint, a textile fiber or a coal face). Again n might be the distance, in units of 100 m, downstream from some permanent feature of a river and $x_j(n), j = 1, 2, 3$, might be measures of the speed and direction of flow of the river (or the discharge and direction of discharge).

The feature of the situations we have in mind that distinguishes them from those found in the classical part of statistics is the fact that $x_j(m)$ and $x_j(n)$ will not be independent, even for $m \neq n$. Indeed, it is almost always true that observations made at equidistant intervals of time form a sample of observations from what was, in principle, a continuously observable phenomenon.† Now it is inconceivable that $x_j(s)$ and $x_j(t)$ should remain independent as s approaches t, so that the lack of independence is forced on us by this most fundamental fact, namely, the (seeming) continuity of natural phenomena. Although, as we have said, most discrete time series may be regarded as samples from a continuous time function, we shall at times

† We use the symbol $x_j(t)$ whenever a continuous time phenomenon is being discussed, but also, sometimes, for a discrete time phenomenon. When the distinction between the two cases has to be emphasized, we use $x_j(n)$ for the discrete time phenomenon.

treat such series as probabilistic phenomena in their own right without imbedding them in a continuous time background. We use the symbol $x(t)$ without subscripts for the vector with components $x_j(t)$. Of course, $x(t)$ may, in particular, be a scalar.

The observable $x(t)$ is to be thought of as an observation on a vector-valued random variable. Thus we imply that for every finite set $t_1, t_2, \ldots, t_r$ of values of t the joint probability distribution of all elements of the vectors $x(t_j)$, $j = 1, \ldots, r$, is prescribed. We might obtain that distribution in another way. We might begin from a set of time points $t_1, t_2, \ldots, t_r, t_{r+1}$ and the corresponding probability distribution of the elements of the $x(t_j)$, $j = 1, \ldots, r + 1$ and obtain the marginal distribution of the elements of the $x(t_j)$, $j = 1, \ldots, r$ from it. If this gives a different distribution from that originally prescribed, then clearly the various prescriptions were inconsistent. We thus naturally impose the requirement that such inconsistencies do *not* occur. When this is done, it has been shown that a probability measure space (sample space) Ω may be constructed on which all the random variables $x_j(t)$ may be defined. Thus on a Borel field, $\mathcal{A}$, of sets in Ω there is defined a probability measure P, in terms of which all probability statements concerning the $x_j(t)$ are made. This space Ω may be thought of as the space of all "histories" or "realizations" of the vector function $x(t)$ and, to emphasize the fact that these random variables are definable, for each fixed j, t, as measurable functions on Ω, we sometimes replace the symbol $x_j(t)$ with $x_j(t, \omega)$. (For a further discussion see Billingsley, 1968, p. 228.) When ω is fixed and t varies, we obtain a particular history. As ω varies we get different histories (of which part of only one may be observed). The family of all such histories together with the associated structure of probabilities we call a stochastic (or random) process, and we use the words "time function" (or "time series" in the discrete case) for a particular history (or "realization") of the stochastic process. Often we do not exhibit the variable ω, replacing, for example, an explicit integration symbolism with respect to the probability measure P on Ω with a symbolism involving the expectation operator $\mathcal{E}$.

We are mainly concerned with situations in which all of the second moments

$$\mathcal{E}\big(x_j(s)\, x_k(t)\big) = \gamma_{j,k}(s, t)$$

are finite. When $j = k$, we write $\gamma_j(s, t)$ for simplicity. We arrange these p^2-quantities, $\gamma_{j,k}(s, t)$, in a symmetric matrix which we call $\Gamma(s, t)$ and we refer to it as a covariance matrix, even though we have not made mean corrections. Moreover, in the continuous time case we almost always assume that $\gamma_{j,k}(s, t)$ is a continuous function of each of its arguments.

In this connection we have the following theorem:

Theorem 1. *In order that the $\gamma_{j,k}(s, t)$ may be continuous, $j, k = 1, \ldots, p$, it is necessary and sufficient that $\gamma_j(s, t)$ be continuous at $s = t$ for $j = 1, \ldots, p$.*

Proof. The necessity is obvious. For the sufficiency we have, using Schwartz's and Minkowski's inequalities,†

$$\begin{aligned}|\gamma_{j,k}(s+u, t+v) - \gamma_{j,k}(s,t)| &= |\mathcal{E}\{x_j(s+u)x_k(t+v)\} - \mathcal{E}\{x_j(s)x_k(t)\}| \\ &= |\mathcal{E}\{(x_j(s+u) - x_j(s))x_k(t+v)\} + \mathcal{E}\{x_j(s)(x_k(t+v) - x_k(t))\}| \\ &\le [\mathcal{E}\{(x_j(s+u) - x_j(s))^2\}\gamma_k(t+v, t+v)]^{1/2} \\ &\quad + [\mathcal{E}\{(x_k(t+v) - x_k(t))^2\}\gamma_j(s,s)]^{1/2}.\end{aligned}$$

If the condition of Theorem 1 is satisfied, then $\gamma_k(t + v, t + v)$ is uniformly bounded in v for v in any finite interval and thus the continuity of the $\gamma_{j,k}(s, t)$ is implied by the condition that, for all s,

$$\lim_{u\to 0} \mathcal{E}\{(x_j(s+u) - x_j(s))^2\} = 0, \qquad j = 1, \ldots, p, \tag{1.1}$$

and, since the left-hand side is

$$\lim_{u\to 0} \{\gamma_j(s+u, s+u) + \gamma_j(s,s) - 2\gamma_j(s+u, s)\},$$

this is implied by the continuity of $\gamma_j(s, t)$ at $s = t$.

We have also shown that the continuity of the $\gamma_{j,k}(s, t)$ is equivalent to (1.1), which is the condition for what is called mean-square continuity. More erratic behavior than this implies is, perhaps, not unthinkable, but presumably it could not be recorded because of the inability of a recording device to respond instantly to a change in the level of the phenomenon being recorded. We put

$$\|x(t)\| = [\mathcal{E}\{x'(t)x(t)\}]^{1/2},$$

where the prime indicates transposition of the vector. Later we shall have occasion to introduce into the theory linear combinations of the $x(t_j)$ with complex coefficients; z being such a combination, we shall then put

$$\|z\| = [\mathcal{E}\{z^*z\}]^{1/2},$$

the star indicating transposition combined with conjugation. This is called the norm of z.

2. DIFFERENTIATION AND INTEGRATION OF STOCHASTIC PROCESSES

Let x_n, $n = 1, 2, \ldots$, be a sequence of random variables for which $\mathcal{E}(|x_n|^2) < \infty$. Then x_n is said to converge in mean square‡ to a random

† See the Mathematical Appendix at the end of this volume.

‡ The notion of mean-square convergence is discussed further in the Mathematical Appendix at the end of this volume.

variable x if

$$\lim_n \mathcal{E}(|x - x_n|^2) = \|x - x_n\|^2 = 0. \tag{2.1}$$

We then write $x_n \to x$. We use the modulus sign here to cover the case in which the random variables x_n are complex-valued, so that each x_n can be considered as a pair of real-valued random variables. The necessary and sufficient condition that an x exist such that (2.1) will hold true is the "Cauchy" condition

$$\lim_{m,n\to\infty} \|x_n - x_m\| = 0. \tag{2.2}$$

The limit x is then uniquely defined in the sense that any two random variables x satisfying (2.1) differ only on a set of measure zero. Moreover, $\mathcal{E}(x\bar{y})$ is a continuous function of x and y, so that if $\|x_n\|$ and $\|y_n\|$ are finite and $x_n \to x$, $y_n \to y$ then $\mathcal{E}(x_n\bar{y}_n) \to \mathcal{E}(x\bar{y})$. (See the Mathematical Appendix.) On the other hand, if $\mathcal{E}(x_n\bar{x}_m)$ converges to a limit independently of the way m and n tend to infinity, the Cauchy condition is satisfied, since $\|x_n - x_m\|^2 = \|x_n\|^2 + \|x_m\|^2 - \mathcal{E}(x_n\bar{x}_m) - \mathcal{E}(x_m\bar{x}_n)$.

We now need to introduce the concepts of differentiation and integration of continuous time processes.

(i) We say that the scalar process $x(t)$ is mean-square differentiable at t if $\delta^{-1}\{x(t+\delta) - x(t)\}$ has a unique limit in mean square as δ approaches zero.

We are now considering a variable tending continuously to zero. However, taking an arbitrary sequence δ_n so that $\delta_n \to 0$ as $n \to \infty$, we see that the following Cauchy condition for mean-square differentiability is necessary:

$$\lim_{\delta_1,\delta_2\to 0} \|\delta_1^{-1}\{x(t+\delta_1) - x(t)\} - \delta_2^{-1}\{x(t+\delta_2) - x(t)\}\| = 0. \tag{2.3}$$

However, this condition is also sufficient, for if δ_n and δ_n' are two sequences converging to zero, we may form from them the combined sequence δ_n'', and (2.3) assures us that $\{x(t+\delta_n'') - x(t)\}/\delta_n''$ also converges in mean square and that the limit is independent of the sequence chosen.

From what was said above concerning convergence in mean square we know that the necessary and sufficient condition for (2.3) to hold is the existence of the limit

$$\lim_{\delta_1,\delta_2\to 0} \mathcal{E}[\delta_1^{-1}\{x(t+\delta_1) - x(t)\}\, \delta_2^{-1}\{x(t+\delta_2) - x(t)\}]$$
$$= \lim_{\delta_1,\delta_2\to 0} \frac{\gamma(t+\delta_1, t+\delta_2) - \gamma(t+\delta_1, t) - \gamma(t, t+\delta_2) + \gamma(t,t)}{\delta_1\delta_2}.$$

For this, in turn, it is sufficient that $\{\partial^2\gamma(s,t)/\partial s\, \partial t\}$ exist and be continuous (e.g., see Goursat and Hedrick, 1904, p. 13). We know that the covariance function of $\dot{x}(t)$, the mean-square derivative, is

$$\lim_{\delta_1,\delta_2\to 0} \mathcal{E}\{\delta_1^{-1}(x(s+\delta_1)-x(s))\,\delta_2^{-1}(x(t+\delta_2)-x(t))\} = \frac{\partial^2\gamma(s,t)}{\partial s\,\partial t}.$$

In the same way

$$\mathcal{E}(x(s)\,\dot{x}(t)) = \frac{\partial\gamma(s,t)}{\partial t}.$$

Theorem 2. *The necessary and sufficient condition that the scalar process $x(t)$ be mean-square differentiable is the existence of the limit*

$$\lim_{\delta_1,\delta_2\to 0} \frac{\gamma(t+\delta_1, t+\delta_2) - \gamma(t+\delta_1, t) - \gamma(t, t+\delta_2) + \gamma(t,t)}{\delta_1\,\delta_2}.$$

If $\dot{x}(t)$ is the mean-square derivative, this has covariance function $\partial^2\gamma(s,t)/\partial s\,\partial t$ and $\mathcal{E}(x(s)\,\dot{x}(t)) = \partial\gamma(s,t)/\partial t$.

(ii) With regard to integration we first consider the definition of integrals to be represented by symbols of the form†

$$\int_{-\infty}^{\infty} x(t)\,m(dt), \tag{2.4}$$

where m is a σ-finite measure adjusted, let us say, so that the corresponding distribution function is continuous from the right. We consider only $x(t)$ which are mean-square continuous and functions for which

$$\iint_{-\infty}^{\infty} \gamma(s,t)\,m(ds)\,m(dt) < \infty. \tag{2.5}$$

We could accomplish the definition of (2.4) by defining first

$$\int_a^b x(t)\,m(dt) \tag{2.6}$$

by considering approximating sums

$$\sum_1^n x(t_j)\{m((s_{j-1}, s_j])\},$$

wherein the points s_j divide the interval $[a, b]$ into subintervals of length less than ϵ and $t_j \in (s_{j-1}, s_j]$. If this sum converges to a unique limit in mean square, as n increases so that ϵ converges to zero, we call this limit the

† We prefer the notation $m(dt)$ to the common one which replaces that by $dm(t)$ for the obvious reason that it is t that is changing, differentially, not necessarily m.

integral of $x(t)$ with respect to $m(t)$ over $[a, b]$ and indicate it by the symbol (2.6). When $x(t)$ is mean-square continuous, the integral (2.6) will exist. Proof of this is obtained by a completely straightforward modification of the well known proof of the existence of the Riemann-Stieltjes integral of a continuous function over a finite interval. Indeed, to obtain the modification we need only to replace the absolute value in the proof with the norm [regarding $x(t)$ as the continuous function]. The integral (2.4) can now be defined as the limit in mean square of a sequence of integrals of the form (2.6) as $a \to -\infty$, $b \to \infty$. A sufficient condition for the existence (and uniqueness) of this limit is the condition (2.5). Indeed, from the definition of (2.6) it follows that

$$\left\| \int_a^b x(t)\, m(dt) \right\|^2 = \iint_a^b \gamma(s, t)\, m(ds)\, m(dt),$$

so that the Cauchy condition becomes, for example, as $b \to \infty$,

$$\lim_{b_1, b_2 \to \infty} \left\| \int_{b_1}^{b_2} x(t)\, m(dt) \right\|^2 = \lim_{b_1, b_2 \to \infty} \iint_{b_1}^{b_2} \gamma(s, t)\, m(ds)\, m(dt) = 0$$

and (2.5) implies that this converges to zero.

We may state these results as follows:

Theorem 3. *If $x(t)$ is mean-square continuous and $m(t)$ is a σ-finite measure, then* (2.6) *is uniquely defined as a limit in mean square of approximating Riemann-Darboux sums. The necessary and sufficient condition that* (2.4) *exist as a limit in mean square of* (2.6) *as $a \to -\infty$, $b \to \infty$, is given by* (2.5).

Of course, the definition (2.4) can be extended to a much wider class of functions than those that are mean-square continuous in the same way as the definition of the Lebesgue integral can be extended from the Riemann integral definition for continuous functions of compact support to the wide class of Lebesgue integrable functions. For a discussion of this problem we refer the reader to Hille and Phillips (1957, Chapter III). We do not need the extensions in this book.

We also wish to define what we mean by

$$\int_{-\infty}^{\infty} g(t)\, \xi(dt), \tag{2.7}$$

where $g(t)$ is now a complex-valued function of t and $\xi(t)$ is the stochastic process. We have in mind, particularly, the situation in which $\xi(t)$ is a process of orthogonal increments, that is, one for which $\xi(t_1) - \xi(t_2)$ and

$\xi(s_1) - \xi(s_2)$ have zero covariance if the intervals $[t_1, t_2]$ and $[s_1, s_2]$ have no common point. This is one situation, however, in which we do not wish to assume that $\xi(t)$ is mean-square continuous. However, it is evidently true that

$$\mathcal{E}([\xi(t) - \xi(t_0)]^2) \tag{2.8}$$

increases monotonically with t for $t \geq t_0$. This quantity (2.8) defines a Lebesgue-Stieltjes measure on $(-\infty, \infty)$, the increment over the interval $(t_0, t]$ being just the expression (2.8). We write $F(t) - F(t_0)$ for that expression, keeping especially in mind the situation, often occurring later, in which $\mathcal{E}\{\xi^2(t)\} \leq a < \infty$, so that $F(t)$ is a distribution function. Again we may define (2.7) for the case in which the integral is over the interval $[a, b]$ as the limit in mean square of a sequence of approximating sums

$$\sum g(t_j)\{\xi(s_j) - \xi(s_{j-1})\},$$

if this limit exists and is uniquely defined. This will certainly be so if $g(t)$ is continuous. Again the proof is straightforward. We may then extend the definition to infinite intervals by a further limiting process. However, a more inclusive and direct procedure can be used (see the appendix to this chapter) which enables (2.8) to be defined for any function $g(t)$ that is measurable with respect to the Lebesgue-Stieltjes measure induced by $F(t)$ and for which

$$\int_{-\infty}^{\infty} |g(t)|^2 F(dt) < \infty.$$

3. SOME SPECIAL MODELS

(i) An important class of scalar time series is that generated by a linear mechanism of the form

$$\sum_0^q \beta(j)\, x(n-j) = \sum_0^s \alpha(k)\, \epsilon(n-k), \qquad \beta_0 = 1, \tag{3.1}$$

wherein the $\beta(j)$ and $\alpha(k)$ are real, of course, and the $\epsilon(n)$ satisfy

$$\mathcal{E}\big(\epsilon(m)\, \epsilon(n)\big) = \sigma^2 \delta_m^{\ n}.\dagger$$

Such a mechanism should often be a reasonable approximation to reality, since the idea that $x(n)$ is determined by immediate past values of itself, together with past disturbances, is an appealing one. The linearity is a convenient mathematical fiction only. Experience suggests that a model of the form (3.1) will fit a wide range of data. When $s = 0$, (3.1) is said to be

† We shall always assume that $\epsilon(n)$ has these covariance properties unless we indicate otherwise. Of course $\delta_m^{\ n}$ is "Kronecker's delta."

an autoregressive relation. When $s > 0$, the terminology "mixed autoregression and moving average" is sometimes used. When $q = 0$, we speak of a (finite) moving average.

Evidently, if $x(n)$ is given q initial values $x(-q), \ldots, x(-1)$, (3.1) serves to define it thereafter in terms of the $\epsilon(n)$. We wish to examine the nature of the formula expressing $x(n)$ in terms of $\epsilon(n)$, $\epsilon(n-1), \ldots$. We consider the homogeneous equation, obtained by replacing the right-hand term of (3.1) with zero, and seek for solutions of the form z^n, which procedure leads us to the characteristic equation

$$\sum_0^q \beta(j) z^{-j} = 0. \tag{3.2}$$

A solution $b_u z_u^{\,n}$ corresponds to each nonrepeated root z_u. If a root z_u is repeated p_u times, it is easily verified that $b_{u,j} n^j z_u^{\,n}$ for $j = 0, 1, \ldots, p_u - 1$ are all solutions. If z_u is complex, evidently a conjugate corresponds to each of these solutions. Thus we obtain the general solution of the homogeneous equation as

$$\sum_{u=1}^{r} \sum_{j=0}^{p_u-1} b_{u,j} n^j z_u^{\,n}, \tag{3.3}$$

where the q constants $b_{u,j}$ are sufficient to satisfy q initial conditions. Of course, we could also write (3.3) as a sum of terms of the form

$$n^j \rho_u^{\,n}(b'_{u,j} \cos \theta_u n + b''_{u,j} \sin \theta_u n), \tag{3.4}$$

where $z_u = \rho_u \exp i\theta_u$ and $b_{u,j} = \frac{1}{2}(b'_{u,j} - ib''_{u,j})$ if $\theta_u \neq 0, \pi$. (For $\theta_u = 0$, π, of course, $b_{u,j} = b'_{u,j}$ and $b''_{u,j} = 0$.)

The $b_{u,j}$ in (3.3) may be chosen so that (3.3) takes any prescribed values at q time points; for example, at the points $n = -q + k$, $k = 0, \ldots, q - 1$. Indeed, if that were not so, the q sequences $n^j z_u^{\,n}$ would be linearly dependent when evaluated at these q values, which implies that a nonnull polynomial of degree not greater than $q - 1$ has q zeros. This is impossible. Provided no z_u has unit modulus, we may find a sequence $f(n)$ such that $|f(n)|$ converges to zero exponentially as $|n| \to \infty$ and which satisfies the equation

$$\sum_0^q \beta(j) f(n-j) = \delta_0^{\,n}.$$

Indeed, under this condition $(\sum \beta(j) z^j)^{-1}$ has a valid Laurent expansion in an annulus containing the unit circle in its interior. If

$$\sum_{-\infty}^{\infty} f(n) z^n$$

is this expansion, then, since $\sum \beta(j)z^j \sum f(n)z^n \equiv 1$ for z in the annulus, $f(n)$ satisfies the required relation. It evidently converges to zero exponentially as we have said.

Theorem 4. *If $\epsilon(n)$ is a sequence of random variables with covariance $\mathcal{E}(\epsilon(m)\,\epsilon(n)) = \delta_m{}^n\sigma^2$ and $x(-q), \ldots, x(-1)$ are prescribed initial values, there exists a unique random sequence, $x(n)$, which satisfies* (3.1) *and takes these initial values. If no z_u has unit modulus, this solution is of the form*

$$x(n) = \sum_{u=1}^{r}\sum_{j=0}^{p_u-1} b_{u,j}n^j z_u{}^n + \sum_{k=0}^{s}\alpha(k) \sum_{v=-\infty}^{\infty} f(n-v)\,\epsilon(v-k).$$

Proof. It is evident that $x(n)$ is uniquely defined once the initial values are given in terms of these initial values and the $\epsilon(n)$ sequence; for example, if $\beta(q) \neq 0$, then

$$x(-q-1) = -\beta(q)^{-1}\sum_{1}^{q}\beta(j)x(-j) + \sum_{0}^{s}\alpha(k)\epsilon(-q-1-k)$$

and the $x(n)$ for all $n < -q$ may be defined by an iteration of this process. The sum

$$\sum_{v=-M}^{N} f(n-v)\,\epsilon(v-k)$$

certainly converges in mean square as M, N increase, since the $|f(n)|$ decrease exponentially with $|n|$. Thus the expression for $x(n)$ is well defined. Moreover,

$$\sum_{k=0}^{s}\alpha(k)\sum_{v=-\infty}^{\infty} f(n-v)\,\epsilon(v-k)$$

may be verified to satisfy (3.1). Indeed, we have

$$\sum_{0}^{q}\beta(j)\left\{\sum_{v=-\infty}^{\infty} f(n-j-v)\,\epsilon(v-k)\right\} = \sum_{v=-\infty}^{\infty}\left\{\sum_{0}^{q}\beta(j)f(n-v-j)\right\}\epsilon(v-k)$$
$$= \epsilon(n-k),$$

since the term in curly brackets is null for $n - v \neq 0$, whereas for $v = n$ it is unity. Since we know that the $b_{u,j}$ may be chosen to satisfy any q initial conditions, the proof is complete.

Putting

$$\lambda(j) = \sum_{k=0}^{s}\alpha(k)f(j-k),$$

we may rewrite the solution as

$$\text{(3.5)}\quad x(n) = \sum_{u}\sum_{j} n^j\rho_u{}^n(b'_{u,j}\cos\theta_u n + b''_{u,j}\sin\theta_u n) + \sum_{-\infty}^{\infty}\lambda(j)\,\epsilon(n-j).$$

For a solution, $x(n)$, of the form

$$\sum_{-\infty}^{\infty} \lambda(j)\, \epsilon(n-j) \tag{3.6}$$

we have, let us say,

$$\mathcal{E}\big(x(n)\, x(m)\big) = \sigma^2 \sum_{-\infty}^{\infty} \lambda(j)\, \lambda(j + |m-n|) = \gamma(m-n).$$

A series $x(n)$ whose covariances depend only on $(m-n)$ is said to be second-order (or wide sense) stationary.†

The most important case of Theorem 4 is that in which, for all u, $\rho_u < 1$, for then $f(n) = 0$, $n < 0$; that is, $\lambda(j) = 0, j < 0$. In this case the stationary solution (3.6) of (3.1) is an infinite moving average of past $\epsilon(n)$. This is the case likely to be of interest, for when we "model" some real world system by means of (3.1) we usually think of the $\epsilon(n)$ as disturbances affecting the system for the first time at time n. In this case also the solution of the homogeneous equation approaches zero as $n \to \infty$, so that all solutions approach the form (3.6).

The case in which, for certain u, $\rho_u = 1$ has also attracted attention in both economics and systems engineering (see Orcutt, 1948; Box and Jenkins, 1962). Let us factor the polynomial $\sum \beta(j) z^{q-j}$ into two factors, the first of which accounts for all the zeros of unit modulus and the second accounts for all remaining zeros. Let the degrees of the polynomials be q_1 and q_2, respectively, so that $q = q_1 + q_2$. Let $\beta'(j)$ and $\beta''(j)$ be the coefficients of the polynomials, with $\beta'(0) = \beta''(0) = 1$. We may find a stationary solution to

$$\sum_{0}^{q_2} \beta''(j)\, y(n-j) = \epsilon(n) \tag{3.7}$$

of the form in (3.6).

The first polynomial may be written in the form

$$\sum_{0}^{q_1} \beta'(j) z^{q-j} = \prod_{u=1}^{q_1} (z - e^{i\theta_u}),$$

where, if $\theta_u \neq 0$, π occurs, then so must $-\theta_u$, since the $\beta'(j)$ are real. Call S_1 the operator that replaces a sequence such as $y(n)$ by

$$y_1(n) = S_1\, y(n) = \sum_{j=-q_1}^{n} y(j) e^{i\theta_1(n-j)}, \qquad n \geq -q_1.$$

† We see that $\gamma(m, n) = \gamma(0, n-m)$ in this stationary case and we have, somewhat confusingly, put $\gamma(n-m) = \gamma(0, n-m)$. A similar notation is used elsewhere below. Of course, in this scalar case we also have $\gamma(n-m) = \gamma(m-n)$.

Then

$$y_1(n) - e^{i\theta_1} y_1(n-1) = y(n), \qquad n \geq -q_1 + 1.$$

Now defining S_u in terms of exp $i\theta_u$, as for S_1, we may form $y_2(n) = S_2\, y_1(n)$ and see that it satisfies

$$y_2(n) - (e^{i\theta_1} + e^{i\theta_2})\, y_2(n-1) + e^{i(\theta_1+\theta_2)}\, y_2(n-2) = y(n), \quad n \geq -q_1 + 2.$$

Repeating the operation q_1 times, we see that

$$\left(\prod_{u=1}^{q_1} S_u\right) y(n)$$

is a solution of (3.1) for $n \geq 0$. Thus we see that

$$\text{(3.8)} \qquad \sum_u \sum_j n^j \rho_u{}^n (b'_{u,j} \cos \theta_u n + b''_{u,j} \sin \theta_u n) + \left(\prod_{u=1}^{q_1} S_u\right) y(n), \qquad n \geq 0$$

may, by suitable choice of the $b_{u,j}$, be made to satisfy (3.1) for all $n \geq 0$ and to satisfy, at $n = -q, -q+1, \ldots, -1$, any q initial conditions. There will now be no stationary solution to (3.1). The general situation may be understood from a simple case. We put

$$\text{(3.9)} \qquad \Delta x(n-1) = x(n) - x(n-1)$$

and, considering

$$\Delta x(n-1) = \epsilon(n),$$

we obtain the solution

$$x(n) = x(-1) + \sum_0^n \epsilon(j), \qquad n \geq 0.$$

For this case we have

$$\mathcal{E}\{(x(n) - x(-1))(x(m) - x(-1))\} = \sigma^2 \min(m, n).$$

Thus the correlation between $x(n)$ and $x(m)$ tends to unity if m and n both increase in such a fashion, for example, that $(n - m)$ remains constant. Thus a realization of this process will show a relatively smooth appearance. For a discussion of the nature of the behavior of these realizations we refer the reader to the second chapter of Cox and Miller (1965).

(ii) We now consider the case in which $x(n)$ is a vector. We write the model as

$$\text{(3.10)} \qquad \sum_0^q B(j)\, x(n-j) = \sum_0^s A(k)\, \epsilon(n-k), \qquad B(0) = I_p,$$

wherein the $B(j)$ are square matrices,† but we do not restrict the $A(k)$. We assume that the $\epsilon(n)$ satisfy

$$\mathcal{E}(\epsilon(n)\,\epsilon'(m)) = \delta_m{}^n G. \tag{3.11}$$

We use the same terminology (autoregression, moving average, etc.) as in the scalar case.

Now we seek for solutions of the homogeneous system of the form $b(u)\,z_u{}^n$, where $b(u)$ is a vector that satisfies

$$\left[\sum_0^q B(j)\,z_u^{q-j}\right] b(u) = 0$$

and z_u satisfies

$$\det\left[\sum_0^q B(j)\,z_u^{q-j}\right] = 0. \tag{3.12}$$

If z_u is a multiple root, we may be able to find more than one solution $b(u)$ corresponding to the same z_u. We call these $b(i, u)$. However, in general we shall also have to adjoin additional terms of the form $n^j z_u{}^n b(i, u)$. The theory now proceeds along "readymade" lines so that we may state Theorem 4′, the proof of which we have inserted as part of the appendix to this chapter.

Theorem 4′. *If* $\epsilon(n)$, $n = 0, \pm 1, \ldots$, *is a sequence of random vectors satisfying* (3.11), *then, given* q *initial values,* $x(-q), \ldots, x(-1)$, *the solution to* (3.10), *is uniquely defined. If no zero,* z_u, *of* (3.12) *lies on the unit circle, the solution to* (3.10) *is of the form*

$$x(n) = \sum_u \sum_i \sum_j c(i, j, u)\,z_u{}^n n^j b(i, u) + \sum_{k=-\infty}^{\infty} \Lambda(k)\,\epsilon(n - k),$$

where the elements $\lambda_{ij}(k)$ *of the* $\Lambda(k)$ *converge to zero exponentially as* $|k| \to \infty$.

Once again, if, for all μ, $|z_u| < 1$, the first term in the expression for $x(n)$ eventually decays to zero and we obtain a solution of the form

$$\sum_0^\infty \Lambda(j)\,\epsilon(n - j),$$

which every solution of (3.10) then eventually approaches. For this solution we have, let us say,

$$\mathcal{E}(x(m)\,x'(n)) = \sum_0^\infty \Lambda(j)\,G\Lambda'(n - m + j) = \Gamma(n - m).$$

† More general formulations are possible (and important, particularly in economics), but if the relation uniquely determines $x(n)$ in terms of its past and the $\epsilon(n - k)$, the system must be reducible to the form used here.

Then we can easily see that $\Gamma(n - m) = \Gamma'(m - n)$. When the covariance matrices of the process $x(n)$ depend only on $(n - m)$, we once again say that $x(n)$ is second-order stationary.†

(iii) We can now set up a model based on a differential equation, much as has been done in examples (i) and (ii). We will not attempt generality here but will discuss a simple example which leads to a model important in some applications. It is customary to speak of a process (in continuous time) $\epsilon(t)$ for which $\mathcal{E}(\epsilon(s)\,\epsilon(t)) = \delta_s^{\,t}$ as white noise. If $\epsilon(s)$ and $\epsilon(t)$ are actually independent the term pure white noise is sometimes used. Such a process evidently can never be observed for the definition implies an irregularity for a realization which is virtually unimaginable. The model which earlier cases suggest is

$$\dot{x}(t) + \beta(t)\,x(t) = a(t)\,\epsilon(t). \tag{3.13}$$

As we shall see, this is unacceptable and indeed that is fairly obvious for (3.13) implies that $x(t)$ has increments, $x(t + \delta t) - x(t)$, which have variance which is $O(\delta t^2)$. In any case, proceeding formally and putting

$$\lambda(t) = \exp\left\{-\int_0^t \beta(\tau)\,d\tau\right\},$$

we obtain the general solution as

$$b\,\lambda(t) + \int_0^t a(\tau)\,\epsilon(\tau)\exp\left\{-\int_\tau^t \beta(u)\,du\right\}d\tau. \tag{3.14}$$

Unfortunately the treatment is not mathematically acceptable for $\epsilon(t)$ is far too irregular to be integrable in any accepted sense. The integral in (3.14) can, however, be given a meaning in the form

$$\int_0^t \exp\left\{-\int_\tau^t \beta(u)\,du\right\}\xi(d\tau),$$

where $\xi(t)$ is a process of orthogonal increments for which

$$\mathcal{E}(\xi^2(t)) = \int_0^t a^2(\tau)\,d\tau, \qquad t \geq 0.$$

If $\xi(t)$ is actually Gaussian and $a(t) \equiv 1$ then $\xi(t)$ is called a Brownian motion process.†

Then (3.14) becomes

$$x(t) = b\,\lambda(t) + \int_0^t \frac{\lambda(t)}{\lambda(\tau)}\,\xi(d\tau), \tag{3.15}$$

† See the footnote on page 12. We again have put $\Gamma(n) = \Gamma(0, n)$ in this stationary case.

† This is often called the Wiener process.

which is to be interpreted as the solution of the integral equation

$$x(t) + \int_0^t \beta(\tau)\, x(\tau)\, d\tau = \xi(t), \qquad t \geq 0. \tag{3.16}$$

If $\beta(t) \equiv \beta$, then (3.15) becomes

$$x(t) = b\, \lambda(t) + \int_0^t \lambda(t-\tau)\, \xi(d\tau), \qquad \lambda(t) = e^{-\beta t},$$

which becomes just $\xi(t) + b$ for $\beta = 0$. If $\beta > 0$ and

$$\int_{-\infty}^t \lambda^2(t-\tau)\, a^2(\tau)\, d\tau < \infty,$$

the solution approaches

$$\int_{-\infty}^t \lambda(t-\tau)\, \xi(d\tau) \tag{3.17}$$

in mean square. If, finally, $a^2(t) \equiv \sigma^2$, we see that (3.17) has covariance function

$$\sigma^2 \int_0^\infty \lambda(\tau)\, \lambda(t-s+\tau)\, d\tau,$$

so that (3.17) is second-order stationary.†

Evidently we could increase the order of the differential equation (3.13) and achieve a variety of solutions which, for the constant coefficient case, would parallel those studied under (i). Systems of equations could also be studied as in (ii).

(iv) Robinson (1962) calls a time function, $w(t)$, $-\infty < t < \infty$, a wavelet if

(a)
$$\int_{-\infty}^\infty w^2(t)\, dt < \infty,$$

(b)
$$w(t) = 0, \qquad t < 0,$$

Then for $\xi(t)$ having orthogonal increments, with $\mathcal{E}(\xi^2(dt)) = dt$,

$$\int_{-\infty}^t w(t-\tau)\, \xi(d\tau) \tag{3.18}$$

is a stationary process. Evidently $w(t) = \lambda(t)$, $t \geq 0$, with $\lambda(t)$ as in (iii), is a wavelet. The interpretation of (3.18), leading to this terminology, is of a superposition of waves, propagated in one direction at successive time

† Again see the footnote on page 12.

instants, all having the same shape but having amplitudes† $\xi(d\tau)$ (for the wave propagated at time τ). More generally we could consider a time function $w(t)$ satisfying (a) but not (b) when

$$\int_{-\infty}^{\infty} w(t-\tau)\,\xi(d\tau) \tag{3.19}$$

is still well defined as a second-order stationary process but for which an interpretation as a superposition of waves now involves the implausible requirement (for time series) that the waves are propagated in both directions through time. We often call (3.19) a moving average, adding the adjective "one-sided" if $w(t)$ is indeed a wavelet. The one-sided case is of fundamental importance in prediction theory as we shall see. This same (wavelet) terminology carries over to the discrete time case in an obvious fashion. These definitions also extend to vector cases; for example, we call a $(p \times q)$ matrix-valued function $W(t)$ a wavelet if

(a) $$\int_{-\infty}^{\infty} W(t)\,W^*(t)\,dt < \infty,$$

(b) $$W(t) = 0, \qquad t < 0.$$

The integral in (a) is to be interpreted as a matrix of integrals, taking the elements of $W(t)\,W^*(t)$ individually. If $\xi(t)$ is a vector process (of q components) of orthogonal increments, with $\mathcal{E}\big(\xi(dt)\,\xi^*(dt)\big) = G\,dt$,

$$\int_{-\infty}^{t} W(t-\tau)\,\xi(d\tau)$$

is a vector stationary process. Again it has an interpretation as a superposition of the effects of vector waves, propagated in one direction at each point of time, with the only variation from time to time being due to the chance variation in $\xi(dt)$. For the discrete time, vector, case (3.10) provides an example of such a system for a particular form of $W(n) = \Lambda(n)$. Again, also, we may consider two-sided averages

$$\int_{-\infty}^{\infty} W(t-\tau)\,\xi(d\tau),$$

when (a) alone holds, but again also the usefulness of the physical interpretation is lost because the wave needs to be propagated in both directions through time.

(v) We have earlier considered the case of a discrete time, second-order, process for which $\Delta^q\,x(n)$ is a finite moving average. An analogous definition

† Actually $|\xi(d\tau)|$ is more properly called the amplitude.

may be given for continuous time and, following Yaglom (1962), we shall say that $x(t)$ has stationary qth order increments if

$$\Delta_h^q x(t) = \sum_0^q \binom{q}{j}(-)^{q-j} x(t+jh)$$

is stationary for every positive h.

If $x(t)$ (scalar) is q times, continuously, differentiable then, since this qth derivative is the mean square limit of $h^{-q}\Delta_h^q x(t)$, with covariance function which is

$$\frac{\partial^{2q}}{(\partial s)^q (\partial t)^q} \gamma(s, t) = \lim_{h \to 0} \mathcal{E}\{h^{-2q} \Delta_h^q x(s) \Delta_h^q x(t)\},$$

we see that $x^{(q)}(t)$ is stationary since its covariance function evidently depends only on $s - t$. However, the case in which this derivative does not exist is also interesting as the example of a Brownian motion process shows, for such a process (evidently) has stationary first-order increments but is not differentiable.

We close this section with a comment. It is sometimes implied that an equation of the form

$$x(n) = \rho x(n-1) + \epsilon(n), \qquad \rho > 1$$

has no solution which is stationary. This is not so, as the solution $x(n) = \sum_0^\infty \rho^j \epsilon(n+j)$ shows. What is true, of course, is that there is no stationary solution involving past and present values of $\epsilon(n)$ only. Often this is the only type of solution which is physically meaningful. Thus when, in future, we say that $x(n)$ is a stationary solution of an equation of the form of (3.10), we mean a stationary solution expressed in terms of $\epsilon(m)$, $m \leq n$, so that when we use this terminology we imply that all zeros of (3.12) lie inside the unit circle.

4. STATIONARY PROCESSES AND THEIR COVARIANCE STRUCTURE

By a strictly stationary vector process $x(t)$ we mean one for which, for all N, $t_1, \ldots, t_N$ and h (these being integers in the discrete time case, but only the first need be an integer in the continuous time case) the distributions of $x(t_1), \ldots, x(t_N)$ and of $x(t_1 + h), \ldots, x(t_N + h)$ are the same. If the $x(t)$ have finite mean square, this means that

$$\mathcal{E}\big(x(s)\, x'(t)\big) = \mathcal{E}\big(x(0)\, x'(t-s)\big) = \Gamma(0, t-s),$$

and, as explained earlier, we shall call this $\Gamma(t - s)$. As we are going to base our statistical treatment very substantially on second-order properties, this last condition, which we have previously called second-order stationarity will be of importance. (However, we shall need more than this for some purposes.) Evidently

$$\mathcal{E}\big(x(t)\,x'(s)\big) = \Gamma'(t-s) = \Gamma(s-t).$$

In particular, for the scalar case, $\gamma(s, t)$ is now a function only of $|s - t|$. *We now see also, from Theorem 1, that for mean square continuity it is necessary and sufficient that the $\gamma_j(t)$ should be continuous at $t = 0$.*

A very important range of statistical techniques, to be discussed in Part Two of this book, is based on the following result.

Theorem 5. *If $x(n)$ is a discrete time stationary vector process generated by* (3.10) *and* (3.11) *and with all zeros of* (3.12) *inside of the unit circle then the covariance sequence of the $x(n)$ satisfies*

$$(4.1)\qquad \sum_{j=0}^{q} B(j)\,\Gamma(j-u) = \sum_{k=\max(0,u)}^{s} A(k)\,G\Lambda'(k-u), \qquad -\infty < u \le s,$$

$$= 0, \qquad s < u.$$

The proof follows from the relation

$$\sum_0^q B(j)\,\mathcal{E}\{x(n-j)\,x'(n-u)\} = \sum_0^s A(k)\,\mathcal{E}\{\epsilon(n-k)\,x'(n-u)\}.$$

The left side is the left side of (4.1) by definition and the right side follows from the fact that

$$x(n) = \sum_0^\infty \Lambda(j)\,\epsilon(n-j),$$

as shown in Section 3.

It is evident from (4.1) that under the conditions of Theorem 5 the $\Gamma(n)$ will decay exponentially with n just as $\Lambda(n)$ does.

It is often convenient to work in terms of the scale free quantities

$$\rho_{jk}(n) = \frac{\gamma_{jk}(n)}{\{\gamma_j(0)\,\gamma_k(0)\}^{1/2}},$$

which we call serial correlations.† If $j = k$, we call these autocorrelations and use the notation $\rho_j(n) = \gamma_j(n)/\gamma_j(0)$.

A direct solution of the equations (4.1) is sometimes useful in theoretical work. A first set of $(q + 1)$ of these systems of equations (those for u going from 0 to q) involve only $\Gamma(0), \Gamma(1), \ldots, \Gamma(q)$ (or their transposes). Thus

† Again we have used a slightly inappropriate terminology since no mean corrections have been made.

these can be solved for these initial values which serve to determine the solution of the remaining relations ($u > q$). Let us illustrate by the simple case in which $p = 1, q = 2, s = 0$. Then $\epsilon(t)$ may as well be taken as a scalar and, putting $\alpha_0 = 1$, we take its variance as σ^2. Then

$$\gamma(0) + \beta(1)\,\gamma(1) + \beta(2)\,\gamma(2) = \sigma^2,$$
$$\gamma(1) + \beta(1)\,\gamma(0) + \beta(2)\,\gamma(1) = 0,$$
$$\gamma(2) + \beta(1)\,\gamma(1) + \beta(2)\,\gamma(0) = 0,$$

whose solution is

$$\rho(1) = \frac{\gamma(1)}{\gamma(0)} = -\frac{\beta(1)}{1 + \beta(2)},$$
$$\rho(2) = \frac{\gamma(2)}{\gamma(0)} = \frac{\beta^2(1)}{1 + \beta(2)} - \beta(2),$$
$$\gamma(0) = \left(1 + \beta(1)\,\rho(1) + \beta(2)\,\rho(2)\right)^{-1}\sigma^2.$$

Let us assume that the discriminant $\beta^2(1) - 4\beta(2)$ is negative. Then the zeros of the characteristic equation are complex and we must have

$$\rho(n) = \frac{a^n \cos(\theta n + \phi)}{\cos\phi},$$

wherein

$$a = [\beta(2)]^{1/2}, \qquad \cos\theta = -\frac{\beta(1)}{2[\beta(2)]^{1/2}}, \qquad \tan\theta\tan\phi = 1 - \frac{\rho(1)}{a\cos\theta}.$$

Thus the $\rho(n)$ oscillate in a damped fashion, the damping depending only on $\beta(2)^{1/2}$.

In this simple case it is easy to interpret the covariance structure of the process but in general that is difficult. Even if it is known that the process is autoregressive, but of unknown order, it is not easy to see from the $\Gamma(n)$ what the order is. (Of course, in practice the dominant problem is usually that introduced by sampling fluctuations in the estimates of the $\Gamma(n)$ but we leave the discussion of this until Part Two of this volume.) This particular problem can, however, be overcome by using coefficients of partial association which we now discuss for the scalar case.

Consider the $m + 1$ dimensional real linear space $\mathcal{X}$, spanned by the random variables $x(j, \omega)$, $j = n, n - 1, n - 2, \ldots, n - m$. Thus $\mathcal{X}$ consists of all real linear combinations

$$\sum_{j=0}^{m} a_j\, x(n - j),$$

where we once more drop the ω from our notation for simplicity. We may now consider the regression of $x(n)$ on the $x(n - j)$, $j = 1, \ldots, m$, that is, the projection of $x(n)$ on the m dimensional subspace of $\mathfrak{X}$ spanned by $x(n-j), j = 1, \ldots, m$. We put

$$P_m = [\rho(j-k)], \qquad p_m = (\rho(j)),$$

where by this notation we mean that $\rho(j-k)$ is in row j, column k of the matrix P_m while the vector p_m has $\rho(j)$ in the jth place. Now the vector of regression coefficients† is $P_m^{-1}p_m$. The regression coefficient for $x(n-m)$ is thus

$$\frac{1}{\det(P_m)} \sum_{i=1}^{m} (-)^{m+i} \rho(i) \det(P_m^{(i,m)}),$$

where $P_m^{(i,m)}$ is the matrix obtained from P_m by deleting the ith row and mth column. However, expanding $\det(P_{m+1}^{(1,m+1)})$ by the elements of the first column

$$\det(P_{m+1}^{(1,m+1)}) = \sum_{i=1}^{m} (-)^{i+1} \rho(i) P_m^{(i,m)}$$

so that the regression coefficient is

$$\rho(m \mid 1, \ldots, m-1) = \frac{(-)^{m+1} \det(P_{m+1}^{(1,m+1)})}{\det(P_m)}. \tag{4.2}$$

This coefficient, $\rho(m \mid 1, \ldots, m-1)$, is called the mth coefficient of partial autocorrelation. If the $x(n)$ are generated by an autoregression‡ of order m, then $\rho(m \mid 1, \ldots, m-1) = -\beta(m)$. Indeed then

$$x(n) = -\sum_{1}^{m} \beta(j)\, x(n-j) + \epsilon(n)$$

and, since the $x(n-j), j \geq 1$, are orthogonal to $\epsilon(n)$, the projection of $x(n)$ on $x(n-1), \ldots, x(n-m)$ is just $-\sum \beta(j)\, x(n-j)$. Thus $\rho(m \mid 1, \ldots, m-1)$ will be zero if $\beta(m)$ is zero.

The fact that $\rho(m \mid 1, \ldots, m-1)$ is both a regression coefficient and a partial correlation may seem strange to readers familiar with the classical theory of regression. It is due, of course, to the very special form of the matrix P_m. To investigate this further we introduce the coefficient

$$R_m^{\ 2} = p_m' P_m^{-1} p_m. \tag{4.3}$$

R_m is the coefficient of multiple correlation between $x(n)$ and the $x(n-j)$, $j = 1, \ldots, m$. Indeed the projection of $x(n)$ on these latter has mean square

† For the regression formulas and results used here without proof see Cramér (1946, Chapter 23).

‡ Unless we say otherwise we shall assume that the zeros of $\sum \beta(j) z^{-j}$ lie inside the unit circle when we speak of our autoregression.

(i.e., "regression mean square")

$$p'_m P_m^{-1} \Gamma_m P_m^{-1} p_m, \qquad \Gamma_m = \gamma(0) P_m$$

since Γ_m is the covariance matrix of the $x(n-j), j = 1, \ldots, m$. Thus the ratio of the regression mean square to the total mean square (namely, $\gamma(0)$) is (4.3).

Theorem 6. *If* R_m^2 *is defined by* (4.3) *and* $\rho(m \mid 1, \ldots, m-1)$ *by* (4.2), *then*

$$(1 - R_m^2) = \prod_{j=1}^{m} (1 - \rho^2(j \mid 1, \ldots, j-1))$$

and $\rho(m \mid 1, \ldots, m-1)$ *is the partial correlation of* $x(n)$ *with* $x(n-m)$ *after the effects of* $x(n-1), \ldots, x(n-m+1)$ *have been removed by regression.*

Proof. We establish the formula by showing that

$$(1 - R_m^2) = (1 - R_{m-1}^2)(1 - \rho^2(m \mid 1, \ldots, m-1)).$$

In turn this follows from a classical formula of regression theory (Cramér, 1946, p. 319) if we show that $\rho(m \mid 1, \ldots, m-1)$ is the partial correlation as stated in Theorem 6. Let $\hat{x}(n)$ be the regression of $x(n)$ on $x(n-1), \ldots, x(n-m+1)$. (Thus the coefficients of these $(m-1)$ variables are comprised by the elements of $P_{m-1}^{-1} p_{m-1}$.) Let $\hat{x}(n-m)$ be the regression of $x(n-m)$ *on the same variables.* Then the square of the partial correlation is, by definition, the ratio of the mean square of the regression of $x(n) - \hat{x}(n)$ on $x(n-m) - \hat{x}(n-m)$ to the mean square of $x(n) - \hat{x}(n)$. The latter we know to be $\gamma(0)(1 - R_{m-1}^2)$. The former is $\rho^2(m \mid 1, \ldots, m-1)$ multiplied by the mean square of $x(n-m) - \hat{x}(n-m)$ since $\rho(m \mid 1, \ldots, m-1)$ is, we know, the regression coefficient. Thus our result will follow from the fact that the mean square of $x(n-m) - \hat{x}(n-m)$ is also $\gamma(0)(1 - R_{m-1}^2)$ or, which is the same thing, R_{m-1}^2 is also the multiple correlation between $x(n-m)$ and $x(n-m+1), \ldots, x(n-1)$. That this is so is simply a consequence of the fact that Γ_m (or P_m) is not altered when the orders of rows and columns are simultaneously reversed.

Thus the $\rho(m \mid 1, \ldots, m-1)$ provide not only a criterion as to whether the process is autoregressive of this or that order but also a form of decomposition of the variance of $x(n)$ into components due to the influence of the past. As we shall see later† the statistics which estimate these partial autocorrelations also have simple properties of statistical independence. Thus a graph, for example, of the $\rho(m \mid 1, \ldots, m-1)$ is easier to comprehend and is more informative than one of the $\rho(j)$ themselves. Nevertheless, it must be emphasized that the usefulness of such a description depends largely on the

† See Section VI. 3.

supposition that $x(n)$ is near to being generated by an autoregression of reasonable order. Consider, for example, the case where

$$x(n) = a \cos n\theta + b \sin n\theta + y(n), \qquad y(n) = \rho y(n-1) + \epsilon(n), \qquad |\rho| < 1,$$

where a and b are independent random variables, independent also of all $\epsilon(n)$, with unit variance. Then $\gamma(n) = \cos n\theta + \sigma^2\rho^n$ and the $\rho(m \mid 1, \ldots, m-1)$ are quite complicated and would be extremely difficult to interpret if they were presented without any further indication of the nature of the process.

5. HIGHER MOMENTS

A very great deal of the analysis of time phenomena is accomplished through moments of the first and second order and we have so far discussed only these. However, if the process generating $x(t)$ is strictly stationary, then, of course,

$$\mathcal{E}\{x_j(t_1)\, x_k(t_2)\, x_l(t_3) \cdots x_r(t_m)\} = \gamma_{j,k\ldots}(t_1, t_2, \ldots, t_m)$$

is a function only of the configuration of the points, $t_1, t_2, \ldots, t_m$, which are involved and not of the points themselves. By this we mean that

$$\gamma_{j,k\ldots}(t_1, t_2, \ldots, t_m) = \gamma_{j,k\ldots}(0, t_2 - t_1, \ldots, t_m - t_1).$$

Similarly, we may say that $x(t)$ is stationary to the mth order if the moments up to those of order m obey this rule. We shall use the symbol $k_{j,k,\ldots,r}(t_1, t_2, \ldots, t_m)$ for the corresponding cumulant so that this is the cumulant of the mth order between $x_j(t_1), x_k(t_2), \ldots, x_r(t_m)$. Some of the subscripts, $j, k, \ldots$, may be repeated. Of course the moments may be expressed in terms of the cumulants by well-known formulas and we mention here only that

$$\begin{aligned}(5.1)\quad \mathcal{E}\{x_j(t_1)\, x_k(t_2)\, x_l(t_3)\, x_m(t_4)\} &= \gamma_{jk}(t_2 - t_1)\, \gamma_{lm}(t_4 - t_3) + \gamma_{jl}(t_3 - t_1)\, \gamma_{km}(t_4 - t_2) \\ &\quad + \gamma_{jm}(t_4 - t_1)\, \gamma_{kl}(t_3 - t_2) + k_{j,k,l,m}(t_1, t_2, t_3, t_4).\end{aligned}$$

We observe the obvious fact that for any moment or cumulant the argument, t_j, corresponding to repeated subscripts, may be permuted and the moment or cumulant remains unchanged.

6. GENERALIZED RANDOM PROCESSES†

It is a classic fact, first established rigorously by Wiener (1923), that the realizations of a Brownian motion are not differentiable, almost everywhere.

† This topic is special and will not be used in most of the remainder of the book. It can therefore be omitted if the reader so desires.

Nevertheless, it is intuitively appealing to think of $\xi(t)$, the Brownian motion process, as being got from $\dot{\xi}(t)$ by integration, where $\dot{\xi}(t)$ is a process satisfying the relation

$$\mathcal{E}(\dot{\xi}(s)\,\dot{\xi}(t)) = \delta_s^{\,t}.$$

It is not surprising that in fact $\dot{\xi}(t)$ can be given a meaning by introducing the concept of generalized random process (GRP) in a way closely related to that in which the notion of generalized function is introduced (see the Mathematical Appendix to this book for a survey of what is needed of this concept of generalized function). We follow Gel'fand and Vilenkin (1964).

We consider the space K of infinitely differentiable functions, $\phi(t)$, whose support is compact (i.e., which vanish off a closed, bounded set). If $\dot{\xi}(t)$ were already meaningfully defined, then we could form

$$\int_{-\infty}^{\infty} \dot{\xi}(t)\,\phi(t)\,dt, \qquad \phi \in K,$$

and, formally, integrating by parts this is

$$-\int_{-\infty}^{\infty} \xi(t)\,\phi'(t)\,dt \tag{6.1}$$

because of the vanishing of $\phi(t)$ at the end points of a suitable interval. The latter expression is certainly well defined. This leads us to begin by defining a generalized random process via a generalization of this kind of expression and to define the derivative of a Brownian motion by reversing the process which led us to (6.1) from $\dot{\xi}(t)$. Thus we define a GRP as a law that allots to elements $\phi \in K$, random variables $\Phi(\phi)$ in such a fashion that (a) $\Phi(\phi)$ is a linear functional of ϕ; that is, α, β being real numbers $\Phi(\alpha\phi_1 + \beta\phi_2) = \alpha\Phi(\phi_1) + \beta\Phi(\phi_2)$, ϕ_1, $\phi_2 \in K$; (b) $\Phi(\phi)$ is a continuous functional of ϕ. By this is meant the following: if $\phi_{1,n}, \phi_{2,n}, \ldots, \phi_{r,n}$ are any r sequences of elements of K, all of which functions are supported by the same compact set (which may, however, depend on the r sequences considered) and which converge uniformly to functions $\phi_1, \phi_2, \ldots, \phi_r$ (which thus belong also to K), then the joint distribution of $\Phi(\phi_{1,n}), \Phi(\phi_{2,n}), \ldots, \Phi(\phi_{r,n})$ converges to the joint distribution of the random variables $\Phi(\phi_1), \ldots, \Phi(\phi_r)$ (in the usual sense of pointwise convergence of a sequence of distribution functions at every point of continuity of the limiting distribution function). It is, of course, required once more, that the joint distributions of the $\Phi(\phi)$ be compatible in the sense that the prescribed distribution of any set be the marginal distribution got from any more inclusive set.†

† See Gel'fand and Vilenkin (1964, Chapter IV, Section 2), for a discussion of the construction of a probability space on which all of the random variables $\Phi(\phi)$ may be thought of as being defined.

The mean-square continuous random processes, as well as such random processes as processes whose increments are mutually orthogonal, are imbedded in the space of GRP through the definition

$$\Phi_x(\phi) = \int_{-\infty}^{\infty} x(t)\, \phi(t)\, dt. \tag{6.2}$$

The physical motivation for these ideas is partly that given above and partly the notion that what we observe is necessarily some functional of the underlying process since observation has to be made through some apparatus. Thus ϕ in (6.2), may be thought of as characterizing the apparatus. The linearity of the functional is, of course, something of a restriction, and is determined by mathematical convenience, but the concept is still very general if only because $x(t)$ in (6.2) could itself be a highly nonlinear function of some even more basic random phenomenon.

We now define

$$\Phi'(\phi) = -\Phi(\phi'),$$

as suggested earlier. Now the properties of the space K show that any GRP may be differentiated to give a new GRP.

We now define a second-order stationary GRP. We consider Φ for which

$$\gamma_\Phi(\phi, \psi) = \mathcal{E}\{\Phi(\phi)\, \Phi(\psi)\} < \infty, \qquad \phi, \psi \in K.$$

Now put $\phi_{(s)}(t) = \phi(s + t)$. Then Φ is said to be second-order stationary if

$$\gamma_\Phi(\phi_{(s)}, \psi_{(s)}) = \gamma_\Phi(\phi, \psi),$$

for all $s \in (-\infty, \infty)$ and $\phi, \psi \in K$.

It is now evident that Φ' is also second-order stationary and has covariance function which is $\gamma_\Phi(\phi', \psi')$. We conclude with an example. Consider the functional

$$\gamma\{\phi, \psi\} = \iint\limits_0^\infty \phi(s)\, \psi(t) \min(s, t)\, ds\, dt \tag{6.3}$$

and the associated generalized random process which has this covariance functional and a joint distribution for $\Phi(\phi_j), j = 1, \ldots, k$, which is Gaussian with zero mean values and covariance matrix $\gamma(\phi_j, \phi_k)$. This may be identified with the Brownian motion random process $\xi(t)$ through the formula

$$\Phi(\phi) = \int_0^t \phi(t)\, \xi(t)\, dt.$$

Now Φ' has a Gaussian distribution also, obviously, and covariance functional

$$\int_0^\infty \phi(t)\, \psi(t)\, dt.$$

This may most easily be seen by expressing (6.3) in the form

$$\int_0^\infty\left\{\int_0^t \phi(s)\,ds - \int_0^\infty \phi(s)\,ds\right\}\left\{\int_0^t \psi(s)\,ds - \int_0^\infty \psi(s)\,ds\right\}dt.$$

Thus Φ' has the covariance functional of the derivative of a Brownian motion and thus the extension to generalized random processes has called into being a random process which otherwise would not exist. It is evident that Φ' is stationary.

EXERCISES

1. Let $x(n) = x(n-m) + \epsilon(n)$, where $x(n)$ is scalar. Show that

$$x(n) = \sum_0^{m-1} a_j e^{i2\pi jn/m} + \sum_{j=0}^{[n/m]} \epsilon(n - mj), \qquad n \geq 0$$

for suitable a_j, where these may be adjusted to satisfy m initial conditions.

2. If $x(n)$ is stationary and is generated by

$$\sum_0^q \beta(j)x(n-j) = \epsilon(n),$$

show that

$$\text{var}\,(\epsilon(n)) = \frac{(\det \Gamma_{q+1})}{(\det \Gamma_q)}.$$

3. Determine the form of the correlogram for

$$x(n) = \sum_{j=0}^{q} \{\alpha(j)\cos n\lambda_j + \beta(j)\sin n\lambda_j\}, \qquad \lambda_j = \frac{2\pi j}{q},$$

where $\alpha(j)$, $\beta(j)$, $j = 0, \ldots, q$ are random variables for which the only nonzero covariances are

$$\mathcal{E}\{\alpha(j)^2\} = \mathcal{E}\{\beta(j)^2\} = f_j.$$

4. Let $x(t)$ be mean square continuous and put

$$y(t) = \int_0^t x(\tau)\,d\tau, \qquad t \geq 0$$

$$= -\int_t^0 x(\tau)\,d\tau, \qquad t \leq 0.$$

Prove that $y(t)$ is mean square differentiable and that $\dot{y}(t) = x(t)$. (See Mann, 1953, Theorem 1.4.)

5. The characteristic functional of a GRP, Φ, is defined as $L(\phi) = \mathcal{E}\{\exp i\Phi(\phi)\}$. Prove that if Φ is Gaussian (so that the joint distributions of the $\Phi(\phi)$, $\phi \in K$, are Gaussian) then $L(\phi) = \exp -\frac{1}{2}\gamma(\phi, \phi)$.

APPENDIX

1

Hilbert Spaces† of Random Variables, and Integration with Respect to a Process of Orthogonal Increments

We consider a scalar random process $x(t, \omega)$ for which $\mathcal{E}(x(t)^2) < \infty$, $-\infty < t < \infty$. We first take $x(t)$ to be mean square continuous. Each $x(t, \omega)$ thus belongs to the space $L_2(\Omega, \mathcal{A}, P)$ of all square integrable, complex-valued functions with respect to P over Ω. Let $\mathcal{H}$ be the closed subspace of $L_2(\Omega, \mathcal{A}, P)$ spanned by the $x(t, \omega)$. Thus $\mathcal{H}$ consists of all finite complex-linear combinations

$$x = \sum_{j=1}^{N} \alpha_j x(t_j), \tag{1}$$

together with the limit in mean square of sequences x_n, of such linear combinations when these limits exist. The inner product in $\mathcal{H}$ we indicate by (x, y), $x, y \in \mathcal{H}$. Of course, $(x, y) = \mathcal{E}(x\bar{y})$, $x, y \in \mathcal{H}$ and in particular

$$\big(x(s), x(t)\big) = \gamma(s, t). \tag{2}$$

Then $\mathcal{H}$ is separable. Indeed the $x(t)$ for rational t constitute a denumerable set which, by mean square continuity, may be used to approximate arbitrarily closely to any x of the form of (1).

A subset $\mathcal{M} \subset \mathcal{H}$ is a (closed) subspace if it is a linear space which is closed with respect to mean square convergence, i.e., with respect to the norm corresponding to the inner product. The set of all elements $x \in \mathcal{H}$ which are orthogonal to all elements of $\mathcal{M}$, which we call $\mathcal{M}^\perp$, is also a closed subspace and each $z \in \mathcal{H}$ can be uniquely decomposed in the form $z = x + y$, $x \in \mathcal{M}$, $y \in \mathcal{M}^\perp$. The operation which replaces z by x is called the perpendicular projection onto $\mathcal{M}$ (and similarly for the operation replacing z by $y \in \mathcal{M}^\perp$). It is of course just the generalization of the notion of regression introduced in Section 4.

The definitions just given would not have been altered in any essential way if we had taken the initial random variables, $x(t)$, to be complex valued.‡ We now consider the special case in which $x(t)$, which we now call $\xi(t)$, has orthogonal increments, that is,

$$\mathcal{E}\{[\xi(t_2) - \xi(s_2)][\overline{\xi(t_1) - \xi(s_1)}]\} = 0, \qquad t_2 > s_2 \geq t_1 > s_1.$$

We put

$$F(t) - F(s) = \mathcal{E}(|\xi(t) - \xi(s)|^2),$$

† See the Mathematical Appendix at the end of this volume for a summary of the part of Hilbert space theory which we need.

‡ The existence of Ω, $\mathcal{A}$, P is evident, for a complex-valued scalar process may be regarded as a vector process of two components.

where $F(t)$ is an everywhere finite, nondecreasing, real-valued function.

It is easy to see that $\xi(t)$ has, for example, a limit from the right at each point. Indeed

$$\lim_{t_1,t_2\downarrow t} \mathcal{E}\{|\xi(t_2) - \xi(t_1)|^2\} = \lim_{t_1,t_2\downarrow t} F(t_2) - F(t_1) = 0,$$

where the notation signifies that the limit is taken as t_1 and t_2 decrease to t. Thus any sequence $\xi(t_j)$, $t_j \geq t$, for which the t_j decrease to t, has a limit in mean square, which may well be $\xi(t)$ but need not be if $\xi(t)$ "jumps" at t. Of course, the limit, which we call $\xi(t+)$, is uniquely defined in any case since, if t_j and s_j were two sequences decreasing to t,

$$\lim_{j\to\infty} \{|\xi(t_j) - \xi(s_j)|\} = \lim_{j\to\infty} |F(t_j) - F(s_j)| = 0.$$

In most cases we will be able to modify our definition of $\xi(t)$, without any consequences for our final results, so that $\xi(t)$ is continuous from the right, that is, so that $\lim_{t\downarrow 0} \xi(s + t) = \xi(s)$, the limit being in mean square of course. Then $F(t)$ is continuous from the right. We shall assume this has been done in what follows. We now do not assume that $\xi(t)$ is mean square continuous. However, if we form $\mathcal{H}$ from the $\xi(t)$ as we formed it previously from the $x(t)$ this space is still separable. This follows from the fact that $F(t)$ can have at most a denumerable sequence of jumps so that the $\xi(t)$ for the points of jump in $F(t)$ together with the $\xi(t)$ for rational t which are not points of jump constitute a denumerable set dense in $\mathcal{H}$.

Now consider the functions $\phi(t)$ defined by

$$(3)\qquad \begin{aligned} \phi(t) &= 0, & t &\leq t_0, \\ &= c_j, & t_{j-1} &< t \leq t_j \qquad j = 1, \ldots, N, \\ &= 0, & t_N &< t, \end{aligned}$$

and establish the correspondence

$$(4)\qquad \phi(t) \leftrightarrow \phi = \int \phi(t)\, \xi(dt) = \sum_1^N c_j\{\xi(t_j) - \xi(t_{j-1})\}.$$

On the left-hand side we regard the $\phi(t)$ as elements of the closed linear space $L_2(F)$, of functions square integrable with respect to $F(dt)$, while on the right we have random variables belonging to the Hilbert space $\mathcal{H}$ defined by the $\xi(t)$. Of course $L_2(F)$ is itself a Hilbert space and we also have, for two such functions,

$$(5)\qquad \int \phi_1(t)\, \overline{\phi_2(t)}\, F(dt) = (\phi_1, \phi_2),$$

as is easily verified.

Thus this correspondence preserves the inner product in the two spaces. Thus the correspondence may be extended to include *all* functions square integrable with respect to F, for if $\phi_n(t)$ is a sequence of functions of the type just described, which satisfies

$$\lim_{m,n\to\infty} \int |\phi_m(t) - \phi_n(t)|^2\, F(dt) = 0,$$

then $\phi_m(t)$ converges, in the norm of $L_2(F)$, to a function $\phi(t)$ square integrable with respect to F and (5) ensures that the corresponding sequence ϕ_m converges to an element of $\mathcal{H}$. Moreover, the step functions (3) are "dense" in $L_2(F)$ in the sense that all square integrable functions can be got from mean-square convergent sequences of them. (Indeed we can, in particular, restrict the points, t_j, of discontinuity to be rational so that we see that $L_2(F)$ has a dense subset which is denumerable so that $L_2(F)$ is "separable.") Thus the correspondence (4) enables us to define a random variable, which we indicate by the symbol

$$\int \phi(t)\, \xi(dt),$$

for every function $\phi(t)$ belonging to $L_2(F)$.

These considerations also serve to define the meaning of

$$\int \Phi(t)\, \xi(dt), \tag{6}$$

where now $\xi(t)$ is a vector process of orthogonal increments, i.e.,

$$\mathcal{E}([\xi(t_2) - \xi(s_2)][\xi(t_1) - \xi(s_1)]^*) = 0, \qquad t_2 > s_2 \geq t_1 > s_1.$$

Of course, $\Phi(t)$ is a matrix with as many columns as there are rows in ξ while the element $\phi_{ij}(t)$ of $\Phi(t)$ is required to belong to $L_2(F_j)$, where

$$F_j(t) - F_j(s) = \mathcal{E}\{|\xi_j(t) - \xi_j(s)|^2\}.$$

Indeed (6) is defined as a vector whose typical component is

$$\sum_j \int \phi_{ij}(t)\, \xi_j(dt)$$

and the individual summands are already defined.

2

Proof of Theorem 4′

In Section 3 we introduced the model (3.10), which led us to consider the solutions of the system of equations

$$\left[\sum_0^q B(j)\, z^{q-j}\right] b = 0. \tag{1}$$

If z_u satisfies

$$\det\left[\sum_0^q B(j)\, z_u^{q-j}\right] = 0, \tag{2}$$

we call the system of solutions of (1), $b(i, u)$, $i = 1, \ldots, s_u$. The existence of these solutions of (1), and their multiplicities as eigenvectors, derive from the theory of equivalence of matrices whose elements are polynomials in the indeterminate, z. (See, for example, Macduffee, 1956, p. 40 or Bôcher, 1907, Chapter XX). It is there asserted that there exist two matrices, P, Q, whose elements are polynomials in z and whose determinants are constants (not zero) for which the matrix in the left-hand term of (1), which we call $C(z)$, satisfies

$$PC(z)Q = \begin{bmatrix} h_1(z) & & & & \\ & h_2(z) & & & \\ & & \cdot & & \\ & & & \cdot & \\ & & & & \cdot \\ & & & & & h_p(z) \end{bmatrix} = H,$$

where the off-diagonal elements are zero, $h_i(z)$ is a polynomial in z of degree p_i, and h_i divides h_{i+1} for each i. (In general there would be zero terms down the diagonal but this is impossible for $C(z)$, since we have assumed $B(0)$ to be the unit matrix.) The $h_i(z)$ are called invariant factors. If z_u is a zero of $h_i(z)$ but not of $h_{i-1}(z)$, then recalling that P^{-1} is also a matrix of polynomials in z we see that we can obtain $s_u = (p - i + 1)$ solutions, $b(i, u)$, $i = 1, \ldots, s_u$, of $C(z_u)\, b = 0$, namely, those given by inserting this value, z_u, of z in P, C, and Q and taking the last $(p - i + 1)$ columns of the resulting Q as the $b(i, u)$. These solutions are evidently linearly independent since the determinant of Q is a nonzero constant. Thus to each z_u we have associated s_u solutions of $C(z_u)\, b = 0$, namely the $b(i, u)$ defined above. Let $p_{i,u}$ be the number of times the factor $(z - z_u)$ occurs in the invariant factor corresponding to $b(i, u)$. Then we construct the system of vectors:

$$b(i, u)\, z_u^{\,n}\, n^j, \qquad \begin{aligned} u &= 1, \ldots, s, \\ i &= 1, \ldots, s_u, \\ j &= 0, \ldots, (p_{iu} - 1), \end{aligned}$$

there being s distinct zeros z_u. We observe that there are precisely as many of these vectors as the degree, pq, of the determinant of $C(z)$ as a polynomial in z. We now form the expression

$$c(n) = \sum_u \sum_i \sum_j c(i, j, u)\, z_u^{\,n}\, n^j\, b(i, u), \tag{3}$$

where the $c(i, j, u)$ are to be adjusted so that this vector satisfies q initial conditions, for $n = -1, \ldots, -q$. The vector $c(n)$ evidently satisfies

$$\sum_0^q B(j)\, c(n - j) = 0. \tag{4}$$

(The fact that $z_u^{\,n}\, n^j\, b(i, u)$ satisfies (4) for $j > 0$ may be verified by differentiating both sides, of $C(z)\, Q = P^{-1} H$, j times and putting $z = z_u$). The only thing remaining to be checked is the existence of $c(i, j, u)$ satisfying given initial conditions and this amounts to verifying that the matrix, with a kth set of p rows ($k = 1$,

$2, \ldots, q$) having as elements in the (i, j, u)th column

$$z_u^{-k}\,(-k)^j\, b(i, u),$$

has nonzero determinant. But if this determinant were zero one could choose non-null $c(i, j, u)$ so that $c(n) = 0$, $n = -1, \ldots, -q$, and this is manifestly impossible since $c(n)$ has elements which are polynomials in n of degree $(q - 1)$ at most.

We may now complete the proof of Theorem 4′. Let $Q^{-1}(z^{-1}) = \sum Q_j z^{-j}$ and $P(z^{-1}) = \sum P_j z^{-j}$. Then, introducing the new variables $y(n) = \sum Q_j x(n - j)$ and replacing the right-hand side of (3.10) with

$$\sum_j P_j \sum_k A(k)\, \epsilon\,(n - k - j),$$

we have reduced the proof of the existence of a unique solution of (3.10) to that for a system of scalar equations, with generating functions $h(z)$ for the left-hand sides. Thus only the second part of the proof remains and we now assume that no solution of (2) lies on the unit circle. Then $(\sum B(j)z^j)^{-1}$ has a Laurent expansion converging in an annulus containing the unit circle, and calling this

$$\sum_{-\infty}^{\infty} F(n) z^n$$

we are led to consider

$$\sum_{u=-\infty}^{\infty} F(n - u)\, A(k)\, \epsilon(u - k).$$

This expression is well defined as a limit in mean square, since the elements of $F(n)$ converge to zero exponentially with n; but now, since

$$\sum_0^q B(j)\, F(n - j) = \delta_0^n I_p,$$

then

$$\sum_0^q B(j) \left\{ \sum_{j=-\infty}^{\infty} F(n - u - j)\, A(k)\, \epsilon(u - k) \right\} = A(k)\, \epsilon(n - k)$$

and

$$\sum_{k=0}^{s} \sum_{u=-\infty}^{\infty} F(n - u)\, A(k)\, \epsilon(u - k)$$

is a solution of (3.10). Putting

$$\Lambda(j) = \sum_{k=0}^{s} F(j - k)\, A(k),$$

we may express this in the form

$$\sum_{-\infty}^{\infty} \Lambda(j)\, \epsilon(n - j). \tag{5}$$

Combining (3) and (5), we obtain Theorem 4′.

CHAPTER II

The Spectral Theory of Vector Processes

1. INTRODUCTION

This chapter is concerned with the Fourier analysis of time series. These Fourier methods are intimately linked with the notion of stationarity so that this kind of stochastic process will be of central interest, although departures from stationarity will also be considered. It is best, perhaps, to begin by explaining why Fourier methods play such an essential part in the theory. The basic condition of stationarity which we have imposed is the requirement, for a scalar time function, that $\gamma(s, t) = \gamma(t - s)$. Thus we are considering a *restricted* class of covariance functions, which have a symmetry which is that of the group of translations of the real line, i.e., $\gamma(s, t) = \gamma(s + \tau, t + \tau)$. This leads us to ask if we can replace the time function $x(t)$ by some new function, $\hat{x}(t)$, let us say, which is to be a linear function of the $x(t)$ and for which the covariance function will be of a simpler form, i.e., diagonal, so that $\hat{\gamma}(s, t) = 0$, $s \neq t$. (This seems implausible since this $\hat{\gamma}(s, t)$ is certainly not continuous but we shall leave the details of the explanation until after this heuristic introduction.) In example (iii) of Chapter I, Section 3 we have already discussed such a transformation for we there considered a representation, available for a special class of stationary time functions, of the form

$$x(t) = \int_0^\infty \lambda(t - \tau)\, \xi(d\tau). \tag{1.1}$$

Thus we have expressed $x(t)$ linearly in terms of a new random process $\xi(t)$ which, while not stationary has, at least, stationary increments which are, moreover, uncorrelated, so that $\xi(dt)/dt$ could, very loosely, be thought of as occupying the place of our $\hat{x}(t)$. However, this representation (1.1) does *not* provide us with what we want, for we cannot realize it unless we know not merely that $\gamma(s, t) = \gamma(t - s)$ but also $\gamma(t)$ in its entirety. What we are seeking is a transformation which will diagonalize $\gamma(s, t)$ and which is known *a priori* once we know that $\gamma(s, t) = \gamma(t - s)$ (without knowing $\gamma(t)$ itself). It is not surprising that this transformation is accomplished by Fourier methods for, in some appropriate sense which we shall not here make explicit,† exp $it\lambda$ is an eigenfunction of the operator $U(\tau)$ which acts via $U(\tau)f(t) = f(t+\tau)$ and which therefore describes the stationarity of the process.

Before going on to discuss this procedure which we have foreshadowed let us discuss an example, in some ways simpler and more general, which illustrates what we have said. Thus we consider a mean square continuous scalar random process, $x(t)$, whose covariance function, $\gamma(s, t)$, is known and we, at first, confine ourselves to the consideration of an interval, $[0, T]$. Then, by Mercer's theorem (Riesz and Nagy, 1956, p. 245),

$$\gamma(s, t) = \sum_0^\infty \mu_i \phi_i(s)\, \phi_i(t),$$

where the $\phi_i(s)$ are eigenvectors of the kernel $\gamma(s, t)$,

$$\int_0^T \gamma(s, t)\, \phi_i(t)\, dt = \mu_i\, \phi_i(s),$$

and

$$\int_0^T \phi_i(s)\, \phi_j(s)\, ds = \delta_i^j.$$

The double series for $\gamma(s, t)$ converges uniformly, in s and t, to its limit. We now form the series

$$\sum_0^\infty \phi_i(t) \int_0^T x(s)\, \phi_i(s)\, ds = \sum_0^\infty \alpha_i\, \phi_i(t).$$

The α_i are random variables (their definition as limits in mean square of approximating sums is justified in Section 1 of the Appendix to Chapter I). They satisfy

$$\begin{aligned}\mathcal{E}(\alpha_i\alpha_j) &= \int_0^T\!\!\int \mathcal{E}\{x(s)\, x(t)\}\, \phi_i(s)\, \phi_j(t)\, ds\, dt \\ &= \int_0^T\!\!\int \gamma(s, t)\, \phi_i(s)\, \phi_j(t)\, ds\, dt = \mu_i \delta_i^j.\end{aligned}$$

Thus they are "orthogonal." Moreover,

$$\mathcal{E}(x(t)\alpha_i) = \int_0^T \gamma(s, t)\, \phi_i(s)\, ds = \mu_i\, \phi_i(t),$$

from which it follows that

$$\mathcal{E}\left[\left\{x(t) - \sum_0^N \alpha_i\, \phi_i(t)\right\}^2\right] = \gamma(t, t) - \sum_0^N \mu_i\, \phi_i^2(t),$$

† See Section 10.

which, as we know, converges uniformly to zero. Thus we have

$$x(t) = \sum_0^\infty \alpha_i \, \phi_i(t)$$

in the sense of mean-square convergence of approximating sums. These results continue to hold true, for example, for the case where $[0, T]$ is replaced by $(-\infty, \infty)$ if $\gamma(s, t)$ is square integrable over the plane, the integrals over $[0, T]$ being replaced by integrals over $(-\infty, \infty)$. The double series for $\gamma(s, t)$ will now converge only in mean square so that

$$\lim_{N\to\infty} \iint_{-\infty}^{\infty} \left[\gamma(s, t) - \sum_0^N \mu_i \, \phi_i(s) \, \phi_i(t)\right]^2 ds \, dt = 0,$$

though if $\gamma(s, t)$ is continuous the convergence will be uniform on any finite interval.

We have here obtained a diagonalization of the covariance matrix (the α_i are the new variables) but one of limited usefulness, for it requires a complete knowledge of $\gamma(s, t)$. For an extension of this domain of ideas we refer the reader to Parzen (1961).

In Section 2 we enunciate and prove the basic theorems relating to the Fourier analysis of $x(t)$. In section 3 we discuss the relationship of this with the corresponding Fourier analysis for discrete time series. Linear filters are discussed in Section 4 and some special models are given in Section 5. Nonlinear filters and some forms of departure from stationarity which still permit the use of spectral methods are dealt with in Sections 6–11 together with some further spectral theories.

2. THE SPECTRAL THEOREMS FOR CONTINUOUS-TIME STATIONARY PROCESSES

We shall first prove the

Theorem 1. If $\Gamma(s, t) = \Gamma(t - s)$ *is the covariance matrix of a second-order stationary vector process, then*

$$\Gamma(t) = \int_{-\infty}^{\infty} e^{it\lambda} F(d\lambda) = \int_0^\infty \{\cos t\lambda C(d\lambda) + \sin t\lambda Q(d\lambda)\}, \tag{2.1}$$

where $F(\lambda)$ *is a matrix whose increments,* $F(\lambda_1) - F(\lambda_2)$, $\lambda_1 \geq \lambda_2$, *are Hermitian non negative. The function* $F(\lambda)$ *is uniquely defined if we require in addition that* (i) $\lim_{\lambda\to-\infty} F(\lambda) = 0$, (ii) $F(\lambda)$ *is continuous from the right. It is called the spectral distribution matrix. The matrices* $C(\lambda)$ *and* $Q(\lambda)$ *are* *real and, respectively, symmetric and skew symmetric and $C(\lambda_1) - C(\lambda_2)$ is nonnegative, $\lambda_1 \geq \lambda_2$.*

The matrix $F(\lambda)$ is called the spectral distribution matrix while $C(\lambda)$ and $Q(\lambda)$ are called, respectively, the co- and quadrature spectral distribution matrices. The right-hand term of (2.1) is, of course, a matrix of integrals with respect to the complex-valued Lebesgue-Stieltjes measures induced by the elements $F_{jk}(\lambda)$ of $F(\lambda)$.

In the proof of Theorem 1 we make use of the classical "uniqueness and continuity theorems for characteristic functions" (e.g., see Cramér, 1946), which asserts that a sequence of probability distributions converges to a proper probability distribution at all points of continuity of the latter if and only if the corresponding sequence of characteristic functions converges, pointwise, to a function continuous at the origin, which is then the characteristic function of the limit distribution, and that the characteristic function uniquely determines its distribution function subject to the convention that, say, the latter be continuous from the right. We shall use these theorems in cases where all the distribution functions increase to the same finite positive number which is not necessarily unity, but clearly the theorems quoted above still apply.

To prove this theorem we first form the scalar process

$$x_\alpha(t) = \alpha^* x(t)$$

where α is a fixed vector of p complex constants. We put

$$\gamma_\alpha(t) = \mathcal{E}\big(x_\alpha(s)\, \overline{x_\alpha(s+t)}\big) = \alpha^* \Gamma(t)\alpha.$$

Then we have the basic

Lemma 1. *The function $\gamma_\alpha(t)$ is nonnegative definite in the sense that, for any n, $t_1, \ldots, t_n$ and complex constants $\beta_1, \ldots, \beta_n$*

$$\sum_{j=1}^{n} \sum_{k=1}^{n} \bar{\beta}_j \beta_k \gamma_\alpha(t_j - t_k) \geq 0. \tag{2.2}$$

This follows from the simple evaluation

$$\sum_{j=1}^{n} \sum_{k=1}^{n} \bar{\beta}_j \beta_k \gamma_\alpha(t_j - t_k) = \sum_{j=1}^{n} \sum_{k=1}^{n} \bar{\beta}_j \beta_k\, \mathcal{E}\big(x_\alpha(t_k)\, \overline{x_\alpha(t_j)}\big)$$

$$= \mathcal{E}\left\{\left|\sum_j \beta_j\, x_\alpha(t_j)\right|^2\right\} \geq 0.$$

For the remainder of the proof of Theorem 1 let us first put

$$\gamma_\alpha^{(T)}(t) = \left(1 - \frac{|t|}{T}\right)\gamma_\alpha(t), \qquad |t| \leq T,$$
$$= 0 \qquad |t| \geq T.$$
$$f_\alpha^{(T)}(\lambda) = \int_{-\infty}^{\infty} e^{-it\lambda}\gamma_\alpha^{(T)}(t)\, dt.$$

Since

$$f_\alpha^{(T)}(\lambda) = \int_{-T}^{T} e^{-it\lambda}\left(1 - \frac{|t|}{T}\right)\gamma_\alpha(t)\, dt = \frac{1}{T}\int_0^T\int_0^T e^{-i(s-t)}\gamma_\alpha(s-t)\, ds\, dt,$$

we see that $f_\alpha^{(T)}(\lambda) \geq 0$, for the right-hand term may be approximated arbitrarily closely by a sum of the form (2.2) (with $\beta_j = \exp(it_j\lambda)\delta t$, the t_j being equidistant). We now form the Cesaro mean of $f_\alpha^{(T)}(\lambda)$, which is

$$\frac{1}{2\pi}\int_{-M}^{M}\left(1 - \frac{|\lambda|}{M}\right) f_\alpha^{(T)}(\lambda)e^{it\lambda}\, d\lambda \tag{2.3}$$

and which converges, as M increases, to $\gamma_\alpha^{(T)}(t)$. (See Section 2 of the Mathematical Appendix). Since (2.3) is a characteristic function (the factor multiplying $\exp it\lambda$ in the integrand being positive) we see that $\gamma_\alpha^{(T)}(t)$, being the limit of a convergent sequence of characteristic functions, and continuous for $t = 0$, is also a characteristic function. Now allowing T to increase we see that $\gamma_\alpha(t)$ is also the limit of a sequence of characteristic functions and since it also is continuous for $t = 0$ we have established that it also is a characteristic function, i.e.,

$$\gamma_\alpha(t) = \int_{-\infty}^{\infty} e^{it\lambda} F_\alpha(d\lambda).$$

The uniqueness of $F_\alpha(\lambda)$, under the conditions (i) and (ii) in Theorem 1 then follows from the uniqueness theorem for characteristic functions.

Now choose α to have unity in the jth place and zero elsewhere. Then we obtain

$$\gamma_j(t) = \int_{-\infty}^{\infty} e^{it\lambda} F_j(d\lambda).$$

If we take

$$x_{\alpha(1)}(t) = x_j(t) + x_k(t),\ x_{\alpha(2)}(t) = x_j(t) + ix_k(t),$$

then, since

$$\tfrac{1}{2}[\gamma_{\alpha(1)}(t) - \gamma_j(t) - \gamma_k(t)] + \frac{i}{2}[\gamma_{\alpha(2)}(t) - \gamma_j(t) - \gamma_k(t)] = \gamma_{j,k}(t),$$

we obtain

$$\gamma_{j,k}(t) = \int_{-\infty}^{\infty} e^{it\lambda} F_{jk}(d\lambda),$$

where

$$F_{jk}(\lambda) = \tfrac{1}{2}[F_{\alpha(1)}(\lambda) - F_j(\lambda) - F_k(\lambda)] + \frac{i}{2}[F_{\alpha(2)}(\lambda) - F_j(\lambda) - F_k(\lambda)].$$

We call $F(\lambda)$ the matrix with entries $F_{jk}(\lambda)$. Since $\gamma_{jk}(t) = \gamma_{kj}(-t)$, it is evident that $F(\lambda) = F^*(\lambda)$, for the latter matrix satisfies

$$\int_{-\infty}^{\infty} e^{it\lambda} F^*(d\lambda) = \overline{\int_{-\infty}^{\infty} e^{-it\lambda} F'(d\lambda)} = \Gamma'(-t) = \Gamma(t).$$

Since the proof of the theorem shows that $\alpha^* F(\lambda)\alpha = F_\alpha(\lambda)$, it follows that $F(\lambda)$ is Hermitian and has nonnegative definite increments. Since

$$\Gamma(-t) = \int_{-\infty}^{\infty} e^{-it\lambda} F(d\lambda) = \Gamma'(t) = \int_{-\infty}^{\infty} e^{it\lambda} F'(d\lambda),$$

we see that $F'(-\lambda) = \Gamma(0) - F(\lambda)$ at all points of continuity, λ, or, equivalently,

$$F(\lambda_1) - F(\lambda_2) = F'(-\lambda_2-) - F'(-\lambda_1-),\ \lambda_1 \geq \lambda_2, \tag{2.4}$$

where by $F'(-\lambda_1-)$, for example, we mean $\overline{\lim}_{\lambda<-\lambda_1} F'(\lambda)$. We shall paraphrase (2.4) by writing $F(d\lambda) = F'(d(-\lambda))$, which is, of course, $\overline{F(d(-\lambda))}$.

We put, using $\mathfrak{R}$ for "real part of" and $\mathscr{I}$ for "imaginary part of,"

$$\begin{aligned} C_{jk}(d\lambda) &= 2\mathfrak{R}\{F_{jk}(d\lambda)\}, & \lambda > 0,\\ &= F_{jk}(d\lambda), & \lambda = 0,\\ Q_{jk}(d\lambda) &= -2\mathscr{I}\{F_{jk}(d\lambda)\}, & \lambda \geq 0. \end{aligned}$$

This defines two real-valued functions of bounded variation, $C_{jk}(\lambda)$ and $Q_{jk}(\lambda)$. We drop a subscript when they are equal, as in Chapter 1, Section 1. Then $Q_j(\lambda)$ is evidently null while $C_j(\lambda)$ has nonnegative increments. The relations $F(\lambda) = F^*(\lambda)$, $F(d\lambda) = F'(d(-\lambda))$ show that $C(d\lambda)$ is a symmetric matrix and may be defined so as to be an even function of λ while $Q(d\lambda)$ is a skew symmetric matrix and may be defined so as to be an odd function of λ. Thus

$$\begin{aligned} \Gamma(t) = {}& \frac{1}{2}\int_0^{\infty} \cos t\lambda\, C(d\lambda) + \frac{i}{2}\int_0^{\infty} \sin t\lambda\, C(d\lambda) - \frac{i}{2}\int_0^{\infty} \cos t\lambda\, Q(d\lambda)\\ &+ \frac{1}{2}\int_0^{\infty} \sin t\lambda\, Q(d\lambda) + \frac{1}{2}\int_0^{\infty} \cos t\lambda\, C(d\lambda) - \frac{i}{2}\int_0^{\infty} \sin t\lambda\, C(d\lambda)\\ &+ \frac{i}{2}\int_0^{\infty} \cos t\lambda\, Q(d\lambda) + \frac{1}{2}\int_0^{\infty} \sin t\lambda\, Q(d\lambda)\\ = {}& \int_0^{\infty} \cos t\lambda\, C(d\lambda) + \int_0^{\infty} \sin t\lambda\, Q(d\lambda). \end{aligned}$$

This completes the proof of Theorem 1.

The matrix $F(\lambda)$ decomposes, according to the Lebesgue decomposition, into three parts,

$$F(\lambda) = F^{(1)}(\lambda) + F^{(2)}(\lambda) + F^{(3)}(\lambda).$$

Here $F^{(1)}(\lambda)$ is absolutely continuous (with respect to Lebesgue measure on the line) so that

$$F^{(1)}(\lambda) = \int_{-\infty}^{\lambda} f(\lambda)\, d\lambda,$$

where $f(\lambda)$ *is a matrix, with entries* $f_{j,k}(\lambda)$, *called the spectral density matrix. Putting* $f(\lambda) = \frac{1}{2}\big(c(\lambda) - iq(\lambda)\big)$ *we call* $c(\lambda)$ *and* $q(\lambda)$ *the co- and quadrature spectral density matrices. The* $f_{j,k}(\lambda)$, *for* $j \neq k$, *are called the "cross spectral density functions."* The component $F^{(2)}(\lambda)$ increases only by jumps, at a finite or denumerable set of points in $(-\infty, \infty)$, which has no limit point (other than $\pm\infty$). The matrix $F^{(3)}(\lambda)$ consists of elements which are continuous, with derivatives which vanish almost everywhere, with respect to Lebesgue measure. It will be difficult to allot to them a physical meaning and we shall often assume that this third component is missing. Indeed, as we shall see below, insofar as this stationary model is an adequate approximation to the complexity which is reality, it is often (but perhaps not always) true that $F^{(1)}(\lambda)$ is the only component that can be expected to be present. We discuss this further below.

We must now explain the physical meaning of the result in Theorem 1. In order to do this let us first consider a somewhat artificial special case. Let us assume that $x_j(t) = x_j(t + 2k\pi)$, $j = 1, \ldots, p$, $k = 0, \pm 1, \ldots$; so that we are observing a periodic random function. For example, $x_j(t)$, for $-\pi < t \leq \pi$, might correspond to a measurement made around a parallel of latitude, at longitude t, which measurement we have merely continued on periodically. Then also $\gamma_{jk}(t) = \gamma_{jk}(t + 2k\pi)$. We shall write $x_j(\phi)$ and $\gamma_{jk}(\phi)$ for these two functions considered as functions defined on $-\pi < \phi \leq \pi$.

Theorem 1′

$$\Gamma(\phi) = \sum_{-\infty}^{\infty} \Delta(n) e^{in\phi} = \sum_{0}^{\infty} \{A(n) \cos n\phi + B(n) \sin n\phi\},$$

where the Fourier series converge absolutely.

Proof. Since $\Gamma(t)$ is a continuous periodic function of t we have

$$\Gamma(t) = \lim_{N \to \infty} \sum_{-N}^{N} \Delta(n)\left(1 - \frac{|n|}{N}\right) e^{int},$$

$$\Delta(n) = \frac{1}{2\pi} \int_{-\pi}^{\pi} \Gamma(t) e^{-int}\, dt$$

(see the Mathematical Appendix). Since $\Gamma(t)$ satisfies the conditions of Theorem 1 then (2.1) holds for it. Let $F_N(\lambda)$ increase only at the points n, $|n| \leq N$, and there by $\Delta(n)(1 - |n|/N)$. Then by the uniqueness and continuity theorems for characteristic functions $F_N(\lambda) \to F(\lambda)$. Thus the first formula in Theorem 1′ holds. Putting $\Delta(n) = \frac{1}{2}(A(n) - iB(n))$, $n \neq 0$, $\Delta(n) = A(n)$, $n = 0$ the second part follows.

We shall indicate typical elements of these matrices, by $\delta_{j,k}(n)$, $\alpha_{j,k}(n)$, $\beta_{j,k}(n)$. It is interesting to observe that this Fourier series converges absolutely, on account of Theorem 1. This result would not be true for an arbitrary continuous function but is a manifestation of the nonnegative definite properties of $\Gamma(\phi)$.

Theorem 2′

$$x_j(\phi) = \sum_{-\infty}^{\infty} \zeta_j(n) e^{-in\phi} = \sum_{0}^{\infty} \{\xi_j(n) \cos n\phi + \eta_j(n) \sin n\phi\}, \qquad j = 1, \ldots, p \tag{2.5}$$

where the series converges in mean square and

$$\zeta_j(n) = \frac{1}{2\pi} \int_{-\pi}^{\pi} x_j(\phi) e^{in\phi}\, d\phi = \tfrac{1}{2}(\xi_j(n) + i\eta_j(n)), \qquad n \neq 0, \tag{2.6}$$

$$= \xi_j(n), \qquad n = 0.$$

The covariances of the $\xi(n)$ and $\eta(n)$ are

$$\begin{cases} \mathcal{E}(\xi_j(m)\, \xi_k(n)) = \mathcal{E}(\eta_j(m)\, \eta_k(n)) = \delta_m^{\ n}\ \alpha_{j,k}(n) \\ \mathcal{E}(\xi_j(m)\, \eta_k(n)) = -\mathcal{E}(\eta_j(m)\, \xi_k(n)) = \delta_m^{\ n}\ \beta_{j,k}(n) \end{cases} \tag{2.7}$$

so that $\mathcal{E}(\zeta_j(m)\, \overline{\zeta_k(n)}) = \delta_m^{\ n}\, \delta_{j,k}(n)$.

Proof. We put†

$$\xi_j(n) = \frac{1}{\pi} \int_{-\pi}^{\pi} x_j(\phi) \cos n\phi\, d\phi, \qquad n = 1, 2, \ldots,$$

$$= \frac{1}{2\pi} \int_{-\pi}^{\pi} x_j(\phi)\, d\phi, \qquad n = 0.$$

$$\eta_j(n) = \frac{1}{\pi} \int_{-\pi}^{\pi} x_j(\phi) \sin n\phi\, d\phi, \qquad n = 1, 2, \ldots,$$

$$= 0, \qquad n = 0.$$

† See Section 1 of the Appendix to Chapter I for a discussion of the meaning of these integrals. The development which follows is of course just an exemplification of the results relating to continuous covariance functions on finite intervals which were mentioned in Section 1 of this chapter, though here we discuss the vector case.

The formulas (2.7) then follow; for example for $m, n \neq 0$

$$\mathcal{E}(\xi_j(m)\,\xi_k(n)) = \pi^{-2}\iint_{-\pi}^{\pi} \gamma_{jk}(\phi_2 - \phi_1) \cos m\phi_1 \cos n\phi_2 \, d\phi_1 \, d\phi_2$$

$$= \pi^{-2}\int_{-\pi}^{\pi} \gamma_{jk}(\phi)\left\{\int_{-\pi}^{\pi} \cos m\psi \cos n(\phi + \psi)\, d\psi\right\} d\phi,$$

which gives the required result.

Thus, if we consider the partial sums

$$S_j(N) = \sum_{n=0}^{N} \{\xi_j(n) \cos n\phi + \eta_j(n) \sin n\phi\},$$

it becomes apparent that

$$\mathcal{E}[\{x_j(\phi) - S_j(N)\}^2] = \sum_{N+1}^{\infty} \alpha_j(n) \rightarrow 0,$$

the last result following from the fact that $\sum_0^\infty \alpha_j(n) = \gamma_j(0)$, the series converging absolutely. We have used the easily checked fact that

$$\mathcal{E}[\{\xi_j(n) \cos n\phi + \eta_j(n) \sin n\phi\} x_j(\phi)] = \alpha_j(n).$$

Thus the right-hand term of (2.5) is established and the expression for $x_j(\phi)$ in terms of the $\zeta(n)$ follows immediately.

The formula (2.5) thus expresses $x(\phi)$ as a Fourier series with coefficients which are random variables with an especially simple covariance structure. The transformation from the $x_j(\phi)$ to the $\xi_j(n)$ and $\eta_j(n)$, namely, (2.6), involves no special knowledge of the covariance function of the $x_j(\phi)$. We have thus reached the form of decomposition which was sought after in Section 1 of this chapter since we have represented $x(\phi)$ as a sum of orthogonal random variables by means of a transformation independent of the covariance function. If we had begun from the supposition that $x(t)$ was periodic with period $2T$, we should have obtained a representation of the form (2.7) with, however, $e^{-in\phi}$ replaced by $(\pi/T)\exp(-itn\pi/T)$. Allowing T to increase this suggests that in the general, nonperiodic case we shall have a representation (spectral representation) of the form

(2.8) $$x(t) = \int_{-\infty}^{\infty} e^{-i\lambda t} z(d\lambda)$$

obtained from (2.5) by identifying λ with $n\pi/T$ and $z(d\lambda)$ with $(\pi/T)\zeta(n)$. Since

$$\mathcal{E}(\zeta(m)\zeta(n)^*) = 0, \qquad m \neq n,$$

we expect $z(\lambda)$, when defined so as to be continuous from the right, to be a vector, random, process of orthogonal increments, i.e.,

$$\mathcal{E}\{(z(\lambda_1) - z(\lambda_2))(z(\lambda_3) - z(\lambda_4))^*\} = 0, \qquad \lambda_1 > \lambda_2 \geq \lambda_3 > \lambda_4,$$

and to have

$$\mathcal{E}\{(z(\lambda_1) - z(\lambda_2))(z(\lambda_1) - z(\lambda_2))^*\} = F(\lambda_1) - F(\lambda_2), \qquad \lambda_1 > \lambda_2.$$

We paraphrase this last relation by writing it as

$$\mathcal{E}\{z(d\lambda)\, z(d\lambda)^*\} = F(d\lambda).$$

These statements are justified by

Theorem 2. *If $x(t)$ is generated by a second-order stationary process with spectral distribution matrix $F(\lambda)$ then $x(t)$ satisfies (2.8) where the right-hand side is the limit in mean square of a sequence of approximating Riemann-Stieltjes sums and $z(\lambda)$ is a complex-valued process of orthogonal increments with $\mathcal{E}(z(\lambda)\, z(\lambda)^*) = F(\lambda)$. Defining $z(\lambda)$ to be continuous in mean square from the right it is then uniquely determined neglecting a set in Ω of probability measure zero.*

The proof of the theorem is given in Section 2 of the Appendix to this chapter. The nature of the (essentially) unique determination of $z(\lambda)$ by $x(t)$ is given in Theorem 3.

Theorem 3. *We have, for any two points $\lambda_2 > \lambda_1$ which are continuity points of $F(\lambda)$*

$$F(\lambda_2) - F(\lambda_1) = \lim_{T\to\infty} \frac{1}{2\pi} \int_{-T}^{T} \Gamma(t)\, \frac{e^{-i\lambda_2 t} - e^{-i\lambda_1 t}}{-it}\, dt$$

and

$$z(\lambda_2) - z(\lambda_1) = \underset{T\to\infty}{\text{l.i.m}}\, \frac{1}{2\pi} \int_{-T}^{T} x(t)\, \frac{e^{i\lambda_2 t} - e^{i\lambda_1 t}}{it}\, dt.$$

(By l.i.m we mean the limit in mean square of the sequence of random variables.)

The proof is again given in Section 2 of the Appendix to this chapter.

Theorem 4. *If $x(t)$ is as in Theorem 2 we have the alternative, real, representation*

$$x(t) = \int_0^\infty \{\cos \lambda t\, \xi(d\lambda) + \sin \lambda t\, \eta(d\lambda)\} \tag{2.9}$$

where $\xi(\lambda)$ and $\eta(\lambda)$ are real processes of orthogonal increments with also $\mathcal{E}\{(\xi(\lambda_1) - \xi(\lambda_2))(\eta(\lambda_3) - \eta(\lambda_4))\} = 0$ if the intervals $[\lambda_2, \lambda_1)$, $[\lambda_4, \lambda_3)$ do not overlap. The nonzero covariances are of the form

$$\begin{aligned}
\mathcal{E}\{\xi(d\lambda)\, \xi'(d\lambda)\} &= F(d\lambda) = C(d\lambda), \qquad \lambda = 0,\\
\mathcal{E}\{\xi(d\lambda)\, \xi'(d\lambda)\} &= \quad \mathcal{E}\{\eta(d\lambda)\, \eta'(d\lambda)\} = 2\mathcal{R}(F(d\lambda)) = C(d\lambda), \qquad \lambda \neq 0,\\
\mathcal{E}\{\xi(d\lambda)\, \eta'(d\lambda)\} &= -\mathcal{E}\{\eta(d\lambda)\, \xi'(d\lambda)\} = 2\mathcal{I}(F(d\lambda)) = Q(d\lambda),\\
& \qquad \mathcal{E}\{\eta(d\lambda)\, \eta'(d\lambda)\} = 0 = Q(d\lambda), \qquad \lambda = 0.
\end{aligned}$$

Proof. We define the two real processes, $\xi(\lambda)$, $\eta(\lambda)$, of orthogonal increments as follows. If $z(\lambda)$ jumps at $\lambda = 0$ we make $\xi(0)$ equal to this jump. Thereafter we have

$$\xi(\lambda) - \xi(0) = 2\mathcal{R}\big(z(\lambda) - z(0)\big), \qquad \lambda \geq 0.$$

We put $\eta(0) = 0$ and

$$\eta(\lambda) = 2\mathcal{I}\big(z(\lambda) - z(0)\big).$$

Since $x(t)$ is real we see from Theorem 3 that at each pair of continuity points λ_1, λ_2

$$\overline{z(\lambda_2) - z(\lambda_1)} = \underset{T\to\infty}{\mathrm{l.i.m}} \frac{1}{2\pi} \int_{-T}^{T} x(t) \frac{e^{-i\lambda_2 t} - e^{-i\lambda_1 t}}{-it}\, dt$$
$$= z(-\lambda_1) - z(-\lambda_2).$$

Then we may rewrite (2.8) as (2.9). (This is the analog of (2.5).)

We have

$$\mathcal{E}\{z(d\lambda)\, z(d\lambda)'\} = \mathcal{E}\{z(d\lambda)\, z\big(d(-\lambda)\big)^*\} = 0, \qquad \lambda \neq 0,$$

since $z(\lambda)$ has orthogonal increments. Thus

$$\tfrac{1}{4}[\mathcal{E}\{\xi(d\lambda)\, \xi(d\lambda)'\} - \mathcal{E}\{\eta(d\lambda)\, \eta(d\lambda)'\}]$$
$$+ \frac{i}{4}[\mathcal{E}\{\eta(d\lambda)\, \xi(d\lambda)'\} - \mathcal{E}\{\xi(d\lambda)\, \eta(d\lambda)'\}] = 0, \qquad \lambda \neq 0.$$

Since also

$$\mathcal{E}\{z(d\lambda)\, z(d\lambda)^*\} = F(d\lambda),$$

we see that $\xi(\lambda)$ and $\eta(\lambda)$ are processes of orthogonal increments with also $\mathcal{E}\big(\xi(d\lambda_1)\, \eta'(d\lambda_2)\big) = 0$, $\lambda_1 \neq \lambda_2$ and the only nonvanishing variances and covariances are as stated in Theorem 4. This completes the proof.

The interpretation of the co- and quadrature spectra† is clarified by the following considerations. Let us return for the moment to the circular example and formula (2.5) and consider what lead or lag of the component $\xi_j(n) \cos n\phi + \eta_j(n) \sin n\phi$ relative to $\xi_k(n) \cos n\phi + \eta_k(n) \sin n\phi$ $(j \neq k)$ will maximize the square of the correlation between the two components. If we alter the argument ϕ in the first expression to $\phi + \tau$ we obtain the covariance

$$\alpha_{jk}(n) \cos n\tau - \beta_{jk}(n) \sin n\tau = \cos\big(n\tau + \theta_{jk}(n)\big)\{\alpha_{jk}^{\,2}(n) + \beta_{jk}^{\,2}(n)\}^{1/2},$$
$$\theta_{jk}(n) = \arctan\left\{\frac{\beta_{jk}(n)}{\alpha_{jk}(n)}\right\}.$$

† Properly the term "spectrum" means the set of points in $(-\infty, \infty)$ at which $F(\lambda)$ increases but it is also commonly used to mean the actual function itself and we use this customary terminology.

The square of the correlation is evidently maximized for $\tau = -n^{-1}\theta_{jk}(n)$ the maximized correlation being† $\sigma_{jk}(n)$, where

$$\sigma_{jk}^{2}(n) = \frac{\alpha_{jk}^{2}(n) + \beta_{jk}^{2}(n)}{\delta_j(n)\,\delta_k(n)} = \frac{|\delta_{jk}(n)|^2}{\delta_j(n)\,\delta_k(n)}.$$

The quantity $\sigma_{jk}(n)$ is called the coherence (or sometimes coherency) and is an intrinsic measure of the strength of association between the two components at "wavenumber" n. The quantity $\theta_{jk}(n)$ describes the lead or lag necessary to bring the two components into most perfect agreement in a mean square sense, averaging over all realizations. We shall call it the "phase".

We now define analogous quantities in the general case. In the situation where $F_j(\lambda)$ and $F_k(\lambda)$ are absolutely continuous we naturally put

$$\sigma_{jk}^{2}(\lambda) = \frac{|f_{jk}(\lambda)|^2}{f_j(\lambda)\,f_k(\lambda)}; \qquad \theta_{jk}(\lambda) = \arctan\frac{q_{jk}(\lambda)}{c_{jk}(\lambda)}. \tag{2.10}$$

In general we need merely to find a distribution function with respect to which $F_j(\lambda)$, $F_k(\lambda)$ and $F_{jk}(\lambda)$ are all absolutely continuous and define $\sigma_{jk}^2(\lambda)$ in terms of the Radon-Nikodym derivations with respect to this. The distribution function $F_j(\lambda) + F_k(\lambda)$ will evidently suffice for this purpose.‡ Thus we may form the functions

$$\tilde{f}_{jk}(\lambda) = \frac{dF_{jk}(\lambda)}{d(F_j(\lambda) + F_k(\lambda))}, \qquad \tilde{f}_j(\lambda) = \frac{dF_j(\lambda)}{d(F_j(\lambda) + F_k(\lambda))},$$

and in terms of these define $\sigma_{jk}^2(\lambda)$, $\theta_{jk}(\lambda)$. The distribution function $F_j + F_k$ is a natural one to use as any set of λ values which has zero measure with respect to it is of no interest since it contains no spectral mass for either of the two series being related. The quantities $\sigma_{jk}^2(\lambda)$ and $\theta_{jk}(\lambda)$ can now be interpreted in terms of the approximation to $x(t)$ by a periodic process with a long period, T. We will further discuss their meaning later.

To summarize, we have represented $x(t)$ in the form

$$x_j(t) = \int_0^\infty \{\cos\lambda t\cdot\xi_j(d\lambda) + \sin\lambda t\cdot\eta_j(d\lambda)\},$$

where the only nonzero covariances which subsist are of the form

$$\begin{aligned}\mathcal{E}\{\xi_j(d\lambda)\,\xi_k(d\lambda)\} &= \mathcal{E}\{\eta_j(d\lambda)\,\eta_k(d\lambda)\} = C_{jk}(d\lambda),\\ \mathcal{E}\{\xi_j(d\lambda)\,\eta_k(d\lambda)\} &= -\mathcal{E}\{\eta_j(d\lambda)\,\xi_k(d\lambda)\} = Q_{jk}(d\lambda),\end{aligned}$$

† A more common notation is $\rho_{jk}(n)$ but we need ρ for another purpose in this book.

‡ For a more complete discussion see Section 2 of the Mathematical Appendix. The absolutely continuous case is so much the more important that we do not dwell upon the other case here.

whence we obtain

$$\gamma_{jk}(t) = \mathcal{E}\big(x_j(s)\, x_k(s+t)\big) = \int_0^\infty \{\cos \lambda t\, C_{jk}(d\lambda) + \sin \lambda t\, Q_{jk}(d\lambda)\}.$$

The meaning of C_{jk} and Q_{jk} can best be understood in terms of the coherence and phase which in the absolutely continuous case are given by

$$\sigma_{jk}^2(\lambda) = \frac{c_{jk}^2(\lambda) + q_{jk}^2(\lambda)}{c_j(\lambda)\, c_k(\lambda)}, \qquad \theta_{jk}(\lambda) = \arctan \frac{q_{jk}(\lambda)}{c_{jk}(\lambda)},$$

the first describing the strength of association in the sense of the maximum correlation to be achieved by rephasing one of the two $x_j(t)$ while the former describes the phase change required to achieve this.

Thus we have represented $x_j(t)$ as a "sum" (integral) of oscillating components whose phase and amplitude are determined by the random variables $\xi_j(d\lambda)$, $\eta_j(d\lambda)$ and have described the simple covariance structure of these random variables.

3. SAMPLING A CONTINUOUS-TIME PROCESS. DISCRETE TIME PROCESSES

We must now discuss the important problem that arises when a continuous time process is observed only at a set of points which constitute a sample from all of those possible, that is, of all points in the real line. Needless to say most time functions will be studied in this way, if only because of the importance of digital computers. Of course, it is possible to have a series which is, by its very nature, one in discrete time, for example, a series of experiments, conducted once a day. We shall consider this situation also, but shall commence from the case of a sampled continuous-time function because that seems the most important situation and because it motivates the result for the discrete time case. There are two kinds of problem that need to be considered. One is to say what can be known about the spectrum of the full process from the sample. A closely related problem is that of determining how best to estimate the value of $x(t)$ at some nonsampled time point t. We discuss the second question in Chapter III, where we deal with problems of prediction and interpolation, and confine ourselves to the first problem here.

The simplest situation is that where $x(t)$ is observed periodically at the points $x(\Delta n)$. We can at best know only $\Gamma(\Delta n)$, $n = 0, \pm 1, \ldots$. The question whether we can know even this, will be discussed later when we study ergodic theory but certainly, insofar as only second-order quantities are used, at most $\Gamma(\Delta n)$ can be known. Then

$$\Gamma(\Delta n) = \int_{-\infty}^{\infty} e^{i\Delta n\lambda}\, F(d\lambda)$$
$$= \sum_{-\infty}^{\infty} \int_{-\pi/\Delta+2\pi j/\Delta}^{\pi/\Delta+2\pi j/\Delta} e^{in\Delta\lambda}\, F(d\lambda)$$
$$= \int_{-\pi/\Delta}^{\pi/\Delta} e^{in\Delta\lambda}\, F^{(\Delta)}(d\lambda),$$

where

(3.1) $$F^{(\Delta)}(\lambda) = \sum_{-\infty}^{\infty}\left\{F\left(\lambda + \frac{2\pi j}{\Delta}\right) - F\left(\frac{\pi(2j-1)}{\Delta}\right)\right\},$$

where the second term in each summand on the right is adjoined purely to give each $F_{jk}^{(\Delta)}(\lambda)$ finite total variation. In case $F(\lambda)$ is absolutely continuous we have

$$\Gamma(n\Delta) = \int_{-\pi/\Delta}^{\pi/\Delta} e^{in\Delta\lambda} f^{(\Delta)}(\lambda)\, d\lambda.$$

(3.2) $$f^{(\Delta)}(\lambda) = \sum_{-\infty}^{\infty} f\left(\lambda + \frac{2\pi j}{\Delta}\right),$$

so that, for $\lambda \in [0, \pi/\Delta]$,

(3.3) $$c^{(\Delta)}(\lambda) = c(\lambda) + \sum_{1}^{\infty}\left\{c\left(\frac{2\pi j}{\Delta} + \lambda\right) + c\left(\frac{2\pi j}{\Delta} - \lambda\right)\right\},$$

(3.4) $$q^{(\Delta)}(\lambda) = q(\lambda) + \sum_{1}^{\infty}\left\{q\left(\frac{2\pi j}{\Delta} + \lambda\right) - q\left(\frac{2\pi j}{\Delta} - \lambda\right)\right\},$$

The relations (3.1), (3.2), (3.3), and (3.4) are usually described by saying that the frequencies λ, $2\pi j/\Delta \pm \lambda$, $\lambda \in [0, \pi/\Delta]$, are "aliased," the frequency λ being the "principal alias." This terminology is due to J. W. Tukey who took it over from the terminology used in connection with fractionally replicated experimental designs where an entirely analogous problem arises. The relation (3.3) can be described geometrically by considering the graph of $c_{jk}(\lambda)$, which is then folded, concertina fashion, at the points $2\pi j/\Delta$, $j = 1, 2, \ldots$. Then the values of $c_{jk}(\lambda)$ at all points which are aliased with a certain principal alias, λ_0, fall on top of each other and $c_{jk}^{(\Delta)}(\lambda_0)$ is obtained by adding the $c_{jk}(\lambda)$ at these aliased points. There is a similar geometric description of (3.4). It is evident that all we can know from the $\Gamma(n\Delta)$ is $F^{(\Delta)}$, so that two different F giving rise to the same $F^{(\Delta)}$ are indistinguishable. In the a.c. case, since $f_{jk}(\lambda)$ is zero at infinity, provided Δ is taken small enough, aliasing effects will be small also. This is, of course, intuitively obvious. In interpreting estimated spectra of discrete time series aliasing effects must be kept in mind. Thus in interpreting a peak in a spectrum observed from a discrete record one must bear in mind the possibility that the peak in an underlying continuous time series is at an aliased frequency; for example, if the series observed is sea level at midday, the observed spectrum $f^{(\Delta)}(\lambda)$ will have a marked peak at low frequencies due to tidal effects, for the tides may have a period of about 12 hr 25 min, which, if our time unit is 1 day and $\Delta = 1$, produces a peak falling at an angular frequency, $48\pi/(12.416) = 4\pi - 0.041$, which will thus alias with the frequency 0.041.

If we consider the process $x^{(\Delta)}(n) = x(\Delta n)$ we see from Theorem 2 that

$$x^{(\Delta)}(n) = \int_{-\pi/\Delta}^{\pi/\Delta} e^{-i\lambda\Delta n}\, z^{(\Delta)}(d\lambda),$$

wherein

$$z^{(\Delta)}(\lambda) = \sum_{-\infty}^{\infty}\left\{z\left(\lambda + \frac{2\pi j}{\Delta}\right) - z\left(\frac{(2j-1)\pi}{\Delta}\right)\right\},$$

so that

$$\mathcal{E}\{z^{(\Delta)}(\lambda)(z^{(\Delta)}(\lambda))^*\} = F^{(\Delta)}(\lambda).$$

As mentioned in Section 1, one might begin from a process $x(n)$ observed at intervals which for convenience we now take to be unity.

Theorem 1″. *If $\Gamma(m, n) = \Gamma(n - m)$ is the covariance function of a second-order stationary vector process in discrete time then*

$$\Gamma(n) = \int_{-\pi}^{\pi} e^{in\lambda}\, F(d\lambda) = \int_{0}^{\pi} \{\cos n\lambda\, C(d\lambda) + \sin n\lambda\, Q(d\lambda)\},$$

where $F(\lambda)$, $C(\lambda)$, and $Q(\lambda)$ are as described in Theorem 1, *save that now $F(-\pi)$ is null.*

The proof of this theorem is effectively the same as that of Theorem 1. Thus we replace $x(n)$ with $x_\alpha(n) = \alpha^*\, x(n)$, α being an arbitrary vector of complex numbers, we show that $\gamma_\alpha(n) = \alpha^*\, \Gamma(n)\alpha$ is a nonnegative definite sequence, that is,

$$\sum_j \sum_k \bar{\beta}_j \beta_k \gamma_\alpha(n_j - n_k) \geq 0$$

for any set of complex numbers β_j and any N integers n_j, and then form

$$F_\alpha^{(N)}(\lambda) = \frac{1}{2\pi} \sum_{-N}^{N} \gamma_\alpha(n) \frac{e^{in\pi} - e^{-in\lambda}}{in} \left(1 - \frac{|n|}{N}\right),$$

taking $\gamma_\alpha(0)(\pi + \lambda)$ as the term for $n = 0$. We then show that $F_\alpha^{(N)}(\lambda)$ is nondecreasing and that $F_\alpha^{(N)}(-\pi) = 0$, $F_\alpha^{(N)}(\pi) = \gamma_\alpha(0)$ and observe that

$$\int_{-\pi}^{\pi} e^{in\lambda} F_\alpha^{(N)}(d\lambda) = \gamma_\alpha(n)\left(1 - \frac{|n|}{N}\right), \qquad |n| \le N,$$
$$= 0 \qquad |n| > N.$$

The right-hand side converges to $\gamma_\alpha(n)$ for each n, as $N \to \infty$. Since we are concerned with a sequence of distribution functions,† $F_\alpha^{(N)}(\lambda)$, on a finite interval and the exp $in\lambda$ are functions which are dense (in the sense of uniform convergence) in the space of all continuous functions on $(-\pi, \pi]$ then $F_\alpha^{(N)}(\lambda)$ converges at each continuity point to a distribution function $F_\alpha(\lambda)$ which has $\gamma_\alpha(n)$ for its Fourier coefficients. (See Billingsley, 1968, theorems 1.3 and 2.1.) $F_\alpha(\lambda)$ is uniquely defined if we require it to be continuous from the right. The remainder of the proof does not differ from that of Theorem 1 and, in particular, $C(\lambda)$ and $Q(\lambda)$ are defined in the same fashion.

Theorem 2″. *If $x(n)$ is generated by a second-order stationary vector process in discrete time then*

$$x(n) = \int_{-\pi}^{\pi} e^{-in\lambda} z(d\lambda) = \int_{0}^{\pi} \{\cos t\lambda\, \xi(d\lambda) + \sin t\lambda\, \eta(d\lambda)\},$$

where $z(\lambda)$, $\xi(\lambda)$, $\eta(\lambda)$ are vector processes of orthogonal increments on $(-\pi, \pi]$ with the same covariance properties as for the corresponding processes in Theorems 2 and 4.

The proof of this theorem is the same as that for Theorem 2 and is given in Section 1 of the Appendix to this chapter.

Theorem 3″. *At any point $\lambda_2 > \lambda_1$ of continuity of $F(\lambda)$ we have*

$$F(\lambda_2) - F(\lambda_1) = \lim_{N\to\infty} \frac{1}{2\pi} \sum_{-N}^{N}{}' \Gamma(n) \frac{e^{-in\lambda_2} - e^{-in\lambda_1}}{-in}$$

$$z(\lambda_2) - z(\lambda_1) = \operatorname{l.i.m}_{N\to\infty} \frac{1}{2\pi} \sum_{-N}^{N}{}' x(n) \frac{e^{in\lambda_2} - e^{in\lambda_1}}{in},$$

where the symbol $\sum'$ indicates that for $n = 0$ we take the summand to be $(\lambda_2 - \lambda_1)\Gamma(0)$ in the first formula and $(\lambda_2 - \lambda_1)x(0)$ in the second.

† The distribution functions all have total variation $\gamma_\alpha(0)$ which need not be unity, but this fact makes no difference. Here we are once again using the uniqueness and continuity theorems for characteristic functions (see Billingsley, 1968, p. 51), but now for distribution functions defined on a finite interval $(-\pi, \pi)$ so that the characteristic function is determined in its entirety by its value at the integers.

We have placed the proof in Section 3 of the Appendix to this chapter. Of course l.i.m has the same meaning here as in Theorem 3.

The physical interpretation of these theorems need not be dwelt on further here, since it is already contained in the discussion we have given for the continuous-time case, the only difference being the restriction of the spectrum to the range $[-\pi, \pi]$.

There are other modes of sampling $x(t)$ than that which we have so far discussed and for completeness we shall say something about these here, though from the point of view of applications they seem less important. Perhaps the most likely is that called by Shapiro and Silverman (1960) "jittered periodic sampling." Thus we consider the case in which we observe at points $n + u_n$, where u_n are identically and independently distributed (i.i.d.) random variables, the relevant case being that in which u_n has a variance that is small compared with unity. The situation held in mind is that in which jitter in some recording device results in observations being made at points slightly different from those intended. Now put

$$y(n) = x(n + u_n)$$

and we easily see that

$$\begin{aligned}\tilde{\Gamma}(n) &= \mathcal{E}\big(y(m)\, y'(m + n)\big)\\ &= \int_{-\infty}^{\infty} \Gamma(n + u)\, P(du),\end{aligned}$$

where P is the probability distribution of $u_{m+n} - u_n$. Thus $y(n)$ is a stationary discrete sequence. The last integral is

$$\int_{-\infty}^{\infty}\int_{-\infty}^{\infty} e^{i(n+u)\lambda}\, F(d\lambda)\, P(du) = \int_{-\infty}^{\infty} e^{in\lambda}\, |\phi(\lambda)|^2\, F(d\lambda),$$

where $\phi(\lambda)$ is the characteristic function of the u_n. This is therefore

$$\int_{-\pi}^{\pi} e^{in\lambda} \tilde{F}(d\lambda)$$

$$\tilde{F}(\lambda) = \sum_{-\infty}^{\infty} [|\phi(\lambda + 2\pi j)|^2\, F(\lambda + 2\pi j) - |\phi((2j - 1)\pi)|^2\, F((2j - 1)\pi)].$$

It is $\tilde{F}$, at best, which is all we can know from $y(n)$. If $F(\lambda)$ does not increase outside $[-\pi, \pi]$, then evidently we can obtain $F(\lambda)$ from $\tilde{F}(\lambda)$ if we know P and therefore ϕ. Otherwise the observed spectrum will need to be interpreted with care.

Shapiro and Silverman (1960) also consider the situation in which we observe at points t_n so that $(t_n - t_{n-1})$ are i.i.d. (nonnegative) random variables (i.e., the t_n form the realization of a renewal process) with characteristic function which we again call $\phi(\lambda)$. We now have, putting $y(n) = x(t_n)$,

$$\tilde{\Gamma}(n) = \mathcal{E}\{y(m)\, y'(m+n)\} = \int_{-\infty}^{\infty} \Gamma(t) P^{(n)}(dt),$$

where $P^{(n)}$ is the distribution function of $t_n - t_0$. In the same way as before we obtain

$$\tilde{\Gamma}(n) = \int_{-\infty}^{\infty} \phi^n(\lambda)\, F(d\lambda), \qquad n \geq 0, \tag{3.5}$$

so that once more $y(n)$ is a stationary sequence.

We now ask whether it is possible that the sequence $\tilde{\Gamma}(n)$, which is all that can be known so far as second-order quantities are concerned, determines $F(\lambda)$ uniquely, subject to the conventions concerning its value at $-\infty$ and its right continuity. In other words, is it possible to avoid aliasing by sampling in the way described, with a suitable $\phi(\lambda)$. An answer is given by Theorem 5 due to Shapiro and Silverman (1960), the proof of which we omit.

Theorem 5. *A sufficient condition that the spectral distribution function, $F(\lambda)$, should be determined uniquely by* (3.5) *is the condition that for each pair (λ_1, λ_2); $\lambda_1 \neq \lambda_2$; $\lambda_1, \lambda_2 \in (-\infty, \infty)$, $\phi(\lambda_1) \neq \phi(\lambda_2)$ (i.e., $\phi(\lambda)$ should be univalent on $(-\infty, \infty)$).*

An important case is that in which the t_n are generated by a Poisson process with rate of occurrence of events μ. Then $\phi(\lambda) = (1 + i\lambda/\mu)^{-1}$. Now the sufficiency condition in Theorem 5 is evidently satisfied. The fact that the result of the theorem now holds is easily established in this Poisson case† for, taking $\mu = 1$ to make the notation simple, we see that the set, A, of complex linear combinations of the functions $(1 \pm i\lambda)^{-n}$, $n = 0, 1, \ldots,$ form not only a linear space over the complex field but also an algebra in the sense that the product of two such complex, linear combinations is again a complex, linear combination of the $(1 \pm i\lambda)^{-n}$. Indeed $(1 + i\lambda)^{-1}(1 - i\lambda)^{-1} = \frac{1}{2}\{(1 + i\lambda)^{-1} + (1 - i\lambda)^{-1}\}$ from which the result follows. Thus A is an algebra of continuous functions, all vanishing at infinity, closed under complex conjugation, and having the property that there is some element of A taking different values at any prescribed pair of points $\lambda_1 \neq \lambda_2$. Then (Yosida, 1965, p. 10) A is dense, in the sense of uniform convergence, in the space $C(-\infty, \infty)$ of all continuous functions vanishing at infinity. Thus, if the $\tilde{\Gamma}(n)$ did not uniquely determine $F(d\lambda)$ so that two such functions satisfied (3.5), we should have, for the matrix valued function, M, say, which was the

† This result is not needed in the sequel so that the reader may well skip the proof.

difference of these two

$$\int_{-\infty}^{\infty} \phi^n(\lambda)\, M(d\lambda) = \int_{-\infty}^{\infty} \overline{\phi^n(\lambda)}\, M(d\lambda) \equiv 0, \qquad n \geq 0$$

and it follows from what has just been said that

$$\int_{-\infty}^{\infty} a(\lambda)\, M(d\lambda) = 0, \qquad a \in C(-\infty, \infty).$$

However this is impossible unless $M(d\lambda) \equiv 0$ (Yosida, 1965, p. 119) so that uniqueness is established.

The problem of actually determining F in terms of the $\tilde{\gamma}(n)$ will not be discussed here. It is not clear how important this result is in practice. From the point of view of a *designed* sampling it would appear to be somewhat troublesome and one might prefer to use a periodic sampling pattern with a very small sampling interval. If $F(\lambda)$ is a.c. this will effectively avoid the aliasing problem as formulas (3.3) and (3.4) show (since $c(\lambda)$ and $q(\lambda)$ will be very small for λ large).

Another form of sampling which could possibly arise is "periodic non-uniform sampling" (Freeman, 1965, p. 77) where the sample points are of the form $t_j + n\Delta, j = 1, \ldots, N, n = 0, \pm 1, \ldots$, and where $0 \leq t_j < \Delta$. This form of sampling arises, for example, with some types of market, which are open at m irregular arrangement of "days" over a trading period (a week, a month, a year).† We may now know $\Gamma(n\Delta + t_j - t_k); j, k = 1, \ldots, N$. We need only consider the set S of those pairs (j, k) for which the corresponding differences $(t_j - t_k)$ do *not* differ from each other by an integral multiple of Δ. Any pair (t_n, t_0) which has a $t_n - t_0$ differing from one of those in S by $k\Delta$ evidently provides no new information. Let us index the pairs (j, k) in S by an index, m, running from 1 to M and call p_m the value of the mth of the $t_j - t_k$. It is easy to see that $M \geq N$ and, considering the scalar a.c. case for simplicity,

$$\gamma(n\Delta + p_m) = \int_{-\pi/\Delta}^{\pi/\Delta} e^{in\Delta\lambda} \sum_{u=-\infty}^{\infty} \exp\left[ip_m\left(\lambda + \frac{2\pi u}{\Delta}\right)\right] f\left(\lambda + \frac{2\pi u}{\Delta}\right) d\lambda,$$
$$m = 1, \ldots, M.$$

To indicate the nature of the aliasing involved, now let us assume that $f(\lambda)$ is null outside of the interval $[-M\pi/\Delta, M\pi/\Delta]$. Then we may observe

$$\sum_{[\frac{1}{2}(M+1)]}^{[M/2]} \exp\left[ip_m\left(\lambda + \frac{2\pi u}{\Delta}\right)\right] f\left(\lambda + \frac{2\pi u}{\Delta}\right), \qquad m = 1, \ldots, M,$$
$$0 \leq \lambda \leq \frac{\pi}{\Delta},$$

† An example is a futures market for wool in Australia.

where $[x]$ is the largest integer not greater than x. These lead to M equations for the M quantities $f(\lambda + 2\pi u/\Delta)$, $u = -[\frac{1}{2}(M+1)], \ldots, [\frac{1}{2}M]$ with determinant having in row m, column u the quantity $\exp\{ip_m(\lambda + 2\pi u/\Delta)\}$. This determinant is easily seen to be

$$\prod_{m=1}^{M} e^{ip_m\lambda} \prod_{l>k\geq 1}^{M} \{e^{i2\pi p_k/\Delta} - e^{i2\pi p_l/\Delta}\},$$

which is not equal to zero by assumption. Thus we are in the same position as if we had observed at points (Δ/M) apart, though we have observed at points averaging only Δ/N apart. If the t_i are very irregularly spaced, M may be expected to be large relative to N and this sampling avoids aliasing to a much greater degree than does strictly periodic sampling. This provides an intuitive explanation for the phenomenon discovered in relation to sampling according to a Poisson process.

4. LINEAR FILTERS

Before commencing the study of linear filters we wish to introduce some notions connected with the definition of infinite integrals. These notions are discussed more fully in the Mathematical Appendix. Thus in the scalar case we are concerned with expressions (Fourier transforms)

$$\int_{-\infty}^{\infty} \alpha(t)e^{it\lambda}\,dt. \tag{4.1}$$

If $|\alpha(t)|$ is integrable this may be taken to be a Lebesgue integral. However, a more relevant case from our point of view is that (not overlapping completely the one just described) in which we define (4.1) as the limit, in mean square as $T \to \infty$ with weighting $F(d\lambda)$ [or, for short, as the limit in mean square (F)] of the integrals, assumed to exist for all $0 \leq T < \infty$,

$$\int_{-T}^{T} \alpha(t)e^{it\lambda}\,dt.$$

Thus we consider cases in which the function of λ, indicated by (4.1), satisfies

$$\lim_{T\to\infty} \int_{-\infty}^{\infty} \left| \int_{-\infty}^{\infty} \alpha(t)e^{it\lambda}\,dt - \int_{-T}^{T} \alpha(t)e^{it\lambda}\,dt \right|^2 F(d\lambda) = 0.$$

The necessary and sufficient condition for this to be so is that

$$\lim_{S,T\to\infty} \int_{-\infty}^{\infty} \left| \int_{-T}^{T} \alpha(t)e^{it\lambda}\,dt - \int_{-S}^{S} \alpha(t)e^{it\lambda}\,dt \right|^2 F(d\lambda) = 0.$$

A similar definition may be adopted for integrals (Fourier-Stieltjes transforms) of the type

$$\int_{-\infty}^{\infty} e^{it\lambda}\, m(dt),$$

where m is a function of bounded variation on every finite subinterval of $(-\infty, \infty)$. In connection with vector processes we proceed as follows. In the first place we introduce the symbolism

$$\int_{-\infty}^{\infty} X(\lambda)\, F(d\lambda)\, X^*(\lambda)$$

to mean the matrix of integrals of which a representative element is

$$\sum_{k,l} \int_{-\infty}^{\infty} x_{ik}(\lambda)\, \overline{x_{jl}(\lambda)}\, F_{kl}(d\lambda).$$

We are concerned only with the case in which $F(\lambda)$ has Hermitian non-negative definite increments and say that $X(\lambda)$ is square integrable (F) [or belongs to $L_2(F)$] if

$$\operatorname{tr}\left\{\int_{-\infty}^{\infty} X(\lambda)\, F(d\lambda)\, X^*(\lambda)\right\} < \infty.$$

Now we define

$$\int_{-\infty}^{\infty} A(t)e^{it\lambda}\, dt, \qquad \int_{-\infty}^{\infty} e^{it\lambda}\, M(dt) \tag{4.2}$$

as before. Thus, if we put

$$X_{S,T}(\lambda) = \int_{-T}^{T} A(t)e^{it\lambda}\, dt - \int_{-S}^{S} A(t)e^{it\lambda}\, dt$$

and

$$\lim_{S,T\to\infty} \operatorname{tr}\left\{\int_{-\infty}^{\infty} X_{S,T}(\lambda)\, F(d\lambda)\, X^*_{S,T}(\lambda)\right\} = 0,$$

we say that the first member of (4.2) is the limit in mean square (F) of the expressions

$$\int_{-T}^{T} A(t)e^{it\lambda}\, dt.$$

The second member of (4.2) is defined in the same way.

Throughout this section we use these definitions of Fourier transforms (or Fourier-Stieltjes transforms) of functions $\alpha(t)$, $A(t)$, $m(t)$, and $M(t)$ and consider only the functions that have Fourier (Fourier-Stieltjes) transforms in this sense. *Of course, $x(t)$ throughout the section is second-order stationary.*

We now wish to discuss certain (special) linear operators which produce a new stationary vector random process from the original one. These operators are called linear filters. We proceed by first dealing with the most important cases of filters and then give a more general discussion, which we shall continue in Section 11 of this chapter. Many specific examples of filters are given in this chapter and in Chapter III.

(i) Integral Operators

We consider first the operation which replaces $x(t)$ by

$$y(t) = \int_{-\infty}^{\infty} A(s)\, x(t-s)\, ds,$$

where $A(s)$ is such that

$$h(\lambda) = \int_{-\infty}^{\infty} A(t) e^{it\lambda}\, dt$$

exists in the mean-square sense defined above. More generally, we may consider

$$y(t) = \int_{-\infty}^{\infty} M(ds)\, x(t-s),$$

where by this we mean the vector of integrals

$$\sum_k \int_{-\infty}^{\infty} x_k(t-s)\, m_{jk}(ds)$$

and $M(s)$ is a matrix of functions of bounded variation in every finite interval and for which

$$h(\lambda) = \int_{-\infty}^{\infty} e^{is\lambda}\, M(ds)$$

exists in the sense of mean-square convergence with respect to F defined above.

Theorem 6. *The random process $y(t)$ is well defined as a limit in mean square of integrals over finite intervals, as the intervals increase to $(-\infty, \infty)$, when and only when $h(\lambda)$ exists as a limit in mean square (F). When this is so $y(t)$ is stationary with covariance function $\Gamma_y(t)$ satisfying*

$$\Gamma_y(t) = \int_{-\infty}^{\infty} e^{it\lambda}\, h(\lambda)\, F(d\lambda)\, h(\lambda)^*.$$

Proof. A typical element of $y(t)$ is of the form

$$\sum_k \int_{-\infty}^{\infty} x_k(t-s)\, m_{jk}(ds),$$

so that we must consider the mean square convergence as, $a, b \to \infty$ of

$$\int_{-a}^{b} x_k(t-s)\, m_{jk}(ds).$$

From Theorem 3 in Chapter I we know that a necessary and sufficient condition for this is the existence of

$$\iint_{-\infty}^{\infty} \gamma_k(u-v)\, m_{jk}(du)\, m_{jk}(dv);$$

but

$$\iint_{-a}^{b} \gamma_k(u-v)\, m_{jk}(du)\, m_{jk}(dv) = \iint_{-a}^{b} \int_{-\infty}^{\infty} e^{i(u-v)\lambda}\, F_k(d\lambda)\, m_{jk}(du)\, m_{jk}(dv)$$

$$= \int_{-\infty}^{\infty} \left| \int_{-a}^{b} e^{iu\lambda}\, m_{jk}(du) \right|^2 F_k(d\lambda),$$

so that the necessary and sufficient condition for mean-square convergence becomes the condition that $h_{jk}(\lambda)$ exist as a limit in mean square (F_k). Considering all j, k this is equivalent to the existence of the second member of (4.2) as a limit in mean square (F). The covariance function is given by

$$\Gamma_y(t-s) = \mathcal{E}\big(y(s)\, y(t)'\big) = \iint_{-\infty}^{\infty} M(du)\, \mathcal{E}\big(x(s-u)x'(t-v)\big)\, M'(dv)$$

$$= \iint_{-\infty}^{\infty} M(du) \int_{-\infty}^{\infty} e^{i(t-s+u-v)\lambda}\, F(d\lambda)\, M'(dv)$$

$$= \int_{-\infty}^{\infty} e^{i(t-s)\lambda}\, h(\lambda)\, F(d\lambda)\, h(\lambda)^*,$$

so that the $y(t)$ sequence is stationary with the spectrum as stated and the proof is completed.

Thus the action of the filter is very simply described in spectral terms. This is especially so in the scalar a.c. case when $f(\lambda)$ is merely replaced by $|h(\lambda)|^2 f(\lambda)$. In the vector a.c. case $f(\lambda)$ is replaced by $h(\lambda)\, f(\lambda)\, h(\lambda)^*$. The case where $M(t)$ increases only by jumps covers the situation where

$$y(t) = \sum_{-\infty}^{\infty} A_k x(t-t_k),$$

since A_k may be taken as the jump in $M(t)$ at the typical point t_k. In the case where $M(t)$ is a.c. so that $M(ds)$ is replaced by $A(s)\, ds$ *the function* $A(s)$

is sometimes called the matrix impulse response function of the filter since if an impulse function (i.e., a δ-function at the origin; see Section 3 of the Mathematical Appendix) is inserted in place of $x(t)$ in $y(t)$ then the output is just $A(t)$. *The function $h(\lambda)$ is called the matrix frequency response function of the filter.* The meaning of the term "frequency response function" is made more apparent by considering the spectral representation of $y(t)$. We have, in the general case,†

$$y(t) = \int_{-\infty}^{\infty} M(ds) \int_{-\infty}^{\infty} e^{-i(t-s)\lambda} z(d\lambda) = \int_{-\infty}^{\infty} e^{-it\lambda} \left\{ \int_{-\infty}^{\infty} e^{is\lambda} M(ds) \right\} z(d\lambda)$$
$$= \int_{-\infty}^{\infty} e^{-it\lambda} h(\lambda)\, z(d\lambda).$$

Thus the process of orthogonal increments associated with $y(t)$ is obtained from that associated with $x(t)$ by mere multiplication of the component at (each) frequency λ by the frequency response function $h(\lambda)$.

(ii) Differential Operators

We next consider operators of the form

$$\sum_0^p A_k \frac{d^k}{dt^k},$$

where A_k is a matrix. We assume that

$$\operatorname{tr}\left\{ \int_{-\infty}^{\infty} \lambda^{2p} F(d\lambda) \right\} < \infty. \tag{4.3}$$

We now put

$$h(\lambda) = \sum_0^p A_k(-i\lambda)^k. \tag{4.4}$$

Theorem 7. *The necessary and sufficient condition that $x(t)$ be p times mean square differentiable is the condition* (4.3). *In this case*

$$y(t) = \sum_0^p A_k \frac{d^k}{dt^k} x(t)$$

is stationary with covariance function

$$\Gamma_y(t) = \int_{-\infty}^{\infty} e^{it\lambda} h(\lambda) F(d\lambda) h(\lambda)^*,$$

where $h(\lambda)$ is given by (4.4).

† We leave a fuller justification of these rearrangements to the Appendix to this chapter. (See the proof of Theorem 9.)

We know from Theorem 2, Chapter I that the necessary and sufficient condition for mean-square differentiability is, in the stationary scalar case, the existence of the limit

$$\lim_{\delta_1,\delta_2\to 0} \frac{\gamma(\delta_2-\delta_1)-\gamma(-\delta_1)-\gamma(\delta_2)+\gamma(0)}{\delta_1\delta_2}$$

$$= \lim_{\delta_1,\delta_2\to 0} \int_{-\infty}^{\infty} (\delta_1\delta_2)^{-1}\{e^{i(\delta_2-\delta_1)\lambda} - e^{-i\delta_1\lambda} - e^{i\delta_2\lambda} + 1\}\, F(d\lambda)$$

$$= \lim_{\delta_1,\delta_2\to 0} \int_{-\infty}^{\infty} \delta_1^{-1}(e^{-i\delta_1\lambda}-1)\,\delta_2^{-1}(e^{i\delta_2\lambda}-1)\, F(d\lambda).$$

The integrand has modulus

$$4\left|\frac{\sin\tfrac{1}{2}\delta_1\lambda\,\sin\tfrac{1}{2}\delta_2\lambda}{\delta_1\delta_2}\right| < \lambda^2,$$

so that by dominated convergence the condition is sufficient. On the other hand, let us take $\delta_1 = \delta_2$ when we obtain

$$\lim_{\delta\to 0}\int_{-\infty}^{\infty} 4\delta^{-2}\sin^2\tfrac{1}{2}\delta\lambda\, F(d\lambda) = \lim_{\delta\to 0}\int_{-\infty}^{\infty} \frac{\sin^2\frac{1}{2}\delta\lambda}{(\frac{1}{2}\delta\lambda)^2}\lambda^2\, F(d\lambda)$$

$$> \lim_{\delta\to 0}\int_{-a}^{a} \frac{\sin^2\frac{1}{2}\delta\lambda}{(\frac{1}{2}\delta\lambda)^2}\lambda^2\, F(d\lambda) = \int_{-a}^{a}\lambda^2\, F(d\lambda),$$

again by dominated convergence. Thus if the condition of the theorem is not satisfied the limit does not exist and the necessary condition for mean-square differentiability of $x(t)$ is violated. Using induction the condition (4.3) for the existence of the pth derivative in the stationary scalar case is seen to be necessary and sufficient. The extension to the vector case is trivial.

Now if $x(t)$ is p times mean-square differentiable then we have

$$\mathcal{E}(y(s)\,y(t)') = \sum_0^p A(k)\,\mathcal{E}\left\{\left(\frac{d^k}{ds^k}x(s)\right)\left(\frac{d^l}{dt^l}x(t)\right)'\right\} A(l)'$$

$$= \sum\sum_0^p A(k)\frac{\partial^{k+l}}{\partial s^k\,\partial t^l}\Gamma(t-s)\,A(l)'$$

by Theorem 2, Chapter I. This is

$$\int_{-\infty}^{\infty} e^{i(t-s)\lambda}\sum\sum_0^p A(k)(i\lambda)^l(-i\lambda)^k\, F(d\lambda)A(l)' = \int_{-\infty}^{\infty} e^{i(t-s)\lambda}\, h(\lambda)\, F(d\lambda)\, h(\lambda)^*.$$

The reason for the insertion of the factor $(-i)^k$ in the term $(-i)^k\lambda^k$ is found when the spectral representation is considered for now we have†

$$y(t) = \int_{-\infty}^{\infty} \left(\sum_0^p A_k \frac{\partial^k}{\partial t^k}\right) e^{-it\lambda} z(d\lambda) = \int_{-\infty}^{\infty} e^{-it\lambda}\, h(\lambda)\, z(d\lambda).$$

(iii) The typical filter in the discrete time situation is of the form

$$y(n) = \sum_{-\infty}^{\infty} A(m)\, x(n-m) \tag{4.5}$$

with frequency response function

$$h(\lambda) = \sum_{-\infty}^{\infty} A(m) e^{im\lambda},$$

which describes the action of the filter via‡

$$y(n) = \int_{-\pi}^{\pi} e^{-in\lambda}\, h(\lambda)\, z(d\lambda),$$

$$\Gamma_y(n) = \int_{-\pi}^{\pi} e^{in\lambda}\, h(\lambda)\, F(d\lambda)\, h(\lambda)^*.$$

Theorem 8. *The necessary and sufficient condition that* (4.5) *exist as a limit in mean square of partial sums is the condition* $\operatorname{tr}\left(\Gamma_y(0)\right) < \infty$. *Then* $y(n)$ *is stationary with covariance function and spectral representation as shown.*

We omit the proof which follows the same lines as Theorem 6 and 7 (but is simpler).

In case $F(d\lambda) = cI_p$ where c is a constant, the condition of the theorem becomes

$$\sum_{-\infty}^{\infty} A(n)\, A(n)^* < \infty,$$

which is equivalent to

$$\sum_{-\infty}^{\infty} \|A(n)\|^2 < \infty,$$

where (see the Mathematical Appendix) $\|A\|$ is the norm of the matrix A, which we may take to be the positive square root of the greatest eigenvalue of AA^*.

† Again we give a fuller justification of the spectral representation in the Appendix to this chapter (see the proof of Theorem 9.)

‡ See the previous footnote.

(iv) *It is convenient to indicate the operation of replacing $x(t)$ by $y(t)$ by means of a filter by the notation $y(t) = Ax(t)$.* Thus A may stand, in particular, for any of the filtering operations defined above.

It is fairly obvious that if we replace $x(t)$ by $u(t) = Ax(t)$ and then $u(t)$ by $y(t) = Bu(t)$, where A and B are both linear filters, then

$$h^{(BA)}(\lambda) = h^{(B)}(\lambda)\, h^{(A)}(\lambda),$$

where the superscipts indicate to which filter the response function corresponds. In particular, for scalar processes, linear filters commute. In any case, the calculation of a response function of a product of filters can be broken down into the calculation of the individual response functions. This rather trivial observation is important in practice.

Not all linear filters are of the type described above. Of course, this statement begs the question "what is a linear filter?" One answer is "an operation which replaces†

$$x(t) = \int_{-\infty}^{\infty} e^{-it\lambda} F(d\lambda)$$

by

$$y(t) = \int_{-\infty}^{\infty} e^{-it\lambda} h(\lambda)\, F(d\lambda),$$

where $h(\lambda)$ satisfies

$$\int_{-\infty}^{\infty} h(\lambda)\, F(d\lambda)\, h(\lambda)^* < \infty\text{''}.$$

We propose a different definition below, but let us accept this for the moment. It certainly includes everything previously defined as a filter. We give examples in our next chapter of filters, according to this definition, which are not of the type (i), (ii), or (iii) above.

In relation to this definition the following question arises. Let $x(t)$ and $y(t)$ be jointly stationary so that the vector process with components all those of $x(t)$ plus all those of $y(t)$ is stationary. Let us call $F_x(\lambda)$ the matrix spectral distribution function for $x(t)$, $F_y(\lambda)$ that for $y(t)$ and $F_{xy}(\lambda)$ the matrix of cross spectra, so that F_{xy} has as (j, k)th element the cross spectrum between $x_j(t)$ and $y_k(t)$. Now we have the following theorem (Rozanov, 1967).

Theorem 9. *Let $x(t)$ and $y(t)$ be jointly stationary. Then the necessary and sufficient condition that $y(t)$ be got from $x(t)$ by a filter, with response $h(\lambda)$ is the condition that*

$$F_y(\lambda) = h(\lambda)F_x(\lambda)h(\lambda)^*,$$
$$F_{xy}(\lambda) = F_x(\lambda)h(\lambda)^*.$$

† We deal with the continuous time case. Of course the definition applies to the discrete time case also, restricting t to be an integer and the integrations to the interval $[-\pi, \pi]$.

The proof is simple and we have put it in the Appendix to this chapter.

The definition of a filter which we have just given does not seem satisfactory for one would, in the first place, prefer a definition which enabled one, in principle at least, to recognize a filter from its action on the actual $x(t)$ process rather than through an examination of its spectrum. One then hopes to prove that a linear filter is completely described by its response function. In the second place it does not generalize to provide a definition of a nonlinear filter, for such a filter will not be described by a response function.† A more appropriate definition seems to be the following one. Consider an arbitrary random variable x which is a linear combination of $x_{j_1}(t_1), x_{j_2}(t_2), \ldots, x_{j_n}(t_n)$. By $U(t)x$ we mean the same linear combination of $x_{j_1}(t_1 + t), \ldots, x_{j_n}(t_n + t)$. We assume that the random variables Ax and $AU(t)x$ are well defined as limits (in mean square) of random variables of the form of x, that A be linear in the $x_j(t_j)$ and we require that $AU(t)x = U(t)Ax$ for all such x and all t. We choose to call such an operator a filter. We shall give a discussion of such operators in Section 11. (Section 11 is more advanced and is not needed in the remainder of the book.) We shall show that such an operator A does in fact define a filter in our previous sense and moreover, that it does generalize to the nonlinear case. A filter (linear or otherwise) is, according to this definition, a time invariant operator which produces an observed vector $y(t)$ from some (perhaps hidden) vector stationary process $x(t)$. However, through the remainder of the book (outside of Section 11 of this chapter) we shall work with the simpler definition used in Theorem 7.

We now consider the physical meaning of a filter. Of course, we have earlier described this, in one sense, when we said that it was to be a time invariant operator. However, now we wish to describe its action on the spectrum. We consider the scalar case, first. We put

$$h(\lambda) = |h(\lambda)|\, e^{i\theta(\lambda)}$$

and call $|h(\lambda)|$ the gain and $\theta(\lambda)$ the phase of the filter. So far as the spectrum is concerned in this scalar case, assuming absolute continuity, we have changed from $f_x(\lambda)$ to $|h(\lambda)|^2 f_x(\lambda) = f_y(\lambda)$. The action of the filter is to this extent described by the degree to which the component at frequency λ has been modified in amplitude. Frequencies for which the gain is relatively very small are greatly reduced in importance ("filtered out"). By a suitable choice of a filter we may thus accentuate such frequencies as we may wish to. If the gain is, substantially, nonzero only over an interval of frequencies then

† We shall discuss nonlinear filters in Section 7 of this chapter. We consider these only in relation to strictly stationary processes.

the filter is sometimes called a "band-pass" filter, that is, one which substantially passes only frequencies in a certain band or interval. According to the location of the interval(s) the terms "low pass," "medium pass," and "high pass" are also used.

If $y(t) = Ax(t)$ and $x(t)$ are both scalar then evidently the coherence between these two series is unity independently of the filter A. However, the cross spectrum is evidently $h(\lambda)\,F(d\lambda)$ so that the phase change at frequency λ is $\theta(\lambda)$.

The simplest case mathematically is that where $\theta(\lambda) = \lambda\tau$. This is evidently just the phase for the filter $x(t) \to x(t-\tau)$. If $\theta(\lambda)$ is smooth near λ_0, we may put

$$\theta(\lambda) \doteq \theta(\lambda_0) + (\lambda - \lambda_0)\theta'(\lambda_0) = \alpha + \lambda\theta'(\lambda_0),$$

and locally the effect of the filter, so far as phase is concerned, is approximately the resultant of a filter which has response exp $i\alpha$ and one which has response exp $i\lambda\theta'(\lambda_0)$. *The quantity* $\theta'(\lambda_0)$ *is sometimes called the group delay*. It can be thought of as the delay to which the component coming from a narrow band of frequencies around λ_0 has been subjected.

In the vector case we may put (Halmos, 1947, p. 138)

$$h(\lambda) = k(\lambda)p(\lambda),$$

where $p(\lambda)$ is Hermitian nonnegative definite and $k(\lambda)$ effects an isometric transformation of the p-dimensional vector space in which each $x(t)$ lies into the q-dimensional vector space in which $y(t) = Ax(t)$ lies. By saying that k is an isometry we mean that $k(\lambda)x$ has the same length as x for all x. The matrix $p(\lambda)$ is uniquely defined as the (nonnegative) square root of $h^*(\lambda)h(\lambda)$ while $k(\lambda)$ is uniquely defined on the range of $p(\lambda)$ by $k(\lambda)y = h(\lambda)x$, $y = p(\lambda)x$, but may be chosen arbitrarily on the orthogonal complement of this range (but so as to be an isometry). We are led to call $p(\lambda)$ the gain of the filter. In the absolutely continuous case the spectral density is got from

$$p(\lambda)\,f(\lambda)\,p^*(\lambda)$$

by means of the action of a filter with response function $k(\lambda)$ and since after an appropriate change of basis in the component spaces at frequency λ, for the y and x processes, $k(\lambda)$ reduces to a matrix whose elements are null off the diagonal commencing at the top left-hand corner and are either 0 or of modulus unity down that diagonal we see that the filter with response $k(\lambda)$ is one which acts only to alter the phase of the components. This justifies the use of the term "gain" for $p(\lambda)$.

5. SOME SPECIAL MODELS

We now discuss some special models and their associated spectra.

(i) If $x(n)$ is a sequence of uncorrelated random vectors, that is,

$$\mathcal{E}\big(x(n)\,x'(m)\big) = \Gamma(0)\,\delta_n{}^m, \tag{5.1}$$

then evidently $F(d\lambda) = (2\pi)^{-1}\Gamma(0)\,d\lambda$. The spectrum is then said to be uniform since all frequencies are equally important. The term "white noise" is also used, as explained earlier, by an analogy with white light, in which no subinterval of the visible spectrum predominates.

(ii) If

$$x(n) = \sum_{-\infty}^{\infty} A(j)\,\epsilon(n-j), \qquad \sum_{-\infty}^{\infty} \|A(j)\|^2 < \infty \tag{5.2}$$

where $\epsilon(n)$ is as in (i), then by the results of Section 4,

$$f_x(\lambda) = \frac{1}{2\pi}\left(\sum_{-\infty}^{\infty} A(j)e^{ij\lambda}\right)\Gamma(0)\left(\sum_{-\infty}^{\infty} A^*(j)e^{-ij\lambda}\right). \tag{5.3}$$

This process $x(n)$ will be well defined as a limit in mean square as the diagonal elements of $f_x(\lambda)$ are square integrable (with respect to $d\lambda$).

Alternatively, if we are given $x(n)$ with a.c. spectrum and spectral density function $f_x(\lambda)$ we can always find a Hermitian matrix valued function (nonnegative definite), $\sqrt{f(\lambda)}$, so that

$$f(\lambda) = \left(\sqrt{f(\lambda)}\right)\left(\sqrt{f(\lambda)}\right)^*.$$

Then each element of $\sqrt{f(\lambda)}$ is square integrable and thus can be expanded in a mean-square convergent (weighting $d\lambda$) Fourier series. Thus

$$\sqrt{f(\lambda)} = \frac{1}{\sqrt{2\pi}}\sum_{-\infty}^{\infty} A(j)e^{ij\lambda},$$

where

$$A(j) = \frac{1}{\sqrt{2\pi}}\int_{-\pi}^{\pi} \sqrt{f(\lambda)}e^{-ij\lambda}\,d\lambda.$$

It follows (see Section 5 in the Appendix to this chapter for a proof) that

$$x(n) = \sum_{-\infty}^{\infty} A(j)\,\epsilon(n-j), \qquad \sum_{-\infty}^{\infty} \|A(j)\|^2 < \infty, \tag{5.4}$$

where the $\epsilon(n)$ are as in (i), with $\Gamma(0) = I_p$. Thus the class of series of the form (5.4) is the class of stationary processes with absolutely continuous spectra.

As we shall see in Chapter III (see also example (iii) in this section), provided such an $x(n)$ is not perfectly (linearly) predictable from its past then the "moving average," (5.4), is one-sided, i.e.,

$$x(n) = \sum_0^\infty A(j)\,\epsilon(n-j). \tag{5.5}$$

In the case where both $x(n)$ and $\epsilon(n)$ are scalar, when we use α_j in place of $A(j)$ in (5.5), we may interpret (5.5) as in example (iv) of Chapter I, Section 3. The sequence $\epsilon(n-j)\,\alpha_0$, $\epsilon(n-j)\,\alpha_1$, $\epsilon(n-j)\,\alpha_2, \ldots$, is to be thought of as a wave-propagated in one direction through time and initiated at the time point $(n-j)$. The wave form is determined by the sequence α_j and $\epsilon(n-j)$ is a random variable which determines its amplitude (and sign). The observed value, $x(n)$, is a superposition of the effects of all waves propagated at times up to n. Occasionally the appearance of the sequence, $x(n)$, will be not unlike that of the sequence α_j (or $-\alpha_j$), since occasionally a very large $\epsilon(n-j)$ will be followed and preceded by much smaller values. In particular if the α_j constitute a weakly damped harmonic oscillation the appearance of $x(n)$ may show a regularity near to that of a sine curve, locally, but of course the phase and amplitude of the oscillation will change from stretch to stretch as a dominant oscillation propagated at one point of time is gradually damped to unimportance and another dominant oscillation takes over. In this situation, where the α_j resemble a damped sine curve, the quantity $\sum \alpha_j \exp(i\lambda)$ will naturally be relatively large in modulus near to the frequency of that oscillation and a large part of the variance of $x(n)$ will be due to frequencies near that one. This exemplifies the place of the spectrum in explaining the structure of the series.

(iii) We consider next the case in which the spectrum of a discrete time series is a.c. and

$$f(\lambda) = \frac{1}{2\pi}\sum_{-q}^{q} \Gamma(j)e^{-ij\lambda}, \qquad \Gamma(j) = \Gamma'(-j), \qquad \Gamma(q) \neq 0,$$

that is, $f(\lambda)$ is a trigonometric polynomial.

We deal first with the scalar case. We consider

$$w(\zeta) = \frac{1}{2\pi}\sum_{-q}^{q} \gamma(j)\zeta^{-j}.$$

This expression has $2q$ zeros, counting each with its appropriate multiplicity. Let ζ_k be a zero whose modulus is not unity. Then

$$\frac{1}{2\pi}\sum_{-q}^{q} \gamma(j)\bar{\zeta}_k^{\,j} = \frac{1}{2\pi}\sum_{-q}^{q} \gamma(j)\bar{\zeta}_k^{\,-j} = \overline{w(\zeta_k)} = 0.$$

so that $\bar{\zeta}_k^{-1}$ is also a zero. Let there be $r \le q$ such pairs of zeros taking each pair as often as its multiplicity. Then there are $2s = 2q - 2r$ zeros of unit modulus, which we call $\exp i\theta_j$, $j = 1, \ldots, 2s$. Then

$$f(\lambda) = \frac{\gamma(q)}{2\pi} e^{-iq\lambda} \prod_1^r (e^{i\lambda} - \zeta_k)(e^{i\lambda} - \bar{\zeta}_k^{-1}) \prod_1^{2s} (e^{i\lambda} - e^{i\theta_j})$$

$$= \prod_1^r (e^{i\lambda} - \zeta_k)(e^{-i\lambda} - \bar{\zeta}_k)\left\{\frac{\gamma(q)}{2\pi} \prod_1^r (-\bar{\zeta}_k)^{-1} e^{-is\lambda} \prod_1^{2s} (e^{i\lambda} - e^{i\theta_j})\right\},$$

from which it follows that the bracketed factor is real and nonnegative. Thus the θ_j must occur in pairs for this bracketed factor being nonnegative and zero at $\lambda = \theta_j$ (and infinitely differentiable) it must have a derivative which is zero at $\lambda = \theta_j$. Now let us number the θ_j so that one member of each pair occurs with a subscript between 1 and s (inclusive) and put $\zeta_{k+r} = \exp i\theta_k$, $k = 1, \ldots, s$. Then

$$f(\lambda) = \frac{\alpha(q)^2}{2\pi} \left| \prod_1^q (e^{i\lambda} - \zeta_k)\right|^2 = \frac{1}{2\pi}\left|\sum_0^q \alpha(j) e^{ij\lambda}\right|^2, \tag{5.6}$$

putting

$$\alpha(q)^2 = \frac{1}{2\pi} \gamma(q) \prod_1^r (-\bar{\zeta}_k)^{-1} \prod_1^s (-e^{i\theta_j}) > 0.$$

Since $f(\lambda)$ is an even function the complex ζ_k occur in conjugate pairs so that if $\bar{\zeta}_k$ along with ζ_k is given an index between 1 and q (inclusive) the $\alpha(j)$ will be real. We thus have proved the major part of the following theorem, due to Fejér and Riesz (see Achieser, 1956).

Theorem 10. *If*

$$f(\lambda) = \frac{1}{2\pi} \sum_{-q}^{q} \gamma(j) e^{-ij\lambda} \ge 0,$$

then it may be represented in the form (5.6). *If* $\gamma(j) = \gamma(-j)$ *so that* $f(\lambda)$ *is an even function of* λ *the* $\alpha(j)$ *may be taken to be real. A canonical representation of the form* (5.6) *may be obtained by choosing* $\alpha(0)$ *to be real and the* ζ_k *to lie on or outside of the unit circle. In this case the* $\alpha(j)$ *are necessarily real if* $f(\lambda)$ *is even.*

Only the last two statements need proof. We may evidently reverse the roles of $\bar{\zeta}_j$ and $\bar{\zeta}_j^{-1}$, if $|\zeta_j| \ne 1$ so that we may assume $|\zeta_j| \ge 1$. Choosing $\alpha(0)$ to be real merely eliminates a factor of modulus unity from $\sum \alpha(j) \exp(ij\lambda)$. The representation is evidently now unique. Evidently also $\bar{\zeta}_j$ must be chosen as a zero of $\sum \alpha(j) z^j$ along with $\bar{\zeta}_j$ which shows that the $\alpha(j)$ are necessarily real.

A more general case is that in which $f(\lambda)$ is a ratio of trigonometric polynomials, that is, "rational." Thus

$$f(\lambda) = \frac{1}{2\pi} \frac{p(e^{i\lambda})}{q(e^{i\lambda})}.$$

We may eliminate any factor common to the numerator and denominator and, agreeing that this has been done, it now must be true that

$$(5.7) \qquad p(e^{i\lambda}) \geq 0, \qquad q(e^{i\lambda}) > 0, \qquad \lambda \in [-\pi, \pi].$$

Thus both numerator and denominator may be factored in the way just described so that

$$f(\lambda) = \frac{1}{2\pi} |u(e^{i\lambda})|^2,$$

where u is a ratio of polynomials in $\exp i\lambda$. In particular we may choose this so that $u(z)$ is holomorphic within the unit circle and has no zeros there since we may choose the canonical factor for the numerator in this fashion and for the denominator may, in virtue of (5.7), choose the canonical factor to be nonzero in *or on* the unit circle.

Following Rozanov (1967), we shall now discuss the vector case. We begin by assuming that the determinant of $f(\lambda)$ is not identically zero. It is also easier to begin with the case where $f(\lambda)$ is composed of elements which are rational trigonometric expressions. Let us call $W(\zeta)$ the matrix function obtained by replacing $\exp i\lambda$ by ζ in the expression for $f(\lambda)$. Now it is easy to see that we may find a matrix function $H(\zeta)$, whose elements are rational functions of ζ, so that

$$W(\zeta) = H(\zeta)\, D(\zeta)\, \tilde{H}(\zeta),$$

where $D(\zeta)$ is diagonal and $\tilde{H}$ is obtained from H by transposing, conjugating coefficients, and replacing ζ by ζ^{-1}. Indeed the first column of $W(\zeta)$ may be made null below the first element by premultiplying by the matrix, $H_1^{-1}(\zeta)$, which has units in the main diagonal $-w_{11}(\zeta)^{-1}\, w_{1j}(\zeta)$ in the jth place $(j > 1)$ in the first column and zeros elsewhere.†

H_1^{-1} has unit determinant so that H_1 is of the same form. Evidently postmultiplication by $\tilde{H}_1^{-1}$ accomplishes the same task for the first row. The proof may now be completed by induction. The elements of $D(\zeta)$ are individually of the same nature as those discussed for the scalar case so that $D(\zeta)$ admits a factorization, $D = U\tilde{U}$ where U is diagonal with elements which are rational functions of ζ which are holomorphic within the unit circle. We may now find a scalar function $c(\zeta)$ so that $c(\zeta)\, H(\zeta)$ also has this

† The elements $w_{11}(\zeta)$ cannot be null as $f(\lambda)$ is positive definite a.e.

property (by choosing c as a product of all linear factors from the denominators of elements of H which are zero within the unit circle). Now $c(\zeta)\,\tilde{c}(\zeta)$ is again real and nonnegative on the unit circle and thus $c(\zeta)\,\tilde{c}(\zeta) = b(\zeta)\,\tilde{b}(\zeta)$, where $b(\zeta)$ is a polynomial in ζ which has no zeros within the unit circle. Now we have

$$\begin{aligned} W(\zeta) &= \{c(\zeta)\,H(\zeta)\,b(\zeta)^{-1}\,U(\zeta)\}\{\tilde{U}(\zeta)\,\tilde{b}(\zeta)^{-1}\,\tilde{H}(\zeta)\,\tilde{c}(\zeta)\} \\ &= A(\zeta)\,\tilde{A}(\zeta), \end{aligned}$$

where $A(\zeta)$ is holomorphic within the unit circle and is composed of rational functions of ζ.

Now let us specialize to the case where $f(\lambda)$ is composed of elements which are trigonometric polynomials,

$$f(\lambda) = \frac{1}{2\pi}\sum_{-q}^{q}\Gamma(j)e^{-ij\lambda}.$$

In the factorization $W(\zeta) = A(\zeta)\,\tilde{A}(\zeta)$ we may assume $A(0)$ nonsingular by agreeing to cancel any factor ζ in $c(\zeta)$ or $U(\zeta)$ with the factor ζ^{-1} in $\tilde{c}(\zeta)$, $\tilde{U}(\zeta)$. Now expanding the elements of $A(\zeta)$ in a power series convergent within the unit circle we see that if a coefficient matrix A_j for $j > q$ is not null then at $\zeta = 0$ $A(\zeta)\,\tilde{A}(\zeta)$ has a pole of higher order than q. Since this is manifestly impossible, we see that

$$\frac{1}{2\pi}\sum_{-q}^{q}\Gamma(j)e^{-ij\lambda} = \frac{1}{2\pi}\left\{\sum_{0}^{q} A_je^{ij\lambda}\right\}\left\{\sum_{0}^{q} A_je^{ij\lambda}\right\}^*.$$

We now show how we may obtain a canonical factorization. Let ζ_1 be a zero of the determinant of $A(\zeta)$ which lies within the unit circle. Now (Macduffee, 1956, p. 78) we may find unitary matrices U_1, V_1 so that $V_1A(\zeta_1)U_1$ is diagonal with zeros in, say, the first r_1 places down that diagonal and not elsewhere. We now introduce the diagonal matrix $\Delta_1(\zeta)$ which has

$$\frac{\bar{\zeta}_1\zeta - 1}{\zeta - \zeta_1} \tag{5.8}$$

in the first r_1 places on the main diagonal and unity elsewhere in that diagonal. On the unit circle (5.8) is of unit modulus so that Δ_1 is unitary on the unit circle. Then $B_1(\zeta) = A(\zeta)\,U_1\,\Delta_1(\zeta)$ is again a matrix whose elements are polynomials in ζ and

$$f(\lambda) = B_1(e^{i\lambda})\,B_1^*(e^{i\lambda}),$$

but B_1 differs from A in that the factor $(\zeta - \zeta_1)^{r_1}$ in the determinant has been replaced by $(\zeta - \bar{\zeta}_1^{-1})^{r_1}$. Proceeding in this way we may achieve a factorization of $f(\lambda)$ (into a product of polynomials of degree q) having determinant which is not zero within the unit circle. If $A(0)$ is prescribed this factorization is essentially unique for if $W(\zeta) = A(\zeta)\,\tilde{A}(\zeta) = B(\zeta)\,\tilde{B}(\zeta)$ were two such factorizations we would have $\{B^{-1}(\zeta)\,A(\zeta)\}\{\tilde{A}(\zeta)\,\tilde{B}^{-1}(\zeta)\} = I_p$ and $U_0(\zeta) = B^{-1}(\zeta)\,A(\zeta)$ is a rational function of ζ whose determinant is identically unity (since it is holomorphic within the unit circle, is unity at $\zeta = 0$ and is of unit modulus on the circle). Thus on the unit circle U_0 is unitary with unit determinant. But since $U_0(\zeta)$ is holomorphic within the circle then on the circle U_0 is composed only of nonnegative powers of $\exp(i\lambda)$. But $U_0^{-1} = U_0^*$ and U_0^* on the circle is composed only of nonpositive powers of $\exp(i\lambda)$. Thus U_0 is a constant unitary matrix. Thus $A(\zeta) = B(\zeta)U_0$. We have shown that, $A_0(\zeta)$ being a particular canonical factor (i.e., holomorphic and without zeros in the unit circle), we have for *any* factorization of $f(\lambda)$

$$f(\lambda) = \{A_0(e^{i\lambda})\,U(e^{i\lambda})\}\{A_0(e^{i\lambda})\,U(e^{i\lambda})\}^*,$$

where $U(\zeta)$ is unitary on the unit circle and holomorphic within it $\big(U$ would be made up of factors of the type $U_0\Delta_1^{-1}(\zeta)U_1^{-1}\cdots\big)$.

It is easily seen that the factorization is unique if in addition to requiring all zeros of $\det\,(\sum A_j\zeta^j)$ on or outside of the unit circle we require A_0 to be Hermitian positive definite. Thus we have established the following result:

Theorem 10′. *If*

$$f(\lambda) = \frac{1}{2\pi}\sum_{-q}^{q}\Gamma(j)e^{-ij\lambda} \geq 0$$

has a determinant not identically zero, we may represent $f(\lambda)$ in the form

$$\frac{1}{2\pi}\left\{\sum_0^q A_j e^{ij\lambda}\right\}\left\{\sum_0^q A_j e^{ij\lambda}\right\}^*,$$

where A_0 is Hermitian positive definite and all zeros of $\det \sum A_j z^j$ *lie on or outside of the unit circle. The representation is then unique and if $f(\lambda)$ is a spectral density matrix* $\big($*i.e.,* $f(\lambda) = \overline{f(-\lambda)}\big)$ *then the A_j are real. All other representations of $f(\lambda)$ as a product AA^*, with A having elements which are polynomials in* $\exp i\lambda$, *are of the form*

$$\frac{1}{2\pi}\left\{\sum_0^q A_j e^{ij\lambda}\right\}U(e^{i\lambda})\;U^*(e^{i\lambda})\left\{\sum_0^q A_j e^{ij\lambda}\right\},$$

where U is unitary for almost all λ and $U(z)$ is holomorphic within the unit circle.

Of course, Theorem 10′ includes Theorem 10 as a special case.

The case in which $\det\{f(\lambda)\} \equiv 0$ is easily reduced to that just considered, for let us say that the rank of $W(\zeta)$ is $r(\zeta) \leq r \leq p$ and that $r(\zeta_0) = r$. Then there is a certain principal minor of $W(\zeta_0)$ which is not zero, and, since this determinant is a polynomial in ζ and ζ^{-1}, it follows that this principal minor is zero only at a finite number of points and $W(\zeta)$ is of rank r at all points except these.

Then we can find a matrix P whose elements are polynomials in ζ and whose determinant is a constant so that $P(\zeta)W(\zeta)\tilde{P}(\zeta)$ has zeros outside of the first r rows and columns.† Then we need consider only the matrix in the first r rows and columns of

$$P(e^{i\lambda})f(\lambda)P^*(e^{i\lambda})$$

to which the previous considerations apply.

Just as in the last example, a given factorization leads to a representation of $x(n)$ in the form

$$x(n) = \sum_0^q A(j)\,\epsilon(n-j), \qquad \mathcal{E}\big(\epsilon(n)\,\epsilon(m')\big) = \delta_n^{\ m} G.$$

We have dealt with this problem of factorization in some detail here for the following reason. In Chapter III we shall establish the existence of factorizations of this type, but with $q = \infty$, under much more general conditions. These theorems are much more difficult to prove and some readers may wish to omit the proof. The applications of the theorems are to problems of prediction and smoothing. In practice, insofar as these techniques are used at all, we prescribe $f(\lambda)$ as of the form discussed here, since it will be estimated from numerical data. (The more general result can hardly be important on prior physical grounds, since $x(n)$ will be sampled from a continuous time process.) Thus the theorems of the present section will cover all important cases.

(iv) Let the stationary process $x(n)$ be generated by the relation (3.10) of Chapter I and there be no zeros of det $\big(\sum B(j)z^j\big)$ on the unit circle.‡ Then density matrix of $\sum B(j)\,x(n-j)$ is

$$\frac{1}{2\pi}\left(\sum_0^s A(k)e^{ik\lambda}\right) G \left(\sum_0^s A(k)e^{ik\lambda}\right)^*$$

by the filtering formula. Thus the spectral density of $x(n)$ is $(2\pi)^{-1}\,AGA^*$, where

$$A(\lambda) = \left(\sum_0^q B(j)e^{ij\lambda}\right)^{-1}\left(\sum_0^s A(k)e^{ik\lambda}\right)$$

by a further use of the filtering formula.

‡ Since there exists a pair of matrices, $P(\zeta)$, $Q(\zeta)$, with element polynomials in ζ and constant, nonzero determinants, so that $P(\zeta)\,W(\zeta)\,Q(\zeta)$ has zero elements in the last $(p-r)$ rows and columns, then $P(\zeta)\,W(\zeta)$ has zeros in the last $(p-r)$ rows; but then so has $P(\zeta)\,W(\zeta)\,\tilde{P}(\zeta)$. $W(\zeta)\,\tilde{P}(\zeta) = [P(\zeta^{-1})\,W(\zeta^{-1})]'$ and it has zeros in the last $(p-r)$ columns; $P(\zeta)\,W(\zeta)\,\tilde{P}(\zeta)$ has such zeros also. Thus the result is established.

† This condition could be relaxed, but we shall not discuss that here.

In the scalar case, and for $s = 0$, we see that

$$f(\lambda) = \frac{\sigma^2}{2\pi}\left|\sum_0^q \beta(j)e^{ij\lambda}\right|^{-2} \qquad \beta(0) = 1$$
$$= \frac{\sigma^2}{2\pi}\left(\left|\prod_1^q (e^{i\lambda} - \zeta_j)\right|^2\right)^{-1}$$

If ζ_u is nearly of unit modulus, say $\zeta_u = \rho_u \exp i\theta_u$, then when λ is near to θ_u, $f(\lambda)$ will be large. We showed, for the case where all ζ_u are less than 1 in modulus, that

$$\sum_0^q \beta(j)\, x(n-j) = \epsilon(n)$$

has a solution of the form

$$x(n) = \sum_0^\infty \lambda(j)\,\epsilon(n-j),$$

where the λ_j are of the form

$$\sum_u \sum_k n^k \rho_u{}^j (b'_{u,k} \cos \theta_u j + b''_{u,k} \sin \theta_u j).$$

If one particular ρ_u is near to unity in modulus and is much larger than the others, this will be near to

$$\lambda(j) = \rho_u{}^j (b'_u \cos \theta_u j + b''_u \sin \theta_u j)$$

and $x(n)$ will consist of a superposition of damped sinusoidal oscillations with frequency θ_u. It is not surprising therefore that $f(\lambda)$ has a peak at $\pm\theta_u$.

(v) The model

$$\sum_0^p B(j) \frac{d^j}{dt^j} x(t) = y(t), \tag{5.9}$$

where $y(t)$ has spectral density matrix $f_y(\lambda)$, leads to the spectral density matrix for $x(t)$, which is a stationary solution of (5.9),

$$\left\{\sum_0^p B(j)(-i\lambda)^j\right\}^{-1} f_y(\lambda) \left\{\sum_0^p B_j(j)(-i\lambda)^j\right\}^{*-1}$$

if we assume that

$$\det\left\{\sum_0^p B(j)z^j\right\}$$

has no zeros on the imaginary axis.

(vi) In the continuous time vector case where $x(t)$ has an absolutely continuous spectrum we may write

$$f(\lambda) = \{\sqrt{f(\lambda)}\}\{\sqrt{f(\lambda)}\}^*$$

where $\sqrt{f(\lambda)}$ is Hermitian nonnegative definite, and putting

$$A(t) = \frac{1}{\sqrt{2\pi}} \int_{-\infty}^{\infty} \sqrt{f(\lambda)} e^{-it\lambda}\, d\lambda,$$

we have

$$f(\lambda) = \frac{1}{2\pi} \left(\int_{-\infty}^{\infty} A(s) e^{is\lambda} \right) \left(\int_{-\infty}^{\infty} A(s) e^{is\lambda} \right)^{*}.$$

We shall discuss this result further in Section 5 of the Appendix to this chapter where we shall also show that

$$x(t) = \int_{-\infty}^{\infty} A(t-s)\, \xi(ds) \tag{5.10}$$

where $\xi(s)$ is a process of orthogonal increments having $\mathcal{E}(\xi(ds)\, \xi^*(ds)) = ds/2\pi$. Again we would hope for a one-sided representation, so that $A(s)$ may be taken as null for $s < 0$, for then we would, for example, in the scalar case, give (5.10) an intuitively acceptable interpretation as a linear superposition of waves $\alpha(t-s)\, \xi(ds)$, all having the same wave form, propagated at time s with amplitude and sign determined by the random variable $\xi(ds)$. We shall see that this is a very plausible model in Chapter III.

(vii) We have so far considered cases where the phenomenological model generating $x(t)$ is prescribed or the spectrum is prescribed. A third possibility is the prescription of the covariance sequence. Consider for example the case where

$$\Gamma(t) = e^{-A|t|}\Gamma$$

where the eigenvalues of A have positive real parts and Γ is a positive definite symmetric matrix. Then since $\Gamma(t) = \Gamma'(-t)$ we must have $A\Gamma = \Gamma A'$ (as can be seen by expanding the exponential in a power series and taking $|t|$ arbitrarily small). Consider the matrix

$$f(\lambda) = \frac{1}{2\pi} \int_{-\infty}^{\infty} e^{-A|t|}\Gamma e^{-i\lambda t}\, dt = \frac{1}{2\pi} \{(A + i\lambda I)^{-1} + (A - i\lambda I)^{-1}\}\Gamma.$$

(See the exercises at the end of this chapter.) This is

$$\frac{1}{2\pi}(A + i\lambda I)^{-1}\, 2A(A - i\lambda I)^{-1}\, \Gamma = \frac{1}{2\pi}(A + i\lambda I)^{-1}\, 2A\Gamma(A' - i\lambda I)^{-1}.$$

However, $2A\Gamma$ is symmetric and positive definite. Indeed if $P\Gamma P' = I$ then $PAP^{-1} = PAP^{-1}P\Gamma P' = P\Gamma A'P' = P'^{-1}A'P'$. Then the eigenvalues of PAP^{-1} are real (since it is symmetric), and since they are the eigenvalues of A they are real and positive. But $PAP^{-1} = P(A\Gamma)P'$ so that $A\Gamma$ is symmetric with positive eigenvalues, which proves what is required. Putting $2A\Gamma = G$ we have the factorization

$$f(\lambda) = \frac{1}{2\pi}(A + i\lambda I)^{-1} G(A + i\lambda I)^{*-1}. \tag{5.11}$$

Since

$$\frac{1}{2\pi}\int_{-\infty}^{\infty} (A + i\lambda)^{-1} e^{it\lambda}\, d\lambda = e^{-At}, \quad t \geq 0,$$

$$= 0, \qquad t < 0,$$

then

$$f(\lambda) = \frac{1}{2\pi}\left\{\int_0^{\infty} e^{-At} e^{-it\lambda}\, dt\right\} G \left\{\int_0^{\infty} e^{-At} e^{-it\lambda}\, dt\right\}^*.$$

Correspondingly we may write

$$x(t) = \int_t^{\infty} e^{A(t-s)}\, \xi(ds), \tag{5.12}$$

where $\mathcal{E}(\xi(ds)\, \xi(ds)^*) = G ds$ and $\xi(s)$ has orthogonal increments. Of course, (5.11) also leads to the model

$$x(t) + A\,\dot{x}(t) = \epsilon(t),$$

where $\epsilon(t)$ is white noise with $\mathcal{E}(\epsilon(s)\, \epsilon^*(t)) = (2\pi)^{-1} G\delta_s^{\,t}$. As we have seen this model has associated mathematical difficulties but (5.12) is clearly a valid reinterpretation of it.

(viii) A model of some importance in textile research is that where a yarn is postulated to be made up of overlaying fibers of length l, the left end points of the fibers lying at points t_j along the yarn, which are the points at which the events of a Poisson process, with parameter μ, occur. Thus the chance that j points occurs in an interval $(s, s + t)$ is

$$e^{-\mu t}\frac{(\mu t)^j}{j!}.$$

The number of yarn fibers overlaying a point t is then

$$x(t) = \int_0^{\infty} \alpha(l, t - s)\, \xi(ds),$$

where $\alpha(l, t)$ is unity for $0 \leq t \leq l$ and is otherwise zero and $\{\xi(s) - \xi(0)\}$ is the number of left end points in the interval $(0, s]$. Then $\xi(s)$ is a process of independent increments with var $(\xi(ds)) = \mu\, ds$. It is convenient in this case to make a mean correction and we then see that the spectral density function of the mean corrected quantity is

$$\frac{\mu}{2\pi}\left|\int_0^\infty \alpha(l, t - s)e^{is\lambda}\,ds\right|^2 = \frac{\sin^2 \frac{1}{2}l\lambda}{(\frac{1}{2}\lambda)^2}\frac{\mu}{2\pi},$$

which is concentrated at low frequencies to a degree dependent upon how large l is. This agrees with intuition, of course, since if l is large the yarn should be relatively even and thus low frequencies should predominate.

A somewhat more plausible model is obtained if the length of the fiber with end point at t_j is considered to be itself a random variable, l_j, independent for different j and independent also of the t_k. We call

$$P(x) = P\{0 \le l \le x\}$$

and may assume that there is no jump in $P(x)$ at the origin. Then

$$x(t) = \int_0^\infty \int_0^\infty \alpha(l, t - s)\,\xi(dl, ds),$$

where $\{\xi(l, s) - \xi(l, 0)\}$ is the number of left end points of fibers of lengths not greater than l lying in the interval $(0, s]$. We assume once more that the locations of the left end points are generated by a Poisson process. It is easily seen that $\xi(l_1, s) - \xi(l_2, s)$, $l_1 > l_2$, is generated by Poisson process with parameter

$$\mu \int_{l_2}^{l_1} P(dl).$$

Moreover, the processes corresponding to two intervals $[l_2, l_1)$, $[l_4, l_3)$ which do not overlap are independent. (For a discussion see Cox and Miller, 1965, p. 155.) This can be seen from the fact that, superimposing two independent processes with these parameters, the correct resulting process is obtained. Thus $x(t)$ is stationary with spectral density (after mean correction)

$$f(\lambda) = \frac{\mu}{2\pi}\int_0^\infty \left|\int_0^\infty \alpha(l, t - s)e^{is\lambda}\,ds\right|^2 P(dl)$$

$$= \frac{\mu}{2\pi}\int_0^\infty \frac{\sin^2 \frac{1}{2}l\lambda}{(\frac{1}{2}\lambda)^2} P(dl).$$

Since

$$\int_{-\infty}^\infty \frac{\sin^2 \theta}{\theta^2}\,d\theta = \pi,$$

the variance is

$$\mu\int_0^\infty l\,P(dl),$$

which depends only on μ and the mean of the distribution of l and is, in fact, the same as the mean of $x(t)$.

A model related to that discussed above is usually called a "random telegraph signal." Here once more we consider a sequence of points t_j generated by a Poisson process with parameter μ and a process $x(t)$ which takes only the values ± 1, switching values as a point t_j is passed.† Then it is easily seen that

$$\mathcal{E}\big(x(s)\,x(s+t)\big) = \exp -2\mu\,|t|.$$

Indeed the value $+1$ is taken for $x(s)\,x(s+t)$ if and only if there is an even number of t_j in $(s, s+t]$, $t \geq 0$, the value -1 being taken if there is an odd number of t_j in this interval. Thus

$$\gamma(t) = e^{-\mu t}\sum_0^\infty \frac{(-\mu t)^j}{j!} = e^{-2\mu t}, \qquad t \geq 0.$$

Thus the spectral density is

$$\frac{2\mu}{\pi\{4\mu^2 + \lambda^2\}}.$$

This example shows the degree to which the spectrum may not uniquely define the probabilistic behavior of the process, for the graph of a Gaussian process with the covariance function $\exp(-2\mu\,|t|)$ would look very different from that of this random telegraph signal.

Processes of the type considered in this example are called "point processes" (see Bartlett, 1955, p. 78), that is, processes at which any nondeterministic change in the state of the process takes place only through jumps at a discrete set of points. We have reduced the Poisson process to the form of a stationary process with finite variance by associating a function ($\alpha(l, t)$, for example) with each jump. The process $\xi(t)$ is not itself stationary within the content of our theory (though see the later section on generalized processes). Of course, we need not restrict ourselves to the Poisson process, but we shall not pursue the matter further here (see Bartlett, 1963). In the cases discussed the probability distribution of $x(t)$, in its entirety, is known once μ, α, and P are known. This makes the spectral theory, which deals only with second-order properties, seem somewhat inappropriate. This is not necessarily so, however, for the models proposed *are* only models and reality will never be so simple. Thus the study of the covariance function, say, of the diameter of a yarn may be a useful statistical device, made even more so when the simple models discussed above (or variants of them) are held in mind so that an illuminating interpretation of empirical results may be given.

† We define $x(t)$ to be continuous from the right.

(ix) We now give an example designed to illustrate the meaning of the spectral representation (Theorem 2) by the construction of a realistic model which is near to that discussed in Theorem 2′ but does not lead to a periodic phenomenon. Let

$$x_j(t) = \sum_{n=0}^{\infty} \{\xi_j(n, t) \cos 2\pi nt + \eta_j(n, t) \sin 2\pi nt\} + u_j(t) = w_j(t) + u_j(t),$$

where for all j, k, s, t

$$\mathcal{E}\{\xi_j(m, s)\, \xi_k(n, t)\} = \mathcal{E}\{\eta_j(m, s)\, \eta_k(n, t)\}$$
$$= \mathcal{E}\{\xi_j(m, s)\, \eta_k(n, t)\} = 0 \qquad m \neq n$$

and for all j, k, s, t, n

$$\mathcal{E}\{u_j(s)\, \xi_k(n, t)\} = \mathcal{E}\{u_j(s)\, \eta_k(n, t)\} = \mathcal{E}\{\xi_j(n, s)\, \eta_j(n, t)\} = 0,$$
$$\mathcal{E}\{\xi_j(n, s)\, \xi_k(n, s + t)\} = \mathcal{E}\{\eta_j(n, s)\, \eta_k(n, s + t)\} = \alpha_{j,k}(n, t)$$
$$\mathcal{E}\{\xi_j(n, s)\, \eta_k(n, s + t)\} = -\mathcal{E}\{\eta_j(n, s)\, \xi_k\{(n, s + t)\} = \beta_{j,k}(n, t),$$

whereas for each j

$$\sum_0^{\infty} \alpha_j(,n\ 0) < \infty.$$

Then the first component $w_j(t)$ is well defined as a limit in mean square. We assume that all spectral functions are absolutely continuous and call them $c_{jk}(n, \lambda)$ (for $\xi_j(n, t)$ with $\xi_k(n, t)$ and $\eta_j(n, t)$ with $\eta_k(n, t)$) and $q_{jk}(n,\lambda)$ (for $\xi_j(n, t)$ with $\eta_k(n, t)$). Then the time functions

$$(5.13) \qquad \{\xi_j(n, t) \cos 2\pi nt + \eta_j(n, t) \sin 2\pi nt\}, \qquad j = 1, 2, \ldots, p,$$

have covariance sequences

$$\alpha_{jk}(n, t) \cos 2\pi nt + \beta_{jk}(n, t) \sin 2\pi nt,$$

so that the cross-spectral density functions for these components are

$$\tfrac{1}{2}\{c_{jk}(n, \lambda - 2\pi n) + c_{jk}(n, \lambda + 2\pi n) + iq_{jk}(n, \lambda - 2\pi n) - iq_{jk}(n, \lambda + 2\pi n)\}.$$

If we put

$$f_{jk}(n, \lambda) = \tfrac{1}{2}\big(c_{jk}(n, \lambda) - iq_{jk}(n, \lambda)\big) = \overline{f_{jk}(n, -\lambda)} = \overline{f_{jk}(-n, \lambda)},$$

the spectral densities for the components (5.13) are

$$f_{jk}(n, \lambda + 2\pi n) + f_{jk}(n, \lambda - 2\pi n).$$

Thus the spectral density matrix of the $x(t)$ vector is, putting $f(u, \lambda)$ for that of the vector $u(t)$,

$$f(\lambda) = \sum_{n=-\infty}^{\infty} f(n, \lambda + 2\pi n) + f(u, \lambda)$$

(for $n = 0$, of course, we put $\eta_j(n, t) \equiv 0$ so that $q_{jk}(0, \lambda) \equiv 0$ and $f_{jk}(0, \lambda) = c_{jk}(0, \lambda)$).

An interesting case is that in which, for each (n, j), $f_j(n, \lambda)$ is concentrated at the origin (i.e., "near a δ-function at 0," let us say) for then $f(\lambda)$ will be the sum of $f(u, \lambda)$ and a sequence of near δ-functions at $2\pi n$, $n = 0, \pm 1, \ldots$, with the "mass" at $2\pi n$ decreasing as $|n| \to \infty$. Correspondingly $\xi_j(n, t)$ and $\eta_j(n, t)$ will change only very slowly (if the concentration of $f_j(n, \lambda)$ is very marked) so that over appreciable periods of time $w_j(t)$ will behave like a periodic function with period unity. Of course, if each $\xi_j(n, t)$ becomes independent of t then $f(n, \lambda)$ is truly a δ-function and $x(t)$ is just the sum of a periodic function (with a mean square convergent Fourier series) and a "noise" component $u(t)$. In this situation $x(t)$ has a spectral distribution matrix, $F(\lambda)$, with an absolutely continuous part with density matrix $f(u, \lambda)$, together with jumps at $2\pi n$, $n = 0, \pm 1, \ldots$. It is fairly evident that from one finite record, $0 \le t \le T$, it will be almost impossible to distinguish the case where $F(\lambda)$ jumps from that where the $f(n, \lambda)$ are merely *very* near to δ-functions. It is something of a commonplace that natural phenomena which are at first sight perfectly periodic are on closer examination seen not to be so (e.g., the rotation of the earth). Thus perfectly periodic phenomena seem to appear nowhere in nature. Nevertheless *no* mathematical model can represent nature in all of its complexity and some natural phenomena are so near to periodic that the model of perfect periodicity is adequate by standards which, if they were maintained everywhere in science, would leave us happy indeed. The consideration of spectra containing jumps is not a useless activity.

The contrary case to the one we have discussed in relation to the $f(n, \lambda)$ is that in which the spectral density function is nearly constant. Of course, it cannot strictly be constant, since its integral, which is the variance, must be finite. However, if $x(t)$ is scalar and

$$\begin{aligned} f(\lambda) &\equiv \frac{1}{2T}, \qquad |\lambda| \le T, \\ &\equiv 0 \qquad\quad |\lambda| > T. \end{aligned}$$

then

$$\gamma(t) = \frac{1}{2T} \int_{-T}^{T} e^{it\lambda}\, d\lambda = \frac{\sin tT}{tT}.$$

If T is very large, then $\gamma(t)$ quickly descends from its value (unity) at $t = 0$ to a very low value so that observations only a relatively small distance apart are almost uncorrelated.

(x) Finally, we discuss an example which illustrates a filter which removes the zero frequency precisely from the spectrum and otherwise has almost the simplest response function imaginable.

Let the output of the filter be

$$x(t)\cos\alpha + \frac{1}{\pi}\int_{0+}^{\infty}\frac{x(t-s)-x(t+s)}{s}\,ds\sin\alpha.$$

We may write this as

$$x(t)\cos\alpha + \frac{1}{\pi}\int_{-\infty}^{\infty}\frac{x(s)}{t-s}\,ds\sin\alpha,$$

it being understood that the infinite integral is taken as a Cauchy principal value as we have indicated in the preceding expression. In order that the filter may be well defined as a limit in mean square it is necessary that the frequency response of the filter in the second term, say, truncated at S, namely,

$$\int_{0+}^{S}\frac{e^{is\lambda}-e^{-is\lambda}}{s}\,ds,$$

converge as S increases in mean square with weighting $F(d\lambda)$. However, this is obviously so, since the Cauchy principal value

$$\int_{-S}^{S}\frac{\sin s\lambda}{s}\,ds$$

is well known to converge to π, for $\lambda > 0$, 0 for $\lambda = 0$, $-\pi$ for $\lambda < 0$, and it does so, as is evident from a consideration, say, of a graph of $x^{-1}\sin x$, boundedly. The mean-square convergence then follows from dominated pointwise convergence. Thus the filter we have constructed has response

$$\begin{aligned}\cos\alpha + i\sin\alpha, &\qquad \lambda > 0,\\ \cos\alpha, &\qquad \lambda = 0,\\ \cos\alpha - i\sin\alpha, &\qquad \lambda < 0,\end{aligned}$$

and we have constructed a filter with the required response function. Taking the case $\alpha = \pi/2$ first, we obtain the Hilbert transform (or Stieltjes transform)

$$x(t) \to \hat{x}(t) = \frac{1}{\pi}\int_{-\infty}^{\infty}\frac{x(s)}{t-s}\,ds.$$

Thus this precisely removes the zero frequency and does not affect the modulus of any other. Correspondingly $x(t)\cos\alpha + \hat{x}(t)\sin\alpha$ modifies the component at zero frequency by the factor $\cos\alpha$ and changes the modulus of no other. For the discrete time case we obtain precisely the same result if we define

$$\hat{x}(n) = \frac{1}{\pi}\sum_{-\infty}^{\infty}{}' x(n-m)\left\{\frac{1}{m}(\cos\pi m - 1)\right\},$$

where the prime indicates that $m^{-1}(\cos \pi m - 1)$ is to be taken as zero at $m = 0$.†

The Hilbert transform is sometimes used to define the "envelope" of a process this envelope being put equal to

$$|x(t) - i\hat{x}(t)|.$$

The intuitively held concept here is perceived by taking $x(t)$ to have all of its mass in a very narrow band around λ_0, that is,

$$x(t) \doteq ze^{i\lambda_0 t} + \bar{z}e^{-i\lambda_0 t}.$$

Then

$$\hat{x}(t) \doteq ize^{i\lambda_0 t} - i\bar{z}e^{-i\lambda_0 t}, \qquad x(t) - i\hat{x}(t) = 2ze^{i\lambda_0 t},$$

whose modulus is $2\,|z|$, which is the amplitude of the sine curve $x(t)$. Thus for narrow-band processes, for which the envelope will vary with t, this envelope can be thought of as a smooth line drawn through the maxima of the nearly sinusoidal oscillation $x(t)$, which justifies the use of the term envelope.

As a less obvious example of a filter with unit gain consider the operator

$$x(t) \to u(t) = \sum_0^\infty a^j x(t - j) \to y(t) = au(t) - u(t - 1), \qquad |a| < 1,$$

which has the response function

$$\frac{a - e^{i\lambda}}{1 - ae^{i\lambda}},$$

whose modulus is unity. The phase is now

$$\theta(\lambda) = \arctan \left\{ \frac{(a^2 - 1) \sin \lambda}{2a - (a^2 + 1) \cos \lambda} \right\},$$

which is not proportional to λ. Thus

$$y(t) = \int_{-\infty}^{\infty} e^{-it\lambda} e^{i\theta(\lambda)} z(d\lambda).$$

The group delay $\theta'(\lambda)$ is in this case easily seen to be

$$\frac{1 - a^2}{1 + a^2 - 2a \cos \lambda},$$

which is, for a near to 1, concentrated at $\lambda = 2k\pi$.

† The implication that this type of transform is useful for removing a trend (see Granger and Hatanaka, 1964, p. 146) seems wrong. It is designed to remove in a very precise fashion only the zero frequency. Indeed, it does not seem to have any practical value.

6. SOME SPECTRAL THEORY FOR NONSTATIONARY PROCESSES

We discuss some cases of nonstationary phenomena in this section for which some kind of spectral theory results. The first case discussed is the only one that will be of any importance later in the book and the remaining subsections of this section may therefore be omitted.

(i) We begin with the consideration of discrete time processes which are asymptotically stationary. The basic idea here is that though $\Gamma(m, n)$ may not be a function only of $(n - m)$ yet we may have an *observed* covariance function, such as $N^{-1} \sum_1^N x(n)\, x'(n + m)$, converging to a limiting value, in some appropriate sense, dependent only upon m and not on the initial time point considered. The most important of these theories appears to be due to Grenander (1954) who following Wiener (1933, Chapter IV) considers vector discrete time series for which he requires that the following conditions (*which we shall call Grenander's conditions*) should hold, almost surely,

$$\text{(6.1a)} \quad \lim_{N\to\infty} d_j^2(N) = \lim_N \sum_{n=1}^N x_j^2(n) = \infty, \qquad j = 1, \ldots, p$$

$$\text{(6.1b)} \quad \lim_{N\to\infty} \left\{\frac{x_j^2(N)}{d_j^2(N)}\right\} = 0, \qquad j = 1, \ldots, p$$

$$\text{(6.1c)} \quad \lim_{N\to\infty} \frac{\{\sum_{m=1}^N x_j(m)\, x_k(m + n)\}}{\{d_j(N)\, d_k(N)\}} = \rho_{jk}(n), \qquad j, k = 1, \ldots, p.$$

By (c) we mean that the limit exists, with probability 1. In the exercises to this chapter we indicate how to prove that a necessary condition that a sequence of nonnegative numbers $x_j^2(N)$ satisfy (b) above is that

$$\lim_{N\to\infty} \frac{x_j^2(N)}{b^N} = 0, \qquad b > 1$$

so that the $x_j(n)$ sequence must increase more slowly than exponentially. Moreover, if $\overline{\lim}_N \{x_j^2(N)/d_j^2(N)\} > 0$ we can extract a subsequence $x_j(n_k)$ which increases at least exponentially with k. Cases of exponential rates of increase will always need special consideration and the condition (b) is designed to exclude them. It also ensures that $\rho_{jk}(n)$ is free from end effects so that, for example,

$$\lim_{N\to\infty} \frac{\{\sum_{m=1}^{N-n-q} x_j(m + q)\, x_k(m + q + n)\}}{\{d_j(N)\, d_k(N)\}} = \rho_{jk}(n).$$

We call $R(n)$ the matrix with entries $\rho_{jk}(n)$. If we form $\rho_\alpha(n) = \alpha^* R(n)\alpha$ where α is a vector of complex numbers then $\rho_\alpha(n)$ forms a nonnegative definite sequence in the sense of the first part of the proof of Theorem 1″. (We leave the reader to check this by evaluating

$$\sum_1^r \sum z_i \bar{z}_j \rho_\alpha(n_i - n_j)$$

as the limit as $N \to \infty$ of the corresponding expressions replacing the $\rho_{jk}(n)$ by the left-hand member of condition (c) of Grenander's conditions.) Thus Theorem 1″ applies to the $R(n)$ and we have

Theorem 11. *If the $x_j(n)$ satisfy Grenander's conditions and $R(n)$ is the matrix with entries $\rho_{jk}(n)$ then*

$$R(n) = \int_{-\pi}^{\pi} e^{in\lambda} M(d\lambda)$$

where $M(\lambda)$ is a matrix whose increments, $M(\lambda_2) - M(\lambda_1)$, $\lambda_2 \geq \lambda_1$, are Hermitian nonnegative, and which is uniquely defined if it is required to be continuous from the right and null at $-\pi$.

Of course, once again at all points of continuity λ_1, λ_2

$$M(\lambda_2) - M(\lambda_1) = \{\overline{M(-\lambda_1) - M(-\lambda_2)}\}.$$

As an example of a sequence fulfilling Grenander's conditions we may take $x_j(n) = n^{j-1}, j = 1, \ldots, p$. It is easily seen that

$$\lim_{N\to\infty} \left\{(j+1) \sum_{m=1}^{N} \frac{m^j}{N^{j+1}}\right\} = 1,$$

so that for these $x_j(n)$, (a), (b), and (c) of (6.1) hold and

$$\rho_{jk}(n) = \frac{2(jk)^{1/2}}{j+k-1}.$$

Thus $M(\lambda)$ increases only at the origin, the increase being $R(0)$. Here the description of the $x(n)$ sequence through $M(\lambda)$ is a very poor one and all "detail" is lost. It is, however, important for some purposes.

A further example is $x_{2j}(n) = \cos \theta_j n$, $x_{2j-1}(n) = \sin \theta_j n$, $j = 1, \ldots, p$, $\theta_j \neq \theta_k$, $j \neq k$. Now again (a), (b), and (c) of (6.1) are satisfied and $R(n)$ decomposes into p blocks down the main diagonal (the remainder of the matrix being null) the jth block being of two rows and columns and of the form

$$\begin{bmatrix} \cos \theta_j n & \sin \theta_j n \\ -\sin \theta_j n & \cos \theta_j n \end{bmatrix}.$$

Thus $M(\lambda)$ increases only at the points $\pm\theta_j$, having jumps at $\pm\theta_j$ which are null save for the jth block down the main diagonal and for which the jth blocks are, at θ_j and $-\theta_j$, respectively,

$$\frac{1}{2}\begin{bmatrix} 1 & -i \\ i & 1 \end{bmatrix}, \quad \frac{1}{2}\begin{bmatrix} 1 & i \\ -i & 1 \end{bmatrix}.$$

In Chapter IV we shall see that under by no means onerous conditions the sample autocovariances of a stationary vector process will converge with probability 1. In that case the stationary vector process will also satisfy (a), (b), and (c) of (6.1) and now, of course, $\rho_{jk}(n) = \gamma_{jk}(n)/\{\gamma_j(0)\,\gamma_k(0)\}^{1/2}$ so that $M(\lambda)$ has entries $m_{jk}(\lambda) = F_{jk}(\lambda)/\{\gamma_j(0)\,\gamma_k(0)\}^{1/2}$, $F(\lambda)$ being the spectral distribution matrix for $x(n)$. It is not difficult to see that if $x(n)$ is a vector process of the type presently considered, $y(n)$ is of the same dimension and with entries n^j and $z(n)$ is of the same dimension and with entries $\cos n\theta_j$, $\sin n\theta_j$ then $x(n) + y(n) + z(n)$ again fulfills the conditions (a), (b), and (c). Much more general examples than this can be constructed, of course, but this is sufficient to show that Grenander's conditions admit an $x(n)$ sequence of a rather general type.

The considerations of the present subsection can easily be extended to deal with processes in continuous time. Their usefulness, however, arises mainly in connection with discrete time and the use of digital methods and we confine ourselves to the foregoing.

(ii) Loève (1960, p. 474) calls a (scalar) process harmonizable if

$$x(t) = \int_{-\infty}^{\infty} e^{-it\lambda}\, z(d\lambda),$$

wherein

$$\mathcal{E}\big(z(\lambda_1)\,\overline{z(\lambda_2)}\big) = F(\lambda_1, \lambda_2)$$

and F is of bounded variation in the plane. Probably a more fruitful approach is to allow $z(\lambda)$ itself to depend upon t so that

$$y(t) = \int_{-\infty}^{\infty} e^{-it\lambda}\, z_t(d\lambda). \tag{6.2}$$

Of course it is implied that $z_t(\lambda)$ changes slowly with time for otherwise the representation would both be rather vacuous and make the determination of the corresponding time-dependent spectrum difficult. A model of the type now being considered arises of course in connection with the modulation of a signal. The simplest case is amplitude modulation, where we observe $y(t) = a(t)\,x(t)$ and $x(t)$ is stationary. Now $z_t(d\lambda) = a(t)\,z_x(d\lambda)$. Of course, if $a(t)$ itself is chosen to be stationary and independent of $x(t)$, then $y(t)$

is once more stationary with

$$\gamma_y(t) = \mathcal{E}\big(y(s)y(s+t)\big) = \gamma_x(t)\{\gamma_a(t) + \mu^2\},$$

where μ is the mean of $a(t)$, which it is now unreasonable to take as zero. If $x(t)$ and $a(t)$ have a.c. spectrum, then

$$\gamma_x(t)\gamma_a(t) = \int_{-\infty}^{\infty} e^{it\lambda}\left\{\int_{-\infty}^{\infty} f_a(\lambda - \theta) f_x(\theta)\, d\theta\right\} d\lambda$$

and $y(t)$ has spectrum $\mu^2 f_x(\lambda) + f_x * f_a(\lambda)$, where by $f_x * f_a$ we mean the convolution of these two functions, i.e., the inner integral on the right. Thus, if $x(t)$ has a narrow band spectrum concentrated at $\pm\lambda$ and $a(t)$ has a narrow-band spectrum at the origin, then $y(t)$ will have a spectrum with a peak at the origin and at $\pm\lambda$, the latter being somewhat wider than for $x(t)$.

A second example is that of a frequency modulated sine wave, say

$$y(t) = \alpha \cos\big(\theta a(t) + \phi\big),$$

where ϕ is a random variable uniformly distributed over $[-\pi, \pi)$. If $z_t(d\lambda) = \frac{1}{2}\alpha \exp\{i\theta\big(a(t) + t\big)\}\, z_x(d\lambda)$ and $z_x(d\lambda)$ increases only at $\pm\theta$, the increment being $\exp \pm i\phi$ then again $y(t)$ is of the form (6.2). If ϕ is independent of $a(t)$ then

$$\mathcal{E}\big(y(s)\, y(s+t) \mid a(t)\big) = \tfrac{1}{2}\alpha^2 \cos\big(\theta(a(s+t) - a(s))\big),$$

where on the left we mean the conditional expectation given the function $a(t)$. Now once again $x(t)$ *may* be stationary, as would be the case if $a(t)$ is a random process independent of ϕ and has strictly stationary first-order increments (see Chapter I, Section 3, example (v)) for then, taking expectation with respect to $a(t)$,

$$\mathcal{E}\big(y(s)\, y(s+t)\big) = \tfrac{1}{2}\alpha^2 \mathcal{R}\big(\phi_t(\theta)\big),$$

where $\phi_t(\theta)$ is the characteristic function of $a(s+t) - a(s)$. For example, we might have

$$a(t) = \int_0^t u(s)\, ds, \qquad t \geq 0$$

$$= \int_{-t}^0 u(s)\, ds, \qquad t \leq 0,$$

where $u(s)$ is itself a stationary Gaussian process. (See Wainstein and Zubakov, 1962, p. 322 ff.) Now

$$\mathcal{E}\big(y(s)\, y(s+t)\big) = \tfrac{1}{2}\alpha^2 \exp\left\{-\tfrac{1}{2}\theta^2 t \int_{-t}^{t} \gamma_u(v)\left(1 - \frac{|v|}{t}\right) dv\right\}.$$

We shall discuss this example further in Section 7.

One should not try to pretend that all problems associated with non-stationarity can be avoided by the introduction of a suitable stationary model. In some cases the system is evolving so rapidly that this is certainly impossible. However, when such rapid evolution occurs, it is often true that a special model has to be considered and modifications of the type considered in this subsection will not be suitable either (since z_t could not be considered to be changing slowly with t). In the contrary case in which the evolution may be very slow the introduction of a stationary model may be rather precious and may not lead to useful techniques.

(iii) In Chapter I, Section 3, we introduced the concept of a process with stationary increments. As shown there, if $x(t)$ has stationary qth order increments *and* is q times mean square differentiable then $x^{(q)}(t)$ is stationary. Thus

$$x^{(q)}(t) = \int_{-\infty}^{\infty} e^{-it\lambda} z(d\lambda)$$

and therefore

$$x^{(q-1)}(t) = \int_{-\infty}^{\infty} \left\{\frac{1 - e^{-it\lambda}}{i\lambda}\right\} z(d\lambda) + x_0,$$

where x_0 is a (vector) random variable. Continuing we obtain

$$x(t) = \int_{-\infty}^{\infty} \left[\sum_{1}^{q} \left\{\frac{t^{q-j}}{(q-j)!\,(i\lambda)^j} (-)^{j-1}\right\} + (-)^q \frac{e^{-it\lambda}}{(i\lambda)^q}\right] z(d\lambda) + \sum_{0}^{q-1} x_j t^j. \tag{6.3}$$

We may introduce the new process of orthogonal increments with

$$\zeta(d\lambda) = \frac{(-)^q z(d\lambda)}{(i\lambda)^q}, \qquad \lambda \neq 0,$$

and rewrite (6.3) in the form

$$\int_{R_0} \left\{e^{-it\lambda} - \sum_{0}^{q-1} \frac{t^j}{j!} (-i\lambda)^j\right\} \zeta(d\lambda) + \sum_{0}^{q-1} x_j t^j,$$

where R_0 is the complement of the origin in $(-\infty, \infty)$. Now $G(\lambda) = \mathcal{E}(|\zeta(d\lambda)|^2)$ satisfies

$$\int_{-\infty}^{\infty} \lambda^{2q}\, G(d\lambda) < \infty.$$

We shall return to these considerations and deal with the case in which $x(t)$ is not q times mean-square differentiable in Section 9 below (see also Doob, 1953, p. 551). It is evident that $x(t)$ of the form (6.3) is of the type of the slowly evolving processes discussed under (ii) above. In a sense they are borderline cases since they are on the very verge of stationarity. This is evident in the case $q = 1$, for if we consider

$$x(t) = \int_0^t e^{-\beta(t-\tau)}\,\xi(d\tau), \qquad \beta > 0,\ \mathcal{E}\big(\xi(dt)^2\big) = \sigma^2\,dt$$

then $x(t)$ is stationary. However, as β tends to zero, this ceases to be so and $x(t)$ approaches a process of stationary increments.

7. NONLINEAR TRANSFORMATIONS OF RANDOM PROCESSES

We saw in Section 4 that for linear filters a simple and intuitively very acceptable picture of the effect of the filter can be given through the frequency response function. Many transformations of random processes are, however, nonlinear. The problem of describing their action is one of great complexity and no systematic theory comparable to that for the linear filters is conceivable. To begin with, the action of the filters will no longer be describable purely in forms of second-order quantities. This immediately elevates the Gaussian processes to a premier place. In principle, any process whose higher moments are specified can be handled. In practice the manipulations become exceedingly complex in the non-Gaussian case. *We shall, therefore, restrict ourselves to the Gaussian case.*

We consider mainly instantaneously acting nonlinear filters, that is, transformations of the form

$$x(t) \to g\big(x(t)\big) = y(t).$$

The most obvious expressions from which to begin are the polynomials in $x(t)$. However, because of the Gaussian assumption it is preferable to equivalently use the Hermite polynomials,†

$$H_n(x) = (-)^n e^{\frac{1}{2}x^2}\,(d^n/dx^n e^{-\frac{1}{2}x^2}).$$

The first few are

$$H_0 \equiv 1,\ H_1 = x,\ H_2 = x^2 - 1,\ H_3 = x^3 - 3x,\ H_4 = x^4 - 6x^2 + 3, \ldots .$$

These satisfy‡

$$\frac{1}{\sqrt{2\pi}}\int_{-\infty}^{\infty} H_m(x)H_n(x)e^{-\frac{1}{2}x^2}\,dx = n!\,\delta_m^{\ n},$$

as is fairly easily established by a repeated process of partial integration. Moreover, if x and y are jointly Gaussian with zero mean, unit variance and

† This definition differs from the more customary one in which $\exp(-x^2)$ replaces $\exp(-\frac{1}{2}x^2)$.

‡ They are easily calculated iteratively from the formula $H_{n+1}(x) = xH_n(x) - H_n'(x)$.

correlation ρ we have the somewhat surprising result,

$$\mathcal{E}\{H_m(x)\, H_n(y)\} = \delta_m{}^n \rho^n n!$$

(See the exercises to this chapter.)

This enables us to describe the action of an instantaneous filter of the form of a polynomial function, p, of $x(t)$ on the spectrum. Thus we put

$$y(t) = p(x(t)) = \sum_0^d a_m\, H_m(x(t)),$$

where

$$m!\, a_m = \frac{1}{\sqrt{2\pi}} \int p(x)\, H_m(x) e^{-\frac{1}{2}x^2}\, dx. \tag{7.1}$$

Then assuming $x(t)$ to have zero mean and unit variance, as we evidently may do by changing from an initial to a new polynomial function, we have

$$\mathcal{E}(y(s)\, y(s+t)) = \sum_0^d a_m{}^2 \{\gamma(t)\}^m\, m!, \qquad \gamma(0) = 1.$$

Let us consider first the case in which $x(t)$ has a.c. spectrum with density $f(\lambda)$. Then

$$\{\gamma(t)\}^m = \int_{-\infty}^{\infty} f^{*m}(\lambda) e^{it\lambda}\, d\lambda \qquad m > 0, \tag{7.2}$$

where by f^{*m} we mean the m-fold convolution of $f(\lambda)$ with itself. (The above considerations apply equally as well to discrete as to continuous time, but in the latter case we continue $f(\lambda - \theta)$ periodically, outside of $[-\pi, \pi]$, so as to define the convolution). For $m = 0$ we interpret $f^{*m}(\lambda)$ as $\delta(\lambda)$, the Dirac delta function.† We now have

$$f_y(\lambda) = \sum_0^d a_m{}^2\, m!\, f^{*m}(\lambda). \tag{7.3}$$

The introduction of $\delta(\lambda)$ can, of course, be avoided by replacing $y(t)$ by $y(t) - \mathcal{E}(y(t)) = y(t) - a_0$ (since it is easily seen that $\mathcal{E}(H_m(x)) = 0$, $m > 0$ when x is Gaussian with zero mean).

Thus, for example, $x^2(t) - 1$ has spectrum $2f^{*2}(\lambda)$; that is,

$$2\int_{-\infty}^{\infty} f(\lambda - \theta) f(\theta)\, d\theta.$$

Of course, this theorem has a wide extension and we may first relax the requirement that $x(t)$ have a.c. spectrum. The spectral distribution function

† The Dirac delta function is discussed in the Mathematical Appendix. Any use of it in this section is quite trivial and should cause no problems.

of $y(t)$ is now

$$\sum_0^d a_m^2 m!\, F^{*m}(\lambda),$$

where F^{*m} is the m-fold convolution of the distribution function F and we now take $F^{*0}(\lambda)$ to be the Heaviside unit function, which is zero for $\lambda < 0$ and 1 for $\lambda \geq 0$. Thus

$$F^{*2}(\lambda) = \int_{-\infty}^{\infty} F(\lambda - \theta)\, F(d\theta).$$

In the second place, we need not restrict ourselves to polynomial functions. Indeed, if $p(x)$ is square integrable with respect to $\exp(-\frac{1}{2}x^2)$, we may put

$$y(t) = \sum_0^\infty a_m H_m(x(t)),$$

where a_m is given by (7.1) above and the infinite series converges† to $y(t)$, in mean square with weighting $\exp(-\frac{1}{2}x^2)$. Correspondingly

$$\gamma_y(t) = \sum_0^\infty a_m^2\, m!\, \{\gamma(t)\}^m$$

(the series converging absolutely) and we obtain

(7.4) $$F_y(\lambda) = \sum_0^\infty a_m^2\, m!\, F^{*m}(\lambda).$$

In the third place we may extend our considerations to instantaneous nonlinear functions of a vector-valued random variable. The considerations here are, however, somewhat more complicated and we defer them to Section II in this chapter, in which we also further discuss noninstantaneous filters. We now discuss the results (7.3) and (7.4) by means of examples.

It is not as easy to interpret these results as the filtering theorem for linear filters. A case that indicates the nature of the situation is one in which $F(\lambda)$ has jumps at points $\pm\lambda_1, \pm\lambda_2, \ldots, \pm\lambda_k$, $\lambda_j \geq 0$, $j = 1, \ldots, k$. It is then easy to see that $F^{*2}(\lambda)$ will have jumps at all points λ that are sums of frequencies from the set of jumps of $F(\lambda)$. By induction we see that $F_y(\lambda)$ will have jumps at all frequencies of the form

(7.5) $$\sum_1^k n_j \lambda_j,$$

† We are asserting the completeness of the sequence of functions $H_n(x) \exp -\frac{1}{4}x^2$, that is, the fact that they span the space of all square integrable functions on the real line. For a proof see Wiener (1933, p. 64).

where n_j are integers. Thus (Munk and Cartwright, 1966) if the gravitational forces causing the tides have six basic frequencies $\lambda_1, \ldots, \lambda_6$, corresponding to the frequencies of the earth's rotation, of the moon's orbital motion, of the earth's orbital motion around the sun, of the lunar and solar perigree and of the regression of lunar nodes then the nonlinear transmission of these effects to the oceans can be expected to produce all integral linear combinations of them, of the form of (7.5). Of course if $F(\lambda)$ is a.c. but $f(\lambda)$ has relatively sharp peaks at points λ_j then $f_y(\lambda)$ can be expected to have, somewhat less sharp, peaks at the frequencies given by (7.5).

We next consider an example. The simplest model for a frequency modulated signal represents the output of the transmitter as of the form

$$y(t) = \alpha \cos\big(\theta t + \phi + \beta x(t)\big)$$

As mentioned above (see the last part of Section 6, example (ii)) if ϕ is uniformly distributed over $[-\pi/2, \pi/2]$ independently of $x(t)$, then

$$\mathcal{E}\big(y(s)\, y(s+t)\big) = \tfrac{1}{2}\alpha^2 \mathcal{E}\big[\cos\{\theta t + \beta\big(x(t+s) - x(s)\big)\}\big]$$

so that if $x(t)$ is Gaussian with stationary increments then $y(t)$ is stationary also. We, more specially, take $x(t)$ to be stationary with $\gamma(0) = 1$ and to have a.c. spectrum. The allocation of such stochastic properties to ϕ seems reasonable and, because of the presence of β, taking $\gamma(0) = 1$ does not matter. Now

$$\exp\{iax(t) + \tfrac{1}{2}a^2\} = \sum_0^\infty \frac{(ia)^m}{m!} H_m\big(x(t)\big),$$

as follows immediately from the definition of $H_m(x)$, by expanding $\exp(zx - \tfrac{1}{2}z^2)$ in a power series. Thus

$$y(t) = \alpha e^{-1/2\beta^2} \sum_0^\infty \frac{a_m(t)}{m!} H_m\big(x(t)\big),$$

$$a_m(t) = \mathcal{R}\{e^{i(\theta t+\phi)}(i\beta)^m\}.$$

Thus

$$\mathcal{E}\big(y(s)\, y(s+t)\big) = \tfrac{1}{2}\alpha^2 e^{-\beta^2} \cos\theta t \sum_0^\infty \frac{\{\beta^2\gamma(t)\}^m}{m!}$$

$$= \tfrac{1}{2}\alpha^2 e^{-\beta^2(1-\gamma(t))} \cos\theta t$$

Correspondingly, of course, we have

$$f_y(\lambda) = \tfrac{1}{2}\alpha^2 e^{-\beta^2} \sum_{m=0}^\infty \beta^{2m}\{f^{*m}(\lambda+\theta) + f^{*m}(\lambda-\theta)\}\frac{1}{m!}$$

since $\gamma(t)^m \exp i\theta t$ is the Fourier transform of $f^{*m}(\lambda) * \delta(\lambda - \theta) = f^{*m}(\lambda - \theta)$. If $f(\lambda)$ is substantially concentrated at the origin so that the modulation is by means of a slowly varying $x(t)$ then $f_y(\lambda)$, apart from the "spectral line" at $\lambda = 0$, which will be removed by mean correction, will be substantially concentrated at the frequencies $\pm n\theta$ and if β is small it will be substantially concentrated at $\pm\theta$, as is to be expected.

In cases in which the derivatives of $g(x)$, from some early derivative onward, are expressible solely in terms of the derivatives of δ-functions then the coefficients in (7.4) are quite easy to obtain. Thus we have, putting $\phi(x)$ for the standard normal density function, $(\sqrt{2\pi})^{-1} \exp -\frac{1}{2}x^2$,

$$m!\, a_m = \int_{-\infty}^{\infty} g(x)\, H_m(x)\, \phi(x)\, dx = (-)^m \int_{-\infty}^{\infty} g(x)\, \phi^{(m)}(x)\, dx$$
$$= \int_{-\infty}^{\infty} g^{(m)}(x)\, \phi(x)\, dx,$$

by integration by parts, assuming $g(x)$ to be well behaved at $\pm\infty$. From some m onward this will involve only expressions of the form

$$\int_{-\infty}^{\infty} \delta^{(k)}(x-a)\, \phi(x)\, dx = (-)^k \phi^{(k)}(a) = H_k(a)\, \phi(a).$$

As an example let

$$g(x) = \begin{cases} -a, & x \le -a, \\ x, & -a \le x \le a, \\ a, & a \le x. \end{cases}$$

Then a_0 is easily seen to be zero and

$$a_1 = \int_{-a}^{a} \phi(x)\, dx.$$

Since

$$g^{(m)}(x) = \delta^{(m-2)}(x+a) - \delta^{(m-2)}(x-a),$$

then

$$a_m = H_{m-2}(-a)\, \phi(-a) - H_{m-2}(a)\, \phi(a),$$

which is zero for m even and is $-2\phi(a)\, H_{m-2}(a)$ for m odd. Thus

$$f_y(\lambda) = \left\{\int_{-a}^{a} \phi(x)\, dx\right\}^2 f(\lambda) + 4 \sum_{1}^{\infty} \frac{\{\phi^{(2j-1)}(a)\}^2}{(2j+1)!} f^{*(2j+1)}(\lambda).$$

The result (7.4) may be extremely difficult to interpret qualitatively and if $g(\cdot)$ is a simple enough function one might hope for some simpler expression. One technique of interest is due to Price (see Deutsch, 1962, p. 16). The result in our formulation is as follows. We consider two generalized† functions $g(x)$, $h(y)$ of two jointly Gaussian random variables with zero means,

† The mathematics needed for what follows is surveyed in the Mathematical Appendix.

unit variances, and correlation ρ. Of course, the restrictions on the means and variances are of no consequence. Though there is some problem with the imbedding of functions which are not integrable on every finite interval in the space of generalized functions this will not be important here for the formulas we are about to derive will be used only when, after a few differentiations, g and h are reduced to linear combinations of derivatives of δ-functions.

We have

Theorem 12. *Let g, h be generalized functions and x, y be jointly Gaussian with zero mean, unit variance and correlation ρ. Then*

$$\frac{d^k}{d\rho^k}\,\mathcal{E}\{g(x)\,h(y)\} = \mathcal{E}\{g^{(k)}(x)\,h^{(k)}(y)\}.$$

The proof is no more than an application of Plancherel's theorem but we have placed it in the Appendix to this chapter.

Two examples show how the theorem may be applied. We consider $h(x) \equiv g(x) = x,\ x \geq 0;\ \ =0,\ x \leq 0$. We take $x = x(s)$, $y = x(s+t)$, $\gamma(0) = 1$.

Then $g''(x) = \delta(x)$ and

$$\frac{d^2}{d\gamma(t)^2}\,\mathcal{E}(y(s)\,y(s+t)) = \frac{1}{2\pi\sqrt{1-\gamma^2(t)}}$$

by the theorem. We also have

$$\left[\frac{d}{d\gamma(t)}\,\mathcal{E}\{y(s)\,y(s+t)\}\right]_{\gamma(t)=0} = \iint_0^\infty xy\,\frac{1}{2\pi}\,e^{-\frac{1}{2}(x^2+y^2)}\,dx\,dy = \frac{1}{2\pi}$$

$$\mathcal{E}\{y(s)\,y(s+t)\}_{\gamma(t)=0} = \tfrac{1}{4}.$$

Thus

$$\gamma_y(t) = \frac{1}{2\pi}\,\{\gamma(t)\arcsin\gamma(t) + \sqrt{1-\gamma^2(t)} - 1 + \gamma(t)\} + \tfrac{1}{4}.$$

Mean correction of $y(t)$ eliminates the term $\frac{1}{4}$. The formula for the spectrum of $y(t)$ in terms of that of $x(t)$ will be very complicated. Of course one can expand $g(\cdot)$ in Hermite polynomials and use (7.4) and indeed this may give a simpler, more easily interpreted, result but one may also prefer to work in terms of the simple closed expression for the covariance function.

A second example is that of a "clipped" signal where $y(t)$ is ± 1 according as $x(t)$ is ≥ 0 or < 0. Taking $h(x) \equiv g(x)$, we have $h' \equiv g' = 2\delta(t)$, and, putting $\gamma(0) = 1$ as before,

$$\frac{d}{dt}\gamma_y(t) = \frac{4}{2\pi\sqrt{1-\gamma^2(t)}}, \qquad [\gamma_y(t)]_{\gamma(t)=0} = 0;$$

thus

$$\gamma_y(t) = \frac{2}{\pi}\arcsin \gamma(t).$$

A similar technique may be used to determine the covariance function of a quantized signal, that is, one of the form

$$y(t) = n, \qquad a_n \leq x_t < a_{n+1}, \qquad n = 0, \pm 1, \ldots,$$

but the calculations are, of course, more complicated.

Both techniques extend to the discovery of cross spectra and cross covariances between instantaneous nonlinear filters of Gaussian processes.

So far we have discussed only instantaneous nonlinear filters. Of course, these may be combined with results for non-instantaneous linear filters, if these latter do not interact with the nonlinear filter. Thus, if

$$y(t) = \alpha \cos\left(\theta t + \phi + \beta z(t)\right), \qquad z(t) = \int_0^\infty \alpha(s)\, x(t-s)\, ds,$$

where θ, ϕ, and $x(t)$ are as in the first example discussed, and $z(t)$ is also stationary, then

$$f_y(\lambda) = \tfrac{1}{2}\alpha^2 e^{-\beta^2} \sum_{m=0}^{\infty} \frac{\beta^{2m}}{m!} \tfrac{1}{2}\{f_z^{*m}(\lambda+\theta) + f_z^{*m}(\lambda-\theta)\},$$

where

$$f_z(\lambda) = \left| \int_0^\infty \alpha(s) e^{is\lambda} \right|^2 f_x(\lambda).$$

Of course the result is even simpler if the linear filter follows the nonlinear one. (*One needs to note of course that two filters, of which one at least is nonlinear, need not commute.*)

8. HIGHER ORDER SPECTRA

In the last section we discussed the spectra of nonlinear functions of Gaussian processes. A situation which may arise is that where neither the nature of the nonlinear filter nor the underlying Gaussian process, $x(t)$, is observed but one wishes to obtain some idea of the nonlinear mechanism producing the observed time function. Hasselmann et al. (1963) provide an example where nonlinearities in the equations of (ocean) wave propagation produce slightly non-Gaussian records from a Gaussian input. One might hope to study the nature of the mechanism through the higher moments. To motivate these ideas consider an example in which a scalar, continuous-time process is of the form

$$y(t) = \sum \alpha_j\, x(t + s_j) + \sum\sum \alpha_{j,k}\, x(t + t_j)\, x(t + t_k), \tag{8.1}$$

and we assume that $x(t)$ is Gaussian with zero mean and unit variance. We take the sums to be finite.† Then

$$y(t) = \int_{-\infty}^{\infty} a^{(1)}(\lambda)e^{-it\lambda}\, z_x(d\lambda) + \iint_{-\infty}^{\infty} a^{(2)}(\lambda_1, \lambda_2)e^{-it(\lambda_1+\lambda_2)} z_x(d\lambda_1)\, z_x(d\lambda_2)$$

$$= \int_{-\infty}^{\infty} e^{-it\lambda}\, z_y(d\lambda),$$

wherein

$$z_y(\lambda) = \int_{-\infty}^{\lambda} a^{(1)}(\lambda_1)\, z_x(d\lambda_1) + \int_{\lambda_1+\lambda_2\leq\lambda} a^{(2)}(\lambda_1, \lambda_2)\, z_x(d\lambda_1)\, z_x(d\lambda_2)$$

and

$$a^{(1)}(\lambda) = \sum a_j e^{-is_j\lambda}, \qquad a^{(2)}(\lambda_1, \lambda_2) = \sum\sum a_{j,k} e^{-i(t_j\lambda_1+t_k\lambda_2)}.$$

The random function $z_y(d\lambda)$ has orthogonal increments because of the Gaussian nature of $x(t)$. Let us take the spectrum of $x(t)$ to be a.c. We can clearly require $a^{(2)}(\lambda_1, \lambda_2)$ to be a symmetric function without loss of generality. Now, if the second component in (8.1) is small compared with the first, $y(t)$ has a spectrum close to $|a^{(1)}(\lambda)^2|\, f_x(\lambda)$, whereas

$$\mathcal{E}\{y(s)\, y(s + t_1)\, y(s + t_2)\}$$

$$= \iiint_{-\infty}^{\infty} \exp i\{s(\lambda + \lambda_1 + \lambda_2) + t_1\lambda_1 + t_2\lambda_2\}\mathcal{E}\{z_y(d\lambda)\, z_y(d\lambda_1)\, z_y(d\lambda_2)\}.$$

If the left-hand side is independent of s, as is the case for our model, then the expectation on the right-hand side can contribute to the integral only if $\lambda + \lambda_1 + \lambda_2 = 0$. This assertion is independent of the Gaussian assumption for the underlying process or the polynomial nature of $a^{(1)}$ and $a^{(2)}$ and depends only on the $y(t)$ process being stationary to the third order. However, in general, we have no reason for believing $\mathcal{E}\{z_y(d\lambda)\, z_y(d\lambda_1)\, z_y(d\lambda_2)\}$ to

† See Section 11 for a discussion of the generality of multinomial functions of a realization as representations of noninstantaneous, nonlinear filters.

be a reasonable function. Reverting to the model we are using for illustration, let us evaluate this expectation. It is, for $\lambda_1 + \lambda_2 \neq 0$,

$$2d\lambda_1\, d\lambda_2\{a^{(1)}(\lambda_1)a^{(1)}(\lambda_2)\overline{a^{(2)}(\lambda_1, \lambda_2)}f_x(\lambda_1)f_x(\lambda_2)$$
$$+ a^{(1)}(\lambda_1)\overline{a^{(1)}(\lambda_1 + \lambda_2)}a^{(2)}(\lambda_1 + \lambda_2, -\lambda_1)f_x(\lambda_1)f_x(\lambda_1 + \lambda_2)$$
$$+ a^{(1)}(\lambda_2)\overline{a^{(1)}(\lambda_1 + \lambda_2)}a^{(2)}(\lambda_1 + \lambda_2, -\lambda_2)f_x(\lambda_1)f_x(\lambda_1 + \lambda_2)\}.$$

Thus we can hope to discover the nature of $a^{(2)}$ from $f_x(\lambda)$ and the function

$$(8.2) \quad \beta(\lambda_1, \lambda_2)\, d\lambda_1\, d\lambda_2 = \mathcal{E}\{z_y(d\lambda)\, z_y(d\lambda_1)\, z_y(d\lambda_2)\}, \qquad \lambda + \lambda_1 + \lambda_2 = 0$$

to which it is the main contributor. This latter function is called the bispectrum. Because of the possibility of permuting the arguments in (8.2) and because of the reality of $y(t)$ we have

$$\beta(\lambda_1, \lambda_2) = \beta(\lambda_2, \lambda_1) = \beta(\lambda_1, -\lambda_1 - \lambda_2) = \overline{\beta(-\lambda_1, -\lambda_2)}$$

for all λ_1, λ_2. In our case we therefore have

$$(8.3) \quad \mathcal{E}\{y(s)\, y(s + t_1)\, y(s + t_2)\} = h_y(t_1, t_2) = \iint_{-\infty}^{\infty} e^{i(t_1\lambda_2 + t_2\lambda_2)}\beta(\lambda_1, \lambda_2)\, d\lambda_1\, d\lambda_2.$$

Of course, we have not proved that such a representation exists in general nor even that we can write

$$(8.4) \qquad h_y(t_1, t_2) = \iint_{-\infty}^{\infty} e^{i(t_1\lambda_1 + t_2\lambda_2)} B(d\lambda_1, d\lambda_2),$$

since we have not established that the right-hand side of (8.2), as a function of (λ_1, λ_2) is a reasonable function, say of bounded variation. Brillinger (1965), to whom the reader may refer also for a definition of higher order spectra (polyspectra as he calls them), discusses conditions under which the representation (8.4) exists.

It is in many ways preferable to avoid the asymmetry inherent in the choice of the one among $\lambda, \lambda_1, \lambda_2$ which is to be eliminated. We may illustrate this by considering the general case for a vector process. It is perhaps preferable to replace higher moments by the corresponding cumulants. (The third-order moments and cumulants are identical.) Let us recall the notation introduced in Chapter I, Section 5, namely,

$$(8.5) \qquad k_{j(1),j(2),\ldots,j(m)}(t_1, \ldots, t_m),$$

for the cumulant between $x_{j(1)}(t_1), x_{j(2)}(t_2), \ldots, x_{j(m)}(t_m)$ it being understood that some of the $j(k)$ may be equal. The only cumulant with which we shall be concerned later in this book is the fourth (see Chapter I, Equation

(5.1)) which naturally arises in connection with the sampling properties of second moments. We shall then replace (8.5) by the less cumbersome notation $k_{ijkl}(s, t, u, v)$, meaning by this the fourth cumulant between $x_i(s)$, $x_j(t)$, $x_k(u)$, $x_l(v)$. We may now write

$$k_{ijkl}(t_1, t_2, t_3, t_4) = \iiiint\limits_{\Sigma\lambda_j=0} \exp i\left(\sum_1^4 t_j\lambda_j\right) B_{ijkl}(d\lambda_1, d\lambda_2, d\lambda_3, d\lambda_4), \tag{8.6}$$

where B describes the fourth cumulant behavior of $z_i(d\lambda_1)$, $z_j(d\lambda_2)$, $z_k(d\lambda_3)$, $z_l(d\lambda_4)$. Again the question of the existence of such a representation arises. Of course this will certainly be so if $k_{ijkl}(o, s, t, u)$ is absolutely integrable as a function of the three arguments, s, t, u. Then also B will be absolutely continuous with derivative which we may call $\beta_{ijkl}(\lambda_1, \lambda_2, \lambda_3, \lambda_4)$. Since β is null off the line $\sum \lambda_j = 0$ we may eliminate one of the three variables. Thus we may put

$$\beta_{ijkl}(\lambda_1, \lambda_2, \lambda_3, \lambda_4) = {}_l\beta_{ijk}(\lambda_1, \lambda_2, \lambda_3), \qquad \sum \lambda_j = 0.$$

This is what we have done in the third-order case in defining $\beta(\lambda_1, \lambda_2)$. That could better have been called ${}_1\beta(\lambda_1, \lambda_2)$, the prescript referring to the variable eliminated from the notation. (This notation is due to Brillinger and Rosenblatt, 1967a, who deal in detail with spectra for higher order cumulants.) However, once the vector case is considered it seems easier to use the full symbolism of (8.5) and (8.6).

We shall not take this subject any further since uses of higher order spectra in data analysis have been very few so far and none appear to have involved anything beyond the bispectrum. As Brillinger says, experience in the classical part of statistics suggests that these concepts will be of limited use and will be replaced, rather, by the consideration of special models when the need for them appears to arise.

9. SPECTRAL THEORY FOR GRP†

(i) In Chapter I, Section 6 we introduced the concept of a generalized random process (GRP). We recall the definition of the Fourier transform of a generalized function as

$$\hat{\Phi}(\phi) = \Phi(\hat{\phi}), \qquad \phi \in K,$$

where $\hat{\phi}$ is the (ordinary) Fourier transform of ϕ (see the Mathematical Appendix). We now consider a stationary GRP, Φ. We shall hold Φ fixed

† This topic is special and will not be used other than in certain sections elsewhere in the book.

and thus shall suppress this part of the notation and shall write $\gamma(\phi, \psi)$ for the covariance functional.

Theorem 1‴

$$\gamma(\phi, \psi) = \int_{-\infty}^{\infty} \hat{\phi}(\lambda)\overline{\hat{\psi}(\lambda)}\, F(d\lambda),$$

where $F(\lambda)$ defines a positive-tempered measure, i.e., a measure such that for some $p \geq 0$

$$\int_{-\infty}^{\infty} (1 + \lambda^2)^{-p}\, F(d\lambda) < \infty. \tag{9.1}$$

This theorem, whose proof we omit (referring the reader to Gelfand and Vilenkin, 1964) is the true analog of Theorem 1. If Φ is in fact an ordinary stationary process, with realizations indicated, as usual, by $x(t)$, then

$$\Phi(\phi) = \int_{-\infty}^{\infty} \phi(t)\, x(t)\, dt.$$

Thus

$$\begin{aligned}
\mathcal{E}\{\Phi(\phi)\overline{\Phi(\psi)}\} &= \iint_{-\infty}^{\infty} \phi(s)\, \overline{\psi(t)}\, \gamma(s - t)\, ds\, dt \\
&= \int_{-\infty}^{\infty} \left\{\int_{-\infty}^{\infty} \phi(t)e^{it\lambda}\, dt\right\}\left\{\int_{-\infty}^{\infty} \overline{\psi(t)}e^{-it\lambda}\, dt\right\} F(d\lambda) \\
&= \int_{-\infty}^{\infty} \hat{\phi}(\lambda)\, \overline{\hat{\psi}(\lambda)}\, F(d\lambda).
\end{aligned}$$

Now F will satisfy (9.1) with $p = 0$. If we differentiate $x(t)$ then as shown in Chapter I, Section 6 we obtain for Φ' a covariance function $\gamma(\phi', \psi')$ where γ corresponds to the initial Φ. Now

$$\gamma(\phi', \psi') = \int_{-\infty}^{\infty} \hat{\phi}(\lambda)\, \overline{\hat{\psi}(\lambda)}\, \lambda^2 F(d\lambda).$$

Thus the new spectral function has differential $\lambda^2\, F(d\lambda)$ which does not necessarily define a totally finite measure on the real line. However, (9.1) is still satisfied which is all that Theorem 1‴ requires.

Theorem 2 also has an extension.

Theorem 2‴. *If Φ is a GRP, then*

$$\Phi = \int_{-\infty}^{\infty} e^{-i\lambda t} z(d\lambda),$$

where $z(\lambda)$ is a process of orthogonal increments with

$$\mathcal{E}\{|z(d\lambda)|^2\} = F(d\lambda)$$

and equality is to be interpreted in the sense that

$$\Phi(\phi) = \int_{-\infty}^{\infty} \hat{\phi}(\lambda)\, z(d\lambda), \qquad \phi \in K.$$

Again we refer to Gelfand and Vilenkin (1964) for the proof.

This immediately shows the scope of the extension of the theory for the class of processes considered is coterminous with the extension one makes in going from Theorem 2 to Theorem 2‴ by dropping the assumption that F be totally finite and replacing it by the weaker condition (9.1). In particular, the output of any linear filter made up from differential operators has been called into being.

We conclude this section by returning to the discussion of processes with stationary increments of the qth order. It is evident that these may be approached through the present apparatus for we can think of such a process as being got by repeated integration from a GRP. Thus we may as well begin from a GRP with stationary increments of the qth order, by which we mean that $\Delta_h{}^q\Phi$ is a stationary GRP for every h where $\Delta_h\Phi(\phi) = \Phi(\phi(t+h) - \phi(t))$, and, of course, $\Delta_h{}^q\Phi = \Delta_h\{\Delta_h^{q-1}\Phi\}$. This may be shown to be equivalent to asserting that $\Phi^{(q)}$ is a stationary GRP. Gelfand and Vilenkin (1964, p. 266) show that a GRP of this form has covariance functional γ satisfying

$$(9.2)\quad \gamma(\phi, \psi) = \int_{R_0} \left\{\hat{\phi}(\lambda) - \alpha(\lambda)\sum_0^{q-1} \hat{\phi}^{(j)}(0)\frac{\lambda^j}{j!}\right\}\overline{\left\{\hat{\psi}(\lambda) - \alpha(\lambda)\sum_0^{q-1} \hat{\psi}^{(j)}(0)\frac{\lambda^j}{j!}\right\}}\, F(d\lambda)$$
$$+ \mathrm{a}_{2q}\alpha_q\bar{\beta}_q + \sum_{j=0}^{q-1}\alpha_j\overline{L_j(\psi)} + \sum_{j=0}^{q-1}\bar{\beta}_j L_j(\phi) + \sum_{j,k=0}^{q-1}\sum c_{jk}\alpha_j\bar{\beta}_k.$$

This calls for some explanation! To begin with by R_0 we mean the set consisting of the line with the origin removed while $\alpha(\lambda)$ is an entire analytic function in the space Z (see the Mathematical Appendix, Section 4) such that $\lambda = 0$ is a zero of order $(q-1)$ for $\alpha(\lambda) - 1$. We have

$$\alpha_j = \int \phi(t)t^j\, dt, \qquad \beta_j = \int \psi(t)t^j\, dt, \qquad j = 0, \ldots, q,$$

whereas $\mathrm{a}_{2q} \geq 0$. The L_j are linear functionals on K. The function $F(\lambda)$ defines a positive-tempered measure for which

$$(9.3)\qquad \int_{0<\lambda<1} |\lambda|^{2q}\, F(d\lambda) < \infty.$$

The last term in (9.2) can be understood by reference to (6.3) if we put $c_{jk} = \mathcal{E}(x_j\bar{x}_k)$. The functional $L_j(\phi)$ in the case of (6.3) is

$$\mathcal{E}\{\bar{x}_j\Phi(\phi)\}.$$

The function $\alpha(\lambda)$ has to be introduced so that the factors under the integral sign will converge to zero (appropriately) as $\lambda \to \infty$ and thus will belong to Z (on which the positive tempered measures define linear functionals). The term $a_{2q}\bar{\beta}_q\alpha_q$ derives from the fact that we have introduced a factor λ^{-2q} into $F(d\lambda)$ so as to make (9.3) hold (see Section 6), *after removing any jump at the origin in* λ^{2q} $F(d\lambda)$. This jump is a_{2q} and, since

$$\lim_{\lambda\to 0} \lambda^{-q}\left\{\hat{\phi}(\lambda) - \sum_{j=0}^{q-1} \hat{\phi}^{(j)}(0)\frac{\lambda^j}{j!}\right\} = \hat{\phi}^{(q)}(0),$$

whereas

$$\lim_{\lambda\to 0} \lambda^{-q}(\alpha(\lambda) - 1)\sum_{j=0}^{q-1} \hat{\phi}^{(j)}(0)\frac{\lambda^j}{j!} = \text{const } \hat{\phi}(0),$$

then absorbing the contribution from this last expression into the last term in (9.2) (i.e., redefining some of the c_{jk}) we obtain the expression (9.2).

It is evident that we shall have a corresponding representation of Φ in the form

$$\Phi = \int_{R_0}\left\{e^{-it\lambda} - \alpha(\lambda)\sum_0^{q-1}\frac{t^j}{j!}(-i\lambda)^j\right\}\zeta(d\lambda) + \sum_0^q x_j t^j, \tag{9.4}$$

where $\mathcal{E}\{|\zeta(d\lambda)|^2\} = F(d\lambda)$ and where, again, this expression (9.4) has to be interpreted in the same way as in Theorem 2‴ that is, as

$$\Phi(\phi) = \int_{R_0}\left\{\hat{\phi}(\lambda) - \alpha(\lambda)\sum_0^{q-1}\hat{\phi}^{(j)}(0)\frac{\lambda^j}{j!}\right\}\zeta(d\lambda) + \sum_0^q x_j\alpha_j.$$

The generalized processes of the form (9.4) constitute therefore rather a general type and may be thought of as obtained by repeated integrations from a GRP got itself, perhaps, by repeated differentiation of an initial (proper) random process. The procedure is closely related therefore to that leading to Equation 3.9 in Chapter I.

10. SPECTRAL THEORIES FOR HOMOGENEOUS RANDOM PROCESSES ON OTHER SPACES†

So far we have considered vector random processes $x(t)$ in which t varies over the real line or some subset of it. As mentioned at the beginning of Chapter I, in some applications t would in fact be a space variable, for

† This section is special and can be omitted.

example, distance downstream from some fixed point on a river where the vector $x(t)$ might have three components corresponding to the "velocity" of the river. Alternatively, $x(t)$ might be measurements made along a line across a field, across the bed of the ocean and so on. In these circumstances one might wish also to consider time variation so that one will observe a realization of a random process $x(v, t)$, let us say, where v and t vary independently upon the line so that x is a function upon the plane. For that matter one might commence from the case where, with each point v of a plane surface (e.g., a field, or the ocean), an observable is associated. We shall use v again to indicate the point in the plane so that now v needs two coordinates to "name" it. Again time variation might be present so that $x(v, t)$ could be considered. Now the argument of $x(\cdot)$ varies over three-dimensional Euclidean space. Evidently we need to consider a random process $x(v)$ defined upon n-dimensional Euclidean space, R^n. We shall separate t from the other coordinates of $x(\cdot)$ when we need to emphasize its special nature as the time coordinate but begin by considering the case where this is not done. We assume that

$$\mathcal{E}\big(x(v_1)x'(v_2)\big) = \Gamma(v_1, v_2)$$

is finite but shall confine ourselves to the analog of a stationary process for which

$$\Gamma(v_1 + v, v_2 + v) = \Gamma(v_1, v_2),$$

where by $v_1 + v$ we mean, of course, the point got from v_1 by translation through v or, equivalently, in coordinates the sum of the vectors v_1 and v. As before we put

$$\Gamma(v_1, v_2) = \Gamma(0, v_2 - v_1) = \Gamma(v_2 - v_1).$$

We then say that $x(v)$ is a "homogeneous random process." We shall use v both for the point v and the vector with components the coordinates of the point v. We do the same with the point and vector θ we immediately introduce. *We take $\Gamma(v)$ to be continuous. Thus $x(v)$ is mean square continuous.*

We say that $z(\theta)$, $\theta \in R^n$, is a process of orthogonal increments if for any bounded measurable sets S_1, S_2

$$\mathcal{E}\left\{\int_{S_1} z(d\theta) \int_{S_2} z(d\theta)\right\} = 0, \qquad S_1 \cap S_2 = \varnothing.$$

We omit the proof of the following theorem because it is a rather simple elaboration of Theorems 1 and 2 in Chapter II.

Theorem 13. *If $x(v)$ is a homogeneous random process on R^n with covariance function $\Gamma(v)$, then*

$$\Gamma(v) = \int_{R^n} e^{i(v,\theta)} F(d\theta), \tag{10.1}$$

where θ is a vector in n-dimensional Euclidean space and (v, θ) is the inner product of v and θ. F is a matrix valued function defined upon R^n which is composed of functions of bounded variation and for which

$$\int_S F(d\theta)$$

is always Hermitian nonnegative definite. Correspondingly we have

$$x(v) = \int_{R^n} e^{-i(v,\theta)} z(d\theta), \tag{10.2}$$

where $z(\theta)$ is a function of "orthogonal increments" with

$$\mathcal{E}\{z(d\theta)\, z(d\theta)^*\} = F(d\theta).$$

Of course (10.2) may be represented in real form, which we exhibit for the case $n = 2$ with u and t as the two coordinates. We take x scalar also and have

$$x(u, t) = \iint_0^\infty \{\cos(ku + \lambda t)\, \xi(dk, d\lambda) + \sin(ku + \lambda t)\, \eta(dk, d\lambda)\}.$$

We have called k, λ the coordinates of θ. Now the only nonvanishing covariances are

$$\begin{aligned}\mathcal{E}\{|\xi(d\theta)|^2\} = \mathcal{E}\{|\eta(d\theta)|^2\} &= 4F(d\theta),\ \lambda \neq 0, k \neq 0\\ &= 2F(d\theta),\ \lambda \neq 0, k = 0 \text{ or } \lambda = 0, k \neq 0,\\ \mathcal{E}\{|\xi(d\theta)|^2\} = F(d\theta),\ \theta = 0.&\end{aligned}$$

One would usually call λ the frequency and k the wavenumber. The nature of a "fundamental particle"

$$\{\cos(ku + \lambda t)\, \xi(d\theta) + \sin(ku + \lambda t)\, \eta(d\theta)\}$$

is apparent.

In the type of situation last discussed it is natural to consider Γ as a function of $(v_2 - v_1)$ (i.e., when v is two-dimensional with a space and time coordinate). However, when v varies over a plane surface, such as a field this seems too general for one would then expect that Γ will depend only upon the Euclidean distance, $|v_2 - v_1|$, between the two points, or at least this would often be very plausible. We then say that $x(v)$ is homogeneous and isotropic. Consider first the two-dimensional case. Put $(v, \theta) = r\lambda \cos(\psi - \phi)$, where (r, ϕ) and (λ, ψ) are the polar coordinates of v and θ. Then, from the definition of the Bessel functions J_l (Whittaker and Watson, 1946, p. 362), we have

$$e^{-i(v,\theta)} = \sum_{-\infty}^{\infty} J_l(r\lambda) e^{il(\psi-\phi-\pi/2)},$$

so that

$$x(v) = \sum_{-\infty}^{\infty} e^{-il\phi} \int J_l(r\lambda) e^{il(\psi-\pi/2)}\, z(d\theta),$$

which we write as

$$\sum_{-\infty}^{\infty} e^{-il\phi} \int_0^{\infty} J_l(r\lambda)\, \zeta^{(l)}(d\lambda), \tag{10.3}$$

where $\zeta^{(l)}(d\lambda)$ is the contribution to the integral from $\exp\{il(\psi - \pi/2)\}\, z(d\theta)$ in the annular region, A, defined by the circles with radii λ and $\lambda + d\lambda$. Evidently

$$\mathcal{E}\{\zeta^{(l)}(d\lambda_1)\, \overline{\zeta^{(m)}(d\lambda_2)}\} = 0 \qquad \lambda_1 \neq \lambda_2$$

but also

$$\mathcal{E}\{\zeta^{(l)}(d\lambda)\, \overline{\zeta^{(m)}(d\lambda)}\} = \int_A e^{i(l-m)\psi}\, F(d\theta).$$

However,

$$\begin{aligned}\gamma(v) &= \int e^{i(v,\theta)}\, F(d\theta) \\ &= \int e^{ir\lambda\cos(\psi-\phi)}\, F(d\theta) \\ &= \sum_{-\infty}^{\infty} e^{-il\phi} \int J_l(r\lambda) e^{il(\psi-\pi/2)}\, F(d\theta)\end{aligned}$$

is a function only of r, which implies that

$$\int J_l(r\lambda) e^{il(\psi-\pi/2)} F(d\theta) = 0, \quad l \neq 0,$$

which implies that F is a function only of λ. Thus we obtain

$$\mathcal{E}\{\zeta^{(l)}(d\lambda)\, \overline{\zeta^{(m)}(d\lambda)}\} = \delta_l^{\,m}\, H(d\lambda),$$

where

$$\gamma(r) = \int_0^{\infty} J_0(r\lambda)\, H(d\lambda). \tag{10.4}$$

Let us rewrite the covariance formula (10.4) and the formulas for the covariances of the $\zeta^{(l)}$ in a form suitable for the case where $x(v)$ is, for each v, a vector of p components, namely,

$$\Gamma(r) = \int_0^\infty J_0(r\lambda)\, H(d\lambda), \tag{10.4$'$}$$

$$\mathcal{E}\{\zeta^{(l)}(d\lambda)\, \zeta^{(m)}(d\mu)^*\} = \delta_l^{\ m}\delta_\lambda^{\ \mu}\, H(d\lambda). \tag{10.5}$$

Then we may state the following theorem, whose proof we have outlined for the scalar case. (We omit all details of the extension to the vector case. These are not difficult.)

Theorem 13′. *If $x(v)$ is a vector, homogeneous, isotropic random process on the plane then it has the representation* (10.3) *where $\zeta^{(l)}(\lambda)$ are vector processes of orthogonal increments with covariances satisfying* (10.5). *Here $H(\lambda)$ is a real symmetric matrix with nonnegative definite increments, uniquely defined (up to an additive constant matrix) by the requirement that it be continuous from the right. The covariance function, $\Gamma(r)$, is related to $H(\lambda)$ by* (10.4)′. *The expression* (10.3) *may be rewritten in the equivalent real form*

$$x_k(v) = \sum_0^\infty \int_0^\infty J_l(r\lambda)\{\cos l\phi\, \xi_k^{(l)}(d\lambda) + \sin l\phi\, \eta_k^{(l)}(d\lambda)\} \tag{10.3$'$}$$

where the only nonvanishing covariances are

$$\mathcal{E}\{\xi_j^{(l)}(d\lambda)\, \xi_k^{(l)}(d\lambda)\} = \mathcal{E}\{\eta_j^{(l)}(d\lambda)\, \eta_k^{(l)}(d\lambda)\} = 2H_{jk}(d\lambda).$$

It is to be observed that no "quadrature spectrum" occurs. This is because $\mathcal{E}\big(x(v)\, x(w)'\big) = \mathcal{E}\big(x(w)\, x(v)'\big)$ since the covariance function depends only upon the distance. If experience does not verify this hypothesis then we must return to the representation (10.2).

Example after example with the same basic structure can be constructed. Thus one might consider the case where observations are made at points u, t where $u \in R^2$ and t is the time variable. (Thus $v \in R^3$ but we distinguish the time and space components.) We are led to assume that

$$\mathcal{E}\big(x(u, s)\, x(w, t)\big) = \Gamma(r, t - s),$$

where $r = |u - w|$. Now introducing the polar coordinates (r, ϕ) for u we obtain the representation

$$x(u, t) = \int_0^{2\pi} \int_0^\infty \int_{-\infty}^\infty \exp i\big\{r\lambda(\phi - \psi) - t\mu\big\}\, z(d\mu, d\lambda, d\psi), \tag{10.3$''$}$$

wherein

$$\mathcal{E}\big\{z(d\mu, d\lambda, d\psi) \cdot z(d\mu', d\lambda', d\psi')^*\big\} = \delta_\mu^{\ \mu'}\delta_\lambda^{\ \lambda'}\delta_\psi^{\ \psi'}\, F(d\mu, d\lambda)\, \frac{d\psi}{2\pi}.$$

This expression represents $x(u, t)$ as a linear superposition of wave forms with frequency μ and wavenumber λ propagated in the direction ψ, with all directions given μ, λ contributing equally to the variance and with phase and amplitude determined by the vector process of orthogonal increments, z. The covariance function satisfies

$$\Gamma(r, t) = \int_{-\infty}^{\infty}\int_{0}^{\infty} e^{-it\mu} J_0(r\lambda)\, F(d\mu, d\lambda). \tag{10.4$''$}$$

If R^3 replaces R^2 analogous formulas are obtained with $J_l(r\lambda) \exp(-il\phi)$ replaced by $Y_l^m(\theta, \phi) J_{l+\frac{1}{2}}(r\lambda)$ where

$$Y_l^m(\theta, \phi) = \frac{1}{(2\pi)^{1/2}} e^{im\phi} P_l^m(\cos\theta)$$

is a spherical harmonic and P_l^m is the normalized associated Legendre function. It is evident that these examples must be special cases of some general theory. That theory is necessary first because there are so many other cases which can arise that it is economical to treat them in a unified fashion, and second because without the general theory the particular cases become bewildering as a profusion of special functions is introduced. It is impossible in the short space available here to deal fully with the general theory. We refer the reader to Hannan (1965a) and, especially, Yaglom (1962) for a more complete treatment.

To try to understand the nature of the situation we begin with the special case of the sphere S^2, in three-dimensional space. Thus v is a point on S^2 and $x(v)$ is a random variable for each v. Moreover, taking the scalar case at first, $\gamma(v, w) = \mathcal{E}\big(x(v)\, x(w)\big)$, $v, w \in S^2$, is a continuous function of v and w. We now introduce $O_+(3)$ the group of (3×3), real, orthogonal matrices with determinant 1, i.e., the group of rotations. We consider $O_+(3)$ as a group of rotations about the center of S^2 and write gv for the point into which $v \in S^2$ is taken by $g \in O_+(3)$. To avoid confusion we shall, in this section use e for the identity operator in $O_+(3)$. *Our essential requirement is that* $\gamma(gv, gw) = \gamma(v, w)$, $v, w \in S^2$, $g \in O_+(3)$. In this case it is evident that γ depends only on the great circle distance between v and w. As we have previously done we use the symbol $\gamma(v)$ for the function $\gamma(v_0, v)$, v_0 being, say, the north pole. This prescribes $\gamma(v, w)$ in its entirety via $\gamma(v, w) = \gamma(v_0, gw)$, $gv = v_0$. This function, $\gamma(v)$, may be considered as a function on the group $O_+(3)$ by introducing the function $\tilde{\gamma}(g) = \gamma(v)$ if $gv_0 = v$. We observe that if $k \in O_+(2)$, the subgroup of $O_+(3)$ leaving v_0 fixed (i.e., the group of rotations in the horizontal plane), then, since $\gamma(v_0, v) = \gamma(kv_0, kv) = \gamma(v_0, kv)$, we have $\gamma(kv) = \gamma(v)$; but, if $gv_0 = v$, then also $gkv_0 = v$ (since $kv_0 = v_0$). Thus $\tilde{\gamma}(kgk') = \tilde{\gamma}(g)$, $k, k' \in O_+(2)$. Such a function $\tilde{\gamma}(g)$ is said to be biinvariant (under $O_+(2)$). Our $\tilde{\gamma}(g)$ is a special biinvariant function since it is also nonnegative definite, i.e., the matrix with entries, $\gamma(v_i, v_j)$, $i, j = 1, \ldots, N$, is nonnegative definite for all choices of v_i, $i = 1, \ldots, N$. Now, appropriately modified, Theorem 1 continues to hold for such nonnegative definite biinvariant functions. (Of course biinvariance is rather trivial for stationary time series since now the place of $O_+(3)$ is taken by the additive group of reals and the place of $O_+(2)$ is taken by the subgroup leaving t_0 fixed, i.e., the trivial group consisting of one element, the real number 0.) This theorem, in its general form, is usually called Bochner's theorem since Bochner proved it first in the form of Theorem 1. What we wish to do is to approach the theorem in another way which exhibits its meaning clearly.

We introduce $\mathcal{H}$, the Hilbert space spanned by the random variables, $x(v)$, with inner product

$$\big(x(v), x(w)\big) = \gamma(v, w),$$

in precisely the same way as in the Appendix to Chapter I. Now we consider the operators in $\mathcal{H}$ which are defined by

$$U(g)\, x(v) = x(gv).$$

Since

$$\begin{aligned}\big(U(g) \textstyle\sum \alpha_j\, x(v_j), U(g) \sum \beta_k\, x(w_k)\big) &= \textstyle\sum\sum \alpha_j\, \bar{\beta}_k \big(x(gv_j), x(gw_k)\big)\\ &= \textstyle\sum \alpha_j \bar{\beta}_k\, \gamma(gv_j, gw_k) = \sum \alpha_j \bar{\beta}_k\, \gamma(v_j, w_k)\\ &= \big(\textstyle\sum \alpha_j\, x(v_j), \sum \beta_k\, x(w_k)\big),\end{aligned}$$

we see that $U(g)$ leaves the inner product invariant between any two sets of linear combinations of the $x(v)$. It follows immediately that, since $U(g)$ has norm which is unity on this set, we may extend the definition of $U(g)$ to all of $\mathcal{H}$ by putting $U(g)x = \lim_n U(g)x_n$, for $x_n \to x$ and $U(g)$ will continue to leave the inner product invariant, so that it is unitary.† Moreover, the correspondence

$$g \to U(g)$$

is a "representation" of $O_+(3)$ by unitary operators in $\mathcal{H}$ in the sense that it is a homomorphism, that is, $e \to I$, $g_1 g_2^{-1} \to U(g_1)\, U^{-1}(g_2)$ and the correspondence is continuous. By the latter we mean that if g_1 approaches g (in the sense that its axis of rotation and angle of rotation approach those of g) then $\| U(g_1) - U(g) \|$ approaches zero. (This is a consequence of the continuity of $\gamma(v, w)$.) We may visualize $U(g)$ by introducing an orthonormal basis, ϕ_j $j = 1, 2, \ldots$ say, in $\mathcal{H}$ so that $U(g)$ can be exhibited as a matrix

† We refer the reader, again, to the Mathematical Appendix for a survey of the concepts needed in this section.

$u_{i,k}(g)$ where

$$U(g)\phi_j = \sum_1^\infty u_{k,j}(g)\phi_k \tag{10.6}$$

The essential fact that we now state (for a proof see, for example, Naimark, 1964) is that if the ϕ_j are suitably chosen the matrix, $[U(g)]$, let us say, is reduced to a particularly simple form. Indeed $\mathcal{H}$ decomposes into subspaces† $\mathcal{H}^{(\lambda)}$, $\lambda = 0, 1, 2, \ldots$ which are such that $U(g)\mathcal{H}^{(\lambda)} \subset \mathcal{H}^{(\lambda)}$, $g \in O_+(3)$, so that $\mathcal{H}^{(\lambda)}$ is an "invariant subspace," and $\mathcal{H}^{(\lambda)}$ is itself irreducible so that it contains no nontrivial proper subspace having the same property. Thus $U(g)$ can be considered as an operator in $\mathcal{H}^{(\lambda)}$ and in $\mathcal{H}^{(\lambda)}$ we call it $U^{(\lambda)}(g)$. Thus $U^{(\lambda)}(g)$ gives again a unitary representation of $O_+(3)$ and is said to be irreducible. Two other important things are true. In the first place each $\mathcal{H}^{(\lambda)}$ is finite-dimensional and is indeed of dimension $(2\lambda + 1)$. In the second place each occurs at most once. This latter property is described by saying that $U(g)$ is "multiplicity-free." We will discuss it further, and the reason for it, in the next section. The finite dimensionality of the $\mathcal{H}^{(\lambda)}$ and the simple form of decomposition of $U(g)$ is special and is entirely due to the fact that the sphere S^2 and, equivalently, $O_+(3)$ are compact topological spaces. Thus the matrix form $[U(g)]$ is now easily understood. We choose our orthonormal basis, ϕ_j, so that ϕ_1 spans $\mathcal{H}^{(0)}$, ϕ_2, ϕ_3, ϕ_4 span $\mathcal{H}^{(1)}, \ldots$. Within each $\mathcal{H}^{(\lambda)}$ the ϕ_j will later be determined to bring $[U^{(\lambda)}(g)]$ to a simple form but in any case $[U(g)]$ becomes an infinite matrix composed of an infinite sequence of blocks down the diagonal, with the λth block, $\lambda = 0, 1, 2, \ldots$, being of $(2\lambda + 1)$ rows and columns. We may write this as

$$\mathcal{H} = \sum_{\oplus} \mathcal{H}^{(\lambda)}, \quad U(g) = \sum_{\oplus} U^{(\lambda)}(g), \quad [U(g)] = \sum_{\oplus} [U^{(\lambda)}(g)]$$

Each subspace $\mathcal{H}^{(\lambda)}$, as already said, is irreducible under $O_+(3)$ but it will not be irreducible under the smaller group $O_+(2)$. However, there is just one, one-dimensional, subspace of $\mathcal{H}^{(\lambda)}$ for every λ, which is such that $O_+(2)$ acts in that subspace as the trivial identity operator. We choose, conventionally, the $(\lambda + 1)$th of the $(2\lambda + 1)$ orthonormal vector spanning $\mathcal{H}^{(\lambda)}$ as the vector spanning this one-dimensional space. Thus $[U^{(\lambda)}(k)]$, $k \in O_+(2)$ has zero elements in its center row and column save for unity on the main diagonal. *However, this is only so for* $k \in O_+(2)$. The other elements in this center row are determined when the remaining orthonormal basic vectors in $\mathcal{H}^{(\lambda)}$ are chosen. They are called spherical functions.‡ This may always be done so that the elements in this center column are of the form

$$\left(\frac{4\pi}{2\lambda + 1}\right)^{1/2} Y_\lambda{}^n(\theta, \phi), \tag{10.7}$$

† More properly all of these $\mathcal{H}^{(\lambda)}$ occur in the direct decomposition *at most* once.

‡ The element in the center row and column is called a zonal spherical function.

where θ is the latitude (measured from zero at the north pole to π at the south pole) and ϕ is the longtitude of the point into which the north pole is taken by the rotation g. Now $x(v_0)$ must be of the form

$$\sum_\lambda \alpha(\lambda)\, z^{(0)}(\lambda)$$

(where $z^{(0)}(\lambda)$ lies in the one-dimensional subspace of $\mathcal{H}^{(\lambda)}$ left fixed by $O_+(2)$, and is normalized to have unit norm), since $U(k)\, x(v_0) = x(v_0)$. Thus, if $gv_0 = v$, we obtain immediately from (10.6)

$$x(v) = U(g)\, x(v_0) = \sum_\lambda \alpha(\lambda)\, U^{(\lambda)}(g)\, z^{(0)}(\lambda) \tag{10.8}$$

$$= \sum_{n=-\lambda}^{\lambda} \sum_{\lambda=0}^{\infty} Y_\lambda{}^n(\theta, \phi)\, \zeta^{(n)}(\lambda),$$

$$\zeta^{(n)}(\lambda) = \alpha(\lambda)\left(\frac{4\pi}{2\lambda + 1}\right)^{1/2} z^{(n)}(\lambda),$$

and $z^{(n)}(\lambda)$ is the nth orthonormal vector ($n = -\lambda, -\lambda + 1, \ldots, \lambda$) chosen in the space $\mathcal{H}^{(\lambda)}$ to give $U^{(\lambda)}(g)$ the matrix form in which the middle column is given by (10.7). Thus

$$\mathcal{E}\{|\zeta^{(n)}(\lambda)|^2\} = |\alpha(\lambda)|^2\left(\frac{4\pi}{2\lambda + 1}\right) = H(\lambda),$$

let us say, and of course the $\zeta^{(n)}(\lambda)$ are orthonormal.

Moreover, we obtain

$$\mathcal{E}(x(v)\, x(v_0)) = \sum_{\lambda=0}^{\infty} Y_\lambda{}^0(\theta, \phi)\, H(\lambda)\left(\frac{2\lambda + 1}{4\pi}\right)^{1/2}$$

$$= \frac{1}{2\pi} \sum_{\lambda=0}^{\infty} P_\lambda(\cos\theta)\, H(\lambda)\left(\frac{2\lambda + 1}{2}\right)^{1/2}$$

if $P_\lambda(\cos\theta)$ is the (ordinary) normalized Legendre polynomial of the order λ. It is, up to a constant factor, the zonal spherical function.

The vector case is not essentially more difficult and may be handled by the device introduced in the proofs of Theorems 1 and 2. Thus we have outlined the proof of the following.

Theorem 13″. *If $x(v)$ is a vector homogeneous random process on the sphere then $x(v)$ satisfies* (10.8), *where the vectors $\zeta^{(n)}(\lambda)$ satisfy*

$$\mathcal{E}\{\zeta^{(m)}(\lambda)\, \zeta^{(n)}(\mu)^*\} = \delta_m{}^n \delta_\lambda{}^\mu\, H(\lambda),$$

$$m, n = -\lambda, -\lambda + 1, \ldots, \lambda; \qquad \lambda, \mu = 0, 1, \ldots.$$

and $H(\lambda)$ is a nonnegative symmetric matrix such that, r being the great circle distance from v_1 to v_2,

$$\Gamma(r) = \mathcal{E}(x(v_1)x(v_2)') = \frac{1}{2\pi}\sum_{\lambda=0}^{\infty} P_\lambda(\cos r)\, H(\lambda)\left(\frac{2\lambda+1}{2}\right)^{1/2}.$$

The spectral representation (10.8) is more complicated than that given by Theorem 13 because for each λ value there are $(2\lambda + 1)$ contributions to the sum. It is simpler, on the other hand, because the sum is discrete. The first fact is due to the noncommutativity of the group $O_+(3)$; the second is due, as we have said, to the fact that this group is compact.

The theory for the stationary processes on the real line may be developed in a corresponding fashion. Once more we construct $\mathcal{H}$. Once more we have a group, that of the additive reals, so that

$$U(s)\, x(t) = x(t + s),$$

and now to accord with the usual conventions we have $U(s)\, U(t) = U(s + t)$. Again this leads to a unitary representation in $\mathcal{H}$. Once again we consider the irreducible unitary representations of the group which now are of the simplest kind, the $\mathcal{H}^{(\lambda)}$ being one-dimensional and the $[U^{(\lambda)}(s)]$ being of the form $\exp(i\lambda s)$ with $\lambda \in (-\infty, \infty)$. The description of the decomposition of an arbitrary unitary representation into irreducible components is, however, more complex because of the (only) local compactness of the group. It can be described as follows. We can represent $\mathcal{H}$, isomorphically with respect to the action of the group, as a family of functions $L_2(\mu)$ square integrable with respect to a measure function μ on the real line $\lambda \in (-\infty, \infty)$. The action of $U(t)$ in $L_2(\mu)$ is of the form

$$y(\lambda) \in L_2(\mu) \to U(s)\, y(\lambda) = U^{(\lambda)}(s)\, y(\lambda) = e^{-i\lambda s}\, y(\lambda). \tag{10.9}$$

It is fairly evident that the only essential thing about μ is its support, that is, we can replace μ by any measure μ_1 which is such that μ and μ_1 are mutually absolutely continuous. This will merely change $y(\lambda)$ to $y(\lambda)(d\mu/d\mu_1)$, and since $(d\mu/d\mu_1)$ will be *nonzero* a.e. (μ) nothing essential has changed. We choose μ so that the function representing $x(0)$, say $x_0(\lambda)$, is unity a.e. μ. We can do this since $x_0(\lambda)$, for a given initial μ_1, must be nonzero a.e. μ_1 since if that were not so all representatives, $\exp(-i\lambda s)\, x_0(\lambda)$ would be zero on one and the same set and the mapping of $\mathcal{H}$ onto $L_2(\mu_1)$ would not give an isomorphic mapping of the initial representation $U(s)$, onto the new representation described by (10.9). Thus $\mu(d\lambda) = x_0(\lambda)\, \mu_1(d\lambda)$ is a suitable new measure and we now have $x(0)$ being mapped onto the function identically unity. Thus $x(t) = U(t)\, x(0)$ must be mapped onto the function $x_t(\lambda) = \exp(-it\lambda)\, x_0(\lambda) = \exp -it\lambda$. If we replace μ by the distribution function on the real line which it defines then the isomorphism between $\mathcal{H}$ and $L_2(F)$ is generated in its entirety by the correspondence

$$x(t) \leftrightarrow e^{-it\lambda}, \tag{10.10}$$

which we have used in Section 1 of the Appendix to this chapter to establish Theorem 2.

This formalism is often replaced by another one which superficially looks more like that used in the case of S_2. We call $E(\theta)$ the operator in $L_2(F)$ which projects onto the subspace of $L_2(F)$ of functions zero for $\lambda > \theta$. It is easy to see that $E(\theta)$ is indeed a perpendicular projection. The family of projections $E(\theta)$ has the properties

(i) $$E(-\infty) = 0, \qquad E(\infty) = I,$$

(ii) $$E(\lambda)E(\theta) = E(\theta), \qquad \lambda \geq \theta,$$

(iii) $$\lim_{\delta \downarrow 0} \|E(\lambda + \delta) - E(\lambda)\| = 0,$$

which characterize what is usually called a spectral family. Thus we are led to write $U(s)$ in the form

$$U(s) = \int_{-\infty}^{\infty} e^{-i\lambda s}\, E(d\lambda) \tag{10.11}$$

so that

$$U(s)\, x(0) = \int_{-\infty}^{\infty} e^{-i\lambda s}\, E(d\lambda)\, x(0) = \int_{-\infty}^{\infty} e^{-i\lambda s}\, z(d\lambda),$$

where $z(d\lambda) = E(d\lambda)\, x(0)$. This is only another way of writing (10.9) and is to be preferred only insofar as it aids intuition through the analogy between (10.11) and the formula for the spectral decomposition of a unitary operator in a finite-dimensional vector space.

The case in which the space where v lies is R_2 and the group of symmetries is $I_0(R_2)$, the group of Euclidean motions, exhibits some of the full complexity, since now the group is both noncompact (locally compact) and noncommutative. The group $I_0(R_2)$ of transformations we now consider is of the form

$$v \to kv + x, \tag{10.12}$$

where k is an element of $O_+(2)$ and x is, of course, a vector of two components. This group is evidently noncommutative and noncompact (since it contains a subgroup, $v \to v + x$, which is topologically R_2 itself). Again we construct $\mathcal{H}$ and again we have a representation of $I_0(R_2)$, which is unitary, generated by

$$x(v) \to U(g)\, x(v) = x(gv), \qquad g \in I_0(R_2),$$

where gv is the point into which v is taken by g. Once again we consider the decomposition of $U(g)$ into irreducible components. In this case, however, these irreducible unitary representations, $U^{(\lambda)}(g)$ are either infinite dimensional or map the group element defined by (10.12) onto an irreducible unitary representation $U(k)$ of $O_+(2)$. The latter do not arise in our further considerations, since they are not of "class one"; that is, they do not contain that single one-dimensional subspace in which $U(k)$, $k \in O_+(2)$ acts as the identity operator. The remaining, infinite-dimensional, unitary representations are indexed by $\lambda \in [0, \infty)$. The elements in the "center column" of an irreducible representation are, for a suitable choice of orthonormal basis in $\mathcal{H}^{(\lambda)}$, of the form

$$e^{-il\phi}J_l(r\lambda), \tag{10.13}$$

where r, ϕ are the polar coordinates of the point into which the origin goes under the particular group element being considered. The decomposition of the representation can be defined by an extension of the notion already introduced. We consider a measure μ on $[0, \infty)$ and with each $\lambda \in [0, \infty)$ associate a Hilbert space $\mathcal{H}^{(\lambda)}$. Then our representation U is equivalent to a representation obtained as follows. The representation space is realized as a family of functions $x(\lambda)$ where $x(\lambda) \in \mathcal{H}^{(\lambda)}$ and†

$$\int_0^\infty \|x(\lambda)\|_\lambda^2 \, \mu(d\lambda) < \infty,$$

where by $\|x(\lambda)\|_\lambda^2$ we mean the squared norm of $x(\lambda)$ as an element of $\mathcal{H}^{(\lambda)}$. The inner product between two elements $x(\lambda)$, $y(\lambda)$ is, of course,

$$\int_0^\infty (x(\lambda), y(\lambda))_\lambda \, \mu(d\lambda),$$

where $(x(\lambda), y(\lambda))_\lambda$ is the inner product in $\mathcal{H}^{(\lambda)}$.

Now our representation is equivalent to that given by

$$U(g)\, x(\lambda) = U^{(\lambda)}(g)\, x(\lambda), \qquad g \in I_0(R_2), \qquad \lambda \in [0, \infty).$$

Again the only fundamental aspect of μ is its support and again we may choose μ so that $x(v_0)$ (v_0 the origin of coordinates in R_2) is of a simple form, viz., of the form $x_0(\lambda)$ where $x_0(\lambda)$, for each λ, belongs to the one-dimensional subspace in which $O_+(2)$ acts as the identity operator. Then if $gv_0 = v$

$$x(v) \to U(g)\, x_0(\lambda) = U^{(\lambda)}(g)\, x_0(\lambda) = \sum_l e^{-il\phi}J_l(r\lambda)\, x_l(\lambda), \tag{10.14}$$

where $x_l(\lambda)$ varies over an orthonormal basis for $\mathcal{H}^{(\lambda)}$ so chosen as to make the appropriate elements of the matricial form of $U(g)$ equal to (10.13).

† Of course it is implied that $\|x(\lambda)\|^2$ is measurable so that the integral is defined.

Here, of course, r, ϕ are as before. Now

$$\mathcal{E}\big(x(v)\,x(v_0)\big) = \int_0^\infty J_0(r\lambda)\,H(d\lambda),$$

calling H the distribution function defined by μ. Thus we have obtained (10.4)′. Now (10.3) is just another way of writing (10.14).

The general pattern is now evident. A topologized space V is given and a group G acts as a transformation group on V transitively (so as to take v_0 to any $v \in V$). The group leaving v_0 fixed, K, is compact. G is of a sufficiently special kind to permit a constructive, simple, representation theory (e.g., G is the group of symmetries of a globally symmetric space, V, as in the examples considered. See Helgason, 1962). The formulas we then need are obtained via the representation theory for G. Considerable detail concerning these is given in Vilenkin (1968).

Two points only need mention before we close this section. In the first place we may put spaces, V, together as topological products with the group of symmetries of the product space being the direct product of the corresponding groups. Then Λ, the set over which λ varies, will be the corresponding product space and the "spherical functions" are correspondingly products. Thus in the case of $R_2 \times R$ with $I_0(R_2) \otimes R$ as the group of symmetries we have spherical functions, using θ, λ for coordinates in Λ,

$$e^{-i\theta t}e^{-il\phi}J_l(r\lambda),$$

and the formula given in (10.3)″ are obtained.

In the second place all of the results extend to the vector case in a quite straightforward fashion which does not deviate from that used to treat the case for Theorems 1 and 2 in Section 2.

11. FILTERS, GENERAL THEORY†

We wish in this section to give a rather general account of the theory of filters as well as some account of the general concept of a filter. The first subsection, though rather technical, is straightforward and concerns nonlinear noninstantaneous filters and is based on Wiener (1958). In the second subsection an attempt is made to indicate the generality and central importance of the concept of a filter.

(i) If the action of a nonlinear filter is to be discussed, even through the spectrum of the output of the filter, then one needs to prescribe the whole probability structure of the filtered process. This leads us to prescribe the initial process $x(t)$ as a strictly stationary vector process of p components.

† This section is special and can be omitted.

We consider the probability space $(\Omega, \mathcal{A}, P)$ of all "histories" of the process $x(t)$, so that $x(t) = x(t, \omega)$ is composed of elements which are random variables over $(\Omega, \mathcal{A}, P)$. We then consider $L_2(\Omega, \mathcal{A}, P)$ which is the space of functions on Ω, square integrable with respect to P. We shall call this L_2 for short. This space L_2 is spanned by the functions $\chi(\omega)$

$$\begin{cases} \chi(\omega) = 1, & \omega \in \{\omega \mid (x_{j(k)}(t_k, \omega), k = 1, \ldots, n) \in B\}, \\ \quad\;\; = 0, & \text{otherwise}, \end{cases} \tag{11.1}$$

where by $(x_{j(k)}(t_k, \omega), k = 1, \ldots, n)$ we mean the point in R^n with these coordinates and B is a Borel set in R^n. (Thus $\chi(\omega)$ is the indicator function of a cylinder set in Ω which has a finite dimensional Borel set as base.) If $x(t)$ is Gaussian so that each $x_j(t, \omega) \in L_2$ then L_2 is spanned by multinomials in the $x_j(t, \omega)$ since any indicator function of a Borel set B in R^n may be approximated in mean square by such a multinomial.

We commence our study of nonlinear filters by assuming $x(t)$ to be Gaussian and scalar and we define our filter to be a function

$$y(t) = f(\{x(t)\}), \qquad \mathcal{E}(y(t)^2) < \infty, \tag{11.2}$$

which allots to $x(t)$ an element $y(t) \in L_2$. We mean by this that $y(t + s) = f(\{x(t + s)\})$. Equivalently we might describe this as follows. In L_2 there is defined the operator $U(s)$ by means of $U(s)\, x(t) = x(t + s)$. This definition extends to all of L_2 first by

$$U(s) \sum a(j_1, \ldots, j_k)\, x(t_1)^{p_1}\, x(t_2)^{p_2} \cdots x(t_k)^{p_k} = \sum a(j_1, \ldots, j_k)\, x(t_1 + s)^{p_1} \cdots x(t_k + s)^{p_k}$$

and then by continuity, so that if x_n is a sequence of such multinomials converging in mean square to x then $U(s)x = \lim_n U(s)x_n$. (The limit always exists.)† Then $U(s)$ is for all s a unitary operator in L_2, by stationarity. Indeed $U(s)$ is a unitary representation of the group of translations of the real line. (See Section 10.) Now let A be a linear operator in L_2 with "reasonable" properties. We interpret these to be as folllows.

(a) A has domain $\mathfrak{D}(A)$ dense in L_2. If A is defined for all multinomials, which we certainly need, this must be so.

(b) A is closed. We need this also for if x_n is a sequence of elements which converges to x and Ax_n converges to y we shall certainly want $x \in \mathfrak{D}(A)$ and $Ax = y$. We cannot require more and say that A is bounded for differential operators will not be bounded.

We may now define a filter to be such an operator which also satisfies

$$AU(s) \supseteq U(s)A, \qquad s \in (-\infty, \infty),$$

† See the Mathematical Appendix for a survey of the theory needed here and below.

where this means that $x \in \mathfrak{D}(A)$ implies $U(s)x \in \mathfrak{D}(A)$ and $AU(s)x = U(s)\,Ax$. We now put $y(t) = Ax(t) = AU(t)\,x(0) = U(t)\,Ax(0)$, *assuming* $x(0) \in \mathfrak{D}(A)$. This is the formal and precise definition corresponding to (11.2).

We now attempt to describe in some detail how the action of A on $x(t)$ may be described, at least for sufficiently simple operators A. We recall that we have, from Section 5, example (vi), *if* $F(\lambda)$ *is a.c.*,

$$x(t) = \int_{-\infty}^{\infty} \alpha(t - s)\, \xi(ds),$$

where $\xi(t)$ satisfies

$$\mathcal{E}\{(\xi(t) - \xi(s))^2\} = |t - s|, \qquad \mathcal{E}(\xi(t)) = 0.$$

Moreover, $\xi(t)$ is Gaussian, for as we show in the Appendix to this chapter, $\xi(t)$ is the limit in mean square of a sequence of expressions which are finite linear combinations of the $x(t)$, each of which is Gaussian. (Thus $\xi(t)$ is a standard Brownian notion on $(-\infty, \infty)$.) Now we may approximate to $y(t)$ in mean square by expressions of the form

$$\text{(11.3)} \quad a(0) + \int_{-\infty}^{\infty} a_1(t - s)\, \xi(ds) + \iint_{-\infty}^{\infty} a_2(t - s_1, t - s_2)\, \xi(ds_1)\xi(ds_2) + \cdots.$$

Indeed this is the same as saying that we may approximate to $y(t)$ by a sequence of multinomial functions of the random variable $x(t)$,

$$\text{(11.4)} \quad a(0) + \sum_j a_1'(j) \int_{-\infty}^{\infty} \alpha(t_j - s)\, \xi(ds)$$
$$+ \sum_{j,k}\sum a_2'(j, k) \iint_{-\infty}^{\infty} \alpha(t_j - s_1)\, \alpha(t_k - s_2)\, \xi(ds_1)\, \xi(ds_2) + \cdots.$$

However, we should observe that in (11.3) not only must $a_2(u, v)$ be square integrable on the plane but also $a_2(u, u)$ must be integrable and so on.

To achieve a unique form of decomposition we proceed as follows. We call V_ν the linear subspace of $L_2(\Omega, \mathcal{A}, P)$ spanned by all expressions of the form

$$\text{(11.5)} \qquad \int_{-\infty}^{\infty} \cdots \int \alpha(s_1, \ldots, s_\nu)\, \xi(ds_1) \cdots \xi(ds_\nu).$$

Of course V_0 is just the space of constants. In (11.5) not only is $\alpha(s_1, \ldots, s_\nu)$ square integrable over ν dimensional space (with respect to Lebesgue measure) but also all integrals of the type

$$\int \cdots \int \alpha(s_1, s_1, t_1, s_2, s_2, \ldots, t_2 \ldots)\, \alpha(u_1, t_1, u_1, u_2, u_2, \ldots, t_2 \ldots)$$
$$ds_1\, ds_2 \cdots dt_1\, dt_2 \cdots du_1\, du_2 \cdots$$

must be finite. By this we mean that we have broken the indices $s_1, \ldots, s_\nu$, $u_1, \ldots, u_\nu$ up into ν pairs in an arbitrary way and have indicated pairs occurring in the first $\alpha(\cdot)$ by s_j, in the second by u_j, and those which have one member in each set of arguments by t_j. The fact that all such expressions as these need to be finite follows from the property of Gaussian variables with zero mean, that an expectation of a product of an odd number of them is zero and for an even number is found as the sum of the products of expectations of pairs, taken in every possible way. It may be observed also that there is no loss of generality in considering all such expressions as (11.5) to have a kernel, $\alpha(\cdot)$, which is a symmetric function of its arguments since (11.5) is clearly not changed by a permutation of the subscripts. Let $\mathcal{H}_\nu$ be the projection of V_ν onto

$$\left(\bigcup_{j=0}^{\nu-1} V_j\right)^{\perp}$$

Thus $\mathcal{H}_\nu$ is the part of V_ν orthogonal to $V_j, j < \nu$. Then

$$\sum_{\oplus} \mathcal{H}_\nu = L_2(\Omega, \mathcal{A}, P).$$

Thus we may write

$$y(t) = \sum_0^\infty G_\nu\big(y(t)\big) = \sum_0^\infty g_\nu(t), \tag{11.6}$$

where $G_\nu\big(y(t)\big)$ is the projection of $y(t)$ onto $\mathcal{H}_\nu$. The components $G_\nu\big(y(t)\big)$ are orthogonal random processes, for different ν, so that

$$\mathcal{E}\{G_\nu\big(y(s)\big)\,\overline{G_\mu\big(y(t)\big)}\} = 0, \qquad \mu \neq \nu.$$

(It may be convenient to allow coefficient functions in (11.4) to be complex, hence the conjugation.) Thus the computation of the spectrum of $y(t)$ is reduced to that of computing the spectrum of an individual $G_\nu\big(y(t)\big)$. Unfortunately this, even in simple cases, is a task of formidable difficulty. To understand the nature of the problem take the case of formula (11.4) where only the third term in that expansion is present. The space $\mathcal{H}_0 = V_0$ is that of constants. If $x(t)$ has mean value zero, which it costs us nothing to assume, then $V_0 \perp V_1$ and $\mathcal{H}_1 = V_1$. We must now make

$$\iint_{-\infty}^{\infty} a_2(t - s_1, t - s_2)\, \xi(ds_1)\, \xi(ds_2) \tag{11.7}$$

orthogonal to V_0 and V_1. Since the expectation of this expression is

$$\int_{-\infty}^{\infty} a_2(t - s, t - s)\, ds,$$

we achieve the first objective by subtracting this expression from (11.7). Moreover, (11.7) is orthogonal to $\mathcal{H}_1$ since a product of (11.7) with a typical element of $\mathcal{H}_1$ involves taking expectations with respect to an odd number (i.e., 3) of random variables. Thus we are left with the problem of finding the spectrum of (11.7), after mean correction. The autocovariances are easy to find and are

$$(11.8) \qquad 2\iint_{-\infty}^{\infty} a_2(t - s_1, t - s_2)\, a_2(s_1, s_2)\, ds_1\, ds_2.$$

Since a_2 is square integrable on the plane, we have

$$a_2(u, v) = \frac{1}{2\pi}\int_{-\infty}^{\infty} e^{i(u\theta_1 + v\theta_2)}\, \hat{a}_2(\theta_1, \theta_2)\, d\theta_1\, d\theta_2$$

in the sense of mean-square convergence, where

$$\hat{a}_2(\theta_1, \theta_2) = \frac{1}{2\pi}\int_{-\infty}^{\infty} e^{-i(u\theta_1 + v\theta_2)}\, a_2(u, v)\, du\, dv.$$

Thus by Plancherel's theorem (11.8) is

$$2\iint_{-\infty}^{\infty} \hat{a}_2(\theta_1, \theta_2)\, \hat{a}_2(-\theta_1, -\theta_2)\, e^{-it(\theta_1 + \theta_2)}\, d\theta_1\, d\theta_2 = \int_{-\infty}^{\infty} e^{-it\lambda}\, \hat{a}(\lambda)\, d\lambda,$$

$$\hat{a}(\lambda) = 2\int_{-\infty}^{\infty} \hat{a}_2(\theta_1, \lambda - \theta_1)\, \hat{a}_2(-\theta_1, -\lambda + \theta_1)\, d\theta_1,$$

which is $2\int |\hat{a}_2(\theta, \lambda - \theta)|^2\, d\theta$ if a_2 is real.

In general the projection of the typical νth term in (11.2) onto $\mathcal{H}_\nu$ will be of the form

$$(11.9) \qquad g_\nu(t) = \int_{-\infty}^{\infty} \cdots \int a_\nu(t - s_1, t - s_2, \ldots, t - s_\nu)$$
$$\times\, \xi(ds_1) \cdots \xi(ds_\nu) + b_{\nu-1},$$

where $b_{\nu-1}$ is a collection of similar terms with at most $(\nu - 1)$ of the s_j occurring; but (11.9) is evidently then orthogonal to $b_{\nu-1}$ so that the serial covariances of (11.9) are of the form

$$\mathcal{E}\left\{\int_{-\infty}^{\infty} \cdots \int a_\nu(t - s_1, \ldots, t - s_\nu)\, \xi(ds_1) \cdots \xi(ds_\nu)\, g_\nu(s)\right\}.$$

Now we have to identify the s_j occurring in both factors in pairs in all possible ways and add. If we identify any two s_j in the first factor, we get an expression of lower order for the first factor after taking the expectation with respect to the $\xi(ds_j)$ for those two, and when we take expectations over other pairs the overall expectation vanishes. Thus we are left with

$$\nu!\int\cdots\int a_\nu(t-s_1,\ldots,t-s_\nu)\,a_\nu(s-s_1,\ldots,s-s_\nu)\,ds_1\cdots ds_\nu,$$

since there are $\nu!$ possible ways of identifying pairs of variables from the two factors. Thus calling

$$\hat{a}_\nu(\theta_1,\ldots,\theta_\nu)=\frac{1}{(2\pi)^{\nu/2}}\int e^{-iu'\theta}\,a_\nu(u_1,\ldots,u_\nu)\,du_1\cdots du_\nu,$$

where u and θ are vectors of ν components, we have the spectral density of $g_\nu(t)$ as

$$\nu!\int_{S(\lambda)}\cdots\int|\hat{a}_\nu(\theta_1,\ldots,\theta_\nu)|^2\,d\theta_1\cdots d\theta_\nu, \tag{11.10}$$

where the integration is over the plane $S(\lambda)$ on which $\sum\theta_\nu=\lambda$ (we have taken a_ν to be real).

Formally, we have reduced the calculation of the spectrum to that of the discovery of the expansion (11.6) together with the evaluation of expressions such as (11.10). The problem of determining the $g_\nu(t)$ may be aided by the following considerations.

We introduce an arbitrary family, $\phi_j(t), j=0,1,\cdots$, of functions orthonormal with respect to Lebesgue measure on the real line and which span L_2.† We put

$$u_j(t)=\int_{-\infty}^{\infty}\phi_j(t-s)\,\xi(ds),$$

so that the $u_j(t)$, $j=0,1,\ldots$, then span $\mathcal{H}_1$ and are independent and Gaussian with zero mean and unit variance. We next form

$$H_m^{(\nu)}(j,t)=\prod_{k=1}^{r}H_{m_k}(u_{j_k}(t)),\qquad j_1<j_2<\cdots<j_r,\qquad \sum_k m_k=\nu,$$

wherein H_u is the uth Hermite polynomial. Here we use m, j as symbols for the ordered multiplets $\{m_k\}$, $\{j_k\}$. Then

$$\begin{aligned}\mathcal{E}\{H_m^{(\mu)}(j,t)\,H_n^{(\nu)}(k,t)\}&=0\quad\text{unless}\quad j=k,\ m=n,\\ &=\prod_j m_j!,\qquad j=k,\ m=n,\end{aligned}\tag{11.11}$$

† The choice of an appropriate family will depend on the form of $y(t)$ as a functional of the $x(t)$ process.

since the Hermite polynomials in different $u_j(t)$ are clearly independent and (11.11) breaks up into products of the form

$$\mathcal{E}\{H_{m_i}(u_j(t))\, H_{n_i}(u_j(t))\}.$$

If $\mu \neq \nu$, then $m_i \not\equiv n_i$ and the result follows. If $\mu = \nu$ and $j \neq k$, the result follows again, since some such expectation must vanish, and if $\mu = \nu$, $j = k$, $m \neq n$ again it is easy to see that this happens. Also

$$(11.12) \qquad \mathcal{E}\{H_m^{(\mu)}(j, s)\, H_n^{(\nu)}(k, t)\} \equiv 0, \qquad \mu \neq \nu,$$

by the argument just given, but for $\mu = \nu$, $s \neq t$ the situation is not so simple as in (11.11). In any case the $H_m^{(\nu)}(j, t)$ lie in $\mathcal{H}_\nu$ and for any fixed t they span $\mathcal{H}_\nu$. Thus

$$g_\nu(t) = \sum_{m,j} a^{(\nu)}(m, j)\, H_m^{(\nu)}(j, t),$$

where the summation is over all m, j satisfying $j_1 < j_2 < \cdots$ and $\sum m_k = \nu$ and

$$\begin{aligned} a^{(\nu)}(m, j) &= \frac{\mathcal{E}\{y(t)\, H_m^{(\nu)}(j, t)\}}{\mathcal{E}[\{H_m^{(\nu)}(j, t)\}^2]} \\ &= (\Pi m_j!)^{-1}\, \mathcal{E}\{y(t)\, H_m^{(\nu)}(j, t)\}, \end{aligned}$$

which is independent of t because of stationarity. We now need

$$\mathcal{E}\{H_m^{(\nu)}(j, s)\, H_n^{(\nu)}(k, t)\}, \qquad s \neq t.$$

As in the argument that leads to (11.10), this is

$$\begin{aligned} \mathcal{E}\{(\prod_i u_{j_i}(s))^{m_i} (\prod_i u_{k_i}(t))^{n_i}\} &= \sum \prod \mathcal{E}\{u_j(s)\, u_k(t)\} \\ &= \sum \prod \{\phi_j * \phi_k(t - s)\}, \end{aligned}$$

where the summation is over all possible pairings of factors from the first product with factors from the second and the product is over pairs (j, k) in such a pairing. The complexity of the general situation is evident.

If only a small number of the ϕ_j are involved, the calculations may be manageable, but then a direct approach, not introducing the ϕ_j, may be just as easy or easier.

The extension of these ideas to the vector case or the discrete-time case is in principle straightforward. We shall not consider the matter further here.

(ii) The definition of a filter given in subsection (i) extends to the general case in which $x(t)$ is strictly stationary but not Gaussian and a vector process. Thus L_2 is defined as before, and $U(s)$ is defined in L_2 by means of the definition of $U(s)\,\chi(\omega)$, where $\chi(\omega)$ is defined by (11.1) and where by $U(s)\,\chi(\omega)$ we mean the indicator function of the cylinder set $\{\omega \mid (x_{j(k)}(t_k + s, \omega),\ k = 1, \ldots, n) \in B\}$. Again $U(s)$ is a unitary representation of the group of translations of the real line, and again we assume A to be a closed linear operator with $\mathfrak{D}(A)$ dense in L_2 and *define A to be a filter if $AU(s) \supseteq U(s)A$*, $s \in (-\infty, \infty)$. If $x_j \in L_2, j = 1, \ldots, p$, then we put

$$y_j(t) = U(t)\,Ax_j = AU(t)x_j, \qquad j = 1, \ldots, p$$

and the vector process $y(t)$ is strictly stationary with finite variance. Heuristically we may regard A as corresponding to measuring apparatus and the x_j as corresponding to those underlying aspects of the system being studied which determine the state of that system at time $t = 0$ and which determine the measurements $y_j(0)$ made at $t = 0$.† In this formalism "filter" and "strictly stationary process with finite variance" are almost equivalent concepts. It is evident that the definition of a filter extends to the general case of a strictly stationary homogeneous random process with finite variance (see Section 10) but we omit further discussion here.

The space L_2 will contain $\mathcal{H}$ as a closed linear subspace if the $x_j(t)$ have finite variance. (Here $\mathcal{H}$ is the subspace spanned by linear combinations of the $x_j(t_k)$.) If A leaves $\mathcal{H}$ invariant it may be said to be a linear filter. In this case we are led to consider A as an operator on $\mathcal{H}$. (It is again closed and with domain dense in $\mathcal{H}$.)‡ *Now if we consider A as defined only on $\mathcal{H}$ we may be more general and it is necessary only to assume that $x(t)$ is second order stationary.* We assume first that $x(t)$ is scalar and that $x(0) \in \mathfrak{D}(A)$ (hence $x(t) = U(t)\,x(0) \in \mathfrak{D}(A)$). Then, as in the previous section [see (10.10)], we may establish the correspondence

$$x(t) \leftrightarrow e^{-it\lambda}, \tag{11.13}$$

where the $\exp(-it\lambda)$ are elements of $L_2(F)$, F being the spectral distribution function of $x(t)$. Then $U(s)$ acts in $L_2(F)$ by means of $U(s)x = \exp(-is\lambda)x$, $x \in L_2(F)$. Now $y(0) = Ax(0)$ is mapped by (11.13) onto a function $h(\lambda)$ in $L_2(F)$.

$$Ax(0) = y(0) \leftrightarrow h(\lambda);$$

but now

$$Ax(t) = y(t) = U(t)\,Ax(0) \leftrightarrow e^{-it\lambda}\,h(\lambda).$$

Thus the action of the filter A is completely described by means of the response function $h(\lambda)$.

In the vector case we may proceed as follows. Let $f(\lambda)$ be the matrix of Radon-Nikodym derivatives of $F(\lambda)$ with respect to $m(\lambda) = \operatorname{tr} F(\lambda)$. Let $\mathcal{H}^{(\lambda)}$ be, for each λ, a vector space with basic vectors $e_j(\lambda), j = 1, \ldots, p$ and

† For a discussion of related concepts see Mackey (1968) p. 159.

‡ It may be shown that the closedness of A follows from its other properties. The discussions may also be generalized to the case $x(t) \notin \mathfrak{D}(A)$. (See Hannan, 1967a.)

inner product given by $f(\lambda)$, so that the inner product is

$$(e_j(\lambda), e_k(\lambda))_\lambda = f_{jk}(\lambda).$$

Then $L_2(m)$ is the space of all measurable functions of λ, $u(\lambda)$, $v(\lambda)$ whose values lie in the $\mathcal{H}^{(\lambda)}$ and with inner product

$$(u, v) = \int_{-\infty}^{\infty} (u(\lambda), v(\lambda))_\lambda \, m(d\lambda).$$

Now we may establish the isomorphism between $\mathcal{H}$ and $L_2(m)$ by means of

$$x_j(t) \leftrightarrow e^{-it\lambda} e_j(\lambda).$$

On the right the $e_j(\lambda)$ are elements of $L_2(m)$. Moreover, $U(t)$ is represented in $L_2(m)$ by means of

$$U(t)e_j(\lambda) = e^{-it\lambda} e_j(\lambda).$$

Now

$$y_j(0) = Ax_j(0) \leftrightarrow \sum_k h_{kj}(\lambda)\, e_k(\lambda);$$

since any $Ax_j(0) \in \mathcal{H}$ and thus corresponds to a function $u_j(\lambda)$ which must be expressible as shown since the $e_k(\lambda)$ from a basis. Thus

$$y_j(t) = U(t)\, Ax_j(0) \leftrightarrow \sum_k h_{kj}(\lambda) e^{-it\lambda} e_k(\lambda)$$

and the action of the filter A is described entirely by the matrix response function $h(\lambda)$.

Conversely, if A is a linear filter as defined in Section 4, example (iv) then $\mathfrak{D}(A)$ is certainly dense in $\mathcal{H}$ (since the $x(t) \in \mathfrak{D}(A)$). It may be seen that A is closed as follows. Let $m(\lambda) = \operatorname{tr}(F(\lambda))$ and $f(\lambda)$ be the matrix of Radon-Nikodym derivations of $F(\lambda)$ with respect to $m(\lambda)$. Let $E(\lambda)$ be the Hermitian idempotent which projects onto the range of $f(\lambda)$. We may assume $h(\lambda)\, E(\lambda) = h(\lambda)$ since replacing $h(\lambda)$ by $h(\lambda)\, E(\lambda)$ does not alter the filter (by Theorem 9). Let $h^{(1)}(\lambda)$ satisfy $h(\lambda) f(\lambda) = f(\lambda)\, h^{(1)}(\lambda)^*$. Then $h^{(1)}(\lambda) f(\lambda) h^{(1)}(\lambda)^* = f(\lambda)\, h(\lambda)^* f(\lambda)^{-1}\, h(\lambda) f(\lambda)$. Let us temporarily drop the λ argument for convenience. This last matrix function is integrable for

$$\operatorname{tr}\{fhf^{-1}h^*f\} \le p\|fhf^{-1}h^*f\| \le p\|fhf^{-\frac{1}{2}}\|\, \|f^{-\frac{1}{2}}h^*f\|$$

$$\le p\{\operatorname{tr}(fhf^{-\frac{1}{2}})\}^2 = p\{\operatorname{tr}(f^{\frac{1}{2}}h)\}^2 = p \operatorname{tr}(h^*fh).$$

Thus $h^{(1)}(\lambda)$ defines a filter, that is,

$$A_1 x_j(t) = \int_{-\infty}^{\infty} e^{-it\lambda} \sum_k h_{jk}^{(1)}(\lambda)\, z_k(d\lambda)$$

if well defined for all j, t; but now if A corresponds to $h(\lambda)$, then

$$\mathcal{E}\big(Ax_j(s), x_k(t)\big) = \int_{-\infty}^{\infty} e^{i(t-s)\lambda} \sum_l h_{jl}(\lambda)\, F_{lk}(d\lambda)$$

$$= \int_{-\infty}^{\infty} e^{i(t-s)\lambda} \sum_l F_{jl}(d\lambda)\, \overline{h_{kl}^{(1)}(\lambda)} = \big(x_j(s), A_1 x_k(t)\big).$$

Since the $x_j(t)$ are dense in $\mathcal{H}$ then $A_1 = A^*$ (the adjoint of A) on their common domain and this common domain is certainly dense in $\mathcal{H}$. Thus A^* has domain dense in $\mathcal{H}$ and thus A has a closed linear extension (see Riesz and Nagy, 1956, p. 305). Thus we may take the operator A to be closed.

Thus we have established the following theorem

Theorem 14. *If A is a closed linear operator in $\mathcal{H}$ with $\mathfrak{D}(A)$ dense in $\mathcal{H}$, $AU(t) \supseteq U(t)A$ and $x(0) \in \mathfrak{D}(A)$ then the action of A is described by a matrix response function, i.e.,*

$$y_j(t) = Ax_j(t) = \int_{-\infty}^{\infty} e^{-it\lambda} \sum_k h_{jk}(\lambda)\, dz_k(\lambda), \tag{11.14}$$

where, if $h(\lambda)$ is the matrix with entries $h_{jk}(\lambda)$, then

$$\int_{-\infty}^{\infty} h(\lambda)\, F(d\lambda)\, h(\lambda)^* < \infty. \tag{11.15}$$

Conversely, if A satisfies (11.14) *and* (11.15), *then it defines a linear operator in $\mathcal{H}$ of the above type.*

Once again this theorem extends to homogeneous random processes on a class of spaces† including all of those discussed in Section 10. It is of interest to ask why a linear filter is described by a frequency response function for these spaces. The scalar case illustrates a reason. It is because $U(g)$ is multiplicity-free. (See Section 10 for a definition of this term in the case of S^2.) In general, by saying that $U(g)$ is multiplicity-free we mean that all bounded linear operators commuting with all $U(g)$ commute with each other. The reason, in turn, why $U(g)$ is multiplicity-free is found in a geometric property of those spaces. For S_2 it is that for any v_1, v_2 there is a g so that $gv_1 = v_2$, $gv_2 = v_1$. In R, for example, reflection in the origin has to be adjoined for this to be so but the need for this adjunction does not affect the result. These spaces, and their generalizations (globally symmetric spaces) are particularly appropriate for Fourier methods and for the theories with which this book is concerned. For a discussion of these ideas we refer the reader to Hannan (1965).

† This class includes all globally symmetric spaces. See Helgason (1962) for their definition. For a discussion of the general case for a filter see Hannan (1965, 1967, 1969).

EXERCISES

1. Let $u(t)$ and $v(t)$ be scalar random processes with zero mean and covariance structure

$$\mathcal{E}(u(s)u(s+t)) = \mathcal{E}(v(s)v(s+t)) = \sigma^2 \rho^{|t|}, \qquad |\rho| < 1,$$

$$\mathcal{E}(u(s)v(t)) \equiv 0.$$

Show that

$$x(t) = u(t) \cos \theta t + v(t) \sin \theta t$$

is stationary and determine its spectral density.

2. Let $x(t)$ (scalar) be observed only at the points $t = 0, 1, 2, \ldots$ and be then replaced by

$$y(k) = \frac{1}{n} \sum_{j=0}^{n-1} x(k + mj), \qquad k = 0, \pm 1, \pm 2, \ldots.$$

Determine the nature of the spectrum of $y(k)$. (As a motivation we mention that one might estimate a periodic component of period m by such a device as this.)

3. We introduce the notation

$$[k]x(n) = \sum_{j=0}^{k-1} x(n+j).$$

Determine the frequency response functions of the following filters

(a) $x(n) \to x(n) - \frac{1}{24}[2][12]\, x(n-6)$

(b) $x(n) \to \Delta^p x(n)$

(c) $x(n) \to x(n) - (350)^{-1}[5]^2[7]\{2x(n-7) + x(n-6)$
$+ x(n-8) - x(n-4) - x(n-10)\}$.

Each of two stationary series $x(n)$, $y(n)$ is filtered by one of the above three filters (not necessarily the same one for the two series). For each of the nine possible combinations determine the effects on the coherence and phase between the two series. Graph the gain of the filter (c).

4. Let $x(n) = \cos n\xi + \sin n\eta$, where ξ and η are independent random variables with the same *probability* distribution function $F(x)$ (which is symmetric about the origin). Show that $x(n)$ is stationary and determine its spectrum.

5. Prove that, the eigenvalues of the matrix A having positive real parts,

$$\int_{-\infty}^{\infty} e^{-A|t|} e^{-i\lambda t}\, dt = \{(A + i\lambda I)^{-1} + (A - i\lambda I)^{-1}\}.$$

(This may be accomplished by reducing A to the canonical form by the similarity transformation $PAP^{-1} = \Sigma_{\oplus} (\lambda_j E_j + N_j)$, where the typical element of this direct sum is composed from an idempotent E_j and a nilpotent N_j for which $E_j N_j = N_j = N_j E_j$ and which is of maximal rank for such a nilpotent. The individual direct summands may be separately considered and the result obtained by reversing the similarity transformation.)

6. Let $a(n)$ be a sequence of nonnegative numbers. Show that

$$\lim_N \frac{a(N)}{\sum_1^N a(n)} = 0 \tag{1}$$

only if

$$\lim_n \frac{a(n)}{b^n} = 0, \qquad b > 1.$$

Show also that if this condition (E.1) is not satisfied there is a subsequence, $a(n_k)$, so that for some $b > 1$

$$\frac{a(n_k)}{b^k} \to 0.$$

Relate these to the results of subsection (i) of Section 6.

7. Let Γ_q be the matrix of qp rows and columns having the (m, n)th block of p rows and columns constituted by $\Gamma(n - m)$. (Thus $\gamma_{ij}(n - m)$ occurs in row $(m - 1)p + i$, $(n - 1)p + j$). Show that the necessary and sufficient condition that the $\Gamma(n)$ satisfy

$$\Gamma(n) = \int_{-\pi}^{\pi} e^{in\lambda}\, F(d\lambda),$$

with $F(\lambda)$ as in Theorem 1″, is the condition that Γ_q be nonnegative definite for all q.

8. Let

$$y(t) = \cos\left(\theta\left[\int_{-\infty}^{\infty} \alpha(s)\, x(t - s)\, ds\right]^2 + \phi\right),$$

where $x(t)$ is Gaussian and has a.c. spectrum with spectral density $f(\lambda)$ and

$$\int_{-\infty}^{\infty} \alpha(s) e^{is\lambda}\, ds$$

is square integrable with respect to $f(\lambda)$. Take ϕ to be uniformly distributed on $[-\frac{1}{2}\pi, \frac{1}{2}\pi]$ and to be independent of $x(t)$. Obtain an expression for the spectrum of $y(t)$.

9. Let x and y be Gaussian with zero mean, unit variance, and correlation ρ. Show that

$$\mathcal{E}\{H_m(x)\, H_n(y)\} = \delta_m{}^n \rho^n n!$$

(Take $m \le n$. Put $z = x - \rho y$ so that y and z are independent. Then $H_m(x) = H_m(z + \rho y)$. Expanding this in powers of y observe that only the term in y^n contributes to the expectation and that only when $m = n$. The result then follows.)

APPENDIX

1 *Proof of Theorem* 2

We begin by considering once more $x_\alpha(t) = \alpha * x(t)$ with covariance function

$$\gamma_\alpha(t) = \int_{-\infty}^{\infty} e^{it\theta}\, F_\alpha(d\theta). \tag{1}$$

The $x_\alpha(t)$ are random variables, i.e., measurable functions defined upon the measure space $(\Omega, \mathcal{A}, P)$ of all realizations of our process. They are, moreover, square integrable with respect to P (i.e., they have finite mean square) and thus belong to $L_2 = L_2(\Omega, \mathcal{A}, P)$. The subspace of L_2 spanned by the $x_\alpha(t)$, $t \in R$, we call $\mathcal{H}_\alpha$. In $\mathcal{H}_\alpha$ the inner product is, of course, defined through

$$(x_\alpha(s), x_\alpha(t)) = \mathcal{E}(x_\alpha(s)\,\overline{x_\alpha(t)}).$$

We also consider the Hilbert† space, $L_2(F_\alpha)$ of all complex-valued functions square integrable with respect to F_α, with the inner product

$$(\phi_1(\theta), \phi_2(\theta)) = \int_{-\infty}^{\infty} \phi_1(\theta)\overline{\phi_2(\theta)}\, F_\alpha(d\theta).$$

We now map $\mathcal{H}_\alpha$ into $L_2(F_\alpha)$ by means of the correspondence

$$x_\alpha(t) \leftrightarrow e^{-it\theta}, \tag{2}$$

which, from (1), preserves the inner product. Among the images of the elements of $\mathcal{H}_\alpha$ are those functions of the type

$$\begin{aligned} e_\lambda(\theta) &= 1, & -\infty < \theta \le \lambda \\ &= 0, & \lambda < \theta < \infty. \end{aligned}$$

To show this consider the function $f_{a,\epsilon}(\theta)$, which is periodic with period $2a$ and is defined as follows for $\epsilon > 0$, $a > |\lambda|$, and ϵ sufficiently small.

$$\begin{aligned} f_{a,\epsilon}(\theta) &= \epsilon^{-1}(\theta + a + \epsilon), & -a - \epsilon \le \theta \le -a \\ &= 1, & -a \le \theta \le \lambda \\ &= 1 - \epsilon^{-1}(\theta - \lambda), & \lambda \le \theta \le \lambda + \epsilon \\ &= 0, & \lambda + \epsilon \le \theta \le a - \epsilon. \end{aligned}$$

Then

$$\lim_{a\to\infty} \lim_{\epsilon\to 0} \int \{e_\lambda(\theta) - f_{a,\epsilon}(\theta)\}^2 F_\alpha(d\theta) < \lim_{a\to\infty} \int_{|\theta|\ge a} F_\alpha(d\theta) = 0,$$

so that it is sufficient to establish that $f_{a,\epsilon}(\theta)$ lies in the image of $\mathcal{H}_\alpha$. But this is a continuous periodic function so that the Cesaro sum of its Fourier series converges uniformly to the function and thus also in mean square (F_α). Thus we have established what we wish, for these Cesaro sums certainly belong to the image of $\mathcal{H}_\alpha$. This shows also that (2) maps $\mathcal{H}_\alpha$ onto $L_2(F_\alpha)$.

Now call $z_\alpha(\lambda)$ the element of $\mathcal{H}_\alpha$ which corresponds to $e_\lambda(\theta)$. Since

$$\int_{-\infty}^{\infty} [e_{\lambda_1}(\theta) - e_{\lambda_2}(\theta)][e_{\lambda_3}(\theta) - e_{\lambda_4}(\theta)]\, F_\alpha(d\theta) = 0, \qquad \lambda_1 > \lambda_2 \ge \lambda_3 > \lambda_4,$$

we see, from the fact that (2) preserves the inner product, that $z_\alpha(\lambda)$ has orthogonal increments. Evidently by the same argument $z_\alpha(d\lambda)$ has mean square $F_\alpha(d\lambda)$.

† See the Mathematical Appendix for a discussion of Hilbert space and for a discussion of the mathematical developments in general needed for this section.

Now consider

$$e_{A,N}(\theta) = \sum_1^N e^{it\lambda_j}\{e_{\lambda_j}(\theta) - e_{\lambda_{j-1}}(\theta)\},$$

where the λ_j constitute a set of $(N + 1)$ points of subdivision of $[-A, A]$, with $\lambda_0 = -A$, $\lambda_N = A$ and N so large that

$$|e^{-it\theta} - e^{-it\lambda_j}| < \epsilon, \qquad \theta \in [\lambda_{j-1}, \lambda_j], \qquad j = 1, \ldots, N,$$

and A so large that

$$\gamma_\alpha(0) - F_\alpha(A) + F_\alpha(-A) < \epsilon.$$

Note that $e_{A,N}(\theta)$ is zero for $\theta \notin (-A, A]$. Then

$$\int_{-\infty}^{\infty} |e^{-it\theta} - e_{A,N}(\theta)|^2 F_\alpha(d\theta) \le \int_{|\theta|>A} F_\alpha(d\theta) + \epsilon^2 \sum_1^N \int_{\lambda_{j-1}}^{\lambda_j} F_\alpha(d\theta) \le \epsilon + \gamma_\alpha(0)\epsilon^2.$$

Thus we may approximate arbitrarily closely to $\exp -it\theta$ by $e_{A,N}(\theta)$, in mean square with weighting F_α. Correspondingly we may approximate arbitrarily closely, in the sense of the norm in $\mathcal{H}_\alpha$, to $x_\alpha(t)$ by means of

$$\sum_1^N e^{-it\lambda_j}\{z_\alpha(\lambda_j) - z_\alpha(\lambda_{j-1})\}$$

and we have proved that

$$x_\alpha(t) = \int_{-\infty}^{\infty} e^{-it\lambda}\, z_\alpha(d\lambda)$$

in the sense that the right-hand side is the limit in mean square of a sequence of approximating Riemann-Stieltjes sums.

Allowing α to vary as in the proof of Theorem 1, we obtain Theorem 2 save for the uniqueness of the determination of $z(\lambda)$, which is established as part of the proof of Theorem 3, to which we immediately turn.

2 *Proof of Theorem* 3

Let us put

$$F(\lambda_2) - F(\lambda_1) = \int_{-\infty}^{\infty} \phi(\lambda)\, F(d\lambda)$$

where $\phi(\lambda)$ is unity in the interval $[\lambda_1, \lambda_2]$ and is otherwise zero. This has the Fourier transform

$$\hat{\phi}(t) = \frac{e^{-i\lambda_2 t} - e^{-i\lambda_1 t}}{-it} = \int_{-\infty}^{\infty} \phi(\lambda) e^{-i\lambda t}\, d\lambda.$$

Moreover,

$$\lim_{T\to\infty} \frac{1}{2\pi} \int_{-T}^{T} \hat{\phi}(t) e^{i\lambda t}\, dt = \phi(\lambda), \qquad \lambda \ne \lambda_1, \lambda_2, \tag{3}$$

since, for example, if the factor $1 - |t|/T$ is inserted this becomes the Cesaro mean of the Fourier series of $\phi(\lambda)$ which converges uniformly to $\phi(\lambda)$ in every interval of continuity of that function, while it is easily checked that

$$\frac{1}{T}\int_{-T}^{T} |t|\hat{\phi}(t)e^{i\lambda t}\,dt \to 0.$$

Moreover, the convergence to $\phi(\lambda)$ is bounded, by the same argument. Thus calling $\phi_T(\lambda)$ the expression under the limit sign on the left of (3), we have, remembering that λ_1, λ_2 are points of continuity of $F(\lambda)$,

$$\begin{aligned} F(\lambda_2) - F(\lambda_1) &= \lim_{T\to\infty} \int_{-\infty}^{\infty} \phi_T(\lambda)\,F(d\lambda) \\ &= \lim_{T\to\infty} \frac{1}{2\pi}\int_{-T}^{T} \frac{e^{-i\lambda_2 t} - e^{-i\lambda_1 t}}{-it}\left(\int_{-\infty}^{\infty} e^{-i\lambda t}F(d\lambda)\right) dt, \end{aligned}$$

which is the expression we require.

In the same way

$$z(\lambda_2) - z(\lambda_1) = \int_{-\infty}^{\infty} \phi(\lambda)\,z(d\lambda)$$

and evidently $\phi_T(\lambda)$ converges to $\phi(\lambda)$ in mean square with weighting $\operatorname{tr}\{F(d\lambda)\}$. Thus the right-hand side is

$$\begin{aligned} \underset{T\to\infty}{\text{l.i.m.}} \int_{-\infty}^{\infty} \phi_T(\lambda)z(d\lambda) &= \underset{T\to\infty}{\text{l.i.m.}}\, \frac{1}{2\pi}\int_{-T}^{T} \frac{e^{i\lambda_2 t} - e^{i\lambda_1 t}}{it}\left\{\int_{-\infty}^{\infty} e^{-i\lambda t}z(d\lambda)\right\}dt \\ &= \underset{T\to\infty}{\text{l.i.m.}}\, \frac{1}{2\pi}\int_{-T}^{T} \frac{e^{i\lambda_2 t} - e^{i\lambda_1 t}}{it}\, x(t)\,dt, \end{aligned}$$

as required. (For the justification of the reversal of the order of integration see theorem 9 proved in Section 4 below.)

This shows that $z(\lambda_2) - z(\lambda_1)$ is uniquely determined, with probability 1, for each pair of continuity points λ_1, λ_2. Thus $z(\lambda)$ is uniquely determined for each λ, with probability 1, by the requirement that it be continuous in mean square from the right.

3 *Proof of Theorem* 3″

The proof is almost the same as for Theorem 3. We introduce $\phi(\lambda)$ as before, now with $|\lambda_1|, |\lambda_2| \le \pi$. Now we need form $\hat{\phi}(t)$ only for t an integer, however. Thus (3) is replaced by

$$(3)' \qquad \lim_{N\to\infty} \frac{1}{2\pi}\sum_{-N}^{N} \hat{\phi}(n)e^{i\lambda n} = \phi(\lambda), \qquad \lambda \ne \lambda_1, \lambda_2.$$

In the same way as in the proof of Theorem 3 we obtain

$$\begin{aligned} F(\lambda_2) - F(\lambda_1) &= \lim_{N\to\infty} \int_{-\pi}^{\pi} \frac{1}{2\pi}\sum_{-N}^{N} \hat{\phi}(n)e^{i\lambda n}\,F(d\lambda) \\ &= \lim_{N\to\infty} \frac{1}{2\pi}\sum_{-N}^{N} \int_{-\pi}^{\pi} \frac{e^{-in\lambda_2} - e^{-in\lambda_1}}{-in}\,F(d\lambda) \\ &= \lim_{N\to\infty} \frac{1}{2\pi}\sum_{-N}^{N}{}' \,\Gamma(n)\,\frac{e^{-in\lambda_2} - e^{-in\lambda_1}}{-in}, \end{aligned}$$

as in the statement of the theorem. The proof of the second part of the theorem is a simple paraphrase of the proof of that part of Theorem 3.

4 *Proof of Theorem 9*

We follow Rozanov (1967). Let

$$(4)\qquad y(t) = \int_{-\infty}^{\infty} e^{-it\lambda}\, h(\lambda)\, z(d\lambda), \qquad x(t) = \int_{-\infty}^{\infty} e^{-it\lambda}\, z(d\lambda)$$

with

$$\int_{-\infty}^{\infty} h(\lambda)\, F(d\lambda)\, h(\lambda)^* < \infty.$$

Then the $y_j(t)$ are well defined as elements of the Hilbert space $\mathcal{H}$ which is the subspace of $L_2(\Omega, \mathcal{A}, P)$ spanned by the $x_j(t)$. Moreover, $y(t)$ is a stationary process with spectrum

$$F_y(\lambda) = \int_{-\infty}^{\lambda} h(\theta)\, F(d\theta)\, h(\theta)^*$$

and the matrix of cross spectra of the $y_j(t)$ and $x_k(t)$ is

$$(5)\qquad F_{xy}(\lambda) = \int_{-\infty}^{\lambda} F(d\theta) \cdot h(\theta)^*.$$

All of these statements may be proved by considering sequences of approximating Riemann-Stieltjes sums, $y_N(t)$, $x_N(t)$, approximating to $y(t)$ and $x(t)$ and using the fact that $\mathcal{E}(y_N(t)y_N(t)')$ and $\mathcal{E}(y_N(t)x_N(t)')$ converge to $\mathcal{E}(x(t)y(t)')$ and $\mathcal{E}(y(t)x(t)')$ if $y_N(t)$ and $x_N(t)$ converge in mean square to $y(t)$, $x(t)$. We leave the details to the reader. We thus need to prove only the necessity of the condition of Theorem 9. Thus let $w(t)$ be a stationary process with $F_w(\lambda) = F_y(\lambda)$, $F_{xw}(\lambda) = F_{xy}(\lambda)$. We need to establish that $w(t)$ and $y(t)$ are identical. Let $\mathcal{H}'$ be the Hilbert space spanned by all $w_j(t)$ and $x_k(t)$ (i.e., the subspace of $L_2(\Omega', \mathcal{A}', P')$ spanned by these where $(\Omega', \mathcal{A}', P')$ is the measure space corresponding to the vector process composed of the measurements $w_j(t)$, $x_k(t)$ at time t). Then $\mathcal{H} \subset \mathcal{H}'$, as a closed subspace. Now consider the mapping T of $\mathcal{H}$ into $\mathcal{H}'$ defined by

$$(6)\qquad \begin{cases} Ty_j(t) = w_j(t), \\ Tx_k(t) = x_k(t). \end{cases}$$

Then T may be extended to be a linear operator defined on all of $\mathcal{H}$ and is moreover isometric in the sense that $(Tu, Tv) = (u, v)$, where $u, v \in \mathcal{H}$ and the inner product is the inner product in $\mathcal{H}'$; but T is the identity operator on $\mathcal{H}$ as the second part of (6) shows. Thus $Ty_j(t) = y_j(t)$ and $w_j(t) = y_j(t)$ and Theorem 9 is established.

This theorem establishes the validity of the formulas of sections (4(i), 4(ii), 4(iii) of the form (4); for example, in the case of 4(ii) we have

$$h(\lambda) = \sum_0^p A_k(-i\lambda)^k,$$

and we already know (see Theorem 7) that $F_y(d\lambda) = h(\lambda)\,F(\lambda)\,h(\lambda)^*$. Thus we need only to check (5). However, this follows from Theorem 2 in Chapter I, for that theorem asserts that

$$\mathcal{E}(x(s)\dot{x}(t)') = \frac{\partial}{\partial t}\Gamma(t-s) = \frac{\partial}{\partial t}\int_{-\infty}^{\infty} e^{i(t-s)\lambda}\,F(d\lambda)$$
$$= \int_{-\infty}^{\infty}(i\lambda)e^{i(t-s)\lambda}F(d\lambda),$$

the validity of the differentiation under the integral sign following as in the proof of Theorem 7. Thus

$$\mathcal{E}(x(s)\,y(t)') = \int_{-\infty}^{\infty} e^{it\lambda}\,F(d\lambda)\,h(\lambda)^*$$

and (5) is established.

5 *Proof of Formulas* (5.4) *and* (5.10)

We first prove

Theorem 15. *Let $x(t)$ be a stationary vector process with a.c. spectrum having spectral density matrix $f(\lambda)$. Let $f(\lambda) = \phi(\lambda)\,\phi(\lambda)^*$, where $\phi(\lambda)$ is a $(p \times q)$ dimensional, matrix of measurable functions. Then,*

(7) $$\phi(\lambda) = \int_{-\infty}^{\infty} A(s)e^{is\lambda}\,ds,\quad \int_{-\infty}^{\infty} \operatorname{tr}\{A(s)A(s)^*\}\,ds < \infty$$

and

(8) $$x(t) = \int_{-\infty}^{\infty} A(t-s)\,\xi(ds),\qquad \mathcal{E}(\xi(ds)\,\xi(ds)^*) = I_q\,ds,$$

where $\xi(s)$ is a vector process of orthogonal increments.

Proof. We wish to show that we may write $\phi(\lambda) = p(\lambda)k(\lambda)$ where (i) $p(\lambda) = (\phi(\lambda)\phi(\lambda)^*)^{1/2} = f(\lambda)^{1/2}$ and by these we mean the uniquely defined, Hermitian nonnegative, square root† of the $f(\lambda)$ and (ii) $k(\lambda)^*$ maps the range of $p(\lambda)$ (i.e., the range of $f(\lambda)$) isometrically onto the range of $\phi(\lambda)^*$ and satisfies $k(\lambda)k(\lambda)^* = I_p$.

To this purpose we define $k(\lambda)^*$ for any vector y (of complex numbers) in the range of $f(\lambda)$ as $k(\lambda)^*y = z$ if $p(\lambda)x = y$ and $\phi(\lambda)^*x = z$. Then $y^*k(\lambda)\,k(\lambda)^*y = x^*\phi(\lambda).\phi(\lambda)^*x = x^*p(\lambda)^2x = y^*y$ and $k(\lambda)^*$ is isometric on the range of $f(\lambda)$. If the dimension of the range of $f(\lambda)$ is less than p then we may define $k(\lambda)^*$ on the orthogonal complement of that vector space arbitrarily but so that $k(\lambda)\,k(\lambda)^* = I_p$. Thus, if p_0 is the dimension of the range of $f(\lambda)$ and $q > p_0$, we choose $q - p_0$ orthonormal vectors u_j orthogonal to the range of $f(\lambda)$ and $q - p_0$ orthonormal vectors v_j orthogonal to the range of $\phi(\lambda)^*$ and put $k(\lambda)^*u_j = v_j$.

† See the Mathematical Appendix.

Now if $p(\lambda)$ is the generalized inverse† of $p(\lambda)$ then $p(\lambda)\,p(\lambda)^{-1} = E(\lambda)$ projects onto the range of $f(\lambda)$. The matrix functions $k(\lambda), p(\lambda), E(\lambda), p^{-1}(\lambda)$ are all measurable (i.e., the choice of $k(\lambda)$ may be made so that $k(\lambda)$ is measurable). Now

$$\int_{-\infty}^{\infty} e^{-it\lambda}\phi(\lambda)\,k(\lambda)^*\,p(\lambda)^{-1}\,z(d\lambda)$$

is, almost surely, identical with $x(t)$. This follows immediately from Theorem 9 since $E(\lambda)f(\lambda) = f(\lambda)\,E(\lambda) = f(\lambda)$. Put

$$\tilde{\xi}_1(\lambda) = \int_0^{\lambda} k(\theta)^*\,p(\theta)^{-1}\,z(d\theta), \qquad \lambda \geq 0; \qquad = -\int_{\lambda}^{0} k(\theta)^*\,p(\theta)^{-1}\,z(d\theta), \qquad \lambda \leq 0.$$

The integrals are well defined, element by element, by precisely the same construction as was used in Section 1 of the Appendix to Chapter I. Now $\tilde{\xi}_1(\lambda)$ is a vector process of orthogonal increments with

$$\mathcal{E}\{\tilde{\xi}_1(d\lambda)\,\tilde{\xi}_1(d\lambda)^*\} = \tilde{E}(\lambda)\,d\lambda, \qquad \tilde{E}(\lambda) = k(\lambda)^*\,E(\lambda)\,k(\lambda).$$

Then $\tilde{E}(\lambda)$ is also an Hermitian idempotent. We wish to construct a vector process of orthogonal increments with

$$\mathcal{E}\{\tilde{\xi}(d\lambda)\,\tilde{\xi}(d\lambda)^*\} = I_q\,d\lambda. \tag{9}$$

For this purpose we introduce a vector process $\tilde{\xi}_2(\lambda)$, of orthogonal increments, with

$$\mathcal{E}\{\tilde{\xi}_2(d\lambda)\,\tilde{\xi}_2(d\lambda)^*\} = \{I_q - \tilde{E}(\lambda)\}\,d\lambda, \qquad \mathcal{E}\{\tilde{\xi}_2(d\lambda)\,\tilde{\xi}_1(d\mu)^*\} \equiv 0.$$

It may be that no such process, $\tilde{\xi}_2(\lambda)$, can be constructed over $(\Omega, \mathcal{A}, P)$. However, in any case there exists a probability space $(\Omega_2, \mathcal{A}_2, P_2)$ over which a $\tilde{\xi}_2(\lambda)$, satisfying the first of these relations, is defined. We then form the product space‡ $\Omega \times \Omega_2$ with product measure, $P \times P_2$, defined on the Borel field $\mathcal{A} \times \mathcal{A}_2$. Then on $\Omega \times \Omega_2$ we have defined the vector process of orthogonal increments, $\tilde{\xi}(\lambda) = \tilde{\xi}_1(\lambda) + \tilde{\xi}_2(\lambda)$, which satisfies (9). As we shall see, the introduction of Ω_2 is of no real consequence and is a device introduced purely to simplify the statement of the theorem.

Now we may put

$$x(t) = \int_{-\infty}^{\infty} e^{-it\lambda}\phi(\lambda)\,\tilde{\xi}(d\lambda). \tag{10}$$

Again the validity of the identification follows from Theorem 9. Put

$$\xi(t) - \xi(s) = \int_{-\infty}^{\infty} \frac{e^{-it\lambda} - e^{-is\lambda}}{-2\pi i\lambda}\,\tilde{\xi}(d\lambda).$$

This expression is well defined, element by element, by the procedure of Section 1 of the Appendix to Chapter I since the integrand, $\psi_{s,t}(\lambda)$ let us say, is square

† See the Mathematical Appendix.

‡ For the definitions of product space, product measure, see Billingsley (1968, p. 224).

integrable with respect to Lebesgue measure. (We give it the value $(t - s)$ at $\lambda = 0$.) Also

(11) $$\mathcal{E}\{(\xi(t) - \xi(s))(\xi(v) - \xi(u))^*\} = I_q \int_{-\infty}^{\infty} \psi_{s,t}(\lambda)\overline{\psi_{u,v}(\lambda)}\, d\lambda$$

But

$$\psi_{s,t}(\lambda) = \frac{1}{2\pi}\int_{-\infty}^{\infty} \{e_t(\theta) - e_s(\theta)\}e^{-i\theta\lambda}\, d\theta$$

where $e_t(\theta)$ is as defined in the first appendix to this chapter (now with t replacing λ). Thus by Plancherel's theorem (see the Mathematical Appendix) (11) is

$$I_q \int_{-\infty}^{\infty} \{e_t(\theta) - e_s(\theta)\}\{e_v(\theta) - e_u(\theta)\}\, d\theta,$$

which is null if $(s, t]$, $(u, v]$ do not overlap and if $(s, t] = (u, v]$ is $I_p(t - s)$; $\xi(t)$ is a vector process of orthogonal increments satisfying the second part of (8).

Now the functions $\chi_{s,t}(\theta) = e_t(\theta) - e_s(\theta)$, for all s, t, are square integrable (with respect to Lebesgue measure) and dense in the space of all such square integrable functions (in the sense of mean square convergence with respect to Lebesgue measure). Thus the functions $\psi_{s,t}(\lambda)$, have, as functions of λ, precisely the same properties, by Plancherel's theorem. Thus we may approximate arbitrarily closely to the right side of (10) by an expression

(12) $$x_N(t) = \sum A_j \int_{-\infty}^{\infty} e^{-it\lambda}\psi_{s_j t_j}(\lambda)\, \tilde{\xi}(d\lambda)$$

for which

(13) $$\sum A_j \psi_{s_j, t_j}(\lambda)$$

approximates arbitrarily closely to $\phi(\lambda)$. Indeed this latter matrix is certainly composed of square integrable elements since $\operatorname{tr}(\phi(\lambda)\phi(\lambda)^*) = \operatorname{tr}(f(\lambda))$ which is integrable. The mean-square error of approximation is

$$\mathcal{E}\{(x_N(t) - x(t))'(x_N(t) - x(t))\}$$

$$= \operatorname{tr}\int_{-\infty}^{\infty} \{\textstyle\sum A_j \psi_{s_j, t_j}(\lambda) - \phi(\lambda)\}\{\textstyle\sum A_j \psi_{s_j, t_j}(\lambda) - \phi(\lambda)\}^*\, d\lambda,$$

which may thus be made arbitrarily small; but (12) is

(14) $$\sum A_j \int_{-\infty}^{\infty} \{e_{t_j+t}(s) - e_{s_j+t}(s)\}\, \xi(ds)$$

from the definition of $\xi(s)$. Now, since $\phi(\lambda)$ is composed of square integrable elements, it may be written in the form (7) with

$$A(s) = \frac{1}{2\pi}\int_{-\infty}^{\infty} e^{-is\lambda}\phi(\lambda)\, d\lambda,$$

and from Plancherel's theorem, since (13) approximates arbitrarily closely to $\phi(\lambda)$,

$$\sum A_j\{e_{t_j+t}(-s) - e_{s_j+t}(-s)\} = \sum A_j\{e_{t_j}(t-s) - e_{s_j}(t-s)\}$$

approximates arbitrarily closely to $A(t-s)$. Thus (14) converges in mean square by the same argument used in relation to (12) to

$$\int_{-\infty}^{\infty} A(t-s)\,\xi(ds)$$

and the theorem is established.

If we take $\phi(\lambda) = f(\lambda)^{1/2}$, where this is the Hermitian, nonnegative, square root of $f(\lambda)$, we obtain formula (5.10). The formula (5.4) follows from Theorem 15′ in the same way as (5.10) followed from Theorem 15. We omit the proof of Theorem 15′, since it is essentially the same as that for Theorem 15 but simpler.

Theorem 15′. *Let $x(n)$ be a stationary, discrete time, vector process with a.c. spectrum with density matrix $f(\lambda) = \phi(\lambda)\,\phi(\lambda)^*$, where $\phi(\lambda)$ is as in Theorem* 15. *Then*

$$(7)' \qquad \phi(\lambda) = \sum_{-\infty}^{\infty} A(j)e^{ij\lambda}, \qquad \sum_{-\infty}^{\infty} \operatorname{tr}(A(j)\,A(j)^*) < \infty,$$

and

$$(8)' \qquad x(n) = \sum_{-\infty}^{\infty} A(j)\,\epsilon(n-j), \qquad \mathcal{E}(\epsilon(m)\,\epsilon(n)') = \delta_m^{\ n} I_q.$$

6 *Proof of Theorem* 12

Let $\phi_\rho(x, y)$ stand for the normal density corresponding to x and y. The expression

$$\mathcal{E}\{g(x)\,h(y)\} = \int_{-\infty}^{\infty} g(x)\,h(y)\phi_\rho(x, y)\,dx\,dy$$

may be written

$$(1) \qquad g \times h(\phi_\rho(x, y)),$$

where by $g \times h$ we mean the generalized function (of two variables), defined by g and h on the space S, of all infinitely differentiable functions of two variables, which decrease, together with all of their derivatives, faster than any power of $\sqrt{(x^2+y^2)}^{-1}$ as $\sqrt{x^2+y^2} \to \infty$. Thus† $g \times h \in S'$. Then $\phi_\rho(x, y)$ certainly belongs to S. It is in this sense (1) in which, in general, we interpret the expression $\mathcal{E}(g(x)\,h(y))$ for g and h generalized functions, since one cannot allot a value to a generalized function at a point. Now by definition

$$g \times h(\phi_\rho) = (4\pi^2)^{-1}\hat{g} \times \hat{h}(\hat{\phi}_\rho),$$

where the "hats" indicate the taking of Fourier transforms. Thus, using θ_1, θ_2 for the variables on which $\hat{\phi}_\rho$ depends,

$$\frac{d^k}{d\rho^k} g \times h(\phi_\rho) = (4\pi^2)^{-1}\frac{d^k}{d\rho^k}\hat{g} \times \hat{h}(\hat{\phi}_\rho) = (4\pi^2)^{-1}\hat{g} \times \hat{h}(\theta_1^{\ k}\theta_2^{\ k}(-)^k\hat{\phi}_\rho).$$

† See the Mathematical Appendix.

The carrying out of the differentiations under the integral sign is justified if, as $\delta \to 0$,

$$u_\delta^{(j)} = \frac{\hat{\phi}_{\rho+\delta}^{(j)} - \hat{\phi}_\rho^{(j)}}{\delta}, \qquad j < k$$

converges to $\hat{\phi}_\rho^{(j+1)}$ in the topology of S, that is,

$$\lim_{\delta\to 0} \max_{\theta_1,\theta_2} \{1 + (\theta_1^2 + \theta_2^2)^{1/2}\}^r \frac{\partial^m}{\partial\theta_1^{m_1}\,\partial\theta_2^{m_2}} (u_\delta^{(j)} - \hat{\phi}_\rho^{(j+1)}) = 0, \qquad m = m_1 + m_2.$$

This is easily checked for $\hat{\phi}_\rho$.

Now for $f \in S$, then†

$$(-)^{2k}(2\pi)^2(g^{(k)} \times h^{(k)})(f) = \hat{g} \times \hat{h}((i\theta_1)^k(i\theta_2)^k \hat{f}),$$

so that

$$\frac{d^k}{d\rho^k} g \times h(\phi_\rho) = (4\pi^2)^{-1}\hat{g} \times \hat{h}(\theta_1^k\theta_2^k(-)^k\hat{\phi}_\rho) = g^{(k)} \times h^{(k)}(\phi_\rho)$$

as required.

† See the Mathematical Appendix.

CHAPTER III

Prediction Theory and Smoothing

1. INTRODUCTION

We shall in this chapter give an account of the Wiener-Kolmogoroff theory of prediction. We follow this with an account of the simpler theory of interpolation. We then discuss signal extraction, including both the classical theory and the later theories, associated with the name of Kalman, which commence from a structurally simple model (a vector autoregressive model) but balance this by allowing the autoregressive constants to be time-dependent. These later methods have been widely used. We conclude with a discussion of some more *ad hoc* smoothing theories.

The mathematically most difficult part of the present chapter is that contained in the general theory of prediction. The physical importance of this generality is doubtful for the spectral density function, $f(\lambda)$, will in practice be a relatively simple function, such as a polynomial in $\exp i\lambda$ and $\exp -i\lambda$, and for such functions (in the discrete time case) the basic results have already been obtained in Chapter II, Section 5, subsection (iii). The general theory is, nevertheless, given here for a variety of reasons. In the first place, it gives an insight into the structure of stationary processes. It also has an undeniable mathematical appeal and, finally, because of the generality it makes later subsidiary results easy to prove since regularity conditions do not have to be checked as they would if a more special theory was developed. The theory of prediction and signal extraction, nevertheless, has wide application and to make this theory available to a reader who wishes to omit the more difficult parts we commence, in the next section, by dealing with the prediction theory for the case of a rational spectral density. The reader who wishes to do so may then omit Sections 3, 4, and 5, which deal with that theory for the general spectral density.

2. VECTOR DISCRETE-TIME PREDICTION FOR RATIONAL SPECTRA

We begin with a vector discrete-time stationary process $x(n)$. We seek for that linear function, $\hat{x}^{(\nu)}(n)$, of the $x(n-j)$, $j \geq \nu$ which best approximates to $x(n)$ in the mean square sense, i.e., for which

$$\|x(n) - \hat{x}^{(\nu)}(n)\|^2 = \mathcal{E}\{(x(n) - \hat{x}^{(\nu)}(n))'(x(n) - \hat{x}^{(\nu)}(n))\}$$

is a minimum. By saying that $\hat{x}^{(\nu)}(n)$ is to be linear in the $x(n-j)$ we mean that it is an expression of the form

$$\hat{x}_N(n) = \sum_{\nu}^{N} A_{N,j}x(n-j), \qquad N \geq \nu \tag{2.1}$$

or is the limit in mean square of a sequence of such expressions, so that $\lim_{N\to\infty} \|\hat{x}^{(\nu)}(n) - \hat{x}_N(n)\| = 0$. Thus

$$\begin{aligned}\lim_{N\to\infty} \|x(n) - \hat{x}_N(n)\| &\leq \|x(n) - \hat{x}^{(\nu)}(n)\| + \lim_{N\to\infty} \|\hat{x}_N(n) - \hat{x}^{(\nu)}(n)\| \\ &= \|x(n) - \hat{x}^{(\nu)}(n)\|.\end{aligned} \tag{2.2}$$

Since the left-hand limit cannot be less than the minimized value, the two extreme members are equal. This shows that the same sequence of sets of matrices, $A_{N,j}, j = \nu, \ldots, N$, will provide a minimizing sequence for any n, as our notation has already implied. Let

$$h_N(e^{i\lambda}) = \sum_{\nu}^{N} A_{N,j}x(n-j)$$

and h_ν be such that†

$$\lim_{N\to\infty} \operatorname{tr}\left\{\int_{-\pi}^{\pi} (h_N - h_\nu)\, F(d\lambda)(h_N - h_\nu)^*\right\} = 0.$$

Then we must have

$$\hat{x}_N(n) = \int_{-\pi}^{\pi} e^{-in\lambda}\, h_N(e^{i\lambda})\, z(d\lambda), \qquad \hat{x}^{(\nu)}(n) = \int_{-\pi}^{\pi} e^{-in\lambda}\, h_\nu(e^{i\lambda})\, z(d\lambda),$$

since this $\hat{x}^{(\nu)}(n)$ certainly satisfies $\lim \|\hat{x}^{(\nu)}(n) - \hat{x}_N(n)\| = 0$.

We shall often suppress any reference to ν in $x^{(\nu)}$, h_ν and shall use the notation $\hat{x}(n)$, $\hat{x}_N(n)$, h_N, h for simplicity since for the moment we shall keep ν fixed. From what has been said we may as well take $n = 0$ since the response functions h_N, h do not depend on n.

We now specialize to the case where $F(d\lambda) = f(\lambda)\, d\lambda$ and $f(\lambda)$ is a rational function of $\exp \pm i\lambda$ with $\det\{f(\lambda)\} \neq 0$, a.e. Then

$$f(\lambda) = \frac{1}{2\pi}\, \Phi_0(e^{i\lambda})\, \Phi_0(e^{i\lambda})^*,$$

† It is now convenient to express the response functions h, h_N, as functions of $\exp i\lambda$ rather than of λ, as in Chapter II. As in the following expression we shall drop the argument variable $\exp i\lambda$ when it is felt that there will be no confusion as a result.

where $\Phi_0(z)$ is rational in z, holomorphic in and on the unit circle (i.e., in a region containing the unit circle) and has a determinant without zeros in the unit circle. Indeed

$$f(\lambda) = \left\{\frac{1}{2\pi}\sum_{-q}^{q} \Delta(j)e^{-ij\lambda}\right\}\left(\sum_{-r}^{r} \delta(j)e^{-ij\lambda}\right)^{-1},$$

where the second factor is the least common multiple of all the denominators of elements of $f(\lambda)$. Using Theorems 10 and 10′ in Chapter II the representation of $f(\lambda)$ results. Putting $G = \Phi_0(0)\,\Phi_0(0)^*$ (where $\Phi_0(0) = \Phi_0(z)$, $z = 0$), we have the representation

$$f(\lambda) = \frac{1}{2\pi}\,\Phi(e^{i\lambda})\,G\Phi(e^{i\lambda})^*, \tag{2.3}$$

where now $\Phi(z)$, when expanded in a power series, has leading coefficient matrix which is the unit matrix. As we shall see, G may be defined as the matrix of one-step prediction errors for $x(n)$. The matrix $\Phi(z)$ may be equivalently and uniquely defined as being holomorphic in and on the unit disc, satisfying $(2\pi)^{-1}\,\Phi G\Phi^* = f$ and with $\Phi(0) = I_p$ for we know from Theorem 10′ in Chapter II that any other matrix $B(z)$ satisfying these conditions must satisfy $B(z) = \Phi(z)\,U(z)$, where $U(z)$ is unitary on the unit circle and holomorphic within it. If $\Phi(0) = B(0) = I_p$, then $U(0)$ is the unit matrix and it follows as in Section 5(ii) of Chapter II† that $U(z)$ must be the unit matrix for all z. This characterization is mentioned here for in the case where $f(\lambda)$ is not a rational function it still defines the factorization which we require.

We introduce the following notation. If $B(z)$ is a matrix of functions each of which has a valid Laurent expansion in an annulus, then we use $[B(z)]_+$ to mean that we retain only the terms from that expansion for nonnegative powers of z. $[B(z)]_-$ is formed similarly taking only negative powers, so that $B(z) = [B(z)]_+ + [B(z)]_-$ in the region in which the Laurent expansion holds. We now prove Theorem 1.

Theorem 1. *Let $f(\lambda)$ be rational with det $(f(\lambda)) \neq 0$, a.e. Then there exists a unique factorization* (2.3) *in which $\Phi(z)$ is holomorphic within the unit circle and satisfies $\Phi(0) = I_p$. Then det $\{\Phi(z)\}$ is never zero within the unit circle. The response function h of the filter giving the best linear ν-step predictor, $\hat{x}^{(\nu)}(n)$, is*

$$h_\nu(e^{i\lambda}) = e^{i\nu\lambda}[e^{-i\nu\lambda}\,\Phi(e^{i\lambda})]_+\Phi^{-1}(e^{i\lambda}) \tag{2.4}$$

† Since $U^*(\exp i\lambda) = U(\exp i\lambda)^{-1}$ is the boundary value of a function holomorphic outside of the unit disc and inside the unit disc it must be a constant unitary matrix and thus the unit matrix since $U(0) = I_p$.

and the covarience matrix of the prediction errors is

$$(2.5)\quad \mathcal{E}\{(x(n) - \hat{x}^{(\nu)}(n))(x(n) - \hat{x}^{(\nu)}(n))'\} = \frac{1}{2\pi}\int_{-\pi}^{\pi} [e^{-i\nu\lambda}\,\Phi(e^{i\lambda})]_-\, G[e^{-i\nu\lambda}\,\Phi(e^{i\lambda})]_-^*\, d\lambda,$$

which is G for $\nu = 1$.

Proof. Since we are keeping ν fixed we once more drop the ν subscript. We seek to minimize

$$\operatorname{tr}\int_{-\pi}^{\pi} \{I - h(e^{i\lambda})\} f(\lambda)\{I - h(e^{i\lambda})\}^*\, d\lambda$$

$$= \operatorname{tr}\frac{1}{2\pi}\int_{-\pi}^{\pi} \{\Phi(e^{i\lambda})\sqrt{G} - h(e^{i\lambda})\,\Phi(e^{i\lambda})\sqrt{G}\} \times \{\Phi(e^{i\lambda})\sqrt{G} - h(e^{i\lambda})\,\Phi(e^{i\lambda})\sqrt{G}\}^*\, d\lambda$$

$$= \operatorname{tr}\left[\frac{1}{2\pi}\int_{-\pi}^{\pi} \{a(e^{i\lambda}) + b(e^{i\lambda})\}\{a(e^{i\lambda}) + b(e^{i\lambda})\}^*\, d\lambda\right]$$

$$a(e^{i\lambda}) = [e^{-i\nu\lambda}\,\Phi(e^{i\lambda})]_-\sqrt{G},$$

$$b(e^{i\lambda}) = \{[e^{-i\nu\lambda}\,\Phi(e^{i\lambda})]_+ - e^{-i\nu\lambda}\, h(e^{i\lambda})\,\Phi(e^{i\lambda})\}\sqrt{G}.$$

Now the matrix a can be expressed as a matrix of polynomials in terms of negative powers only of $\exp i\lambda$. Replacing b by b_N where b_N contains h_N in place of h we see that b_N can be expanded in an absolutely convergent series of *nonnegative* powers of $\exp i\lambda$ since this is true of $\exp(-i\nu\lambda)\, h$ and Φ. Thus the integral of $ab_N{}^*$ vanishes. Since

$$\left\|\int a(b^* - b_N^*)\, d\lambda\right\| \le \int \|a\|\, \|b^* - b_N^*\|\, d\lambda \le \left\{\int \|a\|^2\, d\lambda \int \|b^* - b_N^*\|^2\, d\lambda\right\}^{1/2} \to 0$$

we see that the integral of (ab^*) vanishes also. Thus we must minimize

$$\operatorname{tr}\left[\frac{1}{2\pi}\int (aa^* + bb^*)\, d\lambda\right].$$

However, aa^* does not depend upon h and we must minimize $\operatorname{tr}\int bb^*\, d\lambda$. This is zero for h given by (2.4) so that in that case the prediction error-matrix is $(2\pi)^{-1}\int aa^*\, d\lambda$, which is (2.5). The function $z^{-\nu}h(z)$, where h is given by (2.4), is holomorphic within the unit circle since $\Phi(z)$ has determinant without zeros there, and is evidently square integrable with respect to $f(\lambda)$ on the circle. Thus there is a sequence h_N of the form defined above such that $\|\hat{x}(n) - \hat{x}_N(n)\|$ converges to zero, where $\hat{x}(n)$ is got from h and $\hat{x}_N(n)$ from h_N. Thus the theorem is established.

To bring the point home that h_N is of the required form we indicate here how h may be synthesized. We point out, before doing this, that the predictions $\hat{x}^{(\nu)}(n)$ will be obtained iteratively once an initial prediction is made and moreover, in practice, this first prediction may often be got by some "rough and ready" method rather than by $\hat{x}^{(1)}(n)$, for obvious reasons. Thus the synthesis may not be of great practical importance. In the examples at the end of this section, we show how these iterations are performed. We now return to the synthesis of h. We need to show how $z^{-\nu}h(z)$ may be made up, as the z transform,† of a filter operating only on the present and past of the sequence. Evidently $z^{-\nu}h(z)$ may be represented as the product of two factors, one of which is a matrix polynomial in z, the other the reciprocal of a polynomial, $p(z)$, in z which may have zeros on the unit circle but has none within it. The first factor defines a filter of the required form, directly. The result of this filtering is to produce a new spectral density with respect to which $p(\exp i\lambda)^{-1}$ is square integrable. We decompose $p(z)^{-1}$ into partial fractions. The only terms causing concern are those corresponding to zeros on the unit circle since a partial fraction of the form $(\exp i\lambda - z_0)^{-1}$, $|z_0| > 1$, has the expansion

$$-z_0^{-1} \sum_0^\infty z_0^{-j} e^{ij\lambda},$$

which converges exponentially and therefore in mean square with respect to any rational function and thus defines a filter of the required kind. Thus we are left with the problem of producing a synthesis of the filter with response

$$\left\{\frac{1}{1 - e^{i(\lambda-\theta)}}\right\}^q I_p.$$

We take $\theta = 0$, $q = 1$ for simplicity and consider

$$\left\{(1 - e^{i\lambda})^{-1} - \sum_0^N \left(1 - \frac{j}{N}\right) e^{ij\lambda}\right\} I_p = e^{i\lambda}\left\{\frac{1 - e^{iN\lambda}}{N(1 - e^{i\lambda})^2}\right\} I_p. \tag{2.6}$$

Since $(1 - e^{iN\lambda})/\{N(1 - e^{i\lambda})\}$ is bounded by unity in absolute value and $(1 - e^{i\lambda})^{-1} I_p$ is square integrable with respect to the new density it is evident that (2.6) converges in mean square to zero and

$$\left\{\sum_0^N \left(1 - \frac{j}{N}\right) e^{ij\lambda}\right\} I_p$$

† The response function of a filter considered as a function of z within a region where this function is well defined is spoken of as the z transform of the filter.

constitutes an approximating sequence which accomplishes the desired synthesis. If $\theta \neq 0$, $q > 1$ we see that we may replace $[1 - \exp i(\lambda - \theta)]^{-q}$ by

$$\left\{\sum_0^N \left(1 - \frac{j}{N}\right) e^{ij(\lambda-\theta)}\right\}^q,$$

which is again of the form we require.

If $f(\lambda)$ does not satisfy $\det\{f(\lambda)\} \neq 0$ a.e. then we may always find a matrix $P(z)$ whose elements are polynomials in z and whose determinant is unity, so that

$$P(e^{i\lambda})\, f(\lambda)\, P(e^{i\lambda})^*$$

is identically null outside of the first r (say) rows and columns and for which the determinant of the matrix in the first r rows and columns is nonzero, a.e. (see the end of subsection (iii) of Chapter II, Section 5). This means that we may change from $x(n)$ to a new r dimensional process $y(n)$, by a reversible, autoregressive transformation (reversible in the sense that we may regain $x(n)$ as a finite linear combination of the $y(n-j), j \geq 0$). Thus the prediction problem for $x(n)$ is turned into one for $y(n)$, which is solved by the previous considerations.

Theorem 2.15′† shows us that $x(n)$ may be put in the form

(2.7)

$$x(n) = \sum_0^\infty A(j)\,\epsilon(n-j), \qquad \sum_0^\infty A(j)\, e^{ij\lambda} = \Phi(e^{i\lambda}), \qquad \mathcal{E}(\epsilon(m)\,\epsilon(n)') = \delta_m{}^n G,$$

wherein

$$\epsilon(n) = \int_{-\pi}^{\pi} e^{-in\lambda}\left(\sum_0^\infty A(j)\, e^{ij\lambda}\right)^{-1} z_x(d\lambda) = \int_{-\pi}^{\pi} e^{-in\lambda}\, z_\epsilon(d\lambda).$$

Since, in case $\nu = 1$,

$$h = \left(\sum_1^\infty A(j)\, e^{ij\lambda}\right)\left(\sum_0^\infty A(j)\, e^{ij\lambda}\right)^{-1},$$

then

$$\hat{x}(n) = \int_{-\pi}^{\pi} e^{-in\lambda}\left(\sum_1^\infty A(j)\, e^{ij\lambda}\right) z_\epsilon(d\lambda) = \sum_1^\infty A(j)\,\epsilon(n-j).$$

Thus $\epsilon(n)$ is the one-step prediction error.

The decomposition (2.7) is an aspect of the Wold decomposition theorem which we prove in Section 3. Before closing this, however, we shall discuss some examples. These mainly relate to the case $p = 1$, so that $f(\lambda)$, $h_\nu(\lambda)$, and so on, are all scalar-valued functions. In this case we shall use σ^2 for G,

† See Section 5 of the Appendix to Chapter II.

so that σ^2 is the 1-step prediction error and is the variance of $\epsilon(n)$. We then also replace the notation $A(j)$ for the matrices in (2.7) by $a(j)$.

(i) If $f(\lambda)$ is rational we may put (2.3) in the form

$$\frac{c\,|\sum_0^s \alpha(j)\,e^{ij\lambda}|^2}{2\pi\,|\sum_0^q \beta(j)\,e^{ij\lambda}|^2},$$

where $\sum \alpha(j)z^j$ has no zeros in the circle and $\sum \beta(j)z^j$ has no zeros on or in the circle, the $a(j)$ and $b(j)$ are real and $a(0) = b(0) = 1$. Then evidently $c = \sigma^2$ and

$$\frac{\sum_0^s \alpha(j)z^j}{\sum_0^q \beta(j)z^j} = \sum_0^\infty a(j)z^j.$$

(ii) If $s = 0$ in (i), then

$$\sum_0^\infty a(j)z^j = \left(\sum_0^q \beta(j)z^j\right)^{-1}$$

and

$$h_1(z) = \sum_0^q \beta(j)z^j\left\{\left(\sum_0^q \beta(j)z^j\right)^{-1} - 1\right\}$$

$$= 1 - \sum_0^q \beta(j)z^j = -\sum_1^q \beta(j)z^j.$$

Then the one-step predictor is

$$\hat{x}^{(1)}(n) = -\sum_1^q \beta(j)\,x(n-j),$$

as is, in other ways, obvious.

Since

$$h_\nu(z) = \left(\sum_0^q \beta(j)z^j\right)\left\{\left(\sum_0^q \beta(j)z^j\right)^{-1} - \sum_0^{\nu-1} a(j)z^j\right\},$$

we observe that

$$h_\nu(z) - h_{\nu+1}(z) = \left(\sum_0^q \beta(j)z^j\right) z^\nu\, a(\nu).$$

Thus

$$\hat{x}^{(\nu)}(n) = \hat{x}^{(\nu+1)}(n) + a(\nu)\sum_0^q \beta(j)\,x(n-\nu-j),$$

which enables the predictor of $x(n)$ to be updated as new data comes to hand. We also have

$$\hat{x}^{(\nu)}(n) = -\sum_1^q \beta(j)\,\hat{x}^{(\nu-j)}(n-j); \qquad \hat{x}^{(\nu)}(n) = x(n), \qquad \nu < 0,$$

which enables $\hat{x}^{(\nu)}(n)$ to be calculated, for each ν, from the predictions made up to the previous time point. Indeed,

$$\sum_0^q \beta(j)\, \hat{x}^{(\nu-j)}(n-j)$$

is generated by the filter with z transform

$$\sum_0^q \beta(j)\, h_{\nu-j}(z) z^j = \left[\sum_0^q \left\{\beta(j) z^j \sum_{\nu-j}^{\infty} a(u) z^u\right\}\right]\left[\sum_0^q \beta(j) z^j\right],$$

and the first factor has as coefficient of z^w, $w \geq \nu > 1$, the expression

$$\sum_0^{\min(q,w)} \beta(j)\, a(w-j) = \frac{1}{2\pi}\int_{-\pi}^{\pi} \sum_0^q \beta(j) e^{ij\lambda}\left(\frac{1}{\sum_0^q \beta(j) e^{ij\lambda}}\right) e^{-iw\lambda}\, d\lambda,$$

which is null.

(iii) Let $x(n) = \epsilon(n) + \epsilon(n-1)$, $\mathcal{E}(\epsilon(m)\,\epsilon(n)) = \delta_m^{\ n}\sigma^2$. Then

$$f(\lambda) = \frac{\sigma^2}{2\pi}|1 + e^{i\lambda}|^2, \qquad \sum_0^{\infty} a(j) z^j = 1 + z$$

$$h_1(z) = \frac{z}{1+z}.$$

Now $h_1(z)$ has a pole on the unit circle and cannot be written in the simple "closed" form $\sum_1^{\infty} \beta(j) z^j$. As we show below, the proof of Theorem 1

$$h_N(z) = \sum_1^N (-)^{j-1}\left(1 - \frac{j}{N}\right) e^{ij\lambda}$$

defines a sequence of one-step predictors which converges appropriately to $h_1(z)$. This can also be seen by evaluating the vector of coefficients of regression of $x(n)$ on $x(n-1), \ldots, x(n-N)$, which are obtained from

$$\begin{bmatrix} 2 & 1 & 0 & 0 & . & . & . & 0 \\ 1 & 2 & 1 & 0 & . & . & . & 0 \\ 0 & 1 & 2 & 1 & . & . & . & 0 \\ . & . & . & . & . & . & . & . \\ & . & . & . & . & . & . & . \\ & & . & . & . & . & . & . \\ 0 & 0 & 0 & 0 & . & . & . & 2 \end{bmatrix}^{-1} \begin{pmatrix} 1 \\ 0 \\ 0 \\ . \\ . \\ . \\ 0 \end{pmatrix}.$$

It is easily checked that the first column of the inverse matrix has

$$(-)^{j-1}\left(1 - \frac{j}{N}\right)$$

in the jth place.

It is intrinsically difficult to say *just* when $h(z)$ can be given in the simple closed form $\sum_1^\infty \beta(j)z^j$ (i.e., when $x(n)$ is generated by an infinite autoregression) for we are concerned in the expansion of h_1, on the unit circle, in mean square with respect to the function, $f(\lambda)$, with respect to which the exp $in\lambda$ are not orthogonal.

This function $h_1(z)$, provides an example of the response function of a filter which is not of the form $\sum_{-\infty}^{\infty} a(j)\, x(n-j)$.

(iv) We return to (i). Now consider

$$h_\nu(z) = \frac{\sum_0^q \beta(j)z^j}{\sum_0^s \alpha(j)z^j}\left[\frac{\sum_0^s \alpha(j)z^j}{\sum_0^q \beta(j)z^j} z^{-\nu}\right]_+ z^\nu$$

$$= -\left[z^{-\nu}\left\{1 - \left(\sum_0^q \beta(j)z^j\right)^{-1}\right\}\sum_0^s \alpha(j)z^j\right]_+ \left\{\frac{\sum_0^q \beta(j)z^j}{\sum_0^s \alpha(j)z^j}\right\} z^\nu$$

$$+ \left[z^{-\nu}\sum_0^s \alpha(j)z^j\right]_+ \left\{\frac{\sum_0^q \beta(j)z^j}{\sum_0^s \alpha(j)z^j}\right\} z^\nu$$

$$= \left\{-\sum_1^q \left(\beta(j)\left[z^{-\nu+j}\frac{\sum_0^s \alpha(j)z^j}{\sum_0^q \beta(j)z^j}\right]_+ z^\nu + \sum_\nu^s \alpha(j)z^j\right\}\frac{\sum_0^q \beta(j)z^j}{\sum_0^s \alpha(j)z^j}.$$

However

$$1 - h(z) = \sum_0^q \beta(j)z^j \Big/ \sum_0^s \alpha(j)z^j$$

so that

$$h_\nu(z) = -\sum_0^q \beta(j)z^j h_{\nu-j}(z) + \left\{\sum_\nu^s \alpha(j)z^j\right\}\{1 - h(z)\}$$

and thus

$$\hat{x}^{(\nu)}(n) = -\sum_1^q \beta(j)x^{(\nu-j)}(n-j) + \sum_1^s \alpha(j)\{x(n-j) - \hat{x}^{(1)}(n-j)\}$$

which enables $\hat{x}^{(\nu)}(n)$ to be computed iteratively. Using the $a(j)$ defined in paragraph (i), above, we also have

$$h_\nu(z) - h_{\nu+1}(z) = a(\nu)z^\nu\left\{\sum_0^q \beta(j)z^j \Big/ \sum_0^s \alpha(j)z^j\right\}$$

so that

$$\hat{x}^{(\nu)}(n) = \hat{x}^{(\nu+1)}(n) + a(\nu)\{x(n-\nu) - \hat{x}^{(1)}(n-\nu)\}$$

which enables the estimate $\hat{x}^{(\nu+1)}(n)$ to be updated as the next observation, $x(n-\nu)$, comes to hand.

The first of these iterative formulae is valid even if $\sum \beta(j)z^j$ has zeros on the unit circle. It provides best predictors giving an exact representation of a deterministic component. (See Theorem 11 below.)

(v) We illustrate the vector case by the extension of (iv). Let

$$f(\lambda) = (2\pi)^{-1} B(e^{i\lambda})^{-1} A(e^{i\lambda})\, GA^*(e^{i\lambda})\, B^*(e^{i\lambda})^{-1}, \quad A(0) = B(0) = I_p.$$

Then for ν-step prediction we need

$$\begin{aligned} z^\nu[z^{-\nu}B^{-1}A]_+A^{-1}B &= -z^\nu[z^{-\nu}(I_p - B^{-1})A]_+A^{-1}B + z^\nu[z^{-\nu}A]_+A^{-1}B \\ &= -z^\nu[z^{-\nu}(B - I_p)B^{-1}A]_+A^{-1}B + z^\nu[z^{-\nu}A]_+A^{-1}B \\ &= \textstyle\sum_1^q B(j)z^j h_{\nu-j}(z) + z^\nu[z^{-\nu}A]_+A^{-1}B. \end{aligned}$$

Now, almost exactly as for the case $p = 1$, treated in paragraph (iv) above, the second term is

$$\sum_1^s A(j)z^j\{1 - h(z)\}$$

and thus

$$\hat{x}^{(\nu)}(n) = -\sum_1^q B(j)x^{(\nu-j)}(n-j) + \sum_\nu^s A(j)\{x(n-j) - \hat{x}^{(1)}(n-j)\}.$$

The formula which enables $\hat{x}^{(\nu)}(n)$ to be updated is obtained, again in the same way as in (iv), as

$$\hat{x}^{(\nu)}(n) = \hat{x}^{(\nu+1)}(n) + C(\nu)\{x(n-\nu) - \hat{x}^{(1)}(n-\nu)\}$$

where

$$B^{-1}A = \sum_0^\infty C(j)z^j.$$

3. THE GENERAL THEORY FOR STATIONARY, DISCRETE-TIME, SCALAR PROCESSES†

We call $\mathcal{M}_n$ the Hilbert space which is spanned by the $x(m)$, $m \le n$. Thus $\mathcal{M}_n$ contains all finite linear combinations, $\sum a_j x(n_j)$, $n_j \le n$, as well as their limits in mean square. Clearly $\mathcal{M}_m \subset \mathcal{M}_n$, $m \le n$. We put $\mathcal{M}_{-\infty} = \bigcap_{-\infty}^{\infty} \mathcal{M}_n$. $\mathcal{M}_\infty$ we call $\mathcal{H}$, as before. The most direct prediction problem is that of approximating to $x(n)$ by an element of $\mathcal{M}_{n-1}$. This approximation is got by projecting $x(n)$ perpendicularly onto $\mathcal{M}_{n-1}$, the resulting projection being called $\hat{x}(n)$. We put $\epsilon(n) = x(n) - \hat{x}(n)$ so that $\epsilon(n) \perp \mathcal{M}_{n-1}$. Clearly $\hat{x}(n)$ got in this way minimizes the mean-square error. This minimized error we call σ^2 so that $\mathcal{E}(\epsilon(n)^2) = \sigma^2$. Our notation implies that σ^2 is independent of n which is evidently so by stationarity.

If $\sigma^2 = 0$ we say that the $x(n)$ process is purely deterministic. This is a special use of the term since we are restricting ourselves to linear prediction.

† For most purposes of the remainder of the book this section may be omitted.

(We illustrate this in example (ii) of this section.) We commence by proving an important theorem due to Wold (1938). This is

Theorem 2 (Wold Decomposition Theorem)

$$x(n) = \sum_0^\infty a(j)\,\epsilon(n-j) + v(n) = u(n) + v(n),$$

where

(i) $$\mathcal{E}\{\epsilon(m)\,\epsilon(n)\} = \delta_m^{\,n}\sigma^2,$$

(ii) $$\mathcal{E}\{\epsilon(m)\,v(n)\} \equiv 0,$$

(iii) $v(n)$ is purely deterministic.

Proof. The $\epsilon(n)$ are as defined above. Since, for $m < n$, $\epsilon(n) \perp \mathcal{M}_m \subset \mathcal{M}_{n-1}$ and $\epsilon(m) \in \mathcal{M}_m$ we have $\epsilon(n) \perp \epsilon(m)$, $m < n$ and (i) is proved. Project $x(n)$ on the subspace of $\mathcal{H}$ spanned by the $\epsilon(m)$, $m \leq n$. Then this projection is

$$u(n) = \sum_0^\infty a(j)\,\epsilon(n-j), \qquad a(j) = \sigma^{-2}\,\mathcal{E}\big(x(n)\,\epsilon(n-j)\big)$$

since the $\epsilon(m)$ are orthogonal. Note that $a(0) = 1$. Put $v(n) = x(n) - u(n)$. Evidently $v(n) \perp \epsilon(m)$, $m \leq n$. But $v(n) \in \mathcal{M}_n$ and $\epsilon(m) \perp \mathcal{M}_n$, $m > n$ so that (ii) is proved. Let $\mathcal{V}_n$ be spanned by the $v(m)$, $m \leq n$. Then $v(n) \in \mathcal{M}_n = \mathcal{M}_{n-1} \oplus [\epsilon(n)]$ where by $[\epsilon(n)]$ we mean the one dimensional subspace of $\mathcal{H}$ spanned by $\epsilon(n)$. Thus $v(n) \in \mathcal{M}_{n-1}$. But the projection of $v(n)$ on $\mathcal{V}_{n-1}$ is the projection of $v(n)$ on $\mathcal{M}_{n-1}$ since the latter space is made up of $\mathcal{V}_{n-1}$ and an orthogonally complementary space spanned by the $\epsilon(m)$, $m \leq n-1$. Thus $v(n)$ belongs to the space on which it is projected and its prediction error is zero. This proves (iii).

Of course if $\sigma^2 = 0$ the $u(n)$ component is missing in the Wold decomposition. Since $v(n) \in \mathcal{M}_{n-j}$ for all j it follows that $v(n) \in \mathcal{M}_{-\infty}$. On the other hand, $v(n) \in \mathcal{M}_{-\infty}$ implies $v(n)$ perfectly predictable so that the two are equivalent. If $\mathcal{M}_{-\infty}$ is null then evidently $x(n)$ is purely nondeterministic in the sense that it contains no deterministic component. The orthogonal complement of $\mathcal{M}_{n-\nu}$ in $\mathcal{M}_n$ is spanned by $\epsilon(n), \epsilon(n-1), \ldots, \epsilon(n-\nu+1)$ and the prediction error for $x(n)$ in terms of $x(n-j)$, $j \geq \nu$ is

$$\sum_0^{\nu-1} a(j)\,\epsilon(n-j)$$

with mean-square error

$$\sigma^2 \sum_0^{\nu-1} a(j)^2.$$

We use this later.

We next establish the basic theorem:

Theorem 3

$$\sigma^2 = \exp\left[\frac{1}{2\pi}\int_{-\pi}^{\pi} \log\{2\pi f(\lambda)\}\,d\lambda\right].$$

If $\log f(\lambda)$ is not integrable, this can be only because the integral diverges to $-\infty$, since $\log f(\lambda) \le f(\lambda)$. When this is so, we interpret the right side of the formula as zero and the theorem continues to hold. This remarkable formula is due to Szego in the a.c. case, with the extension to the general case due to Kolmogoroff. We substantially follow Helson and Lowdenslager (1958), whose approach is clearly exhibited in Hoffman (1962) and Helson (1964).

Proof. We have to minimize

$$\int_{-\pi}^{\pi} |1 - h_N(e^{i\lambda})|^2 \, F(d\lambda), \tag{3.1}$$

where h_N is a polynomial for which $h_N(0) = 0$. These polynomials are, however, dense, in the sense of uniform convergence, in the space, A_0, of functions, h, continuous on the boundary of the circle, holomorphic within it and with zero mean value, that is, satisfying

$$\int_{-\pi}^{\pi} h(e^{i\lambda}) \, d\lambda = 0.$$

Indeed every such function, h, is, on the unit circle, the uniform limit of the Cesaro sum of its Fourier series† and such a Cesaro sum defines a polynomial function of the required form which evidently converges uniformly to h within the circle (by the maximum modulus principle). Thus we may replace (3.1) by the corresponding expression with h_N varying over A_0; since if h_N converges to h uniformly, it converges to it in mean square (with weighting $F(d\lambda)$). Thus we are led to consider‡

$$\inf_{h \in A_0} \int |1 - h|^2 \, F(d\lambda) \tag{3.2}$$

(We shall often omit arguments of functions under integral signs and limits of integration where there can be no confusion.) If $g \in A_0$, $h \in A_0$ then $gh \in A_0$. When speaking of functions in A_0 we may consider them either as functions on the circle or the disk, since as the former they completely determine their properties as the latter.

We first consider the a.c. case and show that

$$\inf_{h \in A_0} \int \exp\{\mathfrak{R}(h)\} f \, d\lambda = \exp\left[\frac{1}{2\pi} \int \log(2\pi f) \, d\lambda\right]. \tag{3.3}$$

† See the Mathematical Appendix, Section 3, for a survey of the relevant theorems.

‡ We shall often omit the limits of integration in the remainder of this section and in sections 5 and 6 since these are always $-\pi$ and π.

We use repeatedly below the inequality between the arithmetic and geometric means, and emphasize that this inequality holds for the mean value of any integrable function with respect to any probability measure (see Hewitt and Stromberg, 1965, p. 202). Applying this to $2\pi f \exp \mathcal{R}(h)$, obtain

$$\exp\left[\frac{1}{2\pi}\int \log 2\pi f\, d\lambda\right] \le \inf_{h\in A_0} \int f \exp \mathcal{R}(h)\, d\lambda,$$

the left side being the geometric mean of the function since h has zero mean value. We shall, in Lemma 1 proved below, show that the right side is not altered if the infinium is taken over all real integrable h with zero mean value. If $\log f$ is integrable, then $g = (2\pi)^{-1}\int \log (2\pi f)\, d\lambda - \log 2\pi f$ satisfies

$$\int fe^{g}\, d\lambda = \exp\left[\frac{1}{2\pi}\int \log (2\pi f)\, d\lambda\right]$$

and the infinium is attained, proving (3.3). If $\log f$ is not integrable we take $\log (2\pi f + \epsilon)$, $\epsilon > 0$, which gives

$$\inf_{h\in A_0} \int \exp \{\mathcal{R}(h)\}(f + \epsilon)\, d\lambda = \exp\left[\frac{1}{2\pi}\int \log (2\pi f + \epsilon)\, d\lambda\right]$$

and letting ϵ tend to zero we obtain (3.3), in general.

Now if $g \in A_0$ so does $h = (1 - \exp g)$ since A_0 is closed under the simple algebraic operations and with respect to uniform convergence. (In this lies the usefulness of A_0.) Also

$$\exp \{2\mathcal{R}(g)\} = |1 - h|^2.$$

Thus

$$\exp\left[\frac{1}{2\pi}\int \log (2\pi f)\, d\lambda\right] \ge \inf_{h\in A_0} \int |1 - h|^2 f\, d\lambda,$$

the inequality holding, since we are now taking the infinium over the possibly wider class of functions of the form $|1 - h|^2$, $h \in A_0$ in place of $\exp \mathcal{R}(g)$, $g \in A_0$. However, if $2\pi f$ is taken to be of the form $|1 - h|^2$, we have

$$\exp\left[\frac{1}{2\pi}\int \log |1 - h|^2\, d\lambda\right] \ge \inf_{g\in A_0} \frac{1}{2\pi}\int |1 - h - g + hg|^2\, d\lambda \ge 1$$

(the last inequality following since $h - g + hg \in A_0$). Thus $\log |1 - h|^2$ is integrable and we may put $|1 - h|^2 = k \exp p$, where p is real and integrable with zero mean value and $k \ge 1$. Thus, since p can be the real part of some $g \in A_0$, we derive

$$\int f|1 - h|^2\, d\lambda = k\int fe^{p}\, d\lambda \ge \inf_{g\in A_0} \int fe^{\mathcal{R}(g)}\, d\lambda = \exp\left[\frac{1}{2\pi}\int \log 2\pi f\, d\lambda\right]$$

and the inequality has been reversed and the theorem established in the a.c. case. We now turn to the general case.

If (3.2) is null, then it is evidently also so with $F(d\lambda)$ replaced by $f\,d\lambda$ (where f is the derivative of the a.c. part of F) and $\exp[(2\pi)^{-1}\int \log 2\pi f\,d\lambda] = 0$, so that the theorem holds. Thus we may take the infinium in (3.2) to be positive. Let $h_0 \in L_2(F)$ be the projection of 1 (the function identically unity) on to the closure, S, of A_0 in $L_2(F)$. Then $(1 - h_0) \perp S$ and thus to all elements of A_0. Let $h_n \in A_0$ and h_n converge as an element of $L_2(F)$, to h_0. Then $(1 - h_n) \exp im\lambda \in A_0$, $m > 0$, and converges in $L_2(F)$ to $(1 - h_0) \exp im\lambda$.

$$\int |1 - h_0|^2 e^{im\lambda} F(d\lambda) = \lim_n \int \overline{(1 - h_0)}(1 - h_n)e^{im\lambda} F(d\lambda) = 0, \qquad m > 0,$$

since $(1 - h_0) \perp A_0$ and $(1 - h_n) \exp im\lambda \in A_0$. Since $|1 - h_0|^2$ is real,

$$\int |1 - h_0|^2 e^{im\lambda} F(d\lambda) = 0, \qquad m = \pm 1, \pm 2, \ldots,$$

and $|1 - h_0|^2 F(d\lambda)$ must be a constant multiple of Lebesgue measure. Now h_0 is square integrable with respect to f and is the limit in mean square (weighting f) of a sequence of elements of A_0 and

$$\int \overline{(1 - h_0)} hf\,d\lambda = \int \overline{(1 - h_0)} h\,F(d\lambda) = 0, \qquad h \in A_0,$$

(since $(1 - h_0)$ vanishes on the support of the jump and singular part of F). Thus h_0 is the minimizing function when $F(d\lambda)$ is replaced by $f(\lambda)\,d\lambda$, and since we have shown that only f contributes to σ^2 we have established our theorem, save for the lemma used above. Before proving that we prove Theorem 4:

Theorem 4. *If $x(n)$ is not purely deterministic (i.e., $\sigma^2 > 0$) and $F = F^{(1)} + F^{(2)} + F^{(3)}$ is the Lebesgue decomposition of F, $F^{(1)}$ with derivative f, then f is the spectral density of $u(n)$ (in Theorem 2) and $F^{(2)} + F^{(3)}$ is the spectral function of $v(n)$.*

This means that

$$f(\lambda) = \frac{\sigma^2}{2\pi}\left|\sum_0^\infty a(j)e^{ij\lambda}\right|^2.$$

Note. It should be observed that the theorem says that $F^{(2)} + F^{(3)}$ is the spectral function of $v(n)$ only when $\log f$ is integrable.

Proof. From the proof of the last part of Theorem 3 we know that

$$|1 - h_0|^2 f = \frac{\sigma^2}{2\pi}, \qquad \text{a.e. } (d\lambda)$$

and that $|1 - h_0|^2$ is null on the support of $F^{(2)} + F^{(3)}$. Now, from Theorem 2,

$$\epsilon(n) = \int e^{-in\lambda}\phi(\lambda)\ z_x(d\lambda) = \int e^{-in\lambda}\ \phi(\lambda)\left\{\left(\sum_0^\infty a(j)e^{ij\lambda}\right) z_\epsilon(d\lambda) + z_v(d\lambda)\right\}$$

where $\phi(\lambda) = (1 - h_0)$ is the response function of the filter giving $\epsilon(n)$ in terms of the $x(n)$. Thus $\phi \neq 0$, a.e. $(d\lambda)$ but is null almost everywhere with respect to the spectral function of $v(n)$ and is $\{\sum a(j) \exp ij\lambda\}^{-1}$ a.e. (with respect to Lebesgue measure). Therefore $v(n)$ has a singular spectral function, and since the Lebesgue decomposition is unique the theorem is established.

Before going on to prove the last basic theorem of this section we establish the lemma used above.

Lemma 1

$$\inf_{h \in A_0} \int f e^{\mathcal{R}(h)}\, d\lambda$$

is not altered when the infinium is taken over all integrable functions with zero mean value.

Proof. (Helson and Lowdenslager, 1958.) We assume $\log f$ to be integrable since the remaining case is proved by introducing $f + \epsilon$, $\epsilon > 0$, as in the proof of Theorem 3. We put $g(\lambda) = f(\lambda)[\exp\{(2\pi)^{-1}\int \log f\, d\lambda\}]^{-1}$ so that $\log g$ has zero mean value. Then put $\log g = u - v$, where u is the nonnegative part of $\log g$ and v is the negative part, i.e., $u = \log g$ where $g \geq 1$ and is otherwise zero. It is easy to construct a sequence of bounded nonnegative functions, u_n, v_n, which increase, pointwise a.e. to u, v, respectively. Then

$$\lim_n \int u_n\, d\lambda = \int u\, d\lambda, \qquad \lim_n \int v_n\, d\lambda = \int v\, d\lambda.$$

Then for each n there is an m so that

$$\int u_n\, d\lambda \leq \int v_m\, d\lambda$$

We then multiply v_m by a constant, $c_n \leq 1$, so that this inequality becomes equality and rename the modified function v_n. As n increases $c_n \to 1$ since $\int u\, d\lambda = \int v\, d\lambda$. Now $0 \leq u_n \leq u$, $0 \leq v_n \leq v$, $\int u_n\, d\lambda = \int v_n\, d\lambda$ and u_n increases monotonically to u while v_n converges pointwise to v. For λ such that $g(\lambda) \leq 1$ we have $u(\lambda) = 0$, that is, $u_n(\lambda) = 0$ so that, since $\exp(v_n - v) \leq 1$, for such λ we have $\exp\{u - u_n - v + v_n\} \leq 1$. For λ such that $g(\lambda) \geq 1$ we have $v = 0$ and thus $v_n = 0$ and $\exp(u - u_n) = \exp(\log g - u_n) \leq g$. Thus $0 \leq \exp\{u - u_n - v + v_n\} \leq \max(1, g)$. Thus by dominated convergence

$$\lim_n \frac{1}{2\pi} \int \exp(v_n - u_n) g \, d\lambda = \lim_n \frac{1}{2\pi} \int \exp\{u - u_n - v + v_n\} \, d\lambda = 1$$

Thus, multiplying by $\exp[(2\pi)^{-1} \int \log f \, d\lambda]$, we have

$$\text{(3.4)} \qquad \inf \int f e^{\psi} \, d\lambda \leq \exp\left[\frac{1}{2\pi} \int \log f \, d\lambda\right]$$

where the infinium is over all bounded measurable functions ψ with zero mean value (to which class $v_n - u_n$ belongs). Since we already know that equality holds even for the infinium over all integrable ψ with zero mean value, it holds also for this class of ψ just described. However, a bounded measurable ψ with zero mean value is boundedly the limit a.e. of the Cesaro means of its Fourier series† so that by dominated convergence the infinium on the left in (3.4) is not changed when ψ varies over real trigonometric polynomials with zero mean value and thus when the infinium is taken over $\mathfrak{R}(h)$, $h \in A_0$, as we wished to show.

We now give an important characterization of the functions $g(z)$ for which $f(\lambda) = |g|^2$ is the spectral density of a purely nondeterministic $x(n)$ process.

Theorem 5. *Let $\log f(\lambda)$ be integrable where $f(\lambda)$ is a spectral density. If*

$$f(\lambda) = |g(e^{i\lambda})|^2$$

then $g(z)$ uniquely decomposes as

$$g(z) = e^{i\alpha} B(z) C(z) Q(z),$$

where

$$B(z) = z^p \prod_1^\infty \left[\frac{\bar{a}_n}{|a_n|} \frac{(a_n - z)}{(1 - \bar{a}_n z)}\right]^{p_n}, \qquad a_n \neq 0, |a_n| < 1,$$

$$C(z) = \exp\left[-\int_{-\pi}^{\pi} \frac{e^{i\lambda} + z}{e^{i\lambda} - z} \mu(d\lambda)\right]$$

$$Q(z) = \exp\left[\frac{1}{4\pi} \int_{-\pi}^{\pi} \frac{e^{i\lambda} + z}{e^{i\lambda} - z} \log f(\lambda) \, d\lambda\right].$$

Here μ is a positive singular measure. The p_n are positive integers. The function $Q(z)$ is uniquely defined by the requirement that it be holomorphic

† Though this is true we need here only the much more easily proved fact that a subsequence of the sequence of Cesaro means converges boundedly, a.e., to ψ. See the Mathematical Appendix.

within the unit circle and satisfy

$$|Q(e^{i\lambda})|^2 = f(\lambda), \quad Q(0) = \exp\left[\frac{1}{4\pi}\int_{-\pi}^{\pi} \log f(\lambda)\, d\lambda\right] = \frac{\sigma}{\sqrt{2\pi}}$$

A function of the form of $Q(z)$ is called an outer function. The functions $B(z)$ and $C(z)$ are inner functions, the former being called a Blaschke product. They are holomorphic within the unit circle and bounded in modulus by unity there and of unit modulus almost everywhere on the unit circle. This may be taken to be the definition of an inner function and we show that such a function is of the form $B(z)C(z) \exp i\alpha$.

For the proof we substantially follow Hoffman (1962).

We have

$$|Q(re^{i\lambda})|^2 = \exp\left[\frac{1}{2\pi}\int_{-\pi}^{\pi} P_r(\lambda - \theta) \log f(\theta)\, d\theta\right], \qquad r < 1,$$

where $P_r(\lambda)$ is Poisson's kernel (see the Mathematical Appendix). Once more, using the inequality between the arithmetic and geometric means (this time with respect to the weight function $(2\pi)^{-1} P_r(\lambda - \theta)$), we see that

$$\frac{1}{2\pi}\int |Q|^2\, d\lambda \le \frac{1}{4\pi^2}\iint P_r(\lambda - \theta) f(\theta)\, d\theta\, d\lambda = \frac{1}{2\pi}\int f(\theta)\, d\theta,$$

so that $Q(r \exp i\lambda)$ is square integrable for every r with integral uniformly bounded and thus defines a function square integrable on the unit circle (whose Fourier coefficients are the coefficients in the power series expansion of $Q(z)$). Moreover, since the Abel sum for $\log f(\lambda)$ converges pointwise almost everywhere† to $\log f(\lambda)$,

$$|Q(e^{i\lambda})|^2 = f(\lambda) \quad \text{a.e.}$$

We now make use of Lemma 2, whose proof we give below, to show that, if $g(z)$ is any function holomorphic within the unit circle, for which $f = |g|^2$, then

$$|g(re^{i\lambda})|^2 \le |Q(re^{i\lambda})|^2$$

through the inequality in the series of relations

$$\log |g(re^{i\lambda})| = \log\left|\frac{1}{2\pi}\int g(e^{i\theta})\, P_r(\lambda - \theta)\, d\theta\right| \le \frac{1}{2\pi}\int \log |g(e^{i\theta})|\, P_r(\lambda - \theta)\, d\theta$$
$$= \log |Q(re^{i\lambda})|.$$

† Again we may avoid using this theorem for, $\log f$ being integrable, the Abel sum converges in the L_1 mean and therefore in measure to $\log f$ and the result we require follows.

Thus $u(z) = g(z)/Q(z)$ is holomorphic within the unit circle, of unit modulus on it, and bounded by unity in modulus within it. Thus it is an inner function, and we now need to prove the formula for the decomposition of such a function. If $u(z)$ has a zero of order p at the origin, we divide by z^p, and if the resulting function is not positive for $z = 0$ we divide by exp $i\alpha$ to make it so. We again call the new function $u(z)$. Let a_n, $n = 1, 2, \ldots$, be the remaining zeros of $u(z)$ within the unit circle, the nth having multiplicity p_n. Consider first

$$v_n(z) = \prod_1^n \left(\frac{z - a_k}{1 - \bar{a}_k z}\right)^{p_k}.$$

Now $u(z)/v_n(z)$ is holomorphic within the circle and bounded in modulus by unity. Thus $0 < u(0) \leq |v_n(0)| = \prod_1^n |a_k|^{p_k}$. Thus the infinite product, $\prod_1^\infty |a_k|^{p_k}$ converges since it certainly does not diverge to $+\infty$ since $|a_k| < 1$. Now

$$B_n(z) = \prod_1^n \left\{\frac{\bar{a}_k}{|a_k|}\frac{a_k - z}{1 - \bar{a}_k z}\right\}^{p_k}$$

satisfies

$$\overline{B_n(e^{i\lambda})} = \{B_n(e^{i\lambda})\}^{-1},$$

so that

$$\frac{1}{2\pi}\int |B_n - B_m|^2 \, d\lambda = \frac{1}{2\pi}\int 2\{1 - \Re(B_n\overline{B_m})\} \, d\lambda$$

$$= 2\left[1 - \Re\left\{\frac{1}{2\pi}\int \frac{B_n}{B_m} \, d\lambda\right\}\right] = 2\left\{1 - \Re\left(\frac{B_n(0)}{B_m(0)}\right)\right\}$$

$$= 2\left(1 - \prod_{m+1}^n |a_k|^{p_k}\right), \qquad n > m.$$

This shows that the sequence B_n converges in mean square on the unit circle to a function B which must be of unit modulus a.e. on the unit circle since a subsequence of the B_n will converge pointwise a.e. (Halmos, 1950, p. 93). Putting

$$B(re^{i\lambda}) = \frac{1}{2\pi}\int P_r(\lambda - \theta)\, B(e^{i\theta}) \, d\theta, \qquad 0 \leq r < 1,$$

we see that

$$|B_n(re^{i\lambda}) - B(re^{i\lambda})| \leq \frac{1}{2\pi}\int P_r(\lambda - \theta)\, |B_n(e^{i\theta}) - B(e^{i\theta})| \, d\theta$$

$$\leq \left[\int |B_n(e^{i\theta}) - B(e^{i\theta})|^2 \, d\theta\right]^{1/2},$$

so that $B_n(z)$ converges to $B(z)$ uniformly on every closed disk within the unit circle and there is of modulus not greater than unity. Thus $B(z)$ is an inner function of the special form called a Blaschke product.

We next form $k(z) = u(z)/B(z)$, which is holomorphic within the unit circle, without zeros there and positive at the origin. It is of unit modulus a.e. on the circle and thus of not greater than unit modulus within it. We put $k(z) = \exp\{-h(z)\}$ where $h(z) = v(z) + iw(z)$ is holomorphic within the circle and $v(z) \geq 0$ (since $|k(z) \leq 1$). Then

$$v(re^{i\lambda}) = \frac{1}{2\pi}\int P_r(\lambda - \theta)\,\mu(d\theta), \tag{3.5}$$

where μ is a positive measure. Indeed the Abel sum of the Fourier series of the harmonic function $v(z)$ is

$$v_r(\lambda) = \frac{1}{2\pi}\int P_r(\lambda - \theta)\,v(e^{i\theta})\,d\theta \geq 0$$

and

$$\int v_r(\lambda)\,d\lambda = \int v(e^{i\lambda})\,d\lambda.$$

The a.c. measures on $[-\pi, \pi]$ with densities $v_r(\lambda)$ have uniformly bounded total variation. Thus, by Helly's theorem, we may find a convergent subsequence with limit a positive measure μ. Moreover, as s varies over the subsequence,

$$\lim_s \frac{1}{2\pi}\int P_r(\lambda - \theta)\,v_s(\theta)\,d\theta = \frac{1}{2\pi}\int P_r(\lambda - \theta)\,\mu(d\theta),$$

since P_r is continuous; but the left side also converges to $v(r \exp i\lambda)$, establishing (3.5). Now

$$\int \frac{e^{i\theta} + z}{e^{i\theta} - z}\,\mu\,(d\theta)$$

is holomorphic in the unit circle and its real part is $v(z)$. Up to an additive imaginary constant it must be $h(z)$, and this constant must be zero (mod $2\pi i$), since $h(0)$ is real (mod $2\pi i$). Thus (mod $2\pi i$)

$$h(z) = \int_{-\pi}^{\pi} \frac{e^{i\theta} + z}{e^{i\theta} - z}\,\mu(d\theta).$$

Since $|k(\exp i\lambda)| = 1$, a.e., $|v(\exp i\lambda)| = 0$, a.e.; but

$$\lim_{r\to 1} v(re^{i\lambda}) = \lim_{r\to 1}\frac{1}{2\pi}\int P_r(\lambda - \theta)\,\mu(d\theta) = \frac{1}{2\pi}\frac{d\mu}{d\lambda}, \quad \text{a.e.}$$

(see the exercises to this chapter), so that μ is singular.

This completes the proof of the theorem save for the uniqueness assertion. The factors $\exp i\alpha$, $B(z)$, $Q(z)$ are evidently uniquely defined so that $C(z)$ is uniquely defined also. It is easily seen that $B(0)$ and $C(0)$ are real and less than unity unless $B(z)$ and $C(z)$ are identically unity. Thus, if $g(z)$ satisfies $g(0) = \sigma/\sqrt{2\pi}$, then $g(z) = Q(z)$, which completes the proof.

Lemma 2. *If $g(z)$ is integrable on the unit circle and analytic within it then*

$$(3.6)\quad \frac{1}{2\pi}\int \log |g(e^{i\theta})|\, P_r(\lambda - \theta)\, d\theta \geq \log \left|\frac{1}{2\pi}\int g(e^{i\theta})\, P_r(\lambda - \theta)\, d\theta\right|.$$

Proof. (Hoffman, 1962.) If $g \in A$ (see the Mathematical Appendix), we can find $h \in A$ so that

$$\frac{1}{2\pi}\int |g - h|\, P_r(\lambda - \theta)\, d\theta < \epsilon, \quad \frac{1}{2\pi}\int g\, P_r(\lambda - \theta)\, d\theta = \frac{1}{2\pi}\int h\, P_r(\lambda - \theta)\, d\theta;$$

for example, the nth Cesaro mean for g will satisfy the first inequality for n large enough, and, multiplying this by a constant, near to unity for n large enough, the second equation may be satisfied also. Also $(2\pi)^{-1}\int \log \{|h| + \epsilon\}\, P_r(\lambda - \theta)\, d\theta$, $\epsilon > 0$, will be near to the left side of (3.6) when h is as described and $|g|$ is replaced by $|g| + \epsilon$ (by dominated convergence). Thus we shall have proved the result if we prove it for $h \in A$. Now $\log(|h| + \epsilon)$ is a real continuous function on the circle and can be approximated uniformly by the real part u of $p = u + iv \in A$ (since Cesaro sums for a real continuous function are real and belong to A). Put $k = \exp(-p) \in A$. Then $|k| = \exp(-u)$ and $|hk| = |h| \exp(-u)$, which is less than $\exp \epsilon$ if

$$u - \epsilon < \log \{|h| + \epsilon\} < u + \epsilon.$$

Now for $g_i \in A$ the integral with weight function $(2\pi)^{-1} P_r(\lambda - \theta)$ is multiplicative, as is easily seen; that is,

$$\frac{1}{2\pi}\int g_1 g_2 P_r(\lambda - \theta)\, d\theta = \frac{1}{2\pi}\int g_1 P_r(\lambda - \theta)\, d\theta\, \frac{1}{2\pi}\int g_2 P_r(\lambda - \theta)\, d\theta,$$

so that

$$\left|\frac{1}{2\pi}\int h P_r(\lambda - \theta)\, d\theta\right| \left|\frac{1}{2\pi}\int k P_r(\lambda - \theta)\, d\theta\right| = \left|\frac{1}{2\pi}\int hk P_r(\lambda - \theta)\, d\theta\right| < e^{\epsilon}.$$

Thus

$$\log\left|\frac{1}{2\pi}\int h P_r(\lambda - \theta)\, d\theta\right| + \log\left|\frac{1}{2\pi}\int k P_r(\lambda - \theta)\, d\theta\right| < \epsilon;$$

but

$$\frac{1}{2\pi}\int kP_r(\lambda-\theta)\,d\theta = \frac{1}{2\pi}\int e^{-p}P_r(\lambda-\theta)\,d\theta = \exp\left\{-\frac{1}{2\pi}\int pP_r(\lambda-\theta)\,d\theta\right\}$$

because of the multiplicative nature of the integral. The modulus of the last expression is

$$\exp\left[-\frac{1}{2\pi}\int uP_r(\lambda-\theta)\,d\theta\right],$$

so that

$$\log\left|\frac{1}{2\pi}\int hP_r(\lambda-\theta)\,d\theta\right| - \frac{1}{2\pi}\int uP_r(\lambda-\theta)\,d\theta < \epsilon,$$

and, since $u < \epsilon + \log\{|h| + \epsilon\}$,

$$\frac{1}{2\pi}\int uP_r(\lambda-\theta)\,d\theta < \epsilon + \frac{1}{2\pi}\int(\log|h| + \epsilon)\,P_r(\lambda-\theta)\,d\theta,$$

so that

$$\log\left|\frac{1}{2\pi}\int hP_r(\lambda-\theta)\,d\theta\right| < 2\epsilon + \frac{1}{2\pi}\int\{\log|h| + \epsilon\}\,P_r(\lambda-\theta)\,d\theta,$$

which, since ϵ is arbitrary, proves what we require.

We can now finish our discussion of the optimal predictor.

Theorem 6. *Let $x(n)$ be purely non deterministic and let $h_\nu(z)$ be the z transform of the filter giving the best linear predictor of $x(n+\nu)$, $\nu > 0$, from $x(n-j)$, $j \geq 0$. Then*

$$h_\nu(z) = \frac{\sum_\nu^\infty a(j)z^j}{\sum_0^\infty a(j)z^j} \tag{3.7}$$

and the $a(j)$ may be had from

$$\frac{\sigma}{\sqrt{2\pi}}\sum_0^\infty a(j)z^j = \exp\left\{c(0) + 2\sum_0^\infty c(n)\,z^n\right\}$$

$$c(n) = \frac{1}{4\pi}\int_{-\pi}^{\pi} e^{-in\lambda}\log f(\lambda)\,d\lambda, \qquad \frac{\sigma}{\sqrt{2\pi}} = \exp c(0).$$

Proof. We know that $(\sigma/\sqrt{2\pi})\sum_0^\infty a(j)z^j$ is holomorphic within the circle, has squared modulus $f(\lambda)$ a.e. on the circle, and is $\sigma/\sqrt{2\pi}$ at $z = 0$. Thus it is $Q(z)$. Moreover,

$$\begin{aligned} c(0) + 2\sum_1^\infty c(n)z^n &= \frac{1}{4\pi}\int\left\{1 + 2\sum_1^\infty z^n e^{in\lambda}\right\}\log f(\lambda)\,d\lambda \\ &= \frac{1}{4\pi}\int\frac{e^{i\lambda}+z}{e^{i\lambda}-z}\log f(\lambda)\,d\lambda = \log Q(z). \end{aligned}$$

Since $Q(z)$ has no zeros within the circle, $h_\nu(z)$ is holomorphic within it and it is trivially verified that it satisfies

$$\int |1 - h_\nu|^2 f(\lambda)\, d\lambda = \sigma^2 \sum_0^{\nu-1} a(j)^2,$$

which is the correct prediction error. Thus it is the z transform of the optimal filter. Indeed, we already know that

$$x(n) = \sum_0^\infty a(j)\epsilon(n-j) = \int_{-\pi}^{\pi} e^{-in\lambda}\left(\sum_0^\infty a(j)e^{ij\lambda}\right) z(d\lambda),$$

so that, since $\sum a(j) \exp ij\lambda \neq 0$, a.e.,

$$z_\epsilon(d\lambda) = \left(\sum_0^\infty a(j)e^{ij\lambda}\right)^{-1} z(d\lambda).$$

Thus

$$\int_{-\pi}^{\pi} e^{-in\lambda} h_\nu(e^{i\lambda})\, z(d\lambda) = \sum_\nu^\infty a(j)\, x(n-j).$$

We now deal with some examples of a theoretical nature that use and illustrate the theory of this section.

Examples

(i) As a first example we prove a result due to Grenander (see Grenander and Rosenblatt, 1957, p. 103) which shows the close relation between the spectrum and the eigenvalues of the matrix Γ_N, having $\gamma(m-n)$ in row m, column n, $m, n = 1, \ldots, N$. We consider the case where $f(\lambda) \le (2\pi)^{-1}M < \infty$. Let $\lambda_{j,N}$ be the eigenvalues of Γ_N. We have

$$\sum_{m,n=1}^{N}\sum \gamma(m-n) z_m \bar{z}_n = \int_{-\pi}^{\pi} \left|\sum_1^N z_m e^{im\lambda}\right|^2 f(\lambda)\, d\lambda \le M \sum_1^N |z_m|^2.$$

so that $|\lambda_{j,N}| \le M$. Let α be real with $|\alpha| < M^{-1}$ and $g(\lambda) = (2\pi)^{-1} \times \{1 + 2\pi\alpha f(\lambda)\}$. Then $0 < g(\lambda) < \pi^{-1}$ and is an even function of λ. It may therefore be considered as a spectral density of a purely nondeterministic scalar process. Let us call that process $y(n)$. Now let σ_N^2 be defined by

$$\sigma_N^2 = \inf \mathcal{E}\left\{\left(y(n) - \sum_1^N \beta(j) y(n-j)\right)^2\right\},$$

where we take the infimum over all $\beta(j)$. This may be evaluated in the same way as in Exercise 2 to Chapter I, in terms of the covariance matrix of the $y(j), j = 0, \ldots, N$. (See also Cramer, 1946, p. 305.) This covariance matrix is of course $I_N + \alpha\Gamma_N$, which we call T_N, so that using that exercise

$$\sigma_N^2 = \frac{\det(T_{N+1})}{\det(T_N)}.$$

where

$$\det(T_N) = \prod_{j=1}^{N} (1 + \alpha\lambda_{j,N}).$$

Now σ_N^2 decreases with N and converges to

$$\sigma^2 = \exp\left\{\frac{1}{2\pi}\int_{-\pi}^{\pi} \log\{2\pi g(\lambda)\}\, d\lambda\right\}$$

by Theorem 3. Thus

$$\frac{1}{2\pi}\int_{-\pi}^{\pi} \log\{1 + 2\pi\alpha f(\lambda)\}\, d\lambda = \lim_{N\to\infty} \log \sigma_N^2 = \lim_{N\to\infty} N^{-1}\sum_{n=1}^{N} \log \sigma_n^2$$

$$= \lim_{N\to\infty} N^{-1}\{\log(\det(T_{N+1})) - \log(\det(T_1))\} = \lim_{N\to\infty} N^{-1}\sum_{1}^{N} \log(1 + \alpha\lambda_{j,N}).$$

Let us put

$$m_{N,p} = \frac{1}{N}\sum_{1}^{N} \lambda_{j,N}^p.$$

Now

$$\log(1 + x) = \sum_{1}^{\infty} (-)^{p-1} \frac{x^p}{p}$$

and this converges uniformly for $|x| < 1 - \delta$, $\delta > 0$. Now

$$\lim_{N\to\infty} \frac{1}{N}\sum_{1}^{N} \log(1 + \alpha\lambda_{j,N}) = \lim_{N\to\infty} \frac{1}{N}\sum_{1}^{N}\left\{\sum_{1}^{\infty}(-)^{p-1}\frac{\alpha^p\lambda_{j,N}^p}{p}\right\}$$

$$= \lim_{N\to\infty} \sum_{1}^{\infty} (-)^{p-1}\frac{\alpha^p}{p} m_{N,p},$$

and since $\alpha < M^{-1}$, $2\pi f(\lambda) \le M$ we may also expand $\log(1 + 2\pi\alpha f(\lambda))$ in the logarithmic series and integrate term by term, so that

$$\lim_{N\to\infty} \sum_{1}^{\infty} (-)^{p-1}\frac{\alpha^p}{p} m_{N,p} = \sum_{1}^{\infty} (-)^{p-1}\frac{\alpha^p}{p}\frac{1}{2\pi}\int_{-\pi}^{\pi}\{2\pi f(\lambda)\}^p\, d\lambda.$$

Now the left-hand side is bounded uniformly by

$$\sum \frac{(\alpha M)^p}{p} < \sum \frac{1}{p}\frac{1}{2^p} < \infty$$

if $\alpha < (2M)^{-1}$. Thus by dominated convergence we may take the limit under the summation sign on the left and identifying powers of α in the two expressions we have proved the following theorem.

Theorem 7. *If $f(\lambda)$ is uniformly bounded and $\lambda_{j,N}$, $j = 1, \ldots, N$ are the eigenvalues of Γ_N then*

$$\lim_{N\to\infty} N^{-1} \sum_1^N \lambda_{j,N}^p = \frac{1}{2\pi}\int_{-\pi}^{\pi} \{2\pi f(\lambda)\}^p \, d\lambda, \qquad p = 0, 1, \ldots.$$

This theorem suggests that the quantities $2\pi f(\omega_j - \pi)$, $\omega_j = 2\pi j/N$, $j = 0, 1, \ldots, N-1$, in an average sense, provide good approximations to the eigenvalues of Γ_N, when N is large as we see by approximating the integral in the theorem by an approximating sum. Otherwise put, the expression on the right may be regarded as the pth moment of the function $2\pi f(\cdot)$ of a random variable, X, uniformly distributed in $(-\pi, \pi]$. The theorem asserts that all of the moments of the sequence of sets of observations $\{\lambda_{j,N}, j = 1, \ldots, N\}$ converge to those of $2\pi f(X)$.

(ii) In Exercise 4 of Chapter II we saw that if $x(n) = \cos n\xi + \sin n\eta$ with ξ and η independent with the same (symmetric) probability distribution, F, then $x(n)$ is stationary with F as its spectral distribution function. Then if F is not singular and has $\log f$ integrable $x(n)$ is not perfectly predictable *by a linear operation on the past.* It is evidently perfectly predictable by a nonlinear operation, however.

(iii) We conclude this section with some discussion of the meaning of the inner functions in Theorem 4, following Robinson (1962). In the first place for each of these factors the group delay (see Chapter II, Section 4) is nonnegative. Indeed this is obvious for exp $i\alpha$ and z^p and for the other factors we have, respectively,

$$\frac{1}{i}\frac{d}{d\lambda}\log\left\{\frac{a_n}{|a_n|}\left(\frac{a_n - e^{i\lambda}}{1 - \bar{a}_n e^{i\lambda}}\right)\right\} = \frac{1 - |a_n|^2}{|1 - a_n e^{-i\lambda}|^2} > 0$$

$$\frac{1}{i}\frac{d}{d\lambda}\left\{-\int \frac{e^{i\theta} + e^{i\lambda}}{e^{i\theta} - e^{i\lambda}}\mu(d\theta)\right\} = \int \frac{2\{1 - \cos(\theta - \lambda)\}}{|e^{i\theta} - e^{i\lambda}|^4}\mu(d\theta) \geq 0$$

Thus the filter with response Q is the "minimum delay filter" for the group delay of a product of filters is additive so that this is minimized (subject to the gain being fixed) when the inner factors are absent.

Consider next a filter with response

$$\frac{\sigma}{\sqrt{2\pi}}\sum_0^\infty \beta(j)e^{ij\lambda}, \qquad \frac{\sigma^2}{2\pi}\left|\sum_0^\infty \beta(j)e^{ij\lambda}\right|^2 = f(\lambda)$$

Then

(3.8) $$\sum_0^\infty |\beta(j)|^2 = \sum_0^\infty a(j)^2, \qquad \sum_0^p |\beta(j)|^2 \leq \sum_0^p a(j)^2,$$

Indeed, we may put $x(n)$ in the form

$$x(n) = \sum_0^\infty \beta(j)\,\eta(n-j), \qquad \mathcal{E}\big(\eta(n)\,\eta(m)\big) = \delta_m{}^n\sigma^2.$$

Then $\mathcal{M}_{n-1}$ is certainly contained in the space spanned by the $\eta(n-j)$, $j \geq 1$ so that $|\beta(0)|^2\sigma^2$, the mean square of the deviation of $x(n)$ from the projection on the elements of this last space, is not greater than $\sigma^2 = a(0)^2\sigma^2$. The general result follows in the same way. If we think of the two filters as having the same input then the optimal filter gives as much weight as possible (the gain being fixed) to the most recent inputs. The general one-sided filter with this gain is got from the optimal filter by the adjunction of inner factors which serve to delay the inputs, as we have seen.

4. THE GENERAL THEORY FOR STATIONARY, CONTINUOUS-TIME, SCALAR PROCESSES†

We shall not go into full details here but shall indicate how the results of Section 3 lead to the result for the continuous-time case. We follow Doob (1953) and Hoffman (1962). For an alternative approach see Helson (1964).

The basic device is the mapping of the closed unit disk on to the left half plane (together with the imaginary axis) which is accomplished by

$$w = \frac{z-1}{z+1}, \qquad z = \frac{1+w}{1-w}.$$

This maps the boundary ($z = \exp i\lambda$) onto the boundary ($w = i\tau$) and we have $\tau = \tan \frac{1}{2}\lambda$ so that

$$\frac{1}{2\pi}\frac{d\lambda}{d\tau} = \frac{1}{\pi(1+\tau^2)}$$

A basic result of the last section was that which asserted that $f(\lambda)$ is the square of the modulus of a function holomorphic within the unit circle when and only when its logarithm is integrable. The analogous theorem for the half plane is due to Paley and Wiener (1934) and asserts that the spectral density $f(\tau)$, $-\infty < \tau < \infty$, satisfies

$$f(\tau) = \frac{1}{2\pi}\left|\int_0^\infty \alpha(s)e^{is\tau}\,ds\right|^2$$

when and only when

$$\int_{-\infty}^{\infty} \frac{\log f(\tau)}{1+\tau^2}\,d\tau < \infty. \tag{4.1}$$

† This section may be omitted.

The proof follows from the relation between the spaces H_j $(j = 1, 2)$ of the unit disk and of the left half plane†, the latter consisting of functions which are holomorphic in the half plane and integrable or square integrable on its boundary. We use the same symbols, H_j, for both spaces since the argument symbols (w or z, etc.) will indicate with which space we are concerned. Now it is easy to see that $g(w) \in H_j$ if and only if

$$h(z) = (1 + z)^{-2/j}\, g\left(\frac{z - 1}{z + 1}\right) \in H_j.$$

If $f(\tau)$ satisfies the condition of the Paley-Wiener theorem, then $\phi(\lambda) = f(\tan \frac{1}{2}\lambda)$ is such that $\log \phi(\lambda)$ is integrable, so that $f(\tau) = |g(i\tau)|^2$ and $g(w) \in H_2$ (since it is certainly holomorphic within the half-plane, being of the form

$$h\left(\frac{1 + w}{1 - w}\right)$$

and is square integrable on the boundary since $f(\tau)$ is integrable).

Now we put

$$a(s) = \frac{1}{\sqrt{2\pi}} \int_{-\infty}^{\infty} g(i\tau)e^{-is\tau}\, d\tau$$

and need to show that $a(s) = 0$, $s < 0$, for the sufficiency part of the proof. It is sufficient to assume $g(w) \in H_1 \cap H_2$ for otherwise we consider

$$[1 - \{w/(1 + w)\}^n]\, g(w)$$

in its place, which lies in $H_1 \cap H_2$ for $n > 0$ and converges in mean square to g. On this assumption there is an $h \in H_1$ so that

$$g = h\left(\frac{1 + w}{1 - w}\right)(1 - w)^{-2} \in H_1$$

and for $s < 0$

$$\frac{1}{\pi} \int_{-\infty}^{\infty} g(i\tau)e^{-is\tau}\, d\tau = \frac{1}{\pi} \int_{-\infty}^{\infty} \frac{1 + i\tau}{1 - i\tau}\, e^{-is\tau}(1 - i\tau)^2 \frac{g(i\tau)}{1 + \tau^2}\, d\tau$$

$$= \frac{1}{2\pi} \int_{-\pi}^{\pi} h_1(e^{i\lambda})e^{i\lambda}\, d\lambda = 0,$$

since

$$h_1 = h(z) \exp\left(-s\, \frac{z - 1}{z + 1}\right) \in H_1.$$

† See the Mathematical Appendix.

(This would not necessarily be true for $s > 0$ since then there would be a pole at $z = -1$.)

The necessity is easily established since if $f(\tau)$ is of the required form then $f(\tau) = |g(i\tau)|^2$ where $g(w)$ is certainly in H_2 of the half plane and then $f(\tau)$ transforms into $\phi(\lambda) = |h|^2$, $h \in H_2$, so that $\log \phi(\lambda)$ is integrable whence the necessity follows immediately.

It now follows that, when $\log f(\tau)$ satisfies (4.1),

$$x(t) = \int_0^\infty \alpha(t + s)\xi(ds),$$

where $\xi(s)$ is, as usual, a process of orthogonal increments with $\mathcal{E}(\xi(ds)^2) = ds$.

Now, following Doob (1953), we wish to make a discrete parameter process, $y(n)$, correspond to the continuous parameter process, $x(t)$, in such a way that the subspace $\mathcal{M}_0$, say, of $\mathcal{H}_x$ (i.e., the Hilbert space spanned by the $x(t)$) which is spanned by the $x(t)$ for $t \leq 0$ is mapped onto the subspace, $\mathcal{N}_0$ say, of the Hilbert space $\mathcal{H}_y$ (spanned by the $y(n)$) which is spanned by the $y(n)$ for $n \leq 0$. Since $\mathcal{H}_x$ and $\mathcal{H}_y$ are both separable Hilbert spaces there are infinitely many isomorphic mappings of one into the other but we want more, as has just been said. We assume that $y(n)$ has spectral distribution function $G(\lambda) = F(\tan \frac{1}{2}\lambda)$ where $F(\tau)$ is the spectral distribution function for $x(t)$. Now $\mathcal{M}_0$ is isomorphic to the subspace of $L_2(F)$ spanned by $\exp -it\tau$, $t \geq 0$ and $\mathcal{N}_0$ is isomorphic to the subspace of $L_2(G)$ spanned by $\exp -in\lambda$, $n \geq 0$. To establish the correspondence we may consider mapping $L_2(F)$ onto $L_2(G)$ and we thus consider the correspondence

$$\frac{1 + i\tau}{1 - i\tau} \leftrightarrow \exp i\lambda. \tag{4.2}$$

Now

$$\frac{1 - i\tau}{1 + i\tau} = -1 + \frac{2}{1 + i\tau} = -1 + 2\int_{-\infty}^0 e^{-it\tau}e^t\,dt,$$

so that $\exp -i\lambda$, and thus $\exp -in\lambda$ for all $n \geq 0$ corresponds to an element in $\mathcal{M}_0$, since the last integral may be approximated in mean square $(F(d\tau))$ by a linear combination of the $\exp -it\tau$, $t \geq 0$, and thus belongs to $\mathcal{M}_0$. The image of $\mathcal{N}_0$ is contained in $\mathcal{M}_0$. Conversely,

$$\exp wt = \exp \{t(z - 1)/(z + 1)\}$$

is holomorphic for $|z| > 1$ and has modulus <1 if $t < 0$. It can be expanded in a series of nonpositive powers of z. On the unit circle it is

$$\exp it(\tan \tfrac{1}{2}\lambda) = \exp it\tau,\ t < 0,$$

the bounded pointwise limit (a.e.) of exp wt, $t < 0$. Thus exp $it\tau$, $t < 0$, can be approximated boundedly and thus in mean square $(F(d\tau))$ by the image of an element of $\mathcal{N}_0$ so that the image of $\mathcal{N}_0$ is contained in $\mathcal{M}_0$. Thus the two are mapped onto each other by (4.2). Since $x(t)$ is purely deterministic (by definition) iff $\mathcal{M}_0 = \mathcal{M}_{-\infty}$ and evidently $\mathcal{M}_{-\infty}$ corresponds to $\mathcal{N}_{-\infty}$ (by the same type of proof as just given)†, we have‡ Theorem 3:

Theorem 3′. *The process $x(t)$ is purely nondeterministic if and only if $F(\tau)$ is a.c. and* (4.1) *holds.*

(We say that $x(t)$ is purely nondeterministic if $\mathcal{M}_{-\infty}$ is null.)

It is easiest next to establish the analog of Theorem 5. Indeed we omit the proof for the theorem is obtained from Theorem 5 by easy substitutions.

Theorem 5′. *Let $x(t)$ be nondeterministic with spectral density*

$$f(\tau) = \left|\int_0^\infty \alpha(s)e^{is\tau}\,ds\right|^2 = |g(i\tau)|^2. \tag{4.3}$$

Then, putting $w = \sigma + i\tau$,

$$g(w) = e^{i\alpha}\,B(w)\,C(w)\,Q(w),$$

$$B(w) = \left(\frac{1+w}{1-w}\right)^p \prod_n \left\{\frac{1-\bar{\beta}_n{}^2}{|1-\beta_n{}^2|}\,\frac{w-\beta_n}{w+\bar{\beta}_n}\right\}^{p_n},$$

$$C(w) = e^{\rho/w}\exp\left[-\int_{-\infty}^{\infty}\frac{1-i\tau w}{i\tau - w}\,\nu(d\tau)\right],$$

$$Q(w) = \exp\left[\frac{1}{2\pi}\int_{-\infty}^{\infty}\log f(\tau)\,\frac{1-i\tau w}{i\tau - w}\,\frac{d\tau}{1+\tau^2}\right]. \tag{4.4}$$

wherein the β_n are the zeros of $g(w)$ in the left half-plane, ρ is nonpositive and ν is a singular measure.

Note: The factor exp (ρ/w) comes from the (possible) jump in $\mu(d\lambda)$ at $\pm\pi$. As before Q is uniquely defined by the properties of being holomorphic in the left half-plane and satisfies

$$Q(-1) = \exp\frac{1}{2\pi}\int_{-\infty}^{\infty}\log f(\tau)\,\frac{d\tau}{1+\tau^2}.$$

† See the Exercises to this chapter.

‡ We number the theorems in this section with a number which corresponds to that of the analogous theorem of the preceding section.

We may also put

$$Q(w) = \int_0^\infty a(s)e^{sw}\,ds,$$

for w in the left half-plane, with the integral converging in mean square on the boundary.

We now may consider the problem of minimizing

$$\int_{-\infty}^{\infty} |e^{-it\tau} - h(i\tau)|^2\, F(d\tau),$$

where $h(w)$ is to be holomorphic in the left-hand plane. This is the spectral form of the mean-square error of prediction of $x(t)$ from $\mathcal{M}_0$. If F is a.c. and nondeterministic, then we may rewrite this as

$$\int_{-\infty}^{\infty} |e^{-it\tau}\, Q(i\tau) - h(i\tau)\, Q(i\tau)|^2\, d\tau$$

and by rearranging this as in the argument used in Section 1 we see that

$$h(i\tau) = [e^{-it\tau}Q(i\tau)]_+\, Q(i\tau)^{-1},$$

where

$$[e^{-it\tau}Q(i\tau)]_+ = \int_t^\infty e^{is\tau}e^{-it\tau}\, a(s)\, ds = \int_0^\infty e^{is\tau}\, a(s+t)\, ds.$$

The prediction error is

$$\int_{-\infty}^{\infty} |[e^{-it\tau}Q(i\tau)]_-|^2\, d\tau,\ [e^{-it\tau}Q(i\tau)]_- = \int_0^t e^{i\tau(s-t)}\, a(s)\, ds,$$

which is

$$\int_0^t a^2(u)\, du.$$

The proof that the singular part of F contributes nothing to the prediction error hardly differs from that given in Section 3 for the discrete time case and we omit it. We now prove Theorem 2′ after which we can state Theorem 4′ and conclude with Theorem 6′.

Theorem 2′

$$x(t) = u(t) + v(t) = \int_{-\infty}^{t} a(t-s)\xi(ds) + v(t), \tag{4.5}$$

where

(i) $\xi(s)$ *has orthogonal increment with* $\mathcal{E}(\xi(ds)^2) = ds$

(ii) $\mathcal{E}(u(s)v(t)) \equiv 0$
(iii) *$v(t)$ is purely deterministic, $u(t)$ is purely nondeterministic.*

We decompose $\mathcal{H}$ into $\mathcal{M}_{-\infty} \oplus \mathcal{M}$ where $\mathcal{M}$ is, thus, the orthogonal complement of $\mathcal{M}_{-\infty}$ in $\mathcal{H}$. We let $u(t)$ be the projection of $x(t)$ onto $\mathcal{M}$ and $v(t)$ its projection onto $\mathcal{M}_{-\infty}$. The translation operator $U(\tau)$, $U(\tau)x(t) = x(t+\tau)$, commutes with the projection onto $\mathcal{M}_{-\infty}$ since evidently $U(\tau)\mathcal{M}_{-\infty} = \mathcal{M}_{-\infty}$ and thus it commutes also with the projection onto $\mathcal{M}$ and thus $u(t)$ and $v(t)$ are stationary. Also the $u(t)$ span $\mathcal{M}$ and the $v(t)$ span $\mathcal{M}_{-\infty}$ and evidently (ii) holds. Let $\mathcal{V}_0$ be the space spanned by $v(t)$, $t \le t_0$. Then $\mathcal{V}_0 \subset \mathcal{M}_{t_0}$, and thus $\mathcal{V}_0 = \mathcal{M}_{-\infty}$ since $\mathcal{M}_{-\infty} \subset \mathcal{M}_{t_0}$, $\mathcal{V}_0 \subset \mathcal{M}_{-\infty}$ and $\mathcal{M}_{t_0} = \mathcal{V}_0 + \mathcal{U}_0$, where $\mathcal{U}_0$ is spanned by $u(t)$, $t \le t_0$. Thus $v(t)$ is deterministic. On the other hand, the remote past of $u(t)$, say $\mathcal{U}_{-\infty}$, satisfies $\mathcal{U}_{-\infty} \subset \mathcal{M}_{-\infty}$ and $\mathcal{U}_{-\infty} \subset \mathcal{M}$ so that $\mathcal{U}_{-\infty}$ is null and $u(t)$ is purely nondeterministic. Thus $u(\tau)$ has a.c. spectrum satisfying (4.1) and the representation (4.5) follows from Theorem 15 in the appendix to Chapter II.

The prediction error is evidently

$$\int_s^{t+s} a^2(s-u)\,du = \int_0^t a^2(u)\,du,$$

which shows that our identification of $a(s)$ with the Fourier transform of $Q(i\tau)$ is justified.

Now we have, as in Section 3,

Theorem 4′. *If $x(t)$ is not purely deterministic and $F(\tau)$ has the Lebesgue decomposition $F = F^{(1)} + F^{(2)} + F^{(3)}$ then the derivative of the a.c. component $F^{(1)}$ is the spectral density of $u(t)$ in Theorem 1′ and $F^{(2)}(t) + F^{(3)}(t)$ is the spectrum of $v(t)$.*

Theorem 6′. *Let $x(t)$ be purely nondeterministic and let $h_t(i\tau)$ be the response function of the filter giving the best predictor of $x(s+t)$ from observations up to time s. Then*

$$h_t(i\tau) = \frac{[e^{-it\tau}Q(i\tau)]_+}{Q(i\tau)},$$

where $Q(w)$ is defined by (4.4). We also have

$$\sum_0^\infty \frac{(1+w)^k}{k!}\int_0^\infty e^{-s}s^k a(s)\,ds = \exp\left\{a_0 + 2\sum_1^\infty a_j\left(\frac{1+w}{1-w}\right)^j\right\}, \qquad \mathcal{I}(w) < 0,$$

$$a_j = \frac{1}{2\pi}\int_{-\infty}^\infty (\log\sqrt{f(\tau)})\left(\frac{1-i\tau}{1+i\tau}\right)^n \frac{d\tau}{1+\tau^2}$$

Example

Let $f(\tau) = (1/2\pi)\,|p(i\tau)|^{-2}$, where $p(w)$ is a (real) polynomial with no zeros in the left half-plane. Then evidently $Q(w) = p(w)^{-1}$. Thus

$$a(s) = \frac{1}{2\pi}\int_{-\infty}^{\infty} \frac{1}{p(i\tau)}\, e^{-is\tau}\, d\tau, \qquad s \geq 0$$
$$= \sum_j - \frac{\Pi(-\beta_j)}{p(0)}\left(\prod_{k\neq j}\frac{1}{(\beta_j - \beta_k)}\right) e^{-\beta_j s},$$

taking the zeros β_j of $p(w)$ in the right half-plane to be simple. Thus

$$h_t(i\tau) = -\sum_j e^{-\beta_j t} \prod_{k\neq j}\left\{\frac{i\tau - \beta_k}{\beta_j - \beta_k}\right\}$$

and correspondingly the predictor of $x(s + t)$ from observations up to time s is

$$-\left[\sum_j e^{-\beta_j t} \prod_{k\neq j}\left\{\frac{1}{(\beta_j - \beta_k)}\left(\frac{d}{ds} - \beta_k\right)\right\}\right] x(s)$$

5. VECTOR DISCRETE-TIME PREDICTION†

The vector case of the general prediction theory is considerably more difficult than the scalar case due to the fact that $f(\lambda)$ may be singular but not null. Nevertheless, a considerable part of the theory has been carried over due to the work of Wiener and Masani (1957, 1958), Helson and Lowdenslager (1958) and Rozanov (1967). We restrict ourselves to the discrete time case. For the continuous time case see Robertson (1968).

We reintroduce $\mathcal{H}$, the Hilbert space spanned by $x_j(n)$, $j = 1, \ldots, p$, $n = 0, \pm 1, \ldots$, and call $\mathcal{M}_n$ the closed subspace spanned by $x_j(m)$, $j = 1, \ldots, p, m \leq n$. We obtain the best ν-step linear predictor of $x(n + \nu)$ by projecting the components of that vector on $\mathcal{M}_n$. The error of prediction for $\nu = 1$ we call $\epsilon(n + 1)$. Then, as before,

$$\mathcal{E}\{\epsilon(m)\,\epsilon(n)'\} = \delta_m^{\ n} G. \tag{5.1}$$

Putting

$$A(j) = \mathcal{E}\{x(n)\,\epsilon(n - j)'\}G^{-1}, \qquad j > 0 \tag{5.2}$$

(where G^{-1} is the generalized inverse of G) we have

$$A(j)G = \mathcal{E}\{x(n)\,\epsilon(n - j)'\},$$

since, with probability 1, $\epsilon(n - j)$ is orthogonal to the null space of G. Thus $A(j)\epsilon(n - j)$ is the projection of $x(n)$ in the space spanned by $\epsilon(n - j)$,

† This section may be omitted.

since

$$\mathcal{E}[\{x(n) - A(j)\,\epsilon(n-j)\}\,\epsilon(n-j)'] = 0.$$

Thus we may form

$$u(n) = \sum_0^\infty A(j)\,\epsilon(n-j), \qquad \sum_0^\infty A(j)G\,A^*(j) < \infty. \tag{5.3}$$

The last relation follows from Bessel's inequality. We put $v(n) = x(n) - u(n)$ and have the following theorem:

Theorem 2″ (Wold Decomposition Theorem)

If $x(n)$ is a stationary vector process of p components, then $x(n) = u(n) + v(n)$, where $u(n)$ is given by (5.1), (5.2), *and* (5.3), *$v(n)$ is deterministic and* $\mathcal{E}\big(v(n)\,\epsilon(m)'\big) \equiv 0$.

The proof is essentially the same as for the scalar case. If only $u(n)$ occurs we say that $x(n)$ is purely nondeterministic.

We follow Helson and Lowdenslager (1958) with respect to the initial part of the frequency domain analysis of the prediction theory. The quantity to be minimized in the one-step prediction problem is

$$\operatorname{tr}\left\{\int_{-\pi}^{\pi}(I - \sum A_N(j)e^{ij\lambda})\,F(d\lambda)(I - \sum A_N(j)e^{ij\lambda}),\right.$$

since this is the sum of squares of the prediction errors for the components of $x(0)$ from a prediction formula using $x_j(m), j = 1, \ldots, p, m = -1, \ldots, -N$. Minimization is over all sequences $A_N(j)$, of course. We consider instead the (only apparently) more general problem of evaluating

$$\inf \operatorname{tr}\int\{A(0) - h_N(e^{i\lambda})\}\,F(d\lambda)\{A(0) - h_N(e^{i\lambda})\}^*, \tag{5.4}$$

where $h_N = \sum A_N(j) \exp ij\lambda$ and $A(0)$ varies over all matrices with unit determinant.

We recall that if B is Hermitian nonnegative there is a unique Hermitian matrix H so that $B = \exp H$ (this being defined via the exponential series). We call $H = \log B$

Theorem 3″. *The infinium in* (5.4) *is*

$$\exp\left[\frac{1}{2\pi p}\int_{-\pi}^{\pi}\operatorname{tr}\,\{\log 2\pi f(\lambda)\}\,d\lambda\right]. \tag{5.5}$$

We again interpret (5.4) and (5.5) as zero when $\operatorname{tr}\big(\log f(\lambda)\big)$ is not integrable. Holding $A(0)$ fixed exactly as before we may take the infinium over A_0, the matrix valued functions holomorphic within the unit circle and continuous on its boundary and with mean value zero. Then almost precisely as in the scalar case we may show that, h being the projection of $A(0) \in L_2(F)$ (see the Mathematical Appendix) on to the closure of A_0 in $L_2(F)$ then the singular part of F contributes nothing to the infinium and

$$(A(0) - h)f(\lambda)(A(0) - h)^* = C \quad \text{a.e.}$$

where h and C depend upon $A(0)$, of course. *We thus consider the a.c. case from now on.* We establish the formula

$$\text{(5.6)} \qquad \exp\left[\frac{1}{2\pi p}\int \operatorname{tr}(\log 2\pi f)\, d\lambda\right] = \inf \frac{1}{2\pi p}\int \operatorname{tr}(e^{\psi} 2\pi f)\, d\lambda,$$

where ψ varies over Hermitian matrices which are polynomials with $\int \operatorname{tr} \psi \, d\lambda = 0$. Now

$$\exp\left[\frac{1}{2\pi p}\int \operatorname{tr}(\log 2\pi f)\, d\lambda\right] = \exp\left[\frac{1}{2\pi p}\int \log\det(2\pi f)\, d\lambda\right]$$
$$\leq \frac{1}{2\pi}\int \{\det(2\pi f)\}^{1/p}\, d\lambda$$
$$\leq \frac{1}{2\pi p}\int \operatorname{tr}(2\pi f)\, d\lambda,$$

where we have used $\operatorname{tr}\log B = \log\det B$ and the inequality between geometric and harmonic means (once for sums, of the eigenvalues of $2\pi f$, and once for integrals). Putting† $W = (\exp \frac{1}{2}\psi)\, 2\pi f\, (\exp \frac{1}{2}\psi)$, we obtain

$$\exp\left[\frac{1}{2\pi p}\int \operatorname{tr}\log W\, d\lambda\right] \leq \frac{1}{2\pi p}\int \operatorname{tr} W\, d\lambda$$

and since $\operatorname{tr}\log W = \log\{\det(\exp\psi)\det(2\pi f)\}$ we obtain (from the fact that $\operatorname{tr}\psi$ has zero mean value)

$$\text{(5.7)} \qquad \exp\left[\frac{1}{2\pi p}\int \operatorname{tr}\log 2\pi f\, d\lambda\right] \leq \frac{1}{2\pi p}\int \operatorname{tr}(e^{\psi} 2\pi f)\, d\lambda.$$

Now as before it is possible to show that the infinium of the right side is not altered when ψ is allowed to be any matrix with integrable trace (which is Hermitian and has $\operatorname{tr}\psi$ with zero mean value). Indeed if we define the positive part of an Hermitian matrix by replacing the eigenvalues by their positive parts and leaving the eigenvectors unchanged then the proof of Lemma 1 of Section 3 is almost unchanged. The same trick then allows us

† We define W in the fashion following so that it shall be Hermitian.

to show that equality holds in (5.6), for when tr log $2\pi f$ is integrable†

$$\psi_0 = p^{-1}\left(\int \operatorname{tr} \log 2\pi f \, d\lambda\right) I - \log 2\pi f$$

is admissible and produces equality in (5.7).

We now need a different argument to that used to establish the corresponding theorem of Section 3, since $A = \exp B$ does not imply $AA^* = \exp(B + B^*)$ unless A commutes with A^*. To this effect we prove the following lemma.

Lemma 3. *If the infinium in* (5.4) *is positive then* $f(\lambda)$ *factorizes as*

$$f(\lambda) = \frac{1}{2\pi} \Phi(e^{i\lambda})\Phi^*(e^{i\lambda})$$

where Φ *is holomorphic within the unit circle with*

$$\Phi(z) = \sum_0^\infty C(j)z^j, \qquad \det\{C(0)\} \neq 0.$$

Proof. Let h be the element of the closure of A_0 in $L_2(F)$, which minimizes

$$\int (I + h) f (I + h)^* \, d\lambda.$$

Then $(I + h) f (I + h)^* = C$, *where* C *is nonsingular*, for otherwise we can find $A(0)$ with $\det\{A(0)\} = 1$ so that

$$\operatorname{tr}\bigl(A(0)C\,A(0)^*\bigr) = \int \operatorname{tr}\bigl(A(0) + A(0)h\bigr) f \bigl(A(0) + A(0)h\bigr)^* \, d\lambda$$

is as small as is desired. (Choose the first row a^* of $A(0)$ so that $a^*Ca = 0$. Choose the other rows to be linearly independent and linearly independent of a^*. By multiplying them by a suitable constant the trace may be made as small as desired, whereas we may multiply a^* by another nonzero constant to give $A(0)$ unit determinant without affecting the trace.) Putting $A = C^{-1/2}$ (the positive square root being taken) we have $(A + Ah) f (A + Ah)^* = I$ so that $f = (A + Ah)^{-1} (A + Ah)^{*-1}$. Since $(A + Ah)^{-1}$ is evidently composed of elements square integrable $(d\lambda)$ and

$$\begin{aligned}\int (A + Ah)^{-1} e^{in\lambda} \, d\lambda &= \int f(A + Ah)^* e^{in\lambda} \, d\lambda \\ &= \int f(I + h)^* e^{in\lambda} \, d\lambda \, A^* = 0, \qquad n = 1, 2, \ldots,\end{aligned}$$

we have established the lemma.

† Otherwise we proceed as in the proof of Theorem 3 in Section 3.

Let ψ be as in (5.6). Then $\exp -\psi$ has eigenvalues bounded away from zero and if f is replaced by $\exp -\psi$ the infinium in (5.4) must be positive so that $\exp -\psi = BB^*$, where

$$B = \sum_0^\infty B(n)e^{in\lambda}, \qquad \det B(0) \neq 0.$$

We now show that $\det B(0) = 1$. Since $BB^* = \operatorname{tr}(\exp -\psi)$, we see that the elements of B are uniformly bounded and thus integrable, and, since

$$\det B = \det B(0) + \sum_1^\infty c(n)e^{in\lambda},$$

we obtain from the inequality between geometric and arithmetic means

$$\frac{1}{2\pi}\int \log|\det B|^2\,d\lambda \leq \log|\det B(0)|^2.$$

From the construction of B in the lemma $\exp\psi = D^*D$ where $D^{-1} = A + Ah$ and we similarly have for D

$$\frac{1}{2\pi}\int \log|\det D|^2\,d\lambda \leq \log|\det A|^2.$$

Since $B(0)A = I$ adding corresponding sides of the two inequalities we see that they must be equalities. Thus, by assumption,

$$0 = \frac{1}{2\pi}\int \log(\det e^{\psi})\,d\lambda = \log|\det A|^2,$$

so that, A having a positive real determinant since it is Hermitian, its determinant must be unity as must also be that of $B(0)$. Thus we know that, ψ being as in (5.6),

$$\frac{1}{2\pi p}\int \operatorname{tr}(e^{\psi}2\pi f)\,d\lambda = \frac{1}{2\pi p}\int \operatorname{tr}(D2\pi f D^*)\,d\lambda, \tag{5.8}$$

where $D = A + H$ with A having unit determinant while

$$H = \sum_1^\infty H(j)e^{ij\lambda}.$$

Since we have shown that $\exp(2\pi p)^{-1}\int \operatorname{tr}(\log 2\pi f)\,d\lambda$ is the infinium of all expressions of the form of the left side of (5.8), we see that we can take the infinium over the apparently wider class on the right and obtain

$$\inf \frac{1}{2\pi p}\int \operatorname{tr}\{A(0) + h\}2\pi f\{A(0) + h\}^*\,d\lambda \leq \exp\frac{1}{2\pi p}\int \operatorname{tr}(\log 2\pi f)\,d\lambda.$$

Just as in the proof for the scalar case the opposite inequality may be got from (5.8) itself and Theorem 3″ is thus established.

The theorem can be stated in the equivalent form:

Theorem 3‴

$$\det(G) = \exp \frac{1}{2\pi} \int_{-\pi}^{\pi} \log\{\det 2\pi f(\lambda)\}\, d\lambda.$$

We leave the proof as an exercise.

If $\det(G) > 0$ we say that the process is of full rank. Of course $\det(G)$ may be zero and the process purely nondeterministic as G may not be null.

Theorem 4″. *If $x(n)$ is of full rank, then*

$$f_u(\lambda) = f(\lambda), \qquad F_v(\lambda) = F^{(2)}(\lambda) + F^{(3)}(\lambda).$$

We need only show that $F_v(\lambda)$ is singular, the result then following from the uniqueness of the Lebesgue decomposition. The result is due to Wiener and Masani (1957).

We first establish that $\det(A + B)/\mathrm{tr}(A + B) \geq \det(A)\,\mathrm{tr}(A)$ if A and B are Hermitian nonnegative. We may evidently assume that A is diagonal and then see that $\det(A + B) \geq \det A + \sum b_{j,j}A^{j,j}$ (where $A^{j,j}$ is the indicated cofactor). Since $\mathrm{tr}\, A \geq a_{j,j}$ we obtain $\mathrm{tr}(A)\det(A + B) \geq \mathrm{tr}(A)\det(A) + \mathrm{tr}(B)\det(A) = \det(A)\,\mathrm{tr}(A + B)$, as required. Thus

$$\frac{\det(f)}{\mathrm{tr}(f)} \geq \frac{\det(f_u)}{\mathrm{tr}(f_u)},$$

so that

$$\det(f) \geq \det(f_u)\left\{1 + \frac{\mathrm{tr}\big(d/d\lambda\, F_v(\lambda)\big)}{\mathrm{tr}(f_u)}\right\}.$$

Thus

$$\int \log(\det 2\pi f)\, d\lambda \geq 2\pi \log(\det G) + \int \log\left(1 + \frac{\mathrm{tr}\big(d/d\lambda\, F_v(\lambda)\big)}{\mathrm{tr}\, f_u}\right) d\lambda\,.$$

Thus the second term is null and since $\mathrm{tr}\, f_u$ is finite a.e. then $\mathrm{tr}\big(d/d\lambda F_v(\lambda)\big)$ is null a.e. and F_v being Hermitian this establishes the result.

This Theorem 4″ may fail if $x(n)$ is not of full rank (see Masani (1959a)).

Theorem 5 has not been fully generalized so far, for the form of an "outer" function has not been completely described. Thus $Q(z) = (\sum A(j)z^j)G^{1/2}$ has not been expressed in a closed form in terms of f and indeed this has not even been done for $Q(0)^2 = G$. The basic work here has been done by Masani (1959b, 60) using mathematical results due to Potapov (1955).

See also Helson (1964). However, we have the complete analog of Theorem 1 in the full rank case, which we call Theorem 1″, though we have stated no Theorem 1′.

Theorem 1″. *If $x(n)$ is purely nondeterministic and of full rank then Theorem 1 holds in its entirety where*

$$\Phi(e^{i\lambda}) = \sum_{0}^{\infty} A(j) e^{ij\lambda}$$

is uniquely defined by the requirement that $\Phi(z)$ be holomorphic within the unit circle, satisfy $f(\lambda) = (2\pi)^{-1}\Phi G \Phi^*$, and $\Phi(0) = I_p$.

The proof of this theorem is of almost precisely the same form as that of Theorem 1, except for the uniqueness, which we now establish. Consider $\det \Phi(z)$. This is certainly nonzero almost everywhere on the circle since this is true of $\det\big(f(\lambda)\big)$. Put $R(z) = \Phi(z)^{-1}\Psi(z)G^{1/2}$, where $\Psi(z)$ satisfies the same conditions as does $\Phi(z)$. Then on the circle $RR^* = 2\pi\Phi^{-1}f\Phi^{*-1} = G$ so that $R = G^{1/2}S$ where S is unitary, from the polar decomposition formula for a matrix. Moreover, $G^{1/2}S$ is holomorphic within the circle for consider

$$G(z) = \exp\left\{\frac{1}{2\pi}\int \frac{e^{i\lambda}+z}{e^{i\lambda}-z}\log(\det\Phi)\,d\lambda\right\}.$$

Then, just as in the proof of Theorem 3, $|G(z)| \geq |\det\big(\Phi(z)\big)|$; but $G(0) = \det\Phi(0) = 1$ so that $G(z)$, having no zeros in the circle, $\det\big(\Phi(z)\big)/G(z)$ is holomorphic within the circle and has modulus ≤ 1. Thus by the maximum modulus principle it is identically unity and $\det\big(\Phi(z)\big)$ is holomorphic and without zeros in the circle. Thus $R(z)$ is holomorphic within the circle and thus so is $S(z)$. But $S(0) = I_p$ and now the proof is completed as for Theorem 3; namely, since $S^* = S^{-1}$, then $S(\exp i\lambda)^{-1}$ is both holomorphic within and outside of the circle and thus is a constant matrix which must be I_p. Thus $R(z) = G^{1/2}$, that is, $\Phi(z) = \Psi(z)$ and uniqueness is established.

The problem remains of actually constructing the $A(j)$, hence $\Phi(z)$, which, in principle, was solved by Theorem 5 for the scalar case. However, the construction of an algorithm to determine the $A(j)$ does not seem important in practice, for the reasons given in the introduction, and we shall discuss the problem no further here.

6. PROBLEMS OF INTERPOLATION

Problems of interpolation are simpler than prediction problems since there are fewer restrictions on the filter giving the interpolation. We consider first the interpolation of a missing value in a discrete time series and then interpolation between the sampled points of a continuous time function. For other work on interpolation we refer the reader to Grenander and Rosenblatt (1957), Rozanov (1967), and Bonnet (1965), whose treatments we have also used here.

Once more we seek that linear combination $\hat{x}(n)$ of the $x(n-j)$, $j \neq 0$, which minimizes the error of interpolation $\|x(n) - \hat{x}(n)\|^2$. Just as in Section 2, we introduce the response functions

$$h_N(e^{i\lambda}) = \sum_{-N}^{N}{}' A_N(j)e^{ij\lambda}, \tag{6.1}$$

where the prime indicates that the term for $j = 0$ is omitted. Now we seek for a response function† h such that

$$\lim_{N\to\infty} \operatorname{tr}\left[\int_{-\pi}^{\pi}(h - h_N)\, F(d\lambda)(h - h_N)^*\right] = 0$$

and

$$\left[\int_{-\pi}^{\pi}(I_p - h)\, F(d\lambda)(I_p - h)^*\right]$$

is minimized. Then

$$\hat{x}(n) = \int_{-\pi}^{\pi} e^{-in\lambda} h(e^{i\lambda})\, z(d\lambda).$$

We put

$$\Sigma = \mathcal{E}\{(x(n) - \hat{x}(n))(x(n) - \hat{x}(n))'\}'.$$

Evidently we may take $n = 0$ without any loss of generality. If $F(\lambda)$ is not a.c., we know from Section 5 that the singular part of F corresponds to a perfectly predictable process and thus one which may be perfectly interpolated. This leads us to treat the a.c. case. *We assume that there is no non-null vector* α *such that* $\alpha' x(n) \equiv 0$, *almost surely*. This is clearly a costless assumption, since we could otherwise reduce the dimension of $x(n)$.

Theorem 8. *Let $x(n)$ satisfy the above assumption and have a.c. spectrum and let $f(\lambda)^{-1}$ be the generalized inverse of $f(\lambda)$. The necessary and sufficient condition that Σ be nonsingular is the condition that $f(\lambda)^{-1}$ be integrable. Then the response function of the optimal interpolating filter is*

$$h = I_p - \left\{\frac{1}{2\pi}\int_{-\pi}^{\pi} f(\lambda)^{-1}\, d\lambda\right\} f^{-1}(\lambda) \tag{6.2}$$

† Again we often omit the argument of h.

and

$$\Sigma = \left[\frac{1}{2\pi}\int_{-\pi}^{\pi}\{2\pi f(\lambda)\}^{-1}\right]^{-1}.$$

Proof. Evidently for each pair of vectors α, β of complex numbers we must have

$$\mathcal{E}\{\alpha^*\{x(0) - \hat{x}(0)\}\, x(n)'\beta\} = 0, \qquad n \neq 0.$$

This is

$$\alpha^*\int (I_p - h) f(\lambda)e^{in\lambda}\, d\lambda\beta = 0, \qquad n \neq 0. \tag{6.3}$$

This implies that

$$(I_p - h)f = C,$$

where C is a constant matrix. Thus

$$(I_p - h) = Cf^{-1}.$$

This solution is not unique, but any solution differs from it by a matrix which, when multiplied on the right by f, is annihilated and thus leads to the same Σ. Moreover, (6.3) also shows that

$$\int_{-\pi}^{\pi}(I_p - h)fh^*\, d\lambda = 0,$$

since h is a limit in mean square of expressions of the form of (6.1). Thus

$$\Sigma = \int (I_p - h)f\, d\lambda = 2\pi C,$$

which shows that $C = C^* = \bar{C}$. Now, assuming the integrability of f^{-1} for the first time, we have

$$\Sigma = C\int_{-\pi}^{\pi} f^{-1}ff^{-1}\, d\lambda = C\int_{-\pi}^{\pi} f^{-1}\, d\lambda C$$

and

$$C = C\left\{\frac{1}{2\pi}\int_{-\pi}^{\pi} f(\lambda)^{-1}\, d\lambda\right\}\cdot C, \tag{6.4}$$

of which a solution is

$$C = \left\{\frac{1}{2\pi}\int_{-\pi}^{\pi} f(\lambda)^{-1}\, d\lambda\right\}^{-1}. \tag{6.5}$$

Again any other solution of (6.4) leads to the same Σ, for, if we write

$$A = \frac{1}{2\pi}\int_{-\pi}^{\pi} f(\lambda)^{-1}\, d\lambda,$$

we have to investigate the solutions of $X = XAX$. If $X = C + D$, where C is given by (6.5), we may assume that $ED = DE = EDE = D$, where $E = AA^{-1} = A^{-1}A$, for EXE certainly leads to the same Σ as X. Then we obtain

$$C + D = (C + D)\,A(C + D) = CAC + CAD + DAC + DAD,$$

that is,

$$D = ED + DE + DAD,$$

that is,

$$-D = DAD = (-D)A(-D),$$

and $X = C - D_1$, where D_1 also solves $X = XAX$, but now $XAX = CAC$ as required.

Thus we may take C as given by (6.5) and the theorem results, save for the assertion concerning the nonsingularity of Σ. If $f(\lambda)^{-1}$ is integrable then certainly Σ is nonsingular for otherwise there must be a vector α, so that

$$\int \alpha' f^{-1}(\lambda)\alpha\, d\lambda = 0, \qquad \alpha'\alpha = 1.$$

Taking α as the first row of an orthogonal matrix P this implies that $Pf(\lambda)P'$ must have null elements, for all λ, in the first row and column, which implies that $\alpha' x(n) \equiv 0$, almost surely. On the other hand, if Σ is nonsingular then since $\int (I_p - h)\, f(I_p - h)\, d\lambda = \int (4\pi)^{-2}\, \Sigma\, f^{-1}\, \Sigma\, d\lambda$ we see that $\Sigma\, f^{-1}\Sigma$ is integrable and thus so must f^{-1} be integrable. This completes the proof.

If $x(n)$ is scalar then when $f(\lambda)^{-1}$ is not integrable the interpolation error is zero and conversely (as may be seen by considering $f(\lambda) + \epsilon$, $\epsilon > 0$, and allowing ϵ to decrease). In the scalar case Σ is the harmonic mean of $2\pi f(\lambda)$ and thus could be called that in general. One thus has the interesting array of results for estimation of $x(0)$. In the case where we have no information on future or past we put $x(0) = 0$ and have an error matrix $\Gamma(0)$.

Information used	Error matrix
Nil	Arithmetic mean of $2\pi f$
Past	Geometric mean of $2\pi f$
Past and future	Harmonic mean of $2\pi f$

We now consider the problem of interpolating the value $x(t)$ of a scalar continuous time process, from observations $x(n)$, $n = 0, \pm 1, \ldots$. We again consider only the a.c. case, without loss of generality. Thus we have to minimize

$$\int_{-\infty}^{\infty} |e^{-it\lambda} - h_t(\lambda)|^2 f(\lambda)\, d\lambda, \qquad 0 < t < 1,$$

where h_t is to belong to the part of $L_2(f\,d\lambda)$ spanned by $\exp in\lambda, n = 0, \pm 1, \ldots$.

Thus we must have

$$\int_{-\infty}^{\infty} \{e^{-it\lambda} - h_t\} e^{-in\lambda} f\,d\lambda = 0, \qquad n = 0, \pm 1, \ldots$$

and, since $h_t(\lambda) = h_t(\lambda + 2k\pi)$,

$$\int_{-\pi}^{\pi} \left\{ \sum_{-\infty}^{\infty} e^{-it(\lambda+2k\pi)} f(\lambda + 2k\pi) - h_t(\lambda) \sum_{-\infty}^{\infty} f(\lambda + 2k\pi) \right\} e^{in\lambda}\,d\lambda \equiv 0.$$

Thus

$$h_t(\lambda) = \frac{\sum_{-\infty}^{\infty} f(\lambda + 2k\pi) e^{-it(\lambda+2k\pi)}}{\sum_{-\infty}^{\infty} f(\lambda + 2k\pi)}, \qquad -\pi \le \lambda \le \pi,$$

whose modulus is not greater than unity. The interpolation error is

$$\int_{-\infty}^{\infty} \left| 1 - \frac{\sum_{-\infty}^{\infty} f(\lambda + 2k\pi) e^{-it2k\pi}}{\sum_{-\infty}^{\infty} f(\lambda + 2k\pi)} \right|^2 f(\lambda)\,d\lambda.$$

If this error is to be zero then $h_t(\lambda) \exp it\lambda$ must be unity a.e. $(f\,d\lambda)$. But, when λ is increased by $2j\pi$, $h_t(\lambda) \exp it\lambda$ is merely multiplied by $\exp(i2j\pi t)$ so that f must be zero outside of $[-\pi, \pi]$. Conversely, when this is so the error is evidently zero so that this is the rather obvious necessary and sufficient condition that any $x(t)$ can be exactly known when only the $x(n)$ are observed. If the observation points had been Δn, $\Delta > 0$, we consider $y(t) = x(\Delta t)$ having spectrum $f_y(\lambda) = \Delta^{-1} f_x(\lambda/\Delta)$ and thus obtain the so-called sampling theorem.

Theorem 9. *If $x(t)$ has a.c. spectrum, f, the necessary and sufficient condition that any value $x(t)$ can be exactly known from a linear interpolation based on $x(\Delta n)$, $n = 0, \pm 1, \ldots$, is that f be null outside of $[-\pi/\Delta, \pi/\Delta]$.*

The interpolation formula is easily written down when the interpolation error is zero:

$$h_t = \frac{e^{-it\lambda} f(\lambda)}{f(\lambda)} = e^{-it\lambda}, \qquad |\lambda| < \pi.$$

Then, if

$$h_t(\lambda) = \sum_{-\infty}^{\infty} h_j e^{-ij\lambda},$$

we have

$$h_j = \frac{1}{2\pi} \int_{-\pi}^{\pi} e^{i(j-t)\lambda}\,d\lambda = \frac{\sin \pi(j - t)}{\pi(j - t)}$$

and

$$\hat{x}(t) = \sum_{-\infty}^{\infty} \frac{\sin \pi(j - t)}{\pi(j - t)} x(j),$$

a formula due to Shannon.

7. SMOOTHING AND SIGNAL MEASUREMENT

We now consider the case where we observe a signal, $s(n)$, with noise, $x(n)$, added and wish to extract $s(n)$ from $y(n) = s(n) + x(n)$ using observations up to $n + \mu$, where now μ may be positive. Though the treatment given below is over general in the sense that the mathematical generality is irrelevant from the viewpoint of what one can conceivably know, it is at this stage no more difficult to be general than to be particular. Again, as we said in the introduction, rather special, nonstationary, models have lately become important. Nevertheless, the results of this section still appear of great value and can be used to suggest techniques which can be used in a more *ad hoc* spirit than that motivating the narrowly specified model which gives rise to them.

(i) The scalar, discrete time case.

Our model is

$$y(n) = s(n) + x(n), \qquad \mathcal{E}\big(s(n)\, x(m)\big) \equiv 0,$$

where $s(n)$ and $x(n)$ are stationary. Only $y(n)$ is observed. We choose the estimator† $\hat{s}(n)$ of $s(n)$, using observations to $n + \mu$, so as to minimize $\mathcal{E}\{(s(n) - \hat{s}(n))^2\}$. Just as in previous sections we shall have

$$\hat{s}(n) = \int_{-\pi}^{\pi} e^{-in\lambda}\, h(e^{i\lambda}) z_y\,(d\lambda), \qquad \int_{-\pi}^{\pi} h\ F_y(d\lambda)\ h^*\ d\lambda < \infty,$$

and we must determine h. We may consider only the case where $y(n)$ is purely nondeterministic since in principle the deterministic component can be perfectly predicted so that for it the signal extraction problem becomes one for $\mu = \infty$, which is a simpler problem with which we deal below.

Theorem 10. *If $y(n)$ is purely nondeterministic then the response function h_μ of the optimal extractor $s^{(\mu)}(n)$ of the signal $s(n)$ in $y(n) = s(n) + x(n)$, using $y(m)$, $m \leq n + \mu$, is*

$$h_\mu(e^{i\lambda}) = e^{-i\mu\lambda}\, g(e^{i\lambda})^{-1} \left[\frac{e^{i\mu\lambda} f_s(\lambda)}{g(e^{-i\lambda})}\right]_+, \tag{7.1}$$

† We suppress notational references to μ for the moment.

where

(7.2) $$f_y(\lambda) = g(e^{i\lambda})\, g(e^{-i\lambda})$$

and $g(z)$ is holomorphic within the unit circle and without zeros there. The mean square error is

$$\int_{-\pi}^{\pi} \frac{f_s(\lambda) f_x(\lambda)}{f_y(\lambda)}\, d\lambda + \int_{-\pi}^{\pi} \left| \left[\frac{e^{i\mu\lambda} f_s(\lambda)}{g(e^{-i\lambda})} \right]_- \right|^2 d\lambda.$$

When $\mu = \infty$, whether $y(n)$ is purely nondeterministic or otherwise,

$$h_\infty = \frac{dF_s(\lambda)}{dF_y(\lambda)}$$

and the mean square error is

$$\int_{-\pi}^{\pi} \frac{dF_s}{dF_y} F_x(d\lambda).$$

Proof. We omit the subscript μ during the proof since this is held constant. The mean-square error is

$$\mathcal{E}\left[\left\{\int_{-\pi}^{\pi} e^{-in\lambda} z_s(d\lambda) - \int_{-\pi}^{\pi} e^{-in\lambda} h(e^{i\lambda})\, z_y(d\lambda)\right\}^2\right] = \int_{-\pi}^{\pi} \{f_s + |h|^2 f_y - 2\mathfrak{R} f_s\}\, d\lambda.$$

The factorization (7.2) certainly exists by Theorems 5 and 3 of Section 3. Inserting this into the last expression and rearranging we get

(7.3) $$\int \left\{ \left(f_s - \frac{f_s^2}{f_y} \right) + \left| \frac{f_s}{\bar{g}} - hg \right|^2 \right\} d\lambda.$$

Since $f_s/f_y < 1$, the expression f_s^2/f_y is integrable. Since this is so the expression $f_s/\bar{g}$ is square integrable. Now exp $i\mu\lambda h$ is to be the limit in mean square [with respect to $f_y(\lambda)$] of expressions

$$\sum_0^N a_{N,j} e^{ij\lambda},$$

since observations up to $n + \mu$ only are to be used. Thus, since g also is a limit in mean square of such expressions, to minimize (7.3) we must minimize†

$$\int \left| \frac{f_s}{\bar{g}} - hg \right|^2 d\lambda = \int \left| \frac{f_s}{\bar{g}} e^{i\mu\lambda} - e^{i\mu\lambda} hg \right|^2 d\lambda,$$

† The technique used to find the solution here (and in Theorem 1) is an aspect of the Wiener-Hopf technique.

and the best we can do is to put

$$e^{i\mu\lambda} hg = \left[\frac{e^{i\mu\lambda} f_s}{\bar{g}}\right]_+ . \tag{7.4}$$

Now

$$\int_{-\pi}^{\pi} \left| \left[\frac{e^{i\mu\lambda} f_s}{\bar{g}}\right]_+ \right|^2 d\lambda \le \int_{-\pi}^{\pi} \frac{f_s^2}{|g|^2 \, d\lambda} < \infty$$

Moreover, just as in the proof of Theorem 6

$$\int_{-\pi}^{\pi} e^{-in\lambda} e^{-i\mu\lambda} g^{-1}(e^{i\lambda}) \, z_y(d\lambda) = \epsilon(n + \mu),$$

where $\epsilon(n)$ is the innovation sequence for the purely nondeterministic process $y(n)$. Since the right-hand side of (7.4) is of the form

$$\sum_0^\infty \beta(j) e^{ij\lambda}, \qquad \sum_0^\infty \beta(j)^2 < \infty,$$

(7.1) defines a filter

$$\sum_0^\infty \beta(j) \, \epsilon(n + \mu - j)$$

of the required form and thus is the optimal filter since it makes the mean-square error as small as is possible. The formula for the mean-square error follows immediately from (7.3) since

$$f_s - \frac{f_s^2}{f_y} = \frac{f_s f_y - f_s^2}{f_y} = \frac{f_s f_x}{f_y} .$$

In the case $\mu = \infty$, without any special assumptions concerning F_y, we have to minimize

$$\int \{F_s(d\lambda) + |h_\infty|^2 F_y(d\lambda) - 2\Re(h_\infty) F_s(d\lambda)\}$$

$$= \int \left\{ \frac{dF_s}{dF_y} - 2\Re(h_\infty) \frac{dF_s}{dF_y} + |h_\infty|^2 \right\} F_y(d\lambda)$$

$$= \int \left\{ \left| h_\infty - \frac{dF_s}{dF_y} \right|^2 + \frac{dF_s}{dF_y} - \left| \frac{dF_s}{dF_y} \right|^2 \right\} F_y(d\lambda)$$

which is clearly minimized by

$$h_\infty = dF_s(\lambda)/dF_y(\lambda)$$

which is certainly square integrable with respect to F_y. The mean-square error is

$$\int \left(1 - \frac{dF_s}{dF_y}\right) F_s(d\lambda) = \int \frac{dF_x}{dF_y} F_s(d\lambda) = \int \frac{dF_s}{dF_y} F_x(d\lambda)$$

as required, and the proof is completed.

The quantity $R = dF_s/dF_x$, *insofar as it is defined, is called the signal to noise ratio.* Loosely phrased the optimal filter for $\mu = \infty$ may be expressed as $(R/(1 + R))$ so that the optimal filter has a simple physical interpretation in terms of that ratio.

Even if we can assume f_s, f_y known and can accomplish the canonical factorization of f_y the reduction of (7.3) to a form where signal extraction can be digitally performed (for example, its expansion in a mean square convergent Fourier series if it admits such an expansion, as will be the case if $f_y \geq a > 0$) may be very difficult. For these kinds of reason the formulas have been little used except in simple cases. The usefulness of such formulations as these, as has seen indicated above, lies partly in the fact that they give an attractively neat mathematical solution to the problem but mainly in the fact that the optimal solution provides a standard against which less systematic (but more practically useful) procedures may be measured. A study of the solution also leads to an understanding of the problem.

As an example let us consider

$$f_s = \frac{\sigma_\epsilon^2}{2\pi(1 + \rho^2 - 2\rho \cos \lambda)}, \qquad 0 \leq \rho < 1$$

$$f_x = \gamma_x(0)/2\pi.$$

Then

$$f_y = \frac{1}{2\pi} \frac{\gamma_x(0)\{1 + \rho^2 - 2\rho \cos \lambda\} + \sigma_\epsilon^2}{(1 - \rho e^{i\lambda})(1 - \rho e^{-i\lambda})}$$

To factorize the numerator we need to consider

$$\{\gamma_x(0)(1 + \rho^2) + \sigma_\epsilon^2\} - \rho\gamma_x(0)z - \rho\gamma_x(0)z^{-1}$$
$$= -\rho\gamma_x(0)z^{-1}\{z^2 - z(\rho + \rho^{-1})(1 + \theta) + 1\},$$
$$\theta = \frac{\sigma_\epsilon^2}{\gamma_x(0)(1 + \rho^2)}.$$

This has zeros at

$$\tfrac{1}{2}\{(\rho + \rho^{-1})(1 + \theta)\} \pm [\tfrac{1}{4}\{(\rho + \rho^{-1})(1 + \theta)\}^2 - 1]^{1/2},$$

from which the one with less than unit modulus is obtained by taking the minus sign. Call this β. Then

$$f_y = \frac{\sigma^2}{2\pi} \frac{(1 - \beta e^{i\lambda})(1 - \beta e^{-i\lambda})}{(1 - \rho e^{i\lambda})(1 - \rho e^{-i\lambda})}, \qquad \sigma^2 = \frac{\rho\gamma_x(0)}{\beta}.$$

Thus

$$h_\mu = e^{-i\mu\lambda}\frac{2\pi(1-\rho e^{i\lambda})}{\sigma^2(1-\beta e^{i\lambda})}\left[e^{i\mu\lambda}\frac{(1-\rho e^{-i\lambda})}{(1-\beta e^{-i\lambda})}\frac{\sigma_\epsilon^{\,2}}{2\pi(1-\rho e^{i\lambda})(1-\rho e^{-i\lambda})}\right]_+$$

$$= \left(\frac{\rho-\beta}{\rho}\right)(\rho e^{i\lambda})^{-\mu}\left(\frac{1}{1-\beta e^{i\lambda}}\right); \qquad \mu \le 0$$

$$= \left(\frac{\rho-\beta}{\rho}\right)e^{-i\mu\lambda}\frac{1}{1-\beta e^{i\lambda}}\left\{e^{i\mu\lambda}+\beta\left(\frac{e^{i\mu\lambda}-\beta^\mu}{e^{i\lambda}-\beta}\right)(1-\rho e^{i\lambda})\right\}, \qquad \mu \ge 0.$$

The first formula follows from the fact that

$$[e^{i\mu\lambda}\{(1-\beta e^{-i\lambda})(1-\rho e^{i\lambda})\}^{-1}]_+ = \left(\frac{\rho^{-\mu}}{1-\rho\beta}\right)\frac{1}{1-\rho e^{i\lambda}}, \qquad \mu \le 0,$$

together with the fact that

$$\beta^2 - \beta\left(\rho + \rho^{-1} + \frac{\sigma_\epsilon^{\,2}}{\rho\gamma_x(0)}\right) + 1 = 0,$$

so that

$$(1-\rho\beta)\left(\frac{\rho-\beta}{\rho}\right) = \beta^2 + 1 - \beta(\rho+\rho^{-1}) = \frac{\beta\sigma_\epsilon^{\,2}}{\rho\gamma_x(0)} = \frac{\sigma_\epsilon^{\,2}}{\sigma^2}.$$

On the other hand,

$$[e^{i\mu\lambda}\{(1-\beta e^{-i\lambda})(1-\rho e^{i\lambda})\}^{-1}]_+$$
$$= \left[\sum_0^{\mu-1}\beta^j e^{i(\mu-j)\lambda} + \{\beta^\mu(1-\rho\beta)\}\right](1-\rho e^{i\lambda})^{-1}, \qquad \mu > 0,$$

and rearranging this we obtain the second formula. We take $\mu = 0$ to discuss the nature of the result. We always have $|\beta| < |\rho|$ since $\beta + \beta^{-1} = (\rho + \rho^{-1})(1 + \theta)$, $\theta > 0$. As $\theta \to 0$ clearly $\beta \to \rho$ and as $\theta \to \infty$ then $\beta \to 0$. The "bandwidth"† of the filter, which depends only upon β, and decreases as it increases, thus depends, given ρ, only upon θ. As $\theta \to 0$ the noise becomes dominant except in a very narrow range about the zero frequency, where the signal power is largest. Thus β is taken as large as possible and the bandwidth becomes narrow. As $\theta \to \infty$ the bandwidth becomes wider as we can afford to do so since over a wider band the signal is appreciable relative to the noise. The calculations, once $\sigma_\epsilon^{\,2}$, ρ and $\gamma_x(0)$ are known, are simple. We now use $\hat{s}^{(\mu)}(n)$ for the smoothed value of $s(n)$ using observations to $n + \mu$. Then (by simple rearrangements which we

† "Bandwidth" is a loosely defined concept (see Chapter V). Here we mean the width of the band over which the gain of the filter is appreciably different from zero.

leave as an exercise)

$$\hat{s}^{(0)}(n+1) = \beta\hat{s}^{(0)}(n) + \left(1 - \frac{\beta}{\rho}\right) y(n+1),$$

$$\hat{s}^{(\mu+1)}(n) = \frac{\beta}{\rho}\,\hat{s}^{(\mu)}(n) + \rho^{-\mu-1}\left(1 - \frac{\beta}{\rho}\right) y(n+\mu+1), \qquad \mu < 0$$

$$= \hat{s}^{(\mu)}(n) + (\beta-\rho)\beta^{\mu}\{\hat{s}^{(0)}(n+\mu+1) - y(n+\mu+1)\}, \quad \mu \geq 0.$$

Thus, in particular, having initially computed $\hat{s}^{(0)}(n_0)$, we obtain $\hat{s}^{(0)}(n_0+j)$, $j \geq 1$, iteratively, which may be "updated" to give $\hat{s}^{(\mu)}(n_0+j)$, $\mu \geq 0$, by the third relation.

(ii) We now turn to the vector case. If $\mu < \infty$ we shall again assume $y(n)$ to have a.c. spectrum and be purely nondeterministic but for $\mu = \infty$ this is not needed. *We shall, for this case $\mu = \infty$, introduce the matrices f_s, f_y, of derivatives of F_s, F_y, with respect to $m(\lambda) = \operatorname{tr}(F_y(\lambda))$.* Thus, for example,

$$\int_S F_y(d\lambda) = \int_S f_y(\lambda)\, m(d\lambda)$$

for all Lebesgue measurable sets S. We shall use f_y^{-1} for the generalized inverse of f_y.

We then have the following theorem, whose proof we omit, since it follows precisely the same pattern as the proof of Theorem 10.

Theorem 10′. *Let $y(n) = s(n) + x(n)$ be a vector process which is purely nondeterministic and of full rank. Then the optimal filter for extracting $s(n)$ from $y(m)$ using $m \leq n + \mu$ has response function*

$$h_\mu = e^{-i\mu\lambda}[e^{i\mu\lambda} f_s C^*(e^{i\lambda})^{-1}]_+ C(e^{i\lambda})^{-1},$$

where $C = \Phi G^{1/2}$ and Φ and G are given by Theorem 1″. The error covariance matrix is

$$\int_{-\pi}^{\pi} \{f_s f_y^{-1} f_x + [e^{i\mu\lambda} f_s C^{*-1}]_- [e^{i\mu\lambda} f_s C^{*-1}]_-^*\}\, d\lambda.$$

In case $\mu = \infty$, and with no special assumptions concerning the spectrum of $y(n)$, we have

$$h_\mu = f_s f_y^{-1}$$

with error covariance matrix

$$\int_{-\pi}^{\pi} f_s f_y^{-1} f_x\, m(d\lambda).$$

(iii) We now consider an example based on a nonstationary model but one which is on the verge of stationarity. (See Chapter I, Section 3 and Chapter

II, Section 6, subsection (iii). The considerations of that last section could undoubtedly be applied to give a general development covering the results of this example. However, we shall proceed more conventionally here.) We consider the scalar case and construct a model for an evolving seasonal pattern in a monthly economic time series.† Thus we take $s(n)$ as

$$s(n) = \sum_1^6 \{\alpha_j(n) \cos \lambda_j n + \beta_j(n) \sin \lambda_j n\} = \sum_1^6 s_j(n), \qquad \lambda_j = \frac{2\pi j}{12},$$

where $\alpha_j(n)$ and $\beta_j(n)$ are totally unrelated to each other and to $x(n)$, that is

$$\mathcal{E}\big(x(n)\, s(n+m)\big) \equiv \mathcal{E}\big(\alpha_j(n)\, \beta_k(n+m)\big) \equiv 0.$$

A suitable model for $\alpha_j(n)$, $\beta_j(n)$ is of the form‡

$$\alpha_j(n) = \rho_j \alpha_j(n-1) + \epsilon_j(n),\ \beta_j(n) = \rho_j \beta_j(n-1) + \eta_j(n),$$

where

$$\mathcal{E}\big(\epsilon_j(m)\, \epsilon_k(n)\big) = \mathcal{E}\big(\eta_j(m)\, \eta_k(n)\big) = \delta_j^k \delta_m^n \sigma_j^2; \qquad \mathcal{E}\big(\epsilon_j(m)\, \eta_k(n)\big) \equiv 0.$$

Then putting $\zeta_j(n) = \frac{1}{2}\{\alpha_j(n) - i\beta_j(n)\}$, $j = 1, \ldots, 5$ $\zeta_6(n) = \alpha_6(n)$ we may write

$$s(n) = \sum_{-5}^{6}{}' \zeta_j(n) e^{in\lambda_j},$$

where the prime indicates that the term for zero lag is omitted. In fact the ρ_j would need to be very near to unity if the model is to be realistic. Indeed

$$\mathcal{E}\{s(m)\, s(m+n)\} = \sum{}' \{\rho_j^{\,n} \sigma_j^2 (1-\rho_j^2)^{-1}\} e^{-in\lambda_j},$$

so that for $n = 0 \pmod{12}$ we have an autocorrelation

$$\frac{\sum' \rho_j^{\,n} \sigma_j^2 (1-\rho_j^2)^{-1}}{\sum' \sigma_j^2 (1-\rho_j^2)^{-1}} \le \max_j (\rho_j^{\,n}).$$

If all ρ_j are 0.95 then for $n = 60$ this is (nearly) e^{-3} so that over 5 years the seasonal pattern would probably change entirely. This is probably too radical. Thus we are led to take $\rho_j \equiv 1$. Then $\zeta_j(n) = \zeta_j(n-1) + \xi_j(n)$ where $\xi_j = \frac{1}{2}(\epsilon_j - i\eta_j) \exp in\lambda_j$. Now we may put $s(n)$ in the form

$$s(n) = \sum_{-5}^{6}{}' \zeta_j(n) e^{i\lambda_j n} = \sum_{-5}^{6}{}' \left\{ \int \frac{1 - e^{-in\lambda}}{e^{i\lambda} - 1}\, e^{in\lambda_j}\, z_j(d\lambda) + \zeta_j(0) e^{i\lambda_j n} \right\},$$

† In practice the data would need to be prefiltered to remove the dominant low-frequency component, for example, by subtracting a centered 12 months moving average. This will affect $s(n)$ but the effect will usually be slight.

‡ A more general model would put $\alpha_j(n) = \rho_j \alpha_j(n-1) + \tau_j \beta_j(n-1) + \epsilon_j(n)$, $\beta_j(n) = -\tau_j \alpha_j(n-1) + \rho_j \beta_j(n-1) + \eta_j(n)$. This case is covered by the general model studied below (if $\rho_j^2 + \tau_j^2 \le 1$) but seems inappropriate in a seasonal variation context as if $\tau_j \ne 0$ the spectrum of $s_j(n)$ is not concentrated at $\pm\lambda_j$.

where z_j corresponds to the (complex) stationary process $\xi_j(n)$. We rewrite this in the form

$$s(n) = \sum_{-5}^{6}{}' \left[\int \left\{ \frac{e^{-in\lambda}}{1 - \kappa_j e^{i\lambda}} - \kappa_j^{\,n}(1 - \kappa_j e^{i\lambda})^{-1} \right\} z_j(d\lambda) + \kappa_j^{\,n}\zeta_j(0) \right]$$

and we have put $\kappa_j = \exp i\lambda_j$. We find it convenient to write this in the form

$$s(n) = \int e^{-in\lambda}\, \phi(e^{i\lambda})^{-1}\, z_\xi(d\lambda) + p(n), \tag{7.5}$$

where

$$z_\xi(d\lambda) = \sum{}' \left\{ \prod_{k \neq j} (1 - \kappa_k e^{i\lambda}) \right\} z_j(d\lambda),$$

$$\phi(e^{i\lambda}) = \prod_j (1 - \kappa_j e^{i\lambda}),$$

$$p(n) = -\sum{}' \left\{ \int (1 - \kappa_j e^{i\lambda})^{-1}\, z_j(d\lambda) + \zeta_j(0) \right\} \kappa_j^{\,n}.$$

Both terms on the right side of (7.5) have infinite variance and should not be separated in this way. As we shall see, however, both terms will occur only in a filtered form giving them finite variance. The model is evidently susceptible to generalization. We, more generally, take z_ξ to be a process of orthogonal increments with $\mathcal{E}\{|z_\xi|^2\} = f_\xi\, d\lambda$. We take $\phi(z)$ to be a polynomial in z with zeros outside *or on* the unit circle.† The expression $p(n)$ will be of the form

$$\sum_j \sum_{k=1}^{p_j - 1} c(j, k)\kappa_j^{\,n} n^k, \tag{7.6}$$

where p_j is the multiplicity of the zero κ_j^{-1} of $\phi(z)$. We put

$$k(\lambda) = f_\xi(\lambda) + \phi(e^{i\lambda})\, \overline{\phi(e^{i\lambda})}\, f_x(\lambda)$$

and assume that $\log k(\lambda)$ is integrable. We put

$$k(\lambda) = c(e^{i\lambda})\, \overline{c(e^{i\lambda})}$$

and proceed to consider the minimization of $\mathcal{E}\{[s(n) - \hat{s}(n)]^2\}$, where $\hat{s}(n)$ uses observations to time $n + \mu$ and where

$$s(n) - \hat{s}(n) = \int e^{-in\lambda_j}\{1 - h\}\phi^{-1}\, z_\xi(d\lambda)$$
$$- \int e^{-in\lambda} h\, z_x(d\lambda) + p(n) - \hat{p}(n).$$

† The case of zeros inside the circle also needs discussion but we are not able to give that here.

Here h is the response function of a yet to be determined filter. The function $z^\mu h(z)$ is to be holomorphic within the circle and $\hat{p}(n)$ is the output of the filter from the input $p(n)$. Now, if we neglect the term $p(n) - \hat{p}(n)$ and choose h to minimize the mean square of the remaining term, we shall show that the term $p(n) - \hat{p}(n)$ is identically zero, that is, h necessarily reproduces a polynomial of the form (7.6). This is a desirable feature, for it means, in this context of seasonal adjustment, that a strictly periodic seasonal component is reproduced. Our model and its treatment is more general than it might appear to be, since the seasonal component can consist of a stochastic part plus a strictly periodic "trend" part. Thus we proceed to consider the minimization of

$$\int \{|1 - h|^2 |\phi|^{-2} f_\xi + |h|^2 f_x\} \, d\lambda. \tag{7.7}$$

under somewhat more general conditions. We point out that Theorem 11 below gives, in particular, a solution to the problem of extracting a signal from noise when the signal is nonstationary but of the type discussed below Theorem 4 of Chapter I. In particular, taking $f_x \equiv 0$, the theorem gives a solution also to the prediction problem for such a signal.

Theorem 11. *Let f_x and f_ξ be spectral densities and let ϕ be square integrable with respect to f_x on the boundary of the unit circle and holomorphic in its interior. Let $k = f_\xi + \phi\bar{\phi} f_x$ be such that $\log k$ is integrable and let $k = c\bar{c}$ where c is holomorphic and without zeros within the unit circle. The minimum of the expression* (7.7), *where $h \exp i\mu\lambda$ is to be holomorphic within the unit circle, is obtained for*

$$h_\mu = 1 - e^{-i\mu\lambda}[e^{i\mu\lambda} f_x \bar{\phi} \bar{c}^{-1}]_+ c^{-1}\phi - e^{-i\mu\lambda}[e^{i\mu\lambda} qc]_- c^{-1}\phi, \tag{7.8}$$

where $q(z) \equiv 0$, $\mu \geq 0$, and for $\mu < 0$ $q(z)$ is a polynomial of degree $-\mu - 1$ with coefficients q_j satisfying

$$\sum_0^l q_{l-j}\phi_j = \delta_0^{\,l}, \qquad l = 0, \ldots, -\mu - 1,$$

and the ϕ_j are the Fourier coefficients of ϕ. The mean-square error is

$$\int_{-\pi}^{\pi} (k^{-1} f_\xi f_x + b_0 \bar{b}_0) \, d\lambda, \qquad b_0 = [e^{i\mu\lambda}(f_x \bar{\phi} \bar{c}^{-1} - qc)]_-. \tag{7.9}$$

If all zeros κ_j of ϕ lie inside the circle, (7.8) *becomes*

$$h_\mu = e^{-i\mu\lambda}[e^{i\mu\lambda} \phi^{-1} f_\xi \bar{c}^{-1}]_+ c^{-1}\phi. \tag{7.8$'$}$$

The filter h annihilates a polynomial of the form of $p(n)$, given by (7.6).

Proof. As usual we shall drop the subscript μ. The expression (7.7) may be rearranged into the form

$$\int k^{-1} f_\xi f_x + |\phi^{-1} f_\xi \bar{c}^{-1} - h\phi^{-1} c|^2 \, d\lambda,$$

so that we must minimize

$$\int |\phi^{-1} f_\xi \bar{c}^{-1} - h\phi^{-1} c|^2 \, d\lambda. \tag{7.10}$$

The quantity $k^{-1} f_\xi f_x$ is integrable, since $|k^{-1} f_\xi|$ is less than unity. An h (which we call h_1) making (7.10) finite for $\mu \geq 0$ is unity, as is evident from (7.7). If $\mu < 0$, this is not available and to find such an h we solve

$$1 - h_1(z) = q(z)\,\phi(z), \tag{7.11}$$

where $q(z)$ is to be a polynomial in z and $z^\mu h_1(z)$ is to be holomorphic within the circle. Again from (7.7) it is evident that such an h_1 will make (7.7) finite (since the factor $|\phi|^2$ from $|1 - h_1|^2$ will cancel with $|\phi|^{-2}$, whose poles may cause the integral to diverge). To find such an h_1 we have to solve

$$\sum_0^{-\mu-1} q_j z^j \sum_0^\infty \phi_j z^j - 1 = -h_1(z) = -z^{-\mu} \sum_0^\infty h_{1,j} z^j,$$

where q_j, ϕ_j are the coefficients of $q(z)$, $\phi(z)$. Thus we have to solve

$$\sum_0^l q_{l-j}\phi_j = \delta_0^{\,l}, \qquad l = 0, \ldots, -\mu - 1.$$

and because $\phi_0 \neq 0$ [since $\phi(0) \neq 0$] this is clearly uniquely solvable. Thus putting

$$b = \{\phi^{-1} f_\xi \bar{c}^{-1} - h_1 \phi^{-1} c\}, \qquad h_0 = h - h_1,$$

we have to minimize

$$\int |b - h_0 \phi^{-1} c|^2 \, d\lambda,$$

where now b is square integrable. Now h_0 must be square integrable with respect to $|\phi^{-1} c|^2 = |\phi|^{-2} k$ and holomorphic within the circle. We apply the Wiener-Hopf technique to obtain

$$h_0 = [b]_+ \, \phi c^{-1},$$

so that

$$\begin{aligned} h &= [b]_+ \phi c^{-1} + h_1 \\ &= e^{-i\mu\lambda}[\phi^{-1} f_\xi \bar{c}^{-1} - (1 - q\phi)]_+ \phi c^{-1} + 1 - q\phi \\ &= 1 - e^{-i\mu\lambda}[e^{i\mu\lambda} f_x \bar{\phi} \bar{c}^{-1}]_+ c^{-1}\phi - e^{-i\mu\lambda}[e^{i\mu\lambda} q c]_- c^{-1}\phi, \end{aligned}$$

since the only negative power of z in $z^\mu q c\phi$ is z^μ, $\mu < 0$. (When $\mu \geq 0$ the term involving $[\;]_-$ is null); $\{h - 1\}$ is of the form $a\phi$, where a is square integrable with respect to k. Thus, since the filter with response function ϕ annihilates† a polynomial of the form of $p(n)$, we see that $p(n) - \hat{p}(n)$ is annihilated and our initial assertion concerning it is proved true and (7.8) gives us the optimal filter we require.

If the zeros of ϕ, namely, κ_j^{-1}, are outside the unit circle, (7.8) reduces to

$$h = e^{-i\mu\lambda}[\phi^{-1}f_\xi \bar{c}^{-1}]_+ c^{-1}\phi,$$

which is (7.8)′ and agrees with our previous result since $|\phi|^{-2}f_\xi$ is our old f_s and $\phi^{-1}c$ is now the canonical factor for f_y. The formula (7.8)′ follows from the fact that we may now write

$$e^{i\mu\lambda} = e^{i\mu\lambda}\phi^{-1}cc^{-1}\phi = [e^{i\mu\lambda}\phi^{-1}c]_+c^{-1}\phi + [e^{i\mu\lambda}\phi^{-1}c]_-c^{-1}\phi,$$

since $e^{i\mu\lambda}\phi^{-1}c$ now has a valid Laurent expansion on the unit circle. Thus we obtain

$$[e^{i\mu\lambda}(\phi^{-1}c - f_x\bar{\phi}\bar{c}^{-1})]_+c^{-1}\phi + [e^{i\mu\lambda}(\phi^{-1} - q)c]_-c^{-1}\phi,$$

since $\phi^{-1}c - f_x\bar{\phi}\bar{c}^{-1} = (\phi^{-1}\, c\bar{c}\bar{\phi}^{-1} - f_x)\bar{\phi}\bar{c}^{-1} = \phi^{-1}f_\xi\bar{c}^{-1}$ and $\exp i\mu\lambda(\phi^{-1} - q) = h_1\phi^{-1}$ so that the second term is null. However, in general (7.8)′ will not be meaningful since the expression in square brackets will not be square integrable on the circle.

The error variance is, from (7.10),

$$\int \{k^{-1}f_\xi f_x + b_0\bar{b}_0\}\, d\lambda, \qquad b_0 = [e^{i\mu\lambda}(f_x\bar{\phi}\bar{c}^{-1} - qc)]_-.$$

If $\mu \geq 0$ the second term within $[\;]_-$ can be omitted, and as $\mu \to \infty$, we obtain an error variance that is $\int \{f_\xi f_x/(f_\xi + |\phi|^2 f_x)\}\, d\lambda$.

These considerations extend, of course, to the vector case but we refer the reader to Hannan (1967b) for details.

Let us return to the case in which the zeros of ϕ, namely, κ_j^{-1}, are simple and of the form $\exp i\lambda_j$, $\lambda_j = 2\pi j/12$, so as to illustrate the results. Let us take the spectrum of $\epsilon_j(n)$ and $\eta_j(n)$ to be $\sigma_j^2/2\pi$. We also assume $c(z)$ holomorphic and without zeros within a disk containing the unit disk in its interior. This imposes only a mild restriction on f_x, from a practical point of view. Now (7.8)′ is a valid form of (7.8) for z on a circle lying within the unit circle since $z^\mu\ \phi^{-1}(z)\ c(z)$ now has a valid Laurent expansion on such a circle and we may evaluate $h(z)c(z)\phi(z)^{-1}$, which is square integrable on the circle, by evaluating

$$\left[z^\mu\phi^{-1}\left\{\sum{}' \prod_{k\neq j}(1 - \kappa_k z)(1 - \kappa_k^{-1}z^{-1})\frac{\sigma_j^2}{2\pi}\right\}c(z^{-1})^{-1}\right]_+$$

† To see this, consider (7.6), where κ_j is a zero of order p_j of $\phi(z)$.

and taking the boundary value of that function on the unit circle. We need therefore to evaluate

$$\sum_j{}' \frac{\sigma_j^2}{2\pi} \frac{1}{2\pi i} \oint \frac{\zeta^{-\mu}\{\prod_{k\neq j}(1-\kappa_k^{-1}\zeta)\, z^{-1}\}}{(\zeta-\kappa_j)(z^{-1}-\zeta)\, c(\zeta)}\, d\zeta, \qquad 1 < |\zeta| < |z|^{-1}.$$

For $\mu \leq 0$ this is

$$\sum{}' \frac{\sigma_j^2}{2\pi} \frac{\kappa_j^{-\mu}\prod_{k\neq j}(1-\kappa_j/\kappa_k)}{(1-\kappa_j z)\, c(\kappa_j)}$$

so that

$$h(e^{i\lambda}) = \sum_j{}' \left\{\frac{\sigma_j^2}{2\pi} \frac{(\kappa_j e^{i\lambda})^{-\mu} \prod_{k\neq j}(1-\kappa_j/\kappa_k)(1-\kappa_k e^{i\lambda})}{c(\kappa_j)\, c(e^{i\lambda})}\right\}, \qquad \mu \leq 0$$

For $\mu > 0$ the expression is more complicated to write (though still simple in principle). For $\mu = 1$ it is

$$\sum{}' \frac{\sigma_j^2}{2\pi} \left\{\frac{(\kappa_j e^{i\lambda})^{-\mu} \prod_{k\neq j}(1-\kappa_j/\kappa_k)(1-\kappa_k e^{i\lambda})}{c(\kappa_j)\, c(e^{i\lambda})} - \frac{\kappa_j^{-1}\, \phi(e^{i\lambda})}{c(0)\, c(e^{i\lambda})}\right\}.$$

A main problem with the application of such a formula is associated with the calculation of $c(z)$ (assuming that f_ξ, ϕ, and f_x can be prescribed). This will involve the factoring of a high degree polynomial and the nature of the prediction formulas will be such that the iterative formulas needed for calculations will be quite complicated. An approximate procedure appropriate for the case just studied (for the estimation of a seasonal component) is the following. The series may be filtered first so as to eliminate or vastly reduce all components in the signal save that at frequency λ_j; for example, we might prefilter by

$$y(n) \to \frac{1}{12} \sum_0^{11} y(n-k) \cos k\lambda_j, \qquad \lambda_j = \frac{2\pi j}{12},$$

whose response is unity at $\lambda = \pm\lambda_j$ and is zero at $\lambda_k \neq \pm\lambda_j$. Then each component in the signal is separately extracted, treating the other components as if they were not present. An examination of the response of the filter for extracting the signal, appropriate to λ_j, shows that if $\sigma_j^2/f_x(\lambda_j)$ is small (as would be normal) the response of the optimal filter is very concentrated around $\pm\lambda_j$ so that, since the remaining components are substantially reduced by the prefiltering we may treat the spectrum of the noise as constant. For a detailed description together with numerical examples the reader may consult Hannan and Terrell (1969).

8. KALMAN FILTERING

The procedures discussed so far for filtering data, so as to separate a signal from noise, have all been based upon Fourier methods and have correspondingly assumed stationarity or something very near to it. These methods have led to the description of the filter in terms of a frequency response function and a further step of Fourier expansion is needed to obtain the impulse response function. If the filtering is to be done by an analog device the description by means of a frequency response function may be reasonable since one can attempt to synthesize a filter with this response by physical means. However, digital machines have lately become much more important and, of course, the frequency response function is not then a useful characterization of the filter. Two other factors also suggest a modification of the model. The first of these is the recognition that the spectra involved will not be of such a general form as has so far been allowed for. After all, they will either be given by prior physical theory (subject to estimation of constants) or from an estimate from past data. In either case they will certainly be rational, at least. The second factor arises from the possibility of nonstationarity. One cause of this could be the fact that a system takes some time before it settles into something approaching a stationary state. If one has to extract the signal from the very commencement of operation of the system stationary models will not do. In addition, of course, the system may be more fundamentally nonstationary. These considerations make recent work by Kalman (1960, 1963) and Kalman and Bucy (1961) important. We shall give an account for the discrete time vector case only, though it extends to continuous time. The model is of the form

$$\left.\begin{cases} x(n) = \Phi(n-1)\,x(n-1) + \epsilon(n-1) \\ y(n) = H(n)\,x(n) + \eta(n) \end{cases}\right\}. \tag{8.1}$$

Here $x(n)$ and $\epsilon(n)$ are vectors of p components while $y(n)$ and $\eta(n)$ are of q components. The random vectors $\epsilon(n)$, $\eta(n)$ satisfy

$$\mathcal{E}\{\epsilon(n)\epsilon'(m)\} = \delta_m{}^n Q(n), \qquad \mathcal{E}\{\eta(n)\eta(m)'\} = \delta_m{}^n R(n)$$
$$\mathcal{E}\{\epsilon(n)\eta'(m)\} = \delta_m{}^n S(n).$$

Thus $x(n)$ corresponds to our old $s(n)$ and $\eta(n)$ to our old $x(n)$. We use Kalman's notation for ease of reference. In Kalman's original work $\epsilon(n-1)$ is written as $\Gamma(n, n-1)u(n-1)$, where $u(n)$ has the properties of $\epsilon(n)$. This is done because Kalman uses the model in connection with control problems where $u(n)$ is the controllable quantity† and the additional generality

† More generally $u(n)$ might be made up of a control variable, a deterministic time function, and an uncontrolled random time function.

introduced through the matrix Γ is then useful. We do not discuss control theory in this book and from the point of view of signal extraction we may as well absorb Γ into Q and S. The vector x is the transmitted signal to be measured but we observe not x but y which involves the noise η. We assume that Φ, H, Q, R, and S are known. The model is more general than it appears, as we see from the following generalization. Let

$$x_0(n) = \sum_1^r \Phi_k(n)\, x_0(n-k) + f(n) + \epsilon_0(n-1),$$

$$y_0(n) = \sum_1^s H_{0,k}(n)\, x_0(n-k) + u(n),$$

$$u(n) = \sum_1^t D_k(n)\, u(n-k) + \eta(n),$$

wherein $f(n)$ is a *known* deterministic sequence while $\epsilon_0(n)$ has the properties previously ascribed to $\epsilon(n)$. The term $f(n)$ is included to account for systematic effects of the nature of a trend. Again we assume that all of the matrices, including the matrices of variances and covariances are known. (This is not likely to be quite true and we shall discuss it later in this section.) Now $x_0(n)$ is evidently of the form $x_1(n) + g(n)$, where $g(n)$ satisfies

$$g(n) = \sum_1^r \Phi_k(n)\, g(n-k) + f(n),$$

together with r appropriate initial conditions, whereas $x_1(n)$ satisfies the same equation but with $f(n)$ replaced by $\epsilon_0(n-1$.) Thus we may form

$$y_1(n) = y_0(n) - \sum_1^s H_{0,k}(n)\, g(n-k)$$

and reduce ourselves to extracting $x_1(n)$. Thus we now consider the equations

$$x_1(n) = \sum_1^r \Phi_k(n)\, x_1(n-k) + \epsilon_0(n-1),$$

$$y_1(n) = \sum_1^s H_{0,k}(n)\, x_1(n-k) + u(n),$$

$$u(n) = \sum_1^t D_k(n)\, u(n-k) + \eta(n).$$

Then we may form a new set of relations

$$x_1(n) = \sum_1^r \Phi_k(n)\, x_1(n-k) + \epsilon_0(n-1),$$

$$y(n) = y_1(n) - \sum_1^t D_k(n)\, y_1(n-k) = \sum_1^{s+t} H_k(n)\, x_1(n-k) + \eta(n),$$

where the $H_k(n)$ are got in a simple fashion from the $D_k(n)$ and $H_{0,k}(n)$. Thus now we are reduced to

$$x_1(n) = \sum_1^r \Phi_k(n)\, x_1(n-k) + \epsilon_0(n-1),$$

$$y(n) = \sum_1^{s+t} H_k(n)\, x_1(n-k) + \eta(n).$$

Now putting $m = \max\,(r, s+t)$ we form the new "state" vector (i.e., one describing the state of the system being studied)

$$x(n)' = (x_1(n)' \vdots x_1(n-1)' \vdots \ldots \vdots x_1(n-m)').$$

We introduce

$$\Phi(n-1) = \begin{bmatrix} \Phi_1(n) & \Phi_2(n) & \ldots & \Phi_{m-1}(n) & \Phi_m(n) \\ I & 0 & \ldots & 0 & 0 \\ 0 & 0 & \ldots & I & 0 \end{bmatrix},$$

$$\epsilon(n)' = (\epsilon_0(n)' \;\; 0 \;\; \ldots \;\; 0),$$

$$H(n) = [H_1(n) \vdots H_2(n) \vdots \ldots \vdots H_m(n)],$$

where some $\Phi_k(n)$, $H_k(n)$ may be null (e.g., $\Phi_m(n)$ if $r = s + t - 1$). Then we have

$$x(n) = \Phi(n-1)x(n-1) + \epsilon(n-1),$$

$$y(n) = H(n)x(n) + \eta(n),$$

and we have carried ourselves back to Kalman's initial model (8.1), albeit at the expense of a (possibly greatly) increased dimension for $x(n)$. Now the matrices $Q(n)$ and $S(n)$ are certainly singular but this does not affect the validity of the method.

The model is evidently a very useful one, made so by two basic factors. The first of these is the possibility of allowing the matrices Φ, H, Q, R, and S to be time dependent. Of course, they also have to be known a priori. In some applications this is not too troublesome, particularly with respect to Φ and H. For example, in trajectory estimation† Φ is got from the dynamics of the vehicle being tracked while H is got from the properties of the tracking equipment. There is evidently some need for estimation of some of the quantities involved. For example, $f(n)$ will possibly involve some unknowns, and $Q(n)$, $S(n)$ may also be unknown. As has been pointed out to me by Dr. D. B. Duncan these matrices are themselves, to some extent, susceptible to treatment by the same methods so that a model can be built for them which

† My knowledge of this field is largely due to conversations with Dr. D. B. Duncan.

treats them as $x(n)$ is treated above, $y(n)$ being some estimate of $Q(n)$ (say), written as a vector, obtained from the past output. However, we shall not go into that problem here and instead we return to the model (8.1).

The problem now is to obtain the best linear estimator of $x(n)$ from observations $y(0), y(1), \ldots, y(n+\mu)$. We call this estimator $\hat{x}^{(\mu)}(n)$. We consider only the case where $\mu \leq 0$ since the other case is more complicated (though in principle not more difficult). We find it more convenient now to put $\mu = -m$ and change our notation so that we project $x(n+m)$ on the space spanned by $y(0), y(1), \ldots, y(n)$. It is also convenient to adopt the notation $\hat{x}(n+m \mid n)$ in place of $\hat{x}^{(-m)}(n+m)$ because the nonstationarity means that we cannot so easily suppress notational reference to m. We introduce the vector space $\mathcal{H}_n$ spanned by the components of $\eta(j), j \leq n$, $\epsilon(k), k \leq n-1$, together with $x(0)$, the random vector initiating the $x(n)$ sequence. Let $\mathcal{M}_n$, now, be the complex vector space spanned by the elements of $y(0), y(1), \ldots, y(n)$. We shall sometimes, loosely, say that $y(j) \in \mathcal{M}_n$, $j \leq n$, when we mean that all the elements of that vector lie therein. Similarly, we shall write $\epsilon(n) \perp \mathcal{M}_{n-1}$ when we mean that this is true of all elements. Then $\mathcal{M}_n \subset \mathcal{H}_n$.

Theorem 12. *Let $\hat{x}(n+m \mid n)$, $m \geq 1$, be the estimate of $x(n+m)$ got by projecting $x(n+m)$ onto $\mathcal{M}_n$. Then*

$$\hat{x}(n+m \mid n) = \Phi(n+m-1)\hat{x}(n+m-1 \mid n), \qquad m > 1,$$

$$\hat{x}(n+1 \mid n) = \Psi(n)\hat{x}(n \mid n-1) + K(n)y(n), \tag{8.2}$$

$$\Psi(n) = \Phi(n) - K(n)H(n), \tag{8.3}$$

$$K(n) = \{\Phi(n)\Sigma(n)H(n)' + S(n)\}\{H(n)\Sigma(n)H(n)' + R(n)\}^{-1},$$

and $\Sigma(n)$ is got recursively from

$$\Sigma(n+1) = \Phi(n)\Sigma(n)\Phi(n)' + Q(n) - K(n)\,\{H(n)\,\Sigma(n)\,H(n)' + R(n)\}\,K(n)^*. \tag{8.4}$$

Before going on to the proof let us discuss this result. The first relation shows that to know $\hat{x}(n+m \mid n)$ we need only to know $\hat{x}(n+1 \mid n)$. To know this one needs only to know $\hat{x}(n \mid n-1)$ and $\Psi(n)$, and $\Psi(n)$ is known when $K(n)$ and hence $\Sigma(n)$ is known. Now (8.4) shows that $\Sigma(n)$ is known once $K(n-1)$ and $\Sigma(n-1)$ are known. Thus the iterative procedure is complete.

The proof is as follows. Decompose $\mathcal{M}_n$ as $\mathcal{M}_n = \mathcal{M}_{n-1} \oplus \mathcal{V}_n$, where $\mathcal{V}_n$ is evidently spanned by the elements of $z(n) = y(n) - H(n)\,\hat{x}(n \mid n-1)$, since $\eta(n) \perp \mathcal{M}_{n-1}$ and thus the projection of $y(n)$ on $\mathcal{M}_{n-1}$ is

$$H(n)\,\hat{x}(n \mid n-1).$$

Then $z(n) = \eta(n) + H(n)\{x(n) - \hat{x}(n \mid n-1)\}$. Thus since $\epsilon(n) \perp \mathcal{M}_n$ we see that, considering the decomposition $\mathcal{M}_n = \mathcal{M}_{n-1} \oplus \mathcal{V}_n$,

$$\hat{x}(n+1 \mid n) = \Phi(n)\hat{x}(n \mid n-1) + u(n),$$

where $u(n)$ is the projection of $x(n+1)$ on $\mathcal{V}_n$. This is

$$u(n) = \mathcal{E}\{x(n+1)z(n)^*\}[\mathcal{E}\{z(n)z(n)^*\}]^{-1}z(n),$$

assuming for the moment that the matrix being inverted is nonsingular. Thus putting

$$\Sigma(n) = \mathcal{E}\{(x(n) - \hat{x}(n \mid n-1))(x(n) - \hat{x}(n \mid n-1))'\},$$

we have

$$\begin{aligned} u(n) &= \{\Phi(n)\,\Sigma(n)\,H(n)' + S(n)\}\{H(n)\,\Sigma(n)\,H(n)' + R(n)\}^{-1}\,z(n) \\ &= K(n)\,z(n). \end{aligned}$$

We now have

$$\begin{aligned} \hat{x}(n+1 \mid n) &= \Phi(n)\,\hat{x}(n \mid n-1) + K(n)\,z(n) \\ &= \{\Phi(n) - K(n)\,H(n)\}\,\hat{x}(n \mid n-1) \\ &\qquad + K(n)\,\{z(n) + H(n)\,\hat{x}(n \mid n-1)\}, \end{aligned}$$

which from (8.3) and the definition of $z(n)$ is $\Psi(n) + K(n)\,y(n)$, giving (8.2).

Since $x(n+m) = \Phi(n-1)\,x(n+m-1) + \epsilon(n+m-1)$ and $\epsilon(n+m-1) \perp \mathcal{M}_n$, $m > 1$, the first relation of the theorem follows. Thus only (8.4) needs to be established.

To obtain an equation for $\Sigma(n)$ put

$$\begin{aligned} x(n+1) - \hat{x}(n+1 \mid n) &= \{x(n+1) - \hat{x}(n+1 \mid n-1)\} \\ &\qquad - \{\hat{x}(n+1 \mid n) - \hat{x}(n+1 \mid n-1)\}, \end{aligned} \tag{8.5}$$

where the second term is the projection of the first on $\mathcal{M}_n$. This we break into the projection onto $\mathcal{M}_{n-1}$ plus the projection onto $\mathcal{V}_n$. Then

$$\{\hat{x}(n+1 \mid n) - \hat{x}(n+1 \mid n-1)\} = u(n)$$

since

$$\{x(n+1) - \hat{x}(n+1 \mid n-1)\} \perp \mathcal{M}_{n-1}$$

and

$$\hat{x}(n+1 \mid n-1) \in \mathcal{M}_{n-1} \perp \mathcal{V}_n$$

and $u(n)$ is the projection of $x(n+1)$ on $\mathcal{V}_n$. Moreover

$$x(n+1) - \hat{x}(n+1 \mid n-1) = \Phi(n)\,\{x(n) - \hat{x}(n \mid n-1)\} + \epsilon(n)$$

and thus

$$x(n+1) - \hat{x}(n+1 \mid n) = \Phi(n)\,\{x(n) - \hat{x}(n \mid n-1)\} + \epsilon(n) - u(n)$$

and

$$(8.6)\qquad \begin{aligned}\Sigma(n+1) = \Phi(n)\,\Sigma(n)\,\Phi(n)' + Q(n) + \mathcal{E}\big(u(n)\,u(n)'\big) \\ - \mathcal{E}\{\big(x(n+1) - \hat{x}(n+1 \mid n-1)\big)\,u(n)'\} \\ - \mathcal{E}\,\{u(n)\,\big(x(n+1) - \hat{x}(n+1 \mid n-1)\big)'\};\end{aligned}$$

but

$$(8.7)\qquad \begin{aligned}\mathcal{E}\big(u(n)\,u(n)'\big) &= K(n)\,\mathcal{E}\big(z(n)\,z(n)'\big)\,K(n)' \\ &= K(n)\,\{H(n)\,\Sigma(n)\,H(n)' + R(n)\}\,K(n)',\end{aligned}$$

since $z(n) = \eta(n) + H(n)\,\{x(n) - \hat{x}(n \mid n-1)\}$. Moreover,

$$\begin{aligned}&\mathcal{E}\{\big(x(n+1) - \hat{x}(n+1 \mid n-1)\big)\,u(n)'\} \\ &\qquad = \mathcal{E}\{\big(x(n+1) - \hat{x}(n+1 \mid n-1)\big)\,z(n)'\}\,K(n)' \\ &\qquad = \mathcal{E}\{x(n+1)\,z(n)'\}\,K(n)',\end{aligned}$$

since $\hat{x}(n+1 \mid n-1) \in \mathcal{M}_{n-1} \perp \mathcal{V}_n$. The last expression is

$$K(n)\,\mathcal{E}\big(z(n)\,z(n)'\big)\,K(n)'$$

since $K(n)\,z(n)$ is the projection of $x(n+1)$ on $\mathcal{V}_n$. Inserting this and (8.7) in (8.6) establishes (8.4). In the case where $\mathcal{E}\big(z(n)\,z(n)'\big)$ is singular the proof may be easily modified by using the generalized inverse of that matrix. Evidently $K(n)\,z(n)$ is then well defined since $z(n)$ is orthogonal, with probability 1, to the null space of $K(n)$. Of course, if we wish to break $K(n)\,z(n)$ into $K(n)\,y(n)$ and $K(n)\,H(n)\,x(n \mid n-1)$ we must adopt a suitable convention, for example, that we replace $y(n)$ and $H(n)\,x(n \mid n-1)$ by their projections on to the orthogonal complement of the null space of $K(n)$. It does not seem worthwhile going into details here.

Putting

$$\Phi(n+m \mid n) = \prod_{1}^{m-1} \Phi(n+j),$$

we have

$$\hat{x}(n+m \mid n) = \Phi(n+m \mid n)\,\hat{x}(n+1 \mid n).$$

Evidently, from (8.2), $\hat{x}(0)$ (the initiating estimate) enters into $\hat{x}(n+1 \mid n)$ with a coefficient matrix

$$\prod_{1}^{n} \Psi(j),$$

and this determines the stability of the formulas, that is, the degree to which an error in the initial estimate will be propagated into the future.

If the matrices Φ, H, R, S, and Q do not depend on n and Φ has eigenvalues of less than unit modulus, then $\hat{x}$ will eventually become the solution of the signal extraction problem which was discussed in the last section.

9. SMOOTHING FILTERS

Throughout this section we consider only the scalar case.

During the major part of this section we discuss filters designed to reproduce, at least locally, a polynomial of prescribed degree. For want of a better name we have therefore spoken of smoothing filters. Of course, this means that the gain of the filter will be substantially concentrated at $\lambda = 0$. Nevertheless, we may always transfer this point at which the gain concentrates to any other frequency, θ say. For example with a discrete time filter with impulse response $a(m)$ we need merely to replace $a(m)$ by $a(m) \cos \theta m$ to produce a filter with response $\frac{1}{2}\{h(\lambda - \theta) + h(\lambda + \theta)\}$, $h(\lambda)$ being the response of the original filter.

We may obtain filters, of the "smoothing" type we require, by the methods of previous sections, for example by prescribing the signal as having spectrum substantially concentrated at zero frequency. Indeed we have given an example of such a procedure in Section 7. However, in many situations our prior knowledge of the nature of the signal is quite vague (and the same may be true of the noise for that matter). Moreover, it may be required that the filter should not be too "long" so that the impulse response should be nonzero only for $t \in [0, T]$, for example. A filter designed by the method, say, of Section 7 may have impulse response going quickly to zero so that early truncation would not greatly affect the output.

We begin with the discrete time case and consider the construction of a filter of length $2N + 1$,

$$x(n) \to \sum_{-N}^{N} \alpha(j)\, x(n + j)$$

designed so as to reproduce an arbitrary polynomial of degree $d < 2N + 1$ at the time point $n + m$ and which minimizes the variance of the output when the input $x(n)$ has spectrum $f(\lambda)$. Because m is arbitrary the requirement that the range of j go from $-N$ to N is no restriction. As before we call Γ_{2N+1} the matrix of $2N + 1$ rows and columns with $\gamma(j - k)$ in row j column k, where these are the autocovariances of $x(n)$. Let α be the vector with the $\alpha(j)$ as components. Then we must minimize

$$\alpha' \Gamma_{2N+1} \alpha$$

subject to

$$\sum_{-N}^{N} \alpha(j) \left\{ \sum_{0}^{d} a_k (n + j)^k \right\} = \sum_{0}^{d} a_k (n + m)^k, \qquad d < 2N + 1.$$

This condition is equivalent to

$$\sum_{-N}^{N} \alpha(j) j^p = m^p, \qquad p = 0, 1, \ldots, d,$$

since the a_k are arbitrary. We may replace this by the equation

$$\sum_{-N}^{N} \alpha(j)\phi_p(j) = \phi_p(m), \qquad 0 \leq p \leq d,$$

where the $\phi_p(j)$ are the polynomials got by orthonormalizing the $(d + 1)$ sequences $\{j^p, j = -N, \ldots, N\}$ by the Gram-Schmidt process.† Call Φ the matrix with $\phi_p(j)$ in row p column j and $\phi(m)$ the vector with $\phi_p(m)$ in row p. Then we have to minimize

$$\alpha' \, \Gamma_{2N+1}\alpha$$

subject to

$$\Phi\alpha = \phi(m).$$

This leads to the equation

$$\Gamma_{2N+1}\alpha = \Phi'\lambda$$

where λ is a vector of Lagrangean multipliers.

Then, assuming Γ_{2N+1} to be nonsingular, we obtain

$$\phi(m) = \Phi\Gamma_{2N+1}^{-1}\Phi'\lambda$$

that is,

$$\alpha = \Gamma_{2N+1}^{-1}\Phi'(\Phi\Gamma_{2N+1}^{-1}\Phi')^{-1}\,\phi(m)$$

and the minimized variance is

$$\phi(m)'(\Phi\Gamma_{2N+1}^{-1}\Phi')^{-1}\,\phi(m). \tag{9.1}$$

As is well known (and as we shall see in Chapter VII) this is exactly the result one gets by computing the best linear unbiased regression of $x(n)$ on the orthonormalized polynomials and using the estimated regression coefficients to compute the prediction of the polynomial values at time m. We shall, in Chapter VII (Theorem 8), show that under very general conditions on $f(\lambda)$ (e.g., that $f(\lambda)$ is piecewise continuous with no discontinuity at $\lambda = 0$) when d is held fixed and $N \to \infty$ the ratio of the quantity (9.1), to the value which would be obtained if the computations were effected acting as though $f(\lambda)$ is constant, approaches unity. In other words, if N is large relative to d then we shall do almost as well by using a simple regression.

† These sequences or appropriate multiples of them are tabulated in many places (see Owen, 1962, p. 515). We do not imply that $|m| \leq N$.

The smoothed value of $x(n)$ is then

$$\sum_j \sum_k \phi_j(m)\, \phi_j(k)\, x(n+k) = \sum_k x(n+k) \left\{ \sum_j \phi_j(m)\, \phi_j(k) \right\}.$$

The variance of the output is (see Section 2 in Chapter VII)

$$\phi(m)'(\Phi \Gamma_{2N+1} \Phi')\, \phi(m),$$

whose ratio to (9.1) approaches unity as $N \to \infty$ as we have said. The particular case, $m = 0$, is of interest. Then, since $\phi_j(0)$ is zero for j odd and, since $2N + 1$ is odd, $\phi_j(-k) = \phi_j(k)$ for j even, we see that the resulting coefficients

$$a(k) = \sum_j \phi_j(0)\, \phi_j(k)$$

satisfy $a(k) = a(-k)$ and d may always be taken to be even. For various N and d (and $m = 0$) these coefficients are shown in Kendall and Stuart (1966, p. 370). For example with $d = 4$, $N = 6$ the row vector of $a(k)$ is

$$\tfrac{1}{2431}\,(110, -198, -135, 110, 390, 600, 677, \ldots)$$

where we have shown only 7 since the remaining 6 are obtained by symmetry. We shall adopt this convention elsewhere in this section when the filter coefficients have the required symmetry. The response function of the resulting filter is $\sum a(k) \cos k\lambda$, which is also the output when the input is $\cos k\lambda$. Thus $\{1 - \sum a(k) \cos k\lambda\}$ has a zero of order $(d + 2)$ at $\lambda = 0$ (assuming d to be even). This is easily seen by differentiation, for example. Thus

$$\{1 - \sum a(k) \cos k\lambda\} = \{1 - \sum a(k) e^{ik\lambda}\} = (1 - e^{i\lambda})^{d+2}\, q(\lambda),$$

where q is a polynomial in $\exp i\lambda$, $\exp -i\lambda$. It follows that if we take $d = 0$ and iterate $(1 + \frac{1}{2}d)$ times we shall have a response function which also contains the factor $(1 - \exp i\lambda)^{d+2}$ and, therefore, will have a shape near to the origin not unlike that of the optimal filter. The same is true if we repeatedly apply simple averages ($d = 0$) of *different* lengths. It is usually convenient to have the length of the resulting filter an odd number so that an even number of simple averages of even length is always taken. By considerations such as these [see Kendall and Stuart, 1966] several useful formulas have been derived, partly for use in actuarial work. We adopt Kendall's notation $[p]\, x(n)$ for the sum of p successive terms of the $x(n)$ sequence. Two examples are

Spencer's 15-point formula: $\frac{1}{320}\,[4]^2\,[5]\,(-3, 3, 4, \ldots)$,
Spencer's 21-point formula: $\frac{1}{350}\,[5]^2\,[7]\,(-1, 0, 1, 2, \ldots)$.

The output of each filter is regarded as located at the middle term of the (odd) number of terms included in the average. The response functions of these filters are easily obtained. The gains are, respectively,

$$\left|\frac{\sin^2 2\lambda \sin 5\lambda/2}{80 \sin^3 \frac{1}{2}\lambda}\right| |1 + \tfrac{3}{4}\cos \lambda - \tfrac{3}{4}\cos 2\lambda|,$$

$$\left|\frac{\sin^2 5\lambda/2 \sin 7\lambda/2}{175 \sin^3 \frac{1}{2}\lambda}\right| |1 + \cos \lambda - \cos 3\lambda|.$$

Both, and particularly the second, are quite sharp in their shape as the following tabulation of the gain shows

	$\lambda = 0$	$\pi/12$	$\pi/6$	$3\pi/12$	$2\pi/6$	$3\pi/6$	$4\pi/6$	$5\pi/6$	π
Spencer's 15 pt formula	1	0.983	0.809	0.292	0.094	0.000	0.013	0.003	0.000
Spencer's 21 pt formula	1	0.951	0.554	0.080	0.014	0.006	0.003	0.000	0.000

Some indication of the length of the filter required to substantially pass a given narrow band of frequencies and substantially to exclude the remainder of the spectrum can be obtained as follows. (We follow Granger and Hatanaka, 1964.)

We take the band to be passed to be $\lambda \in [-\lambda_0, \lambda_0]$ and study only symmetric filters, thus filters having response

$$\sum_0^N a(k) \cos k\lambda,$$

which we may rewrite in the form

$$\sum_0^N b_j(\cos \lambda)^j.$$

We take $\lambda \in [0, \pi]$ and put $x = \cos \lambda$ so that $x \in [-1, 1]$. We now consider

$$\sum_0^N b_j x^j$$

and require such a polynomial to be small over the range $[-1, x_0]$, where $x_0 = \cos \lambda_0$ and substantial over the range $[x_0, 1]$. We consider in particular the Chebyshev polynomial

$$C_N(x) = \tfrac{1}{2}\{[x + (x^2 - 1)^{1/2}]^N + [x - (x^2 - 1)^{1/2}]^N\},$$

which is

$$\begin{aligned} &\cos N\,(\text{arcos}\, x), && x \in [-1, 1], \\ &\cosh N\,(\text{arcosh}\, x), && |x| > 1. \end{aligned}$$

This polynomial has the following property (Lorentz, 1966, p. 40) that at the end points of the interval $[-1, 1]$ it has the largest possible derivative (in absolute magnitude), namely N^2, for any polynomial in x of degree N which is dominated by unity in $[-1, 1]$. Thus, if we put

$$y = \frac{2x - (x_0 - 1)}{1 + x_0},$$

the polynomial $P_N(x) = C_N(y)$ will be bounded by 1 in modulus in $[-1, x_0]$ and increase as quickly as possible at x_0. If we take $\alpha P_N(x)$, where α is to be the maximum gain of the filter we need, over $[\lambda_0, \pi]$, and λ_0 is small so that x_0 is near to 1 we must have something near to an optimal solution to our problem. Since the ratio $\alpha P_N(1)$ to α is what we are concerned with we take $\alpha = 1$. Now $P_N(1)$ is the response of the filter at $\lambda = 0$, that is, $x = 1$, which is clearly a maximum at $P_N(1)$, which corresponds to $C_N(y_0)$, $y_0 = (3 - x_0)/(1 + x_0) = (3 - \cos\lambda_0)/(1 + \cos\lambda_0) \approx 1 + \frac{1}{2}\lambda_0^2$, if λ_0 is small. Since $\cosh y \approx 1 + \frac{1}{2}y^2$ for y small we see that

$$C_N(y_0) \approx C_N(1 + \tfrac{1}{2}\lambda_0^2) \approx C_N(\cosh\lambda_0) = \cosh N\lambda_0.$$

Thus the ratio of the maximum of our filter at $\lambda = 0$ to its maximum outside $\lambda \in [-\lambda_0, \lambda_0]$ is approximately $\cosh N\lambda_0$. If $\cosh N\lambda_0$ and λ_0 are fixed this gives an equation for N, that is, for the length, $2N + 1$, of the filter with the required properties. For example, if λ_0 is to be $\pi/12$ and $\cosh N\lambda_0$ is to be 10 we have to solve $\cosh(N\pi/12) = 10$, that is, $N\pi/12 = 2.993$, that is, $N = 11.43$ so that the length of the filter must be about 23. This means that we can find a filter with gain which is 10% of the maximum gain at all distances greater than $\pi/12$ from the point of maximum gain and whose length is 23 terms. Similarly if $\cosh N\lambda_0 = (1/0.367)$, $\lambda_0 = \pi/12$ we get $N = 6.34$ so that the length is about 14. A centered 12-term moving average will achieve this ratio at $\pi/12$, also and for that filter the length is 13. The shape of the centered 12-term moving average will not be so sharp as can be seen either from the fact that at $\pi/12$ it is not decreasing as fast as the "optimal" filter considered above or from the location of the nearest zero after λ_0. On the other hand, it eventually becomes very small while the optimal filter has gain which repeatedly comes up to the value at λ_0.

The location of the first zero of the response function of the filter based on the Chebyshev polynomial gives, in some ways, a better indication of the "sharpness" of its shape. This will be at $\arccos y = \pi/(2N)$, that is, $y = \cos(\pi/(2N))$, that is, $x = \cos\lambda = \frac{1}{2}\{\cos(\pi/(2N))(1 + \cos\lambda_0) + \cos\lambda_0 - 1\}$. Then, approximately

$$\cos\lambda = -\frac{1}{4}\frac{\pi^2}{4N^2}(1 + \cos\lambda_0) + \cos\lambda_0.$$

and, if N is large, we reach the approximation

$$(\lambda - \lambda_0)^2 \approx \frac{1}{2}(1 + \cos \lambda_0)\frac{\pi^2}{4N^2};$$

for example, at $\lambda_0 = \pi/12$, $N = 7$ the first zero near to λ_0 is at

$$\lambda_0 + 0.05 \approx \pi/12 + \pi/60.$$

A filter designed to produce a smooth output may be used to remove a "trend," that is, to remove the smooth component from the series, by subtracting the output of the filter from the input. An alternative technique for this purpose in the discrete time case is repeated differencing, that is,

$$x(n) \to \Delta^d\, x(n),$$

where

$$\Delta x(n) = x(n + 1) - x(n).$$

The response of such a filter is $(e^{-i\lambda} - 1)^d$ so that the gain is

$$|2 \sin \tfrac{1}{2}\lambda|^d.$$

There is a considerable literature surrounding the "variate difference method" (see, for example, Tintner, 1940) which has been used to estimate the degree of smoothing required to represent a polynomial trend. We shall not discuss that here, however.

We may consider the problem for continuous time in a similar fashion. Thus we consider filters of the type

$$y(t) \to \int_{-T}^{T} \alpha(s)\, y(t + s)\, ds, \tag{9.2}$$

which have the added property of reproducing exactly $p(t + \mu)$ where $p(t)$ is any polynomial of degree d. In addition we wish to make the output of (9.2) for input with spectrum $f(\lambda)$ have as small a variance as is possible. This means, however, that we must consider not only filters of the form of (9.2) but also their limits in mean square, which may not be of that form. It is fairly evident that the solution to this problem will be achieved by a process of "best linear unbiased" regression for a model of the form

$$y(s) = \sum_0^d \beta_j s^j + x(s), \qquad t - T \le s \le t + T,$$

where $x(s)$ is a stationary process with zero mean and spectrum $f(\lambda)$. We do not wish to discuss the difficulties which arise in connection with the solution of that problem here but refer the reader to Heble (1961) for a treatment of it. Substantially Heble shows that it remains true that as $T \to \infty$ the best linear unbiased procedure and least squares procedures are again equivalent, in the sense that they will give estimates of the β_j whose covariance matrices, after appropriate normalization, approach equality as $T \to \infty$. The least squares procedure is of course the simple one of solving the equations

$$\sum \hat{\beta}_j \int_{-T}^{T} (t+s)^{j+k}\, ds = \int_{-T}^{T} y(t+s)(t+s)^k\, ds, \qquad k = 0, \ldots, d.$$

Then the filter output at t is

$$\sum_0^d \hat{\beta}_j t^j. \tag{9.3}$$

Once again it is convenient to introduce the Legendre polynomials† $P_j(x)$ which satisfy

$$\int_{-1}^{1} P_j(x) P_k(x)\, dx = \delta_j^k \frac{2}{2j+1}, \qquad j, k = 0, 1, \ldots .$$

and the jth of which is a polynomial of degree j. If

$$\phi_j(s) = \left\{\frac{2j+1}{2T}\right\}^{1/2} P_j\left(\frac{s}{T}\right),$$

the $\phi_j(s)$ satisfy

$$\int_{-T}^{T} \phi_j(s)\phi_k(s)\, ds = \delta_j^k.$$

Then we may alternatively write

$$y(s) = \sum_0^d \alpha_j \phi_j(s-t) + x(s), \qquad t - T \le s \le t + T,$$

where the α_j are linear functions of the β_k, $k \le j$, and where now

$$\hat{\alpha}_j = \int_{t-T}^{t+T} y(s)\, \phi_j(s-t)\, ds.$$

so that (9.3) becomes

$$\sum_0^d \phi_j(0) \int_{t-T}^{t+T} y(s)\, \phi_j(s-t)\, ds = \int_{-T}^{T} y(t+s)\, \alpha(s)\, ds, \tag{9.4}$$

$$\alpha(s) = \sum_0^d \phi_j(s)\phi_j(0)$$

To illustrate the closeness of the approximation to the optimal filter by the least squares filter which may hold under favorable circumstances we close this section with an example. This is taken from Blackman (1965) who substantially arrives at the conclusions of the last part of this section.‡

† See, for example, Abramowitz and Stegun (1964, p. 342).

‡ Apparently without recognizing the closeness of his theory to that of regression problems or the relevance of the work of Grenander and Rosenblatt or Heble.

Example

Let $x(t)$ be stationary with a.c. spectrum and density

$$f(\lambda) = \frac{\rho}{\pi(\rho^2 + \lambda^2)}, \qquad \rho > 0, \tag{9.5}$$

and let us take $d = 0$. Thus we are considering almost the simplest case of any interest. Let $h(\exp i\lambda)$ be the response function of the optimal filter in the sense that it is the limit in mean square of a sequence, h_n, of response functions of filters of the form of (9.2) which reproduce a polynomial of degree 0. Thus

$$\lim_{n\to\infty}\int_{-\infty}^{\infty} |h - h_n|^2 f(\lambda)\, d\lambda = 0. \tag{9.6}$$

Let us call σ^2 the minimized variance of output. Then

$$\int_{-\infty}^{\infty} e^{-is\lambda} h(e^{i\lambda}) f(\lambda)\, d\lambda \equiv \sigma^2, \qquad s \in [-T, T]. \tag{9.7}$$

Indeed if $\alpha + \beta = 1$ then $h_1 = \{\alpha h + \beta \exp i\lambda s\}$, $s \in [-T, T]$ is also of the required form and the variance of the output is

$$|\alpha|^2 \int_{-\infty}^{\infty} |h|^2 f(\lambda)\, d\lambda + |\beta|^2 \int_{-\infty}^{\infty} f(\lambda)\, d\lambda + 2\mathfrak{R}\left[\alpha\bar{\beta}\int_{-\infty}^{\infty} h\, e^{-i\lambda s} f(\lambda)\, d\lambda\right].$$

If this is to be at a minimum for $\beta = 0$ then we must have (9.7). On the other hand, if h_1 and h_2 satisfy (9.7) then so does $k = ah_1 + bh_2$ and a, b may be chosen so that

$$\int_{-\infty}^{\infty} e^{-is\lambda} k(e^{i\lambda}) f(\lambda)\, d\lambda \equiv 0, \qquad s \in [-T, T].$$

Now there is a sequence k_n of response functions of filters of the form of (9.2) for which

$$\lim_{n\to\infty} \int |k - k_n|^2 f(\lambda)\, d\lambda = 0.$$

Let $\alpha_n(s)$ be the impulse response corresponding to k_n. Then since

$$\alpha_n(s) \int_{-\infty}^{\infty} k e^{-i\lambda s} f(\lambda)\, d\lambda \equiv 0, \qquad s \in [-T, T]$$

then

$$\int_{-\infty}^{\infty} \bar{k}_n k\, f(\lambda)\, d\lambda \equiv 0,$$

and $k \equiv 0$ almost everywhere with respect to $f(\lambda)$. Thus the solution h of (9.7), which satisfies (9.6), for the h_n being response functions of filters of the form of (9.2), is unique in the sense that any two such solutions differ only on a set over which $f(\lambda)$ integrates to zero (i.e., in the case of $f(\lambda)$ of the form of (9.5), a set of zero Lebesgue measure).

When $f(\lambda)$ is given by (9.5) then the solution to (9.7) is

$$\frac{1}{1+\rho T}\left\{\cos T\lambda + \rho T\,\frac{\sin T\lambda}{T\lambda}\right\},$$

with $\sigma^2 = (1+\rho T)^{-1}$. We leave the reader to check this by substitution in the left side of the expression (9.7).

If we take $\alpha(s) = 2T^{-1}$, corresponding to the least squares regression procedure, we obtain a variance of output

$$\sigma_\infty^2 = \frac{1}{2T}\int_{-2T}^{2T} e^{-\rho|t|}\left(1 - \frac{|t|}{2T}\right) dt = \frac{1}{\rho T}\left\{1 - \frac{1}{2\rho T}(1 - e^{-2\rho T})\right\}$$

The maximum value of σ_∞^2/σ^2 is approximately 1.07 so that little can be lost by using a simple average. Of course this result depends upon the nature of $f(\lambda)$ and the fact that $d = 0$ was being considered.

EXERCISES

1. Let $v(r \exp i\lambda)$ be harmonic within the unit circle and zero almost everywhere upon it and let

$$v(r\,e^{i\lambda}) = \frac{1}{2\pi}\int P_r(\lambda - \theta)\mu(d\theta),$$

where μ is a nonnegative measure on the unit circle. Prove that μ is singular with respect to Lebesgue measure on the circle. *Note:* A proof of this follows from the first corollary on page 38 of Hoffman (1962).

2. Let

$$x(t) = \left(\sum_0^\infty e^{-\rho t_j}\right) - \rho^{-1}\mu, \qquad \rho > 0,$$

where the t_j form the realization of a Poisson process with parameter μ. Determine the best linear predictor of $x(t)$ and discuss its merits.

3. Robinson (1962). Let

$$x(t) = \int_0^\infty \alpha(t-s)\xi(ds)$$

be the canonical factor in the Wold decomposition of a stationary scalar process.

Show that if $\beta(t)$ satisfies

$$\int_0^\infty \beta(t)^2\, dt < \infty, \qquad \int_0^\infty \alpha(t+s)\beta(s)\, ds = 0, \qquad t < 0,$$

then $\beta(t) \equiv 0$, a.e.

4. If $m \geqq 2$, show that

$$\int_\infty^\infty e^{it\tau} \left(\frac{1+i\tau}{1-i\tau}\right)^m d\tau$$

is zero for $t > 0$ and continuous at $t = 0$. Hence show that $\{(1+i\tau)/(1-i\tau)\}^m$ may be approximated arbitrarily closely in mean square (with weighting $F(d\tau)$) by a linear combination of functions $\exp it\tau$ for $t > 0$. Hence show that (in the notation of Section 4) under the mapping $(1+i\tau)/(1-i\tau) \leftrightarrow \exp i\lambda$ the subspace $\mathcal{M}_{-\infty}$ is mapped onto $\mathcal{N}_{-\infty}$. On the other hand $\exp\{t(z-1)/(z+1)\}$ is an entire function for $t > 0$ and is $\exp -t$ at $z = 0$. Use this result to show conversely that the above mapping sends $\mathcal{N}_{-\infty}$ onto $\mathcal{M}_{-\infty}$.

5. Prove Theorem 3‴, using Theorem 3″.

6. Derive the iterative formulas for $\hat{s}^{(0)}(n+1)$ in terms of $\hat{s}^{(0)}(n)$ and $\hat{s}^{(\nu+1)}(n)$ in terms of $\hat{s}^{(\nu)}(n)$ quoted above Theorem 10′.

7. If the filter

$$x(t) \to \int_0^T \alpha(s)x(t-s)\, ds, \qquad \int_0^T |\alpha(s)|\, ds < \infty,$$

predicts exactly a polynomial of degree d with prediction lag μ show that it has response function

$$h(i\tau) = \sum_0^d \frac{(i\tau\mu)^j}{j!} + \lambda^{d+1} h_R(i\tau),$$

where $h_R(z)$ is analytic.

PART TWO

Inference

CHAPTER IV

The Laws of Large Numbers and the Central Limit Theorem

1. INTRODUCTION

We are henceforth going to be concerned with inference problems in time series analysis. This means that we shall be concerned with problems that, we hope, are reasonably near to those of the real world. Needless to say there will have to be compromises. On the one hand, models of the real world, which are known to be false, based at best on models which are evolutive† only because of simple forms of departure from stationarity, will have to be used. On the other hand, it will be impossible to maintain the same generality of treatment so that, for example, it is pointless to consider inference problems only on the supposition that $f(\lambda)$ is integrable. (From this point of view some of the developments of the first part are needlessly general.) This does not mean that a lack of mathematical precision is permissible but merely that additional restrictions will have to be imposed which, although severe from a mathematical viewpoint, will not be so from a practical viewpoint, it is hoped.

We confine our inference theory almost entirely to the discrete time case. This is not by any means necessary but it seems advisable. For unless we do this the space required for a clear treatment will be very large. Moreover, almost all analyses of time series data are now effected by digital computers. In this case sampling of a continuous record is necessary and the discrete time case is relevant. This does not mean that the treatment of continuous time in earlier chapters was unnecessary for usually a continuous-time phenomenon will underlie the data analyzed.

We shall not attempt to give detailed instructions as to how the necessary computations can be effected (though we shall often discuss computational problems and details).

Because we are dealing with problems of the real world we shall, where possible, lower the mathematical level somewhat as compared to the first

† That is, nonstationary.

part in the hope that this part will then be more widely accessible. This will be made easier by the fall off in generality discussed above and because of our restriction to discrete time situations. In addition we shall proceed somewhat more slowly.

2. STRICTLY STATIONARY PROCESSES. ERGODIC THEORY

We recall that a (vector) process is strictly stationary if the joint distribution of $x(n_1 + m), x(n_2 + m), \ldots, x(n_k + m)$ depends only upon $n_1, n_2, \ldots, n_k$ and not upon m, for each $k, n_1, \ldots, n_k$ and m. We shall state without proof here certain theorems of a central nature concerning the strong law of large numbers. We do not prove them because the proofs are long and not easy, are readily available† elsewhere and cannot be simplified, at least by this author.

It will be recalled (Chapter I, Section 1) that the random variables $x_j(n)$ can be regarded as being defined over one and the same measure space $(\Omega, \mathcal{A}, P)$. The class of measurable sets in this space, namely, $\mathcal{A}$, includes all cylinder sets

$$\{\omega \mid (x_{j(k)}(n_k, \omega), k = 1, \ldots, m) \in B\},$$

where $(x_{j(k)}(n_k, \omega), k = 1, \ldots, m)$ is the point in R^m with these coordinates and B is a Borel set in R^m. Indeed $\mathcal{A}$ is the smallest Borel field of sets including all such cylinder sets. For these cylinder sets it is meaningful to define the shift operator $S \to TS$ where, for a set S of the form

$$S = \{\omega \mid x_{j(1)}(n_1, \omega) < a_1, x_{j(2)}(n_2, \omega) < a_2, \ldots, x_{j(m)}(n_m, \omega) < a_m\},$$

we have

$$TS = \{\omega \mid x_{j(1)}(n_1 + 1, \omega) < a_1, \ldots, x_{j(m)}(n_m + 1, \omega) < a_m\}.$$

This definition extends to all sets in $\mathcal{A}$. It is easily seen that T is measure preserving so that S and TS always have the same probability content. Evidently T^{-1} is also well defined, in the same way, and is measure-preserving. *Thus we can define a shift operator, which we also call T, on the space of random variables (measurable functions) over Ω via $Tf(\omega) = f(T^{-1}\omega)$. If $Tf = f$ then we say that f is invariant and similarly for S.*

We observe that $Tx_j(n, \omega) = x_j(n + 1, \omega)$ since, for all a,

$$\{\omega \mid Tx_j(n, \omega) \le a\} = \{\omega \mid x_j(n, T^{-1}\omega) \le a\} = T\{\omega \mid x_j(n, \omega) \le a\}$$
$$= \{\omega \mid x_j(n + 1, \omega) \le a\}.$$

† See, for example, Doob (1953) and Billingsley (1965).

We put

$$\bar{x}_N = N^{-1} \sum_1^N x(n).$$

We now have the ergodic theorem.

Theorem 1. *If* $x(n)$ *is strictly stationary with* $\mathcal{E}\{|x_j(n)|\} < \infty$, $j = 1, \ldots, p$ *then there is a vector,* $\hat{x}$, *of invariant random variables for which* $\mathcal{E}\{|\hat{x}_j|\} < \infty$, *and*

$$\lim_{N\to\infty} \bar{x}_N = \hat{x}, \quad \text{a.s.},$$

$$\mathcal{E}\{x(n)\} = \mathcal{E}(\hat{x}).$$

The only defect in this theorem lies in the fact that $\hat{x}$ is not necessarily a constant with probability 1. Evidently this requirement will be satisfied if the *only* invariant random variables are constants with probability 1 which is the same as saying that the only sets S for which $S = TS$ (in the sense that the two sides differ by a set of probability zero) have probability content zero or 1. *If this is so the process* $\{x(n)\}$ *is said to be ergodic.* (The process and the shift operator, T, are then also sometimes said to be "metrically transitive.")

The property of ergodicity is not observationally verifiable from a single history. This is illustrated by the example of a jump in the spectrum of a scalar process, which produces a component $2\Re\{z(d\lambda_0) \exp -in\lambda_0\}$ in $x(n)$. Now, as we show almost immediately, $|z(d\lambda_0)|$ is an invariant random variable. However, from one history one cannot tell whether it is almost surely a constant. To see that $|z(d\lambda_0)|$ is invariant consider $\lambda_1 < \lambda_0 < \lambda_2$ where λ_1, λ_2 are continuity points of $F(\lambda)$. It follows from Theorem 3″ in Chapter II that

$$T\big(z(\lambda_2) - z(\lambda_1)\big) = \int_{\lambda_1}^{\lambda_2} e^{-i\lambda} z(d\lambda),$$

so that

$$\big\|T\big(z(\lambda_2) - z(\lambda_1)\big) - e^{-i\lambda}\big(z(\lambda_2) - z(\lambda_1)\big)\big\|^2 = \int_{\lambda_1}^{\lambda_2} 4 \sin^2 \tfrac{1}{2}(\lambda - \lambda_0)\, F(d\lambda),$$

converges to zero as λ_1 and λ_2 approach λ_0. Thus

$$Tz(d\lambda_0) = \exp(-i\lambda_0)\, z(d\lambda_0)$$

and $|z(d\lambda_0)|^2$ is invariant as we have said. In the Gaussian case the absence of jumps in the spectrum is also sufficient for ergodicity.† More generally we have the following result. Let

$$x(n) = \sum_{j=-\infty}^{\infty} e^{in\lambda_j} z_j, \quad \mathcal{E}(z_j \bar{z}_k) = \delta_j^k f_j, \quad z_{-j} = \bar{z}, \quad \lambda_{-j} = -\lambda_j, \quad \sum_{-\infty}^{\infty} f_j < \infty.$$

† For a proof of the results cited in the present paragraph we refer the reader to Rozanov (1967).

Then in the first place for $x(n)$ to be ergodic $|z_j|$ must be, almost surely, a constant for all j. (This immediately rules out the Gaussian case.) Thus randomness enters into $x(n)$, essentially, only through a random determination of the phasing of the oscillations. In the second place if S is any finite set of λ_j for which there are integers m_j so that

$$\sum_{\lambda_j \in S} m_j \lambda_j = 0, \qquad (\text{mod } 2\pi)$$

then also we must have

$$\sum_{\lambda_j \in S} m_j \lambda_j = \phi \, (\text{mod } 2\pi), \qquad z_j = |z_j| e^{i\theta_j},$$

where ϕ is almost surely a constant. The necessity is again fairly evident since for

$$\eta = \exp i \left\{ \sum_{\lambda_j \in S} m_j \theta_j \right\}$$

we have

$$T\eta = \exp i \left\{ \sum_{j \in S} m_j(\theta_j + \lambda_j) \right\} = \eta.$$

A weaker condition than ergodicity which implies it and whose meaning is somewhat more apparent is the condition

$$\text{(2.1)} \qquad \lim_{n \to \infty} P(A \cap T^{-n}B) = P(A)P(B), \qquad A, B \in \mathcal{A}$$

If this condition holds then the $\{x(n)\}$ *process is said to be mixing.* That the mixing property implies ergodicity is seen by taking $A = B =$ invariant when we obtain $P(A)^2 = P(A)$ so that $P(A) = 0$ or 1. Thus, loosely, provided events determined by sets of $x(n)$ far apart in time are near to independent then the sample means of our process converge with probability 1 to the true mean. A stronger condition than (2.1) is the "uniform mixing condition" which we now define. We consider all events (i.e., all sets in $\mathcal{A}$) depending only upon $x(n)$ for $n \leq p$. These events constitute a Borel field $\mathcal{B}_p$. Similarly $\mathcal{F}_q$ is to be the Borel field of events determined by $x(n)$, $n \geq q$. *Let there exist a positive function g defined upon the integers so that* $g(n) \to 0$ *as* $|n| \to \infty$ *and*

$$\text{(2.2)} \qquad |P(B \cap F) - P(B)P(F)| < g(q - p), \qquad B \in \mathcal{B}_p,\ F \in \mathcal{F}_q$$

Then $\{x(n)\}$ *is said to satisfy a uniform mixing condition.* Thus we now imply that events separated in time by a distance $(q - p)$ approach independence uniformly in that the departure of the left-hand side of (2.2) from zero depends only upon $(q - p)$ and not on the particular B and F considered. This condition obviously implies (2.1) and thus the ergodicity of $\{x(n)\}$. The uniform mixing condition is rather strong but also rather appealing. For one is inclined to believe that, shall we say, the behavior of the waves that beat upon the shores some age ago is totally unrelated to that of the waves that beat today and that this is true of all phenomena which those waves determined. Whether or not we are justified in thinking this we shall have to make considerable use of this condition, introduced into statistical considerations by M. Rosenblatt, in this book. However, we shall not use it immediately, having introduced it now so as to connect it with the mixing condition (1) and thence with ergodicity.†

If $x(n)$ is strictly stationary and ergodic, consider $x(n)x(n+m)'$. Fixing m and allowing n to vary this constitutes a sequence of matrices of random variables which again constitutes a vector strictly stationary process. We assume that the process is second-order stationary also so that $\Gamma(m) = \mathcal{E}\{x(n)x(n+m)'\}$ is well defined. If $\{x(n)\}$ is ergodic then so is

$$\{x(n)x(n+m)'\}$$

for the shift operator for this process agrees with that for the $\{x(n)\}$ process on the domain in which it is defined (i.e., sets determined by conditions on the $x_j(n)x_k(n+m)$) so that the only invariant sets have measure 0 or 1.

Theorem 2. *If $\{x(n)\}$ is strictly stationary and ergodic and $\mathcal{E}\{|x_j(n)|\} < \infty$, then*

$$\lim_{N\to\infty} \frac{1}{N} \sum_1^N x(n) = \mathcal{E}\big(x(n)\big) \quad \text{a.s.}$$

Also if

$$\mathcal{E}\{x_j(n)^2\} < \infty,$$

then

$$\lim_{N\to\infty} \frac{1}{N} \sum_1^N x(m)x(m+n)' = \Gamma(n) \quad \text{a.s.}$$

In particular we see that Theorem 2 holds if $x(n)$ is mixing. It also holds for any Gaussian stationary process with continuous spectral distribution function. The result extends to other functions of a realization than

$$x(m)x(m+n)'.$$

If the process $x(n)$ is mixing then so is any new process defined by

$$y(n, \omega) = f(T^{-m}\omega)$$

where f is a vector-valued measurable function upon the measure space

† Rozanov (1967) discusses the meaning of this condition (which he calls complete regularity) in relation to Gaussian processes and establishes a number of interesting relations. We use the condition mainly in circumstances where the Gaussian assumption is not made.

$(\Omega, \mathcal{A}, P)$. Indeed if A and B are sets of the form

$$\{\omega \mid (y_{j(k)}(n(k), \omega), k = 1, \ldots, m) \in C\}$$
$$= \{\omega \mid (f_{j(k)}(T^{-n(k)}\omega), k = 1, \ldots, m) \in C\}$$

where C is a Borel set in R^m then (2.1) is immediately seen to hold, while the class of measurable sets $\mathcal{A}_y$ in the measure space $(\Omega_y, \mathcal{A}_y, P_y)$ corresponding to $y(n)$ is determined by such sets A, B. It is not difficult to show that a sequence of independent identically distributed random vectors, $\epsilon(n)$, with zero mean and finite covariance matrix is mixing. (We leave this as an exercise to the chapter.) It follows that if

$$x(n) = \sum_{-\infty}^{\infty} A(j)\epsilon(n-j), \qquad \sum_{-\infty}^{\infty} \|A(j)\|^2 < \infty \tag{2.3}$$

then $x(n)$ is mixing also and thus is ergodic. Processes of the form (2.3) are sometimes called "linear processes." We shall use this term in a more special sense, however. (We know that if $x(n)$ is Gaussian and has absolutely continuous spectrum it is of this form.)

Theorem 3. *If $x(n)$ is given by (2.3) where the $\epsilon(n)$ are independent and identically distributed then $x(n)$ is mixing and therefore ergodic.*

3. SECOND-ORDER STATIONARY PROCESSES. ERGODIC THEORY

The purposes of this section are to prove theorems analogous to those of the previous section but using hypotheses only upon the moments of the second order, so far as is possible. We begin by proving theorems relating to the means, modify these so as to apply to covariance and conclude by discussing certain statistics which are building blocks in the estimation of the spectrum and are basic to many inferential problems.

Theorem 4. *Let $x(n)$ be stationary† with*

$$x(n) = \int_{-\pi}^{\pi} e^{-in\lambda} z(d\lambda), \qquad \Gamma(n) = \mathcal{E}(x(m)x(m+n)') = \int_{-\pi}^{\pi} e^{in\lambda} F(d\lambda).$$

Then

$$\lim_{N\to\infty} \bar{x}_N = z(0) - z(0-),$$

$$\lim_{N\to\infty} \frac{1}{N}\sum_{0}^{N-1} \Gamma(n) = F(0) - F(0-) = \mathcal{E}\{(z(0) - z(0-))(z(0) - z(0-))^*\}.$$

† Throughout the remainder we will continue as earlier in the book to use stationary to mean second-order stationary.

Proof. Let $\phi(\lambda) = 1$ for $\lambda = 0$; $= 0$, $\lambda \neq 0$. Then

$$\lim_{N\to\infty} \int_{-\pi}^{\pi} \left| \frac{1}{N} \sum_{1}^{N} e^{-in\lambda} - \phi(\lambda) \right|^2 F(d\lambda)$$

$$= \lim_{N\to\infty} \int_{-\pi}^{\pi} \left\{ \frac{\sin^2 \frac{1}{2}N\lambda}{N^2 \sin^2 \frac{1}{2}\lambda} + \phi(\lambda) - 2\phi(\lambda) \right\} F(d\lambda) = 0$$

by dominated convergence.

Thus

$$\underset{N\to\infty}{\text{l.i.m}}\, x_N = \underset{N\to\infty}{\text{l.i.m}}\, \frac{1}{N} \sum_{1}^{N} \int_{-\pi}^{\pi} e^{-in\lambda}\, z(d\lambda) = \underset{N\to\infty}{\text{l.i.m}} \int_{-\pi}^{\pi} \left\{ \frac{1}{N} \sum_{1}^{N} e^{-in\lambda} \right\} z(d\lambda)$$

$$= \int_{-\pi}^{\pi} \phi(\lambda)\, z(d\lambda) = z(0) - z(0-)$$

This has the stated covariance matrix. For the last statement we observe that

$$\lim_{N\to\infty} \frac{1}{N} \sum_{0}^{N-1} \Gamma(n) = \lim_{N\to\infty} \int_{-\pi}^{\pi} \frac{1}{N} \sum_{0}^{N-1} e^{in\lambda} F(d\lambda) = \int_{-\pi}^{\pi} \phi(\lambda) F(d\lambda)$$

$$= F(0) - F(0-),$$

by dominated convergence, again.

Theorem 4 shows immediately that if $x(n)$ is a stationary process with mean vector μ then the sample mean vector $\bar{x}_N$ converges in mean square to μ if and only if there is no jump in the spectrum at the origin. This is often called the "statistical ergodic theorem" and is attributed to J. von Neumann, as distinct from the "individual ergodic theorem" of the previous section.

If $x(n)$ is strictly stationary, we know that $\bar{x}_N$ has a limit almost surely. The last theorem tells us that $\bar{x}_N$ converges in mean square to a random variable, $z(0) - z(0-)$. Since both mean square convergence and a.s. convergence imply convergence in probability, $\bar{x}_N$ must converge a.s. to $z(0) - z(0-)$.

Corollary 1. *If $x(n)$ is strictly stationary (as well as second-order stationary) with mean vector 0 then $\bar{x}_N$ converges to 0 a.s. if and only if there is no jump in the spectrum at the origin.*

A completely analogous discussion may be given for the quantity

$$\frac{1}{N} \sum_{1}^{N} x(n) e^{in\lambda}$$

and it may be shown that this converges in mean square and a.s. to the random variable $\{z(\lambda) - z(\lambda-)\}$ and thus it converges in mean square and a.s. to zero if and only if there is no jump in the spectrum at λ.

We now prove a theorem due to Doob (1953) which gives conditions under which convergence in mean square can be replaced by almost sure convergence for second-order stationary processes.

Theorem 5. *Let $x(n)$ be as in Theorem 4 and have $\mathcal{E}(x(n)) \equiv 0$. Then $\bar{x}_N$ converges with probability one to zero if there are constants $K > 0$, $\alpha > 0$, so that*

$$\frac{1}{N^2}\sum_{n,m=0}^{N-1}\sum \gamma_j(n-m) = \frac{1}{N}\sum_{-N+1}^{N-1} \gamma_j(n)\left(1 - \frac{|n|}{N}\right) \le KN^{-\alpha}. \tag{3.1}$$

Clearly the theorem needs to be proved only in the scalar case. The expresssion (3.1) is then the variance of $\bar{x}_N$. Choose $\beta \ge 1$ so that $\beta\alpha > 1$. Then if $N \ge M^\beta$ we have $\mathcal{E}\{\bar{x}_N{}^2\} \le KM^{-\alpha\beta}$. If $\epsilon > 0$ and $N(M)$ is the smallest integer not smaller than M^β, by Chebyshev's inequality

$$\sum_{M=1}^{\infty} P\{|\bar{x}_{N(M)}| \ge \epsilon\} \le K \sum_{M=1}^{\infty} \epsilon^{-2} M^{-\alpha\beta} < \infty.$$

Thus by the Borel-Cantelli lemma only finitely many of the events occur, whose probabilities are summed on the left. Thus $|\bar{x}_{N(m)}| < \epsilon$ with probability 1 for M large enough and since ϵ is arbitrary $\bar{x}_{N(m)}$ converges to zero with probability 1. Moreover,

$$\mathcal{E}\left\{\max_{N(M)\le N \le N(M+1)} \left|\bar{x}_N - \frac{N(M)}{N}\bar{x}_{N(M)}\right|^2\right\} \le N(M)^{-2}\,\mathcal{E}\left\{\left[\sum_{N(M)+1}^{N(M+1)} |x(n)|\right]^2\right\}$$

$$= N(M)^{-2} \sum_{n,m=N(M)+1}^{N(M+1)}\sum \gamma(n-m) \le \frac{\{N(M+1) - N(M)\}^2\gamma(0)}{N(M)^2} \le \frac{K_1}{M^2}$$

since

$$\frac{N(M+1) - N(M)}{N(M)} \le \frac{(M+1)^\beta + 1 - M^\beta}{M_\beta} \le (1 + M^{-1})^\beta - 1 + M^{-1} \le \frac{K}{M}.$$

Again applying Chebyshev's inequality and the Borel-Cantelli lemma we see that $|\bar{x}_N - (N(M)/N)\bar{x}_{N(M)}|$ converges to zero with probability 1 as $N \to \infty$, with $N(M) \le N \le N(M+1)$. Indeed

$$\sum_{m=1}^{\infty} P\left(\max_{N(M)\le N\le N(M+1)} |\bar{x}_N - (N(M)/N)\bar{x}_{(NM)}|^2 \ge \epsilon\right) \le K_1 \sum_{m=1}^{\infty} \epsilon^{-2}M^{-2} < \infty$$

and the result follows as before. Thus for $N(M)$ satisfying the same relation $\max |\bar{x}_N - \bar{x}_{N(m)}|$ also converges to zero with probability 1 (since, for

$N(M) \leq N \leq N(M+1)$, $N(M)/N \to 1$) and we have established what we wish.

Corollary 2. *Let $x(n)$ be as in Theorem 5 with $\gamma_j(n) = O(n^{-\beta})$, $\beta > 0$ as $n \to \infty, j = 1, \ldots, p$, then $\bar{x}_N$ converges to zero with probability* 1.

Proof.

$$N^{-1} \sum_{0}^{N-1} \gamma_j(n) = N^{-1} \sum{}' \gamma(n) + N^{-1} \sum{}'' \gamma(n)$$

where in Σ' the sum is over $n = 0, \ldots, [N^{1-\alpha}]$ and in Σ'' from $([N^{1-\alpha}] + 1)$ to N. Thus the left side is not greater than

$$\gamma_j(0)N^{-\alpha} + (1 - N^{-\alpha}) \overline{\lim_{n > N^{1-\alpha}}} |\gamma(n)| \leq \gamma_j(0)N^{-\alpha} + c(1 - N^{-\alpha})\, N^{(\alpha-1)\beta}$$

If we choose $\alpha > 0$ so that $\beta(1 - \alpha) > 0$ we thus see that

$$N^{-1} \sum_{0}^{N-1} \gamma_j(n) < KN^{-\gamma}, \qquad \gamma > 0$$

and evidently the condition of Theorem 5 is satisfied.

Corollary 3. *Let $x(n)$ be as in Theorem 5 and*

$$\int_0^\delta F_j(d\lambda) = O(\delta^\beta), \qquad \beta > 0, \quad \text{as} \quad \delta \to 0, j = 1, \ldots, p$$

then $\bar{x}_N$ converges to zero with probability 1.

Proof

$$\mathcal{E}\left\{\left(N^{-1} \sum_{1}^{N} x_j(n)\right)^2\right\} = 2\int_0^\pi N^{-1} L_N(\lambda)\, F(d\lambda),$$

where $L_N(\lambda)$ is Fejérs kernel $N^{-1}\{(\sin \frac{1}{2}N\lambda)/(\sin \frac{1}{2}\lambda)\}^2$. The right side is not greater than

$$2\int_A F(d\lambda) + 2N^{\alpha-2} \int_B N^{1-\alpha} L_N(\lambda)\, F(d\lambda), \qquad 0 < \alpha < 1,$$

where A is the interval $[0, N^{-\frac{1}{2}\alpha}]$, and B the interval $[N^{-\frac{1}{2}\alpha}, \pi]$. This expression is not less than

$$K_1 N^{-\frac{1}{2}\alpha\beta} + K_2 N^{\alpha-2};$$

since, for $\lambda \in B$,

$$\frac{\sin^2 \frac{1}{2}N\lambda}{N^\alpha \sin^2 \frac{1}{2}\lambda} \leq \frac{1}{N^\alpha \sin^2 (\frac{1}{2}N^{-\frac{1}{2}\alpha})},$$

which approaches 4 as $N \to \infty$. If we put $\alpha = 4/(2 + \beta)$ we see that the variance of the mean of $x_j(n)$ is dominated by $KN^{-\gamma}$, $\gamma = 2\beta/(2 + \beta) > 0$. Thus the result is established.

The third corollary shows that the spectral density may actually be infinite at the origin and still the sample mean converge with probability 1. It cannot, necessarily, be infinite in an arbitrary way since the density

$$f(\lambda) = |\lambda|^{-1} \log (1 + |\lambda|)$$

does not satisfy the condition of the theorem. However, since a jump in $F(\lambda)$ at $\lambda = 0$ certainly prevents the conclusion of the corollary from holding (as Theorem 4 shows) and the condition on the behavior of $f(\lambda)$ at $\lambda = 0$ may be very radical (e.g., increasing as $\lambda^{-\alpha}$, $0 \leq \alpha < 1$) the result is a rather strong one.

Corollary 4. *If $x(n)$ is wide-sense stationary with zero mean and absolutely continuous spectrum, continuous at the origin, then*

$$\lim_{N\to\infty} N\mathcal{E}(\bar{x}_N \bar{x}_N') = 2\pi f(0).$$

This follows immediately from the fact that N by the covariance matrix for $\bar{x}_N$ is the Nth Cesaro mean of the Fourier series of $2\pi f(\lambda)$. (See the Mathematical Appendix.)

We turn next to the study of the quantities

$$\tilde{C}(n) = N^{-1} \sum_{m=1}^{N} x(m)x(m+n)',$$

which constitute natural estimators for $\Gamma(n)$. We have inserted the tilde since when N observations are available we shall be able to form only

$$C(n) = (N-n)^{-1} \sum_{m=1}^{N-n} x(m)x(m+n)', \qquad n \geq 0,$$

$$C(-n) = C(n)',$$

and we prefer to retain the simpler notation for these quantities. We call the elements of $C(n)$ serial covariances, and autocovariances for diagonal elements.† However, for the present purposes we may as well preserve the simplicity involved in fixing the number of terms in the sum. For any fixed n, or any fixed set of values of this parameter, the ergodic properties of $C(n)$ are the same as those of $\tilde{C}(n)$. If $x(n)$ has finite fourth moment

$$N^{-1}\{x(N-j)x(N+n-j)\}$$

converges in mean square to zero so that the asymptotic expressions for the covariances of the elements of $\tilde{C}(n)$ as well as the mean square convergence properties of the matrix carry over and apply also to $C(n)$.

† Later we shall use the same symbols, $c_{ij}(n)$, $C(n)$ and so on, for these quantities when computed from $x(n) - \bar{x}_N$, that is when mean corrections are made.

We are led to introduce the quantities

$$u_{ij}^{(n)}(m) = x_i(m)\, x_j(m+n) - \gamma_{ij}(n)$$

and to study the ergodic behavior of the elements of $\tilde{C}(n)$, as $N \to \infty$, through these quantities, which evidently have zero mean. We know that $\tilde{c}_{ij}(n)$ converges with probability 1 to $\gamma_{ij}(n)$ if the $x(n)$ process is strictly stationary and ergodic, or, more strongly, mixing but here we wish to investigate mean square convergence and convergence with probability 1 using only the lower moments of the process. Clearly for the former we need the existence of fourth moments for $x(n)$ and this we now assume. *If the fourth moments behave as they would for a stationary sequence, that is, if*

$$\begin{aligned}(3.2)\quad \mathcal{E}\big(x_i(m)x_j(m+n_1)x_k(m+n_2)x_l(m+n_3)\big)\\ \equiv \mathcal{E}\big(x_i(0)x_j(n_1)x_k(n_2)x_l(n_3)\big),\\ m = 0, \pm 1, \ldots; \qquad i, j, k, l = 1, \ldots, p.\end{aligned}$$

then we say that the $x(n)$ process is stationary to the fourth order, and similarly for stationarity to any order. We recall (see Chapter I, Equation 5.1) that (3.2) is

$$\begin{aligned}\gamma_{ij}(n_1)\gamma_{kl}(n_3 - n_2) + \gamma_{ik}(n_2)\gamma_{jl}(n_3 - n_1) + \gamma_{il}(n_3)\gamma_{jk}(n_2 - n_1)\\ + k_{ijkl}(m, m+n_1, m+n_2, m+n_3).\end{aligned}$$

If $x(n)$ is fourth-order stationary, the last term does not depend on m and we may replace it by $k_{ijkl}(0, n_1, n_2, n_3)$. If $i = j = k = l$, we write $k_i(0, n_1, n_2, n_3)$ for simplicity. Of course, if $x(n)$ is Gaussian, the fourth cumulant vanishes.

We now consider the typical covariance of the elements of $\tilde{C}(n)$. We leave it as an exercise to this chapter for the reader to check that the covariance between $\tilde{c}_{ij}(m)$ and $\tilde{c}_{kl}(n)$ is

$$\begin{aligned}(3.3)\quad N^{-1}\sum_{u=-N+1}^{N-1}\left(1 - \frac{|u|}{N}\right)\{\gamma_{ik}(u)\gamma_{jl}(u+n-m) + \gamma_{il}(u+n)\gamma_{jk}(u-m)\\ + (N-|u|)^{-1}\sum_v{}' k_{ijkl}(v, v+m, v+u, v+u+n)\}, \quad i, j, k, l = 1, \ldots, p,\end{aligned}$$

wherein the index v for the summation $\sum'$ runs over values such that $1 \le v + u \le N$.

If

$$(3.4)\qquad x(n) = \sum_{-\infty}^{\infty} A(j)\epsilon(n-j), \qquad \sum_{-\infty}^{\infty} \|A(j)\| < \infty$$

and the $\epsilon(n)$ are independent and identically distributed with finite covariance matrix G, we say that $x(n)$ is generated by a linear process. If we replace

$\sum \|A(j)\| < \infty$ *with* $\sum \|A(j)\|^2 < \infty$, *we say that* $x(n)$ *is generated by a generalized linear process.* This terminology is cumbersome (and somewhat unusual), but some such terminology is needed. In any case, a generalized linear process, and thus a linear process, is mixing and thus ergodic.

We shall now prove the following theorem:

Theorem 6. *Let* $x(n)$ *be second-order stationary. A necessary and sufficient condition that the* $c_{ij}(n)$ *converge in mean square to* $\gamma_{ij}(n)$ *is that* (3.3) *converge to zero. If* $x(n)$ *is fourth-order stationary and* (3.3) *is* $O(N^{-\alpha})$, $\alpha > 0$, *convergence also takes place almost surely. If the fourth-order spectrum of* $x(n)$ *exists and is absolutely continuous, mean-square convergence holds for all* i, j, n *if and only if there are no jumps in the spectrum. If* $x(n)$ *is also strictly stationary with a.c. fourth-order spectrum, the absence of jumps in* $F(\lambda)$ *is necessary and sufficient for almost sure convergence. In particular this condition is necessary and sufficient for both mean square and almost sure convergence in the Gaussian case. If* $x(n)$ *is generated by a generalized linear process, almost sure convergence holds, and if it is generated by a linear process and the fourth cumulants of the* $\epsilon_j(n)$ *are finite then mean square convergence also holds.*

Proof. We observe first that all statements for $c_{ij}(n)$ follow immediately from those for $\tilde{c}_{ij}(n)$ by what was said above. The first assertion is obvious and the sufficiency of the condition that (3.3) be $O(N^{-\alpha})$, $\alpha > 0$, when $x(n)$ is fourth-order stationary follows immediately from Theorem 5, replacing $x(m)$ by $u_{ij}(m)$, for i, j, n fixed. Next take $m = n$, $i = k$, $j = l$. Then (3.3) becomes

$$(3.5)\quad N^{-1}\sum_{-N+1}^{N-1}\left(1-\frac{|u|}{N}\right)\{\gamma_i(u)\gamma_j(u)+\gamma_{ij}(u+n)\gamma_{ji}(u-n)$$
$$+(N-|u|)^{-1}\sum_v{}' k_{ijij}(v, v+n, v+u, v+u+n)\}.$$

If $i = j$ and $m = n$, then the first two terms constitute $2N^{-1}$ by the Cesaro mean of the Fourier series with Fourier coefficients $\gamma_i(u)^2$. These are just the Fourier coefficients of the convolution

$$F_i(\lambda)^{*2} = \int_{-\infty}^{\infty} F_i(\lambda-\theta)F_i(d\theta).$$

Thus these first two terms converge to twice the jump in $F_i(\lambda)^{*2}$ at the origin. (To see this replace $F(\lambda)$ by $F_i(\lambda)^{*2}$ in the proof of Theorem 4.) However, that jump is evidently just the sum of squares of the jumps in $F_i(\lambda)$. When $i \neq j$ but $m = n$ the first two terms involve the jump in $F_i{*}F_j$ at the origin and the jump in the convolution of $F_{ij}\exp-(i2n\lambda)$ with F_{ji} at the origin. Since F_{ij} clearly cannot jump if neither F_i nor F_j does so (because of the Hermitian nonnegative nature of the increments in $F(\lambda)$) then, provided the term involving the fourth cumulant converges to zero, a necessary and sufficient condition for mean-square convergence of all $\tilde{c}_{ij}(n)$ is the condition that there be no jumps in $F(\lambda)$. In the fourth-order stationary case the fourth cumulant term in (3.3) becomes

$$N^{-1}\sum_{u=-N+1}^{N-1}\left(1-\frac{|u|}{N}\right)k_{ijkl}(0, m, u, u+n),$$

which for $m = n$, $i = k$, $j = l$ depends upon the behavior of B_{ijij} along the line $(\lambda_2 + \lambda_3) = 0$ (see Chapter II, Section 8). If B_{ijij} is a.c. certainly the last term will vanish asymptotically. Thus a necessary and sufficient condition for mean square convergence when B_{ijkl} is a.c. is then that there be no jumps in $F(\lambda)$. If $x(n)$ is strictly stationary as well as second-order stationary and the fourth-order spectrum is a.c. then evidently the absence of jumps in $F(\lambda)$ is necessary for almost sure convergence. But it must also be sufficent for $\tilde{c}_{ij}(n)$ converges almost surely by Theorem 1 and since it converges in mean square to $\gamma_{ij}(n)$ it must converge almost surely to that quantity. If $x(n)$ is a generalized linear process then we already know that it is ergodic.†
Also

$$\Gamma(n) = \sum_{-\infty}^{\infty} A(j)GA'(j+n)$$

and

$$\sum_{-\infty}^{\infty}\|\Gamma(n)\| \leq \|G\|\sum_{-\infty}^{\infty}\|A(j)\|)^2 < \infty,$$

and thus $F(\lambda)$ is a.c. and $f(\lambda)$ is continuous (with an absolutely convergent Fourier series). Moreover, if the fourth cumulants of the $\epsilon(n)$ are finite, so are the fourth cumulants of the $x(n)$, and if $x(n)$ is generated by a linear process

$$\sum\sum_{q,r,s}\sum |k_{ijkl}(0, q, r, s)| < \infty.$$

Indeed

$$k_{ijkl}(0, q, r, s) = \sum\sum_{bcde=1}^{p}\sum\sum \kappa_{bcde}\sum_{u} a_{ib}(u)a_{jc}(u+q)a_{kd}(u+r)a_{le}(u+s),$$

where κ_{bcde} is the fourth cumulant between $\epsilon_b(n)$, $\epsilon_c(n)$, $\epsilon_d(n)$, $\epsilon_e(n)$. Thus from (3.3) we see that the $\tilde{c}_{ij}(n)$ converge in mean square to $\gamma_{ij}(n)$.

This completes the proof.

We observe that the condition that there be no jumps in $F(\lambda)$ is equivalent to the condition that

$$\lim_{N\to\infty}\frac{1}{N}\sum_{1}^{N}\gamma_j(n)^2 = 0, \qquad j = 1, \ldots, p,$$

† Because we have not proved Theorem 1 and because of the importance of linear processes in the remainder of the book we have outlined a separate proof of the a.s. convergence of the $c_{ij}(n)$ in the exercises to the chapter.

since $F_j^{*2}(\lambda)$ has a jump at the origin if and only if $F_j(\lambda)$ has jumps and

$$\lim_{N\to\infty} \frac{1}{N} \sum_{-N+1}^{N-1} \gamma_j(n)^2 \left(1 - \frac{|j|}{N}\right) = \lim_{N\to\infty} \int_{-\pi}^{\pi} \frac{\sin^2 \frac{1}{2}N\lambda}{N^2 \sin^2 \frac{1}{2}\lambda} F_j^{*2}(d\lambda),$$

whose convergence to zero is equivalent to the stated condition.

Theorem 6 can hardly be called neat! Some of the conditions are grossly too strong and the justification for them is the difficulty in writing down weaker conditions in a simple form. One's impression is that the requirements of convergence in mean square and with probability 1, when added onto the initial requirement of wide-sense stationarity, should not materially alter one's willingness to adopt the model for an analysis of data.

In the case of a linear process the asymptotic form of the expression (3.3) for the variances and covariances is somewhat neater since the fourth cumulant term approaches

$$\sum\sum_{pqrs}\sum\sum \kappa_{pqrs}\left\{ \sum_{u=-\infty}^{\infty} \alpha_{ip}(u)\alpha_{jq}(u+m) \sum_{u=-\infty}^{\infty} \alpha_{kr}(u)\alpha_{ls}(u+n)\right\}, \tag{3.6}$$

where p, q, r, s run over a set of indices corresponding to the columns of the $A(j)$ while κ_{pqrs} is the fourth cumulant between $\epsilon_p(n)$, $\epsilon_q(n)$, $\epsilon_r(n)$ and $\epsilon_s(n)$ When the $x(n)$ and $\epsilon(n)$ processes are scalar we obtain

$$N \operatorname{var}\{c(n)\} \to 2\pi \int_{-\pi}^{\pi} f^2(\lambda)(1 + e^{i2n\lambda})\, d\lambda + 2\pi\kappa\gamma^2(n),$$

where κ is the fourth cumulant of the $\epsilon(n)$. A similar expression can be obtained for the general linear process but it is of little value and we omit it here.

For more detail concerning the asymptotic covariance properties of the $c_i(n)$ the reader may consult Anderson (1970). These expressions are not of great importance in practice (partly because of the occurrence of fourth-cumulant terms) and we do not further emphasize them here.

We consider now the ergodic behavior of some statistics whose importance in data analysis could not be exaggerated and indeed which form the basis for most analyses. We put†

$$\begin{aligned} w(\lambda) &= \frac{1}{(2\pi N)^{1/2}} \sum_1^N x(n) e^{in\lambda} \\ &= \int_{-\pi}^{\pi} \frac{1}{(2\pi N)^{1/2}} \sum_1^N e^{in(\lambda-\theta)} z\,(d\theta). \end{aligned} \tag{3.7}$$

† The corresponding quantities with $\sqrt{N}$ replaced by N are of interest only in connection with regression problems. They are analogs of $\bar{x}_N$, which we shall also consider again in relation to regression but which we considered here partly for historical reasons, partly because the mean stands in a special place and partly so as to introduce the considerations involved in Theorem 6.

We evidently have

$$\mathcal{E}\big(w(\lambda)w(\lambda)^*\big) = \frac{1}{2\pi}\sum_{-N+1}^{N-1}\left(1 - \frac{|n|}{N}\right)\Gamma(n)e^{-in\lambda}.$$

If $F(\lambda)$ is a.c. this converges to $f(\lambda)$ a.e. (a result which is not proved in the Mathematical Appendix and which we shall not use). We now consider only the a.c. case. If f is additionally continuous at λ then the expression converges boundedly to $f(\lambda)$ and if $f(\lambda)$ is a continuous function the convergence is uniform. When $f(\lambda) = 0$, then $w(\lambda)$ evidently converges in mean square to zero. When $f(\lambda) > 0$, then almost everywhere it cannot converge in mean square to any random variable and, in particular, it cannot so converge where $f(\lambda)$ is continuous (see the exercises to this chapter). Thus the ergodic behavior of these quantities turns out to be rather uninteresting. Their usefulness lies, as we shall see, in constructions of estimates of spectra and cross spectra from functions of sets of them.

The quantity

$$I(\lambda) = w(\lambda)\, w(\lambda)^* \tag{3.8}$$

we call the periodogram. This is an unsatisfactory terminology since the expression is a function of frequency not period. It should also be mentioned that the periodogram is often considered only for scalar processes and, moreover, is then often defined as

$$\frac{2}{N}\left|\sum_{1}^{N} x(n)e^{in\lambda}\right|^2. \tag{3.9}$$

This definition is adopted so that when $x(n)$ is Gaussian and independent with unit variance then, for $\lambda = 2\pi k/N$, $k \neq 0, \pi$, (3.9) will be distributed as chi square with two degrees of freedom. We shall retain the definition (3.8). In Olshen (1967) a number of somewhat special theorems are proved which establish that $I(\lambda)$ does not have an interesting ergodic behavior. We do not discuss this in detail here but later we shall discuss a particular example of this type of result. Thus the periodogram tends to have no asymptotic properties which make it valuable, directly, as an estimate of anything. In Section 4 we show in fact that under rather general conditions the distribution of the matrix $I(\lambda)$ converges to a slightly generalized form of Wishart's distribution with two degrees of freedom ($\lambda \neq 0, \pi$).

We now consider

$$r_N = w(\lambda) - h(\lambda)\, w(\epsilon, \lambda),$$

where $w(\epsilon, \lambda)$ *is constructed from the sequence* $\epsilon(1), \ldots, \epsilon(N)$ *as* $w(\lambda)$ *is, from the* $x(n)$ *sequence*, and†

$$x(n) = \sum_{-\infty}^{\infty} A(j)\,\epsilon(n-j), \qquad h(\lambda) = \sum_{-\infty}^{\infty} A(j)e^{ij\lambda}, \qquad \sum_{-\infty}^{\infty} \|A(j)\|^2 < \infty.$$

We now have Theorem 7 (see Olshen, 1967).

Theorem 7. *If* $x(n)$ *is a.c. with spectral density continuous at* λ *then* $w(\lambda) - h(\lambda)\,w(\epsilon, \lambda)$ *converges in mean square to zero and if* $f(\lambda)$ *is continuous for all* λ *then the mean square error converges to zero uniformly in* λ.

Proof. If f is continuous at the point λ we may choose $h(\lambda)$ to be continuous at that point. (Indeed the positive square root of $f(\lambda)$ satisfies this condition.) Consider now

$$\lim_{N\to\infty} \|w(\lambda) - h(\lambda)w(\epsilon, \lambda)\|^2$$

$$= f(\lambda) + \frac{1}{2\pi} h(\lambda)Gh(\lambda)^* - \lim_{N\to\infty} \mathcal{E}\{ w(\lambda)w(\epsilon, \lambda)^*h(\lambda)^* + h(\lambda)\,w(\epsilon, \lambda)w(\lambda)^*\}$$

where $G = \mathcal{E}(\epsilon(n)\,\epsilon(n)')$, as usual. The second term is also $f(\lambda)$. Then

$$\mathcal{E}\{w(\lambda)w(\epsilon, \lambda)^*\} = \sum_{-\infty}^{\infty} A(j)\frac{1}{2\pi N}\,\mathcal{E}\left\{\sum_{1}^{N} \epsilon(n-j)e^{in\lambda} \sum_{1}^{N} \epsilon'(n)e^{-in\lambda}\right\}$$

$$= \frac{1}{2\pi}\sum_{-N+1}^{N-1} A(j)\left(1 - \frac{|j|}{N}\right)e^{ij\lambda}G.$$

Since this converges to $(1/2\pi)\,h(\lambda)G$ if $h(\lambda)$ is continuous we see that $w(\lambda)$ converges in mean square to $h(\lambda)\,w(\epsilon, \lambda)$ (see the Mathematical Appendix) so that, choosing $h(\lambda)$ to be continuous, we see that the mean-square error converges uniformly to zero.

We say that a sequence x_n of random variables converges in the mean to a random variable x if $\mathcal{E}|x_n - x|$ converges to zero.

We put $I(\epsilon, \lambda) = w(\epsilon, \lambda)w(\epsilon, \lambda)^*$.

Corollary 5. *If* $x(n)$ *is as in Theorem* 7 *the elements of* $I(\lambda) - h(\lambda)I(\epsilon, \lambda)h(\lambda)^*$ *converge in the mean to zero and the mean error converges to zero uniformly in* λ *if* $f(\lambda)$ *is uniformly continuous.*

Indeed

$$\begin{aligned} I(\lambda) = h(\lambda)I(\epsilon, \lambda)h(\lambda)^* = {} & \{w(\lambda) - h(\lambda)w(\epsilon, \lambda)\}\,w(\epsilon, \lambda)^*h(\lambda)^* \\ & + h(\lambda)w(\epsilon, \lambda)\,\{w(\lambda) - h(\lambda)w(\epsilon, \lambda)\}^* \\ & + \{w(\lambda) - h(\lambda)w(\epsilon, \lambda)\}\{w(\lambda) - h(\lambda)\,w(\epsilon, \lambda)\}^* \end{aligned}$$

† For ease of printing it is now simpler to revert to the notation $h(\lambda)$ for what is called $h(\exp i\lambda)$ in Chapter III.

Taking expectations of the norm of this expression and using Schwartz's inequality, we see that the result is bounded by

$$2\|f(\lambda)\|\;\|w(\lambda) - h(\lambda)\,w(\epsilon, \lambda)\| + \|w(\lambda) - h(\lambda)\,w(\epsilon, \lambda)\|,$$

which clearly converges to zero in the manner required.

In the statistical work which follows the $w(\lambda)$ will be computed not at a fixed frequency λ but rather at a set of frequencies

$$\omega_j = 2\pi j/N, j = 0, 1, \ldots, [\tfrac{1}{2}N].$$

In particular we shall want to consider a set of ω_j which cluster around some fixed frequency in which we may be interested or which lie in a narrow band of frequencies in which we are interested. Thus in this section we wish to study the mean square convergence and almost sure convergence properties of quantities such as $w(\omega_j)$ for ω_j values clustering in the band of interest. This means that we are considering the ergodic properties of sequences of regression coefficients of the type

$$\frac{\sum x(n)y^{(N)}(n)}{\sum \{y^{(N)}(n)\}^2},$$

where $y^{(N)}(n)$ depends, as we have shown, on N. For later purposes (see Chapter VII) we shall develop the theory more generally by means of a model for the $y^{(N)}(n)$ of the type satisfying the conditions (3.1) introduced in Chapter II, Section 6. More generally we consider a sequence of vectors $y^{(N)}(n)$, of q components each, for which

(a) $\lim_N d_j^2(N) = \lim_N \sum_1^N \{y_j^{(N)}(n)\}^2 = \infty, \quad j = 1, \ldots, q,$

(b) $\lim_N \left[\dfrac{\{y_j^{(N)}(N)\}^2}{d_j^2(N)}\right] = 0, \quad j = 1, \ldots, q,$

(c) $\lim_N \left\{\dfrac{\sum_{m=1}^N \left(y_j^{(N)}(m) y_k^{(N)}(m+n)\right)}{d_j(N)\,d_k(N)}\right\} = \rho_{jk}(n).$

Here $d_j(N)$ is the positive square root of $d_j^2(N)$. Then just as in Chapter II, Section 6 (see Theorem 11)

$$\rho_{jk}(n) = \int_{-\pi}^{\pi} e^{in\lambda}\, M(d\lambda).$$

We call $R(n)$ the matrix with $\rho_{jk}(n)$ as the typical element and assume that $R(0)$ is nonsingular.

We introduce the matrix Y_N which has $y_j^{(N)}(n)$ in row n column j and the *matrix X_N similarly defined in terms of the $x_j(n)$.* Then we put

$$\hat{B} = (Y'_N Y_N)^{-1} Y'_N X_N. \tag{3.10}$$

The motivation for this definition is clear from the theory of regression for $\hat{B}$ has as jth column the vector of regression coefficients of $x_j(n)$ on the $y_k^{(N)}(n)$, $k = 1, \ldots, q$. Thus, when $q = 1$ and $y^{(N)}(n) \equiv 1$, $\hat{B}$ is $\bar{x}_N.$. *We introduce the diagonal matrix D_N having $d_j(N)$ in the jth place.* In order to state and prove our theorem properly it is necessary to introduce a tensor notation. We take the columns of $\hat{B}$ and successively place them down a column $\hat{\beta}$ so that $\hat{\beta}_{i,j}$ goes into row $(j-1)q + i$, $i = 1, \ldots, q$; $j = 1, \ldots, p$. (Recall that there are p of the x variables, $x_j(n), j = 1, \ldots, p$ and q of the y variables, $y_k(n)$, $k = 1, \ldots, q$.) We remind the reader that the tensor (or Kronecker) product,† $A \otimes B$, of a $p \times p$ matrix A and a $q \times q$ matrix B is the $pq \times pq$ matrix having $a_{ij}b_{kl}$ in row $q(i-1) + k$, column $q(j-1) + l$, $i, j = 1, \ldots, p$; $k, l = 1, \ldots, q$. Thus we may define the matrix

$$\{I_p \otimes R(0)\}^{-1} \int_{-\pi}^{\pi} 2\pi f(\lambda) \otimes M'(d\lambda) \{I_p \otimes R(0)\}^{-1}, \tag{3.11}$$

where by the middle matrix we mean that obtained by integrating the tensor product $f(\lambda) \otimes M'(d\lambda)$ element by element. We have the following theorem which establishes the mean-square convergence to zero of the elements of $\hat{B}$ but also shows what the limiting formula for the covariance matrix is. We discuss these results in relation to the $w(\omega_j)$ in the next section.

Theorem 8. *If $x(n)$ is stationary with a.c. spectrum which is piecewise continuous with no discontinuities at the jumps of $M(\lambda)$ then $(I_p \otimes D_N)\, \mathcal{E}(\hat{\beta}\hat{\beta}') \times (I_p \otimes D_N)$ converges as N increases to* (3.11) *above.*

Proof. We have $\mathcal{E}(\hat{\beta}) = 0$ since $\hat{\beta}$ is linear in the $x_j(n)$. We observe that $Y'_N X_N$ when arranged as a column vector,‡ u, in the same way as $\hat{B}$ was arranged as $\hat{\beta}$, is of the form

$$u = \sum_{n=1}^{N} x(n) \otimes y^{(N)}(n).$$

Then

$$(I_p \otimes D_N)\hat{\beta} = \{(I_p \otimes D_N)(I_p \otimes Y'_N Y_N)^{-1}(I_p \otimes D_N)\}(I_p \otimes D_N^{-1})u.$$

Now the first factor (within braces) is

$$\{I_p \otimes D_N^{-1} Y'_N Y_N D_N^{-1}\}^{-1},$$

† See the Mathematical Appendix, Section 5.

‡ We suppress an explicit reference to N in the symbol u, for notational convenience, though u does depend on N.

and, as $N \to \infty$, this converges to $[I_p \otimes R(0)]^{-1}$ by condition (c) (for $n = 0$). Thus we must prove the theorem for $(I_p \otimes D_N^{-1})u$, with covariance matrix

$$\int_{-\pi}^{\pi} 2\pi f(\lambda) \otimes M'(d\lambda).$$

We consider first the case in which $f(\lambda)$ is continuous and $f(\lambda) > 0$ in $[-\pi, \pi]$ (in which here and below the inequality refers to the usual partial ordering of nonnegative matrices†). If $f^{(1)}(\lambda) \le f(\lambda) \le f^{(2)}(\lambda)$ and u_1, u, u_2 are formed from processes, $x_1(n)$, $x(n)$, $x_2(n)$, with these spectra, then

$$\mathcal{E}(u_1u_1') \le \mathcal{E}(uu') \le \mathcal{E}(u_2u_2').$$

Indeed

$$\text{(3.12)} \qquad \mathcal{E}(uu') = \sum_{m,n}\sum \gamma(n-m) \otimes y^{(N)}(m)y^{(N)}(n)'$$
$$= \int_{-\pi}^{\pi} f(\lambda) \otimes \sum_{m,n}\sum y^{(N)}(m)y^{(N)}(n)'e^{i(n-m)\lambda}\, d\lambda$$
$$= \int_{-\pi}^{\pi} f(\lambda) \otimes J(\lambda)\, J(\lambda)^*\, d\lambda,$$

let us say, where

$$J(\lambda) = \sum_{n=0}^{N} y^{(N)}(n)e^{-in\lambda}.$$

Since

$$f^{(1)}(\lambda) \otimes J(\lambda)\, J(\lambda)^* \le f(\lambda) \otimes J(\lambda)\, J(\lambda)^* \le f^{(2)}(\lambda) \otimes J(\lambda)\, J(\lambda)^*,$$

the result follows. Moreover, if $f^{(2)}(\lambda) - f^{(1)}(\lambda) \le \epsilon I_p$, then

$$(I_p \otimes D_N^{-1})\mathcal{E}(u_2u_2' - u_1u_1')(I_p \otimes D_N^{-1})$$
$$= \int_{-\pi}^{\pi} \{f^{(2)}(\lambda) - f^{(1)}(\lambda)\} \otimes D_N^{-1}J(\lambda)J(\lambda)^*D_N^{-1}\, d\lambda$$
$$\le \epsilon \int_{-\pi}^{\pi} I_p \otimes D_N^{-1}J(\lambda)J(\lambda)^*D_N^{-1}\, d\lambda,$$

which converges to $2\pi\epsilon\{I_p \otimes R(0)\}$, since

$$\int_{-\pi}^{\pi} J(\lambda)\, J(\lambda)^*\, d\lambda = 2\pi Y_N Y_N'.$$

† In particular we use $f(\lambda) > 0$ to mean that $f(\lambda)$ is positive definite and not merely nonnegative.

Thus, if we can find spectra $f^{(i)}(\lambda)$, $i = 1, 2$, which satisfy, for all $\epsilon > 0$,

$$f^{(1)}(\lambda) \leq f(\lambda) \leq f^{(2)}(\lambda), \qquad f^{(2)}(\lambda) - f^{(1)}(\lambda) \leq \epsilon I_p,$$

$$\lim_{N\to\infty} (I_p \otimes D_N^{-1})\mathcal{E}(u_i u_i')(I_p \otimes D_N^{-1}) = \int_{-\pi}^{\pi} 2\pi f^{(i)}(\lambda) \otimes M'(d\lambda),$$

we shall have established what we wish for $\mathcal{E}(uu')$. If $f(\lambda)$ is continuous with $f(\lambda) > 0$, this is easily done, for we may find trigonometric polynomials,

$$f^{(i)}(\lambda) = \sum_n A^{(i)}(n)e^{-in\lambda}, \qquad A^{(i)}(n) = A^{(i)}(-n), \qquad i = 1, 2,$$

that satisfy the first two of these conditions. (To see this we may consider $f(\lambda) \pm \eta I_p$, with $\eta > 0$ chosen so that this matrix is positive definite, and then use the fact that the Cesaro means of the Fourier series for these matrices converge uniformly and are positive definite.) The third is satisfied also for

$$(I_p \otimes D_N^{-1})\mathcal{E}(u_i u_i')(I_p \otimes D_N^{-1}) = \sum_n A^{(i)}(n) \otimes \int_{-\pi}^{\pi} e^{-in\lambda} D_N^{-1} J(\lambda)\, J(\lambda)^* D_N^{-1}\, d\lambda,$$

which converges to

$$2\pi \sum_n A^{(i)}(n) \otimes R(-n) = \int_{-\pi}^{\pi} 2\pi \sum A^{(i)}(n)e^{-in\lambda} \otimes M'(d\lambda)$$

$$= \int_{-\pi}^{\pi} 2\pi f^{(i)}(\lambda) \otimes M'(d\lambda).$$

Thus the formula for the covariance matrix is established when $f(\lambda) > 0$, $\lambda \in [-\pi, \pi]$. If $f(\lambda)$ is continuous, but this is not so, consider $f(\lambda) + \epsilon I_p$, $\epsilon > 0$, and call the associated vector u_ϵ. Then from what has been said above

$$\overline{\lim_{N\to\infty}} (I_p \otimes D_N^{-1})\mathcal{E}(uu')(I_p \otimes D_N^{-1}) \leq \int_{-\pi}^{\pi} 2\pi f(\lambda) \otimes M'(d\lambda) + 2\pi\epsilon\big(I_p \otimes R(0)\big),$$

$$\underline{\lim_{N\to\infty}} (I_p \otimes D_N^{-1})\mathcal{E}(uu')(I_p \otimes D_N^{-1}) \geq \int_{-\pi}^{\pi} 2\pi f(\lambda) \otimes M'(d\lambda) - 2\pi\epsilon\big(I_p \otimes R(0)\big)$$

which establishes what we wish to prove, since ϵ is arbitrary.

If $f(\lambda)$ is merely piecewise continuous, we may find two continuous spectra, $g^{(1)}(\lambda) \leq f(\lambda) \leq g^{(2)}(\lambda)$, so that

$$\int_{-\pi}^{\pi} \{g^{(2)}(\lambda) - g^{(1)}(\lambda)\} \otimes M'(d\lambda) \leq \epsilon I_{pq}; \tag{3.13}$$

for example, if λ_0 is a point of discontinuity and $f(\lambda_0+) > f(\lambda_0-)$, we put

$$\begin{aligned} g^{(1)}(\lambda) &= f(\lambda_0-) + \eta^{-1}(\lambda - \lambda_0)\{f(\lambda_0 + \eta) - f(\lambda_0-)\}, \\ g^{(2)}(\lambda) &= f(\lambda), \end{aligned} \qquad \lambda \in [\lambda_0, \lambda_0 + \eta]$$

If η is chosen small enough, then $g^{(1)}(\lambda) \le f(\lambda) \le g^{(2)}(\lambda)$ over the interval and the contribution from the interval to (3.13) may be made arbitrarily small also since the increase in $M'(\lambda)$ over $[\lambda_0, \lambda_0 + \eta]$ is arbitrarily small with η. Since the theorem holds for $g^{(1)}$ and $g^{(2)}$ by the foregoing proof and ϵ in (3.13) is arbitrary it must hold for $f(\lambda)$ also. This establishes Theorem 8.

In connection with the proof of a form of central limit theorem we are led to modify the condition (b) to the stronger condition

(b)′ Let S_N be a set of M_N indices j_N, $1 \le j_N \le N$. If

$$\lim_{N\to\infty} \frac{M_N}{N} = 0,$$

then also

$$\lim_{N\to\infty} \sum_{n\in S_N} \frac{\{y_j^{(N)}(n)\}^2}{d_j^2(N)} = 0, \qquad j = 1, \ldots, q,$$

uniformly in M_N/N. (Thus the left-hand side may be made smaller than any $\epsilon > 0$ for all S_N, and all j, for which M_N/N is sufficiently small.)

This condition seems to exclude no cases of real importance.† It does exclude certain cases where $d_j(N)$ increases very slowly, for example $y(n) = n^{-1/2}$. It excludes none of the cases explicitly discussed in Chapter II, Section 6. We now have

Theorem 9. *If* $x(n)$ *is stationary with* $\operatorname{tr}\{f(\lambda)\} \le K_1 < \infty$, $\lambda \in [-\pi, \pi]$, *the* $y^{(N)}(n)$ *satisfy* (a), (b)′, (c) *and* $d_j(N)^2 \ge K_2 N^\alpha$, $K_2 > 0$, $\alpha > 0$, $j = 1, \ldots, q$ *then* $\hat{B}$ *converges almost surely to zero.*

Proof. We may take $p = 1$, since $\hat{\beta}_{ij}$ depends only on $x_j(n)$, so that $f(\lambda)$ is scalar. Then

$$\hat{\beta} = D_N^{-1}(D_N^{-1} Y_N' Y_N D_N^{-1})^{-1} D_N^{-1} Y_N' x_N,$$

where x_N is a column vector with $x(n)$ is the nth place. Since $D_N^{-1} Y_N' Y_N D_N^{-1}$ converges to $R(0)$, and $N^{1/2\alpha} D_N^{-1}$ remains bounded, it is sufficient to show that $N^{1/2\alpha} D_N^{-1} Y_N' x_N$ converges almost surely to zero. This reduces our considerations to the case $q = 1$ since we may take each row of this vector separately. Thus taking $p = q = 1$, we must show that

$$u(n) = \frac{\sum_1^N y^{(N)}(n)\, x(n)}{(d(N) N^{1/2\alpha})}$$

† See the exercises to this chapter for more details.

converges almost surely to zero. This has variance

$$\frac{1}{d^2(N)N^\alpha}\sum_{m,n}\sum\int_{-\pi}^{\pi}\exp\{i(n-m)\lambda\}\, y^{(N)}(m)y^{(N)}(n)f(\lambda)\,d\lambda$$
$$\leq K_1N^{-\alpha}\int_{-\pi}^{\pi}\frac{|\sum_1^N y^{(N)}(n)e^{in\lambda}|^2}{d(N)^2}\,d\lambda = 2\pi K_1N^{-\alpha}.$$

Now choosing β so that $\alpha\beta > 1$ and taking $N(M)$ as in the proof of Theorem 5 then precisely as in that theorem we may show that $u(N(M))$ converges almost surely to zero as $M \to \infty$. Let us put $\{d^2(N)N^\alpha\}^{1/2} = c(N)$. Then

$$\text{(3.14)}\quad \mathcal{E}\left\{\max_{N(M)\leq N\leq N(M+1)}\left|u(N) - \left\{\frac{c(N(M))}{c(N)}\right\}u(N(M))\right|^2\right\}$$
$$\leq c(N(M))^{-2}\sum_{N(M)}^{N(M+1)} y(m)y(n)\gamma(n-m) \leq 2\pi K_1\frac{d^2(N(M+1)) - d^2(N(M))}{c(N(M))^2}$$

by the same evaluation as just used. The last expression is

$$N(M)^{-\alpha}\frac{d^2(N(M+1)) - d^2(N(M))}{d^2(N(M))} \leq M^{-\beta\alpha}\left\{\frac{d^2(N(M+1)) - d^2(N(M))}{d^2(N(M))}\right\}$$

Now if we show that the second factor in the last expression converges to zero then we may complete the proof exactly as in Theorem 5 since we shall have shown that (3.14) is $O(M^{-1})$ and at the same time that $c(N(M))/c(N)$, $N(M) \leq N \leq N(M+1)$, converges to unity. However,

$$\frac{d^2(N(M)) - d^2(N(M-1))}{d^2(N(M))} = \frac{\sum_{S_M} y^2(n)}{\sum_1^{N(M)} y^2(n)},$$

which converges to zero by (b)′ since S_M contains $N(M) - N(M-1)$ terms and $\{N(M) - N(M-1)\}/N(M)$ converges to zero.

4. THE CENTRAL LIMIT THEOREM

There is fairly considerable literature concerning the central limit theorem for stationary processes and we mention, in particular, Moran (1947), Diananda (1954), Grenander and Rosenblatt (1957), Hannan (1961a). Two types of situation have been considered. The first of these is that where

$$\text{(4.1)}\qquad x(n) = \sum_{-\infty}^{\infty} A(j)\epsilon(n-j), \qquad \sum^{\infty}\|A(j)\|^2 < \infty$$

and the $\epsilon(n)$ are independent and identically distributed with covariance matrix G. Then $x(n)$ is a generalized linear process. A very general discussion of this situation is given in Eicker (1967), and indeed, in part, his discussion includes the case where the $\epsilon(n)$ are not identically distributed. We shall use his ideas in modifying the results in Hannan (1961a) insofar as our treatment of (4.1) is concerned. The assumptions in (4.1) are by no means satisfactory for there is no very natural way of justifying the assumption of the independence of the $\epsilon(n)$. (If the $x(n)$ are Gaussian with spectrum which is a.c. then (4.1) necessarily follows of course.) Nevertheless it seems that the introduction of notions of independence *somewhere* in the analysis is essential. An alternative type of condition is that introduced in Rosenblatt (1956a, 1961) where $x(n)$ is required to satisfy the uniform mixing condition of Section 2. In some ways this procedure seems preferable for the condition is one which we can justify in the kind of way introduced in Section 2. We shall also discuss a form of central limit theorem of this type in this section.

Rather than deal directly with the central limit theorem for a mean vector we shall immediately discuss the more general case of the matrix $\hat{B}$ (defined by (3.10)) where the $y^{(N)}(n)$ satisfy the conditions (a), (b), and (c) of the previous section. We have

Theorem 10. *Let $y^{(N)}(n)$ satisfy the conditions* (a), (b), *and* (c) *of Section* 3 *and $\hat{B}$ be defined by* (3.10) *where $x(n)$ is generated by a generalized linear process with piecewise continuous spectral density having no discontinuities at jumps in $M(\lambda)$. Let $\hat{\beta}$ be defined in terms of $\hat{B}$ as described after* (3.10). *Then if* (3.11) *is not the null matrix the distribution of $(I_p \otimes D_N)\hat{\beta}$ converges as $N \to \infty$ to the normal distribution with zero mean vector and covariance matrix given by* (3.11).

We have put the proof of their theorem into the Appendix to this chapter not so much because it is difficult as because it is rather tedious. When $y^{(N)}(n)$ is more special we can expect stronger results and in particular this is so for the mean as the following theorem shows.

Theorem 11. *If $x(n)$ is generated by a generalized linear process with spectrum $f(\lambda)$ continuous at $\lambda = 0$, with* tr $(f(\lambda))$ *uniformly bounded and with $f(0)$ not null then $N^{\frac{1}{2}}\,\bar{x}_N$ is asymptotically normal with covariance matrix $2\pi f(0)$.*

Again we have put the proof in the appendix to this chapter.

Now we consider a case of considerable importance in all that follows, which is closely related to the first example of Chapter II, Section 6. We take a fixed frequency λ, $0 \leq \lambda \leq \pi$, and we take $y^{(N)}(n)$ to be as follows.

$$
(4.2)\quad \begin{cases} \lambda \neq 0, \pi, \qquad y_{2u-1}^{(N)}(n) = \cos n\lambda_u, \qquad y_{2u}^{(N)}(n) = \sin n\lambda_u \\[4pt] \qquad\qquad \lambda_u = \dfrac{2\pi j_u}{N}, \qquad u = 1, \ldots, m, \end{cases}
$$

where the j_u are successive integers so chosen that the λ_u are as near as possible to λ. Thus $q = 2m$. For $\lambda = 0$ there will be no sine term for $j_u = 0$ and for $\lambda = \pi$ and N even the same will be true for $j_u = \frac{1}{2}N$. We shall continue to call q the number of indices k. Now for $\lambda \neq 0, \pi$ the matrix $R(n)$ is

$$R(n) = \begin{bmatrix} \cos n\lambda & \sin n\lambda & & & & & \\ -\sin n\lambda & \cos n\lambda & & & & & \\ & & \cos n\lambda & \sin n\lambda & & & \\ & & -\sin n\lambda & \cos n\lambda & & & \\ & & & & \cdot & & \\ & & & & & \cdot & \\ & & & & & & \cdot \end{bmatrix},$$

where only nonzero terms are shown. To see this observe first that $d_j^2(N) \equiv \frac{1}{2}N$ and consider for example

$$\lim_{N\to\infty} \frac{2}{N} \sum_{m=1}^{N} \cos m\lambda_u \sin (m + n)\lambda_u = \lim_{N\to\infty} \sin n\lambda_u = \sin n\lambda$$

Of course we could take $\cos n\lambda_u$ without the corresponding sine term (or the reverse) in which case some blocks down the diagonal of $R(n)$ will be one dimensional (and of the form $\cos n\lambda$). However, we omit separate consideration of this. If $\lambda = 0, \pi$ then $R(n)$ becomes the q rowed unit matrix, I_q, $(\lambda = 0)$ or $\cos n\pi I_q$ $(\lambda = \pi)$. Correspondingly, $M(\lambda)$ has two points of increase (at most) for $\lambda \neq 0, \pi$ these being at $\pm\lambda$, with the increases being jumps of the form

$$V_q = \begin{bmatrix} \frac{1}{2} & \frac{i}{2} & & & & & \\ -\frac{i}{2} & \frac{1}{2} & & & & & \\ & & \frac{1}{2} & \frac{i}{2} & & & \\ & & -\frac{i}{2} & \frac{1}{2} & & & \\ & & & & \cdot & & \\ & & & & & \cdot & \\ & & & & & & \cdot \end{bmatrix} \tag{4.3}$$

at λ, and with jump V_q' at $-\lambda$. For $\lambda = 0, \pi$ the jump is just I_q, at the point λ.

In the case just studied the elements of $\hat{B}$ are just the sums

$$\frac{2}{N}\sum_{1}^{N} x_j(n) \begin{matrix}\cos\\ \sin\end{matrix} n\lambda_u,$$

where the notation indicates that either or both type of sum may occur.

It is easily checked that $\cos n\lambda_u$ and $\sin n\lambda_u$, with λ_u as described, satisfy (a), (b)′, and (c) of Section 3 so that these elements of $\hat{B}$ converge almost surely to zero. We of course are concerned with a central limit theorem after norming by multiplication by D_N.

In the present case $R(0) = I_q$ (and for $\lambda \neq 0, \pi$ we may take $D_N = \frac{1}{2}NI_q$). Thus from Theorem 10 $(I_p \otimes D_N)\hat{\beta}$ has a limiting normal distribution with covariance matrix

$$\text{(4.4)} \qquad \begin{matrix} 2\pi f(\lambda) \otimes V_q + 2\pi f(-\lambda) \otimes \bar{V}_q, & \lambda \neq 0, \pi, \\ 2\pi f(\lambda) \otimes I_q, & \lambda = 0, \pi. \end{matrix}$$

Let us point out in connection with this formula that if we change the order of the elements of $\hat{\beta}$ by putting $\hat{\beta}_{ij}$ in row $(j-1)p + i$ (i.e., taking them in dictionary order with column number first and then row number) we merely change the order of the terms in each tensor product in (4.4).

Theorem 12. *If $x(n)$ is generated by a generalized linear process with spectrum continuous at the point λ and not null there,* $\operatorname{tr}\big(f(\lambda)\big)$ *is uniformly bounded and the $y_k^{(N)}(n)$ are q sequences of the form* (4.2) *then $(I_p \otimes D_N)\hat{\beta}$ has a distribution converging to the multivariate normal distribution with zero mean vector and covariance matrix* (4.4).

Again we have put the proof in the Appendix to this chapter.

This theorem enables us to discuss the distribution of the $w(\lambda_u)$, defined by (3.7). For simplicity of printing let us put

$$\text{(4.5)} \qquad v(u) = w(\lambda_u) = \frac{1}{(2\pi N)^{1/2}}\sum_{1}^{N} x(n)e^{in\lambda_u}, \qquad \lambda_u = \frac{2\pi j_u}{N}$$
$$= a(u) + ib(u), \qquad u = 1, \ldots, m,$$

where $a(u)$ and $b(u)$ are $\sqrt{(N/8\pi)}$ by the adjacent rows of the matrix $\hat{\beta}$, for $\lambda_u \neq 0, \pi$, and $\sqrt{(N/2\pi)}$ times the row of that matrix for $\lambda_u = 0, \pi$. To describe the asymptotic distribution of the $v(u)$ we introduce the complex multivariate normal distribution

$$\text{(4.6)} \qquad \frac{1}{\pi^p \det(\mathbf{\Sigma})} e^{-z^*\mathbf{\Sigma}^{-1}z},$$

where Σ is a $p \times p$ Hermitian nonnegative matrix and z has z_j in row j $j = 1, \ldots, p$. This density has associated with it a differential form consisting of the product of all differentials of real and imaginary parts of each z_j.

Theorem 13. *If $x(n)$ is generated by a generalized linear process with spectrum continuous and not null at λ and the $v(u)$ are defined by* (4.5) *with the λ_u as in* (4.2) *then for $\lambda \neq 0, \pi$ the joint distribution of the $v_j(u)$ converges to that of m independent random vectors with density* (4.6) *and $\Sigma = f(\lambda)$. For $\lambda = 0, \pi$ the joint distribution converges to that of m independent random vectors with the ordinary multivariate normal distribution with the covariance matrix $f(\lambda)$.*

Proof. We commence with some algebraic details, which we shall use again later, that are relevant to the case $\lambda \neq 0, \pi$.

Let Σ be any square matrix of p rows and columns of complex numbers. Put $\Sigma = C + iQ$. Establish the correspondence

$$\Sigma \leftrightarrow \begin{bmatrix} C & Q \\ -Q & C \end{bmatrix}.$$

Then this is an isomorphism of the algebra of all such matrices Σ (considered as an algebra over the *real* field†) with the algebra of all matrices of the second form. (The fact that the algebraic operations correspond is easily checked and the correspondence is evidently one to one.) If Σ is nonsingular, so is the matrix on the right and conversely for $z = x - iy$, $\Sigma z = 0$ implies $(Cx + Qy) = (Qx - Cy) = 0$ and conversely. If Σ is Hermitian, then C is symmetric and Q is skew symmetric so that the image of Σ is symmetric and conversely. (If Σ is unitary then its image is orthogonal and conversely.) If Σ is Hermitian nonnegative, its image is symmetric nonnegative and conversely, since

$$z^*\Sigma z = (x' : y') \begin{bmatrix} C & Q \\ -Q & C \end{bmatrix} \begin{pmatrix} x \\ \cdot\cdot \\ y \end{pmatrix}.$$

Moreover, if Σ is Hermitian and has z as an eigenvector for eigenvalue μ then

$$\begin{pmatrix} x \\ \cdots \\ y \end{pmatrix}, \quad \begin{pmatrix} y \\ \cdots \\ -x \end{pmatrix}$$

are eigenvectors of its image for the same eigenvalue and conversely. In particular, the determinant of the image of Σ is $(\det \Sigma)^2$.

† By saying that it is considered as an algebra over the real field we mean that it is a set of matrices closed under the algebraic operations of multiplication and the formation of linear combinations with real coefficients.

Now, if we take the elements of the vectors

$$g(u) = \begin{pmatrix} a(u) \\ \cdots \\ b(u) \end{pmatrix},$$

Theorem 13 says that in their limiting distribution these vectors are independent for various u. The elements of $a(u)$ and $b(u)$ are obtained from those of $\hat{\beta}$, for $m = 1$, by multiplying the latter by $(4\pi)^{-1/2}$. Moreover, in writing them down the column $g(u)$ we have altered their order by taking all cosine terms first and then all sine terms, compared with their order in $\hat{\beta}$, for $m = 1$. As a result their covariance matrix in their limiting distribution is got by altering the order of the two factors in the tensor products in (4.6), (as explained immediately before Theorem 12) and multiplying by $(4\pi)^{-1}$. Thus we get for the covariance matrix of $g(u)$, in its limiting distribution,

$$\frac{1}{4}\left\{\begin{bmatrix} 1 & i \\ -i & 1 \end{bmatrix} \otimes f(\lambda) + \begin{bmatrix} 1 & -i \\ i & 1 \end{bmatrix} \otimes \overline{f(\lambda)}\right\} = \frac{1}{2}\begin{bmatrix} C(\lambda) & Q(\lambda) \\ -Q(\lambda) & C(\lambda) \end{bmatrix},$$

$$f(\lambda) = \tfrac{1}{2}\big(C(\lambda) - iQ(\lambda)\big).$$

Thus

$$\frac{1}{2}\sum_u g(u)'\left\{\frac{1}{2}\begin{bmatrix} C(\lambda) & Q(\lambda) \\ -Q(\lambda) & C(\lambda) \end{bmatrix}\right\}^{-1} g(u) = \sum_u v(u)^* f(\lambda)\, v(u)$$
$$= \operatorname{tr}\big(f(\lambda)VV^*\big).$$

The Jacobian of the transformation from the elements of $\hat{\beta}$ to those of the $g(u)$ is $(4\pi)^{mp}$. The constant factor in the limiting density function for the elements of $a(u)$ and $b(u)$ is thus

$$(4\pi)^{mp}\left[2\pi^p \det\left\{\pi\begin{bmatrix} C & Q \\ -Q & C \end{bmatrix}\right\}^{1/2}\right]^{-m} = \left[\pi^p \det\left\{\tfrac{1}{2}\begin{bmatrix} C & Q \\ -Q & C \end{bmatrix}\right\}^{1/2}\right]^{-m}$$
$$= \pi^{-mp}\{\det f(\lambda)\}^{-m},$$

as required, and the theorem is established for $\lambda \neq 0, \pi$. For $\lambda = 0, \pi$ the derivation is simple and we leave the details to the reader.

Note. *It may be seen in precisely the same way that if we take a finite set $\lambda, \lambda', \ldots, \lambda^{(t)}$ of points in $[0, \pi]$ and for the* j*th choose the m_j frequencies $\lambda_{j,i}$, $i = 1, \ldots, m_j$, of the form $2\pi j/N$, j an integer, nearest to $\lambda^{(j)}$ then if $f(\lambda)$ is continuous and not null at the points $\lambda^{(j)}$ and $x(n)$ is a generalized linear process the joint distribution of the vectors $w(\lambda_{j,i})$ converges as $N \to \infty$ to that of $\sum m_j$ independent vectors, those belonging to the* j*th set having the* *distribution* (4.8) *with* $\not\Sigma = f(\lambda^{(j)})$, $\lambda^{(j)} \neq 0, \pi$ *and for* $\lambda^{(j)} = 0, \pi$ *the ordinary multivariate normal distribution with covariance matrix* $f(\lambda^{(j)})$.

The distribution with density (4.6) plays a very central part in the inferential theory which follows as is evident from the fact that the quantities $w(\lambda)$ are normalized approximations to $z(d\lambda)$, the orthogonal increments in terms of which $x(n)$ is Fourier-decomposed.

Of course, Theorem 10 goes well beyond the other theorems so that $y(n)$ might itself be a stationary process (with serial covariances which converge with probability 1). These considerations are taken up again when regression problems are considered. The requirement that the $\epsilon(n)$ have identical distributions can also be relaxed as we have indicated earlier.

We now go on to discuss the alternative form of central limit theorem mentioned at the beginning of this section. Here we require the following conditions.

1. The process $x(n)$ is to satisfy the uniform mixing condition, (4.2), of Section 2.
2. The process $x(n)$ is to be stationary to the fourth-order and the fourth-cumulant function $k_{ijkl}(0, n, p, q)$ is to satisfy

$$\sum\sum\sum_{n,p,q=0}^{\infty} |k_{ijkl}(0, n, p, q)| < \infty.$$

3. The sequence $y^{(N)}(n)$ is to satisfy (a) (b)′ and (c) of Section 3.
4. The process $x(n)$ has a.c. spectrum with continuous spectral density and

$$\int f(\lambda) \otimes M'(d\lambda)$$

is not null.

The second condition could be replaced by a similar condition regarding stationarity to the third-order and the third-absolute moment (and no doubt the condition could be weakened further). However, we retain the form of 2. since that condition will be needed later also (and, indeed, has already been used in Theorem 6 of Section 3). If $x(n)$ is strictly stationary then in the case of the analogs of Theorems 11, 12, and 13 we may replace condition 2. by a condition which is necessary and sufficient. Indeed this may be also done for Theorem 10, when $x(n)$ is strictly stationary, but the necessary and sufficient condition is then too complicated to be interesting. The condition we need is based upon the joint distribution function, $G_{N,\lambda}(x)$ of all of the real and imaginary parts of the elements of $w(\lambda)$. Thus, for example, except for the occurrence of the factor $(N/2\pi)^{1/2}$ in the definition of $w(0)$ the distribution $G_{N,0}(x)$ is that of the mean, whose asymptotic distribution we are investigating. The condition we need is

2′. $$\lim_{a\to\infty} \overline{\lim_{N\to\infty}} \int_{\|x\|>a} \|x\|^2 G_{N,\lambda}(dx) = 0.$$

It is interesting to compare the use of the uniform mixing condition with the use of a generalized linear process in proving theorems of the type of 11, 12, and 13. Since a generalized linear process is strictly stationary the difference lies in replacing the generalized linear process formulation by the uniform mixing condition† together with the change from (b) to (b)′ and the introduction of 2′. We now state the theorems, whose proofs are deferred to the Appendix to this chapter.

Theorem 10′. *Under* 1 *to* 4 *above* $(I_p \otimes D_N)\hat{\beta}$ *is asymptotically normal with zero mean vector and covariance matrix* (4.3).

Theorem 11′. *Under* 1, 2 *and the condition that* $x(n)$ *has a.c. spectrum with density continuous at* $\lambda = 0$ *(and not null there) and* $\operatorname{tr}(f(\lambda))$ *is uniformly bounded the conclusion of Theorem* 11 *holds. If condition* 2 *is eliminated but* $x(n)$ *is strictly stationary then a necessary and sufficient condition for this to hold is condition* 2′, *for* $\lambda = 0$,

Theorem 12′. *Under* 1, 2 *and the condition that* $x(n)$ *has a.c. spectrum with spectral density continuous at* λ *(and not null there) and* $\operatorname{tr}(f(\lambda))$ *is uniformly bounded the conclusion of Theorem* 12 *remains true. If* $x(n)$ *is strictly stationary and condition* 2 *is eliminated then a necessary and sufficient condition for this to hold is condition* 2′.

Theorem 13′. *Under the same conditions as for Theorem* 12′ *the conclusion of Theorem* 13 *remains true.*

Note. Needless to say Theorems 11′ and 12′ could well be coalesced into one theorem, but we have chosen to separate out the special case of the mean. These theorems are close to ones given by Rozanov (1967) and in particular he proves that 2′ is a necessary and sufficient condition under the other stated conditions for the case of $\bar{x}_N$.

We now consider the matrix of quantities $C(n)$. We first state the analog of Theorem 11′ since that follows easily from the stated theorem. For that purpose we need to consider a condition analogous to 2′. The simplest

† As mentioned earlier, the $\epsilon(n)$ in the generalized linear process do not need to be identically distributed provided the family of distributions from which they are chosen is suitably restricted (see Eicker, 1965). However, strict stationarity, in the uniform mixing condition situation, may be similarly relaxed.

way to state this seems to be to introduce the quantity

$$y_{N,q} = \sum_{i,j=1}^{p} \sum_{n=1}^{q} [\sqrt{N}\{c_{ij}(n) - \gamma_{ij}(n)\}]^2,$$

with distribution function which we call $G_{N,q}(x)$. Then the condition we need is

2″. $$\lim_{a\to\infty} \overline{\lim_{N\to\infty}} \int_{x>a} x G_{N,q}(dx) = 0.$$

Then we have the following theorem.

Theorem 14′. *If $x(n)$ is strictly stationary and satisfies* 1 *and* 2 *and has a.c. spectrum with continuous spectral density then for any integer $q \geq 0$ the joint distribution of the quantities*

$$\sqrt{N}\{c_{ij}(n_u) - \gamma_{ij}(n_u)\}; \qquad i, j = 1, \ldots, p; \qquad u = 1, \ldots, q$$

converges to the multivariate normal distribution with zero mean vector and covariance matrix as evaluated in Section 3 *if and only if* 2″ *holds.*

This theorem follows immediately from Theorem 11′. Indeed the p^2q sequences $y_{iju}(m) = x_i(m)\, x_j(m + n_u) - \gamma_{ij}(n_u)$, under the conditions of the theorem, constitute a vector stationary process with zero mean and finite variance satisfying the uniform mixing condition and with absolutely continuous spectrum having continuous spectral density. Thus Theorem 11′ may be applied with the vector of p^2q sequences $\sqrt{N}\{\tilde{c}_{ij}(n) - \gamma_{ij}(n)\}$ replacing $\sqrt{N}\,\bar{x}_N$. The necessary and sufficient condition of Theorem 11′ then becomes 2″. save that $\tilde{c}_{ij}(n)$ replaces $c_{ij}(n)$. However,

$$\sqrt{N}\{\tilde{c}_{ij}(n) - \gamma_{ij}(n)\} - \sqrt{N}\{c_{ij}(n) - \gamma_{ij}(n)\}$$

has variance which is $O(N^{-1})$ so that $c_{ij}(n)$ may be used in place of $\tilde{c}_{ij}(n)$ in the condition, 2″ and the theorem. This establishes the theorem.

The corresponding unprimed theorem is the following:

Theorem 14. *If $x(n)$ is generated by a linear process with the $\epsilon(n)$ having finite fourth-cumulant function, κ_{ijkl}, then the quantities*

$$\sqrt{N}\{c_{ij}(n_u) - \gamma_{ij}(n_u)\}; \qquad i, j = 1, \ldots, p; \qquad u = 1, \ldots, q$$

have, asymptotically, a normal distribution with zero mean vector and covariance matrix as evaluated in Section 3.

We give the proof in the Appendix to this Chapter.

Neither of these theorems is very satisfactory. We shall not pursue the matter because of its complexity and because theorems of this type do not seem to be important. This is also because the quantities

$$r_{ij}(n) = c_{ij}(n) \{c_i(0)\, c_j(0)\}^{-1/2}$$

are more important in practice and because the quantities discussed in Theorems 13 and 13′ are so central.

It is of interest to observe that, at least when

$$\sum_{-\infty}^{\infty} |j|\, \alpha(j)^2 < \infty,$$

the distribution of the autocorrelation coefficient

$$r(n) = \frac{c(n)}{c(0)}$$

is asymptotically normal, for a linear process, without any assumption being made concerning the existence of a fourth moment. The reader is referred to Anderson and Walker (1964). In fact the nature of the situation can be seen from the proof of Theorem 10 of Chapter V. The result is somewhat special, for the corresponding result fails to hold in the vector case, namely, if

$$\sum |j|\, \|A(j)\|^2 < \infty$$

for a vector linear process, it seems that

$$r_{jk}(n) = \frac{c_{jk}(n)}{[c_j(0) c_k(0)]^{1/2}}$$

cannot be shown to be normal without assuming finite moments of the fourth order by Anderson and Walker's technique *unless* $\epsilon(n)$ *is a scalar.*

EXERCISES

1. Let $x(n)$ be scalar and be given by

$$x(n) = \alpha \cos n\lambda + \beta \sin n\lambda$$

where λ is irrational and α, β are independent random variables with zero mean and unit variance. Show that $\bar{x}_N$ converges to zero with probability one but that $x(n)$ is not mixing.

2. Let $x(n)$ be strictly stationary with zero mean and spectral distribution function having no jump at the origin. Show that $\bar{x}_N$ converges to zero with probability 1. (Use Theorem 1 and the fact that $\bar{x}_N$ converges in mean square to a limit which is zero with probability 1.)

3. Derive formula (3.3).

4. Show that a process of independent identically distributed random variables is mixing.

5. Show that $y(n) = n^{-1/2}$ satisfies (b) of Section 4 but not (b)′.

6. Let $x(n)$ be scalar and as in Theorem 14. If $r(n) = c(n)/c(o)$ show that any finite set of the $N^{1/2}\{r(n) - \rho(n)\}$ is asymptotically normal with zero mean and covariances, for $r(u)$ with $r(v)$,

$$\sum_{-\infty}^{\infty} \{\rho(m)\rho(m+u-v) + \rho(m+u)\rho(m-v) - 2\rho(m)\,\rho(u)\,\rho(m-v) \\ - 2\rho(m)\,\rho(v)\,\rho(m-u) + 2\rho(m)^2\,\rho(u)\,\rho(v)\}.$$

7. Let $x(n)$ be a sequence of independent identically distributed random variables with zero mean and finite variance. Show that $N^{1/2}r(n)$ is asymptotically normal with zero mean and unit variance. (*Note:* No assumption is here made about the fourth moment.) Show also that the $N^{1/2}r(n)$ are, for any finite set of values of n, jointly asymptotically normal with unit covariance matrix and hence that $N\sum_1^M r(n)^2$ (M fixed) may be used as chi-square with M degrees of freedom to test the independence of $x(n)$.

8. If $x_i(n)$ and $x_j(n)$ are generated by *independent* linear processes show that $N^{1/2}r_{ij}(n)$ is normally distributed with zero mean and variance

$$\sum_{-\infty}^{\infty} \rho_{ii}(m)\,\rho_{jj}(m) = \frac{\int_{-\pi}^{\pi} f_i(\lambda) f_j(\lambda)\,d\lambda}{\left(\int_{-\pi}^{\pi} f_i(\lambda)\,d\lambda \int_{-\pi}^{\pi} f_j(\lambda)\,d\lambda\right)^{1/2}}.$$

(**Note.** There is again no need for an assumption concerning the fourth moment.) Show also that any set of the $\sqrt{N}r_{ij}(n)$ (for a finite set of values of n) are jointly normal with zero mean vector and covariance matrix (for lags n and p)

$$\sum_{m=-\infty}^{\infty} \rho_{ii}(m)\,\rho_{jj}(m+n-p).$$

9. (Olshen, 1967). Let u be an element of a Hilbert space $\mathcal{H}$ and x_n a sequence of elements therein for which $\|x_n\| \le a < \infty$, while (x_m, x_n) converges to zero with n for all fixed m. Show that u may be put in the form

$$u = \sum_1^k \alpha_j x_j + \epsilon + \eta,$$

where η is orthogonal to all x_n and ϵ has arbitrary small norm. Hence show that (u, x_n) converges to zero for any $u \in \mathcal{H}$. Let $w(\lambda)$ be computed from N observations on a scalar sequence $x(n)$. If $x(n)$ is a scalar second order stationary process with a.c. spectrum having $f(\lambda) \le a < \infty$ for all λ show that the sequence $w(\lambda)$ may be taken as x_N in the above result and hence that $w(\lambda)$ converges in distribution to zero as $N \to \infty$. Hence deduce that if $w(\lambda)$ does not converge in mean square to zero it does not converge to any random variable.

10. Give a direct proof not using the ergodicity of $x(n)$ that for a linear process the $c_{ij}(n)$ converge almost surely to the $\gamma_{ij}(n)$.

Hint.

$$C(n) = (N-n)^{-1}\sum_{m=1}^{N-n}\left\{\sum\sum_{-\infty}^{\infty} A(u)\epsilon(m-u)\epsilon'(m+n-v)\,A(v)\right\}.$$

Remove the "squared" terms

$$* \qquad (N-n)^{-1}\sum_{m=1}^{N-n}\sum_{-\infty}^{\infty} A(u)\epsilon(m-u)\epsilon(m-u)'A(n+u)'.$$

Show that the remainder has mean zero and its elements have variance which is $O(N^{-1})$, without any assumptions concerning the fourth moment of the $\epsilon(u)$. Thus show that it converges almost surely to zero as in Theorem 5. Observe that $(N-n)^{-1}\sum \epsilon(m-u)\epsilon'(m-u)$ converges to $G = \mathcal{E}(\epsilon(u)\epsilon(u)')$, almost surely and show that

$$P\left\{\|(N-n)^{-1}\sum_{1}^{N-n}\epsilon(m-u)\epsilon'(m-u)\| \geq \|A(u)\|^{-1}\right\} \leq C\,\|A(u)\|$$

so that almost surely only finitely many of the events in this probability statement occur. Thus using dominated convergence show that * converges to zero almost surely.

11. Show that if the scalar sequence $y^{(N)}(n)$ satisfies (a), (b)′, (c) of Section 3 then

$$\lim_{N\to\infty}\frac{d(N)}{(\log N)^j} = \infty, \qquad j = 0, 1, \ldots$$

Show also that if

$$y^{(N)}(n) = [\{\log (N)\}^{1/2} - \{\log (N-1)\}^{1/2}]^{1/2}$$

that (a), (b)′, (c) are satisfied but

$$\lim_{N\to\infty}\frac{d(N)}{N^{\alpha}} = 0, \qquad \alpha > 0.$$

APPENDIX

1 *Proof of Theorem* 10

We wish to establish the asymptotic normality of the elements of $D_N\hat{B}$ (see (3.10)). Since this is $(D_N^{-1}Y_N'Y_ND_N^{-1})^{-1}(D_N^{-1}Y_N'X_N)$ and the first factor converges almost surely to $R(0)$ we need to establish the central limit theorem only for $D_N^{-1}Y_N'X_N$. We shall have established what we wish if we prove the theorem for an arbitrary linear combination of the elements of $D_N^{-1}Y_N'X_N$. For if that is so and $\phi_N(\theta)$ is the characteristic function for the linear combination then this converges to $\exp(-\frac{1}{2}\theta^2\sigma^2)$ where $\sigma^2 = \alpha'\Sigma\alpha$ and Σ is the limit as N increases of the covariance matrix for the elements of $D_N^{-1}Y_N'X_N$, arranged in some order. (This limit exists as we know from Theorem 8.) Here also α is the vector of coefficients defining the linear combination and $\phi_N(\theta) = \tilde{\phi}_N(\theta\alpha)$ where $\tilde{\phi}_N$ is the joint characteristic function of the elements of $D_N^{-1}Y_N'X_N$. Thus $\tilde{\phi}_N(\theta\alpha)$ converges to $\exp -\frac{1}{2}(\theta^2\alpha'\Sigma\alpha)$ and since α is arbitrary what we wish to prove has been shown. Thus we consider

$$x_N = \sum_{i=1}^{p}\sum_{j=1}^{q} a_{ij}\sum_{n=1}^{N} x_i(n)\left(\frac{y_j^{(N)}(n)}{d_j(N)}\right)$$
$$= \sum_{k=-\infty}^{\infty}\sum_i\sum_j a_{ij}\sum_{n=1}^{N}\alpha_i'(k)\epsilon(n-k)\left(\frac{y_j^{(N)}(n)}{d_j(N)}\right)$$

where $\alpha_i'(k)$ is the ith row of $A(k)$ and

$$x(n) = \sum_{-\infty}^{\infty} A(k)\epsilon(n-k), \qquad \sum_{-\infty}^{\infty}\|A(k)\|^2 < \infty,$$

where the $\epsilon(n)$ are independent and identically distributed with zero mean vector and covariance matrix G. We may rewrite x_N as

$$x_N = \sum_{u=-\infty}^{\infty}\left\{\sum_i\sum_j\sum_{n=1}^{N} a_{ij}\left(\frac{y_j^{(N)}(n)}{d_j(N)}\right)\alpha_i'(n-u)\right\}\epsilon(u) = \sum_{u=-\infty}^{\infty}\eta_{N,u},$$

let us say. Of course $\eta_{N,u}$ depends on the a_{ij} but since we hold these fixed for the moment we do not introduce them into the notation. We have already established in Theorem 8 that the variance of x_N converges to

$$(1) \qquad 2\pi\int_{-\pi}^{\pi}\sum_{ijkl} a_{ij}a_{kl}f_{ik}(\lambda)M_{lj}(d\lambda) < \infty.$$

The $\eta_{N,u}$ are independent random variables with finite variance, and it follows that we can find a sequence k_N, $k_N \to \infty$ as $N \to \infty$, so that, putting

$$y_N = \sum_{u=-k_N}^{k_N}\eta_{N,u},$$
$$\lim_{N\to\infty}\mathcal{E}\{(x_N - y_N)^2\} = 0.$$

Thus we need only to establish the theorem for the y_N sequence. For this it is sufficient to show (see Gnedenko and Kolmogoroff, 1954, p. 109) that for every $\delta > 0$ Lindeberg's condition holds, namely, that

$$(2) \qquad \sum_{u=-k_N}^{k_N}\int_{|z|>\delta} z^2\, F_{N,u}(dz) \to 0,$$

where $F_{N,u}$ is the distribution function of $\eta_{N,u}$. Now consider the special case where

$$\eta_{N,u} = \eta_{N,u}(i,j) = \sum_{n=1}^{N}\left(\frac{y_j^{(N)}(n)}{d_j(N)}\right)\alpha_i'(n-u)\epsilon(u).$$

Following Eicker (1965), we first observe that from (b) of Section 3

$$\lim_{N\to\infty} \max_{1\le n\le N} \frac{|y_j^{(N)}(n)|}{d_j(N)} = 0.$$

Thus there are integers m_N, $m_N \to \infty$, so that

$$\lim_{N\to\infty} m_N \max_{1\le n\le N} \frac{|y_j^{(N)}(n)|}{d_j(N)} = 0.$$

Let u_N be the integer for which

$$\sup_{-k_N\le u\le k_N} \left\| \sum_{n=1}^{N} \left(\frac{y_j^{(N)}(n)}{d_j(N)}\right) \alpha_i(n-u) \right\|$$

is attained. Then

$$\left\| \sum_{n=1}^{N} \left(\frac{y_j^{(N)}(n)}{d_j(N)}\right) \alpha_i(n-u_N) \right\| \le \left\| \sum{}' \left(\frac{y_j^{(N)}(n)}{d_j(N)}\right) \alpha_i(n-u_N) \right\|$$

$$+ \left\| \sum{}'' \left(\frac{y_j^{(N)}(n)}{d_j(N)}\right) \alpha_i(n-u_N) \right\|,$$

where $\sum'$ is over n for which $|n - u_N| > [\frac{1}{2}m_N]$, whereas $\sum''$ is over the remaining n, $1 \le n \le N$. The right side of the last expression is dominated by

$$\left\{ \sum_{v>[1/2m_N]} \|\alpha_i(v)\|^2 \right\}^{1/2} + m_N \sup_v \|\alpha_i(v)\| \max_n \left\| \frac{y_j^{(N)}(n)}{d_j(N)} \right\|$$

which evidently converges to zero uniformly in i, j. Now

$$\sup_{-k_N\le u\le k_N} \mathcal{E}(\eta_{N,u}^2\,(i,j)) \le \mu \sup_{-k_N\le u\le k_N} \left\| \sum_{N=1}^{N} \left(\frac{y_j^{(N)}(n)}{d_j(N)}\right) \alpha_i(n-u) \right\|^2,$$

where μ is the greatest eigenvalue of G.

Now

$$\eta_{N,u} = \sum\sum a_{ij}\eta_{N,u}(i,j)$$

and thus

$$\sup_{-k_N\le u\le k_N} \{\mathcal{E}(\eta_{N,u}^2)\}^{1/2} \le \mu\sum\sum |a_{ij}| \sup_{-k_N\le u\le k_N} \left\| \sum_{n=1}^{N} \left(\frac{y_j^{(N)}(n)}{d_j(n)}\right) \alpha_i(n-u) \right\|,$$

which converges to zero as $N \to \infty$. Now let $q_{N,u}$ be the variance of $\eta_{N,u}$. Then (2) is

$$\sum_{-k_N}^{k_N} q_{N,u}^2 \int_{|x|>\delta(N,u)} x^2 G_{N,u}(dx), \qquad \delta(N,u) = \frac{\delta}{q_{N,u}},$$

$$< C \max_{-k_N\le u\le k_N} \int_{|x|>\delta(N,u)} x^2 G_{N,u}(dx),$$

where C bounds the sum the $q_{N,u}^2$ and $G_{N,u}$ is the distribution function of $q_{N,u}^{-1}\eta_{N,u}$; but $q_{N,u}^{-1}\eta_{N,u}$ is of the form $\psi_{N,u}'\epsilon(\mu)$, $\psi_{N,u}'G\psi_{N,u} \equiv 1$, so that the region for which this is greater in modulus than $\delta(N, u)$ is contained in the region for which $\epsilon'(u)G^{-1}\epsilon(u) \geq \delta^2(N, u)$, since

$$\begin{aligned}\psi'_{N,u}\epsilon(u)\,\epsilon(u)'\psi_{N,u} &= (\psi'_{N,u}G^{1/2})(G^{-1/2}\epsilon(u)\,\epsilon(u)'G^{-1/2})(G^{1/2}\psi_{N,u}) \\ &= \epsilon(u)'G^{-1/2}(G^{1/2}\psi_{N,u}\psi'_{N,u}G^{1/2})G^{-1/2}\epsilon(u) \\ &\leq \epsilon(u)'G^{-1}\epsilon(u).\end{aligned}$$

Thus since $\epsilon(u)'G^{-1}\epsilon(u)$ has a distribution independent of u and $\delta(N, u) \to \infty$, since $(\max_u q_{N,u})^{-1}$ does so, the theorem is established. It is evident from the proof that the $\epsilon(u)$ do not need to be identically distributed but only that

$$\int_{x>\delta(N,u)} x^2\, F_u(dx)$$

should converge to zero uniformly in u where F_u is the distribution function of $\{\epsilon(u)'G^{-1}\epsilon(u)\}^{1/2}$. For further details see Eicker (1965).

2 *Proofs of Theorems* 11 *and* 12

Theorem 11 would follow immediately from Theorem 10 if $f(\lambda)$ were required to be piecewise continuous. However, it may be observed that the piecewise continuity of $f(\lambda)$ was used in Theorem 10 only through its use in Theorem 8, where the asymptotic covariance matrix was established. However, we already know from Corollary 4 that the covariance matrix for $\sqrt{N}\,\bar{x}$ is as required by Theorem 11 when $f(\lambda)$ is continuous at the origin so that this theorem needs no further discussion. The situation is similar for Theorem 12. We consider the case $\lambda \neq 0, \pi$ and leave the details of the other two cases to the reader. Now

$$K_N(\theta) = D_N^{-1} \sum_{m,n=1}^{N}\sum y^{(N)}(m)\, y^{(N)}(n)' e^{i(n-m)\theta}\, D_N^{-1}$$

has typical element

$$\frac{2}{N}\sum_{m}\sum_{n} \begin{matrix}\cos\\ \sin\end{matrix}\, m\lambda_u \begin{matrix}\cos\\ \sin\end{matrix}\, n\lambda_v\, e^{i(n-m)\theta},$$

where the notation indicates the either cosine or sine may be taken in either place and the λ_u are as in (4.2.) For N sufficiently large, when $\lambda \in (0, \pi)$, then also $\lambda_u \in (0, \pi)$, $u = 1, \ldots, m$, since $\lambda \neq 0, \pi$. Then, defining u as at the beginning of the proof of Theorem 8, the covariance matrix we require is

$$(I_p \times D_N^{-1})\, \mathcal{E}(uu')(I_p \otimes D_N^{-1}) = \int_{-\pi}^{\pi} f(\theta) \otimes K_N(\theta)\, d\theta. \tag{3}$$

The elements of the matrix $K_N(\theta)$ are made up linearly of terms of the form

$$\frac{1}{\sqrt{N}}\sum_{1}^{N} e^{-in(\theta \pm \lambda_u)}\, \frac{1}{\sqrt{N}}\sum_{1}^{N} e^{in(\theta \pm \lambda_v)},$$

the coefficients being $\pm 1/2$ or $\pm i/2$. Consider

$$(4)\quad \int_{-\pi}^{\pi} f(\theta)\frac{1}{\sqrt{N}}\sum_{1}^{N} e^{-in(\theta-\lambda_u)}\frac{1}{\sqrt{N}}\sum_{1}^{N} e^{in(\theta\pm\lambda_v)}\,d\theta$$

$$= \int_{-\pi}^{\pi} f(\lambda)\frac{1}{\sqrt{N}}\sum_{1}^{N} e^{-in(\theta-\lambda_u)}\frac{1}{\sqrt{N}}\sum_{1}^{N} e^{in(\theta\pm\lambda_v)}\,d\theta$$

$$+ \int_{-\pi}^{\pi} (f(\theta)-f(\lambda))\frac{1}{\sqrt{N}}\sum_{1}^{N} e^{-in(\theta-\lambda_u)}\frac{1}{\sqrt{N}}\sum_{1}^{N} e^{in(\theta\pm\lambda_v)}\,d\theta.$$

The first integral vanishes unless $\lambda_u = \lambda_v$, since it is

$$2\pi f(\lambda)\frac{1}{N}\sum_{n=1}^{N} e^{in(\lambda_u\pm\lambda_v)} = 2\pi f(\lambda)\delta_u{}^v.$$

The last two factors in the second integral have squared moduli which are $L_N(\theta - \lambda_u)$, $L_N(\theta \pm \lambda_v)$, respectively. (See Section 3 of the Mathematical Appendix.) The proof that the second term converges to zero with N therefore hardly differs from the proof that the Césaro mean of the Fourier series of a function continuous at λ converges to the value of the function at that point (in this case that value is zero), the only additional (slight) complexity arising from the fact that λ_u, λ_v differ by up to $\pi m N$ from λ. Following the proof in Section 3 of the Mathematical Appendix we may still find an interval, of width δ, about λ within which $\|f(\theta) - f(\lambda)\| < \epsilon$ and outside of which $L_N(\theta - \lambda_u) < \epsilon$, $L_N(\theta - \lambda_v) < \epsilon$. Also $L_N(\theta + \lambda_v) < \epsilon$ except within the same interval about $-\lambda$. Taking, for example, the case where $+\lambda_v$ is used in the second factor, the second term in (4) is dominated by

$$\epsilon\int_{|\theta-\lambda|\le\delta} L_N(\theta-\lambda_u)^{1/2}L_N(\theta+\lambda_v)^{1/2}\,d\theta + \int_{|\theta-\lambda|>\delta}\|f(\theta)-f(\lambda)\|\,L_N(\theta-\lambda_u)^{1/2} \times L_N(\theta+\lambda_v)^{1/2}\,d\theta$$

$$\le 2\pi\epsilon + \epsilon^{1/2}\int_{|\theta-\lambda|>\delta}\|f(\theta)-f(\lambda)\|\,L_N(\theta+\lambda_v)^{1/2}\,d\theta$$

$$\le \epsilon\left\{2\pi + \int\|f(\theta)-f(\lambda)\|\,d\theta\right\} + \epsilon^{1/2}\int_{|\theta+\lambda|<\delta}\|f(\theta)-f(\lambda)\|\,L_N(\theta+\lambda_v)^{1/2}\,d\theta$$

$$\le \epsilon\left\{2\pi + \int\|f(\theta)-f(\lambda)\|\,d\theta\right\} + c\epsilon^{1/2} = O(\epsilon^{1/2}),$$

since $f(\theta)$ is continuous at $-\lambda$.

Thus the only contribution to (4) comes when we take $\pm\lambda_v$ as $-\lambda_u$. Similarly, we get a contribution taking $+\lambda_u$ in the place of $-\lambda_u$ and replacing $\pm\lambda_v$ by $+\lambda_u$. Putting these contributions together with the appropriate coefficients, we obtain

$$2\pi f(\lambda)\otimes V_q + \overline{2\pi f(\lambda)\otimes V_q},$$

as required.

It may be observed that even if f was not continuous at the point λ but the right- and left-hand limits $f(\lambda + 0)$ and $f(\lambda - 0)$ existed the theorem would remain true with $f(\lambda)$ replaced by $\frac{1}{2}(f(\lambda+0)+f(\lambda-0))$. If $\lambda = 0$, since $f(-\lambda) = \overline{f(\lambda)}$, this extension becomes vacuous.

3 *Proof of Theorem* 10′ *in Section* 4

Since (b)′ implies (b), see Section 3, the formula for the asymptotic variance of $\hat{\beta}$ is not affected. We thus consider u once more and, following Rosenblatt (1962), put

$$u = \sum_j u(j) + \sum_k v(k),$$

where

$$u(j) = \sum_{S_j} x(n) \otimes y^{(N)}(n), \qquad v(k) = \sum_{T_k} x(n) \otimes y^{(N)}(n).$$

Here

$$S_j = \{n \mid j(a(N)+b(N)) < n \le (j+1)\,a(N) + jb(N)\},$$
$$0 \le j < c(N), \qquad c(N) = [N\{a(N)+b(N)\}^{-1}]$$

where $a(N)$, $b(N)$, $c(N)$ are integers and $a(N), b(N), c(N) \to \infty$, $b(N)/a(N) \to 0$ and

$$T_k = \{n \mid (k+1)\,a(N) + kb(N) < n \le (k+1)(a(N)+b(N))\},$$
$$k = 0, 1, \dots, c(N)-1,$$
$$T_{c(N)} = \{n \mid c(N)(a(N)+b(N)) < n \le N\}.$$

Put $S = \bigcup S_j$, $T = \bigcup T_k$ so that $S \cup T$ is the set of integers from 1 to N. We first show that $(I_p \otimes D_N^{-1}) \sum_k v(k)$ converges in probability to zero. Indeed this has covariance matrix

$$\int_{-\pi}^{\pi} f(\lambda) \otimes \sum\sum_T [\{D_N^{-1} y^{(N)}(m)\, y^{(N)}(n)' D_N^{-1}\} e^{i(n-m)\lambda}]\, d\lambda$$
$$\le 2\pi \sup_\lambda \|f(\lambda)\|\; I_p \otimes \sum_T D_N^{-1} y^{(N)}(n)\, y^{(N)}(n)'\, D_N^{-1}.$$

But since $\{c(N)\,b(N)\}/\{c(N)\,a(N)\} \to 0$ then certainly $c(N)\,b(N)/N \to 0$ and by condition (b)′ the diagonal elements of the nonnegative definite matrix which is the second factor in the tensor product converge to zero and thus the tensor product does so also. Thus we need consider only $(I_p \otimes D_N^{-1}) \sum_j u(j)$. We consider

$$\frac{\sum_u \sum_{S_u} x_i(n)\, y_j^{(N)}(n)}{d_j(N)} = \sum_u \eta_{N,u}(i,j)$$

Now calling $k_i(.,.,.,.)$ the fourth cumulant function of $x_i(n)$

$$\sum_u \mathcal{E}\{\eta_{N,u}^4(i,j)\} = \sum_u \Bigg(\sum_{\substack{m,n,p,q \\ \in S_u}}\sum\sum\sum \{\gamma_i(n-m)\,\gamma_i(q-p) + \gamma_i(p-m)\,\gamma_i(q-n)$$
$$+ \gamma_i(q-m)\,\gamma_i(p-n) + k_i(0, n-m, p-m, q-m)\}\, y_j^{(N)}(m)\, y_j^{(N)}(n)$$
$$\times\, y_j^{(N)}(p)\, y_j^{(N)}(q)/d_j^4(N) \Bigg).$$

This is not greater than

$$(5)\quad 12\pi^2\left\{\sup_\lambda f_i(\lambda)\right\}^2 \sum_u \left\{\frac{\sum_{S_u} y_j^{(N)}(n)^2}{d_j^2(N)}\right\}^2 + \sum\sum\sum\sum\sum_{umnpq} k_i(0, n-m, p-m, q-m) \times \left\{\frac{y_j^{(N)}(m)\, y_j^{(N)}(n)\, y_j^{(N)}(p)\, y_j^{(N)}(q)}{d_j^4(N)}\right\}.$$

Since $a(N)/N \to 0$, then

$$\frac{\sum_{S_u} y_j^{(N)}(n)^2}{d_j^2(N)} \to 0,$$

by condition (b)′. Since also

$$\frac{\sum_u \sum_{S_u} y_j^{(N)}(n)^2}{d_j^2(N)} \le 1,$$

the first term in (5) converges to zero for all i, j. The second term is bounded by

$$\sup_{1\le n\le N}\left\{\frac{y_j^{(N)}(n)}{d_j(N)}\right\}^2 \sum\sum\sum_{abc=-\infty}^{\infty}\left\{|k_i(0, a, b, c)| \sum_u \sum_{m\in S_u}\left|\frac{y_j^{(N)}(m)\, y_j^{(N)}(m+a)}{d_j^2(N)}\right|\right\},$$

where we adopt the convention of putting $y_j^{(N)}(n) = 0$ unless $0 < n \le N$. The last displayed expression is not greater than

$$\sup_{1\le n\le N}\left\{\frac{y_j^{(N)}(n)}{d_j(N)}\right\}^2 \sum\sum\sum_{abc} |k_i(0, a, b, c)|,$$

since, under the just-stated convention,

$$\sum_u \sum_{m\in S_u}\left\{\left|\frac{y_j^{(N)}(m)y_j^{(N)}(m+a)}{d_j^2(N)}\right|\right\} \le \sum_{m=1}^{N}\left|\frac{y_j^{(N)}(m)y_j^{(N)}(m+a)}{d_j^2(N)}\right|,$$

which is not greater than unity by Schwartz's inequality. Because of (b)′ and condition (4.2) of the theorem being proved, the second term in (5) converges to zero and thus the expression (5) does so in its entirety.

Now call

$$(6)\qquad \eta_{N,u} = \sum\sum a_{ij}\eta_{N,u}(i,j)$$

and consider

$$\sum_u \eta_{N,u}.$$

It is this whose asymptotic normality we must establish by the same argument as was used at the beginning of the proof of Theorem 10. Now

$$[\mathcal{E}\{\eta_{N,u}^4\}]^{1/4} \le \sum\sum_{ij} |a_{ij}|\, [\mathcal{E}\{\eta_{N,u}^4(i,j)\}]^{1/4},$$

and thus

$$\sum_u \mathcal{E}\{\eta_{N,u}^4\}$$

converges to zero.

By Liapounoff's form of the central limit theorem (Gnedenko and Kolmogoroff, 1954, p. 103) $\sum_u \eta_{N,u}$ would be normal if the $\eta_{N,u}$ were independent. To show that the central limit theorem holds as if this were the case consider the event

$$E(j, m_j) = \{m_j\, c(N)^{-2} < \eta_{N,j} \le (m_j + 1)\, c(N)^{-2}\}.$$

Then

$$\sum{}' P\Big\{\bigcap_j E(j, m_j)\Big\} \le P\Big\{\sum_j \eta_{N,j} \le x\Big\} \le \sum{}'' P\Big\{\bigcap_j E(j, m_j)\Big\},$$

where $\sum'$ is over all different sets of $c(N)$ integers m_j such that

$$\frac{\sum m_j + c(N)}{c(N)^2} \le x,$$

whereas $\sum''$ is similarly over all such sets for which

$$\frac{\sum m_j}{c(N)^2} \le x.$$

To see this observe that the events $E(j, m_j)$ with the integers m_j varying arbitrarily constitute a subdivision of the whole $c(N)$ dimensional space into nonintersecting rectangle sets. Clearly those events whose probabilities are summed on the left each imply the middle event so that the first inequality holds. On the other hand, each $\eta_{N,j}$ must lie in some half-open interval $(m_j c(N)^{-2}, (m_j + 1)\, c(N)^{-2}]$ and, if the middle event happens, certainly the last displayed condition must hold. The right-hand inequality holds also. Let $F_{j+1,N}$ be the distribution function of $\eta_{N,j}$, $j = 0, \dots, c(N) - 1$. (We number the $F_{j,N}$ this way only to avoid a complicated subscript.) Then

$$F_{1,N}^* F_{2,N}^* \cdots {}^*F_{c(N),N}(x - c(N)^{-1}) \le \sum{}' \prod_j P(E(j, m_j)), \tag{7}$$

since $\sum_j \eta_{N,j} \le x - c(N)^{-1}$ implies that for some set of m_j each of $E(j, m_j)$ will happen.

Similarly

$$\sum{}'' \prod_j P(E(j, m_j)) \le F_{1,N}^* F_{2,N}^* \cdots {}^*F_{c(N),N}(x + c(N)^{-1}). \tag{8}$$

We now show that we may replace $\sum' P(\bigcap_j E(j, m_j))$ by the right-hand side of (7) and similarly $\sum'' P(\bigcap_j E(j, m_j))$ by the left-hand side of (8) in the sense that the difference between the two sets of expressions may be made to approach zero by a suitable choice of $a(N)$, $b(N)$, $c(N)$. This establishes the central limit theorem, since we already know that

$$F_{1,N}^* F_{2,N}^* \cdots {}^*F_{c(N),N}$$

approaches a normal distribution.

In the first place

$$P\left\{\max_j |\eta_{N,j}| > c(N)\frac{k}{\sqrt{\epsilon}}\right\} < \epsilon,$$

for a suitable constant k, for

$$\mathcal{E}\left\{\max_j |\eta_{N,j}|\right\}^2 \le \mathcal{E}\left\{\sum_j |\eta_{N,j}|\right\}^2 \le c(N)^2k^2.$$

In the second place

$$\left|P\left(\bigcap_{j=0}^{c(N)-1} E(j, m_j)\right) - \prod_{j=0}^{c(N)-1} P(E(j, m_j))\right| \le c(N)g(b(N)).$$

(Here g is the function occurring in the uniform mixing condition.) This is so because the sets $E(j, m_j)$ involve the $x(n)$ for $n \in S_j$ each of which is separated from every other by at least $b(N)$ time units. Repeated application of the uniform mixing condition establishes the result. There are $\{2c(N)^3k\epsilon^{-1/2}\}^{c(N)}$ sets of $E(j, m_j)$ which retain $\max |\eta_{N,j}|$ within the bound $c(N)k\epsilon^{-1/2}$. Thus

$$\left|\sum{}'' P\left(\bigcap_j E(j, m_j)\right) - \sum{}'' \prod P(E(j, m_j))\right| \le c(N)\{2c(N)^3k\epsilon^{-1/2}\}^{c(N)} g(b(N)) + \epsilon,$$

as may be seen by breaking up the sample space into that part for which

$$\max_j |\eta_{N,j}| \le c(N)k\epsilon^{-1/2}$$

and its complement. The same result, of course, holds for $\sum'$. If we agree to take $g(x) > (1 + x)^{-1}$, $x \ge 0$ (which costs nothing), we may choose $c(N)$ so that

$$c(N) \le \{-\log g(b(N))\}^{1/2},$$

since

$$\log\left(c(N)\{2c(N)^3k\epsilon^{-1/2}\}^{c(N)}\right) = \alpha c(N) + 3c(N)\log c(N) + \log c(N),$$

which is of smaller order than $c(N)^2$ as $c(N) \to \infty$. Now, if $b(N) = N^{1/2}$, $a(N) = N/c(N) - N^{1/2}$, we see that $b(N) \to \infty, c(N) \to \infty$,

$$a(N) \ge \frac{N}{\{\log(1 + b(N))\}^{1/2}} - N^{1/2} \to \infty,$$

$$\frac{a(N)}{b(N)} = \left(\frac{N^{1/2}}{\{\log(1 + b(N))\}^{1/2}}\right) - 1 \to \infty,$$

and the result is established.

4 *Proof of Theorems* 11′ *and* 12′

If the $y^{(N)}(n)$ are unity, then $\eta_{N,u}$ is

$$\frac{1}{\sqrt{N}}\sum_{S_u} a'\, x(n),$$

and when $x(n)$ is strictly stationary this has a distribution independent of u. We can now replace Liapounoff's condition by Lindeberg's condition. As will be seen we could do this with Theorem 10′ also and obtain a necessary and sufficient condition for that theorem to hold, under the condition of strict stationarity. However, the condition then becomes rather complicated to interpret and we have considered the strict stationarity only in Theorems 11′ and 12′.

The necessity of the condition (4.2′) is immediate, for if F_N is the distribution function of $\sqrt{N}\,\bar{x}_N$ and F is the (normal) distribution function to which this converges we know that

$$\lim_{N\to\infty} \int \|x\|^2 F_N(dx) = \int \|x\|^2 F(dx)$$

by Theorem 8, the integrals being taken over all of p-dimensional Euclidean space. Moreover, for all $a > 0$

$$\lim_{N\to\infty} \int_{\|x\|\le a} \|x\|^2 F_N(dx) = \int_{\|x\|\le a} \|x\|^2 F(dx),$$

if the theorem holds, so that F_N converges to F. Thus

$$\lim_{N\to\infty} \int_{\|x\|>a} \|x\|^2 F_N(dx) = \int_{\|x\|>a} \|x\|^2 F(dx)$$

and condition (4.2′) certainly holds.

The only part of the proof of Theorem 10′ needing consideration is that dealing with the distribution of $\eta_{N,u}$ when these are assumed independent. Then the necessary and sufficient condition that the distribution of

$$\sum_u \eta_{N,u}$$

converge to a limiting normal distribution is (2) of this Appendix, save that now $0 \le y \le c(N) - 1$. In our present case this becomes, since $F_{N,u}$ is independent of u,

$$\lim_{N\to\infty} c(N) \int_{|x|>\delta} x^2 F_{N,u}(dx) = 0.$$

Since $N/a(N)$ converges to $c(N)$, this becomes

$$\lim_{N\to\infty} \int_{|x|>\delta c(N)} x^2\, H_N(dx) = 0,$$

where H_N is the distribution function of

$$\frac{1}{a(N)^{1/2}} \sum_1^{a(N)} a'\, x(n).$$

A sufficient condition for this to hold is

$$\lim_{a\to\infty} \overline{\lim_{N\to\infty}} \int_{|x|>a} x^2\, F_{a,N}(dx),$$

where $F_{a,N}$ is the distribution function of

$$\frac{1}{\sqrt{N}} \sum_1^N a' \, x(n).$$

Since $|a' \sqrt{N} \bar{x}_N|^2 \leq a' a N \bar{x}_N' \bar{x}_N$, we see that the condition (4.2′) for $\lambda = 0$ in turn becomes a sufficient condition that $\sqrt{N} a' \bar{x}_N$ be asymptotically normal for all $a \neq 0$, which establishes the theorem.

For Theorem 12′ the situation is only slightly more complicated. Now we consider

$$u_{N,j} = \left(\frac{2}{N}\right)^{1/2} \sum_{S_j} x(n) \cos n\lambda_N, \qquad v_{N,j} = \left(\frac{2}{N}\right)^{1/2} \sum_{S_i} x(n) \sin n\lambda_N,$$

where λ_N is a multiple of $2\pi/N$ which converges to λ. We take the case $\lambda \neq 0, \pi$ since the first of these other cases has already been dealt with and the second hardly differs from that first one. We assume f continuous at λ. Now $u_{N,j}, v_{N,j}$ have distributions depending on j, even assuming $x(n)$ to be stationary. However, if

$$a_{N,j} = j\{a(N) + b(N)\}$$

and

$$\tilde{u}_{N,j} = \left(\frac{2}{N}\right)^{1/2} \sum_{jS} x(n) \cos \{(n - a_{N,j})\lambda_N\},$$

$$\tilde{v}_{N,j} = \left(\frac{2}{N}\right)^{1/2} \sum_{S_j} x(n) \sin \{(n - a_{N,j})\lambda_N\},$$

then

$$u_{N,j} = \tilde{u}_{N,j} \cos a_{N,j}\lambda_N - \tilde{v}_{N,j} \sin a_{N,j}\lambda_N,$$

$$v_{N,j} = \tilde{v}_{N,j} \cos a_{N,j}\lambda_N + \tilde{u}_{N,j} \sin a_{N,j}\lambda_N.$$

Now $\tilde{u}_{N,j}, \tilde{v}_{N,j}$ have the same distribution as $u_{N,0}, v_{N,0}$ for each j. Again we have only to consider the situation in which each sequence $\{u_{N,j}, v_{N,j}\}$, $j = 0, 1, \ldots, c(N) - 1$, consists of independent components. The condition of the type of (2) of this Appendix for the asymptotic normality of $\sum \eta_{N,j}$ now is

$$\lim_{N\to\infty} \sum_0^{c(N)-1} \int_{|x|>\delta} x^2 \, F_{j,N}(dx) = 0, \qquad \delta > 0,$$

where $F_{j+1,N}$ is the distribution function of

$$\eta_{N,j} = a_j' u_{N,j} + b_j' v_{N,j}$$

and a_j, b_j are arbitrary vectors of real constants. However,

$$\eta_{N,j} = (a_j \cos a_{N,j}\lambda_N + b_j \sin a_{N,j}\lambda_N)' \tilde{u}_{N,j} + (b_j \cos a_{N,j}\lambda_N - a_j \sin a_{N,j}\lambda_N)' \tilde{v}_{N,j}$$

and thus

$$|\eta_{N,j}| \leq (a_j' a_j + b_j' b_j)^{1/2} \, (\tilde{u}_{N,j}' \tilde{u}_{N,j} + \tilde{v}_{N,j}' \tilde{v}_{N,j})^{1/2}.$$

A sufficient condition for asymptotic normality is

$$\lim_{N\to\infty} c(N) \int_{x>\delta} x^2 G_N(dx),$$

where G_N is the distribution function of $\tilde{u}'_{N,j}\tilde{u}_{N,j} + \tilde{v}'_{N,j}\tilde{v}_{N,j}$. Now, proceeding precisely as in the proof of Theorem 11′, we arrive at condition (4.2′) as sufficient. The necessity is established precisely as for Theorem 11′.

5 *Proof of Theorem* 14

We first consider the quantities

$$\hat{x}(n) = \sum_{-K}^{K} A(j)\epsilon(n-j),$$

which are generated by a finite moving average of independent identically distributed random variables. Now the quantities

$$\sqrt{N}\,\hat{c}_{ij}(n) = \frac{1}{\sqrt{N}} \sum_{m=1}^{N} \hat{x}_i(m)\,\hat{x}_j(m+n),$$

for a fixed finite set of n values, may be written in the form

$$\frac{1}{\sqrt{N}} \sum_{1}^{N} y_u(m), \qquad u = 1, 2, \ldots, r$$

where y_u is $\hat{x}_i(m)\,\hat{x}_j(m+n)$ for some choice of i, j, n, and the $y_u(m)$ are finitely dependent in the sense that $y_u(m)$, $y_v(p)$ are independent if $|m-p|$ is larger than some fixed finite integer (which will be $(2K+q)$, where q depends on the values of n involved). Now, of course, the $y_u(m)$ constitute a vector process which certainly is strictly stationary and satisfies the uniform mixing condition. Moreover, the spectrum of this vector process (after mean correction of the y_u) is absolutely continuous with continuous density. Thus the $\sqrt{N}\{\hat{c}_{ij}(n) - \hat{\gamma}_{ij}(n)\}$ are asymptotically normal by Theorem 10′, putting $\hat{\gamma}_{ij}(n) = \mathcal{E}\{\hat{c}_{ij}(n)\}$.

We first prove the following lemma due to Bernstein (1926) which we also use later in the book. We shall refer to it as *Bernstein's Lemma. Let x_N be a sequence of vector valued random variables with zero mean such that for every $\epsilon > 0$, $\zeta > 0$, $\eta > 0$ there exist sequences of random vectors $y_N(\epsilon)$, $z_N(\epsilon)$* so that $x_N = x_N(\epsilon) + z_N(\epsilon)$ *where $y_N(\epsilon)$ has a distribution converging to the multivariate normal distribution with zero mean vector and covariance matrix $\Sigma(\epsilon)$, and*

$$(9) \qquad \lim_{\epsilon\to 0} \Sigma(\epsilon) = \Sigma, \qquad P(z_N(\epsilon)'\,z_N(\epsilon) > \zeta^2) < \eta.$$

Then the distribution of x_N converges to the multivariate normal distribution with covariance matrix Σ.

Proof. If the second half of (9) is true for any $\eta > 0$, $\zeta > 0$ then it is also true for all such η, ζ for $\alpha'\,z_N(\epsilon)$ where α is an arbitrary vector of real numbers and the same goes for the first half of (9) with $\Sigma(\epsilon)$, Σ replaced by $\alpha'\Sigma(\epsilon)\alpha$, $\alpha'\Sigma\alpha$. Thus by the same argument as was used at the commencement of the proof of Theorem 10 it is sufficient to prove the lemma for the scalar case. We then take $\alpha' \mathfrak{F} \alpha = \sigma^2$. Thus, taking $t_1 > t_0$ and ζ sufficiently small, we may define sets E_j as follows:

$$E_1 = \{\omega \mid |z_N(\epsilon)| \leq \zeta\}, \qquad E_2 = \{\omega \mid t_0 + \zeta < y_N(\epsilon) < t_1 - \zeta\},$$

$$E_3 = \{\omega \mid t_0 < x_N < t_1\}, \qquad E_4 = \{\omega \mid t_0 - \zeta < y_N(\epsilon) < t_1 + \zeta\}.$$

Now $P(E_1) \geq 1 - \eta$ by (9) and

$$P(E_3) \geq P(E_1 \cap E_2) \geq P(E_1) + P(E_2) - 1 \geq P(E_2) - \eta,$$
$$P(E_4) \geq P(E_1 \cap E_3) \geq P(E_1) + P(E_3) - 1 \geq P(E_3) - \eta.$$

Thus

$$P(E_2) - \eta \leq P(E_3) \leq P(E_4) + \eta.$$

Also for ϵ and ζ sufficiently small and N sufficiently large

$$P(E_2) + \eta_1 = \frac{1}{\sqrt{2\pi}} \int_{t_0}^{t_1} e^{-\frac{1}{2}x^2/\sigma^2}\, dx = P(E_3) + \eta_2,$$

where $|\eta_1| < \eta$, $|\eta_2| < \eta$. Thus for N sufficiently large

$$P(E_3) = \frac{1}{\sqrt{2\pi}} \int_{t_0}^{t_1} e^{-\frac{1}{2}x^2/\sigma^2}\, dx + \eta_3, \qquad |\eta_3| < 3\eta,$$

and the lemma is established.

Now we return to the proof of Theorem 14. We first observe that we may consider $\tilde{c}_{ij}(n)$ in place of $c_{ij}(n)$ since we already know that $\sqrt{N}\{\tilde{c}_{ij}(n) - c_{ij}(n)\}$ converges almost surely to zero. (See the proof of Theorem 6 in this present chapter.) We put $x_N = \sqrt{N}\{\tilde{c}_{ij}(n) - \gamma_{ij}(n)\}$, $y_N(\epsilon) = \sqrt{N}\{\hat{c}_{ij}(n) - \hat{\gamma}_{ij}(n)\}$. We may, by taking K sufficiently large, make the variance of $y_N(\epsilon)$, in its limiting distribution, as near to that to which the variance of x_N converges as we wish and the same is true of the covariances between the $y_N(\epsilon)$ for different pairs (i, j) in relation to the corresponding covariances for the x_N for the same pairs. Indeed, this follows from (3.3) for the first two terms in the limit, as $N \to \infty$, of that expression depend only on the matrix $f(\lambda)$ in a simple way (see text below (3.6)), and if $f_K(\lambda)$ is the spectrum of $\hat{x}(n)$ then $f_K(\lambda)$ converges uniformly to $f(\lambda)$. The evaluation of the last term in (3.3) given in (3.6) shows that this last term for $\hat{x}(n)$ and $x(n)$ may be made as near to each other as is desired by taking K large enough. Thus, since $z_N(\epsilon) = x_N - y_N(\epsilon)$ has zero mean, then by Chebyshev's inequality we need only to prove that the variance of $z_N(\epsilon)$, namely,

$$N \operatorname{var}\{\tilde{c}_{ij}(n) - \hat{c}_{ij}(n)\}, \tag{10}$$

converges to zero.

Now let us put $z(n) = x(n) - \hat{x}(n)$. Then

$$\sqrt{N}\{\tilde{c}_{ij}(n) - \hat{c}_{ij}(n)\}$$
$$= \frac{1}{\sqrt{N}} \sum_m \{z_i(m)\hat{x}_j(m+n) + \hat{x}_i(m)z_j(m+n) + z_i(m)\, z_j(m+n)\},$$

and we need to prove that each of the three terms coming from this decomposition has variance converging to zero uniformly in N. The first term has mean square

$$N^{-1}\sum_{l,m=1}^{N}\sum \mathcal{E}\Big[\sum_{|t|>K}\sum_{p}\alpha_{ip}(t)\epsilon_p(l-t)\sum_{|u|\le K}\sum_{q}\alpha_{jq}(u)\epsilon_q(l+n-u)$$
$$\times\sum_{|v|>K}\sum_{r}\alpha_{ir}(v)\epsilon_r(m-v)\sum_{|w|\le K}\sum_{s}\alpha_{js}(w)\epsilon_s(m+n-w)\Big],$$

and, after allowing for the mean correction, this gives a variance

$$N^{-1}\sum_{pqrs}\sum\sum\sum\Big[\sum_{l,m}\sum\Big\{\sigma_{pr}\sigma_{qs}\sum_t{}'\sum_u{}''\alpha_{ip}(t)\alpha_{ir}(m-l+t)\alpha_{jq}(v)\alpha_{js}(m-l+v)$$
$$+\sigma_{ps}\sigma_{qr}\sum_t{}'''\sum_u{}'''\alpha_{ip}(t)\alpha_{js}(m+n-l+t)\alpha_{jq}(u)\alpha_{ir}(m-n-l+u)$$
$$+\kappa_{pqrs}\sum_t\alpha_{ip}(t)\alpha_{jq}(t+n)\alpha_{ir}(t+m-l)\alpha_{js}(t+m+n-l)\Big\}\Big],$$

wherein $\sum'$ in each case runs over a range within that defined by $|t| > K$, $\sum''$ over a range within that defined by $|u| \le K$, $\sum'''$ over a range including at most a finite number of terms in the range $t > K$ or $u > K$, and the final sum over t also runs over such a finite range. Only the first sum therefore could fail to go to zero, and this is, for fixed p, q, r, s and omitting the factor $\sigma_{pr}\sigma_{qs}$,

$$\sum_{|\tau|<N-1}\left(1-\frac{|\tau|}{N}\right)\{\textstyle\sum'\alpha_{ip}(t)\,\alpha_{ir}(t+\tau)\}\{\sum''\alpha_{jq}(v)\,\alpha_{js}(v+\tau)\},\tag{11}$$

where modulus is not greater than

$$\sum{}'|\alpha_{ip}(t)|\sum_{-\infty}^{\infty}|\alpha_{ir}(t)|\left\{\sum_{-\infty}^{\infty}|\alpha_{jq}(t)|^2\sum_{-\infty}^{\infty}|\alpha_{js}(t)|^2\right\}^{1/2},$$

which converges to zero as K is increased.

The second term is clearly of the same nature and will go to zero in the same fashion. The third term has a variance which is again of the same nature but with the summations $\sum''$, $\sum'''$, $\sum_t$ replaced by summations of the form of $\sum'$. This is seen to tend to zero, uniformly in N, in the same way since, for example, the term involving the fourth cumulant will be, for fixed p, q, r, s,

$$N^{-1}\kappa_{pqrs}\sum_{lm}\sum\sum{}'\alpha_{ip}(t)\,\alpha_{jq}(t+n)\,\alpha_{ir}(t+m-l)\,\alpha_{js}(t+m+n-l)$$
$$\le\sum{}'|\alpha_{ip}(t)|\,|\alpha_{jq}(t+n)|\sum_{\tau=-\infty}^{\infty}|\alpha_{ir}(t+\tau)\,\alpha_{js}(t+\tau+n)|,$$

which clearly converges to zero. This establishes the theorem.

CHAPTER V

Inference About Spectra

1. INTRODUCTION

There is certainly a wide range of situations in which inference procedures can best be carried out in the time domain, that is, with the initial observations and without Fourier transformation of them. We discuss some of these in Chapter VI. However, the use of the Fourier transformed data occupies a central place, particularly when the amount of data available is large, as is often the case, for example, in the earth sciences. The techniques to be discussed in Chapter VI, based on mixed autoregressive, moving average, models, are of considerable importance when the number of data points available is small (as is often the case in economics), for then the yield to be had from an appropriate investment in a finite parameter model is very high and inevitably that investment will be made. They are also of importance in problems of prediction and control in which they are especially well adapted to iterative, digital techniques. However, in the common situation (in, say, the earth sciences) in which a large amount of data is available to be analyzed, the Fourier methods seem to be called for. The reason for this goes back to the essential symmetry embodied, a priori, in the stationarity assumption (which symmetry conditions appear in some ways to be the only proper foundation for a priori statements and which symmetry manifests itself in the physically intelligible concepts of frequency, phase, and amplitude). If autoregressive, moving-average type models are introduced in this situation (with say 1000 observations), a very large number of parameters will often be needed to explain adequately the variation in the data. In the scalar case, one may have to estimate, for example, 100 autoregressive parameters in place of 100 spectral ordinates. How, then, are the 100 parameters to be interpreted? It would seem natural to study the eigenvibrations of the autoregressive system, but this brings us back to the spectral considerations to which the Fourier methods lead us. To summarize, in situations in which large numbers of data points are available and analysis of the data is required, not necessarily for purposes of prediction or regulation but rather because of a need to examine the sources of variation in the data statistically, the Fourier methods to be discussed in this chapter appear to play a premier part.

2. THE FINITE FOURIER TRANSFORM

As we have just said a central place is to be occupied by the finite Fourier transforms

$$w(\lambda) = \frac{1}{\sqrt{2\pi N}} \sum_{1}^{N} x(n) e^{in\lambda},$$

which will be computed for $\lambda = 2\pi k/N$, $k = 0, 1, \ldots, N$. *It will be convenient to have a special name for these frequencies and we thus put* $\omega_k = 2\pi k/N$, *suppressing reference to the variable, N, since that appears to cause no confusion.*

We have already given some discussion of these quantities in Chapter IV. (See Theorem 6 and the discussion following (3.7), Theorem 13 and 13′, and exercise 9 in Chapter IV.) In case $x(n)$ is generated by a linear process, a rather complete discussion of the properties of $w(\lambda)$ may be given. Thus we consider

$$\begin{aligned} w(\lambda) &= \frac{1}{\sqrt{2\pi N}} \sum_{1}^{N} x(n) e^{in\lambda} \\ &= \frac{1}{\sqrt{2\pi N}} \sum_{1}^{N} e^{in\lambda} \left\{ \sum_{-\infty}^{\infty} A(j) \epsilon(n-j) \right\} \\ &= \sum_{-\infty}^{\infty} A(j) e^{ij\lambda} \frac{1}{\sqrt{2\pi N}} \sum_{1}^{N} \epsilon(n) e^{in\lambda} + \frac{1}{\sqrt{2\pi N}} \sum_{-\infty}^{\infty} A(j) e^{ij\lambda} R_{j,N}(\lambda), \end{aligned}$$

where

$$\begin{aligned} R_{j,N}(\lambda) &= \left\{ \sum_{n=-j+1}^{0} \epsilon(n) e^{in\lambda} - \sum_{n=N-j+1}^{N} \epsilon(n) e^{in\lambda} \right\}, & 0 \le j \le N, \\ &= \left\{ -\sum_{n=1}^{-j} \epsilon(n) e^{in\lambda} + \sum_{N+1}^{N-j} \epsilon(n) e^{in\lambda} \right\}, & -N \le j \le 0, \\ &= \left\{ \sum_{-j+1}^{N-j} \epsilon(n) e^{in\lambda} - \sum_{1}^{N} \epsilon(n) e^{in\lambda} \right\}, & N \le j < \infty, \\ &= \left\{ -\sum_{1}^{N} \epsilon(n) e^{in\lambda} - \sum_{-j+1}^{N-j} \epsilon(n) e^{in\lambda} \right\}, & -\infty < j \le -N. \end{aligned}$$

Now consider the case in which $\epsilon(n)$ has finite absolute moment of order $2k$.

Then

$$(2.1)\quad \frac{1}{\sqrt{2\pi N}} \max_{\lambda} \left\{ \mathcal{E}\left[\left\{ \left(\sum_{-\infty}^{\infty} A(j)e^{ij\lambda} R_{j,N}(\lambda) \right)^{*} \left(\sum_{-\infty}^{\infty} A(j)e^{ij\lambda} R_{j,N}(\lambda) \right) \right\}^{k} \right] \right\}^{\frac{1}{2k}}$$

$$\leq \frac{1}{\sqrt{2\pi N}} \sum_{-\infty}^{\infty} \|A(j)\| \max_{\lambda} \left[\mathcal{E}\{(R_{j,N}(\lambda)^{*} R_{j,N}(\lambda))^{k}\} \right]^{\frac{1}{2k}}$$

$$\leq c_k \sum_{-\infty}^{\infty} \|A(j)\| \min\left(1, \left(\frac{|j|}{N}\right)^{\frac{1}{2}}\right),$$

since as j and N increase

$$\mathcal{E}\{(R_{j,N}(\lambda)^{*} R_{j,N}(\lambda))^{k}\} = O(j^{k}), \qquad j \leq N,$$
$$= O(N^{k}), \qquad j \geq N,$$

because no term involving an $\epsilon_j(n)$ only to the first power contributes to the expectation. Here c_k is a constant depending only upon the absolute moments of the $\epsilon_j(n)$ up to the $(2k)$th. Now we can find m so that

$$\sum_{|j|>m} \|A(j)\| < \frac{\frac{1}{2}\epsilon}{c_k}.$$

Then (2.1) is less than

$$\tfrac{1}{2}\epsilon + \sum_{|j| \leq m} \|A(j)\| \left(\frac{|j|}{N}\right)^{\frac{1}{2}} \leq \tfrac{1}{2}\epsilon + \left(\frac{m}{N}\right)^{\frac{1}{2}} \sum_{j=-\infty}^{\infty} \|A(j)\|,$$

and taking $\sqrt{N}$ greater than $\epsilon^{-1} 2\sqrt{m} \sum_{-\infty}^{\infty} \|A(j)\|$ we see that (2.1) is less than ϵ. Thus, since ϵ is arbitrary,

$$w(\lambda) = \left\{ \sum_{-\infty}^{\infty} A(j)e^{ij\lambda} \right\} w(\epsilon, \lambda) + r_N(\lambda),$$

where $w(\epsilon, \lambda)$ is defined in terms of $\epsilon(n)$ as $w(\lambda)$ was in terms of $x(n)$ and where

$$\lim_{N\to\infty} \max_{\lambda} \mathcal{E}\{(r_N(\lambda)^{*} r_N(\lambda))^{k}\} = 0,$$

provided the $2k$th absolute moment of $\epsilon(n)$ is finite. In particular, $r_N(\lambda)$ converges in probability to zero.

Under the condition

$$\sum_{-\infty}^{\infty} \|A(j)\| \, |j|^{\frac{1}{2}} < \infty$$

a stronger result may be obtained, for now we see from (2.1) that

$$\max_{\lambda} \left[\mathcal{E}\{(r_N(\lambda)^{*} r_N(\lambda))^{k}\} \right]^{1/(2k)} \leq \frac{1}{\sqrt{N}} c_k \sum_{-\infty}^{\infty} \|A(j)\| \, |j|^{\frac{1}{2}} \leq \frac{c_k'}{\sqrt{N}}$$

so that the $2k$th absolute moments of the components of $r_N(\lambda)$ are of order

N^{-k} as $N \to \infty$, uniformly in λ if the $2k$th absolute moments of the components of $\epsilon(n)$ are finite.†

Theorem 1. *If $x(n)$ is a linear process with $\epsilon(n)$ having finite $2k$th absolute moments then*

$$w(\lambda) = \left\{\sum_{-\infty}^{\infty} A(j)e^{ij\lambda}\right\} w(\epsilon, \lambda) + r_N(\lambda),$$

where $r_N(\lambda)$ has $2k$th *absolute moments converging to zero uniformly in λ, and if also $\sum \|A(j)\|\,|j|^{1/2} < \infty$ then $r_N(\lambda)$ has* $2k$th *absolute moments which are of order N^{-k}, uniformly in λ.*

If $x(n)$ has an absolutely continuous spectrum and a representation

$$x(n) = \sum_{-\infty}^{\infty} A(j)\,\epsilon(n-j), \qquad \sum_{-\infty}^{\infty} \|A(j)\| < \infty, \tag{2.2}$$

then $r_N(\lambda)$ converges to zero in mean square, with mean square error converging to zero uniformly in λ.

We use aspects of this theorem later, but at the moment it does not add greatly to our knowledge, for in the remainder of this section we are going to consider inferential procedures based on the distribution (4.6) in Chapter IV for the $v_j(u)$ (i.e., $w_j(\lambda_u)$), and this has been established under somewhat more general conditions. Before doing that, however, we introduce

$$I(\epsilon, \lambda) = w(\epsilon, \lambda)w(\epsilon, \lambda)^*$$

and record the consequent of Theorem 1.

Theorem 2. *If $x(n)$ is generated by a linear process, then*

$$I(\lambda) = \left\{\sum_{-\infty}^{\infty} A(j)e^{ij\lambda}\right\} I(\epsilon, \lambda) \left\{\sum_{-\infty}^{\infty} A(j)e^{ij\lambda}\right\}^* + R_N(\lambda),$$

where, if the $(2k)$th *moment of $\epsilon(n)$ is finite, the* kth *absolute moment of $R_N(\lambda)$ converges to zero with N, uniformly in λ; whereas, if $\sum \|A(j)\|\,|j|^{1/2} < \infty$, this moment is of order $N^{-1/2 k}$, uniformly in λ. If $x(n)$ has absolutely continuous spectrum and* (2.2) *holds true, then $R_N(\lambda)$ has first absolute moment converging to zero.*

† These results may be improved by more careful argument. See Walker (1965). It is easy to see that if $\sum \|A(j)\|\,|j|^{\delta} < \infty$, $\delta > 0$, then the kth absolute moment of $r_N(\lambda)$ is $O(N^{-k\delta})$ as $N \to \infty$.

Corollary 1.† *If $x(n)$ is generated by a linear process with finite fourth moment then, for any fixed λ_1, λ_2,*

$$\begin{aligned}\lim_{N\to\infty} \operatorname{cov}\{I_{pq}(\lambda_1), I_{rs}(\lambda_2)\} &= 0, && \lambda_1 \neq \pm\lambda_2,\\ &= f_{pr}(\lambda) f_{sq}(\lambda), && \lambda_1 = \lambda_2 = \lambda \neq 0, \pm\pi\\ &= f_{pr}(\lambda) f_{sq}(\lambda) + f_{ps}(\lambda) f_{qr}(\lambda), && \lambda_1 = \lambda_2 = 0, \pm\pi.\end{aligned}$$

Moreover, if λ_1, λ_2 are of the form $2\pi j/N$, $j = 0, 1, \ldots, [\frac{1}{2}N]$, this result continues to hold, the neglected term having mean square which is $O(N^{-1})$ uniformly in j if $\sum \|A(j)\|\,|j|^{1/2} < \infty$.

Proof.

$$\operatorname{cov}\left(I_{pq}(\epsilon, \lambda_1), I_{rs}(\epsilon, \lambda_2)\right) = \frac{1}{4\pi^2 N^2}\, \mathcal{E}\Big\{\sum_{mlkj}\sum\sum\sum \epsilon_p(j)\epsilon_q(k)\epsilon_r(l)\epsilon_s(m)$$
$$\times \exp\left[i\{(j-k)\lambda_1 - (l-m)\lambda_2\}\right]\Big\} - \frac{1}{4\pi^2}\sigma_{pq}\sigma_{rs},$$

where, of course, σ_{pq} is the covariance of $\epsilon_p(n)$ and $\epsilon_q(n)$. This expression is

$$\frac{\sigma_{pr}\sigma_{qs}}{4\pi^2}\left\{\frac{\sin \frac{1}{2}N(\lambda_1 - \lambda_2)}{N \sin \frac{1}{2}(\lambda_1 - \lambda_2)}\right\}^2 + \frac{\sigma_{ps}\sigma_{qr}}{4\pi^2}\left\{\frac{\sin \frac{1}{2}N(\lambda_1 + \lambda_2)}{N \sin \frac{1}{2}(\lambda_1 + \lambda_2)}\right\}^2 + \frac{\kappa_{pqrs}}{N},$$

where κ_{pqrs} is the fourth cumulant between $\epsilon_p(n)$, $\epsilon_q(n)$, $\epsilon_r(n)$, $\epsilon_s(n)$. For λ_1, λ_2 of the form $2\pi j/N$ the first two terms are exactly zero unless $\lambda_1 = \lambda_2$ or $\lambda_1 = -\lambda_2$. They also converge to zero for any fixed λ_1, λ_2 unless $\lambda_1 = \lambda_2$ or $\lambda_1 = -\lambda_2$. Thus since $R_N(\lambda)$ in Theorem 2 converges in mean square to zero we may neglect it in evaluating the covariances. Thus for $\lambda_1 = \lambda_2 = \lambda \neq 0, \pm\pi$, for example, using $g_{pk}(\lambda)$ for the element in row p, column k of $\sum A(j) \exp ij\lambda$,

$$\begin{aligned}\lim_{N\to\infty} \operatorname{cov}\{I_{pq}(\lambda), I_{rs}(\lambda)\} &= \sum_{kl\,mn}\sum\sum\sum g_{pk}(\lambda)\,\overline{g_{ql}(\lambda)}\,\overline{g_{rm}(\lambda)}\,g_{sn}(\lambda)\,\sigma_{km}\sigma_{ln}/4\pi^2\\ &= f_{pr}(\lambda)\,\overline{f_{qs}(\lambda)} = f_{pr}(\lambda)\,f_{sq}(\lambda).\end{aligned}$$

If $\sum \|A(j)\|\,|j|^{1/2} < \infty$ then the mean square of $R_N(\lambda)$ is $O(N^{-1})$ and since $\operatorname{cov}\{I_{pq}(\lambda), I_{rs}(\lambda)\} - f_{pr}(\lambda)f_{sq}(\lambda)$ is $O(N^{-1})$ if λ is of the form $2\pi j/N$ the result follows for the case $\lambda_1 = \lambda_2 = \lambda \neq 0, \pm\pi$. The other cases follow similarly.

We now consider inferences based on the $w(\lambda)$. We shall assume that either Theorem 13 or 13′, Chapter IV, holds, so that the $v_j(u)$ have the distribution (4.6). We have first in mind the situation where a frequency λ is chosen for

† In defining covariances of complex quantities, we remind the reader, we shall always conjugate the second expression after the cov symbol.

investigation and one is concerned with a relatively narrow band around λ so that m is small compared to N. Under these conditions the asymptotic situation embodied in the aforementioned Theorems 4.13 and 4.13′ seems the appropriate one. We can be allowed, perhaps, to stress something that is obvious. All that we can prove is a mathematical theorem based on certain premises. Of course the theorem can be more or less useful according to its generality; for example, an extension might conceivably be obtained to Theorems 13 and 13′ in Chapter IV giving a form of asymptotic expansion for the distribution of the $v_j(u)$ of the type of the Berry-Esséen theorem and its consequents. (See Chapter VII of Cramer, 1937.) The applicability of such theorems in particular instances must, of course, be a matter for judgment. If m was not very small relative to N, one might prefer to consider a limiting situation where both m and N increase together with, perhaps $m/N \to 0$. Such a model might also be rather inappropriate of course, and indeed this would be so if m was small. In any case, we shall deal with the case in which m increases indefinitely later.

In connection with distributions to be derived below we shall consistently refer results back to corresponding results in the theory of multivariate analysis, when this can be done. This is thought to be an earlier part of the training of most readers so that it is appropriate to so act. Unfortunately there are limited possibilities for doing this.

If we put, using the notation of Chapter IV, Theorem 13,

$$v_j(u) = \tfrac{1}{2}\{\xi_j(u) + i\eta_j(u)\}, \qquad u = 1, \ldots, m$$

then the asymptotic distribution of the $\xi_j(u)$, $\eta_j(u)$ from (4.6) in Chapter IV becomes when $\lambda \neq 0, \pi$ the multivariate normal distribution with zero mean vector and covariance matrix which has as its only nonzero entries

$$\begin{cases} \mathcal{E}\{\xi_j(u)\xi_k(u)\} = \mathcal{E}\{\eta_j(u)\eta_k(u)\} = 2\Re\big(f_{jk}(\lambda)\big) = c_{jk}(\lambda), \\ \mathcal{E}\{\xi_j(u)\eta_k(u)\} = -\mathcal{E}\{\eta_j(u)\xi_k(u)\} = -2\mathscr{I}\big(f_{jk}(\lambda)\big) = q_{jk}(\lambda), \end{cases} \tag{2.3}$$

where in case $j = k$, of course, $c_j(\lambda) = 2f_j(\lambda)$, whereas $q_j(\lambda) = 0$. For $\lambda = 0, \pi$, $q_{jk}(\lambda)$ is zero for all j, k, since $q_{jk}(\lambda)$ is assumed continuous at λ and changes sign at $0, \pi$. However, for $\lambda = 0$ and, in case N is even for $\lambda = \pi$, there will be one value of u for which $\lambda_u = 0, \pi$ and the corresponding η_j are the constant zero. Correspondingly these v_j are real and have covariance matrices, respectively, $f(0), f(\pi)$ (which are real symmetric matrices). Thus the cases $0, \pi$ have to be treated with care. We shall sometimes omit separate discussion of these cases to save unnecessary elaboration but their treatment follows easily from the case $\lambda \neq 0, \pi$. *We shall not repeatedly insert the qualification "asymptotic" in what goes below.* It is to be understood that in all cases the statistics involved have distributions which are asymptotically obtainable by treating the $v_j(u)$ as having the complex multivariate normal distribution. In all cases the statistics are continuous functions of the $v_j(u)$ (outside of a set whose probability content with respect to the limiting distribution is zero). Thus in each case the asymptotic distribution is validly obtained in this fashion (see, for example, Fisz (1963) p. 184). We shall also often drop the argument variable, λ, when no confusion will result.

The asymptotic multivariate normal density for the $\xi_j(u)$, $\eta_j(u)$, $\lambda \neq 0, \pi$ factors into a product of m factors with the same covariance matrix, this being, for the vector

$$\zeta'(u) = (\xi_1(u), \xi_2(u), \ldots, \xi_p(u), \eta_1(u), \eta_2(u), \ldots, \eta_p(u)),$$

that with covariance matrix

$$W(\lambda) = \begin{bmatrix} c(\lambda) & q(\lambda) \\ -q(\lambda) & c(\lambda) \end{bmatrix}.$$

Since $q(\lambda)$ has zero elements in its diagonal it is evident that $\xi_i(u)$, $\eta_i(u)$ are independent for each i.

It follows fairly easily that the maximum likelihood estimators of the unspecified parameters in (4.6) Chapter IV are, for $\lambda \neq 0, \pi$

$$\text{(2.4)} \qquad \begin{cases} \hat{c}_{jk}(\lambda) = \dfrac{1}{2m} \sum_u \{\xi_j(u)\xi_k(u) + \eta_j(u)\eta_k(u)\}, \\ \hat{q}_{jk}(\lambda) = \dfrac{1}{2m} \sum_u \{\xi_j(u)\eta_k(u) - \eta_j(u)\xi_k(u)\}, \end{cases}$$

where the summation is over the m values of λ_u. Thus

$$\text{(2.5)} \qquad \hat{f}_{jk}(\lambda) = \tfrac{1}{2}\{\hat{c}_{jk}(\lambda) - i\hat{q}_{jk}(\lambda)\} = \frac{1}{m} \sum_u v_j(u)\, \overline{v_k(u)}.$$

In case $\lambda = 0, \pi$ then $q(\lambda) = 0$ and $f_{jk}(\lambda) = \tfrac{1}{2}c_{jk}(\lambda)$. Now one of the λ_u will always be 0, for $\lambda = 0$ (and π for $\lambda = \pi$). Moreover, a mean correction (at least) will always be made to the data, which has the sole effect of making $w(0) = 0$. The frequencies λ_u nearest to $\lambda = 0, \pi$ are now such that the corresponding $v_j(u)$ occur in conjugate pairs, with perhaps one not paired (if m is even). It is now best to consider only those values of u corresponding to frequencies at or to the right of 0 (or at or to the left of π). Thus

$$\hat{c}_{jk}(0) = \tfrac{1}{2}[\tfrac{1}{2}m]^{-1} \sum_{1}^{[\frac{1}{2}m]} \{\xi_j(u)\xi_k(u) + \eta_j(u)\eta_k(u)\},$$

where the sum is over u such that the λ_u are $2\pi u/N$, $u = 1, \ldots, [\tfrac{1}{2}m]$

$$\hat{c}_{jk}(\pi) = \tfrac{1}{2}[\tfrac{1}{2}(m + 1)] \sum_{1}^{[\frac{1}{2}(m+1)]} \{\xi_j(u)\xi_k(u) + \eta_j(u)\eta_k(u)\},$$

where now the sum is over u such that the λ_u are the frequencies $\pi - 2\pi k/N$, $k = 0, 1, \ldots, [\frac{1}{2}(m-1)]$.

If m is small these seem to be inescapable as estimators provided the computations can be effected without too great a cost. Modifications have been suggested and we discuss these later.

Theorem 3. *Let $x(n)$ satisfy the conditions of Theorems* 13 *or* 13′ *in Chapter IV and $v(u)$ be as defined in* (4.5), *with the λ_u, $u = 1, \ldots, m$, as near as possible to a fixed frequency λ. Then the maximum likelihood estimators for $c(\lambda)$, $q(\lambda)$, $f(\lambda)$ got from the limiting distribution of the $v(u)$ are given by* (2.4) *and* (2.5) *above.*

The asymptotic joint distribution of the $\hat{c}_{jk}(\lambda)$, $\hat{q}_{jk}(\lambda)$ is found from the complex Wishart distribution which we discuss in Section 6, but we do not need it in this section.

The most important inferences seem to be founded upon the spectra and the coherence and phase between pairs of series as well as the estimate of the response function of one series in terms of a number of others and corresponding partial and multiple coherences. We deal with these and some more elaborate cases below. There is a limit to the detail that can be given, for all of the complexity of classical multivariate analysis can enter here, and thus a book on this subject would be needed in its own right. We give some indications of the complex multivariate analysis of variance involved in Section 6 because it seems to be conceivably of some importance though so far no use appears to have been made of it.

We repeat again our warning that we shall not always insert "asymptotic" in statements made below but nevertheless that all of these results have been obtained from the limiting distribution described in Theorem 3.

(a) Individual Spectra

The distribution of $\hat{f}_j$ is obtained directly from Theorem 3 but for $\lambda \neq 0, \pi$ it is clear from (2.4) and the remark just previously that it is the distribution of a variance with $2m$ degrees of freedom. Thus $2m\hat{f}_j/f_j$ is distributed as χ_{2m}^2 (chi-square with $2m$ degrees of freedom). The shortest unbiased†

† By the shortest confidence interval we mean one which has prescribed probability (the confidence coefficient) of containing some particular value, f_j, when that is the true value and has the minimum possible probability of containing f_j when it is not the true value. It is not necessarily physically the shortest interval. By saying that the interval is unbiased we mean that the probability that it contains a particular value f_j is greatest when f_j is the true value.

confidence interval, with confidence coefficient $(1 - \alpha)$, for f_j is of the form

$$\frac{2m}{b_{2m}}\hat{f}_j \leq \hat{f}_j \leq \frac{2m}{a_{2m}} f_j,$$

where $[a_{2m}, b_{2m}]$ contains $(1 - \alpha)$ of the probability mass for $\chi_{2m}{}^2$. These numbers, a_n, b_n, are tabulated in Tate and Klett (1959) for $n = 2(1)29$ and for $\alpha = 0.1, 0.05, 0.01, 0.005, 0.001$. They also tabulate values a_{2m}, b_{2m}, to be used in the same way, for the physically shortest unbised interval (for the same parameter values for α, n). A rough-and-ready procedure treats $\ln(\hat{f}_j/f_j)$ as normal with zero mean and variance m^{-1}. Though one would now probably work in terms of $\ln \hat{f}_j$ and $\ln f_j$, for comparison we point out that the corresponding interval for f_j, for $m = 10$, $\alpha = 0.95$, is

$$0.538\hat{f}_j = \hat{f}_j \exp -\left(1.96 \frac{1}{\sqrt{10}}\right) \leq f_j \leq \hat{f}_j \exp\left(1.96 \frac{1}{\sqrt{10}}\right) = 1.859\hat{f}_j,$$

compared with

$$0.531\hat{f}_j \leq f_j \leq 1.978\hat{f}_j$$

for the minimum length interval. The approximate method, in this case, is certainly adequate for most practical purposes. The width of the interval is worth observing, for it extends over a range of $1.5\hat{f}_j$. Since $m = 10$ would not be an uncommon value for this constant, we see already how carefully the observed value must be interpreted. Even if m is increased to 20, using the approximate method, the interval (for $\alpha = 0.05$) is from $0.645\hat{f}_j$ to $1.550\hat{f}_j$.

(b) Coherence

We define the sample coefficient $\hat{\sigma}_{jk}(\lambda)$ as

$$\hat{\sigma}_{jk}(\lambda) = \frac{|f_{jk}(\lambda)|}{\{\hat{f}_j(\lambda)\hat{f}_k(\lambda)\}^{1/2}}.$$

This is the maximum likelihood estimator. *Its distribution has density, for* $\lambda \neq 0, \pm\pi$,

(2.6) $p(x \mid \sigma_{jk})$

$$= \frac{2(1 - \sigma_{jk}{}^2)^m}{(m - 1)!(m - 2)!} x(1 - x^2)^{m-2} \sum_{n=0}^{\infty} \frac{\sigma_{jk}{}^{2n}\{(m + n - 1)!\}^2}{(n!)^2} x^{2n}, \quad 0 \leq x \leq 1.$$

This distribution is that of a multiple correlation, for Gaussian data, for $\lambda \neq 0, \pm\pi$, *of one variable with two others from 2m observations without mean correction.* We shall not prove this result immediately for it follows as a special case from Theorem 5 which establishes the distribution of the multiple coherence. The reason why the distribution is that of a multiple correlation can be seen immediately by considering the three row vectors

$$(2.7)\ (\mathrm{i}) \qquad (\xi_j(1), \xi_j(2), \ldots, \xi_j(m), \eta_j(1), \ldots, \eta_j(m)),$$

$$(2.7)\ (\mathrm{ii}) \qquad (\xi_k(1), \xi_k(2), \ldots, \xi_k(m), \eta_k(1), \ldots, \eta_k(m)),$$

$$(2.7)\ (\mathrm{iii}) \qquad (-\eta_k(1), -\eta_k(2), \ldots, -\eta_k(m), \xi_k(1), \ldots, \xi_k(m)).$$

The last two are orthogonal. Thus the square of the multiple correlation is just

$$\left[\frac{\{\sum_u \xi_j(u)\,\xi_k(u) + \sum_u \eta_j(u)\,\eta_k(u)\}^2}{\sum_u \xi_k(u)^2 + \sum_u \eta_k(u)^2} + \frac{\{-\sum_u \xi_j(u)\,\eta_k(u) + \sum_u \eta_j(u)\,\xi_k(u)\}^2}{\sum_u \xi_k(u)^2 + \sum_u \eta_k(u)^2}\right]$$

$$\times \{\sum \xi_j(u)^2 + \sum \eta_j(u)^2\}^{-1} = \frac{\hat{c}_{jk}^{\,2} + \hat{q}_{jk}^{\,2}}{\hat{f}_j \hat{f}_k} = \hat{\sigma}_{jk}^{\,2}.$$

Thus the formula for the coherence is just that for a multiple correlation as described. When $\sigma_{jk} = 0$ the distribution (2.6) follows immediately since now (2.7)(i) is independent of (2.7)(ii), (2.7)(iii) and it is well known that when this is so the distribution of a multiple correlation is independent of the distribution of the elements of the vectors regressed upon. When $\sigma_{jk} \neq 0$ it is not quite so obvious that the distribution is as shown since (2.7)(ii) and (2.7)(iii) are of a special form with identical random variables appearing in the two rows.

If $\sigma_{jk} = 0$, the density is, for $\lambda \neq 0, \pm\pi$,

$$2(m-1)x(1-x^2)^{m-2},$$

so that the distribution of $(m-1)\hat{\sigma}_{jk}^2(1-\hat{\sigma}_{jk}^2)^{-1}$ has density

$$\left(1 + \frac{x}{m-1}\right)^{-m}, \qquad 0 \leq x < \infty,$$

which is that of Fisher's F with 2 and $2(m-1)$ degrees of freedom. It may be used, of course, to test for the presence of coherence but this is not a procedure which is very likely to be used. It is more likely that a confidence interval for σ_{jk} will be required. These may be got from the extensive tables of the distribution function corresponding to (2.6) prepared by Amos and Koopmans (1963), for example, by using the "equal tail areas" interval

$$(2.8) \qquad a \leq \sigma_{jk} \leq b, \qquad \int_0^a p(x \mid \sigma_{jk})\,dx = \int_b^1 p(x \mid \sigma_{jk})\,dx = \tfrac{1}{2}\alpha,$$

where $p(x \mid \sigma_{jk})$ is defined by (2.6).

We finally consider the case $\lambda = 0, \pm\pi$. Thus the m frequencies $2\pi j/N$ from which $\hat{f}(0)$ is computed are those for $j = 0, \pm 1, \ldots$. We assume that a mean correction is made so that in fact there is no contribution from $j = 0$ since $w(0) = 0$ for mean corrected data. [No other $w(\omega_k)$ is affected.] *Then $\hat{\sigma}_{jk}(0)$ is (asymptotically) distributed as an ordinary sample correlation between two Gaussian variables, with correlation $\sigma_{jk}(0)$, computed from $(m - 1)$ pairs of observations and $\hat{\sigma}_{jk}(\pi)$ is distributed in the same way in relation to $\sigma_{jk}(\pi)$ but with m replacing $(m - 1)$.* To see this, for example, for $\hat{\sigma}_{jk}(0)$ and m odd, consider the two rows of random variables

$$(\xi_j, \ldots, \xi_j(\tfrac{1}{2}(m-1)), n_j(1), \ldots, \eta_j(\tfrac{1}{2}(m-1))),$$
$$(\xi_k(1), \ldots, \xi_k(\tfrac{1}{2}(m-1)), \eta_k(1), \ldots, \eta_k(\tfrac{1}{2}(m-1))).$$

Here $v_j(u) = w(\omega_u)$, $\omega_u = 2\pi u/N$. The square of the correlation between these two rows is

$$\frac{\{\sum(\xi_j(u)\,\xi_k(u) + \eta_j(u)\eta_k(u))\}^2}{\{(\sum \xi_j(u)^2) + \sum\eta_j(u)^2)(\sum \xi_k(u)^2 + \sum\eta_k(u)^2)\}} = \frac{|\hat{f}_{jk}(0)|^2}{\hat{f}_j(0)\hat{f}_k(0)} = \hat{\sigma}_{jk}{}^2(0),$$

and the stated result follows immediately.

We mention here a problem with the estimation of coherence to which we shall return later. This problem arises because the phase angle may be changing rapidly across a band of frequencies. Thus, if f_j, f_k, σ_{jk} are effectively constant but θ_{jk} is changing rapidly, then, since

$$c_{jk} = 2(f_jf_k)^{1/2}\sigma_{jk}\cos\theta_{jk}, \qquad q_{jk} = -2(f_jf_k)^{1/2}\sigma_{jk}\sin\theta_{jk},$$

it can be seen that the average value of c_{jk}, q_{jk} over the band may be small and so may be the average value of σ_{jk}, which is what our procedures are estimating. However, in this section we are considering narrow bands (because m is constant as N increases) and we defer discussion of the problem to Section 7.

(c) Complex Regression Coefficient and Phase

We define

$$\beta_{jk}(\lambda) = \frac{f_{jk}(\lambda)}{f_k(\lambda)}, \qquad f_k(\lambda) \neq 0,$$
$$= 0, \qquad f_k(\lambda) = 0.$$

(We shall often drop the argument variable λ.) This has the maximum likelihood estimator

$$\hat{\beta}_{jk}(\lambda) = \frac{\hat{f}_{jk}(\lambda)}{\hat{f}_k(\lambda)}, \qquad \lambda \neq 0, \pi, = \frac{\hat{c}_{jk}(\lambda)}{\hat{c}_k(\lambda)}, \lambda = 0, \pi.$$

Let us consider the conditional distribution of the $v_j(u)$ for $v_k(u)$ fixed, $\lambda \neq 0, \pi$. This is easily seen to have the density function which is the product of m factors the uth of which is†

$$\text{(2.9)} \qquad \frac{1}{\pi f_j(1 - \sigma_{jk}^2)} \exp\left\{\frac{-|v_j(u) - \beta_{jk}v_k(u)|^2}{f_j(1 - \sigma_{jk}^2)}\right\},$$

that is, the "one-dimensional" complex normal distribution with mean value $\beta_{jk}v_k(u)$ and variance $f_j(1 - \sigma_{jk}^2)$. (We have the easy proof of this as an exercise for this chapter.) It follows immediately that $(\hat{\beta}_{jk} - \beta_{jk})$ has real and imaginary parts which, conditional on $v_k(u)$ fixed, are independently and identically distributed in a normal distribution with mean zero and variance $(2m)^{-1}f_j(1 - \sigma_{jk}^2)/\hat{f}_k$. (Indeed it is easily checked that in the regression of (2.7)(i) on (2.7)(ii) and (2.7)(iii) the regression coefficients are the real and imaginary parts of $\hat{\beta}_{jk}$. The result now follows from classical regression theory.) Moreover, the residual sum of squares is $4m\hat{f}_j(1 - \hat{\beta}_{jk}^2)$ so that $\hat{f}_j(1 - \hat{\sigma}_{jk}^2)$ is an estimator of $f_j(1 - \sigma_{jk}^2)$ which is independent of $\hat{\beta}_{jk}$. This enables us to carry out tests and determine confidence intervals which are "exact," to the extent that the complex multivariate normal distribution is appropriate in the first place. This exactness should not be overstressed though it clearly is a virtue. Thus a $100(1 - \alpha)\%$ confidence region for β_{jk} is obtained as

$$\begin{aligned}\text{(2.10)} \quad |\hat{\beta}_{jk}(\lambda) - \beta_{jk}(\lambda)| &\\ &\leq [\{(m-1)\hat{f}_k\}^{-1}\hat{f}_j(1 - \hat{\sigma}_{jk}^2)F_{2,2m-2}(\alpha)]^{1/2}, \qquad \lambda \neq 0, \pi \\ &\leq [\{\nu\hat{f}_k\}^{-1}\hat{f}_j(1 - \hat{\sigma}_{jk}^2)]^{1/2}t_\nu(\alpha), \\ &\qquad \nu = m-2,\ \lambda = 0;\ \nu = m-1,\ \lambda = \pi.\end{aligned}$$

Here $F_{2,2m-2}(\alpha)$ is the $100\alpha\%$ point for the F distribution with the indicated degrees of freedom, whereas $t_\nu(\alpha)$ corresponds similarly to the t distribution (Student's distribution). In the case of $\lambda = 0$, as we have already said, we assume that $w(0)$ will have been eliminated (made zero) by mean correction. We may also construct confidence intervals for the real and imaginary parts separately, but this does not seem to be something one will wish to do. More importantly we can now derive a confidence interval for the phase angle. Indeed the argument of the complex number β_{jk} is evidently θ_{jk} and that of $\hat{\beta}_{jk}$ is $\hat{\theta}_{jk}$. Thus $\hat{\beta}_{jk} \exp -i\theta_{jk}$ has expectation $|\beta_{jk}|$ and is thus real. Therefore $|\hat{\beta}_{jk}| \sin(\hat{\theta}_{jk} - \theta_{jk})$ has zero expectation. Since it is of the form

$$\mathcal{R}(\hat{\beta}_{jk}) \cos\theta_{jk} - \mathcal{I}(\hat{\beta}_{jk}) \sin\theta_{jk},$$

† There is room for confusion, we realize, in the use of $v_j(u)$ both for the random variables and the "running variable" in the density function but in a subject already overloaded with notation we have sometimes preferred to risk the confusion.

it has variance which is, conditional on the $v_k(u)$ fixed, $(2m)^{-1}f_j(1 - \sigma_{jk}^2)/\hat{f}_k$. Thus putting $s^2 = \hat{f}_j(1 - \hat{\sigma}_{jk}^2)/\{(2m - 2)\hat{f}_k\}$ we see that

$$|\hat{\beta}_{jk}| \sin \frac{\hat{\theta}_{jk} - \theta_{jk}}{s}$$

has the t distribution with $(2m - 2)$ degrees of freedom. Thus a $100(1 - \alpha)\%$ confidence interval for θ_{jk} is given by those θ_{jk} satisfying

$$|\sin(\hat{\theta}_{jk} - \theta_{jk})| \le \frac{s}{|\hat{\beta}_{jk}|} t_{2m-2}(\alpha) = \left\{\frac{1 - \hat{\sigma}_{jk}^2}{\hat{\sigma}_{jk}^2(2m - 2)}\right\}^{1/2} t_{2m-2}(\alpha), \qquad \lambda \ne 0, \pi, \tag{2.11}$$

where $t_{2m-2}(\alpha)$ is the $100\alpha\%$ point for a two-sided test for the t statistic with $2m - 2$ degrees of freedom.

Theorem 4. *Under the same conditions as for Theorem* 3, *asymptotically valid confidence regions for* $\beta_{jk}(\lambda)$ *and for* $\theta_{jk}(\lambda)$, $\lambda \ne 0, \pi$, *may be obtained from* (2.10) *and* (2.11).

In constructing the confidence interval a problem arises. The numbers $\hat{\theta}_{jk}$ can be determined so as to lie anywhere in the interval $(-\pi, \pi]$ so that in principle the θ_{jk} can be so determined also. However if we proceed in this way and use (2.11) we shall obtain a confidence region which, representing θ_{jk}, $\hat{\theta}_{jk}$ as points on the unit circle, will consist of two disjoint pieces, one centered at $\hat{\theta}_{jk}$ and the other centered at the point diametrically opposite to $\hat{\theta}_{jk}$ and of the same length. (Of course the two regions may coalesce, so that the confidence statement becomes vacuous and indeed this will tend to happen if $\hat{\sigma}_{jk}^2$ is not near to unity, to a degree depending on the size of m.) One might be disposed to eliminate the second of these regions, if $\hat{\sigma}_{jk}^2$ is high, on the grounds that $\hat{\theta}_{jk}$ could hardly have taken its observed value if the true value differed by nearly π from it (or equivalently both $\hat{c}_{jk}$ and $\hat{q}_{jk}$ could hardly have wrong signs if $\hat{\sigma}_{jk}^2$ is near to unity). However, this means that the confidence statement is no longer exact. Another procedure is to locate $\hat{\theta}_{jk}$, θ_{jk} in a chosen interval of length π; for example, to take the principal inverse tangent in defining them so that both are put between $(-\pi/2, \pi/2]$. However, there is still a problem, for now to obtain an exact confidence statement one must take the intersection of the region as first defined (involving the two pieces anywhere on the circle) with the a priori chosen region; *that is, we take the part of the interval in* $(-\pi/2, \pi/2]$ *which intersects with the* θ_{jk} *values which satisfy* (2.11).† This *may* again consist of two

† We leave the reader to check that this is an exact confidence region for θ_{jk}, restricted to lie in $(-\pi/2, \pi/2]$.

pieces, one of which (that not centerd at $\hat{\theta}_{jk}$) will tend to be smaller than the other. Probably workers will tend to use the first method and eliminate the region not centered at $\hat{\theta}_{jk}$, at least when both regions are not too large.

The distribution of the phase angle does not itself seem of great interest, except possibly when $\sigma_{jk}(\lambda) = 0$ when it is the distribution of arctan $(\hat{q}_{jk}/\hat{c}_{jk})$, where $\hat{q}_{jk}$ and $\hat{c}_{jk}$ are now independent Gaussian random variables (so far as their limiting distribution is concerned). Thus, if the principal value of the inverse tangent is taken, it is easy to see that $\hat{\theta}_{jk}$, when $\sigma_{jk} = 0$, is uniformly distributed over $(-\pi/2, \pi/2]$.

The confidence interval we have constructed is not by any means unique but it appears (see Neyman, 1954) to have reasonable properties. Indeed, it has been shown to be the shortest unbiased interval (see the footnote on p. 252). Thus it has some virtues and may be reasonably used, therefore, until a better procedure is forthcoming.

(d) Multiple Regression, Coherence, Phase Procedures

The techniques of the previous subsection extend to the case where one variable, $y(n)$ let us say, is to be related to p others $x_1(n), x_2(n), \ldots, x_p(n)$. These are considered to be jointly stationary and we now distinguish one, by calling it $y(n)$, only for notational convenience. We are led to consider first the conditional distribution of the $v_y(u)$ for fixed values of the $v_k(u)$. Here the k subscript refers to the variable x_k, the y subscript to y and the conditional distribution of which we speak is that obtained from the distribution of form (4.6) in Chapter IV for $v_y(u)$ and the $v_k(u)$. We call f_x the *matrix* with f_{jk} in row j column k and f_{xy}, the vector with f_{ky} in row k, this being the cross spectrum between x_k and y. Of course, f_{yx} will be the transposed vector. The conditional distribution has density

$$\text{(2.12)} \qquad \frac{1}{\pi f_y(1-\sigma_{yp}{}^2)} \exp\left\{\frac{-|v_y(u) - \sum_k \beta_k v_k(u)|^2}{f_y(1-\sigma_{yp}{}^2)}\right\}.$$

Here σ_{yp} is what we call the multiple coherence and is defined by

$$\sigma_{yp}{}^2 = f_y^{-1} f_{yx} f_x^{-1} f_{xy},$$

whereas the vector β having β_k in the kth place is of the form

$$\beta = f_x^{-1} f_{xy}.$$

We assume f_x to be nonsingular, though this condition could be relaxed. The multiple coherence is an intrinsic measure of the strength of association between $y(n)$ and the $x_k(n)$ at frequency λ. It has an interpretation analogous to that given for ordinary coherence in Chapter II, Section 3 [above formula (2.10)]. Thus, taking $y(n)$ and the $x_k(n)$ to be had from continuous-time phenomena, we may form

$$x_{\alpha,\theta}(n) = \sum \alpha_k x_k(n - \theta_k).$$

We may now consider the correlation between the component of $y(n)$ and of $x_{\alpha,\theta}(n)$ due to frequencies $\pm\lambda$ (i.e., to small bands about these points). Maximizing this by appropriate choice of the α_k and θ_k we obtain the multiple coherence. To estimate it and the vector β by maximum likelihood we merely replace f_x, f_y, f_{yk} by $\hat{f}_x, \hat{f}_y, \hat{f}_{yk}$. We now have

Theorem 5. *Under the conditions of Theorem* 3 *the asymptotic distribution of* $\hat{\sigma}_{yp}$ *for* $\lambda \neq 0, \pi$, *has density*

$$\frac{2(1 - {\sigma_{yp}}^2)^m}{B(m - p, p)} \hat{\sigma}_{yp}^{2p-1}(1 - {\sigma_{yp}}^2)^{m-p-1} {}_2F_1(m, m; p; {\hat{\sigma}_{yp}}^2 {\sigma_{yp}}^2)$$

For $\lambda = 0$ *the same result holds with p replaced by* $\frac{1}{2}p$ *and m by* $\frac{1}{2}(m - 1)$ (*assuming* $x(n)$ *to have been mean corrected*) *and for* $\lambda = \pi$ *the same result holds with p replaced by* $\frac{1}{2}p$ *and m by* $\frac{1}{2}m$.

Note. Here $B(.\,,.)$ is the beta function and ${}_2F_1$ is the confluent hypergeometric function. This density can also be written in other forms (Anderson, 1958, p. 96, formula (38)). The density function is that of a multiple correlation of one real normal variable with $2p$ other real normal variables when computed from $2m$ observations without mean correction. The proof, which we give immediately, is easy and covers also the case $p = 1$, of course, which we dealt with in formula (2.6). The result is a little surprising. We shall discuss the geometric reason for it in Section 6 below. The proof for $\lambda = 0, \pi$ is the same as that given for (2.6) and we omit it.

Proof. We consider the $(p + 1)$ vectors which written as rows are of the form

$$a_y' = (\xi_y(1), \ldots, \xi_y(m), \eta_y(1), \ldots, \eta_y(m)),$$

$$a_j' = (\xi_j(1), \ldots, \xi_j(m), \eta_j(1), \ldots, \eta_j(m)); \qquad j = 1, \ldots, p,$$

$$b_j' = (-\eta_j(1), \ldots, -\eta_j(m), \xi_j(1), \ldots, \xi_j(m)); \qquad j = 1, \ldots, p.$$

Here ξ_y, η_y correspond to $y(n)$ and ξ_j, η_j to $x_j(n)$. Then $\hat{\sigma}_{yp}$ is the multiple correlation between a_y and the $2p$ vectors a_j, b_j, $j = 1, \ldots, p$. To establish this consider the correspondence established at the beginning of the proof of Theorem 13 in Chapter IV. Now since $\hat{f}_x = \frac{1}{2}(\hat{c}_x - i\hat{q}_x), \hat{f}_{xy} = \frac{1}{2}(\hat{c}_{xy} - i\hat{q}_{xy})$, where $\hat{c}_x$, $\hat{q}_x$ are square matrices and $\hat{c}_{xy}$, $\hat{q}_{xy}$ are column vectors, we see through that correspondence that

$$\hat{\sigma}_{yp}^{\;2} = \tfrac{1}{2}\hat{f}_y^{\;-1}(\hat{c}_{xy}' \vdots \hat{q}_{xy}')\begin{bmatrix} \hat{c}_x & \hat{q}_x \\ -\hat{q}_x & \hat{c}_x \end{bmatrix}^{-1}\begin{pmatrix} \hat{c}_{xy} \\ \cdots \\ q_{xy} \end{pmatrix},$$

which is easily seen to be the multiple correlation just described.

Let us then first consider the distribution of $\hat{\sigma}_{yp}$ conditional on the elements of a_j, b_j, $j = 1, \ldots, p$ being fixed at their observed values. In the conditional distribution a_y has mean vector

$$\sum_{k=1}^{p} (\gamma_k a_k + \delta_k b_k), \qquad \beta_k = \gamma_k + i\delta_k$$

which we obtain immediately from (2.12). The residual variance, that is, the variance of the $\xi_y(u)$, $\eta_y(u)$, $u = 1, \ldots, m$, in the conditional distribution, is $2f_y(1 - \sigma_{yp}^{\;2})$, again from (2.12). The $\xi_y(u)$, $\eta_y(u)$, $u = 1, \ldots, m$, in this conditional distribution are independent and normal. The conditional distribution of $\hat{\sigma}_{yp}^{\;2}$ is now the conditional distribution of the square of a multiple correlation of one normal variable with $2p$ others from $2m$ observations, without mean correction, since once the a_j, b_j are fixed their special nature as vectors of random variables is of no concern. From Anderson (1958, p. 94, formula (31)) the density is

$$\frac{e^{-\frac{1}{2}\theta^2}(1 - x)^{m-p-1}}{(m - p - 1)!} \sum_{j=0}^{\infty} \left\{ \frac{(\frac{1}{2}\theta^2)^j x^{p+j-1}(m + j - 1)!}{(p + j - 1)!} \right\},$$

where

$$\begin{aligned}\theta^2 &= \{2f_y(1 - \sigma_{yp}^{\;2})\}^{-1} \sum\sum \{\gamma_k\gamma_l a_k' a_l + \delta_k\delta_l b_k' b_l + \gamma_k\delta_l a_k' b_l + \delta_k\gamma_l b_k' a_l\} \\ &= 2m\{f_y(1 - \sigma_{yp}^{\;2})\}^{-1}\beta^*\hat{f}_x\beta.\end{aligned}$$

This has expectation

$$\begin{aligned}2m\{f_y(1 - \sigma_{yp}^{\;2})\}^{-1}\beta^* f_x\beta &= 2m\{f_y(1 - \sigma_{yp}^{\;2})\}^{-1} f_{yx} f_x^{\;-1} f_{xy}. \\ &= \frac{2m\sigma_{yp}^{\;2}}{1 - \sigma_{yp}^{\;2}}.\end{aligned}$$

Moreover θ^2, as a function of the random variables $\xi_j(u)$, $\eta_j(u)$, is a constant multiple of

$$\sum_u |\beta^* v(u)|^2$$

and thus is a constant multiple of a variable, $\chi_{2m}^{\;2}$, which is distributed as chi-square with $2m$ degrees of freedom, since the $\beta^* v(u)$ have the (univariate) complex multivariate normal distribution and are independent. Thus

$\theta^2 = \{\sigma_{yp}{}^2/(1-\sigma_{yp}{}^2)\}\chi_{2m}{}^2$. Now inserting this in the conditional distribution for $\hat{\sigma}_{yp}{}^2$ it becomes of precisely the same form as obtains for the corresponding multiple correlation. (See Anderson, 1958, p. 94 once more.) Thus when we integrate over the range of variation of the $\xi_j(u)$, $\eta_j(u)$, $u = 1, \ldots, m$; $j = 1, \ldots, p$, we obtain the same result as for the multiple correlation since in spite of the difference in form of the vectors being held constant in the two cases we are effectively integrating only over the distribution of $\chi_{2m}{}^2$ in both cases. This completes the proof of the theorem.

When $\sigma_{yp}{}^2 = 0$, it follows from what was said in the note preceding the proof that

$$F_{2p,2m-2p} = \left(\frac{\hat{\sigma}_{yp}{}^2}{1-\hat{\sigma}_{yp}{}^2}\right)\frac{m-p}{p}$$

has the F distribution with $2p$ and $2m - 2p$ degrees of freedom. For $\sigma_{yp}{}^2 \neq 0$ there are no extensive tabulations it seems. For $p = 1$, of course, the tables of Amos and Koopmans (1963) may be used. For $p = 2, 4, 8, 16, 32$; $\hat{\sigma}_{yp}{}^2 = 0.0(0.1)1.0$, $(m - p) = 5, 10, 20, 40, 80, 160$, the upper and lower bounds to an equal tail areas confidence interval for $\sigma_{yp}{}^2$ are shown in Groves and Hannan (1968), for confidence coefficients $1 - \alpha$ with $\alpha = 0.05$ and 0.01. In that tabulation the notation used replaces $\hat{\sigma}_{yp}$ by r, $\sigma_{yp}{}^2$ by ϕ and m by M. Thus, for example, for $m = 42$, $p = 2$, $\alpha = 0.05$ and for $\hat{\sigma}_{yp}{}^2 =$ 0.80 the lower and upper bounds to the confidence interval are 0.69 and 0.86. The corresponding interval for σ_{yp} is thus between 0.83 and 0.93. In Khatri (1966) a discussion is given of transformations of $\hat{\sigma}_{yp}{}^2$ to a form in which the distribution is nearly independent of $\sigma_{yp}{}^2$. However, the results do not seem useful from the point of view of giving a confidence interval though they may be used to test an hypothesized value of $\sigma_{yp}{}^2$.

We may define the partial coherence as

$$\sigma^2_{yk\cdot p} = \frac{|\Sigma^{yk}|^2}{(|\Sigma^{yy}|\,|\Sigma^{kk}|)},$$

where Σ is the matrix

$$\Sigma = \left(\begin{array}{c|c} f_y & f_{yx} \\ \hline f_{xy} & f_x \end{array}\right)$$

and $|\Sigma^{yk}|^2$ means the square of the modulus of the determinant obtained by eliminating the first row and the $(k+1)$st column, $|\Sigma^{yy}| = \det(f_x)$ and $|\Sigma^{kk}|$ is the determinant got by eliminating the $(k+1)$st row and column. Now $\hat{\sigma}^2_{yk.p}$ is got by replacing f_y, f_{yx}, f_x by $\hat{f}_y, \hat{f}_{yx}, \hat{f}_x$. It is also the partial multiple correlation of the column a_y with the pair of columns a_k, b_k after removing $a_j, b_j, j \neq k, j = 1, \ldots, p$ by regression. It is an intrinsic measure of association, at frequency λ, between $y(n)$ and $x_k(n)$ after allowing for all linear effects of other $x_j(n)$, including all linear effects from leads and lags of these other $x_j(n)$. We shall leave the reader to check the correctness of the assertion concerning the form of $\hat{\sigma}^2_{yk.p}$. It follows from the correspondence established at the beginning of Theorem 13 in Chapter IV just as in the proof of Theorem 5. We shall also omit the proof of the following theorem which proof follows the same form as that of Theorem 5.

Theorem 6. *Under the conditions of Theorem* 3 *the asymptotic distribution of* $\hat{\sigma}_{yk \cdot p}$ *is that of an ordinary coherence estimated from* $(m - p + 1)$ *frequencies* λ_u *(or from* $m - p$ *in case* $\lambda = 0$*, because of the effect of mean correction).*

It is somewhat more likely that (when $p > 1$) we shall wish to test $\hat{\sigma}_{yk \cdot p}$ for significance since the null hypothesis, that all the association at frequency λ or $y(n)$ with the $x_j(n)$ is due to those for $j \neq k$, may be a priori credible. This test is accomplished by using

$$(m - p)\left\{\frac{\hat{\sigma}^2_{y \cdot kp}}{1 - \hat{\sigma}^2_{yk \cdot p}}\right\}$$

as $F_{2,2m-2p}$.

We recall that

$$\hat{\beta} = \hat{f}_{xx}^{-1}\hat{f}_{xy}.$$

The typical element is

$$\hat{\beta}_k = |\hat{\beta}_k|\, e^{i\hat{\theta}_k}, \qquad \beta_k = |\beta_k|\, e^{i\theta_k}.$$

Now $\hat{\theta}_k$ describes the lead or lag relationship of $y(n)$ with $x_k(n)$ at frequency λ after the linear effects of the other $x_j(n)$ (including all lags and leads) have been removed. We have

Theorem 7. *Under the conditions of Theorem* 3 *asymptotically valid* $100(1 - \alpha)\%$ *confidence regions for* β_k *and, in case* $\lambda \neq 0, \pi$, θ_k *may be obtained from*

$$\begin{aligned} |\beta_k - \hat{\beta}_k| &\le \{(m - p)^{-1}\hat{f}_y(1 - \hat{\sigma}_{yp}{}^2)\,|\hat{f}_{xx}{}^{kk}|\,F_{2p,2m-2p}(\alpha)\}^{1/2}, && \lambda \neq 0, \pi, \\ &\le \{\nu^{-1}\hat{f}_y(1 - \hat{\sigma}_{yp}{}^2)\,|\hat{f}_{xx}{}^{kk}|\}^{1/2} t_\nu(\alpha), && \nu = m - p - 1,\ \lambda = 0, \\ & && \nu = m - p,\ \lambda = \pi, \end{aligned}$$

$$|\sin(\theta_k - \hat{\theta}_k)| \le \frac{s}{|\hat{\beta}_k|}\, t_{2m-2p}(\alpha),$$

where

$$s^2 = \frac{\hat{f}_y(1 - \hat{\sigma}_{yp}{}^2)\,|f_{xx}{}^{kk}|}{2m - 2p}.$$

Of course the same problems arise in connection with the confidence interval for $\hat{\theta}_k$ as arose in connection with phase in relation to Theorem 4 (see the discussion below that theorem).

The techniques we have discussed in this section are all regression and correlation techniques of a classical form, the only complications arising from the special nature of the vectors regressed upon (the regressor vectors we shall say) which affects the non-null distribution. Of course the fact that the β_{jk} are complex causes the introduction of additional detail, for example in connection with the notion of phase. It will readily be seen that the whole apparatus of multivariate analysis of variance is available in the present situation. We defer the discussion of this to Section 6, not because we think these techniques will prove to be unimportant but rather to avoid too lengthy a discussion at this point of techniques which so far have hardly been used.

3. ALTERNATIVE COMPUTATIONAL PROCEDURES FOR THE FFT

We have used FFT as "shorthand" for finite Fourier transform.† The calculation of these is an onerous task if N is really large as N^2 operations (each a multiplication plus an addition) are needed. For this reason the availability of simpler procedures is important. These procedures go back a long way, it seems, but recently have been associated with a paper by Cooley and Tukey (1965). The procedure rests on the following simple result. Let

$$N = \prod_1^s n_i, \qquad N_j = \prod_{i=1}^{s-j} n_{j+i}, \qquad j = 0, 1, \ldots, s-1, \qquad N_s = 1$$

and put, modulo N,

$$n = N \sum_1^s N_{j-1}^{-1} u_j, \qquad u_j = 0, 1, \ldots, n_j - 1,$$

$$k = \sum_{j=1}^s N_j v_j, \qquad v_j = 0, 1, \ldots, n_j - 1.$$

Thus

$$n = u_1 + n_1\big(u_2 + n_2\big(u_3 + n_3(\cdots n_{s-1}u_s)\big)\cdots\big),$$
$$k = v_s + n_s\big(v_{s-1} + n_{s-1}\big(v_{s-2} + n_{s-2}(\cdots n_2 v_1)\big)\cdots\big),$$

and these formulas show how the u_j, v_j are obtained by successive division. Now

$$\frac{1}{\sqrt{2\pi N}} \sum_1^N x(n) e^{in\omega_k} = \frac{1}{\sqrt{2\pi N}} \sum_{u_s} \cdots \sum_{u_1} x(n) \prod_{j=1}^s \exp\left\{ i2\pi u_j \sum_{l=j}^s \frac{v_l N_l}{N_{j-1}} \right\}.$$

† It is also used to mean fast Fourier transform, which we call the Cooley–Tukey technique.

Indeed in forming the product nk we may eliminate all products $NN_{j-1}^{-1}N_l$, $j > l$, since the contribution to $in\omega_k$ is then an integral multiple of $2\pi i$. We now perform s sequences of operations a typical one of which requires N times $2n_j$ real operations, each being multiplication followed by addition so that approximately $2N\sum n_j$ real operations are done in all. Thus we successively form a number of arrays of which the first has the N entries $A_{s+1}(u_1, \ldots, u_s) = x(n)$, this entry being in the place with address $u_1, \ldots, u_s$. Successive arrays are then formed as

$$A_j(u_1, \ldots, u_{j-1}, v_j, \ldots, v_s) = \sum_{u_j} A_{j+1}(u_1, \ldots, u_j, v_{j+1}, \ldots, v_s) \times \exp\left\{i2\pi u_j \sum_{l=j}^{s} \frac{N_l v_l}{N_{j-1}}\right\}, \qquad j = s, s-1, \ldots, 1. \tag{3.1}$$

In A_j the complex entries for $v_j, \ldots, v_s$ corresponding to k and $N - k$, but the same $u_1, \ldots, u_{j-1}$, are conjugates and for N, $\frac{1}{2}N$ are real.

Then

$$\frac{1}{\sqrt{2\pi N}} A_1(v_1, \ldots, v_s) = w(\omega_k).$$

If $N = \prod_1^s n_i^{p_i}$, we require approximately $2N\sum p_i n_i$ real operations to complete the task. One possible way of carrying out these procedures is to drop a small number of observations, so that the N corresponding to those remaining is highly composite, or to take a few more observations so that this is so; for example, if 1539 observations are to hand (1539 was chosen from a table of random numbers) since this is $3^4 \cdot 19$ we might drop 81 to take $N = 3^6 \times 2$. Alternatively, if this was possible, we might add 81 to make $N = 3^4 \cdot 2^2 \cdot 5$. In the first case we should then be required to carry out $2 \times 1458 \times 20 = 58{,}320$ operations and in the second $2 \times 1620 \times 21 = 68{,}040$ operations (with more data, of course). These compare with 2368521 by direct computation. The new procedure thus involves a saving of over 95% of the original work and brings the calculations down to reasonable proportions.

The formulas become especially simple when $n_i \equiv 2$ for then the typical sum on the right hand side of (3.1) is

$$A_{j+1}(u_1, \ldots, u_{j-1}, 0, v_{j+1}, \ldots, v_s) + A_{j+1}(u_1, \ldots, u_{j-1}, 1, v_{j+1}, \ldots, v_s) \cdot \exp\left\{i\pi \sum_{l=j}^{s} \frac{v_l}{2^{l-j}}\right\} \tag{3.2}$$

so that some of the operations called multiplication plus addition are in fact only additions. In this case the number of operations is $2sN = 2N \log_2 N$ and this, as a proportion of N^2, is $\{2 \log_2 N/N\}$. This has led Cooley and

Tukey (1965) to suggest the following procedure. The series $x(n)$ is augmented with zeros at one or both ends so that the new series is of length N', $N \le N' = 2^s < 2N$. This is then analyzed (after a modification which we discuss below) as for a series of N' observations. Thus we do not get $w(\omega_k)$ but instead

$$w\left(\frac{2\pi k}{N'}\right), \qquad k = 0, 1, \ldots, N' - 1.$$

This procedure may require more operations than that first mentioned but may have other advantages from a computational viewpoint, (see Cooley and Tukey (1965)). In case $N = 1539$ we would take $N' = 2048$ and we would then perform $2048 \times 22 = 45056$ operations of multiplication followed by addition. We obtain N' values of $w(\lambda)$ now in place of N but this can be of no advantage as N observations $x(n)$ only were initially observed.

The modification to the initial series at its end points which was mentioned above consists in the introduction of a "fader",† at both ends of the original series, which avoids the abrupt transition from zero to nonzero values. Thus we replace the original series $x(n)$ by $a_N(n)x(n)$, where the scalars $a_N(n)$ are the coefficients of the fader; for example, one might take

$$\begin{aligned} a_N(n) &= \frac{n}{M}, \qquad n \le M \ll N, \\ &= 1, \qquad M \le n \le N - M, \\ &= \frac{N-n}{M}, \qquad N - M \le n \le N, \end{aligned}$$

or

$$\begin{aligned} a_N(n) &= \frac{1}{2}\left(1 - \cos\frac{\pi n}{M}\right), \qquad n \le M \ll N, \\ &= 1, \qquad M \le n \le N - M, \\ &= \frac{1}{2}\left(1 - \cos\pi\frac{N-n}{M}\right), \qquad N - M \le n \le N. \end{aligned}$$

There seems to have been a tendency to make the fader approach unity at each end of the series much less quickly than this, for example, to use

$$a_N(n) = \frac{1}{2}\left(1 - \cos\frac{2\pi n}{N}\right).$$

† The word "taper" is also used with the same meaning.

We introduce

$$\phi_N(\lambda) = \frac{1}{\sqrt{2\pi N}} \sum_{n-1}^{N} a_N(n) e^{in\lambda}$$

and define $\hat{w}(\lambda)$ by

$$\hat{w}(\lambda) = \frac{1}{\sqrt{2\pi N}} \sum a_N(n)\, x(n) e^{in\lambda}. \tag{3.3}$$

It is this that is computed, *for* $\omega_k' = 2\pi k/N'$, by the Cooley Tukey procedure. Then

$$\mathcal{E}\{\hat{w}(\lambda)\, \hat{w}(\lambda)^*\} = \int_{-\pi}^{\pi} |\phi_N(\lambda - \theta)|^2 f(\theta)\, d\theta.$$

The purpose of the fader may be seen from an examination of this formula. It will be desirable to have $|\phi_N(\theta)|^2$ as small as possible for θ far from zero for only then will we be secure from a bias in $\hat{w}(\lambda)\hat{w}(\lambda)^*$ away from $f(\lambda)$ due to a very much larger value of an element of $f(\theta)$ at some point well away from λ. As our examples suggest and we shall later do, we take $a_N(n) = u_N(n/N)$ where $u_N(x)$ is a continuous function of x. Let us for simplicity at this moment take $a_N(n) = u(n/N)$. By "θ far from zero" we mean many units of size $2\pi/N$ away from that point so we are led to rewrite $\mathcal{E}(\hat{w}(\lambda)\hat{w}(\lambda)^*$ as

$$\mathcal{E}\{\hat{w}(\lambda)\hat{w}(\lambda)^*\} = \int_{-N\pi}^{N\pi} N^{-1} |\phi_N(\psi/N)|^2 f(\lambda - \psi/N)\, d\psi$$

where we have put $\lambda - \theta = \psi/N$. Now

$$N^{-\frac{1}{2}}\phi_N(\psi/N) = \frac{1}{\sqrt{2\pi}} \frac{1}{N} \sum_{1}^{N} u\left(\frac{n}{N}\right) e^{in\psi/N}$$

which rapidly approaches

$$\hat{u}(\psi) = \frac{1}{\sqrt{2\pi}} \int_0^1 u(x) e^{ix\psi}\, dx.$$

Now considering $u(x)$ as a function on $(-\infty, \infty)$ which is zero outside of $(-1, 1)$, we see that if that function can be differentiated ν times so that the resulting derivative is a linear combination of integrable functions and δ-functions (but not derivatives of the latter) then $\hat{u}(\psi) = O(|\psi|^{-\nu})$ as $\psi \to \infty$. Thus the higher the order of contact of $u(x)$ to the horizontal axis at 0, 1 the faster $u(\psi)$ decreases as $|\psi|$ increases, that is, the faster $\phi_N(\lambda)$ decreases as $|\lambda|$ increases. For example, for $u(x) \equiv 1$, $|\phi_N(\lambda)|^2$ is

$$\frac{\sin^2 (\frac{1}{2})N\lambda}{2\pi N \sin^2 (\frac{1}{2})\lambda} \tag{4}$$

which decreases as λ^{-2} as $|\lambda|$ increases. If $u(x) = \frac{1}{2}(1 - \cos 2\pi x)$ then $\nu = 3$ and we expect $|\phi_N(\lambda)|^2$ to decrease as $|\lambda|^{-6}$. Indeed,

$$\left| \int_0^1 \tfrac{1}{2}(1 + \cos 2\pi x) e^{i\psi x}\, dx \right|^2 = \frac{16\pi^4 \sin^2 \frac{1}{2}\psi}{\psi^2(4\pi^2 - \psi^2)^2}$$

and this decreases as ψ^{-6}, as it should.

There is, of course, a price to be paid. In the first place, as we shall see, the use of the fader introduces correlation between the $\hat{w}(\omega_k')$ in their asymptotic distribution. However, once $N/N' \ll 1$ there is substantial correlation among the $\hat{w}(\omega_k')$ in any case and indeed, as will be seen from Theorem 8 below, it may well be true that when $N/N' \ll 1$ the use of a fader simplifies the covariance properties of the $\hat{w}(\omega_k')$. In the second place, again as we shall see below, the use of the fader tends to increase the variance. This effect will be minor if the fader is not far from unity over most of its range and, as we know, we can achieve this if we are only concerned with the behavior of $\phi_N(\psi/N)$ a fair way from zero (in units of $1/N$).

Thus the use of a fader may provide a not very costly form of insurance against a bad distortion of $\mathcal{E}\{\hat{w}(\lambda)\hat{w}(\lambda)^*\}$ away from $f(\lambda)$ due to a very large value of $f(\theta)$ at some θ a long way from λ. (The reason why we emphasize this "long way from λ" is that we shall be averaging the $\hat{w}(\omega_k')$ over a range of consecutive ω_k' and will have chosen this range so that, we hope, $f(\theta)$ will not vary much over it. It seems somewhat contradictory then to worry about a discrepant value just outside of this, somewhat arbitrary, range.) Of course the objection may be made that if $f(\theta)/f(\lambda)$ is very large, for some θ, we shall know of it and could prevent its effects, for example, by prefiltering. The choice is one calling for judgment but it is probably often true that the use of a fader will be a cautious procedure with little cost. *

We need to investigate the asymptotic properties of the estimates $\hat{w}(\lambda)$. We consider a situation where $N = [aN']$, $\frac{1}{2} < a \leq 1$ and allow N', which is of the form 2^s, to increase keeping a fixed. Motivated partly by examples given above we put

$$a_N(n) = u_N\left(\frac{n}{N}\right), \tag{3.5}$$

where $u_N(x)$ is a continuous function which satisfies $|u_N(x)| \leq 1$ and

$$\lim_{N\to\infty} u_N(x) \equiv u(x) \qquad 0 < x < 1, \tag{3.5$'$}$$

where $u(x)$ is also continuous.

If $M/N \to 0$ as $N \to \infty$ this is true for the examples given above, with $u(x) \equiv 1$. If $M/N \to \alpha$ then, in the first case, for example, $u(x)$ is $x\alpha^{-1}$ for $x \leq \alpha$, 1 for $\alpha \leq x \leq 1 - \alpha$ and $(1 - x)\alpha^{-1}$ thereafter. The second example is of the same nature. We leave it to the reader to describe the situation for himself. The situation where $u(x) \equiv 1$ is in some ways the more relevant one since apparently as $N \to \infty$ the ratio M/N would tend to zero. (We say "apparently" for obvious reasons!) However, the other case is also relevant since in any actual example M/N will be nonzero and to gain some idea of the situation for N large we have therefore considered the more general case. We now consider the joint distribution of the $\hat{w}(\lambda)$ for m' values $\lambda'_u = 2\pi j_u/N'$ which are nearest to a fixed value λ. We have in mind the relation $m = am'$, which cannot exactly hold, so that the λ'_u cover nearly the same band as the λ_u. Just as in Chapter IV, Section 4 we are led to consider the sequences $a_N(n) \exp in\lambda'_u$. It is easily seen that these (or their real and imaginary parts) satisfy (a), (b), (c) of Chapter IV, Section 4. Indeed, writing $\lambda'(k)$ in place of λ'_k for ease of printing, consider

$$\lim_{N\to\infty} \frac{e^{-in\lambda'(l)} \sum_{m=1}^N a_N(m)\, a_N(m+n) e^{im(\lambda'(k)-\lambda'(l))}}{\sum_1^N a_N(m)^2}. \tag{3.6}$$

This is

$$\frac{e^{-in\lambda} \lim\limits_{N\to\infty} \left\{\dfrac{1}{N} \sum\limits_{m=1}^{N} a_N(m) a_N(m+n) e^{im(\lambda'(k)-\lambda'(l))}\right\}}{\displaystyle\int_0^1 u^2(x)\, dx,} \tag{3.7}$$

since

$$\frac{1}{N}\sum_1^N a_N(m)^2 = \int_0^1 \hat{u}(x)^2\, dx,$$

where $\hat{u}(x)$ is constant, at $a_N(m)$, over the interval $[m/N - (2N)^{-1}, m/N + (2N)^{-1}]$. Since $\hat{u}(x)^2$ converges boundedly to $u(x)^2$ for almost all x the result is established. But similarly the numerator of (3.7) is, apart from the factor $\exp -in\lambda$,

$$\lim_{N\to\infty} \int_0^1 \hat{u}(x)\, \hat{u}\left(x + \frac{n}{N}\right) \exp\left\{i2\pi a x(j_k - j_l)\left(\frac{N}{aN'}\right)\right\} dx$$

and by the same reasoning (and remembering that $j_k - j_l = k - l$) this evidently converges to

$$\phi_{k-l} = \int_0^1 u^2(x) e^{i2\pi a x(k-l)}\, dx.$$

For $u(x) \equiv 1$ this is

$$e^{ia\pi(k-l)}\left\{\frac{\sin \pi a(k-l)}{\pi a(k-l)}\right\},$$

so that (3.6), then, is

$$e^{-in\lambda} e^{ia\pi(k-l)}\left\{\frac{\sin \pi a(k-l)}{\pi a(k-l)}\right\}.$$

We call Φ the matrix with ϕ_{u-v} in row u, column v. It is Hermitian symmetric and for reasonable $u(x)$ (e.g., $u(x) \not\equiv 0$ a.e.) it is nonsingular. It is, of course, the unit matrix for $a = 1$ and $u(x) \equiv 1$. We put $\hat{v}(u) = \hat{w}(\lambda'_u)$ and call $\hat{v}$ the vector with $\hat{v}(u)$ as the uth set (of p elements) down it, so that $\hat{v}_j(u)$ goes in row $(u - 1)p + j$. Then we have the following theorem.

Theorem 8. *Let $\hat{v}$ be as just described with $\hat{w}(\lambda'_u)$ given by* (3.3) *for m' values, λ'_u, of $2\pi k/N'$ nearest to λ and $a_N(n)$ given by* (3.5) *with $u(x)$ continuous and $u(x) \not\equiv 0$, a.e. Then under the conditions on $x(n)$ of either Theorem* 13 *or* 13′ *of Chapter IV, as $N = [aN']$, $\frac{1}{2} < a \leq 1$, increases the distribution of $\hat{v}$ converges to a complex multivariate normal distribution with density*

$$\frac{1}{\pi^{pm'} \det\{\Phi \otimes f(\lambda)\}} e^{-\hat{v}^*(\Phi \otimes f(\lambda))^{-1}\hat{v}}, \qquad \lambda \neq 0, \pi.$$

If, for $\lambda = 0$, $\lambda'_u = 0$ is excluded and for $\lambda = \pi$, $\lambda'_u = \pi$ is not included then the result continues to hold in these cases also.

The proof of this theorem hardly differs from that of Theorems 13 or 13′ in Chapter IV and is omitted.

Theorem 8 *may not be so important as might be thought.* This is because the Cooley-Tukey procedure will be used only when N is very large and then the considerations of Section 4, where m' is allowed to increase with N, may be more relevant. However, the theorem enables us, for a single fixed frequency λ, to carry through the considerations of Section 2 for the present circumstances. Indeed, Φ is a known matrix. (In practice a is taken to be N'/N) so that we may form

$$\tilde{v} = (\Phi^{-1/2} \otimes I_p)\hat{v}. \tag{3.8}$$

Let us call $\tilde{v}(u)$, the vector of p elements, occurring in the same p places in $\tilde{v}$ as $\hat{v}(u)$ does in $\hat{v}$. Thus to get $\tilde{v}(u)$ we first form the p vectors $\hat{v}_j$, having $\hat{v}_j(u)$ in the uth place, $u = 1, \ldots, m'$, then form $\tilde{v}_j = \Phi^{-1/2}\hat{v}_j$ and then $\tilde{v}_j(u)$ is the uth element in $\tilde{v}_j$.

Corollary 2. *If the conditions of Theorem* 8 *hold and $\tilde{v}(u)$, $u = 1, \ldots, m'$, is formed as just described then these have the same asymptotic distributional properties as the $v(u)$ of Theorem* 3 *and all of the results of that section hold with the $\tilde{v}(u)$ replacing $v(u)$ and m' replacing m.*

Though this general result is evidently of some interest it is rather likely that the constant a will be so near to unity and $u(x) \equiv 1$ (or $a_N(n)$ at or near unity for all but a small portion of its values). Then we shall replace the use of

$\tilde{v}(u)$ by the simpler use of $\hat{v}(u)$. The most direct formula is

$$\hat{f}_{jk}(\lambda) = \left\{\frac{1}{m'}\sum_{u=1}^{m'} \hat{v}_j(u)\,\overline{\hat{v}_k(u)}\right\}\Big/\left\{\frac{1}{N}\sum_{n=1}^{N} a_N^{\,2}(n)\right\}. \tag{3.9}$$

(We do not, at present, notationally distinguish this estimate from the other estimate of the same quantity earlier introduced.)

It is of some interest to examine the first two moments of these quantities. Since

$$\frac{1}{N}\sum_{1}^{N} a_N^{\,2}(n) \to \int_0^1 u^2(x)\,dx,$$

and all of the diagonal elements of Φ are this latter quantity it follows that the mean value of the (asymptotic) distribution of $\hat{f}_{jk}(\lambda)$ is $f_{jk}(\lambda)$, as required. We may obtain the variance of this distribution from Theorem 8. Now we consider the corresponding covariance from the limiting distribution of these quantities.

Corollary 3. *The covariances*† *of the* $\hat{f}_{jk}(\lambda)$, *got from the limiting distribution of theorem* 8 *are given by*

$$\text{(3.10)}\quad \operatorname{cov}\left(\hat{f}_{i}\,(\lambda), \hat{f}_{kl}(\lambda)\right)$$

$$= \left\{m'\int_0^1 u^2(x)\,dx\right\}^{-2} \operatorname{tr}(\Phi^2) f_{ik}(\lambda) f_{lj}(\lambda), \qquad \lambda \neq 0, \pi$$

$$= \left\{m'\int_0^1 u^2(x)\,dx\right\}^{-2} \operatorname{tr}(\Phi^2)\{f_{ik}(\lambda) f_{lj}(\lambda) + f_{il}(\lambda)\, f_{kj}(\lambda)\}, \qquad \lambda = 0, \pi.$$

We leave this corollary to be proved as an exercise to this chapter as it follows easily from Theorem 8.

If m' is not small a fair approximation to $\operatorname{tr}\Phi^2$ may be got as follows, namely,

$$\frac{1}{m'}\operatorname{tr}(\Phi\Phi^*) = \sum_{-m'+1}^{m'-1}\left(1 - \frac{|j|}{m'}\right)|\phi_j|^2$$
$$\simeq \frac{1}{a}\int_0^1 u^4(x)\,dx,$$

since the sum is the Cesaro sum of the Fourier series of the function which is $a^{-2}u^4(y/2\pi a)$ on the interval $[0, 2\pi a)$ and is otherwise zero in $[0, 2\pi)$.

† We again remind the reader that in defining covariances for complex random variables we conjugate the second random variable in the pair being covariated.

This suggests the approximate covariance matrix

$$\frac{1}{m} f(\lambda) \otimes f'(\lambda) \frac{\int_0^1 u^4(x)\, dx}{\left(\int_0^1 u^2(x)\, dx\right)^2} \qquad m = am', \qquad \lambda \neq 0, \pm\pi.$$

It is evident that the asymptotic distribution of the matrix $\hat{f}$, estimated by the Cooley-Tukey technique, taking m, m' as fixed, is not that obtained in Section 2. Thus, in particular, in the scalar case, $2m\hat{f}/f$ is not chi-square, asymptotically, even if $u(x) \equiv 1$, unless $a = 1$. Nevertheless, too much should not be made of this. The kth cumulant of $\hat{f}$, by the Cooley-Tukey procedure, from the limiting distribution, is of order m^{-k+1}. Thus, if m is not very small, the use of $\nu\{\hat{f}/f\}$ as chi-square with

$$\text{(3.11)} \qquad \nu = \frac{2m\left(\int_0^1 u^2(x)\, dx\right)^2}{\int_0^1 u^4(x)\, dx}, \qquad m = am',$$

degrees of freedom should give a reasonable approximation. This rule gives a distribution with the correct mean and nearly the correct variance, the nearly being due to the approximation involved in the introduction of the simple factor in the denominator. It should also have approximately the correct shape. Correspondingly we are led to act in relation to the matrix $\hat{f}$ as if it was a matrix having the complex Wishart distribution (see Section 6) but with "degrees of freedom," ν, got from (3.11) instead of by the rule $\nu = 2m$, used for the finite Fourier transform. Further investigation of these approximations is, no doubt, needed. *To the extent that the approximation may be used it makes all procedures very simple for all the results of the previous section may, via this approximation, be carried over to apply to* (3.9) *with the simple replacement of* $2m$ *by* ν, *got from* (3.11).

The computation of the quantity ν for the $u(x)$ corresponding to those suggested above formula (3.3) is quite trivial, of course; for example, the first corresponds to

$$\begin{aligned} u(x) &= \frac{x}{\alpha}, && 0 \leq x \leq \alpha, \\ &= 1, && \alpha \leq x \leq 1 - \alpha, \\ &= \frac{1 - x}{\alpha}, && 1 - \alpha \leq x \leq 1, \end{aligned}$$

where α is M/N. Then

$$\nu = \frac{2m(1 - 4\alpha/3)^2}{1 - 8\alpha/5}.$$

In many cases α would be quite small (say 0.05) and the expression (3.11) becomes $\nu = 2m$.

When a is much less than unity the use of an appropriate fader can certainly make the $|\phi_j|$ converge to zero with j at a faster rate. Thus it, in a sense, makes Φ nearer to the unit matrix. However, there is evidently a limit to what can be achieved in this direction and for a much less than unity it seems that the approximate treatment for the distribution of the statistics defined by (3.9) can be only a rough approximation and the use of Corollary 2 is needed. As we have already said the real usefulness of the Cooley-Tukey procedure comes when N is very large when it is more realistic to construct an asymptotic theory with m increasing with N, as in Section 4. Though the matter deserves further investigation the use of a fader does not seem to be well justified by the desire to validate the approximate distribution theory for (3.9). However, the Corollary 3 and the discussion following it does show that the use of the fader has some disadvantages; for example, in the scalar case the variance is

$$m^{-1} f(\lambda)^2 \int_0^1 u^4(x)\, dx \Big/ \left\{\int_0^1 u^2(x)\, dx\right\}^2$$

and this is minimized when $u(x) \equiv 1$. Thus the use of the fader tends to increase the variance. Its justification appears to be based upon a reduction in bias.

We close this section with two comments. In the first place there is something paradoxical about the result of Theorem 8 for it tells us that we may, by using the frequencies $2\pi k/N'$ lying in a given band of width $2\pi m/N$ about λ, obtain an estimate of $f(\lambda)$ which, in the scalar case is a multiple of $\chi^2_{2m'}$, where $2m' > 2m$. Why then do we not take a arbitrarily small and thus m' arbitrarily large? The answer is of course found in the fact that when a is much less than 1 we may have to take N'/m' much larger than N/m before the asymptotic theories are relevant. Thus an N which, in relation to a given m', makes the result of Theorem 8 a good approximation to the truth would justify the use of an m in Theorem 3 which is at least as big as m'. This question deserves further investigation than we give it here.

The second question which arises in relation to the Cooley-Tukey procedure relates to the relationship between estimates from neighboring bands. In the case of the technique of Section 2 it is evident that we may split the set of m frequencies from which $f(\lambda)$ was estimated into, say, two subsets of m_1 and m_2 and obtain two estimates of $f(\lambda)$ with degrees of freedom $\nu_i = 2m_i$

which are asymptotically independent. In the case of the Cooley-Tukey procedure this is not so, of course as is evident from Theorem 8. Of course if the fader were well chosen and the m' frequencies were split into two subsets of m_1' and m_2', of which the former set lay altogether to the left of the latter and m_1' and m_2' were not small, then the association might be quite weak. We can, indeed, work out what the association will be, at the limiting distribution of the two estimates, for in the same way as for formula (3.11) we obtain

$$(m_1'm_2')^{-1}\left\{\int_0^1 u^2(x)\,dx\right\}^{-2} \operatorname{tr}(\Phi_{12}\Phi_{12}^*) f(\lambda) \otimes f(\lambda)',$$

where now the element in row $(i-1)p+j$, column $(k-1)p+l$ is the covariance between the estimate of $f_{ij}(\lambda)$ from the first set of m' frequencies and the estimate of $f_{kl}(\lambda)$ from the second set while Φ_{12} has m_1 rows and m_2 columns and contains entries ϕ_{u-v}, $u = 1, \ldots, m_1$; $v = m_1 + 1, \ldots, m_1 + m_2$. We again leave this result as an exercise for this chapter. The correlation in the limiting distribution between the two neighboring estimates in the scalar case thus becomes

$$\frac{\operatorname{tr}(\Phi_{12}\Phi_{12}^*)}{[\operatorname{tr}(\Phi_{11}\Phi_{11}^*)\operatorname{tr}(\Phi_{22}\Phi_{22}^*)]^{1/2}}, \tag{3.12}$$

where Φ_{11} and Φ_{22} are matrices of the form of Φ with m_1 and m_2 rows respectively; in the case of Φ_{11}, for example, the rows and columns being those numbered 1 to m_1 in Φ. This shows the nature of the situation fairly clearly. One can easily extend these results to the case of a partition of m into more than two sets.

4. ESTIMATES OF SPECTRA FOR LARGE N AND m

We now consider the problem of spectral estimation under the circumstances which are probably most relevant to many "real life" situations, where N is very large (say above 2000) and the spectral estimates are to be made for a set of bands, in $(0, \pi]$, of equal width, π/M, centered at $\pi j/M$, which are so chosen that N/M is also large. If N/M is large, then a chi-square law for the distribution of a spectral estimate may as well be replaced by a normal law. The asymptotic situation we have in mind, therefore, is one where N and M both increase but so that M/N converges to zero. This, rather than the procedure of Sections 2 and 3, is the model that has been mainly used to substantiate, statistically, spectral estimation procedures. We restrict ourselves to estimators which are quadratic functions of the observations, that is,

$$\hat{f}(\lambda) = \sum\sum b_{uv}(\lambda) x(u)\, x(v)'. \tag{4.1}$$

We assume for the moment that $x(n)$ has zero expectation, since we shall deal with trend corrections later but emphasize that in practice at least mean corrections will need to be made. The most important class of estimates is that for which $b_{uv}(\lambda) = b_{v-u}(\lambda)$. Since the expectation of $x_j(u)x_k(v)$ depends only on $v - u$, this is a natural estimator to use. In fact Grenander and Rosenblatt (1957), show that we may, without any real loss, restrict ourselves to estimators of this form. However, not all of the estimators used in practice are of this form. If we have

$$b_n(\lambda) = b_n e^{-in\lambda},$$

the estimator we are using is of the form

$$\hat{f}(\lambda) = \sum_{-N+1}^{N-1} b_n e^{-in\lambda} \sum{}' x(u)x'(u+n)\},$$

where the sum $\sum'$ is over all u for which both u and $u + n$ lie between 1 and N inclusive. We put $k_n = 2\pi N b_n$ and write this in the form

$$\frac{1}{2\pi} \sum_{-N+1}^{N-1} k_n e^{-in\lambda} \left(1 - \frac{|n|}{N}\right) C(n) \tag{4.2}$$

where $C(n)$ is the matrix defined in Chapter IV, Section 3. Introducing

$$K_N(\lambda) = \frac{1}{2\pi} \sum_{-N+1}^{N-1} k_n e^{-in\lambda},$$

we may write this as

$$\int_{-\pi}^{\pi} I(\theta)\, K_N(\lambda - \theta)\, d\theta, \tag{4.3}$$

where $I(\theta)$ is defined by (3.8) in Chapter IV. Formula (4.2) and (4.3) are basic and (4.3), in particular, shows that estimators of the type we shall mainly consider in this chapter are weighted averages of the "periodogram," $I(\theta)$, with a kernel function (or "spectral window"),† $K_N(\lambda)$, as weight. Since $I(\theta)$ has expectation that converges to $f(\theta)$ under very general conditions, we see that we shall require that $K_N(\lambda)$ concentrates, as N increases, at the origin. Correspondingly the sequence k_n (which depends on N also) will converge to zero but at a decreasing rate as N increases. We shall see, however, that $K_N(\lambda)$ cannot concentrate at $\lambda = 0$ too quickly as $N \to \infty$ for otherwise the variance of (4.2) will not converge to zero.

In an important range of cases we shall have

$$k_n = k\left(\frac{n}{M}\right), \tag{4.4}$$

† The sequence k_n is sometimes called a "lag window." The subject is replete with special terminology.

where $k(x)$ is a continuous, even function with $k(0) = 1$, $|k(x)| < 1$ and

$$\int_{-\infty}^{\infty} k^2(x)\, dx < \infty.$$

We give examples below. As we shall see, however, not all cases can be fitted into this scheme. In the case of the, so called, Bartlett estimator the variation is slight but estimators of the type of those considered in the last section need separate consideration.

It is often more convenient to consider, in place of $K_N(\lambda)$,

$$K(\lambda) = \frac{1}{2\pi} \int_{-\infty}^{\infty} k(x) e^{-i\lambda x}\, dx,$$

which is approximately

$$\frac{1}{2\pi M} \sum_{-N+1}^{N-1} k\left(\frac{n}{M}\right) e^{-i(n/M)\lambda} = \frac{1}{M} K_N\left(\frac{\lambda}{M}\right).$$

We shall now discuss some examples.

Example 1. The FFT

Here we consider

$$\frac{1}{m} \sum_j I(\omega_j),$$

where j varies over m consecutive integers and m increases with N. Now we have

$$\hat{f}^{(F)}(\lambda) = \frac{1}{m} \sum_j \frac{1}{2\pi} \sum_{-N+1}^{N-1} \left(1 - \frac{|n|}{N}\right) C(n) e^{-in\lambda} e^{in(\lambda - \omega_j)},$$

wherein the ω_j values are the m nearest to λ. This is just our formula (2.5) rewritten. Put

$$k_n(\lambda) = \frac{1}{m} \sum_j e^{in(\lambda - \omega_j)} = e^{in\lambda} e^{-in\omega_q} \frac{1 - e^{inm\omega_1}}{m(1 - e^{-in\omega_1})},$$

where ω_q is the first frequency in the sum defining $\hat{f}^{(F)}(\lambda)$. *We shall often omit the λ in the expressions given below since it is held fixed, but k_n does depend upon λ.* This last expression is

$$e^{in(\lambda - \lambda')} \frac{e^{in(\lambda' - \omega_q + \pi/N)} - e^{-in(m\omega_1 - \lambda' + \omega_q - \pi/N)}}{m 2i \sin(n\pi/N)}.$$

We have put

$$\lambda' = \frac{2\pi}{N} \{q + \tfrac{1}{2}(m - 1)\},$$

so that $|\lambda - \lambda'| \le \pi/N$. *(If λ is exactly the midpoint of the band covered by the m of the FFT then $\lambda - \lambda' = 0$.)* Now

$$k_n = e^{in(\lambda-\lambda')}\,\frac{e^{in\pi m/N} - e^{-in\pi m/N}}{m2i\sin(n\pi/N)};$$

that is,

$$k_n = e^{in(\lambda-\lambda')}\,\frac{\sin(nm\pi/N)}{m\sin(n\pi/N)}, \qquad |\lambda - \lambda'| < \frac{\pi}{N}, \qquad |n| \le N-1.$$

Then $\hat{f}^{(F)}(\lambda)$ may be written as

$$\frac{1}{2\pi}\sum_{-N+1}^{N-1}\left(1 - \frac{|n|}{N}\right)k_n C(n)e^{-in\lambda}.$$

In relation to this estimator we are led to introduce the function

$$k(x) = \left\{\frac{\sin\frac{1}{2}\pi x}{\frac{1}{2}\pi x}\right\},$$

putting $2mM = N$, and

$$K(\lambda) = \pi^{-1}, \qquad |\lambda| \le \tfrac{1}{2}\pi; = 0, \qquad |\lambda| > \tfrac{1}{2}\pi.$$

Example 2. The Cooley-Tukey Procedure

Here we have

$$\hat{f}^{(c)}(\lambda) = \frac{1}{m'}\sum_j\left\{\frac{1}{2\pi N}\left(\sum_n x(n)\,a_N(n)e^{in\omega_j'}\right)\left(\sum_n x(n)\,a_N(n)e^{in\omega_j'}\right)^*\right\},$$

where the ω_j' are of the form $2\pi j/N'$ and are the m' nearest such values to λ. This is the same formula as (3.10) *but with $N^{-1}\sum a_N(n)^2 = 1$, which we assume for simplicity and without loss of generality.* The intervention of the fader now makes it impossible to put this, even approximately, in the form (4.4) above. If $a_N(n) \equiv 1$, we obtain, in the same way as for Example 1,

$$\hat{f}^{(c)}(\lambda) = \frac{1}{2\pi}\sum_{-N+1}^{N-1}\left(1 - \frac{|n|}{N}\right)k_n C(n)e^{-in\lambda},$$

$$k_n(\lambda) = e^{in(\lambda-\lambda')}\,\frac{\sin(nm'\pi/N')}{m'\sin(n\pi/N')}, \qquad |\lambda - \lambda'| < \pi/N', \qquad |n| \le N-1.$$

Again we shall often omit λ in the notation for k_n. Putting $2m'M = N'$, we again are led to introduce

$$k(x) = \left\{\frac{\sin\frac{1}{2}\pi x}{\frac{1}{2}\pi x}\right\},$$

$$K(\lambda) = \pi^{-1}, \qquad |\lambda| \le \tfrac{1}{2}\pi; = 0, \qquad |\lambda| > \tfrac{1}{2}\pi.$$

When $a_N(n) \not\equiv 1$ we assume that $a_N(n)$ is uniformly bounded (in N and n) by a finite constant and (as mentioned before) that

$$\frac{1}{N}\sum_{n=1}^{N} a_N(n)^2 \equiv 1.$$

We further assume that there is a continuous function $u(x)$, $0 \leq x \leq 1$, so that if $u_N(x)$ is the function which is $a_N(n)$ over the interval $(n/N - 1/(2N), n/N + 1/(2N)]$ then $u_N(x)$ converges pointwise to $u(x)$. Evidently

$$\int_0^1 u^2(x)\,dx = 1.$$

Then for the Cooley-Tukey estimator with a fader we introduce

$$k(x) = \frac{\sin \frac{1}{2}\pi x}{\frac{1}{2}\pi x}\left\{\int_0^1 u^4(x)\,dx\right\}^{1/2}.$$

Example 3. Truncated Estimate

Here

$$\hat{f}^{(T)}(\lambda) = \frac{1}{2\pi}\sum_{-M}^{M}\left(1 - \frac{|n|}{N}\right)C(n)e^{-in\lambda},$$

$$k(x) \equiv 1, \qquad |x| \leq 1; \equiv 0, \qquad |x| > 1$$

$$K_N(\lambda) = \frac{\sin \dfrac{2M+1}{2}\lambda}{2\pi \sin \frac{1}{2}\lambda},$$

$$K(\lambda) = \frac{\sin \lambda}{\pi\lambda}.$$

This estimate has a $k(x)$ that is not continuous. It has vices that can be seen from $K(\lambda)$ whose first side lobe has maximum near $3\pi/2$ which is $2/(3\pi)$ times the height of the main lobe. Correspondingly, if there is a sharp peak in a spectral density, $f(\lambda)$, say at λ, then as (4.3) shows, there will tend to be a peak in the estimate at points $\pm 3\pi/2M$, $\pm 5\pi/2M, \ldots$ away from it, caused by a coincidence of some side lobe with this peak. This can lead to misinterpretations. Another undesirable feature is the fact that $K(\lambda)$ is not always positive. Thus an estimate of a spectrum need not be positive if this estimator is used. This will tend to occur at a point where a deep trough, going near to zero, occurs in $f(\lambda)$. For these kinds of reason this estimator is rarely used.

Example 4. Bartlett Estimate

Here

$$\hat{f}^{(B)}(\lambda) = \frac{1}{2\pi}\sum_{-M}^{M}\left(1 - \frac{|n|}{M}\right)C(n)e^{-in\lambda},$$

$$k_n = \frac{1 - |n|/M}{1 - |n|/N}.$$

Again, though this is not quite of the form (4.4), we are led to introduce

$$k(x) = (1 - |x|), \qquad |x| \le 1; = 0, \qquad |x| > 1.$$

Then

$$K(\lambda) = \frac{1}{2\pi}\left\{\frac{\sin \frac{1}{2}\lambda}{\frac{1}{2}\lambda}\right\}^2 \ge 0.$$

Example 5. Tukey-Hanning Estimate

This is a commonly used one due to the influence of Tukey. It is obtained as

$$\hat{f}^{(H)}(\lambda) = \tfrac{1}{2}\hat{f}^{(T)}(\lambda) + \tfrac{1}{4}\hat{f}^{(T)}\left(\lambda - \frac{\pi}{M}\right) + \tfrac{1}{4}\hat{f}^{(T)}\left(\lambda + \frac{\pi}{M}\right).$$

Correspondingly,

$$k_n = \tfrac{1}{2}\left(1 + \cos\frac{\pi n}{M}\right), \qquad n = 0, \pm 1, \ldots, \pm M,$$

$$k(x) = \tfrac{1}{2}(1 + \cos \pi x), \qquad |x| \le 1,$$

$$= 0, \qquad x > 1,$$

$$K(\lambda) = \frac{\sin \lambda}{2\pi\lambda}\left(\frac{\pi^2}{\pi^2 - \lambda^2}\right).$$

Of course, $K_N(\lambda)$ can easily be got from that for the truncated estimate by the same "moving average" as was used to get $\hat{f}^{(H)}$ from $\hat{f}^{(T)}$. This $K(\lambda)$ is not always positive but has a first side lobe, centered nearly at $5\pi/2$, with height (negative) which is nearly $8/(105\pi) = 0.021$ relative to the height of the main lobe. This is the reason the advocates of this estimate give in its favor. Of course, by the same argument as was used before in connection with the use of a fader, if we wish $K(\lambda)$ to have its first zero at $\lambda = 2\pi$ and to minimize the average height relative to the height of the main lobe one can hardly do better than the Bartlett estimate. Nevertheless, the largest side lobe for the Bartlett estimate is relatively somewhat higher (105/9 times as great) and in some cases this may be an important consideration (see the discussion of the truncated estimate).

Example 6. Parzen Estimate

It is natural to seek for $K(\lambda)$ amongst the kernels that have been successful in connection with the summability theory of Fourier series. Such a one is the Jackson-de la Vallée Poussin kernel

$$\frac{12}{\pi}\left(\frac{\sin \frac{1}{2}\lambda}{\lambda}\right)^4$$

which is zero, also, for the first time at $\lambda = 2\pi$. It is more concentrated at $\lambda = 0$ than is the case of Fejér's kernel (the Bartlett estimate) as can be seen from the fact that its height relative to the height of the main lobe is $16/\pi^4$ at $\lambda = \pi$ compared to $4/\pi^2$ for Fejér's kernel. For this reason it has been replaced by

$$K(\lambda) = \frac{3}{8\pi}\left(\frac{\sin \lambda/4}{\lambda/4}\right)^4$$

This is not zero at 2π but is small there and has very low side lobes. It is, of course, positive. Correspondingly

$$\begin{aligned} k_N(x) \equiv k(x) &= 1 - 6x^2 + 6\,|x|^3, && |x| \le \tfrac{1}{2}, \\ &= 2(1 - |x|)^3, && \tfrac{1}{2} \le |x| \le 1, \\ &= 0, && 1 \le |x|. \end{aligned}$$

Example 7. Abel Estimate

The remaining two estimates are mentioned only as items of interest and not because they will be used. The first of these puts

$$k_N(x) \equiv k(x) = e^{-a\,|x|}, \qquad 0 < a,$$

$$K(\lambda) = \frac{a}{\pi(a^2 + \lambda^2)}.$$

Example 8. Daniell Estimate

Here we take

$$k_N(x) \equiv k(x) = \frac{\sin \frac{1}{2}\pi x}{\frac{1}{2}\pi x},$$

$$\begin{aligned} K(\lambda) &= \frac{1}{\pi}, && |\lambda| \le \frac{\pi}{2}, \\ &= 0, && |\lambda| > \frac{\pi}{2}. \end{aligned}$$

Both Examples 7 and 8, like 1, require a large computational effort. In fact, it is evident that Examples 8 and 1 are very closely related. There seems no good reason why either 7 or 8 should be used. Other estimators also have been introduced, for example, of the form (4.4) with $k(x) = 1 - x^2$, $|x| \leq 1$; $\equiv 0$, $x \geq 1$.

The subject of the choice of a "spectral window" has, probably, been greatly exaggerated in importance. The standard procedures appear to be Examples 1, 2, 4, 5, and 6 with 2 and 5 being the most commonly used. It seems likely that Example 2 will become dominant. We may as well discuss the sampling properties of such estimates with generality as at least 7 out of the 8 procedures suggested have been used, on occasions, in practice as well as others. We now commence this study.

We remind the reader, in connection with the interpretation of the covariance formula below that in defining covariances of complex quantities we conjugate the second member of the pair being covariated. Now we have

Theorem 9. *Let $x(n)$ be stationary to the fourth order with*

$$\sum\sum\sum_{npq=-\infty}^{\infty} |k_{ijkl}(0, n, p, q)| < \infty, \qquad \sum_{-\infty}^{\infty} |\gamma_{jk}(n)| < \infty, \qquad i, j, k, l = 1, \ldots, p$$

and $\hat{f}$ be of the form (4.2) *above with k_n of the form* (4.4), *with $k(x)$ continuous and uniformly bounded and $k(0) = 1$, or be one of the estimates of the Examples* 1 *to* 8 *above. Then, for $M \to \infty$, $M/N \to 0$,* (*with $M = \frac{1}{2}m^{-1}N$ for Example* 1 *and $\frac{1}{2}(m')^{-1}N'$ for Example* 2)

$$(4.5) \quad \lim_{N\to\infty} \frac{N}{M} \operatorname{cov}\left\{\hat{f}_{ij}(\lambda_1), \hat{f}_{kl}\left(\lambda_2 + \frac{\pi p}{M}\right)\right\} = 0, \qquad \lambda_1 \neq \pm\lambda_2 \ (\mathrm{mod}\ 2\pi),$$

$$= \int_{-\infty}^{\infty} k^2(x) e^{ip\pi x}\, dx\, f_{ik}(\lambda) f_{lj}(\lambda), \qquad \lambda_1 = \lambda_2 = \lambda \neq 0, \pm\pi,$$

$$= \int_{-\infty}^{\infty} k^2(x) e^{ip\pi x}\, dx \{f_{ik}(\lambda) f_{lj}(\lambda) + f_{il}(\lambda) f_{kj}(\lambda)\}, \qquad \lambda_1 = \pm\lambda_2 = 0, \pm\pi.$$

Convergence is uniform in λ_1, λ_2 for all points of the torus $-\pi \leq \lambda_i < \pi$, $i = 1, 2$, for which $|\lambda_1 \pm \lambda_2| \geq \epsilon > 0$, (*mod* 2π), *for any such* ϵ.

We prove this theorem in the Appendix to this chapter.

It is not necessary to state the result for $\lambda_1 = -\lambda_2 = \lambda$ since this becomes the covariance between $\hat{f}_{ij}(\lambda)$ and $\hat{f}_{lk}(\lambda - \pi p/M)$. If we adopt the tensor notation used before, the result, in case $\lambda_1 = \lambda_2 = \lambda \neq 0, \pi$, can be written in the form

$$\left\{\int_{-\infty}^{\infty} k^2(x) e^{ip\pi x}\, dx\right\} f(\lambda) \otimes f(\lambda)'.$$

Theorem 8 is expressed in the form shown so as to emphasize (a) that the estimates at two fixed points λ_1, λ_2, become uncorrelated as $N \to \infty$ but, (b) estimates at points $\pi p/M$ apart are, in general, correlated. In fact estimates are computed at points $\pi p/M$ apart so that this is relevant.

The expression $\int k^2(x) \exp i(p\pi x)\, dx$ is null for $p \neq 0$ for Examples 1, 2, 3, and 8. For Examples 4, 5, and 7 it is as follows.

Example 4

$$\tfrac{2}{3},\ p = 0; \qquad \frac{4}{\pi^2 p^2},\ p \neq 0.$$

Example 5

$$\tfrac{3}{4},\ p = 0; \qquad \tfrac{1}{2},\ p = \pm 1; \qquad \tfrac{1}{8},\ p = \pm 2; \qquad 0,\ |p| > 2.$$

Example 7

$$\frac{4a}{4a^2 + p^2\pi^2}.$$

Example 6 has properties similar to those of Example 4, though the formulas are more complicated. For Example 6, at $p = 0$, the factor $\int k^2(x)\, dx$ is 367/560.

If we take $k(x)$ as appropriate to the truncated formula and $M = N$ then the truncated estimator becomes the periodogram. Of course when $M = N$ Theorem 9 does not hold. We might, nevertheless, be led to put $M = N$ in (4.5), for this $k(x)$, expecting to get the appropriate covariance formula for the periodogram. However, comparing that result with the correct result of Corollary 1 we see that this is not so but (4.5), used in this improper way, gives twice the correct value. (The reason is found in the factor $(1 - |n|/N)$ whose influence is negligible when $M/N \to 0$ but is not when $M = N$.) This paradoxical result suggests that, for the truncated estimator at least, the use of Theorem 9 may tend in practice to give too large a value for the variance.

In relation to these examples it is of interest to show the constant

$$\nu = \frac{2N}{M \int k^2(x)\, dx}.$$

Since $\nu = 2m$ for the FFT it is natural to define m in general as $\frac{1}{2}\nu$, and we shall do that. The quantity ν is determined according to the rule that the degrees of freedom of a quantity with the distribution of a constant by a chi-square variable with ν degrees of freedom is the ratio of twice its expectation squared to its variance. This gives Table 1.

Table 1

Estimator	ν = Degrees of freedom
1. FFT	$2m = N/M$
2. Cooley-Tukey	$(2m'N/N')/\int_0^1 u^4(x)\,dx = (N/M)/\int_0^1 u^4(x)\,dx$
3. Truncated	N/M
4. Bartlett	$3N/M$
5. Tukey-Hanning	$8N/3M$
6. Parzen	$(1120N)/(367M)$
7. Abel	$2aN/M$
8. Daniell	N/M
9. $k(x) = 1 - x^2,\ \lvert x\rvert \le 1;\ 0,\ \lvert x\rvert \ge 1$	$(15N)/(8M)$

One should beware of assuming that an estimator is better if its degrees of freedom are larger, for given M, N. In fact in all cases where ν rises above N/M this is because the neighboring bands are correlated. This is not altogether a bad thing for we expect the true $f(\lambda)$ values to be close together for neighboring bands and the smooth variation in the error of estimate which will result from a high correlation between such bands will avoid sharp changes of a random nature. However, the difficulty is that this increase in ν has been caused by a spread of the spectral window beyond the band, of width π/M, and consequently the estimate at the center of the band is influenced by values of $f(\lambda)$ more than $(\pi/2M)$ away from that center. In the case of the estimate 2 the use of the fader must reduce the degrees of freedom unless $u(x) \equiv 1$. Again one must be careful in one's judgement for we have not yet discussed the bias in our estimates and the use of the fader is designed to reduce this.

As we indicated at the end of the last paragraph, we need also to consider the bias on spectral estimates. One should not over estimate the importance of these bias calculations for in practice workers appear to treat spectral estimators as much as estimates of the average spectral mass over a band as estimators of the spectral density at a point. Nevertheless, some consideration of a concept such as bias is necessary as otherwise we are led to believe that we can allow our degrees of freedom, ν, to increase as rapidly as we like by making M increase slowly with N. In the following we substantially follow Parzen (1957).

Evidently the bias will depend on the smoothness of $f(\lambda)$ and we assume that

$$\sum_{-\infty}^{\infty} |n|^q \,\|\Gamma(n)\| < \infty, \qquad q \ge 0. \tag{4.6}$$

We shall be concerned only in evaluating the dominant term in an asymptotic expansion for the bias and we also assume that, for the same q,

(4.7) $$\lim_{x\to 0} \frac{1-k(x)}{|x|^q} = k_q < \infty.$$

If k_q is finite for some q_0 then it is zero for $q < q_0$. If we have a given theoretical sequence $\Gamma(n)$ and a given $k(x)$ then q may be taken to be the largest number satisfying (4.6) and (4.7) though it is useful to think of q as the largest real number satisfying (4.6) and of $k(x)$ as chosen so that (4.7) is true. For the Cooley-Tukey procedure we need the following condition also

(4.8) $$\lim_{N\to\infty} M^q\left\{1 - \frac{1}{N-|n|}\sum{}' a_N(n)\, a_N(u+n)\right\} = 0$$

for each fixed n. Here the sum is again over all u for which u and $u+n$ lie between 1 and N. Then we have the following theorem

Theorem 10. *Under the conditions of Theorem* 9 *on* $\hat{f}$, M, N *and for* q *satisfying* (4.6), (4.7), *the fader satisfying* (4.8) *and also* $M^q/N \to 0$, $q \geq 1$,

(4.9) $$\lim_{N\to\infty} M^q \mathcal{E}\{\hat{f}(\lambda) - f(\lambda)\} = -\frac{k_q}{2\pi}\sum_{-\infty}^{\infty} \Gamma(n) e^{-in\lambda}\,|n|^q.$$

Proof. We have, for estimators with k_n of the form of (4.4) and also the Truncated estimator,

(4.10) $$M^q\mathcal{E}\{\hat{f}_{jk}(\lambda) - f_{jk}(\lambda)\}$$
$$= \frac{M^q}{2\pi}\left\{\sum_{-N+1}^{N-1} e^{-in\lambda}\left(1 - \frac{|n|}{N}\right)\gamma_{jk}(n)\, k\left(\frac{n}{M}\right) - \sum_{-\infty}^{\infty} e^{-in\lambda}\,\gamma_{jk}(n)\right\}.$$

Now we break (4.10) into three parts, of which the first leads to

$$\left|-\frac{M^q}{2\pi}\sum_{|n|\geq N}\gamma_{jk}(n)e^{-in\lambda}\right| \leq \frac{1}{2\pi}\left(\frac{M}{N}\right)^q \sum_{|n|\geq N} |n|^q\,|\gamma_{jk}(n)| \to 0, \qquad q \geq 0.$$

The second leads to

(4.11) $$\left|\frac{M^q}{2\pi}\sum_{-N+1}^{N-1} k\left(\frac{n}{M}\right)\frac{|n|}{N}\,\gamma_{jk}(n)e^{-in\lambda}\right|.$$

For $q \geq 1$ we see that this is not greater than

$$\frac{cM^q}{N}\sum_{-N+1}^{N-1} |n|\,|\gamma_{jk}(n)| \to 0.$$

For $q < 1$ we see that (4.11) is dominated by

$$cM^qN^{-q}\sum_{-N+1}^{N-1}|\gamma_{jk}(n)|\,|n|^q \to 0, \qquad 0 < q \leq 1.$$

Again the result holds also for $q = 0$, since

$$\left|\frac{1}{2\pi}\sum_{-N+1}^{N-1}k\left(\frac{n}{M}\right)\frac{|n|}{N}\gamma_{jk}(n)\right| \leq \frac{1}{2\pi}\sum_{-N+1}^{N-1}\frac{|n|}{N}|\gamma_{jk}(n)|$$

and both

$$\frac{1}{2\pi}\sum_{-N+1}^{N-1}\left(1-\frac{|n|}{N}\right)|\gamma_{jk}(n)|, \qquad \frac{1}{2\pi}\sum_{-N+1}^{N-1}|\gamma_{jk}(n)|$$

converge to the same limit.

Finally, we have

$$-\frac{M^q}{2\pi}\sum_{-N+1}^{N-1}\left[\left\{1-k\left(\frac{n}{M}\right)\right\}e^{-in\lambda}\gamma_{jk}(n)\right] = -\frac{1}{2\pi}\sum_{-N+1}^{N-1}\frac{1-k(n/M)}{|n/M|^q}|n|^q\gamma_{jk}(n).$$

The condition

$$\lim_{|x|\to 0}\frac{1-k(x)}{|x|^q} = k_q < \infty$$

implies that $\{1 - k(n/M)\}/|n/M|^q$ converges boundedly to k_q for each fixed n which implies the result of the theorem. The same is evidently true of the Bartlett estimator.

For the FFT the considerations are the same but with $k(x)$ replaced by

$$k_N(x) = e^{ix(\lambda-\lambda')M}\frac{\sin\frac{1}{2}\pi x}{m\sin(\frac{1}{2}\pi x/m)}, \qquad |\lambda-\lambda'| \leq \frac{\pi}{N}.$$

The quantity $(\lambda - \lambda')$ is, in fact,

$$\left\{\lambda - \left(j_0 + \frac{m-1}{2}\right)\frac{2\pi}{N}\right\},$$

where $2\pi j_0/N$ is the lowest frequency in the sum defining the estimator. The first two contributions to the bias converge to zero as before if q satisfies (4.6) of Section 4. If the bias is to be $O(M^{-q})$ we must also have

$$\lim_{N\to\infty}\frac{1-k_N(n/M)}{(n/M)^q} = k_q < \infty \tag{4.12}$$

for each fixed n. Consider

$$\lim_{N\to\infty} n^{-q}M^q\left\{1 - \frac{\sin \frac{1}{2}\pi n/M}{m \sin (\frac{1}{2}\pi n/Mm)}\right\},$$

$$\lim_{N\to\infty} -\frac{\pi^2 n^2}{6n^q}\left\{\frac{m/N^3 - m^3/N^3}{m/N} M^q\right\} = \lim_{N\to\infty} -\frac{\pi^2 n^2}{6n^q}\left\{\frac{(1-m^2)M^q}{N^2}\right\},$$

which is $\pi^2/24$ for $q = 2$ and is null for $q < 2$. On the other hand,

$$(4.13) \qquad (1 - e^{ix(\lambda-\lambda')M})\frac{\sin \frac{1}{2}\pi x}{m \sin (\frac{1}{2}\pi x/m)} M^q, \qquad x = \frac{n}{M},$$

will converge to zero as N increases, if M^q/N does so.

For the Cooley-Tukey procedure we put

$$k_N(x) = e^{ix(\lambda-\lambda')M}\frac{\sin \frac{1}{2}\pi x}{m' \sin (\frac{1}{2}\pi x/m')}\frac{1}{N - |n|}\sum{}' a_N(u)\, a_N(u+n),$$

where the sum varies over $N - |n|$ values for which u and $n + u$ be between 1 and N. Ignoring the last factor (4.12) will hold true for $q \leq 2$, once more, with the same value of k_q. Thus for the same result to hold for the actual $k_N(x)$ of the Cooley-Tukey procedure we also need (4.8).

The theorem shows that the bias is $O(M^{-q})$, at most, if q satisfies the stated conditions. In all cases studied $k_0 = 0$ so that the bias at least converges to zero under the conditions of the theorem. Some particular results are exhibited in Table 2. The k_q in the third column is, of course, the value for the largest q for which it is less than ∞.

Table 2 is in some ways rather misleading, for the truncated estimator, which shows up best in it, is known to have unsatisfactory bias-type properties as explained earlier, for its first side lobe is appreciable and can cause

Table 2

Estimator	q for which $k_q < \infty$	k_q
FFT	$q \leq 2$	$k_2 = \pi^2/24$
Cooley-Tukey	$q \leq 2$	$k_2 = \pi^2/24$
Truncated	$q < \infty$	$k_q \equiv 0$
Bartlett	$q \leq 1$	$k_1 = 1$
Tukey-Hanning	$q \leq 2$	$k_2 = \pi^2/4$
Parzen	$q \leq 2$	$k_2 = 6$
Abel	$q \leq 1$	$k_1 = a$
Daniell	$q \leq 2$	$k_2 = \pi^2/6$
$k(x) = 1 - x^2$	$q \leq 2$	$k_2 = 1$

trouble. The asymptotic analysis which we have given does not reveal this type of defect. The bias in the periodogram is

$$\int_{-\pi}^{\pi} \frac{\sin^2 \frac{1}{2}N(\lambda - \theta)}{2\pi N \sin^2 \frac{1}{2}(\lambda - \theta)} \{f(\theta) - f(\lambda)\}\, d\theta,$$

which is $O(\log N/N)$ if $f \in \text{Lip } 1$ near λ. (See the Mathematical Appendix. This condition holds in particular if f is differentiable at λ.) If $q \geq 1$ since $M^q/N \to 0$ we may take $M^{-q} = O(\log N/N)$ by a suitable choice of M, provided $q \geq 1$. However, such choices of M would make the variance much too large and we have given this discussion only to give some indication of what a bound might reasonably be to the rate of decrease of the bias.

If we wish to compare the standard deviation and the bias it is not unreasonable to consider the mean square error, that is, the variance plus the $(\text{bias})^2$. Since the variance is $O(M/N)$ then if the bias is at best $O(M^{-q})$ the order of the mean-square error is made as small as possible by taking

$$M = O(N^{1/(1+2q)}). \tag{4.14}$$

This result is of course somewhat vacuous for the truncated estimator and implies that M/N may converge to zero as fast as is desired without adverse effects on the estimation procedure *provided* $f(\lambda)$ *is smooth enough*, which is misleading in relation to fixed N. For $q = 2$ the formula implies that $M = O(N^{1/5})$. This prescription is of little value in practice, for obvious reasons. In case $q = 2$ the asymptotic bias is expressible as $-M^{-2}k_2 f''(\lambda)$ so that the mean-square error is, when $M = cN^{1/5}$, in the scalar case for $\lambda \neq 0, \pm\pi$,

$$\lim_{N\to\infty} N^{4/5}\, \mathcal{E}[\{\hat{f}(\xi) - f(\lambda)\}^2] = cf^2(\lambda) \int_{-\infty}^{\infty} k^2(x)\, dx + c^{-4} k_2^{\,2} \{f''(\lambda)\}^2,$$

which is minimized for

$$c = \left[\frac{4k_2^{\,2}\{f''(\lambda)\}^2}{f^2(\lambda) \int_{-\infty}^{\infty} k^2(x)\, dx} \right]^{1/5}.$$

Thus $N^{4/5}$ by the mean-square error then approaches

$$2\left(f^2(\lambda)\int_{-\infty}^{\infty} k^2(x)\, dx\right)^{4/5} (2k_2 f''(\lambda))^{2/5}.$$

The considerations already given show that bias and variance work in opposite directions so that if we increase M so as to reduce the former we reduce the rate at which the variance decreases. For a further analysis of this phenomenon the reader may consult Grenander (1951) or Bartlett and Medhi (1955).

One can use these results concerning the variance to obtain the variances of some of the other statistics we have considered. We do not give all of the details corresponding to those in Theorem 9 but only those of interest later.

Corollary 4. *Under the same conditions as for Theorem* 9, *for* $\lambda \neq 0, \pi$ (mod 2π),

$$\lim_{N\to\infty} \frac{N}{M} \operatorname{var}(\hat{c}_{ij}(\lambda)) = 2\int_{-\infty}^{\infty} k^2(x)\,dx\{f_i(\lambda)f_j(\lambda)(1-\sigma_{ij}^2(\lambda)) + \tfrac{1}{2}c_{ij}^2(\lambda)\},$$

$$\lim_{N\to\infty} \frac{N}{M} \operatorname{var}(\hat{q}_{ij}(\lambda)) = 2\int_{-\infty}^{\infty} k^2(x)\,dx\{f_i(\lambda)f_j(\lambda)(1-\sigma_{ij}^2(\lambda)) + \tfrac{1}{2}q_{ij}^2(\lambda)\},$$

$$\lim_{N\to\infty} \frac{N}{M} \operatorname{cov}(\hat{c}_{ij}(\lambda), \hat{q}_{ij}(\lambda)) = \int_{-\infty}^{\infty} k^2(x)\,dx\, c_{ij}(\lambda)\, q_{ij}(\lambda),$$

$$\lim_{N\to\infty} \frac{N}{M} \operatorname{var}(\hat{\theta}_{ij}(\lambda)) = \tfrac{1}{2}\int_{-\infty}^{\infty} k^2(x)\,dx\left\{\frac{1-\sigma_{ij}^2(\lambda)}{\sigma_{ij}^2(\lambda)}\right\}, \qquad \sigma_{ij}(\lambda) \neq 0,$$

and if the spectral window is never negative

$$\lim_{N\to\infty} \frac{N}{M} \operatorname{var}(\hat{\sigma}_{ij}(\lambda)) = \tfrac{1}{2}\int_{-\infty}^{\infty} k^2(x)\,dx\{1-\sigma_{ij}^2(\lambda)\}^2.$$

Note. The frequencies 0, π are of interest here only for $\hat{\sigma}_{ij}(\lambda)$. We omit separate mention of this case for simplicity only. For $c_{ij}(\lambda) = 0$, under suitable conditions on the $x(n)$ process, we shall later show that $\hat{\sigma}_{ij}(\lambda)$ is uniformly distributed, whereas for $\sigma_{ij}(\lambda) \neq 0$ it is asymptotically normal, so that a variance is of interest. The result for the coherence undoubtedly holds under more general conditions but we have not thought it worthwhile to consider them here.

Proof. The first two formulas are easily found since $\hat{c}_{ij}$ and $\hat{q}_{ij}$ are linear functions of $\hat{f}_{ij}$ and $\hat{f}_{ji}$.

So far as the coherence is concerned we consider here only the case where the spectral window is never negative. In this case it is easy to show that $\hat{\sigma}_{ij}(\lambda)$ is bounded by unity. As a result we may, to order M/N, obtain its variance by a simple argument (see Cramer, 1946, p. 352). Thus we have

$$\hat{\sigma}_{ij}(\lambda) = \left\{\frac{\hat{f}_{ij}(\lambda)\hat{f}_{ji}(\lambda)}{\hat{f}_i(\lambda)\hat{f}_j(\lambda)}\right\}^{1/2} = \left(\frac{ab}{cd}\right)^{1/2},$$

let us say. Expanding this function of four variables in its Taylor series about α, β, γ, δ, where $\alpha = \mathcal{E}\{\hat{f}_{ij}(\lambda)\}$, for example, and retaining only the linear terms we obtain

$$\hat{\sigma}_{ij}(\lambda) - \left(\frac{\alpha\beta}{\gamma\delta}\right)^{1/2} = \tfrac{1}{2}\sigma_{ij}^{-1}(\lambda)\left\{(a-\alpha)\frac{\beta}{\gamma\delta} + (b-\beta)\frac{\alpha}{\gamma\delta} - (c-\gamma)\frac{\alpha\beta}{\gamma^2\delta} - (d-\delta)\frac{\alpha\beta}{\gamma\delta^2}\right\} + o((M/N)^{1/2}), \qquad \lambda \neq 0, \pm\pi,$$

from which we obtain,† to the same order, the variance of $\hat{\sigma}_{ij}(\lambda)$,

$$\lim_{N\to\infty} \frac{N}{M} \operatorname{var}\{\hat{\sigma}_{ij}(\lambda)\} = \tfrac{1}{2}\{1 - \sigma_{ij}^{2}(\lambda)\}^2 \int k^2(x)\,dx.$$

In case $\sigma_{ij}(\lambda) \neq 0$ we can similarly obtain the variance of the phase angle

$$\hat{\theta}_{ij}(\lambda) = \arctan \frac{\hat{q}_{ij}(\lambda)}{\hat{c}_{ij}(\lambda)}.$$

We assume $c_{ij}(\lambda) \neq 0$ and that $\theta_{ij}(\lambda)$ (when account is taken of the signs of $q_{ij}(\lambda)$ and $c_{ij}(\lambda)$) lies in the interior of $[-\pi/2, \pi/2]$. Provided $\sigma_{ij}(\lambda) \neq 0$ other cases can be dealt with in a similar way and the same variance is obtained. Expanding the inverse tangent and proceeding as before we obtain

$$\lim_{N\to\infty} \frac{N}{M} \operatorname{var}\left(\hat{\theta}_{ij}(\lambda)\right) = \tfrac{1}{2}\int k^2(x)\,dx \left\{\frac{1}{\sigma_{ij}^{2}(\lambda)} - 1\right\}, \qquad \sigma_{ij}(\lambda) \neq 0.$$

As is to be expected, the variance is zero when $\sigma_{ij}(\lambda)$ is unity.

5. THE ASYMPTOTIC DISTRIBUTION OF SPECTRAL ESTIMATES

We content ourselves here with proving a central limit theorem for our spectral estimators for the case of a linear process. It is possible, without doubt, to establish such a result by using, instead, a uniform mixing condition but we restrict ourselves to the simpler case here. We assume that

$$x(n) = \sum_{-\infty}^{\infty} A(j)\epsilon(n-j), \qquad \sum_{-\infty}^{\infty} \|A(j)\| < \infty, \tag{5.1}$$

where the $\epsilon(n)$ are i.i.d $(0, G)$ and have finite fourth moment. If useful references are to be made we need to be sure that the bias if of lower order than the standard deviation so that after normalization our limiting distribution of $\hat{f}(\lambda)$ will concentrate around $f(\lambda)$. For this reason we add the requirement

† The quantity $[(\alpha\beta)/\gamma\delta]^{1/2}$ differs from $\mathcal{E}(\hat{\sigma}_{ij})$ by quantities of order M/N.

that

$$\lim_{N\to\infty} \sup_{\lambda} \nu^{1/2} \, \|f(\lambda) - \mathcal{E}(\hat{f}(\lambda))\| = 0. \tag{5.2}$$

There is one further unsatisfactory feature of the specification so far. This lies in the fact that we are forced to assume the existence of fourth moments [for $\epsilon(n)$], whereas our inferences relate only to second moments. We can remedy this defect, as we show in the following theorem, if we add to (5.1) the condition

$$\sup_{-\infty < j < \infty} |j| \, \|A(j)\| < \infty \tag{5.3}$$

(which is neither stronger nor weaker than (5.1), as is easily checked).

Theorem 11. *If* $x(n)$ *satisfies* (5.1), (5.2), *and* (5.3), *the necessary and sufficient condition that the distribution of*

$$\nu^{1/2}\left\{\hat{f}_{jk}\left(\lambda_l + \frac{\pi q}{M}\right) - f_{jk}\left(\lambda_l + \frac{\pi q}{M}\right)\right\}; \qquad j, k = 1, \ldots, p;$$
$$q = 1, \ldots, r; \qquad l = 1, \ldots, s,$$

where the λ_l, $l = 1, \ldots, s$, *are arbitrary frequencies, should converge to the complex, multivariate, normal distribution with covariance matrix given by Theorem* 9, *is the condition that*

$$N^{-1}\nu^{1/2} \sum_{1}^{N} \{\epsilon(n)'\epsilon(n) - q\}$$

converge in probability to zero. This is necessarily true if $\epsilon(n)$ *has finite fourth moment but then the result holds when condition* (5.3) *above is deleted.*

We have again put the proof of this theorem in the appendix to this chapter.

It seems probable that the conditions can be relaxed and that something weaker than absolute convergence of the $A(j)$ will suffice. In this connection we refer the reader to Whittle (1964). We have phrased the theorem in the way stated since the joint distribution of estimates from neighbouring bands will be needed.

It is perhaps unnecessary but to avoid possible confusion we point out that the complex multivariate normal distribution discussed here is *not* the true analogue of that occurring in Section 2 but is a limiting form, in a sense, of a product of m copies of (4.6) in Chapter IV, one for each elementary frequency band in the band (of width π/M) in question. In other words, and speaking loosely, we are using a normal distribution to approximate to a chi-square distribution for large ν.

This theorem, to the extent that it is applicable, justifies all of the techniques of Section 2 as asymptotically valid procedures. In all cases the asymptotic theory leads to a normal or chi-square variable in place of t or F variable, for example, but it is not necessary to make these modifications as the distribution used in Section 2 will always give an asymptotically equivalent result to that obtained by the methods of the present section. We illustrate with some examples. We commence with the confidence interval procedure for a spectrum which now becomes very simple, namely,

$$\hat{f}_j\left\{1 + z(\alpha)\left(\frac{2}{\nu}\right)^{1/2}\right\}^{-1} \le f_j \le \hat{f}_j\left\{1 - z(\alpha)\left(\frac{2}{\nu}\right)^{1/2}\right\}^{-1}$$

where $z(\alpha)$ is the $100(1-\alpha)\%$ point for the standard normal distribution, the confidence coefficient being $100\alpha\%$. This procedure is easily seen to be equivalent to that of Section 2(a), and the latter procedure is possibly preferable though a more careful analysis is needed if we are to establish this.

The distribution of $\hat{c}_{jk}$ and $\hat{q}_{jk}$ follows from this theorem also. Thus, being real linear combinations of complex normal variables, they have a real normal distribution. Their variances and covariances are evidently as calculated in Section 4. In particular, when $\sigma_{jk} = 0$, both $\hat{c}_{jk}$ and $\hat{q}_{jk}$ have the same variance, namely,

$$2\int k^2(x)\,dx\, f_j(\lambda) f_k(\lambda).$$

Their ratio $\hat{q}_{jk}/\hat{c}_{jk}$ is distributed as the ratio of two independent normal variables with zero mean and unit variance and thus $\hat{\theta} = \arctan\,(\hat{q}_{jk}/\hat{c}_{jk})$ is uniformly distributed over $(-\pi, \pi]$, on the assumption that we allocate $\hat{\theta}$ over the whole interval by taking account of the signs of $\hat{q}_{jk}$ and $\hat{c}_{jk}$. For the coherence we consider first the case in which $\sigma_{jk} = 0$ and assume $f_j > 0$, $f_k > 0$. Then, for $\lambda \ne 0, \pm\pi$,

$$\frac{\nu}{4}\,\frac{\hat{c}_{jk}^{\,2} + \hat{q}_{jk}^{\,2}}{f_j f_k}$$

is chi-square with two degrees of freedom so that, since $f_j f_k/\hat{f}_j \hat{f}_k$ converges in probability to unity $\nu\hat{\sigma}_{jk}^2$ is distributed as chi-square with two degrees of freedom. For $\lambda = 0, \pm\pi$, $(\nu/2)\hat{\sigma}_{jk}^2$ is distributed as chi-square with one degree of freedom, by the same reasoning. The result obtained in Section 2, when m was fixed, was that $\frac{1}{2}(\nu-2)\hat{\sigma}_{jk}^2/(1-\hat{\sigma}_{jk}^2)$ had the distribution of F with 2 and $(\nu - 2)$ degrees of freedom. This is consistent with the result established, as $\nu \to \infty$, since $\hat{\sigma}_{jk}^2$ converges to zero in probability, whereas $2F_{2,\nu-2}$ converges to chi-square with two degrees of freedom. Even with

$\nu = 60$ there is a fair difference between the significance points for the two distributions (3.15 for $F_{2,60}$, 2.99 for $\frac{1}{2}\chi_2^2$ at 5% level). For Gaussian data we discuss this type of approximation further in Chapter VI, Section 3, but it seems probable in the present limited state of our knowledge that the F approximation is the better one under most circumstances.

In case $\sigma_{jk} \neq 0$, $f_j > 0$, $f_k > 0$, we proceed as follows. Consider, for $\lambda \neq 0, \pm\pi$,

$$\nu^{1/2}\frac{\hat{\sigma}_{jk}^2 - \sigma_{jk}^2}{2\sigma_{jk}} = \nu^{1/2}(\hat{\sigma}_{jk} - \sigma_{jk})\frac{\hat{\sigma}_{jk} + \sigma_{jk}}{2\sigma_{jk}}.$$

Evidently, if the left-hand side has a normal distribution as $\nu \to \infty$ so will $\nu^{1/2}(\hat{\sigma}_{jk} - \sigma_{jk})$ as the remaining factor converges to unity in probability. But

$$\nu^{1/2}(\hat{\sigma}_{jk}^2 - \sigma_{jk}^2) = \nu^{1/2}\left\{\frac{|\hat{f}_{jk} - f_{jk}|^2 + \bar{f}_{jk}(\hat{f}_{jk} - f_{jk}) + f_{jk}(\bar{\hat{f}}_{jk} - \bar{f}_{jk})}{f_j f_k} - |\hat{f}_{jk}|^2\frac{(\hat{f}_j - f_j)\hat{f}_k + (\hat{f}_k - f_k)f_j}{\hat{f}_j\hat{f}_k f_j f_k}\right\}$$

and this is distributed, asymptotically, as

$$\nu^{1/2}\left\{\frac{\bar{f}_{jk}}{f_j f_k}(\hat{f}_{jk} - f_{jk}) + \frac{f_{jk}}{f_j f_k}(\bar{\hat{f}}_{jk} - \bar{f}_{jk}) - \sigma_{jk}^2\frac{\hat{f}_j - f_j}{f_j} - \sigma_{jk}^2\frac{\hat{f}_k - f_k}{f_k}\right\}$$

since, for example, $\nu^{1/2}|\hat{f}_{jk} - f_{jk}|^2$ converges in probability to zero. But this is a linear combination of complex normal components and thus is normal, with variance, calculated from the formula above, as $(1 - \sigma_{jk}^2)^2 4\sigma_{jk}^2$. Thus $\nu^{1/2}(\hat{\sigma}_{jk} - \sigma_{jk})$ is normal with mean zero and variance $(1 - \sigma_{jk}^2)^2$, in agreement with the formula for the variance obtained in Section 4.† Thus, in particular,

$$\frac{\nu^{1/2}(\hat{\sigma}_{jk} - \sigma_{jk})}{1 - \hat{\sigma}_{jk}^2}$$

is normal with zero mean and unit variance.‡ This leads to the confidence interval

$$\hat{\sigma}_{jk} - \frac{1 - \hat{\sigma}_{jk}}{\nu^{1/2}}z(\alpha) \leq \sigma_{jk} \leq \hat{\sigma}_{jk} + \frac{1 - \hat{\sigma}_{jk}^2}{\nu^{1/2}}z(\alpha),$$

where $z(\alpha)$ is now the $100(1 - \alpha)\%$ point for the standard normal distribution.

† There is now no need for the spectral window of the estimator to be positive.

‡ This result can, of course, also be got directly from theorems concerning the distribution of a function of a vector of asymptotically normal random variables. See Rao (1965, p. 321).

For $\lambda = 0, \pm\pi$ the formula needs modification but we leave these relatively unimportant cases as an exercise. Again it may be preferable to use the procedure, based on the tables of Amos and Koopmans (1963), devised in Section 2 for the case of fixed ν (i.e., m). These tabulations cover very large values of ν (indeed up to $\nu = 250$), though for ν very large the two methods will give almost the same result.

These examples will suffice to illustrate the principles involved. We state the conclusion as

Corollary 5. *Under the conditions of Theorem* 11 *and with* ν *determined by Table* 1 *replacing* $2m$, *all of the results of Section* 2 *continue to hold.*

There is a real need for an extension of Theorem 11 to the case in which $x(n)$ is not a linear process but instead satisfies a uniform mixing condition, for a linear model is unsuitable in a situation in which $x(n)$ is essentially positive, and, except under rather special restrictions, we must take (in the scalar case) both α_j and $\epsilon(n)$ nonnegative. When this is so, however, the autocorrelation must be positive for all lags, which is often unsuitable. There is no doubt that Theorem 11 could be extended to deal with this other case, but we leave the matter with that, rather unsatisfactory, statement.

We now discuss some further results due to Woodroofe and Van Ness (1967). In the preceding section we discussed procedures that enable a confidence interval to be allocated to various quantities determined by $f(\lambda)$. However, a problem arises when all values of λ, for which $\hat{f}(\lambda)$ has been computed (i.e., $\lambda = \pi k/M$, $k = 0, \ldots, M$), are to be considered simultaneously for now we are evidently assigning too high a level of confidence to our intervals if it is required that, *simultaneously for all* k, the quantities being estimated are to lie in these intervals. The acuteness of the problem arises because M increases with N so that ultimately an indefinitely large number of confidence statements are simultaneously being made. What is needed is the distribution of something like

$$\max_k \left| \hat{f}\left(\frac{\pi k}{M}\right) - f\left(\frac{\pi k}{M}\right) \right|,$$

for example. It is this that Woodroofe and Van Ness have discussed.

It is evident that stronger conditions than those used at the end of Section 4 will be needed and we assume that

$$x(n) = \sum_{-\infty}^{\infty} A(j)\epsilon(n-j), \qquad \epsilon(n) = \text{i.i.d.}\ (0, I_p),$$

with finite eighth moment and with

$$\sum_{-\infty}^{\infty} \|A(j)\| \, |j|^{\delta} < \infty, \qquad \delta > \tfrac{1}{2}. \tag{5.4}$$

We assume also that

$$\min_{\lambda} \det \{h(\lambda)\} > 0, \qquad h(\lambda) = \sum_{-\infty}^{\infty} A(j) e^{ij\lambda},$$

that

$$M = O(N^{\alpha}), \qquad \alpha < \tfrac{2}{5},$$

and, if $\delta < \frac{3}{4}$, $2\alpha/(3 - 4\delta) > 1$. These constitute modifications of the conditions of Woodroofe and Van Ness which are not (so far as this writer can see) either unequivocally less or more restrictive than are theirs. They are simpler to state however, and if $\delta \geq \frac{3}{4}$ they are particularly simple. They do not seem unduly restrictive, though it is not easy to relate them to real world phenomena. We shall consider estimation procedures of the form of those discussed in Sections 4 and 5. Woodroofe and Van Ness assume that $K_N(\lambda) \geq 0$ and, effectively, that $k(x) = 0$, $|x| < 1$. This excludes procedures number 3 and 5 of Section 4, both of which are of some importance. In fact, if

$$\int_{-\pi}^{\pi} |K_N(\lambda)| \, d\lambda = O(\log N),$$

the condition $K_N(\lambda) \geq 0$ may be omitted. This condition is satisfied by procedures 3 and 5. The requirement $k(x) = 0$, $|x| < 1$ excludes procedures 7 and 8 and, much more importantly 1 and 2. Again these can be included within the scope of the theorem as can any nontruncated type of procedure for which

$$\int x^{\beta} k^2(x) \, dx < \infty, \qquad \beta > \left\{\left(\frac{10}{3}\right)\alpha - 1\right\}.$$

It is probable that this condition is much too strong but suffices for our purpose as β may be taken as close to 1 as is desired for all of 1, 2, 7 and 8 and $(10/3)\alpha - 1 < \frac{1}{3}$. One problem which arises with the Cooley-Tukey estimator is the use of a fader. The fader will also have to behave "properly" as $N \to \infty$ in order that bias effects shall vanish. We leave it for some industrious reader to pursue the details and confine ourselves to the case where no fader is used. To simplify the notation we put

$$\operatorname{var}\left\{\hat{f}_j\left(\frac{\pi k}{M}\right)\right\} = 2\nu^{-1} f_j^{\,2}\left(\frac{\pi k}{M}\right), \qquad k \neq 0, M,$$

with modified formulas in the case $k = 0, M$ in accordance with results given above. Similarly, we put

$$\operatorname{var}\left\{\hat{\sigma}_{ij}\left(\frac{\pi k}{M}\right)\right\} = \nu^{-1}(1 - \sigma_{ij}{}^2)^2, \qquad k \neq 0, M,$$

$$\operatorname{var}\left[\sin\left\{\hat{\theta}_{ij}\left(\frac{\pi k}{M}\right) - \theta_{ij}\left(\frac{\pi k}{M}\right)\right\}\right] = \nu^{-1}\frac{\hat{f}_i^2(\pi k/M)\big(1 - \sigma_{ij}{}^2(\pi k/M)\big)}{f_j{}^2(\pi k/M)\,|\beta_{ij}|^2},$$

$$a_N = \{2\log(2M)\}^{-\frac{1}{2}},$$

$$b_N = \{2\log(2M)\}^{\frac{1}{2}} - \tfrac{1}{2}(2\log 2M)^{-\frac{1}{2}}\{\log(2\log 2M) + \log 2\pi\},$$

$$z_N = a_N x + b_N.$$

We shall not give a proof of the following theorem, whose derivation does not differ markedly from that given in Woodroofe and Van Ness (1967).

Theorem 12. Under the conditions, on $x(n)$ and on the estimation procedures *for spectra and cross spectra, discussed above we have*

$$\lim_{N\to\infty} P\left[\max_{0\le k\le M} \frac{\{|\hat{f}_j(\pi k/M) - f(\pi k/M)|\}}{\{\operatorname{var}(\hat{f}_j(\pi k/M))\}^{\frac{1}{2}}} \le z_N\right] = \exp(-e^{-x}),$$

$$\lim_{N\to\infty} P\left[\max_{0\le k\le M} \frac{\{|\hat{\sigma}_{ij}(\pi k/M) - \sigma_{ij}(\pi k/M)|\}}{\{\operatorname{var}(\hat{\sigma}_{ij}(\pi k/M))\}^{\frac{1}{2}}} \le z_N\right] = \exp(-e^{-x}),$$

$$\lim_{N\to\infty} P\left[\max_{1\le k< M} \frac{\{|\sin(\hat{\theta}_{ij}(\pi k/M) - \theta_{ij}(\pi k/M))|\}}{\operatorname{var}(\sin\{\hat{\theta}_{ij}(\pi k/M) - \theta_{ij}(\pi k/M)\})} \le z_N\right] = \exp(-e^{-x}).$$

In all cases the variances may be replaced by estimates thereof, needless to say, so that the nett effect of the theorem is to institute $\exp -(\exp - x)$ as the function defining the significance level with $a_N x + b_N$ as the corresponding significance point. Some significance points for this distribution are tabulated in Owen (1962, pp. 307–309). It seems that a lower level of confidence will usually be needed when these results are used.

These results are important in principle but surrounded by some doubt in practice, of a greater magnitude than that accorded to results of earlier sections, because of their asymptotic nature. It is known that such extreme value formulas are relevant only in enormously large samples, when the largest of a series of independent and identically distributed random variables is under consideration. Here further approximations are involved and for the relevance of the formulas it is evidently M as much as N whose magnitude is of importance. One conjectures that the formulas are the roughest of approximations only.

We close this section with a reference to some work by Brillinger and Rosenblatt (1967b) dealing with polyspectra. This work effectively carries over to higher order spectra part of the results of Theorems 9, 10, and 11.

6. COMPLEX MULTIVARIATE ANALYSIS†

We shall here give a summary account of some techniques of multivariate analysis available for the analysis of a vector process at a given frequency λ. The account will be summary, for the subject is an immense one and could incorporate all of classical real multivariate analysis and more. It exceeds the scope of real multivariate analysis because when the complex multivariate normal distribution is expressed as a real multivariate normal distribution, for the real and imaginary parts of all the complex random variables involved, it is of a special form. This specialty corresponds to the occurrence of the phenomenon of phase, which has no meaning in classical real multivariate analysis. However, our account will be by no means complete and in particular the statistical treatment of phase in the situation we study in this section needs investigation. We shall assume that our vector process $x(n)$ satisfies the conditions of Theorem 13 or 13′ in Chapter IV and that we are concerned with a particular frequency $\lambda \neq 0, \pi$, *which we hold fixed and do not introduce into our notation.* (We choose $\lambda \neq 0, \pi$, since otherwise the multivariate analysis does reduce to the classical real multivariate analysis.) We thus commence from the density function

$$\phi(f, S) = \frac{1}{\pi^{mp}(\det f)^m} e^{-\operatorname{tr}(f^{-1}S)}, \qquad S = VV^*, \tag{6.1}$$

where V has $v_j(u)$ in row j, column u, $j = 1, \ldots, p$; $u = 1, \ldots, m$. Thus $m^{-1}S = \hat{f}$. We associate with this density the differential form dV, which is the product of the differentials of the real and imaginary parts of all $v_j(u)$.

Theorem 13. *If the complex random variables $v_j(u)$ have the density* (6.1), *then $S = VV^*$ has the density*

$$\left\{\pi^{\frac{1}{2}p(p-1)} \prod_1^p (m-j)!\right\}^{-1} e^{-\operatorname{tr}(f^{-1}S)}(\det S)^{m-p}(\det f)^{-m}, \tag{6.2}$$

which is called the complex Wishart distribution.

Proof. We observe first that, introducing the transformation

$$\tilde{V} = AV, \qquad \tilde{f} = AfA^*, \qquad \tilde{S} = ASA^*,$$

† This section may be omitted.

where A is a nonsingular $p \times p$ matrix of complex numbers,

$$\phi(\tilde{f}, \tilde{S})\, d\tilde{V} = \phi(f, S)\, dV. \tag{6.3}$$

Indeed, $\phi(\tilde{f}, \tilde{S}) = |\det A|^{-2m}\, \phi(f, S)$, and we have merely to check that the Jacobian of the transformation is $|\det A|^{2m}$. If $A = C + iD$, $v(u) = a(u) + i\, b(u)$, $\tilde{v}(u) = \tilde{a}(u) + i\, \tilde{b}(u)$, then, as at the commencement of the proof of Theorem 13 of Chapter IV,

$$\begin{pmatrix} \tilde{a}(u) \\ \cdots \\ \tilde{b}(u) \end{pmatrix} = \begin{bmatrix} C & -D \\ D & C \end{bmatrix} \begin{pmatrix} a(u) \\ \cdots \\ b(u) \end{pmatrix}.$$

We show in the exercises to this chapter how to prove that this transformation has determinant $|\det A|^2$. Given this, the fact that the Jacobian is $|\det A|^{2m}$ follows immediately. Let us write $\psi(S)\, dS$ for the density function of the elements of S, where now dS is the product of the differentials of the p^2 real and imaginary parts of those elements of S lying on or above the main diagonal. (Those below the main diagonal introduce no new real random variables as their real and imaginary parts.) Thus, from (6.1),

$$\psi(S)\, dS = c_1 e^{-\mathrm{tr}(f^{-1}S)} (\det f^{-1})^m g_1(S)\, dS, \tag{6.4}$$

where $g_1(S)$ is obtained from the integration which eliminates the variables in (6.1) over and above those needed to specify S. Of course, $g_1(S)$ does not depend on f. Since (6.3) holds, changing S to $\tilde{S}$ and f to $\tilde{f}$ cannot change $\psi(S)\, dS$ and we are led to rewrite (6.4) as

$$\psi(S)\, dS = c e^{-tr(f^{-1}S)} \{\det (f^{-1}S)\}^m g(S)\, dS,$$

where now $g(\tilde{S})\, d\tilde{S} = g(S)\, dS$. It is known that up to a constant factor there is only one such "invariant differential form" (e.g., see, Helgason, 1962, p. 290). The transformation $S \to \tilde{S}$ is a linear transformation of the p^2 real random variables in S. It has determinant $|\det A|^{2p}$. (We indicate how to prove this in the exercises.) As a result $(\det S)^{-p}\, dS$ is an invariant differential form. Thus

$$\psi(S)\, dS = c e^{-\mathrm{tr}(f^{-1}S)} (\det S)^{m-p} (\det f)^{-m}\, dS$$

and it remains only to evaluate c.

Essentially we follow Goodman (1963), but first we observe that c is independent of f and consider

$$\int \{\det S\}^k e^{-\mathrm{tr}(S)}\, dS,$$

which we may evaluate by integrating over the set of all positive definite S, the set of singular S having probability measure zero. We put $S = T^*T$

where T is upper triangular and has positive diagonal elements. (It is fairly easy to see that S may be written in this form. Uniqueness is easily established as if $S = T_1^*T_1$, $S = T_2^*T_2$, with T_1, T_2 of the required form, then $T_2^{-1}T_1$ is both unitary and triangular with positive diagonal elements and hence the unit matrix.) The Jacobian $\partial S/\partial T$ is

$$\frac{\partial S}{\partial T} = 2^p t_{11}^{2p-1} t_{22}^{2p-3} \cdots t_{pp},$$

where t_{ij} is the typical element of T. This is shown by observing that

$$s_{jk} = \sum_{i=1}^{\min(j,k)} \bar{t}_{ij} t_{ik}; \qquad j \leq k,$$

and ordering the real parameters involved as $s_{11}, \mathcal{R}(s_{12}), \mathcal{I}(s_{12}), \ldots, \mathcal{R}(s_{1p}), \mathcal{I}(s_{1p}), s_{22}, \ldots, \mathcal{I}(s_{p-1,p}), s_{pp}; \quad t_{11}, \mathcal{R}(t_{12}), \mathcal{I}(t_{12}), \ldots, \mathcal{R}(t_{1p}), \mathcal{I}(t_{1p}), t_{22}, \ldots, \mathcal{I}(t_{p-1,p}), t_{pp}$, when the Jacobian becomes the determinant of a triangular matrix and is easily evaluated. This reduces us to

$$\int |\det T|^{2k} \exp -\operatorname{tr}(T^*T) 2^p t_{11}^{2p-1} \cdots t_{pp}\, dT$$
$$= 2^p \int t_{11}^{2(p+k)-1} t_{22}^{2(p+k)-3} \cdots t_{pp}^{2k+1} \exp -\operatorname{tr}(T^*T)\, dT.$$

Since

$$\operatorname{tr}(T^*T) = t_{11}^2 + (|t_{12}|^2 + t_{22}^2) + \cdots + (|t_{1p}|^2 + \cdots + t_{pp}^2),$$

the integral can be expressed as 2^p by a product of integrals, of which p are of the form

$$\int_0^\infty t_{jj}^{2(p+k)-(2j-1)} \exp(-t_{jj}^2)\, dt_{jj} = \tfrac{1}{2}(p+k-j)!$$

and $p(p-1)/2$ of the form

$$\iint_{-\infty}^{\infty} \exp(-|t_{jk}|^2)\, du_{jk}\, dv_{jk} = \pi, \quad t_{jk} = u_{jk} + iv_{jk}, \quad j < k, j, k = 1, \ldots, p,$$

Thus the integral is

$$\pi^{\frac{1}{2}p(p-1)} \prod_1^p (k + j - 1)!$$

Putting $k = m - p$, we see that

$$c = \left\{\pi^{\frac{1}{2}p(p-1)} \prod_1^p (m - j)!\right\}^{-1},$$

as required.

The use of the argument involving the uniqueness of the invariant differential form may be avoided, now that it is known what the answer is, by evaluating the joint characteristic function of the elements of S by using (6.1) and (6.2). These derivations are now relatively straightforward and we leave them to the reader who wishes another proof of the validity of (6.2).

This Theorem 13 is the analog of the results discussed under (a) of Section 2. It seems to be of limited value. We shall not discuss problems of the type of a test for significance of the difference between two spectra because these problems also do not seem physically important. Instead, we proceed to discuss some, apparently important, analogs of those discussed under (b) and (c). For this purpose we introduce a second stationary vector process, $y(n)$, of q components it being assumed that the (row) vector process $(x(n)' : y(n)')$ is stationary and satisfies the conditions of Theorem 13 or 13′ in Chapter IV. Correspondingly we have the partition

$$f = \begin{bmatrix} f_x & f_{xy} \\ f_{yx} & f_y \end{bmatrix},$$

where f_x is the matrix of spectra and cross spectra for $x(n)$, f_y the matrix of spectra and cross spectra for $y(n)$ and $f_{xy} = f_{yx}^*$ contains, as (j, k)th element the cross spectrum betwen $x_j(n)$ and $y_k(n)$. We shall, of course, use $\hat{f}_x$ and so on for the estimates, obtained from $v_{x,j}(u)$, $j = 1, \ldots, p$, $v_{y,k}(u)$, $k = 1, \ldots, q$, $u = 1, \ldots, m$, where the first subscript indicates which vector process these statistics are computed from. We write $v_x(u)$, etc., for the vector of p components, $v_{x,j}(u)$, $j = 1, \ldots, p$. *We assume* $m \geq p + q$. The p sets of m numbers $v_{x,j}(u)$ span a p dimensional plane in C^m (complex m-space) and similarly for the q sets of m numbers $v_{y,j}(u)$. We seek for two new random vectors, one lying in each plane, so that the true coherence between this pair is as large as possible. Having found these we seek for a second pair, incoherent with each member of the first pair so that this new pair has as high a coherence as possible, consistent with their incoherence with the first pair, and so on. The maximized coherences, ρ_i, may be called canonical coherences. There can be at most min (p, q) which are nonzero. The corresponding random vectors with components $\mu_i^* v_x(u)$, $\nu_i^* v_y(u)$, having coherence ρ_i,

$$\rho_i^2 = \frac{|\mu_i^* f_x \nu_i|^2}{\mu_i^* f_x \mu_i \nu_i^* f_{yy} \nu_i}, \qquad i = 1, \ldots, \min(p, q),$$

are called the x and y discriminant functions and the vectors μ_i (and ν_i) are called the vectors of x (and y) discriminant function coefficients. If $p > q$ we may find $p - q$ discriminant functions $\mu_{q+j}^* v_x(u)$, $j = 1, \ldots, p - q$, which are incoherent with all $\nu_k^* v_y(u)$ and similarly for the case $q > p$. These $\mu_{q+j}^* v_x(u)$

are said to correspond to canonical coherences, $\rho_{q+j} = 0$, which are "identically zero." We may normalize the discriminant functions to have unit mean square (so that the denominator of the last expression is unity). Then we have

$$\mathcal{E}(\mu_i^* f_x \mu_j) = \delta_i^{\,j} = \mathcal{E}(\nu_i^* f_y \nu_j),$$
$$|\mathcal{E}(\mu_i^* f_{xy} \nu_j)|^2 = \delta_i^{\,j} \rho_i^{\,2}, \qquad i = 1, \ldots, p; \qquad j = 1, \ldots, q.$$

The following theorem is of a simple algebraic nature and its proof will be omitted:

Theorem 14. *For $p \geq q$ the normalized vectors of discriminant function coefficients are defined, up to a constant factor of modulus unity, the same for each member of the same pair, as the solutions of*

$$\left.\begin{aligned} [f_{xy} f_y^{-1} f_{yx} - \rho_i^2 f_x]\mu_i &= 0, \\ \det\,[f_{xy} f_y^{-1} f_{yx} - \rho_i^2 f_x] &= 0, \end{aligned}\right\}, \qquad i = 1, \ldots, p,$$
$$\rho_1^{\,2} \geq \rho_2^{\,2} \geq \cdots \geq \rho_q^{\,2} \geq 0, \qquad \rho_{q+1}^2 = \rho_{q+2}^2 = \cdots = \rho_p^{\,2} = 0$$
$$\nu_i = \rho_i f_y^{-1} f_{yx} \mu_i, \qquad i = 1, \ldots, q.$$

The ρ_i are the canonical coherences. For $p < q$ we may obtain these parameters by reversing the places of x and y. *We may now define the corresponding sample quantities, $\hat{\mu}_1$, $\hat{\nu}_i$, $\hat{\rho}_i$, by simply replacing f_x, etc., by $\hat{f}_x$, etc.* These have a simple geometric interpretation which we leave for the reader to establish for himself.

The joint asymptotic distribution of the $\hat{\rho}_i$ has been determined by James (1964, formula (112)). We shall deal neither with it nor its derivation here. Instead, we shall deal with some tests of significance for association between the $x(n)$ and $y(n)$ processes at the frequency λ and associated confidence procedures. Let us be more general and consider the situation where three random vector processes, $x(n)$, $y(n)$, $z(n)$ of p, q, r components are simultaneously observed, where the (row) vector process $(x(n)' \,\vdots\, y(n)' \,\vdots\, z(n)')$ satisfies the conditions of Theorem 13 or 13′ in Chapter IV. We now have

$$f = \begin{bmatrix} f_x & f_{xy} & f_{xz} \\ f_{yx} & f_y & f_{yz} \\ f_{zx} & f_{zy} & f_z \end{bmatrix}.$$

We wish to consider a test for association for x with y (at frequency λ) after allowing for any effects of z. Thus we are initially hypothesizing that $f_{xy} - f_{xz} f_z^{-1} f_{zy}$ is null so that $v_x(u) - f_{xz} f_z^{-1} v_z(u)$ is incoherent with $v_y(u)$ and all of the apparent association between x and y is truly due only to their common association with z.

Theorem 15. *The likelihood ratio, got from the asymptotic distribution* (6.1), *for the test of the hypothesis that* $f_{xy} - f_{xz}f_z^{-1}f_{zy}$ *is null is a monotonic function of the "lambda criterion"*†

$$(6.5) \qquad \tilde{\Lambda}(m-r, p, q) = \frac{\det(A-B)}{\det(A)}$$

$$A = \hat{f}_x - \hat{f}_{xz}\hat{f}_z^{-1}\hat{f}_{zx}, \quad B = (\hat{f}_{xy} - \hat{f}_{xz}\hat{f}_z^{-1}\hat{f}_{zy})(\hat{f}_y - \hat{f}_{yz}\hat{f}_z^{-1}\hat{f}_{zy})^{-1}(\hat{f}_{yx} - \hat{f}_{yz}\hat{f}_z^{-1}\hat{f}_{zx}).$$

We have

$$(6.6) \quad \tilde{\Lambda}(m-r, p, q) = \prod_{j=1}^{p}\prod_{k=1}^{q} \tilde{\Lambda}(m-r-j-k+2, 1, 1), \qquad (m-r) \geq p+q,$$

where each $\Lambda(m', 1, 1)$ *is of the form* $(1 - \hat{\rho}^2)$ *where* $\hat{\rho}$ *is an ordinary coherence from* m' *observations. On the null hypothesis the several factors are asymptotically independently distributed as ordinary coherences on the null hypothesis of zero true coherence. Thus*

$$(6.7) \quad P\big((\tilde{\Lambda}(m-r, p, q)) \leq y\big) = y^{m-r+1}\sum_{j=1}^{p}\sum_{k=1}^{q}\left\{y^{-j-k}\prod_{\substack{u,v=1\\(u,v)\neq(j,k)}}^{p,q}\binom{m-u-v+1}{j+k-u-v}\right\},$$

$$0 \leq y \leq 1, \qquad m \geq r+p+q.$$

For m large

$$-m'' \log_e \tilde{\Lambda}(m-r, p, q), \qquad m'' = \frac{2}{pq}\sum_{j=1}^{p}\sum_{k=1}^{q}(m-r-j-k+1),$$

may be approximately treated as chi-square with $2pq$ *degrees of freedom.*

Note. A more accurate approximation to the distribution of Λ may be obtained from the results of Anderson (1958, pp. 207–209). (See also Rao, 1965, p. 472.) In particular we may, using χ_u^2 for a chi-square variable with u degrees of freedom, use the asymptotic expansion of which the first two terms are

$$P(-m'' \ln \Lambda \leq x) = P(\chi_{2pq}^2 \leq x) + \frac{pq(p^2+q^2-2)}{3} m''^{-2} \times \{P(\chi_{2pq+4}^2 \leq x) - P(\chi_{2pq}^2 \leq x)\} + \cdots.$$

Proof. We shall omit the proof that the likelihood ratio is as stated. Now first consider the case where $r = 0$. Let s be the plane through the origin

† We insert a tilde to differentiate them from their real multivariate analogs.

in C^m spanned by the rows of V_x (having $v_{x,j}(u)$ in row j column u). The $\tilde{\Lambda}$ criterion is now

$$\tilde{\Lambda} = \frac{\det\{\hat{f}_x - \hat{f}_{xy}\hat{f}_y^{-1}\hat{f}_{yx}\}}{\det\{\hat{f}_x\}}$$

$$= \prod_i 1(-\hat{\rho}_i^2).$$

The $\hat{\rho}_i$ determine the relative orientation of the plane s and the plane t spanned by the rows of V_y (having entries $v_{y,k}(u)$). The correspondence between V_x and s induces a probability distribution on the space† $G_{p,m}(C)$ of all such planes through the origin. This induced distribution is invariant under the unitary group $U(m)$ since evidently V_x and $V_xU(m)$ have the same distribution. By saying the induced distribution is invariant we mean that the probability measure of any set S of such planes is the same as that of the set gS got from S by the action of any element g of the unitary group $U(m)$, where by gS we mean the set $\{gs;\ s \in S\}$. Now if t_0 is a fixed element of $G_{q,m}(C)$ then the distribution of the $\hat{\rho}_i$ which describe the relative orientation of s and t_0 (s being invariantly distributed in $G_{p,m}(C)$) is independent of the fixed plane (since s and gs have the same distribution and the $\hat{\rho}_i$ are the same for gs and gt_0 as for s and t_0). This means that the distribution of the $\hat{\rho}_i$ is independent of the distribution of the plane t_0, if it is not fixed, provided t_0 is distributed independently of s and in particular t_0 might itself be invariantly distributed in $G_{q,m}(C)$ independently of s. It also shows that the distribution of $\tilde{\Lambda}$ is the same provided one of the two planes is invariantly distributed and independent of the other, which may be arbitrarily distributed. This shows that to find the distribution of $\tilde{\Lambda}(m, p, q)$ we may take V_y as the matrix

$$V_y = [I_q \vdots 0].$$

It is also true that $\tilde{\Lambda}(m, p, q)$ is not affected by any transformation $V_x \to PV_x$ where P is nonsingular. Thus we may take $f_x = I_p$. Now turning to the case $r \neq 0$ we put $E = \{I_m - V_z^*(V_zV_z^*)^{-1}V_z\}$. Then E projects onto the orthogonal complement of the space spanned by the rows of V_z in C^m. Thus when $r \neq 0$ we are relating V_xE and V_yE and V_xE is independent of V_y (since $\mathcal{E}(V_xEV_y^*) = f_{xy} - f_{xz}f_z^{-1}f_{zy} = 0$.) Moreover, for E fixed the rows of V_xE span a plane $s \in G_{p,m-r}(C)$ which is invariantly distributed in that manifold since a unitary transformation leaving the space spanned by the rows of V_z fixed evidently commutes with E. Thus $\tilde{\Lambda}(m - r, p, q)$ has the same distribution as that corresponding to a set of $\hat{\rho}_i$ between a p-dimensional

† As a topological space with a locally Euclidean topology derived from the local coordinatization by means of elements of the defining matrix V_x the space $G_{p,m}(C)$ is an analytic manifold called a Grassman manifold.

invariantly distributed plane in C^{m-r} and the q-dimensional plane in that space spanned by the rows of

$$[I_q \vdots 0],$$

where the invariantly distributed plane is spanned by the rows of a matrix W having the complex normal distribution (with zero mean vector) with I_p as covariance matrix. Thus the effect of $r \neq 0$ is merely to reduce the degrees of freedom and we may as well consider $\tilde{\Lambda}(m, p, q)$. The ratio $\tilde{\Lambda}(m, p, q)$ is of the form

$$\tilde{\Lambda}(m, p, q) = \frac{\det\{\sum_{q+1}^{m} w_i w_i^*\}}{\det\{\sum_1^m w_i w_i^*\}},$$

where w_i is the ith column of W. This is distributed independently of its denominator for clearly it is functionally independent of the elements of the matrix whose determinant is in the denominator since they could be changed to any admissible set of numbers (i.e., any set giving a positive definite Hermitian matrix) by a transformation $W \to PW$, P nonsingular, which we know leaves $\tilde{\Lambda}$ invariant. We leave it to the reader to establish, by means of an appropriate transformation of coordinates,† that these elements of WW^* are distributed independently of the remaining variables needed to specify the distribution of W and which are all that $\tilde{\Lambda}$ can depend on. We put

$$\det\left\{\sum_j^m w_i w_i^*\right\} = S_j$$

and observe that

$$\tilde{\Lambda}(m, p, q) = \prod_1^q \frac{S_{j+1}}{S_j} = \prod_1^q \tilde{\Lambda}(m - j + 1, p, 1).$$

Moreover, each ratio $\tilde{\Lambda}(m - j + 1, p, 1)$ is independent of all of the others. Consider $\tilde{\Lambda}(m - j + 1, p, 1)$, $\tilde{\Lambda}(m - k + 1, p, 1)$, $k > j$. The latter of the two is independent of the elements of the matrix, namely $\sum_k^m w_i w_i^*$, whose determinant constitutes its denominator. It is certainly independent of

$$\sum_j^{k-1} w_i w_i^*, \qquad \sum_{j+1}^{k-1} w_i w_i^2,$$

and of the elements of both

$$\sum_j^{k-1} w_i w_i^* + \sum_k^m w_i w_i^*, \qquad \sum_{j+1}^{k-1} w_i w_i^* + \sum_k^m w_i w_i^*,$$

† This may be done by putting $W = GLA$ where G is unitary, L is square and diagonal, and A has rows which are orthonormal, and is $p \times m$. The rows of A specify the plane which W specifies. The differential form in the elements of W splits into a product of two forms, one of which involves only the elements of A, and since the density fuction involves only $WW^* = GLG$ the result is established.

and thus of $\tilde{\Lambda}(m-j+1, p, 1)$, which is a function only of the elements of these matrices. This establishes the result. Of course, $\tilde{\Lambda}(m', p, 1) = 1 - \hat{\rho}^2$ where $\hat{\rho}$ is a multiple coherence of one complex normal variate with p others of which it is independent and thus $\hat{\rho}^2$ is distributed as a multiple correlation with $2p$ real variables from $2m'$ sets of observations. We may now repeat the process with each factor $\tilde{\Lambda}(m', p, 1) = \tilde{\Lambda}(m', 1, p)$ to arrive at the factorization

$$\tilde{\Lambda}(m, p, q) = \prod_{j=1}^{p} \prod_{k=1}^{q} \tilde{\Lambda}(m - j - k + 2, 1, 1),$$

where each factor is independent of all other factors and is of the form $1 - \hat{\rho}^2$ where $\hat{\rho}$ is now an ordinary coherence from, typically, $m' = m - j - k + 2$ observations. Thus, from (2.6), $(1 - \hat{\rho}^2)$ has density

$$(m' - 1)x^{m'-2}, \qquad 0 \leq x \leq 1,$$

and $-2(m' - 1) \ln (1 - \hat{\rho}^2)$ has the chi-square distribution with two degrees of freedom. Thus, putting $a_{j,k} = m - j - k + 1$

$$\xi = -\ln \tilde{\Lambda}(m, p, q) = \sum_{j=1}^{p} \sum_{k=1}^{q} a_{j,k}^{-1} x_{j,k},$$

where the $x_{j,k}$ are independently and identically distributed with density $\exp -x$. Thus ξ has characteristic function

$$\prod_{j,k} (1 - i a_{j,k}^{-1} \theta),$$

whence the density of ξ is easily obtained as

$$\sum\sum_{j,k} \left\{ e^{-x a_{jk}} a_{jk} \prod_{\substack{(u,v) \\ \neq j,k}} \left(\frac{1}{1 - a_{jk}/a_{uv}} \right) \right\}$$

and the theorem results, save for the last statement. This last result is fairly obvious for we have

$$-m'' \ln \tilde{\Lambda}(m, p, q) = \sum\sum_{j,k} \frac{m''}{2a_{jk}} \{-2a_{jk} \ln \tilde{\Lambda}(m - j - k + 1, 1, 1)\}$$

and for m large relative to p, q the factors $(m''/2a_{jk})$ are near to unity so that the right side is near to the sum of pq independent random variables, each of which is χ_2^2. *However*, (6.7) *would be simple to evaluate.*

Before proceeding further we discuss some special cases which illustrate the general results given above. So far as we are aware these techniques never been used in practice so that the examples we give cannot be given an air of reality. Nevertheless it is believed that these techniques will have applications:

(i) If a deterministic component, $\sum (\alpha_j \cos n\theta_j + \beta_j \sin n\theta_j)$ where α_j and β_j are vectors of p components, is present in $x(n)$ then the mean value of $v_x(u)$ will not be zero especially for those u which correspond to frequencies $2\pi u/N$ clustering around some frequency λ which is near to one of the θ_j. In this circumstance the mean vector for $v_x(u)$ will depend upon u. The example is introduced here purely to show that one might wish to test the hypothesis that the mean value of $v_x(u)$ is zero. In this circumstance we might consider the canonical association of the $v_x(u)$ with the variates $v_y(u) \equiv 1$ (i.e., $q = 1$). As we have seen above the $\tilde{\Lambda}$ criterion will have the distribution derived above on the null hypothesis that $\mathcal{E}(v_x(u)) \equiv 0$ and this will be distributed as $(1 - \hat{\rho}^2)$, where $\hat{\rho}^2$ is a multiple correlation of one real variable with $2p$ other real variables from $2m$ sets of observations, the true correlation being zero. The $\tilde{\Lambda}$ criterion is, of course,

$$1 - \hat{\mu}_x^*(m^{-1}V_xV_x^*)^{-1}\hat{\mu}_x = 1 - \hat{\mu}_x^*\hat{f}^{-1}(\lambda)\hat{\mu}_x,$$

where $\hat{\mu}_x$ is the vector of mean values of the $v_x(u)$. We shall discuss this type of problem in another way later. A more relevant test, perhaps, might be obtained by taking $v_y(u) = \exp iu\lambda$, which would lead to the same distribution for the new test statistic.

(ii) Let us next consider an example where, let us say, p oceanographic "parameters" are to be related to $q + 1$ meteorological parameters and it is believed that all of the relation is due to one known linear combination of the meteorological parameters. In this case $r = 1$. The criterion is constructed by removing, by complex regression, the effect of the known linear combination from all of the oceanographic variables and q linear combinations of the meteorological variables (linearly independent of the specified one). Of course it may also be constructed more simply via (6.5). Then

$$-(2m - p - q - 2)\log_e \tilde{\Lambda}(m - 1, p, q),$$

used as chi square with $2pq$ degrees of freedom tests† whether this is the only true explanatory variable. The overall $\tilde{\Lambda}(m, p, q + 1)$ factors precisely as

$$\tilde{\Lambda}(m, p, q + 1) = \tilde{\Lambda}(m, p, 1)\,\tilde{\Lambda}(m - 1, p, q),$$

where the first factor on the right is that appropriate to testing the association of the prescribed linear combination of the meteorological variables with the oceanographic variables.

(iii) As a slightly more elaborate case we could consider the situation where it was hypothesized that one linear function of the meteorological variables was responsible for all association but also that this affected only a

† Of course, Theorem 15 in its entirety applies and we do not have to use the approximate χ^2_{2pq} distribution.

specified subset of $p_1 < p = p_1 + p_2$ of the oceanographic parameters. Now we should factor $\tilde{\Lambda}$ as

$$\tilde{\Lambda}(m, p, q + 1) = \tilde{\Lambda}(m, p_1, 1)\, \tilde{\Lambda}(m - p_1, p_2, 1)\, \tilde{\Lambda}(m - 1, p, q),$$

where $\tilde{\Lambda}(m, p_1, 1)$ is the criterion for association of the specified linear combination with the specified oceanographic variables, $\tilde{\Lambda}(m - p_1, p_2, 1) = \tilde{\Lambda}(m, p, 1)/\tilde{\Lambda}(m, p_1, 1)$, the numerator being as in (ii), and $\tilde{\Lambda}(m - 1, p, q)$ is also as in (ii). The first factor leads to a χ^2 with $2p_1$ d.f. if the specified linear function is incoherent with the p_1 specified oceanographic variables at the frequency in question, the second to chi square with $2p_2$ d.f. independently of the truth of the first hypothesis if the remaining p_2 variables are independent of that linear combination (save through their relation to the first p_1), whereas the third leads to a χ^2 with $2pq$ d.f. if this linear combination is the sole explanatory variable.

There are two other problems which we mention only briefly here, the first because its solution at a practical level depends at the moment on asymptotic theory (m large) which is somewhat dubious in the present applications and the second, which is concerned with confidence region procedures for discriminant function coefficients, because of the difficulty in intelligibly presenting the results in the present context where the number of real parameters is never less than 2 and could often be 8 or 10.

The first problem arises when having computed the $\hat{\rho}_i$, which we have ordered in magnitude so that $\hat{\rho}_1 \geq \hat{\rho}_2 \geq \cdots$, we observe that only the first $s < \min(p, q)$ are sizable and the remainder are small. One may now wish to test whether there are only s non-null true canonical coherences. If the true discriminant functions were known, say for the x variables, corresponding to the, precisely, s non-null ρ_i then we could proceed as above by forming the criterion $\tilde{\Lambda}(m - s, p - s, q)$. An alternative is to use the sample discriminant functions when the true ones are unknown. From (6.5) this leads to

$$\frac{\tilde{\Lambda}(m, p, q)}{\{\det(\hat{f}_x - \hat{f}_{xz}\hat{f}_z^{-1}\hat{f}_{zx})/\det \hat{f}_x\}},$$

wherein z corresponds to the s variables defined by the first s sample discriminant functions. The denominator is evidently

$$\prod_1^s (1 - \hat{\rho}_i^2),$$

so that our criterion is

$$\frac{\tilde{\Lambda}(m, p, q)}{\prod_1^s (1 - \hat{\rho}_i^2)} = \prod_{s+1}^{\min(p,q)} (1 - \hat{\rho}_i^2). \tag{6.8}$$

The distribution of this may be obtained from the exact distribution of the $\hat{\rho}_i$, $i = 1, \ldots, \min(p, q)$. An approximation may also be got from the asymptotic joint distribution of these for fixed p, q as $m \to \infty$. For the real case we refer the reader to Hsu (1941). The net result is that we may asymptotically treat (6.8) as $\tilde{\Lambda}(m-s,\ p-s,\ q-s)$ provided there are *exactly* s non-null true canonical coherences, that is, we may approximately use

$$(2m - p - q) \ln \left\{ \frac{\tilde{\Lambda}(m, p, q)}{\prod_1^s (1 - \hat{\rho}_i^{\,2})} \right\}$$

as chi-square with $2(p-s)(q-s)$ degrees of freedom.

Finally, we shall discuss very briefly confidence interval procedures for discriminant function coefficients. Consider the case where we wish to find a confidence region for the vector of coefficients ν defining the discriminant function $\xi(u) = \nu^* v_y(u)$ when it is believed that after the elimination of the effects of r variables $v_{z,j}(u)$ there is only one true discriminant function for x with y which has non-null canonical coherence. We normalize ν, say by requiring the first nonzero coefficient to be unity. We may then factor the overall criterion as

$$\begin{aligned} \tilde{\Lambda}(m, p, q + r) &= \tilde{\Lambda}(m, p, r)\, \tilde{\Lambda}(m - r, p, q) \\ &= \tilde{\Lambda}(m, p, r)\, \tilde{\Lambda}_\nu(m - r, 1, q)\, \tilde{\Lambda}_\nu(m - r - 1, p - 1, q), \end{aligned}$$

wherein the first factor is appropriate for testing the association of the x variables and the z variables, the second for testing the association of ξ (assuming ν known for the moment) with the y variables and the last for testing whether there is only one such ξ with non-null canonical coherence. The last factor is, of course, a function of the unknown coefficients (restricted as above) in ξ. Thus the relation

$$P\{\tilde{\Lambda}_\nu(m - r - 1, p - 1, q) \le a\} = \alpha,$$

which is determined from the fact that $\tilde{\Lambda}_\nu(m - r - 1, p - 1, q)$ has a known distribution on the hypothesis only that there is a single non-null canonical coherence, may be used to find a confidence region with confidence coefficient α for the components of ν by finding those ν which satisfy

$$\tilde{\Lambda}_\nu(m - r - 1, p - 1, q) \le a. \tag{6.9}$$

Now we illustrate with the simple case $p = 2$. Then

$$\tilde{\Lambda}_\nu(m - r - 1,\ p - 1,\ q) = \frac{\tilde{\Lambda}(m - r,\ p,\ q)}{1 - \phi}$$

where $\tilde{\Lambda}(m - r, p, q)$ is the overall criterion (independent of ν) for association after removing the effects of the z variables while ϕ is the canonical coherence of $\xi(u)$ with the q of the y variables after removing the effects of the z variables. It is thus of the form

$$\phi = \frac{\nu^* B \nu}{\nu^* A \nu},$$

where A and B are as defined in (6.5).

Then (6.9) leads to

$$\nu^* \left\{ B - \left\{ \frac{a}{1 - \tilde{\Lambda}(m - r - 1, p - 1, q)} \right\} A \right\} \nu \leq 0,$$

so that it is this Hermitian form which defines the region for ν.

Of course more elaborate cases could also be considered.

To conclude we mention that, in accordance with the principle enunciated in Corollary 5, the techniques of this section become applicable in the case when M and N both increase indefinitely but so that $M/N \to 0$, if we replace $2m$ by the degrees of freedom.

7. PRACTICAL SPECTRAL ANALYSIS

We shall not constantly speak of "spectra and cross spectra" in this section but shall use "spectrum" or "spectra" to cover all cases, where there is no confusion.

In order to conform with the notation customarily used in actual Fourier analysis of data we change our notation:

1. We denote by Δ the sampling interval, which we have so far taken to be unity.
2. We shall work in terms of frequency f in cycles per time unit. Thus $\lambda = 2\pi f$.
3. We call $X_{ij}(f) = 2\pi \hat{f}_{ij}(\lambda)$, $\lambda = 2\pi f$.
4. Then $f_n = (2\Delta)^{-1}$ is the "Nyqvist" or folding frequency.
5. The width of what we call a "band" (i.e., a band over which a spectral estimate is made) is now $(2M\Delta)^{-1}$, since this is $M^{-1} \times (2\Delta)^{-1}$.

We assume that we have chosen our spectral estimation procedure once and for all (e.g., as Cooley-Tukey or Tukey-Hanning, etc.). This is clearly what will happen, since availability of a program known to "run" on the computer to be used, personal experience with a certain estimation technique and so on will determine the technique used and not any desire to choose one optimally for the problem to hand.

A number of decisions then have to be made that are reflected in our choice† of Δ, N, and M (or m). In choosing these clearly we will have in mind such factors as

(a) the rate of decrease of $|X_{ij}(f)|$ as f increases;
(b) the cost of increasing N;
(c) the degree of resolution and of accuracy required.

With respect to these last we mean by accuracy what we have measured by variance but we need to define what we mean by resolution. This involves the introduction of the concept of "bandwidth." This can be defined in various ways but since it is the concept which is primarily needed and there is no intention of setting out accurately to measure the bandwidth there is no need for any extensive discussion of various definitions. What we have in mind is the degree to which the true spectrum (in the scalar case, for example) shows narrow tall peaks. The bandwidth of such a local maximum could be defined as one half of the width of the peak at half of its maximum height above the level of the spectrum in its neighbourhood. The bandwidth of the spectrum over a region could be loosely defined as the minimum of these quantities for "sizable" peaks. As we have already indicated, this concept is too loosely defined for measurement. It corresponds closely, of course, to the oscillation of the function over an interval I, namely, the function

$$\omega(\delta) = \sup |f(\lambda_1) - f(\lambda_2)|, \qquad |\lambda_1 - \lambda_2| \leq \delta, \qquad \lambda_1, \lambda_2 \in I,$$

which is related, in turn, to the rate at which the α_j, in the formula for a linear process, decrease to zero. Thus this concept of bandwidth has already entered into our considerations, in a disguised form (see exercise 2). Here we need something looser and better related to intuitive concepts for our discussion of practical problems.

Of course one cannot decide upon an estimation procedure without some prior concept as to what the spectra are like. In many fields this information will exist in the form of a background of experience of similar investigations. Failing this the only course of action is a "pilot" calculation. Blackman and Tukey (1959) describe a "quick and dirty" method for carrying out such a pilot investigation, based on a technique of summing observations in pairs, differencing in pairs and summing the square of the results. Thus

$$\sum \{x(2\Delta n) - x(2\Delta n - \Delta)\}^2 \tag{6.1}$$

† In this book we have not discussed situations in which relatively high frequencies are very important (e.g., audiofrequency work) so that digital methods become too costly. In such situations analog devices using filtering techniques are of prime importance.

may be used (after division by an appropriate constant) to estimate the total spectral mass in the interval $f_{\frac{1}{2}n}$ to f_n. However, we shall not consider this technique here for it seems to have been designed for earlier days of high-speed digital computers. With equipment now available there is little need for such techniques. The great virtue of the FFT and the Cooley-Tukey technique is that it provides the basic initial estimates, of $I(\lambda)$, from which all other calculations can later proceed. An initial "pilot" investigation might consist therefore of nothing other than a first use of these calculations, that is, a first choice of m or m' (possibly varying with frequency). Apart from the use of these techniques it would also seem preferable if they are not available, rather than to use formulas of the type of (6.1), to make an initial estimate of $X(f)$ by using, say, the Tukey-Hanning procedure with M rather large. In any case we assume that some information, of a qualitative kind at least, is available on the basis of which further calculations can be carried through.

We shall deal with the choice of N last, though it will have to be chosen first. The choice of Δ is substantially motivated by the desire to avoid aliasing problems, but of course also questions of cost arise as, for example, it may be very difficult to read a record accurately at very close intervals. A rough rule given by Tukey is to choose f_n as $\frac{3}{2} f_{\max}$ so that

$$\Delta = (3f_{\max})^{-1},$$

where $f_{\max}$ is the "maximum frequency of interest." If there is no spectral mass above $2f_{\max}$ then this means that there is no effect from aliasing below $f_{\max}$.

Before M can be chosen other treatments of the data will have been carried through.

1. If possible the data should be graphed. This is of immense importance, not only in giving some idea of its nature but also (a) to guard against recording errors, and (b) to indicate departures from stationarity. A large recording error can play havoc with the results of spectral analysis. There are no automatic rules, save care, for avoiding them and graphing (followed by close scrutinization of the graph) is one of the best ways of discerning them. We have so far assumed stationarity but it is not uncommon for the nature of the random mechanism to change radically. An important technique for studying this change is the breaking of the series into lengths within each of which, individually, stationarity seems to obtain.

2. Some form of prefiltering might be performed. A common procedure in economics is to prefilter so as to remove "trends," which can roughly be thought of as concentrations of spectral mass at the origin of frequencies. Ultimately the purpose of filtering is to use prior knowledge so as to adjust

the spectrum, before estimation is commenced, so as to resemble as closely as possible a uniform spectrum. It can, to some extent, be viewed as a procedure for using this information to "tailor" the estimation method to fit the data to hand. With FFT and Cooley-Tukey techniques such prefiltering is reduced in importance since these methods give initial estimates based on spectral windows with such narrow bandwidths ($2\pi/N$ or $2\pi/N'$). There is no doubt still a need for prefiltering techniques, particularly to remove a substantial concentration of spectral mass in situations where a precise prescription of the nature of the concentration is not known. Of course the effects of filtering will have to be allowed for after calculations have been completed. In principle this is a simple procedure involving division by the frequency response of the filter. In fact the sequence filter, spectral estimation, division by response function of filter does not give quite the same result as spectral estimation without filtering and indeed, if it did, there would be no purpose in filtering! This subject seems to need some further investigation and filtering should be used cautiously. Needless to say one may also have to account for filtering which has been accomplished, as it were, by nature (e.g., due to the response of a recording device).

3. It will always be necessary to make mean corrections and a more elaborate regression may be needed. We shall discuss the effects of this in detail in Chapter VII. The effect of a mean correction will be substantially concentrated in the estimate at $f = 0$. Here an approximate correction procedure is to divide by $(1 - M/N)$. More generally one may divide the estimate at $f = (k/2M)$ by $\{1 - (r/N)K_N(2\pi f)\}$ in the case where a polynomial trend of degree $(r - 1)$ has been removed by regression. This reduces to the previous rule at $f = 0$ and for $r = 1$.

4. We next mention a problem previously considered in the estimation of cross spectra, that if $\theta(2\pi f)$ is changing rapidly with f then the estimation of coherence will be badly biased downwards. An approximate device is to lag one series relative to the other so as to subtract a term, $f\theta_0$, from the phase. Of course θ_0, determined by the lag, will have to vary from band to band (i.e., with k). Thus if a pilot survey indicates that, as between the ith and jth series, θ may be of the form $a + bf$, with b large, over a band of interest then the jth series should be lagged relative to the ith series by $[b]$ time units before calculations for that band are commenced.

Let B be the bandwidth of the aliased spectrum or of that part of it in which we are interested. We are then led to put $(M\Delta)^{-1} = B$. For bands are of width (in the units of f) $(2M\Delta)^{-1}$ and if we choose this width so that it is half the bandwidth of the spectrum reasonable resolution should be achieved. This puts

$$M = (B\Delta)^{-1} = \frac{3f_{\max}}{B}.$$

The relative error (in the sense of deviation from its expected value) of an estimate of a spectrum is measured by the coefficient of variation

$$V = \left\{\frac{\operatorname{var}(\hat{X}_i)}{X_i^2}\right\}^{1/2} = \left(\frac{M}{N}\right)^{1/2}\left\{\int k^2(x)\,dx\right\}^{1/2}.$$

Approximating the last factor, this leads to

$$N = \frac{4M}{V^2} = \frac{12 f_{\max}}{BV^2},$$

where V is the required order of magnitude of the fractional error, that is, the required coefficient of variation.† Thus given B, $f_{\max}$ and V we have established rough and ready rules which determine Δ, M and N. They are no more than guides and cannot be taken too seriously if only because often they lead to too large a value of N; for example, if in time units of one second

$$f_{\max} = 10, \qquad B = \tfrac{1}{2}, \qquad V = 0.1,$$

then

$$\Delta = \frac{1}{30}, \qquad M = 60, \qquad N = 24{,}000.$$

It is fairly clear here that for most purposes V is too small. If it is changed to 0.3 the N reduces to about 2700 which seems a more reasonable figure. Of course a more exact-seeming approach could be based upon the Theorem 11, of Section 5. In practice this sequence of steps is unlikely to be used explicitly, even as it stands. Indeed the procedure is likely to be partly experimental, particularly if FFT or Cooley-Tukey are used, which lend themselves to some "play" with the choice of M. A rule suggested by G. M. Jenkins is to choose three values of M. A first choice of a low value gives a rough indication of where most spectral mass lies. A second high value gives some idea of what resolution is needed and a compromise is achieved with the third value. In this subject, as elsewhere, experience is the real teacher and that cannot be got from a book!

EXERCISES

1. Let

$$\sum_{-\infty}^{\infty} \|A(j)\|\,|j|^{\delta} < \infty, \qquad \tfrac{1}{4} \le \delta.$$

† The coefficient of variation is the order of magnitude of fractional error which can be expected to occur.

By considering

$$\left\| \sum_{1}^{r} \epsilon(n) e^{in\lambda} \right\|^2 \le \sum_{-r}^{r} \left\| \sum{}' \epsilon(n)\epsilon(n+s) \right\|$$

show that

$$\mathcal{E}\left\{ \max_{\lambda} \|R_{j,N}(\lambda)\|^2 \right\} \le K\{\min(|j|, N)\}^{3/4},$$

hence that

$$\mathcal{E}\left\{ \max_{\lambda} \|r_N(\lambda)\| \right\} = O(N^{-\delta+1/4}), \qquad \tfrac{1}{4} \le \delta \le \tfrac{3}{4},$$

$$= O(N^{-1/2}), \qquad \delta \ge \tfrac{3}{4}.$$

2. If the assumption of Exercise 1 is true, show that

$$\|h(\lambda + \epsilon) - h(\lambda)\| = O(\epsilon^{\delta})$$

as $\epsilon \to 0$ so that $h(\lambda)$ satisfies a Lipshitz condition of order δ. Hence show that $f(\lambda)$ does so also.

3. Show that if the condition of Exercise 1 is satisfied then also

$$\sum_{-\infty}^{\infty} \|\Gamma(n)\| \, |n|^{\delta} < \infty.$$

4. Commence from a complex multivariate normal distribution for the $v_j(u)$, $j = 1, 2$, for fixed u, but with f, the matrix occurring in (4.4) of Chapter IV, a function of u of the form

$$f_u = \begin{bmatrix} f_1 & \rho_{12} e^{i\theta_u} \\ \rho_{12} e^{-i\theta_u} & f_2 \end{bmatrix}, \qquad u = 1, \ldots, m.$$

Treat the $v_j(u)$ as independent, $u = 1, \ldots, m$ and derive the maximum likelihood estimates of f_1, f_2 and $\sigma_{12} = \rho_{12}\{f_1 f_2\}^{-1/2}$. Determine the likelihood ratio test for the hypothesis that θ_u is independent of u.

5. Take $\theta_u = \alpha + \beta u$ in Example 4. Determine the maximum likelihood estimators for $f_1, f_2, \sigma_{12}, \alpha, \beta$ and the likelihood ratio test for $\beta = 0$. Discuss Exercises 4 and 5 in relation to procedures for estimating coherence suggested in Section 6.

6. Establish formulas (2.9), (2.12), (3.10), and (3.12).

7. Consider an estimator of the form

$$* \qquad f(\lambda) = \int_{-\pi}^{\pi} I(\theta)\, K_N(\lambda - \theta)\, d\theta,$$

$$K_N(\lambda) = \frac{1}{2\pi} \sum_{-M}^{M} k_j e^{ij\lambda}.$$

Show that

$$f\left(\frac{2\pi k}{N}\right) = \frac{\pi}{N} \sum_{-N+1}^{N} K_N\left(\frac{2\pi k}{N} - \frac{\pi j}{N}\right) I\left(\frac{\pi j}{N}\right),$$

so that the spectral estimator is exactly expressed, at the points $2\pi k/N$, in terms of a "natural" approximating sum to the integral on the right of *.

8. Let

$$\tilde{A} = \begin{bmatrix} C & D \\ -D & C \end{bmatrix}.$$

Show that $\tilde{A}$ has determinant $|\det(C + iD)|^2$. (Use the polar decomposition of $A = C + iD$ together with the isomorphism established in the proof of Theorem 13 in Chapter IV to reduce the result to that for an Hermitian matrix and a unitary matrix A.)

9. Determine the form of the inverse of $A = C + iD$ in exercise 8 by using the isomorphism referred to in the exercise.

10. Let S be $p \times p$ and Hermitian and $\tilde{S} = PSP^*$. Show that the Jacobian of this linear transformation of the p^2 real parameters in S is $|\det P|^{2p}$. (Consider a general matrix T of complex numbers. The set of all such matrices forms a p^2-dimensional complex vector space and the transformation $T \to PTP$ is $P \otimes P'$ if T is written as a column, taking its elements in a suitable order, as explained in Section 5 of the Mathematical Appendix. The vector space in which T lies decomposes as the orthogonal direct sum of subspaces corresponding to the Hermitian and the skew Hermitian matrices, each of which is left invariant by $P \otimes P'$. Observing that if Z is skew Hermitian it may be written as iY where Y is Hermitian show that we may take A in exercise 8 as $P \otimes P'$ with D null and C operating in the real vector space which may be identified with the Hermitian symmetric matrices. Thus since $\det(P \otimes P') = |\det P|^{2p}$ then by exercise 8 that of A is $|\det P|^{4p}$ and as an operation in the space of Hermitian symmetric matrices $S \to PSP^*$ has Jacobian as stated.)

APPENDIX

1 *Proof of Theorem* 9

In the same way as we obtained (3.3) in Chapter IV, we see that

$$(1) \quad \frac{(N-n)(N-n')}{N} \operatorname{cov}\left(c_{jk}(n'), c_{rs}(n)\right)$$
$$= N^{-1} \sum_{u=-\infty}^{\infty} \{\gamma_{jr}(u)\gamma_{ks}(u+n-n') + \gamma_{js}(u+n)\gamma_{kr}(u-n') + k_{jkrs}(0, n', u, u+n)\}\, \phi_N(u, n', n),$$

wherein for $n' \geq n$

$$\phi_N(u, n', n) = 0, \qquad u \leq -N + n'; = \left(1 - \frac{n'+u}{N}\right), \quad -N + n' \leq u \leq 0$$
$$= 1 - \frac{n'}{N}, \quad 0 \leq u \leq n - n'; = 1 - \frac{n+u}{N}, \quad n - n' \leq u \leq N - n;$$
$$= 0, \; u \geq N - n.$$

We use this expression in the covariance of $\hat{f}_{jk}(\lambda_1)$ with $\hat{f}_{rs}(\lambda_2 + \pi p/M)$, conjugating the second expression in forming the covariance. We commence with the case of estimators of the type (4.2), where k_n is given by (4.4). Then N/M by this covariance decomposes into three terms, of which the first is

$$(2)\quad \frac{1}{4\pi^2 M}\sum_{n,n'}\sum \left\{k\left(\frac{n'}{M}\right)k\left(\frac{n}{M}\right)e^{-i(n'\lambda_1 - n\lambda_2 - n\pi p/M)} \times \sum_{-\infty}^{\infty} \gamma_{jr}(u)\gamma_{ks}(u+n-n')\phi_N(u,n',n)\right\}$$

$$= \frac{1}{4\pi^2}\sum_{v=-2N+2}^{2N-2}\sum_{u=-\infty}^{\infty}\gamma_{jr}(u)\gamma_{ks}(u-v)e^{-iv\lambda_1} \times \left\{\frac{1}{M}\sum_{-N+1}^{N-1}\phi_N(u,n+v,n)\,k\left(\frac{n+v}{M}\right)k\left(\frac{n}{M}\right)e^{-in(\lambda_1-\lambda_2)}e^{in\pi p/M}\right\}.$$

Call the summand in the bracketed expression $L_N(u, v, n)$. It also depends on λ_1, λ_2, p, but these are constants in the present discussion.

We now show that as $N \to \infty$ the bracketed expression converges boundedly to zero for $\lambda_1 \neq \lambda_2$ and otherwise to

$$\int_{-\infty}^{\infty} k^2(x)e^{i\pi px}\,dx.$$

This will show that as $N \to \infty$ (2) converges to zero for $\lambda_1 \neq \lambda_2$ and otherwise to

$$(3)\quad \int_{-\infty}^{\infty} k^2(x)e^{i\pi px}\,dx\,\frac{1}{2\pi}\sum_{-\infty}^{\infty}\gamma_{jr}(u)e^{iu\lambda_2}\,\frac{1}{2\pi}\sum_{-\infty}^{\infty}\gamma_{ks}(v)e^{iv\lambda_2} = \int_{-\infty}^{\infty} k^2(x)e^{i\pi px}\,dx\, f_{jr}(\lambda_2)f_{sk}(\lambda_2)$$

[using dominated convergence and the absolute convergence of $\gamma_{jk}(n)$]. We first observe that the bracketed expression is less in modulus than

$$\left\{\frac{1}{M}\sum k^2\left(\frac{n}{M}\right)\right\}^{1/2}\left\{\frac{1}{M}\sum k^2\left(\frac{n+v}{M}\right)\right\}^{1/2};$$

but, since each of the sums is an approximating sum to the integral,

$$\int_{-\infty}^{\infty} k^2(x)\,dx,$$

of a positive, continuous† function it is evident that they converge to that expression as M increases and thus are uniformly bounded. In the same way, for

† The truncated estimate has a $k(x)$ with a discontinuity but the statement is evidently still true for it.

any $\epsilon > 0$, we may choose T so large that

$$\left|\frac{1}{M}\sum_{|n|>MT} L_N(u, v, n)\right| < \epsilon$$

for M sufficiently large, uniformly in u, v, λ_1, λ_2 and p. Thus we need consider only

$$\frac{1}{M}\sum_{|n|\leq TM} L_N(u, v, n).$$

We choose disjoint intervals E_i whose union is the interval $[-T, T]$, so that the moduli of continuity† of φ_N, $k(x)$ and $\exp i\pi px$ are always less than $\eta > 0$, on E_i. Let there be e_i points, n/M, in E_i. Then, if $\lambda_1 - \lambda_2 \not\equiv 0 \pmod{2\pi}$ for N sufficiently large,

$$(4)\quad \left|\frac{1}{e_i}\sum_{n\in E_i} k\left(\frac{n}{N}\right)k\left(\frac{n+v}{N}\right) e^{in\pi p/M}e^{-in(\lambda_1-\lambda_2)}\phi_N(u, n+v, n)\right| < c_1\eta + \left|\frac{c_2}{e_i}\sum_{n\in E_i} e^{-in(\lambda_1-\lambda_2)}\right|.$$

Here the c_i depend upon the average value of $k(x)$ over E_i, for example. If N, and thus e_i, is chosen sufficiently large the last expression may also be made less than η. This result is uniform in u, v, p and λ_1, λ_2 provided that $|\lambda_1 - \lambda_2| \geq \epsilon > 0$. Thus, then

$$\left|\frac{1}{M}\sum_{|n|\leq MT} L_N(u, v, n)\right| \leq c\sum\left(\frac{e_i}{M}2\eta\right) = 2c\eta T$$

(since the e_i sum, approximately, to $2MT$). When $\lambda_1 - \lambda_2 = 0$, we have

$$\frac{1}{M}\sum_{|n|\leq MT} k\left(\frac{n}{M}\right)k\left(\frac{n+v}{M}\right)e^{in\pi p/M} \to \int_{-T}^{T} k^2(x)e^{i\pi px}\,dx,$$

since it is the approximating sum to that integral. It follows immediately that (2) approaches 0 for $|\lambda_1 - \lambda_2| \neq 0$ and otherwise (3) and that the convegrence is uniform in p and λ in case $\lambda_1 = \lambda_2 = \lambda$ and also the convergence to zero is uniform in p and $|\lambda_1 - \lambda_2|$ provided $|\lambda_1 - \lambda_2| \geq \epsilon > 0 \pmod{2\pi}$. The second term in (1) gives a completely analogous expression to (2) save that $\lambda_1 - \lambda_2$ is replaced by $(\lambda_1 + \lambda_2)$ and when $(\lambda_1 - \lambda_2) = 0 \bmod 2\pi$ we interchange r and s. This gives a contribution which converges to (3) provided $\lambda_1 = -\lambda_2 = \lambda$, uniformly in λ, and to 0 for $\lambda_1 + \lambda_2 \neq 0$, mod 2π, uniformly in $|\lambda_1 + \lambda_2| \geq \epsilon > 0 \pmod{2\pi}$. The third term is dominated by

$$\frac{1}{4\pi^2 M}\sum\sum\sum_{n',u,n=-\infty}^{\infty}\left|k\left(\frac{n}{M}\right)k\left(\frac{n'}{M}\right)k_{jkrs}(0, n', u, u+n)\right| = O(M^{-1}),$$

which converges to zero uniformly in λ_1, λ_2, p.

† The modulus of continuity, $\omega(h)$, of a function $f(x)$ on an interval E is defined as the maximum with respect to x, $x + t \in E$, $|t| \leq h$ of $|f(x + t) - f(x)|$.

This establishes the theorem for if $\lambda_1 = \lambda_2 = \lambda \neq 0, \pi$ only the first term contributes and all convergence is uniform and similarly if $\lambda_1 = -\lambda_2 = \lambda \neq 0, \pi$. If $\lambda_1 = \lambda_2 = 0, \pm\pi$ both terms contribute. However, convergence is not uniform in these neighborhoods because of the lack of uniformity of convergence to zero of the second term in (4) at $\lambda_1 - \lambda_2 = 0 \pmod{2\pi}$ and similar lack of uniformity of convergence at $\lambda_1 + \lambda_2 = 0 \pmod{2\pi}$ for the corresponding term arising from the second term in (1).

We need finally to cover the case where the k_n are not of the type of (4.4). The truncated estimator was covered by the footnote on page 314. The Bartlett estimator varies rather trivially from an estimator of the present type and we leave the details to the reader. The FFT is a little more complicated. We consider the expression that results from taking the first component of (1) in the expression for the covariance which we require. Just as in (2) we obtain

$$\frac{1}{4\pi^2}\sum_{-2N+2}^{2N-2}\sum_{u=-\infty}^{\infty}\gamma_{jr}(u)\gamma_{ks}(u-v)\left\{\frac{1}{M}\sum_{-N+1}^{N-1}L'_N(u,v,n)\right\},$$

where now

$$L'_N(u,v,n) = \phi_N(u,n+v,n)k_{n+v}(\lambda_1)k_n\left(\lambda_2+\frac{\pi p}{M}\right)e^{-in(\lambda_1-\lambda_2)}e^{in\pi p/M}$$

and k_n is given in Example 1 in Section 4. Thus

$$(5)\quad \frac{1}{M}\sum L'_N(u,v,n) = \frac{1}{M}\sum\left\{e^{-i(n+v)\lambda'}e^{in\lambda''}\frac{\sin\dfrac{m(v+n)\pi}{N}\sin\dfrac{mn\pi}{N}}{m^2\sin\dfrac{(v+n)\pi}{N}\sin\dfrac{n\pi}{N}}\phi_N(u,n+v,n)\right\},$$

where

$$\lambda' = \frac{2\pi}{N}\left\{q'+\tfrac{1}{2}(m-1)\right\},\qquad \lambda'' = \frac{2\pi}{N}\left(q''+\tfrac{1}{2}(m-1)\right),$$

and q', q'' are integers such that λ', λ'' are as near as possible to λ_1 and $\lambda_2 + \pi p/M$, respectively.

We break (5) into two halves, of which one is for $|n| < mM$ and the second for $|n| \geq mM$. (They contain almost equally as many terms since $N = 2mM$.) Since

$$|L'_N(u,v,n)| = \phi_N(u,n+v,n)\left|\frac{\sin\dfrac{m(v+n)\pi}{N}\sin\dfrac{mn\pi}{N}}{m^2\sin\dfrac{(n+v)\pi}{N}\sin\dfrac{n\pi}{N}}\right|$$

and $\phi_N(u,n+v,n) < \max\{(1-|n|/N), (1-|n+v|/N)\}$, then, using Schwartz's inequality for the part of the sum over the second half of the range for $n \geq mM$ and putting $N - n = u$ (or $N - n - v = u$), we obtain a bound for that part of the sum (5) as

$$\frac{1}{M}\sum_{u=0}^{mM}\left(\frac{\sin um\pi/N}{m\sin u\pi/N}\right)^2\frac{u}{N}.$$

Choose a so that

$$\frac{1}{M}\sum_{u=[aM-1]}^{mM}\left\{\frac{\sin(um\pi/N)}{m\sin u\pi/N}\right\}^2\frac{u}{N}\leq\frac{1}{M}\sum_{u=[aM-1]}^{mM}\left\{\frac{\sin(um\pi/N)}{m\sin(u\pi/N)}\right\}^2<\epsilon.$$

This is possible because the sum approximates the tail of the integral of an integrable positive continuous function. However, the part of the sum for $u < aM$ is also asymptotically negligible, for this is

$$\frac{1}{N}\left\{\frac{1}{M}\sum_{u=0}^{[aM]}\left\{\frac{\sin(um\pi/N)}{m\sin(u\pi/N)}\right\}^2\frac{u}{M}\right\}$$

and the bracketed expression approximates to an integral in the same way. The same is true for $n \leq -mM$ so that we are left with the first half of the sum (5). The proof that this converges to zero for $(\lambda_1 - \lambda_2) \neq 0$, mod 2π, uniformly for $(\lambda_1 - \lambda_2) > \epsilon$, and otherwise uniformly in $\lambda_1 - \lambda_2 = \lambda$, to

$$\begin{aligned}\int_{-\infty}^{\infty}\frac{\sin^2\frac{1}{2}\pi x}{\frac{1}{2}\pi x^2}e^{ipx\pi}\,dx &= 0, p\geq 1,\\ &= 2, p = 0,\end{aligned}$$

is no different from that given above for the estimators of the form of (4.4). The second and third terms may be treated in the same way and this part of the theorem is established.

As far as the Cooley-Tukey procedure is concerned we observe first that if no fader is used then the only change is in (5), where m is replaced by m' and N by N' so that $m/m' - N/N'$ is as small as possible (consistent with m, N, m', N' being integers and $N \leq N' = 2^s$). The proof of the theorem in this case is not different from that for the FFT. In the case where a fader is used we proceed as follows. We have

$$(6)\quad \nu\operatorname{cov}\left\{f_{jk}^{(c)}(\lambda_1), f_{rs}^{(c)}\left(\lambda_2+\frac{\pi p}{M}\right)\right\} = \nu\frac{1}{4\pi^2 N}\frac{1}{(m')^2}\sum_{\substack{a,b\\=1}}^{m'}\sum\operatorname{cov}\left\{\sum a_N(s)x_j(s)e^{is\omega_a'}\right.$$

$$\left.\cdot\sum a_N(t)x_k(t)e^{-it\omega_a'}, \sum a_N(u)\,x_r\,(u)e^{iu\omega_b'}\sum a_N(v)\,x_s(v)e^{-iv\omega_b'}\right\},$$

where the m' values ω_a', of the form $2\pi\, l/N'$, cluster around λ_1 and similarly for the m' values ω_b' in relation to $\lambda_2 + \pi p/M$. Once more we can express this covariance as a sum of three terms, of which one involves the fourth cumulant function of the $x_j(n)$ and is shown to be asymptotically negligible in the same way as before. We take the term corresponding to the contribution made by taking expectations of the product of the first and fourth sum in the last bracketed expression and expectations of the product of the second and third such sum. This gives a contribution

$$\frac{1}{4\pi^2}\sum_{\substack{u,v\\=N+1}}^{N-1}\sum\gamma_{js}(u)\gamma_{kr}(u+v)e^{i(u-v)(\lambda_2+\pi p/M)}\left[\frac{1}{M}\sum_{w=-N+1}^{N-1}e^{iw(\lambda_1+\lambda_2+\pi p/M)}\right.$$

$$\left.\cdot\left\{e^{i\{w(\lambda_1-\lambda_1')+(v+u-w)(\lambda_2-\lambda_2')\}}\frac{\sin\pi wm'/N'}{m'\sin\pi w/N'}\cdot\frac{\sin\pi(w+v-u)m'/N'}{m'\sin\pi(w+v-u)/N'}\right\}\phi_N'(u,v,w),\right]$$

where

$$\phi'_N(u, v, w) = \frac{1}{N} \sum_s{}' a_N(s)\, a_N(s - w)\, a_N(s + u)\, a_N(v + s - w)$$

and $(\lambda_1 - \lambda_1') < \pi/N'$, $(\lambda_2 - \lambda_2') < \pi/N'$. The sum defining ϕ'_N varies over such s values as keep arguments between 1 and N for each u, v, w. It is easily seen to be dominated by $c(1 - |w|/N)$ and as N increases for each fixed u, v, w it converges to

$$\int_0^1 u^4(x)\, dx.$$

In precisely the same way as before we may show that the contribution to the covariance converges to zero if $\lambda_1 + \lambda_2 \neq 0$, mod 2π. When $\lambda_1 + \lambda_2 = 0$, mod 2π the expression within square brackets converges to

$$\int_{-\infty}^{\infty} k^2(x)\, e^{ip\pi x}\, dx = \int_0^1 u^4(x)\, dx \int_{-\infty}^{\infty} \frac{\sin^2 \frac{1}{2}\pi x}{(\frac{1}{2}\pi x)^2}\, e^{ip\pi x}\, dx,$$

which is null for $p \neq 0$ and is otherwise

$$2\int_0^1 u^4(x)\, dx.$$

The proof of this is again of the same nature as for the FFT. Carrying out the summation over u, v, we obtain, when $\lambda_1 = -\lambda_2 = \lambda$,

$$\int_{-\infty}^{\infty} k^2(x) e^{ip\pi x}\, dx\, f_{js}(\lambda)\, f_{rk}(\lambda).$$

The remaining pairing of the factors in the bracketed expression on the right in (6) gives a contribution

$$\int_{-\infty}^{\infty} k^2(x) e^{ip\pi x}\, dx\, f_{jr}(\lambda)\, f_{sk}(\lambda)$$

if $\lambda_1 = \lambda_2 = \lambda$ and zero otherwise and the theorem is thus established.

2 *Proof of Theorem* 11

We may consider

$$\nu^{1/2}\{\hat{f}(\lambda) - \mathcal{E}(\hat{f}(\lambda))\} \tag{7}$$

because of the condition (5.2) stated in the theorem. We now replace this by

$$\nu^{1/2}\{\hat{f}(\lambda) - \mathcal{E}(\hat{f}(\lambda))\} - Y'_N(\lambda) \tag{8}$$

$$Y'_N(\lambda) = \frac{\nu^{1/2}}{2\pi} \sum_{-N+1}^{N-1} k_n e^{-in\lambda} \left[\sum_{-\infty}^{\infty} A(j) \frac{1}{N} \sum_u{}' \{\epsilon(u)\, \epsilon(u)' - I_p\} A'(j + n) \right],$$

wherein the sum, $\sum'$, is over $1 - j \leq u \leq N - n - j$, for $n \geq 0$, and $n + 1 - j \leq u \leq N - j$, $n \leq 0$. Thus Y'_N contains the "squared terms" in (7). Now if the fourth moment of the $\epsilon(n)$ was finite all elements in this matrix, Y_N', would converge in mean square to zero for *then* the squared norm of Y_N' has expectation not greater than

$$c \cdot (\nu/N)\left\{\sum_{-N+1}^{N-1} |k_n| \sum_{j=-\infty}^{\infty} \|A(j)\|\,\|A(j+n)\|\right\}^2 = O\left(\frac{\nu}{N}\right),$$

where c is a finite constant.† On the other hand, the variance of (8) involves no moments, of the $\epsilon_j(u)$, higher than the second. Thus the covariances of the elements of (8), for different λ values, are the same whether or not $\epsilon(u)$ has finite fourth moments and since when this is so these covariances are asymptotically the same for (7) and (8) these covariances for (8) must asymptotically be those given by Theorem 9.

We next show that Y_N' converges in probability to zero if

$$(\nu^{1/2}/N)\sum{}' \{\epsilon'(u)\epsilon(u) - q\}$$

does so, when we no longer assume that $\epsilon(u)$ has finite fourth moments. We now show that Y_N' may be replaced by Y_N'' which differs from Y_N' only in the replacement of $\sum'$ by a summation over all u, $1 \le u \le N$. Indeed, the expectation of the norm of $Y_N'' - Y_N'$ is not greater than

$$c\frac{\nu^{1/2}}{2\pi N}\sum_{-N+1}^{N-1} |k_n|\left[\sum_{-\infty}^{\infty} \|A(j)\|\,\|A(j+n)\|\,(|n| + 2|j|)\right] \le c\frac{\nu^{1/2}}{N}\sum_{-N+1}^{N-1} |k_n|,$$

since, because of condition (5.3) of the theorem,

$$\sum_{-\infty}^{\infty} \|A(j)\|\,\|A(j+n)\|\,|j| \le c\sum_{-\infty}^{\infty} \|A(j)\| < \infty,\ \sum_{-\infty}^{\infty} \|A(j)\|\,\|A(j+n)\|\,|n|$$
$$\le \sum_{-\infty}^{\infty} \|A(j)\|\,\|A(j+n)\|\,(|j+n| + |j|) < \infty.$$

However,

$$(\nu^{1/2}/N)\sum_{-N+1}^{N-1} |k_n| \to 0;$$

for example, for the estimate based on the finite Fourier transform we have

$$\frac{\nu^{1/2}}{N}\sum_{-N+1}^{N-1} |k_n| = \left(\frac{M}{N}\right)^{1/2}\frac{1}{M}\sum_{-N+1}^{N-1}\left|\frac{\sin nm\pi/N}{m\sin n\pi/N}\right| = O\left(\frac{\log \nu}{\nu^{1/2}}\right).$$

Now, in turn, we may replace Y_N'' with

$$Y_N = \frac{\nu^{1/2}}{2\pi}\left(\sum_{-\infty}^{\infty} A(j)e^{ij\lambda}\right)\left\{\frac{1}{N}\sum_{1}^{N} \epsilon(u)\epsilon(u)' - I_p\right\}\left(\sum_{-\infty}^{\infty} A(j)e^{ij\lambda}\right)^*,$$

† We continue to use c for such a constant but not necessarily always the same one.

for, z being a vector of p components,

$$z^*(Y_N - Y_N'')z = z^* \frac{\nu^{1/2}}{2\pi}\left\{\sum_{-\infty}^{\infty} A(j)e^{ij\lambda}\left(N^{-1}\sum_{1}^{N} \epsilon(u)\epsilon(u)', - I_p\right)\right.$$
$$\left. \times \left(\sum_{-\infty}^{\infty} (1 - k_n)e^{-i(n+j)\lambda} A(n+j)'\right)\right\}z,$$

where we put $k_n = 0$ for $|n| \geq N$. Since $N^{-1}\sum(\epsilon(u)\epsilon(u)' - I_p)$ converges almost surely to the null matrix this is for any $\eta > 0$ almost surely not greater than

$$\eta \frac{\nu^{1/2}}{2\pi} z^* \sum_{-\infty}^{\infty} A(j)e^{ij\lambda}\left\{\sum_{-\infty}^{\infty}(1 - k_n)e^{-i(n+j)\lambda}A(n+j)'\right\}z = \eta z^*[\nu^{1/2}\{\hat{f}(\lambda) - \mathcal{E}(\hat{f}(\lambda))\}]z,$$

which converges to zero because of condition (5.2) of the theorem. Now the sufficiency of the condition that $N^{-1}\sum_1^N(\epsilon(u)\epsilon(u)' - I_p)$ converge in probability to zero for the convergence in probability to zero of Y_N is evident.

We now prove the central limit theorem for (7). In the proof of this theorem we shall use the finiteness of the fourth moments of the $\epsilon_j(u)$ only through the use of the asymptotic formula for the variance of (7). We shall not use condition (5.3) of the theorem. Thus we shall have established what we wish for we have already shown that (8) has the same asymptotic covariance properties as (7) and that, using condition (5.3), Y_N' may be neglected.

We put

$$\begin{aligned} \eta_1(n) &= 0, \qquad & \epsilon(n)'\epsilon(n) \leq A, \\ &= \epsilon(n), \qquad & \epsilon(n)'\epsilon(n) > A, \end{aligned}$$

and $\eta(n) = \eta^{(1)}(n) - \mathcal{E}(\eta^{(1)}(n))$. Then for A sufficiently large

$$\mathcal{E}(\eta^{(1)}(n)'\eta^{(1)}(n)) \leq \epsilon.$$

Now we put

$$x(n) = u(n) + v(n), \qquad v(n) = \sum_{-\infty}^{\infty} A(j)\eta(n-j).$$

Then $u(n)$ and $v(n)$ are both linear processes with zero means and with spectra, respectively,

$$\frac{1}{2\pi} h(\lambda)\, G_1(\epsilon)\, h(\lambda)^*, \qquad \frac{1}{2\pi} h(\lambda)G_2(\epsilon)h(\lambda)^*, \qquad h(\lambda) = \sum_{-\infty}^{\infty} A(j)e^{ij\lambda},$$

where, as $A \to \infty$ so that $\epsilon \to 0$, $G_1(\epsilon) \to G$, $G_2(\epsilon) \to 0$. Consider

$$\begin{aligned} (10) \quad & \nu^{1/2}\{\hat{f}(\lambda) - \mathcal{E}(\hat{f}(\lambda))\} \\ &= \nu^{1/2}\{\hat{f}_u(\lambda) - \mathcal{E}(\hat{f}_u(\lambda))\} + \nu^{1/2}\{\hat{f}_{uv}(\lambda) - \mathcal{E}(\hat{f}_{uv}(\lambda))\} \\ &\quad + \nu^{1/2}\{\hat{f}_{vu}(\lambda) - \mathcal{E}\hat{f}_{vu}(\lambda)\} + \nu^{1/2}\{\hat{f}_v(\lambda) - \mathcal{E}\hat{f}_v(\lambda)\}, \end{aligned}$$

where for our present purposes *only* we use $\hat{f}_{uv}(\lambda)$, for example, to indicate the matrix of cross spectra between an element of the $u(n)$ process and one of the $v(n)$ process.

Now

$$\lim_{N\to\infty} \nu\, \mathcal{E}\{\operatorname{tr}[(\hat{f}_v(\lambda) - \mathcal{E}(\hat{f}_v(\lambda)))(\hat{f}_v(\lambda) - \mathcal{E}(\hat{f}_v(\lambda)))^*]\}$$
$$= \operatorname{tr}\left(f_v(\lambda) f_v(\lambda)^*\right)\int k^2(x)\,dx, \qquad \lambda \neq 0, \pm\pi$$

with this same value doubled for $\lambda = 0, \pm\pi$. This may be made arbitrarily small for ϵ small enough, that is, A large enough. The same is true of the second and third terms on the right side of (10). On the other hand, for ϵ small enough, that is, A large enough, the covariance tensor of the first term on the right in (10) is arbitrarily near to that for the left of (10). Thus by Bernstein's lemma (see the fifth part of the Appendix to Chapter IV) we need only establish the theorem for $u(n)$ replacing $x(n)$ and correspondingly we may assume that all moments of the disturbance, $\eta(n)$, are finite. To avoid excessive notation we now revert to $x(n)$ and $\epsilon(n)$, but assume all moments of the latter finite. In precisely the same way we may restrict ourselves to the situation in which $A(j) = 0$ for $|j| > T$, the part to be neglected now arising from a new $v(n)$ defined as

$$\sum_{|j|>T} A(j)\epsilon(n-j),$$

whose spectrum is

$$\frac{1}{2\pi}\left(\sum_{|j|>T} A(j)e^{ij\lambda}\right)\left(\sum_{|j|>T} A(j)e^{ij\lambda}\right)^*$$

whose norm is not greater than

$$\frac{1}{2\pi}\left\{\sum_{|j|>T} \|A(j)\|\right\}^2,$$

which may be made arbitrarily small.

Thus we have restricted ourselves to a situation in which

$$x(n) = \sum_{|j|\leq T} A(j)\,\epsilon(n-j), \tag{11}$$

where the $\epsilon(n)$ are independent and identically distributed with zero mean and unit covariance matrix and all moments finite.

Now Theorem 1 shows that

$$w(\lambda) = h(\lambda)w(\epsilon, \lambda) + O(N^{-1/2}) \tag{12}$$

where the second term on the right has norm whose expectation is of order $N^{-1/2}$. A repetition of the same type of argument once more shows that the contribution from this term may be neglected; for example, for the FFT we have

$$\nu^{1/2}\{\hat{f}^{(F)}(\lambda) - \mathcal{E}(\hat{f}^{(F)}(\lambda))\} = \nu^{1/2}\left\{\frac{1}{m}\sum w(\omega_j)\, w(\omega_j)^* - \mathcal{E}\left(\hat{f}^{(F)}(\lambda)\right)\right\},$$

where the ω_j are the nearest m to λ, as before. This is

$$= \nu^{1/2}\left\{\frac{1}{m}\sum h(\omega_j)\left(w(\epsilon,\omega_j)\,w(\epsilon,\omega_j)^* - \frac{1}{2\pi}I_p\right)h(\omega_j)^*\right\} + O(M^{-1/2}).$$

Indeed

$$\nu^{1/2}\left\{m^{-1}\frac{1}{2\pi}h(\omega_j)\,Gh(\omega_j)^* - \mathcal{E}(f^{(F)}(\lambda))\right\} = O(\nu^{1/2}N^{-1/2}) = O(M^{-1/2}),$$

since the term within braces is the average of m quantities each of which is of order $N^{-1/2}$, uniformly in j. The term which is $O(N^{-1/2})$ in (12) similarly contributes a component to the whole expression which is $O(M^{-1/2})$. Thus we have reduced ourselves to the consideration of

$$\nu^{1/2}\left\{\frac{1}{m}\sum_j h(\omega_j)\left(w(\epsilon,\omega_j)\,w(\epsilon,\omega_j)^* - \frac{1}{2\pi}I_p\right)h(\omega_j)^*\right\}.$$

The situation is precisely the same for each of the other estimators and we forgo the details. Now finally we are led to consider

$$\nu^{1/2}\,h(\lambda)\{\hat{f}_\epsilon(\lambda) - \mathcal{E}(\hat{f}_\epsilon(\lambda))\}\,h(\lambda)^*. \tag{13}$$

Take, for example, the case of an estimate of the type (4.2) with k_n satisfying (4.4). This leads to an expression of the form

$$\begin{aligned}(14)\quad \nu^{1/2}&\int_{-\pi}^{\pi} h(\theta)\left\{I_\epsilon(\theta) - \frac{1}{2\pi}I_p\right\}h(\theta)^*\,K_N(\lambda-\theta)\,d\theta \\ &= \nu^{1/2}\left[h(\lambda)\int_{-\pi}^{\pi}\left\{I_\epsilon(\theta) - \frac{1}{2\pi}I_p\right\}K_N(\lambda-\theta)h(\lambda)^*\,d\theta\right. \\ &\qquad + \int_{-\pi}^{\pi}\{h(\theta)-h(\lambda)\}\left\{I_\epsilon(\theta) - \frac{1}{2\pi}I_p\right\}h(\theta)^* \\ &\times K_N(\lambda-\theta)\,d\theta + \int_{-\pi}^{\pi}h(\theta)\left\{I_\epsilon(\theta) - \frac{1}{2\pi}I_p\right\}\{h(\theta)-h(\lambda)\}^*\,K_N(\lambda-\theta)\,d\theta \\ &\left. - \int_{-\pi}^{\pi}\{h(\theta)-h(\lambda)\}\left\{I_\epsilon(\theta) - \frac{1}{2\pi}I_p\right\}\{h(\theta)-h(\lambda)\}^*\,K_N(\lambda-\theta)\,d\theta\right].\end{aligned}$$

The last three terms evidently give a contribution converging in probability to zero since, for example, the second gives a matrix whose components are cross spectra evaluated at λ between two processes, both of the simple type at present under consideration, but one of which has null spectrum at λ. Thus we finally have only to consider (13), that is,

$$\nu^{1/2}\{\hat{f}_\epsilon(\lambda) - \mathcal{E}(\hat{f}_\epsilon(\lambda))\},$$

for which we must establish the theorem. The simplest way to do this appears to be to show that the cumulants higher than the second vanish as $N \to \infty$ and the first and second moments converge to those stated in the theorem (for $x(n)$ replaced by $\epsilon(n)$). This will establish the theorem. As far as the first and second moments this has already been done in Section 4 [since $\epsilon(n)$ is just a special case of $x(n)$]. It will be sufficient to establish the convergence to zero of a typical tth order cumulant $t > 2$. For this it is sufficient to examine the cumulants of the components of $I_\epsilon(\lambda)$ and this in turn requires only the examination, for $t \geq 6$, of the even cumulants of the components of $w(\epsilon, \lambda)$. These are obtained from

$$\log \mathcal{E}\left\{\exp i \sum_1^\tau \theta_u w_u(\epsilon,\lambda)\right\} = \sum_{n=1}^N \log \mathcal{E}\left\{\exp i \left(\sum_u \theta_u \frac{1}{\sqrt{2\pi N}} e^{i\lambda_u n} \epsilon_u(n)\right)\right\},$$

wherein u varies over τ integers between 1 and p,

$$= \sum_{n=1}^N \psi\left(\theta_1 e^{i\lambda_1 n} \frac{1}{\sqrt{2\pi N}}, \ldots, \theta_t e^{i\lambda_t n} \frac{1}{\sqrt{2\pi N}}\right),$$

where ψ is the joint cumulant generating function of the $\epsilon_u(n)$, $n = 1, \ldots, \tau$. The coefficient of

$$(-)^t \frac{\theta_1^{t_1}\theta_2^{t_2}\cdots\theta_\tau^{t_\tau}}{t_1!\, t_2! \cdots t_\tau!}, \qquad \sum t_j = t,$$

which is the required cumulant, is

$$\sum_{n=1}^N k_{u_1,\ldots,u_\tau}(t_1, \ldots, t_\tau) e^{i(\Sigma t_j\lambda_j)n}\left(\frac{1}{2\pi N}\right)^{t/2} \leq c\, N^{1-t/2}.$$

Since $t \geq 6$ this is at most $O(N^{-2})$. This means that any tth cumulant between the components of $I_\epsilon(\lambda)$ is at most of order (N^{-2}), $t \geq 3$. A trivial checking now shows that the sufficiency is established.

We now turn to the proof of necessity. Let us put $Z_N = \nu^{1/2}\{\hat{f} - \mathcal{E}(\hat{f})\}$ and $X_N = Z_N - Y_N$ so that $Z_N = X_N + Y_N$. Let $\eta(n)$ be defined as above and put $\tilde{X}_N$ for X_N constructed, using $u(n)$ in place of $x(n)$ and $\tilde{Y}_N = Z_N - \tilde{X}_N$ so that once again $Z_N = \tilde{X}_N + \tilde{Y}_N$ and $\tilde{Y}_N = Y_N - (\tilde{X}_N - X_N)$. Now let v be an arbitrary vector of complex numbers and put $z_N = \tilde{x}_N + \tilde{y}_N = x_N + y_N$, where $x_N = v^* X_N v$, $y_N = v^* Y_N v$ and so on. We know that $\tilde{x}_N$ has a distribution converging to normality with variance which may be made to differ by arbitrarily little from that obtained from Theorem 9 by taking A large enouth. Moreover, $\mathcal{E}((\tilde{x}_N - x_N)^2) < \epsilon$, for any $\epsilon > 0$, for A large enough. We now assume that the distribution of z_N also converges to a normal distribution. Let $\tilde{F}_N$ be the joint distribution of $\tilde{x}_N$, $\tilde{y}_N$ and let us consider a subsequence of values of N along which $\tilde{F}_N$ converges, in the usual sense of (weak) convergence of distributions to $\tilde{F}$. Then $\tilde{F}$ is a proper distribution, since the distributions of z_N and $\tilde{x}_N$ converge for any subsequence. Then, putting

$$\sigma_z^2 = \iint z^2\, d\tilde{F}(x, y), \qquad \tilde{\sigma}_x^2 = \iint x^2\, d\tilde{F}(x, y),$$

we have

$$\sigma_z^2 = \tilde{\sigma}_x^2 + \eta,$$

where η may be made arbitrarily small. Now $\mathcal{E}(\tilde{x}_N y_N) \equiv 0$ for expressing these as quadratic forms in, respectively, the elements of $\epsilon(n) - \eta(n)$ and $\epsilon(n)$ we see that the only products which occur are of the type

$$(\epsilon_i(m) - \eta_i(m))(\epsilon_j(n) - \eta_j(n))(\epsilon_k(p) - 1)(\epsilon_l(p) - 1), \qquad m \neq n,$$

so that either m or n is different from p and since $\epsilon_i(m) - \eta_i(m)$ has zero expectation then $\mathcal{E}(\tilde{x}_N y_N) = 0$. Thus

$$|\mathcal{E}(\tilde{x}_N \tilde{y}_N)| = |\mathcal{E}(\tilde{x}_N(\tilde{x}_N - x_N))|,$$

and this may be made arbitrarily small for N large enough by taking A sufficiently large. Also

$$\mathcal{E}(|\tilde{x}_N \tilde{y}_N|) \leq c_1\, \mathcal{E}(|\tilde{y}_N|) < c,$$

where c_1 depends upon A. Thus we also have, along the sequence N_j for which $\tilde{F}_N$ converges

$$\int xy\, d\tilde{F}(xy) = \lim_{j\to\infty} \mathcal{E}(\tilde{x}_{N_j}\tilde{y}_{N_j}),$$

and this is arbitrarily small if A is large enough. It follows that

$$\tilde{\sigma}_y^{\,2} = \int y^2\, d\tilde{F}(x, y) = \sigma_z^{\,2} - \tilde{\sigma}_x^{\,2} - 2\int xy\, d\tilde{F}(xy)$$

must be arbitrarily small for A large enough. We have already observed that

$$P(|\tilde{x}_N - x_N| > \epsilon)$$

may be made arbitrarily small for A and N large enough. Thus also

$$P(|y_{N_j}| > \epsilon)$$

may be made arbitrarily small for A and j large enough. Thus, since y_{N_j} does not depend on A, it must converge to zero in probability and since this is true for any sequence for which $\tilde{F}_N$ converges the convergence in probability to zero of y_N is established. Also, Y_N converges in probability to zero and so does

$$u(\lambda) = \frac{\nu^{1/2}}{2\pi} \operatorname{tr}\left\{\left(\sum_{-\infty}^{\infty} A(j)e^{ij\lambda}\right)\left(\frac{1}{N}\sum_{1}^{N} \epsilon(u)\,\epsilon(u)' - I_p\right)\left(\sum_{-\infty}^{\infty} A(j)e^{-ij\lambda}\right)^*\right\}.$$

Thus so must

$$\int_{-\pi}^{\pi} u(\lambda)\, d\lambda = \nu^{1/2} \operatorname{tr}\left\{\left(\sum_{-\infty}^{\infty} A(j)^* A(j)\right)\left(\frac{1}{N}\sum_{1}^{N} \epsilon(u)\,\epsilon(u)' - I_p\right)\right\}.$$

We may assume that

$$\sum_{-\infty}^{\infty} A(j)^* A(j)$$

is nonsingular and we see that the condition of the theorem is established as necessary.

CHAPTER VI

Inference for Rational Spectra

1. INTRODUCTION

In the history of the development of time series analysis by statisticians time domain analysis comes first, the name of G. U. Yule occupying a premier place. By time domain analysis, in this connection, is meant the analysis of autoregressive models, the terminology deriving from the fact that the sample autocorrelations are used, and not the sample spectra, for inference purposes. In general it refers to the use of the initial data and not the Fourier transformed data. By 1943 the asymptotic theory for autoregressive systems was substantially tied up in the paper by Mann and Wald (1943). After the war these systems were studied for low order autoregressions, in a more precise fashion on Gaussian assumptions. From the early fifties the "glamour" of the Fourier methods attracted workers away from the time domain analysis and certainly, in some applications, when N is very large, spectral methods occupy a central place. In others, for example in economics, this is not necessarily so for if the amount of data available is small there is a great reward to be obtained from a successful investment in a "finite parameter" model so that such an investment is nearly certain to be made. A great deal depends, of course, upon the end use of the analysis. For some predictive purposes in economics the autoregressive schemes have a great appeal, and the same is true of some problems of control. On the other hand the study of the gust load on an aircraft, for example, leads inevitably to a Fourier decomposition of the record. There seems to have been a return to the use of finite parameter models in recent years, particularly in connection with problems of prediction and control, to which, as we have said, they are well adapted. By a finite parameter model we substantially mean a mixed autoregressive and moving average model and it is on this basis that the present chapter is founded. Not too much emphasis should, however, be placed upon the "time domain" aspect for there is no intrinsic reason why Fourier methods should not be used for these models, and indeed they shall prove useful in our treatment.

2. INFERENCE FOR AUTOREGRESSIVE MODELS. ASYMPTOTIC THEORY

We consider the vector autoregressive model

$$\sum_0^q B(j)\, x(n-j) = \epsilon(n), \qquad B(0) = I_p, \tag{2.1}$$

where the $\epsilon(n)$ are i.i.d $(0, G)$ and the $B(j)$ are square matrices with all zeros of

$$\det\left\{\sum_0^q B(j)z^j\right\}$$

lying outside of the unit circle. Then we know from Theorem 4′ of Chapter I that a stationary solution of this system is of the form

$$x(n) = \sum_0^\infty A(j)\epsilon(n-j), \qquad A(0) = I_p, \tag{2.2}$$

wherein the $\|A(j)\|$ converge exponentially to zero as j increases. Theorem 4′of Chapter I also shows that any other sequence satisfying (2.1) differs from this stationary solution only by the addition of a sequence of terms of the form

$$\xi(n) = \sum_u \sum_i \sum_j c(i, j, u) z_u^n n^j b(i, u), \qquad |z_u| < 1,$$

where the $b(i, u)$ are fixed vectors and the $c(i, j, u)$ are random variables determined so that the solution satisfies q initial conditions. Thus $a^n \xi(n)$ converges almost surely to zero for some $a > 1$. It will easily be seen that this ensures that the inclusion of $\xi(n)$ affects no part of Theorem 1 below. We therefore proceed, in the remainder of the discussion, to assume $x(n)$ stationary.

This model as it stands is not quite acceptable since it implies that $x(n)$ has zero mean. If $x(n)$ has constant mean, μ, then certainly

$$\sum_0^q B(j)\mu = \mathcal{E}(\epsilon(n)),$$

so that if $x(n)$ is replaced by $x(n) - \mu$ then (2.1) holds for the new process with $\{\epsilon(n) - \mathcal{E}(\epsilon(n))\}$ on the right, and in general, if $\mu(n)$ is the mean of $x(n)$ the relation

$$\sum_0^q B(j)\mu(n-j) = \mathcal{E}(\epsilon(n))$$

is implied by (2.1). We prefer to leave our discussion of such problems until Chapter VII where we deal with regression problems in general. However, because it will always be necessary at least to make mean corrections we shall in this chapter discuss estimation procedures on the assumption that the mean of $x(n)$ is constant but not necessarily zero. It is natural to estimate μ by $\bar{x}$. In relation to the estimation of the remaining parameters we introduce the relation

$$\mathcal{E}\left\{\sum_0^q B(j)x(n-j)\big(x(m)-\mu\big)'\right\} = \mathcal{E}\{\epsilon(n)\big(x(m)-\mu\big)'\}, \qquad m \leqslant n,$$

which leads to

$$\sum_0^q B(u)\Gamma(u-v) = \delta_0^{\,v}G, \qquad v \geq 0. \tag{2.3}$$

It is natural to use these, for $v = 0, 1, \ldots, q$, to estimate the $B(j)$and G, by inserting $C(n)$ (see Chapter IV, Section 3)† in place of $\Gamma(n)$.

An alternative approach would be to assume that the $\epsilon(n)$ are Gaussian and to use the method of maximum likelihood. This does not lead to the estimates just discussed. The situation is illustrated by the case $p = q = 1$, when the likelihood becomes

$$\frac{1}{(2\pi\sigma^2)^{\frac{1}{2}(N-1)}} \exp\left[-\frac{1}{2\sigma^2}\sum_2^N \{x(n) - \rho x(n-1) - \mu(1-\rho)\}^2\right]$$
$$\times\ [2\pi\{\sigma^2(1-\rho^2)\}^{-1}]^{-\frac{1}{2}} \exp\left[-\frac{1}{2\sigma^2}(x(1)-\mu)^2(1-\rho^2)\right]$$

where we have, as is customary, replaced $B(1)$ by $-\rho$ and G by σ^2, in this simple case. The equations of maximum likelihood are

$$(1-\hat{\rho})(\bar{x}-\hat{\mu}) + \frac{\hat{\rho}}{N}\left\{x(N) + x(1) - 2\hat{\mu}\right\} = 0,$$

$$\sum_2^N \{\big(x(n)-\mu\big)\big(x(n-1)-\hat{\mu}\big)\} - \hat{\rho}\sum_1^N \big(x(n)-\hat{\mu}\big)^2 \tag{2.4}$$
$$+\ \hat{\rho}\{\big(x(1)-\hat{\mu}\big)^2 + \big(x(N)-\hat{\mu}\big)^2 - \hat{\sigma}/(1-\hat{\rho}^2)\} = 0,$$

$$\hat{\sigma}^2 = \frac{1}{N}\left\{\sum_2^N \{\big(x(n)-\hat{\mu}\big) - \hat{\rho}\big(x(n-1)-\hat{\mu}\big)\}^2 + \big(x(1)-\hat{\mu}\big)^2(1-\hat{\rho}^2)\right\}.$$

These estimates differ from those derived from (2.3) by quantities of order N^{-1}. The magnitude of the deviation also depends on ρ as is evident by considering, for example,

$$\hat{\mu} = \frac{\bar{x} + \hat{\theta}\frac{1}{2}\{x(1) + x(N)\}}{1+\hat{\theta}}, \qquad \hat{\theta} = \frac{2\hat{\rho}}{N(1-\hat{\rho})}.$$

† See the footnote on p. 208. We are here computing the $C(n)$ after mean correction.

The adjustment to $\bar{x}$ will be small unless N is small or $\hat{\rho}$ is near to 1. It is not difficult to write down the equations of maximum likelihood in any case and when N is not large it might sometimes be wise to use these equations for estimation. However, we shall discuss this matter no further here but proceed to the properties of the solutions of (2.3). Later we shall discuss, precisely, under Gaussian assumptions, the distributional properties of estimators close to those obtained from (2.3) or (2.4).

We need to introduce a tensor notation once more. It is possibly preferable to do away entirely with the vector-matrix notation heretofore used and to write (2.3), for $v > 0$, as, for example,

$$\sum_{u=0}^{q} \beta_i^{\,j}(u)\gamma_j^{\,k}(u-v) = 0, \qquad u, v = 1, \ldots, q;\ i, j, k = 1, \ldots, p,$$

where we have put $\beta_{i,j}(u) = \beta_i^{\,j}(u)$, $\gamma_{j,k}(u-v) = \gamma_j^{\,k}(u-v)$ and the usual tensor summation convention has been used. (It is natural to regard $\beta_{i,j}(u)$ as covariant in its first subscript and contravariant in its second.) However, for statisticians and probabilists more used to a vector-matrix notation we preserve this symbolism and rewrite (2.3) instead as†

$$\{I_p \otimes \Gamma_q\}\beta_q = -\gamma_q, \tag{2.5}$$

wherein Γ_q has q^2 blocks, that in the (u, v)th row and column of blocks being $\Gamma(u-v)$, while β_q has $\beta_{i,j}(u)$ in row $(i-1)pq + (u-1)p + j$, whereas γ_q has $\gamma_{i,k}(v)$ in row $(i-1)pq + (v-1)p + k$. Here again $u, v = 1, \ldots, q$; $i, j, k = 1, \ldots, p$. Thus we order the elements in dictionary order, first by row, then by lag and finally by column. We could, of course, have defined β_q by ordering the elements in dictionary order according to lag, row, and then column. By defining γ_q by ordering the elements $\gamma_{ik}(v)$ in this way also, the only change in (2.5) would then have been that $(I_p \otimes \Gamma_q)$ would be changed to $\Gamma_q \otimes I_p$. We make this observation now, for later we shall have cause to alter the ordering.

The matrix $I_p \otimes \Gamma_q$ is assumed to be invertible. This is equivalent to saying that Γ_q is invertible. If Γ_q is singular, then

$$\mathcal{E}\{(\sum_i \sum_u \alpha_i(u)\, x_i(n-u))^2\} \equiv 0$$

for all n and some set of $\alpha_i(u)$ not identically zero; but

$$\sum_i \sum_u \alpha_i(u)\, x_i(n-u) = \sum_u \alpha'(u)\, x(n-u)$$

† The subscripts q are inserted because they are needed later in connection with partial autocorrelation, for example.

in an obvious notation. This implies that

$$\left(\sum_u \alpha(u)e^{iu\lambda}\right)^* f(\lambda)\left(\sum_u \alpha(u)e^{iu\lambda}\right) = 0 \qquad \text{a.e.}$$

However, $f(\lambda)$ has nonnull determinant a.e. and the last displayed relation is impossible.

Our estimation equations become

$$\{I_p \otimes C_q\}\hat{\beta}_q = -c_q, \tag{2.6}$$

where $c_{ij}(n)$ replaces $\gamma_{ij}(n)$ to obtain C_q from Γ_q, c_q from γ_q and, of course, $\hat{\beta}_q$ estimates β_q. Since

$$G = \sum_0^q B(u)\,\Gamma(u), \tag{2.7}$$

we estimate G by

$$\hat{G}_q = \sum_0^q \hat{B}(u)\,C(u). \tag{2.8}$$

Evidently C_q will be nonsingular with probability 1 and we thus have

$$\hat{\beta}_q = -\{I_p \otimes C_q^{-1}\}c_q,$$

which converges with probability one to β_q, since all elements of C_q and c_q so converge to the elements of Γ_q and γ_q as we saw in Theorem 6, Chapter IV. Correspondingly, $\hat{G}$ converges with probability 1 to G. Thus we have established the first part of Theorem 1:

Theorem 1. *If $x(n)$ is generated by* (2.1), *where the $\epsilon(n)$ and $B(j)$ are as stated below that equation, Γ_q, β_q, C_q are defined at and below* (2.5), *$\hat{\beta}_q$ by* (2.6), *and $\hat{G}_q$ by* (2.8), *then $\hat{\beta}_q$, C_q, and $\hat{G}_q$ converge almost surely to β_q, Γ_q, and G, respectively. Moreover, $\sqrt{N}(\hat{\beta}_q - \beta_q)$ has a distribution which converges as $N \to \infty$ to that of a normally distributed vector of random variables with zero mean and covariance matrix $(G \otimes \Gamma_q)^{-1}$.*

Proof. We first consider the case where $\mu = 0$ and no mean correction is made. To this end we write

$$\begin{aligned} c_{ij}(u - v) &= N^{-1}\sum_{m=1}^{N}\{x_i(m - u)x_j(m - v)\} + d_{ij}(u, v) \\ &= \hat{c}_{ij}(u, v) + d_{ij}(u, v), \end{aligned}$$

let us say. Then, for $u - v = n$,

$$d_{ij}(u, v) = \frac{n}{N}\,c_{ij}(n) + N^{-1}\sum_m{}' \pm\{x_i(m)\,x_j(m + n)\},$$

where $\sum'\pm$ accounts for terms omitted from the summation defining $c_{ij}(n)$ but in $\hat{c}_{ij}(n)$, or the converse. It is easily seen that the last term is of the form $N^{-1}(y + y_N)$ where the y_N are a strictly stationary sequence (and come from the terms in $c_{ij}(n)$, but omitted from $\hat{c}_{ij}$, for which $m > N - u$) and y is a fixed random variable. Thus $N^{1/2} d_{ij}(u, v)$ converges in probability to zero since $nN^{-1/2}c_{ij}(n) + N^{-1/2}y$ converges almost surely to zero and $N^{-1/2}y_N$ converges in probability to zero, by Markov's inequality. We now insert in place of $\Gamma(u - v)$ in Γ_q the matrix with entries $\hat{c}_{ij}(u, v)$. We call the resulting matric $\hat{C}_q$. Similarly, we define $\hat{c}_q$ (to replace c_q), using $\hat{c}_{ij}(u, v)$ in place of $c_{ij}(u - v)$. If b_q satisfies

$$(I_p \otimes \hat{C}_q)b_q = -\hat{c}_q,$$

then

$$\sqrt{N}(I_p \otimes \hat{C}_q)(b_q - \hat{\beta}_q) = \sqrt{N}(c_q - \hat{c}_q) + \sqrt{N}(I_p \otimes D)\hat{\beta}_q,$$

where D is defined in terms of the $d_{ij}(u, v)$ as was $\hat{C}_q$ in terms of the $\hat{c}_{ij}(u, v)$. The right-hand side evidently converges to zero in probability so that, since $\hat{C}_q$ is nonsingular with probability 1, $\sqrt{N}(b_q - \hat{\beta}_q)$ converges to zero in probability and we may thus consider b_q in place of $\hat{\beta}$ in studying the limiting distribution. Now

$$\sqrt{N}(I_p \otimes \hat{C}_q)\beta_q$$

is the tensor form of

$$\frac{1}{\sqrt{N}} \sum_1^q B(u) \sum_1^N x(n - u)x'(n - v),$$

$$= \frac{1}{\sqrt{N}} \sum_1^N \epsilon(n)x'(n - v) - \frac{1}{\sqrt{N}} \sum_1^N x(n)x'(n - v), \qquad v = 1, \ldots, q,$$

so that

$$\sqrt{N}(I_p \otimes \hat{C}_q)(b_q - \beta_q) = -e,$$

where e has an entry in row $(i - 1)pq + (v - 1)p + j$ the quantity

$$e_{ij}(v) = \frac{1}{\sqrt{N}} \sum_{n=1}^N \epsilon_i(n)x_j(n - v).$$

Thus

$$\mathcal{E}(e) = 0, \qquad \mathcal{E}(ee') = (G \otimes \Gamma_q),$$

since

$$\mathcal{E}\left\{\frac{1}{\sqrt{N}} \sum_1^N \epsilon_i(m)x_j(m - u) \frac{1}{\sqrt{N}} \sum_1^N \epsilon_k(n)x_l(n - v)\right\} = \sigma_{ik}\gamma_{jl}(u - v),$$

where σ_{ik} is the typical element of G. If e can be shown to be asymptotically normal, then, since $(I_p \otimes \hat{C}_q)$ converges with probability 1 to $I_p \otimes \Gamma_q$, we shall have established that $\sqrt{N}(b_q - \beta_q)$ is asymptotically normal with covariance matrix $(G \otimes \Gamma_q^{-1})$. *Of course, if it is assumed that the* $\epsilon_j(n)$ *have finite fourth moment, the result follows from Chapter IV, Theorem* 14, since $\sqrt{N}c_{ij}(n)$ *in that theorem* can be identified with $e_{ij}(v)$ and $\gamma_{ij}(n)$ *in that theorem* with zero. However, a check through the proof of Theorem 14, Chapter IV, shows that now asymptotic normality holds without the existence of moments higher than the second. Indeed,

$$e_{ij}(v) = \frac{1}{\sqrt{N}} \sum_{l=1}^{p} \sum_{m=0}^{\infty} \alpha_{j,l}(m) \sum_{n=1}^{N} \epsilon_i(n)\, \epsilon_j(n - v - m),$$

where the $\alpha_{j,l}(m)$ decrease to zero exponentially, and are the entries in the matrices $A(m)$ in the representation

$$x(n) = \sum_{0}^{\infty} A(m)\epsilon(n - m).$$

Thus $e_{ij}(v)$ contains no "squared" terms (i.e., no terms of the form $\epsilon_i(n)\epsilon_j(n)$). Now a check through the proof of Theorem 14 (see the Appendix to Chapter IV) shows that the proof goes through also in the present case. We have only, again, to take $x_i(n)$ (in $c_{ij}(n)$ in Theorem 14) as $\epsilon_i(n)$ and $x_j(n)$ as our present $x_j(n - v)$. After truncation of the infinite sum defining $x(n - v)$ at K and the replacement of $x_j(n - v)$ by the truncated sum then $\epsilon_i(n)x_j(n - v)$ satisfies the conditions of Theorem 10′, Chapter IV, being strictly stationary with continuous spectral density. The proof that the effect of truncation may be neglected is of the same form as in Theorem 14, Chapter IV, though simpler since no fourth cumulant terms occur because of the absence of squared terms and only the expansion of $x_j(n - v)$ in terms of the $\epsilon(n)$ needs to be considered, since the other factor is $\epsilon_i(n)$.

If μ is not known to be zero we shall replace $x_j(n)$ by $x_j(n) - \bar{x}_j$. Then μ may be assumed to be zero. Now, temporarily, call $\bar{c}_{ij}(n)$ the quantity $c_{ij}(n)$ when it is computed using the $x_j(n) - \bar{x}_j$. (We use this notation only for the moment and shall subsequently revert to the notation $c_{ij}(n)$ for the quantities computed after mean correction.) Then

$$c_{ij}(n) - \bar{c}_{ij}(n) = \{\bar{x}_i\bar{x}_j(n) + \bar{x}_i(-n)\bar{x}_j - \bar{x}_i\bar{x}_j\},$$

where, for $n \geq 0$, for example,

$$\bar{x}_j(n) = (N - n)^{-1} \sum_{m=1}^{N} x_j(m), \qquad \bar{x}_i(-n) = (N - n)^{-1} \sum_{m=1}^{N-n} x_i(m).$$

Now evidently $c_{ij}(n) - \bar{c}_{ij}(n)$ converges almost surely to zero. Also

$$\sqrt{N}\big(c_{ij}(n) - \bar{c}_{ij}(n)\big)$$

converges in probability to zero since, for example, $\sqrt{N}\bar{x}_i$ is asymptotically normal and $\bar{x}_j(n)$ converges almost surely to zero. This shows that the theorem continues to hold after mean corrections are made and completes the proof of the theorem.

It is not necessarily true that the method of estimation for the $B(j)$ gives us estimates providing an *estimate* of $g(z) = \sum B(u)z^u$ having all of its zeros outside of the unit circle. Indeed in the simplest case, $p = q = 1$, the estimate is

$$\hat{g}(z) = 1 + r(1)z$$

where when, for example, no mean corrections are made,

$$r(1) = \frac{N}{N-1}\left(\frac{\sum_1^{N-1} x(n)\, x(n+1)}{\sum_1^N x(n)^2}\right),$$

and $r(1)$ will be greater than unity if, say, $x(n) = 1$, $n = 1, \ldots, N-1$, $x(N) = \frac{1}{2}$, $N > 3$. However, if we use the quantities $(1 - |n|/N)c_{ij}(n)$ in place of the $c_{ij}(n)$ it will be true that $\hat{g}(z) = \sum \hat{B}(u)z^u$ has determinant with all of its zeros outside of the unit circle. We do not give the proof here but for the case $p = 1$ it follows from exercise 7 to this chapter. (It is conceivable that a zero on the unit circle can occur but this can only be so if the periodogram is zero for $N - q$ of the ω_j, $j = 0, \ldots, N-1$.)

We could also investigate the asymptotic distribution of $\hat{G}$. This will depend upon the fourth moments of the $x(n)$ and though $\sqrt{N}(\hat{G} - G)$ will be asymptotically normal the covariance tensor of this limiting distribution will be complicated and of little use. We therefore leave this aside and go on to discuss some special cases.

(i) *We consider first the case* $p = 1$. Then our formula become

$$\text{(2.9)}\quad \begin{cases} \hat{\beta}_q = -C_q^{-1}c_q, \qquad \hat{\sigma}_q^{\,2} = c(0) - c_q' C_q^{-1} c_q = \dfrac{\det\{C_{q+1}\}}{\det\{C_q\}}, \\ \lim_{N\to\infty} N\mathcal{E}\{(\hat{\beta}_q - \beta_q)(\hat{\beta}_q - \beta_q)\}' = \sigma^2\Gamma_q^{-1}. \end{cases}$$

By C_q we mean the square matrix of q rows with $c(i-j)$ in row i column j. By c_q we mean the column vector of q rows with $c(j)$ in the jth row. We leave the proof of the second last equality as an exercise for this chapter. Correspondingly, of course,

$$\beta_q = -\Gamma_q^{-1}\gamma_q, \qquad \sigma^2 = \gamma(0) - \gamma_q'\Gamma_q^{-1}\gamma_q = \frac{\det\{\Gamma_{q+1}\}}{\det\{\Gamma_q\}}.$$

The analogy with classical regression theory is almost complete, if we consider the problem as one of regressing the vector composed of $x(q+1)$,

$x(q+2), \ldots, x(N)$ on q other vectors (after eliminating the mean), of which a typical one, the jth, is composed of $x(q+1-j), \ldots, x(N-j)$. The only alteration is that in forming the mean corrected cross product, for example between the two vectors just described, we have also used terms of the type $x(n)\,x(n-j)$ for $j+1 \leq n < q+1$. As a consequence the calculation of C_q requires that only q covariances be computed instead of $q(q+1)/2$ as would be the case in a classical regression model. Otherwise, the estimation procedure and the asymptotic distribution of the estimates is precisely the same as would obtain in a classical regression model. It follows that all inference procedures used in classical regression are asymptotically valid here since these depend only upon the distribution of the vector $\hat{\beta}_q$. Of course, the computations are always simplified in accordance with the principle just discussed. Apart from the simplification due to the reduced number of covariances to be calculated the matrix C_q is easily inverted. Indeed

$$C_q^{-1} = \begin{bmatrix} C_{q-1}^{-1} + \hat{\sigma}_{q-1}^{-2} P \hat{\beta}_{q-1} \hat{\beta}_{q-1}' P' & -P \hat{\beta}_{q-1} \hat{\sigma}_{q-1}^{-2} \\ -\hat{\beta}_{q-1}' P' \hat{\sigma}_{q-1}^{-2} & \hat{\sigma}_{q-1}^{-2} \end{bmatrix},$$

where P is the matrix which reverses the order of the elements of a vector (i.e., P has units in the diagonal going from the top right to the bottom left corner and zeros elsewhere). We leave the establishment of this formula as an exercise to this chapter. If an autoregression of order $q-1$ has already been fitted all of the constituents of this inverse are immediately available and the determination of $\hat{\beta}_q$ is a simple matter. We proceed to describe some of these procedures in more detail. Confidence region procedures are so obvious that we omit a description of them. (The procedures, of course, involve the replacement of σ^2 and Γ_q by $\hat{\sigma}^2$ and C_q to form the estimate of the covariance matrix of $\hat{\beta}_q$.)

(a) Estimation of the Spectrum

We are not so likely to wish to estimate the spectrum when we have postulated an autoregressive model but there is some interest in the problem if only because the autoregressive model may have been postulated in order to provide a method of spectral estimation. The spectral estimator will be

$$\hat{f}(\lambda) = \frac{\hat{\sigma}^2}{2\pi} \left| \sum_0^q \hat{\beta}(u) e^{iu\lambda} \right|^{-2}, \qquad \hat{\beta}(0) \equiv 1.$$

Let us consider $\sqrt{N}(\hat{f} - f)/f$, dropping the λ argument for convenience. Put $h(\lambda) = \sum \beta(u) e^{iu\lambda}$, $\hat{h}(\lambda) = \sum \hat{\beta}(u) e^{iu\lambda}$. Then $\sqrt{N}(\hat{f} - f)/f$ differs from $\sqrt{N}\{|\hat{h}|^{-2} - |h|^{-2}\}/|h|^{-2}$ by a quantity which converges to zero in probability since $\hat{\sigma}^2$ converges to σ^2 with probability 1. But

$$\sqrt{N}\,\frac{\{|\hat{h}|^{-2} - |h|^{-2}\}}{|h|^{-2}} = \sqrt{N}\left\{\frac{h - \hat{h}}{h\,|\hat{h}|^2}\,|h|^2 + \frac{\overline{(h - \hat{h})}}{\bar{\hat{h}}}\right\}$$

and this, again, differs by a quantity converging to zero in probability from

$$2\sqrt{N}\mathcal{R}\left\{\frac{h - \hat{h}}{h}\right\}$$

since $(\hat{h}/h)$ converges to unity with probability 1. This is evidently asymptotically normal with zero mean and variance

$$2f(\lambda)\left\{2\pi \sum_{1}^{q}\sum e^{i(u-v)\lambda}\gamma^{u,v}\right\} + 2\sigma^2\mathcal{R}\left\{\frac{\sum\sum_1^q e^{i(u+v)\lambda}\gamma^{u,v}}{\sum\sum_0^q e^{i(u+v)\lambda}\beta_u\beta_v}\right\}$$

where the $\gamma^{u,v}$ are the elements of the matrix inverse to Γ_q. This is a relatively complicated expression reflecting, it appears, the use of a method of estimation not basically designed for spectral estimation.

It is of some interest to evaluate this last expression when the true order of the autoregression, q_0 let us say, is small compared to the assumed order q. To this end we are led to consider the limit as q increases (q_0 being held fixed) of q^{-1} by the last displayed expression. We shall not go into details here but shall refer the reader to exercise 7 of Chapter VII. The techniques there used show that this limit is 2 when $\lambda \neq 0, \pi$ and 4 when $\lambda = 0, \pi$. Thus for q large relative to q_0 we are led to use $\sqrt{(N/q)}(\hat{f} - f)$ as normal with zero mean and variance $2f^2$. Thus q is to be, roughly, identified with the M of our earlier investigations, which is hardly a surprising result. It is not a result which encourages us to use an autoregressive procedure for spectral estimation for, equating q to M, the computing effort involved when M is large seems much greater for the autoregressive than for the procedures earlier discussed.

(b) Order of Autoregression

A more important problem in the present type of application is that of determining the order of the autoregression. A basic step in this is the examination of $\hat{\beta}(q)$ (in an autoregression of order q). We see that

$$\hat{\beta}(q) = -\sum_{1}^{q} \frac{c(j)\,|C_q^{(i,q)}|\,(-)^{j+q}}{\det(C_q)},$$

where we now find it notationally simpler to write $|C_q^{(i,q)}|$ for the *minor* of the element in row i, column q of C_q, rather than $\det(C_q^{(i,q)})$ as we earlier have done.

Then $\hat{\beta}_q$ is easily seen to be $(-)^{q+1}\,|C_q^{(q+1,1)}|/\det(C_q)$ and in accordance with (4.2) in Chapter I we are led to put

$$-\hat{\beta}(q) = \hat{\rho}(q \mid 1, \ldots, q-1),$$

where the right-hand expression is the estimated autocorrelation of $x(n)$, with $x(n-q)$ eliminating $x(n-1), \ldots, x(n-q+1)$ by regression. Now $\sqrt{N}(\hat{\beta}(q) - \beta(q))$ has variance $\sigma^2 \det(\Gamma_{q-1})/\det(\Gamma_q)$. On the null hypothesis that the process is autoregressive of order $q-1$ this is unity. Thus $\sqrt{N}\hat{\beta}(q)$ may be treated as $N(0, 1)$ to test the hypothesis that the order of the autoregression is $(q-1)$. It would be more in line with what is customary in classical least squares analysis to use

$$\frac{\sqrt{N}\hat{\rho}(q \mid 1, \ldots, q-1)}{\sqrt{1 - \hat{\rho}^2(q \mid 1, \ldots, q-1)}},$$

which clearly has the same limiting distribution. This procedure is almost certainly preferable, as we shall later see when we investigate the exact sampling distribution of $\hat{\rho}$ (for small q) but it can also be seen as follows. If we use $\sqrt{N}\hat{\rho}(q \mid 1, \ldots, q-1)$, we are using an estimate (namely, 1) of its variance on the null hypothesis. However, if the null hypothesis is not true, the variance of $\hat{\beta}(q) = -\hat{\rho}(q \mid 1, \ldots, q-1)$ is smaller for it is asymptotically $N^{-1}\sigma^2 \det(\Gamma_{q-1})/\det(\Gamma_q) = \{N^{-1}\det(\Gamma_{q+1})\det(\Gamma_{q-1})\}/\{\det(\Gamma_q)\}^2$. However, we know Theorem 6, Chapter I, that

$$1 - \rho^2(q \mid 1, \ldots, q-1) = \frac{\{\gamma(0)(1 - R_q^2)\}}{\{\gamma(0)(1 - R_{q-1}^2)\}}$$

and from (4.3) in Chapter I

$$\gamma(0)(1 - R_q^2) = \gamma(0) - \gamma_q'\Gamma_q^{-1}\gamma_q = \frac{\det(\Gamma_{q+1})}{\det(\Gamma_q)}$$

and thus

$$\lim_{N\to\infty} N\{\operatorname{var}\hat{\beta}(q)\} = 1 - \rho^2(q \mid 1, \ldots, q-1) = (1 - \beta^2(q)) < 1.$$

In fact, as we shall see later, it may be preferable to use

$$\frac{\sqrt{(N-q)}\,\hat{\rho}(q \mid 1, \ldots, q-1)}{\sqrt{1 - \hat{\rho}^2(q \mid 1, \ldots, q-1)}} \tag{2.10}$$

as Student's t with $(N-q)$ degrees of freedom. We shall later suggest some other approximations. Even if the null hypothesis is true the introduction of the factor in the denominator will tend to increase the value of the test statistic and this is related to the use of Student's t. However, the power of the test will be increased because on the alternative hypothesis the denominator may be substantially reduced.

It is of interest to write out the formula for $\hat{\rho}(q \mid 1, \ldots, q-1)$ for small q. We have

$$q = 1, \qquad \hat{\rho}(1 \mid 0) = r(1)$$

$$q = 2, \qquad \hat{\rho}(2 \mid 1) = |C_3^{(3,1)}|/\det(C_2) = \frac{r(2) - r(1)^2}{1 - r(1)^2}$$

$$q = 3, \qquad \hat{\rho}(3 \mid 1, 2) = \frac{\begin{vmatrix} r(1) & 1 & r(1) \\ r(2) & r(1) & 1 \\ r(3) & r(2) & r(1) \end{vmatrix}}{\begin{vmatrix} 1 & r(1) & r(2) \\ r(1) & 1 & r(1) \\ r(2) & r(1) & 1 \end{vmatrix}}.$$

(c) H_0: Autoregression of Order q
H_1: Autogregression of Order $q + s$

Under (b) we considered $s = 1$. We have in mind here a circumstance where we want a general test of the appropriateness of the model of a qth order autoregression, that is, a "goodness of fit" test. In this circumstance s is likely to be large, probably much larger than q. In classical regression theory we consider $q + s$ "regressor" variates, $x_1, x_2, \ldots, x_q, \ldots, x_{q+s}$ and to test whether the last s add to the explanation of the variation in the dependent variate we form

$$1 - R^2 = \frac{\text{residual sum of squares from regression on } x_1, x_2, \ldots, x_{q+s}}{\text{residual sum of squares from regression on } x_1, x_2, \ldots, x_q}.$$

The analogous statistic in our case is $\hat{\sigma}_{q+s}^2/\hat{\sigma}_q^{\,2}$ so that we define

$$1 - \hat{R}^2(q + s \mid q) = \frac{\hat{\sigma}_{q+s}^2}{\hat{\sigma}_q^{\,2}} = \frac{\{\det(C_{q+s+1})\det(C_q)\}}{\{\det(C_{q+s})\det(C_{q+1})\}}.$$

This is scale free and can be equivalently expressed in terms of the $r(n)$. By analogy with the classical situation we are led to form

$$(2.11) \qquad F_{s,N-q-s} = \frac{\hat{R}^2(q + s \mid q)}{1 - \hat{R}^2(q + s \mid q)} \frac{N - q - s}{s}.$$

We shall leave it to the reader to check that as $N \to \infty$ $N\hat{R}^2(q + s \mid q)$ has a distribution converging to that of chi-square with s d.f. so that

$$1 - \hat{R}^2(q + s \mid q)$$

converges to unity with probability 1. This can be done by expressing $\hat{R}^2(q + s \mid q)$ as a quadratic form in $\hat{\beta}(q + 1), \ldots, \hat{\beta}(q + s)$ with matrix which converges with probability one to the inverse of the covariance matrix of these elements, which establishes the result. This, asymptotically, validates (2.11) but it seems probable, and this will be supported by our later investigations, that the recommended treatment of the statistic as $F_{s,N-q-s}$ is preferable.

An alternative method of testing the hypothesis here considered was put forward in Quenouille (1947) whose connection with the present test was pointed out in Whittle (1951). Since (see Chapter I, Theorem 6)

$$\{1 - R^2(q + s \mid q)\} = \prod_1^s \{1 - \rho^2(q + j \mid 1, 2, \ldots, q + j - 1)\}$$

we shall have approximately, when H_0 is true,†

$$N\hat{R}^2(q + s \mid q) \approx N[1 - \prod_1^s \{1 - \hat{\rho}^2(q + j \mid 1, 2, \ldots, q)\}]$$

$$\approx N \sum_1^s \hat{\rho}^2(q + j \mid 1, 2, \ldots, q),$$

where we have neglected cross product terms which, on H_0, will make a contribution converging to zero in probability. This is effectively Quenouille's statistic, which is, however, computed as follows. We form

$$a_u = \sum_j \hat{\beta}(j)\, \hat{\beta}(u - j), \qquad u = 0, 1, \ldots, 2q$$

where the sum is over such values of $0 \le j \le q$, as make $(u - j)$ also lie in that range. Then we form

$$d_j = (N + q - j)^{1/2} \frac{\sum_0^{2q} a_u r(j - u)}{\sum_0^{2q} a_u r(u)}, \qquad j = q + 1, \ldots, q + s.$$

Then on H_0 as $N \to \infty$

$$\sum_{j=q+1}^{q+s} d_j^2$$

† By $\hat{\rho}(q + j \mid 1, 2, \ldots, q)$ in the following expression we mean the partial autocorrelation between $x(n)$ and $x(n - q - j)$ removing the linear effects of q of the intermediate variates from each of these, these q being the $x(m)$ for the q values of m nearest to each of $x(n)$, $x(n - q - j)$.

is distributed as chi-square with s d.f. We shall not give a detailed justification of the test here.† The interested reader may refer to Grenander and Rosenblatt (1957, p. 106) or Hannan (1960, p. 96). The test is easy to carry out once the qth order autoregression has been formed since we have only to compute the a_u and use these to form a moving average of the $r(n)$. A change in s requires only few additional calculations. Today, however, the calculations required for (2.11) are in any case fairly minor and there seems to be no very good reason why Quenouille's test should be used in place of that founded on (2.11).‡

(ii) *The case where $p > 1$.* Now $\hat{\beta}$ can be seen to correspond to the matrix

$$\hat{B} = \begin{bmatrix} \hat{B}'(1) \\ \hat{B}'(2) \\ \cdot \\ \cdot \\ \cdot \\ \hat{B}'(q) \end{bmatrix} \tag{2.12}$$

as the two quantities called $\hat{\beta}$ and $\hat{B}$ corresponded in Chapter IV, Section 3. (See (3.10) and text immediately following.) The estimate is analogous to that which one would obtain if one estimated B (defined in the same way in terms of the $B(j)$) by least squares regression of each of the sequences $x_j(n)$, $n = 1, 2, \ldots, N$ on the pq sequences $x_j(n-u)$. There is no exact correspondence, for the same type of reason as was discussed in relation to the case $p = 1$, that cross products unused in the least squares regression will in practice be used so that the sum of squares and cross products for the pq vectors regressed upon will have all elements the same down any diagonal. Once again the whole of classical regression theory goes through in a suitably modified and simplified fashion, as a consequence of Theorem 1. One should observe also that the solution of (2.5) is not so formidable a task as it might at first appear, for in fact we may consider each row of the relation

$$x(n) = -\sum_1^q B(u)\, x(n-u) + \epsilon(n)$$

separately, since the set of coefficients for the ith row is, when arranged as a

† A discussion of a generalization of it, for the case $p > 1$, is given in Section 6, which covers the case $p = 1$ also. A careful examination suggests that the power of the test may be lower than that of the test given by (2.11) since the estimate of σ^2 based on the residuals from the qth order autoregression may be too high if the null hypothesis is not true.

‡ This to some extent contradicts the opinion given by this author in an earlier work, (1960) p. 97. The computational situation has changed a lot.

column vector, first according to u and then according to column number,

$$-\Gamma_q^{-1}\gamma_i,$$

where γ_i has $\gamma_{ik}(v)$ in row $\{(v-1)p+k\}$. Correspondingly, the column of estimates is

$$-C_q^{-1}c_i,$$

where c_i has $c_{ik}(v)$ in row $\big((v-1)p+k\big)$. In other words, we have regressed the vector of $N-q$ values $x_i(n)$, $n = q+1, \ldots, N$ on to the pq vectors composed of the $x_j(n-u)$ for $n = q+1, \ldots, N$ obtained by taking $j = 1, \ldots, p$, $u = 1, \ldots, q$, but have used all relevant cross products so as to simplify the form of regression.

The carrying out of standard tests and confidence interval procedures now follows well-established lines. We illustrate first by a test for the hypothesis that all $B(u)$ are null, $u = 1, \ldots, q$. This is to be a test of the null hypothesis that $x(n)$ is serially independent against that of an autoregression of order up to q. A natural statistic to use is

$$\Lambda(N, p, pq) = \frac{\det\{C(0) - \hat{B}'C_q\hat{B}\}}{\det\{C(0)\}},$$

where $\hat{B}$ was defined by (2.12) above. If we were concerned with a classical regression situation of p vectors of N observations on pq vectors, with the p vectors Gaussian and independent over time, then the corresponding statistic Λ under the null hypothesis of no association would have a distribution whose properties are well known and indeed which is fairly well tabulated.† For details we refer the reader to Anderson (1958). A good approximation is obtained by treating

$$-\{N - \tfrac{1}{2}(p+pq)\}\log\Lambda \tag{2.13}$$

as chi-square with p^2q d.f. We shall justify the use of this treatment here, in a sense to be made precise.

Let us note that

$$C(0) - \hat{B}'C_q\hat{B} = \hat{G}_q,$$

since

$$\sum_1^q \hat{B}(u)\,C(u) = \hat{B}'\begin{bmatrix} C(1) \\ C(2) \\ \cdot \\ \cdot \\ \cdot \\ C(q) \end{bmatrix} = -\hat{B}'C_q\hat{B}.$$

† The considerations of the remainder of this section are intimately related to those of Section 6 Chapter V.

Thus

$$\begin{aligned}\Lambda(N, p, pq) &= \frac{\det(\hat{G}_q)}{\det\{\hat{G}_q + (C(0) - \hat{G}_q)\}} \\ &= \frac{\det(\hat{G}_q)}{\det\{\hat{G}_q + \hat{B}'C_q\hat{B}\}} \\ &= \det\{I_p + \hat{G}_q^{-1/2}\hat{B}'C_q\hat{B}\hat{G}_q^{-1/2}\}^{-1}.\end{aligned}$$

Now

$$\begin{aligned}N\{(I_p + \hat{G}_q^{-1/2}\hat{B}'C_q\hat{B}\hat{G}_q^{-1/2}) &- (I_p + G^{-1/2}\hat{B}'\Gamma_q\hat{B}G^{-1/2})\} \\ = N\{(\hat{G}_q^{-1/2} - G^{-1/2})\hat{B}'C_q\hat{B}G^{-1/2} &+ G^{-1/2}\hat{B}'(C_q - \Gamma_q)\hat{B}\hat{G}_q^{-1/2} \\ &+ G^{-1/2}\hat{B}'\Gamma_q\hat{B}(\hat{G}_q^{-1/2} - G^{-1/2})\}\end{aligned}$$

and this converges to the null matrix in probability, on the hypothesis that $B = 0$, since $N^{1/2}\hat{B}$ is asymptotically normal and, for example, $\hat{G}_q^{-1/2} - G^{-1/2}$ converges to zero with probability 1; but

$$N \operatorname{tr}\{G^{-1/2}\hat{B}'\Gamma_q\hat{B}G^{-1/2}\} = N\hat{\beta}'(G \otimes \Gamma_q^{-1})^{-1}\hat{\beta},$$

which is asymptotically distributed as chi-square with p^2q d.f. However,

$$-N \ln \Lambda(N, p, pq) = N \ln \Pi(1 + r_i^2),$$

where the r_i^2 are the eigenvalues of $\hat{G}^{-1/2}\hat{B}'C_q\hat{B}\hat{G}^{-1/2}$, which evidently converge almost surely to zero. Since we have seen that $N\sum r_i^2$ is chi-square with p^2q d.f. and

$$N\sum r_i^2 < N\sum \ln(1 + r_i^2) < N\sum r_i^2 + N\sum r_i^2 \ln(1 + r_i^2),$$

then the middle term is also distributed as chi-square with p^2q d.f. for N large since $N\sum r_i^2 \ln(1 + r_i^2)$ converges in probability to zero because each $\ln(1 + r_i^2)$ converges almost surely to zero. One may conjecture that the use of (2.13) as indicated below that expression will be preferable to the use of $N\sum r_i^2$ but it must be admitted that this is founded only upon an examination (which we give below) of the cases $p = 1$ and q small.

A more relevant test is that for $B(q) = 0$. Here we form

$$\Lambda(N - (q - 1)p, p, p) = \frac{\det(\hat{G}_q)}{\det(\hat{G}_{q-1})}. \tag{2.14}$$

The proof that, as $N \to \infty$, the distribution of

$$-\{N - p(q - 1)\} \ln \Lambda$$

becomes that of chi-square with p^2 d.f. is of precisely the same nature as that just given and is omitted.

More generally one may take

$$\Lambda(N - q_1p, p, q_2p) = \frac{\det(\hat{G}_q)}{\det(\hat{G}_{q_1})}$$

and treat

$$-\{N - q_1p - \tfrac{1}{2}(p + q_2p)\} \ln \Lambda, \qquad q_1 + q_2 = q,$$

as chi-square with p^2q_2 d.f. to test the null hypothesis that $x(n)$ is generated by an autoregression of order q_1 against the alternative that it is of order $q > q_1$. These procedures seem to have been very little used (but procedures closely related to them have been widely used in economics).

All of the results of this section are unsatisfactory since they are asymptotic and no indication of the sample size needed for a good approximation is given. As we have said, some more exact results on Gaussian assumptions will be got below and we shall also then discuss some sampling experiments (and theoretical results) with non-Gaussian data which suggest that the approximations suggested here will be quite accurate provided the effective number of observations, which is the multiplier of $-\log \Lambda$, is reasonably large, say greater than 50 and in case p and q are small tolerable accuracy may hold for much smaller values (down to 20, say). Of course, these results are somewhat tentative.

We have not explicitly discussed confidence region procedures for the elements of B since these procedures are fairly obvious but at the same time difficult to interpret because of high dimensionalities involved. The procedures which may be used are those of the classical theory of multiple systems of regression (e.g., see Anderson, 1958, pp. 210–211), modified to take account of the simpler nature of the situation which now obtains, as has already been indicated many times. We illustrate by a confidence region for $B(q)$.

If in forming $\det(\hat{G}_q)$ in (2.14) we replace $\hat{B}(q)$ by $\{\hat{B}(q) - B(q)\}$ then the distribution of the resulting statistic, Λ_q, let us say, will be as stated, for Λ, for ultimately that distribution reduces to that of N by the quadratic form in the elements of $\{\hat{B}(q) - B(q)\}$ with the inverse of the (estimated) covariance matrix of $\hat{B}(q)$ as matrix of the quadratic form. Thus having determined $\chi(\alpha)$ so that

$$P\{-(N - p(q-1)) \log \Lambda_q < \chi(\alpha)\} = 1 - \alpha,$$

the region defined by the inequality within the braces will define a confidence region of coefficient $(1 - \alpha)$ for $B(q)$.

3. INFERENCE FOR AUTOREGRESSIVE MODELS. SOME EXACT THEORY

In this section we survey some exact results, *obtained on Gaussian assumptions*, concerning the distributions of those coefficients of autocorrelation and partial autocorrelation from which inferences about autoregressive models are made. We consider, in the main, the case $p = 1$, for it is for it that most of the available results have been obtained. Ultimately it is only a matter of energy and the use of computers that is needed to obtain any of the distributions to any required accuracy. In fact, relatively few tabulations have been made and, indeed, most of the work that we discuss below was done more than 10 years ago.† This partly reflects the changed emphasis, since then, because of the importance of Fourier methods, but partly it reflects the fact that exact results are not truly exact since the Gaussian assumptions do not hold true so that beyond a certain point a very accurate determination of the relevant distributions is probably not worth while. As we have said, there has been some return to finite parameter models and perhaps this will later be associated with a further development of the distribution theory—particularly in association with the case $p > 1$.

We first consider the case $p = 1$ and open the discussion of that case by considering the distribution of $r(1)$ *on the assumption that the* $x(n)$ *are N.I.D.* (μ, σ^2). In fact the first derivation of an exact distribution of a statistic of this nature was made by von Neumann (1941) who considered

$$\frac{(N-1)^{-1}\sum_1^{N-1}\{x(n+1)-x(n)\}^2}{N^{-1}\sum_1^N\{x(n)-\bar{x}\}^2} = \frac{\delta^2}{s^2}.$$

(In this section, since N is fixed, we do not need any special notation to indicate the number of observations from which $\bar{x}$ is computed.) This is sometimes called "von Neumann's ratio." It is, of course, close to $2(1 - r(1))$ the difference being due only to different weight being given to the end terms. As may be seen from the likelihood for $q = 1$ given in Section 2 the sufficient statistics for the simple Markoff case are

$$\sum_2^{N-1} x(n)^2, \quad \sum_1^{N-1} x(n+1)x(n), \quad x(1)^2 + x(N)^2, \quad \sum_2^{N-1} x(n), \quad x(1) + x(N),$$

so that inevitably there will be no unique choice of a statistic, even for testing $p = 0$ in such a simple Markoff model. (For a discussion of this

† We mention in particular, in addition to other references given below, Daniels (1956), Dixon (1944), Jenkins (1954a), Koopmans (1942), Leipnik (1947), Madow (1945), and Rubin (1945).

point in greater detail see T. W. Anderson, 1948.) Slightly later R. L. Anderson (1942) considered the "circular" autocorrelation

$$\hat{r}(1) = \frac{\sum_1^{N-1}\{(x(n) - \bar{x})(x(n+1) - \bar{x})\} + \{x(N) - \bar{x}\}\{x(1) - \bar{x}\}}{\sum_1^N \{x(n) - \bar{x}\}^2}.$$

This is of the form

$$\frac{x'PWPx}{x'Px}, \qquad P = I - N^{-1}\tilde{1}\tilde{1}', \tag{3.1}$$

where I is the N rowed unit matrix and $\tilde{1}$ the vector of N components, all unity, while W is the circulant having $\frac{1}{2}$ in the two diagonals above and below the main diagonal and in the top right-hand and bottom left-hand corners and zeros elsewhere. If U is the basic circulant

$$\begin{bmatrix} 0 & 1 & 0 & 0 & \cdots & 0 \\ 0 & 0 & 1 & 0 & \cdots & 0 \\ 0 & 0 & 0 & 1 & \cdots & 0 \\ \cdot & & & & & \\ \cdot & & & & & \\ \cdot & & & & & \\ 0 & 0 & 0 & 0 & \cdots & 1 \\ 1 & 0 & 0 & 0 & \cdots & 0 \end{bmatrix},$$

then $W = \frac{1}{2}(U + U') = \frac{1}{2}(U + U^{-1})$. In general we may consider $r = \{x'PAPx/x'Px\} = u/v$, where A is nonsingular and symmetric and P is idempotent and symmetric, since all of the three statistics mentioned are of this type. The first relevant result is

Theorem 2. *If* $r = x'PAPx/x'Px = u/v$ *where* x *is a vector, having* $x(n)$ *in the* nth *place, of independent Gaussian random variables with zero mean and unit variance,* A *is symmetric and* P *is symmetric and idempotent then* r *is distributed independently of* $x'Px$. *In particular* $\mathcal{E}(r^m) = \mathcal{E}(u^m)/\mathcal{E}(v^m)$.

This observation appears to be due to E. J. G. Pitman. The proof is very easy for we may, by an orthogonal transformation, reduce r to the form $\sum \mu_j\xi_j^2/\sum \xi_j^2$, wherein the μ_j are the eigenvalues of PAP and the number of terms in each sum is the rank of P. By introducing polar coordinates, it is evident to us that this ratio depends only on the polar angles and its denominator is the length of the vector with components ξ_j that is distributed independently of the polar angles. In particular we have

$$\mathcal{E}\{r^m v^m\} = \mathcal{E}(u^m) = \mathcal{E}(r^m)\,\mathcal{E}(v^m),$$

that is,

$$\mathcal{E}(r^m) = \frac{\mathcal{E}(u^m)}{\mathcal{E}(v^m)}.$$

We illustrate by $r(1)$. Now P is as in (3.1) and

$$A = \frac{N}{N-1}\frac{1}{2}\begin{bmatrix} 0 & 1 & 0 & \cdot & \cdot & \cdot & 0 \\ 1 & 0 & 1 & \cdot & \cdot & \cdot & 0 \\ 0 & 1 & 0 & \cdot & \cdot & \cdot & 0 \\ \cdot & \cdot & \cdot & \cdot & \cdot & \cdot & \cdot \\ 0 & 0 & 0 & \cdot & \cdot & 0 & 1 \\ 0 & 0 & 0 & \cdot & \cdot & 1 & 0 \end{bmatrix}.$$

Thus

$$\mathcal{E}(r(1)) = \frac{\operatorname{tr}(PAP)}{\operatorname{tr}(P)} = \frac{\operatorname{tr}(PA)}{(N-1)} = \frac{\operatorname{tr}(A) - N^{-1}\tilde{1}'A\tilde{1}}{N-1}$$

$$= -\frac{1}{2(N-1)^2}2(N-1) = -\frac{1}{N-1}.$$

$$\mathcal{E}(r(1)^2) = \frac{2\operatorname{tr}\{(PAP)^2\} + \operatorname{tr}\{(PAP)\}^2}{N^2-1} = \frac{N^2-3N+3}{(N-1)^3}.$$

Thus

$$\operatorname{var}\{r(1)\} = \frac{(N-2)^2}{(N-1)^3} = (N-1)^{-1}\left\{1 - \frac{1}{(N-1)}\right\}^2$$

$$= \frac{1}{N+1} - \frac{3(N-1)}{\{(N+1)(N-2)\}^2} + O(N^{-5}).$$

Thus $\{r(1) + 1/(N-1)\}$ has the variance, to order N^{-2}, of an ordinary correlation, with mean corrections, from $N+2$ pairs of observations. The eigenvalues of A are $\{N/(N-1)\}\cos(\pi j/(N+1)\}$, $j = 1, 2, \ldots, N$. This result we leave as an exercise. These are not the eigenvalues of PAP, of course, but the result indicates that though the limits within which $r(1)$ lies are not ± 1, as for an ordinary correlation, yet these limits must be near to the true limits for N not very small. Thus the use of $r(1) + (N-1)^{-1}$ as an ordinary correlation from $N+2$ pairs of observations should yield a good approximation, if N is not very small. We shall discuss such approximations below. The situation is simpler for $A = W$ and $P = I - N^{-1}\tilde{1}\tilde{1}'$, for now 1 is an eigenvector of W for eigenvalue unity, the remaining eigenvalues being $\cos(2\pi j/N)$, $j = 1, \ldots, N-1$. Thus $\tilde{r}(1)$ varies between $\cos 2\pi/N$ and -1 (N even) or $-\cos \pi/N$ (N odd). In the same way we may find its mean and variance as, respectively, $-(N-1)^{-1}$ and

$$(N^2 - 3N)/\{(N+1)(n-1)^2\} = (N+2)^{-1} + O(N^{-3}).$$

This suggests the use of $\tilde{r}(1) + (N + 2)^{-1}$ as an ordinary correlation from $(N + 3)$ pairs of observations. There is no real need to do this since significance points of the distribution are tabulated in R. L. Anderson (1942) for small N and we mention it here only to provide evidence for approximate procedures suggested below. In fact, a comparison of the approximate and actual significance points show that the approximation is very good for $N \geq 20$. The same is true for von Neumann's ratio, whose mean is 2 and whose standard deviation is $4(N - 2)/(N^2 - 1)$. If we put $r = (1 - \frac{1}{2}\delta^2/s^2)$ we are led to use this as a correlation from $(N + 3)$ pairs of observations, since its variance is $(N + 2)^{-1} + O(N^{-3})$. Again significance points for the exact distribution are tabulated and we are merely offering evidence for approximate procedures. In this case we show Table 1.

Table 1 COMPARISON OF ACTUAL AND APPROXIMATE SIGNIFICANCE POINTS FOR $r = (1 - \frac{1}{2}\delta^2/s^2)$
$\{x \mid P(r \geq x) = p\}$

N	$p = 0.05$		$p = 0.01$	
	Actual	Approximate	Actual	Approximate
4	0.610	0.669	0.687	0.833
5	0.590	0.622	0.731	0.789
6	0.555	0.582	0.719	0.750
7	0.532	0.549	0.693	0.716
8	0.509	0.521	0.669	0.685
9	0.488	0.497	0.646	0.658
10	0.469	0.476	0.624	0.634
15	0.397	0.400	0.539	0.543
20	0.350	0.352	0.480	0.482

Again the approximation is clearly quite adequate for $N \geq 20$ (and indeed it is difficult to imagine circumstances that would make it not satisfactory for $N \geq 10$). For most purposes it is adequate to use $r(1) + (N - 1)^{-1}$ as a correlation from $(N + 2)$ pairs of observations.

We now derive the distribution of $\tilde{r}(1)$, still assuming independence for the $x(n)$. Indeed, we consider the joint distribution of the $\tilde{r}(j), j = 1, \ldots, q$, where

$$\tilde{r}(j) = \frac{x'PW_jPx}{x'Px}, \qquad W_j = \tfrac{1}{2}(U^j + U^{-j}).$$

The reason for this choice of the $\tilde{r}(j)$ is found in the fact that the W_j commute and are also near to matrices which occur in the numerator for

$r(j)$. Moreover, for a range of important cases P will project onto the orthogonal complement of the space spanned by certain of the eigenvectors of all W_j. Indeed these eigenvectors are the vectors with cos $(2\pi kl/N)$ or sin $(2\pi kl/N)$ in the lth place, the eigenvalue being cos $(2\pi kj/N)$ (for W_j) for both. Thus as k goes from 0 to $[\frac{1}{2}N]$ we get N vectors which correspond to the vectors on which we regress to fit a finite Fourier series to N points. Thus if we fit a Fourier series up to frequency $2\pi K/N$ then since P projects onto the orthogonal complement of the space spanned by the first $2K + 1$ eigenvectors we shall remove cos $(2\pi jk/N)$, $k = 0, 1, \ldots, K$, from the list of possible eigenvalues of PW_jP and the remaining eigenvalues will be cos $(2\pi jk/N)$, $k = K + 1, \ldots, [N/2]$, with each repeated twice save for $k = \frac{1}{2}N$, when N is even, which occurs only once. Since the matrices PW_jP, $j = 1, \ldots, q$, and P all commute if P is as described, which we assume, they may be simultaneously diagonalized by an orthogonal transformation. We add together the pair of squared normal variables which have cos $(2\pi jk/N)$ as coefficient $(2\pi k/N \neq \pi)$ and since the $\tilde{r}(j)$ are scale and origin free we may assume that the resulting random variable has the density exp $-x$ (is a gamma-1 variate), $x \geq 0$. If N is even we shall have a further variate distributed with density $\pi^{-1/2}\, x^{-1/2}$ exp $-x$ (i.e., a gamma-$\frac{1}{2}$ variate). We thus write

$$\tilde{r}(j) = \frac{\sum_{k=0}^{M} \lambda(k, j) w_k}{\sum_{k=0}^{M} w_k}, \qquad j = 1, \ldots, q, \tag{3.2}$$

where $\lambda(k,j) = \cos\{2\pi(K + k)j/N\}$, $k = 1, \ldots, M$, $\lambda(0,j) = (-)^j$ for N even and zero for N odd, w_k is a gamma-1 variate and w_0 is gamma-$\frac{1}{2}$. For N even $N = 2M + 2K + 2$ while for N odd it is $2M + 2K + 1$. The case where $K > 0$ and $q = 1$ has been treated in detail in Anderson and Anderson (1950) where the 5% and 1% significance points are also tabulated for K corresponding to cases relevant to the estimation of seasonal variation. We now have (Watson (1956)).

Theorem 3. *If the w_k in* (3.2) *are independent gamma-*1 *variates, w_0 is a gamma-$\frac{1}{2}$ variate, independent of the other w_k and $\tilde{r}(j)$ is defined by* (3.2) *then the density function of the $\tilde{r}(j)$, $j = 1, \ldots, q$ is given by* (3.6) *below (for R as described below* (3.6) *and with alterations there described for N even). In particular if $\tilde{r}(j) = x'PW_jPx/x'Px$ where x is as in Theorem* 2, $W_j = \frac{1}{2}(U^j + U^{-j})$ *and P projects onto the orthogonal complement of the space spanned by $2K + 1$ vectors corresponding to a regression of $x(n)$ on* cos $2\pi kn/N$, $k = 0, \ldots, K$, *sin $2\pi kn/N$, $k = 1, \ldots, K$, then* (3.6) *is the joint density function of the $\tilde{r}(j)$ for $\lambda(k,j) =$* cos $\{2\pi(K + k)j/N\}$, *$k = 1, \ldots, M$. $\lambda(0,j) = (-)^j$, M even, $= 0$, M odd where $N = 2M + 2K + 2$, N even, $= 2M + 2K + 1$, N odd.*

Proof. It is convenient to put $s(j) = \tilde{r}(j) - \lambda(0,j) = u_j/v$ and to find first the joint characteristic function of the u_j and v. This is

$$\phi(\theta_1, \ldots, \theta_q, \psi) = \int_0^\infty\int \cdots \int_0^\infty \exp i\left(\sum_1^q \theta_j u_j + \psi v\right) \exp -\sum_1^M w_k$$

$$\left[\sqrt{\frac{1}{\pi}} \exp(-w_0) w_0^{-\frac{1}{2}}\right] \prod_0^M dw_j,$$

the last factor (in square brackets) occurring only when N is even. This is (see Rao, 1965, p. 133)

$$\left[\prod_{k=1}^M \left(1 - i\sum_1^q \theta_j\{\lambda(k,j) - \lambda(0,j)\} - i\psi\right)^{-1}\right](1 - i\psi)^{-\frac{1}{2}},$$

where again we include the last factor only for N even. (We shall not repeatedly make this qualification.) This is

(3.3)

$$(1 - i\psi)^{-(M+\frac{1}{2})} \prod_{k=1}^M \left(1 - i\sum_1^q \theta_j \hat{\lambda}(k,j)\right)^{-1}, \qquad \hat{\lambda}(k,j) = \left(\frac{\lambda(k,j) - \lambda(0,j)}{1 - i\psi}\right).$$

We first observe that the matrix of M rows and $q + 1$ columns

$$[1 \vdots \hat{\lambda}(k,j)]$$

having units in the first column and $\hat{\lambda}(k,j)$ in row k column $(j + 1)$, has all minors of order $(q + 1)$ nonsingular, for all save a finite number of values of ψ. Indeed this will be so if it is true of $[1 \vdots \lambda(k,j) - \lambda(0,j)]$ and thus, equivalently, of $[1 \vdots \lambda(k,j)]$. If the contrary were true for this last, then we could find $(q + 1)$ constants α_j so that

$$\sum_0^q \alpha_j \cos\frac{2\pi k_u j}{N} \equiv 0, \qquad 1 \leq k_u \leq [\tfrac{1}{2}N], \qquad u = 1, \ldots, q + 1,$$

with all k_u distinct. This is, however, impossible for it would imply that the trigonometric polynomial $\sum \alpha_j \cos \theta j$, of degree q, is null for $q + 1$ values $2\pi k_u/N$ in $(0, \pi]$, which is impossible unless $\alpha_j \equiv 0$. Thus no more than q factors of the M in the product in (3.3) may be simultaneously zero (except for special values of ψ) and we may expand the product in a series of partial fractions, taking all possible sets of q from the product, namely,

$$\prod_{k=1}^M \{1 - i\sum \theta_j \hat{\lambda}(k,j)\}^{-1} = \sum_{k_1 < k_2 < \cdots < k_q} \cdots \sum \frac{c(k_1, k_2, \ldots, k_q)}{\prod_{u=1}^q \{1 - i\sum_{j=1}^q \hat{\lambda}(k_u, j)\theta_j\}}.$$

Multiply both sides by a typical denominator from the right side and set the

θ_j at those values making all factors in this denominator zero, namely,

$$i\bar{\theta} = [\hat{\lambda}(k_u, j)]^{-1}\tilde{1},$$

where $\bar{\theta}$ has the required value in the kth place. Then

$$c(k_1, \ldots, k_q) = \prod_{k \neq k_1, \ldots, k_q} \{1 + \tilde{1}'[\hat{\lambda}(k_u, j)]^{-1}\hat{\lambda}(k)\}^{-1},$$

where $\hat{\lambda}(k)$ has $\hat{\lambda}(k, j)$ in the jth place. This is easily evaluated as†

$$c(k_1, \ldots, k_q) = \frac{|\hat{\lambda}(k_u, j)|^{M-q}}{\prod_{k \neq k_1, \ldots, k_q} \begin{vmatrix} 1 & \hat{\lambda}'(k) \\ \tilde{1} & \hat{\lambda}(k_u, j) \end{vmatrix}}.$$

This is clearly not changed when $\hat{\lambda}(k, j)$ is replaced by $\tilde{\lambda}(k, j) = \lambda(k, j) - \lambda(0, j)$ and thus we get

$$\phi(\theta_1, \ldots, \theta_q, \psi) = \sum_{k_1 < k_2 < \cdots < k_q} \cdots \sum \frac{c(k_1, \ldots, k_q)}{(1 - i\psi)^{M-q+\frac{1}{2}} \prod_{u=1}^{q} \{1 - i \sum_j \tilde{\lambda}(k_u, j)\theta_j - i\psi\}}.$$

We may now invert the Fourier transform, term by term. The typical term in the sum, replacing the numerator by unity, is evidently the joint characteristic function of

$$(3.4) \qquad \begin{cases} u_j = \sum_{u=1}^{q} \hat{\lambda}(k_u, j) y_u, & u = 1, \ldots, q, \\ v = y_0 = \sum_{u=1}^{q} y_u, \end{cases}$$

where the y_u are gamma-1 and v is gamma $(M - q - \frac{1}{2})$, that is, has the density

$$\frac{1}{(M - q - \frac{1}{2})!} e^{-x} x^{M-q-\frac{1}{2}}.$$

(Here and below we shall use the factorial notation for the gamma function to avoid confusion with our notation for serial covariances.) Thus the typical term in the sum, after inversion of the Fourier transform, gives the contribution, obtainable from the density of the y_u, $u = 0, \ldots, q$, through the linear transformation (3.4),

$$\frac{c(k_1, k_2, \ldots, k_q)}{(M - q - \frac{1}{2})!} e^{-v} y_0^{M-q-\frac{1}{2}} |\hat{\lambda}(k_u, j)|^{-1} \operatorname{sgn} |\hat{\lambda}(k_u, j)|,$$

where sgn (x) is the sign of x.

† In this section we find it convenient to use $|a_{j,k}|$ for the determinant of the matrix $a_{j,k}]$.

From (3.4)

$$y_0 = u'[\tilde{\lambda}(k_u, j)]^{-1}\,\tilde{1} + v,$$

where u has u_j in the jth place. This is

$$\frac{\begin{vmatrix} v & u' \\ \tilde{1} & \tilde{\lambda}(k_u, j) \end{vmatrix}}{|\tilde{\lambda}(k_u, j)|}.$$

Thus we obtain the contribution to the joint density from the typical term in the sum as

$$\frac{\operatorname{sgn}|\tilde{\lambda}(k_u, j)|}{|\tilde{\lambda}(k_u, j)|^{1/2}}\,\frac{e^{-v}}{(M-q-\frac{1}{2})!}\,\frac{\begin{vmatrix} v & u' \\ \tilde{1} & \tilde{\lambda}(k_u, j) \end{vmatrix}^{M-q-1/2}}{\prod\limits_{k\neq k_1,k_2,\ldots,k_q}\begin{vmatrix} 1 & \tilde{\lambda}'(k) \\ \tilde{1} & \tilde{\lambda}(k_u, j) \end{vmatrix}}$$

Changing variables to $s(j), j = 1, \ldots, q$ and v in place of u_j and v and integrating over v from 0 to ∞ we obtain the contribution

$$\text{(3.5)}\qquad \frac{(M-\frac{1}{2})!}{(M-q-\frac{1}{2})!}\,\frac{\operatorname{sgn}|\tilde{\lambda}(k_u, j)|}{|\tilde{\lambda}(k_u, j)|^{1/2}}\,\frac{\begin{vmatrix} 1 & s' \\ \tilde{1} & \tilde{\lambda}(k_u, j) \end{vmatrix}^{M-q-1/2}}{\prod\limits_{k\neq k_1,\ldots,k_q}\begin{vmatrix} 1 & \tilde{\lambda}'(k) \\ \tilde{1} & \tilde{\lambda}(k_u, j) \end{vmatrix}},$$

where s has $s(j)$ in the jth place. (The Jacobian of the transformation is v^q and the integral of $\exp(-v)v^{M-1/2}$ is $(M-\frac{1}{2})!$.) Of course we put (3.5) equal to zero when the elements of s lie outside of their range of permissible variation. We may now revert to our original random variables $\tilde{r}(j)$ and our original constants $\lambda(k_u, j)$ to obtain the joint density as

$$\text{(3.6)}\qquad \frac{(M-\frac{1}{2})!}{(M-q-\frac{1}{2})!}\sum_{(k_1,k_2,\ldots,k_q)\in R}\!\!\cdots\!\sum \frac{\begin{vmatrix} 1 & \tilde{r}' \\ \tilde{1} & \lambda(k_u, j) \end{vmatrix}^{M-q-1/2}}{\prod\limits_{k\neq k_1,k_2,\ldots,k_q}\begin{vmatrix} 1 & \lambda(k) \\ \tilde{1} & \lambda(k_u, j) \end{vmatrix}}\cdot\frac{\operatorname{sgn}\begin{vmatrix} 1 & \lambda'(0) \\ \tilde{1} & \lambda(k_u, j) \end{vmatrix}}{\begin{vmatrix} 1 & \lambda(0)' \\ \tilde{1} & \lambda(k_u, j) \end{vmatrix}^{1/2}}.$$

Here $\tilde{r}$ has $\tilde{r}(j)$ as its jth component. The last factor in the denominator occurs only when N is even and the exponent in the first factor in the numerator can then also be decreased by $\frac{1}{2}$ and $(M-\frac{1}{2})!$, $(M-q-\frac{1}{2})!$ changed to $(M-1)!$, $(M-q-1)!$. The region R over which $k_1, k_2, \ldots, k_q$ vary is such as to include the sets $k_1, k_2, \ldots, k_q$ for which the $s(j)$ can be possible values of

$$\frac{\sum \tilde{\lambda}(k_u, j) y_u}{y_0 + \sum y_u}$$

that is, all sets $k_1, k_2, \ldots, k_q$ for which $s(j)$ lie in the least convex polyhedron with vertices $(0, 0, \ldots, 0)$,

$$\big(\tilde{\lambda}(k_1, 1), \ldots, \tilde{\lambda}(k_1, q)\big), \cdots, \big(\tilde{\lambda}(k_q, 1), \ldots, \tilde{\lambda}(k_q, q)\big).$$

This is the same as saying that we include all $(k_1, \ldots, k_q)$ in R for which $\tilde{r}(1), \ldots, \tilde{r}(q)$ lie in the least convex polyhedron with vertices $\big(\lambda(0, 1), \ldots, \lambda(0, q)\big)$, $\big(\lambda(k_1, 1), \ldots, \lambda(k_1, q)\big), \ldots, \big(\lambda(k_q, 1), \ldots, \lambda(k_q, q)\big)$, which defines R. Thus the joint distribution of the $\tilde{r}(j), j = 1, \ldots, q$, has been found and the proof of Theorem 3 is completed.

It is this distribution, for $q = 1$ and $K = 0$, whose significance points are tabulated in R. L. Anderson (1942) and for $q = 1$ and various special K in Anderson and Anderson (1950). As it stands it is of little value except for the special case $q = 1$ for one will hardly want the distribution of $\tilde{r}(2)$, or $\{\tilde{r}(2) - \tilde{r}(1)^2\}/\{1 - \tilde{r}(1)^2\}$ for that matter, on the hypothesis that the $x(n)$ are independent. What one requires for $q > 1$ is the distribution on the basis of an autoregressive model of order q (or $q - 1$) for $x(n)$. Once the assumption of independence is dropped, however, the problem becomes much more difficult for now the matrix in the exponent of the likelihood function for the $x(n)$ will not commute with the W_j or with any other A_j one might reasonably choose (that is, which can reasonably be used to define autocorrelations). This has led authors to introduce the fictitious assumption of a circularly correlated process. Then it is assumed that $x(n) = x(n + N)$ for $n = 0, 1, 2, \ldots, N - 1$ so that the process is really one on the integers reduced modulo N. It is also assumed that $x(n)$ is stationary in the sense that $\mathcal{E}\big(x(m)x(n)\big) = \tilde{\gamma}(n - m)$ where all integers occurring are reduced modulo N. Thus, if $N = 20$ then $\mathcal{E}\big(x(-5)x(54)\big) = \tilde{\gamma}(19)$. Then the matrix $\tilde{\Gamma}_N$ having $\tilde{\gamma}(n - m)$ in row m column n is a circulant; for example, with $N = 5$ we have the matrix

$$\begin{bmatrix} \tilde{\gamma}(0) & \tilde{\gamma}(1) & \tilde{\gamma}(2) & \tilde{\gamma}(3) & \tilde{\gamma}(4) \\ \tilde{\gamma}(4) & \tilde{\gamma}(0) & \tilde{\gamma}(1) & \tilde{\gamma}(2) & \tilde{\gamma}(3) \\ \tilde{\gamma}(3) & \tilde{\gamma}(4) & \tilde{\gamma}(0) & \tilde{\gamma}(1) & \tilde{\gamma}(2) \\ \tilde{\gamma}(2) & \tilde{\gamma}(3) & \tilde{\gamma}(4) & \tilde{\gamma}(0) & \tilde{\gamma}(1) \\ \tilde{\gamma}(1) & \tilde{\gamma}(2) & \tilde{\gamma}(3) & \tilde{\gamma}(4) & \tilde{\gamma}(0) \end{bmatrix}$$

Thus it may be written in the form

$$\tilde{\Gamma}_N = \sum_0^{N-1} \tilde{\gamma}(j) U^j,$$

where U is the basic circulant defined above. Since evidently $\tilde{\gamma}(N-j) = \tilde{\gamma}(j)$ it is relevant now to consider as null hypothesis the case in which

$$\tilde{\Gamma}_N = \sum_{j=0}^{N-1} \tilde{\gamma}(j) W_j.$$

How are we to choose the $\tilde{\gamma}(j)$? We may discover this by considering the inverse of the covariance matrix Γ_N of N observations from an autoregression of order q. We leave it to the reader to prove, as one of the exercises to this chapter, that when $N > q$ this matrix Γ_N has inverse whose element in the jth diagonal, above the main diagonal, has in the kth row (choosing $\sigma^2 = 1$)

$$\begin{aligned} &\sum_u \beta(u)\beta(u+j), \qquad && 0 \le j \le q, \\ &0, && q < j, \end{aligned}$$

where the sum is for u from 0 to the minimum of $k-1, q-j, N-k-j$. The matrix is symmetric, of course, and is also symmetric about its main diagonal. This leads us to introduce

$$\tilde{\Gamma}_N^{-1} = \sum_{-q}^{q} \delta_j U^j, \qquad \delta_j = \sum_0^{q-j} \beta(u)\beta(u+j) = \delta_{-j}. \tag{3.7}$$

Note that $\delta_q = \beta_q$ and if $\beta_q = 0$ then $\delta_{q-1} = \beta_{q-1}$, and so on. Moreover, $\tilde{\Gamma}_N$ is positive definite for the eigenvalues of $\tilde{\Gamma}_N^{-1}$ are

$$\sum_{-q}^{q} \delta_j e^{ij\omega_k} = \left| \sum_0^q \beta(u) e^{iu\omega_k} \right|^2, \qquad \omega_k = \frac{2\pi k}{N}.$$

We shall discuss later the relevance of this approximation. The likelihood function is now

$$(2\pi)^{-\frac{1}{2}N} \, |\tilde{\Gamma}_N|^{-\frac{1}{2}} \exp -\tfrac{1}{2} \sum_0^q \delta_j x' W_j x, \tag{3.8}$$

so that the statistics $x'W_j x$ are sufficient for δ_j, $j = 0, \ldots, q$.

Correspondingly, the likelihood function for the $x'PW_jPx$, $j = 0, \ldots, q$, where P is as before, is

$$(2\pi)^{-\frac{1}{2}N} \, |\hat{\Gamma}_N|^{-\frac{1}{2}} \exp -\tfrac{1}{2} \sum_0^q \delta_j x' P W_j P x,$$

where $\hat{\Gamma}_N$ is the matrix $\tilde{\Gamma}_N$ considered as the matrix of a transformation on the range of the projection operator P. Indeed, Px has covariance matrix $\sum \delta_j P W_j P$ and $P^2 = P$ so that the exponent is correct, while P commutes with

$\tilde{\Gamma}_N$ so that $P\tilde{\Gamma}_N P = \hat{\Gamma}_N$ where this is as described. Thus $\hat{\Gamma}_N$ has the eigenvalues

$$\sum_{-q}^{q} \delta_j e^{ij\omega_k}, \qquad k = K+1, \ldots, K+M,$$

each repeated twice save when N is even when $\omega_k = \pi$ occurs, once only.

Theorem 4. *If $x(n)$, $n = 1, \ldots, N$ are circularly correlated with covariance matrix given by* (3.7) *then the joint density of the $\tilde{r}(j) = x'PW_jPx/x'Px$, with P as in Theorem* 3, *is obtained from* (3.6) *by adjoining the factor*

$$\left(\sum_{-q}^{q} \delta_j \cos j\pi\right)^{1/2} \sum_{K+1}^{K+M} \left\{\sum_{-q}^{q} \delta_j e^{ij\omega_k}\right\}^{-1} \left\{\sum_{0}^{q} \delta_j \tilde{r}(j)\right\}^{-M-1/2}, \tag{3.9}$$

where we eliminate the factor $-\frac{1}{2}$ from the exponent of the last factor and also the first factor, when N is odd.

Proof. The joint density of the statistics $x'PW_jPx$, $j = 0, \ldots, q$, may be found by adjoining to that discovered in the case $\delta_j = 0$, $j = 1, \ldots, q$, the factor

$$|\hat{\Gamma}_N|^{-1/2} \exp -\tfrac{1}{2}\left\{(\delta_0 - 1)x'Px + \sum_{1}^{q} \delta_j x'PW_jPx\right\}$$
$$= |\hat{\Gamma}_N|^{-1/2} \exp -v\left\{(\delta_0 - 1) + \sum_{1}^{q} \delta_j \tilde{r}(j)\right\}.$$

Integrating with respect to v to obtain (3.5), we must now replace $\exp -v$ in the last displayed expression before (3.5) with

$$|\hat{\Gamma}_N|^{-1/2} \exp -v\left\{\sum_{0}^{q} \delta_j \tilde{r}(j)\right\}.$$

This introduces the factor (3.9) which adjoined to (3.6) gives the joint density when the parent density is (3.7), as stated in the theorem, ending the proof.

The distribution (3.6) as it stands is of little use except in very special cases since apart from the complexity of the formula the circularly defined universe is an unacceptable fiction. This does not hold in the case $q = 1$ of course since now one may consider the distribution of $\tilde{r}(1)$ on the null hypothesis of independence. We proceed now to study the nature of the approximation to the distribution of the $x'PW_jPx$ afforded by the distribution obtained on the assumption that (3.7) is the parent distribution when the *true* parent distribution is that for the corresponding autoregression. To this purpose we evaluate two characteristic functions, using a tilde always to indicate the circular case. We shall restrict our attention to the case where no mean correction of any kind is made and shall compare the distribution of the $x'W_jx$, $j = 0, 1, \ldots, q$ derived on the circular assumption described above, with that of the statistics $x'A_jx$ derived on the corresponding noncircular assumption, where

$$x'A_jx = \sum_{1}^{N-j} x(n)x(n+j), \tag{3.10}$$

so that $r(j) = x'A_jx/x'x$, when no mean corrections are made. The characteristic function of th $x'W_jx$ is, we know,

$$\tilde{\phi}(\theta) = \tilde{\phi}(\theta_0, \theta_1, \ldots, \theta_q) = |I_N - (2i \sum \theta_j W_j)\tilde{\Gamma}_N|^{-\frac{1}{2}}$$

Then

$$\tilde{\psi}(\theta) = \log \tilde{\phi}(\theta) = -\frac{1}{2}\sum_{1}^{N} \log(1 - \tilde{\phi}_k) = \frac{1}{2}\sum_{1}^{\infty}\frac{1}{s}\left(\sum_k \tilde{\phi}_k^{\,s}\right),$$

$$\tilde{\phi}_k = \left\{2i \sum_{0}^{q} \theta_j \cos j\omega_k\right\} 2\pi f(\omega_k), \qquad \omega_k = 2\pi k/N.$$

Then the typical cumulant for the quantities $N^{-1}x'W_jx$ is

(3.11)

$$\tilde{\kappa}(s_0, s_1, \ldots, s_q) = N^{-s}2^{s-1}(s-1)! \sum_{\kappa}\left[\{2\pi f(\omega_k)\}^s \prod_{j=1}^{q} (\cos j\omega_k)^{s_j}\right], \sum_{0}^{q} s_j = s.$$

Now consider the $x'A_jx$, $j = 0, \ldots, q$ on the noncircular hypothesis. This has characteristic function

$$\phi(\theta) = |I_N - (2i \sum \theta_j A_j)\Gamma_N|^{-\frac{1}{2}}$$

so that

$$\psi(\theta) = \log \phi(\theta) = -\frac{1}{2}\sum_{1}^{N} \log(1 - \phi_k) = \frac{1}{2}\sum_{1}^{\infty}\frac{1}{s}\left(\sum_k \phi_k^{\,s}\right)$$

where ϕ_κ is the kth eigenvalue of $(2i \sum \theta_j A_j)\Gamma_N$.

Theorem 5. *Let x be composed of elements $x(n)$, $n = 1, \ldots, N$, which are normal with zero mean. If x has covariance matrix $\tilde{\Gamma}_N$, given by* (3.7), *then the $N^{-1}x'W_jx$ have cumulants given by* (3.11). *If x has covariance matrix Γ_N, corresponding to the autoregression with coefficients $\beta(j)$, $j = 1, \ldots, q$, and $N^{-1}x'A_jx$ is defined by* (3.10), *then these have cumulants given by*

$$\kappa(s_0, s_1, \ldots, s_q) = N^{-s+1}2^{s-1}(s-1)!\frac{1}{2\pi}\int_{-\pi}^{\pi} \{2\pi f(\lambda)\}^s \prod_{j=1}^{q} (\cos j\lambda)^{s_j}\, d\lambda \tag{3.12}$$

$$+ O(N^{-s}), \sum_{0}^{q} s_j = s,$$

where $f(\lambda)$ is the spectral density corresponding to Γ_N.

Proof. We proceed to show that

$$\frac{1}{N}\sum_{k=1}^{N}\phi_k^{\,s} = \frac{1}{2\pi}\int_{-\pi}^{\pi}\{2\pi f(\lambda)2i\sum_0^q(\theta_j\cos j\lambda)\}^s\,d\lambda + O(N^{-1}),$$

whence the result (3.12) follows immediately from the expression for $\psi(\theta)$ just as (3.11) followed from that for $\tilde{\psi}(\theta)$. The proof is taken from Grenander and Szego (1958, p. 221). The $\phi_k/2i$ are the eigenvalues of the product of two matrices, namely $(\sum\theta_jA_j)$ and Γ_N. Both of these are "Toeplitz matrices" in that they have elements which are the same down any diagonal with the element in the jth diagonal above or below the main diagonal the jth Fourier coefficient of an even integrable function, namely for Γ_N, $2\pi f(\lambda)$, while for $\sum\theta_jA_j$ it is $\sum\theta_j\cos j\lambda$. Both of these functions are, in the present instance, infinitely differentiable (since $f(\lambda)$ corresponds to an autoregression). Let $M_N(f_j)$ be such a matrix, corresponding to a function f_j. Thus $\Gamma_N = M_N(2\pi f(\lambda))$, $2i\sum\theta_jA_j = M_N(2i\sum\theta_j\cos j\lambda)$. Our assertion at the beginning of this proof will follow if we can show that

$$\lim_{N\to\infty} N^{-1}\operatorname{tr}\{M_N(f_1)M_N(f_2)\cdots M_N(f_p) - M_N(f_1f_2\cdots f_p)\} = O(N^{-1}).$$

For taking $p = 2s$ and s of the f_j as $2\pi f(\lambda)$ and s as $2i\sum\theta_j\cos j\lambda$ we see that

$$N^{-1}\operatorname{tr}\left(M_N(f_1)\cdots M_N(f_p)\right) = N^{-1}\operatorname{tr}\left\{\left(2i\sum_j\theta_jA_j\Gamma_N\right)^s\right\} = N^{-1}\sum\phi_k^{\,s},$$

whereas

$$N^{-1}\operatorname{tr}\{M_N(f_1f_2\cdots f_p)\} = \frac{1}{2\pi}\int_{-\pi}^{\pi}\left\{2\pi f(\lambda)2i\sum_0^q\theta_j\cos j\lambda\right\}^s d\lambda.$$

Now, putting $m_{ij}(f_k)$ for the typical element of $M_N(f_k)$ we see that

$$S_N = \operatorname{tr}\left(M_N(f_1)\cdots M_N(f_p)\right) = \sum m_{n_1n_2}(f_1)m_{n_2n_3}(f_2)\cdots m_{n_pn_1}(f_p) \tag{3.13}$$

where the sum is for $1 \le n_j \le N$ for each j. Let us now replace the range $1 \le n_j \le N$ by $1 \le n_j < \infty$, for $j = 2, 3, \ldots, p$ and call the resulting sum T_N. We shall show that $T_N - S_N$ is $O(1)$ and this proves what we wish since, for example,

$$\begin{aligned}\sum_{n_2=1}^{\infty} m_{n_1n_2}(f_1)m_{n_2n_3}(f_2) &= \sum_{n_2=1}^{\infty}\left\{\frac{1}{2\pi}\int_{-\pi}^{\pi}f_1(\lambda)e^{i(n_1-n_2)\lambda}\,d\lambda\,\frac{1}{2\pi}\int_{-\pi}^{\pi}f_2(\lambda)e^{i(n_2-n_3)\lambda}\,d\lambda\right\}\\ &= \frac{1}{2\pi}\int_{-\pi}^{\pi}f_1(\lambda)f_2(\lambda)e^{i(n_1-n_3)\lambda}\,d\lambda = m_{n_1n_3}(f_1f_2),\end{aligned}$$

since the $f_j(\lambda)$ are real functions, so that T_N is $\operatorname{tr}\{M_N(f_1f_2\cdots f_p)\}$. Thus we need only to prove that $S_N - T_N$ is $O(1)$. This difference will reduce to the sum of a finite number of terms of the form (3.13) where $1 \le n_1 \le N$, at least one subscript runs from n to infinity and the remainder from 1 to ∞. The case in which only one runs over the full range is the only relevant one, for by the reduction just used to show that T_N is $\operatorname{tr}\{M_N(f_1 f_2 \cdots f_p)\}$ we may reduce each of the others to that one, with fewer than p subscripts involved. Thus we need to consider

$$\sum m_{n_1 n_2}(f_1) m_{n_2 n_3}(f_2) \cdots m_{n_q n_1}(f_q), \qquad 2 \le q \le p, \tag{3.14}$$
$$1 \le n_1 \le N; \qquad N \le n_j \le \infty; \qquad j = 2, 3, \ldots, q.$$

Now if k and l are arbitrary integers and n is an integer then either $|n - k|$ or $|n - l|$ is not less than $\frac{1}{2}|k - l|$. Also

$$m_{n_1 n_2}(f_j) = \{1 + (n_1 - n_2)^2\}^{-1} O(1),$$

where the bound of $O(1)$ may depend on f_j but not on n_1, n_2. Indeed this would be true if f_j was only twice continuously (or boundedly) differentiable. It is evidently true for any product of the f_j (under the same condition). Thus for f, g linear combinations of products of the f_j

$$\begin{aligned} \sum_n m_{kn}(f) m_{nl}(g) &= O(1) \sum_{n=N}^{\infty} \{1 + (n - k)^2\}^{-1} \{1 + (n - l)^2\}^{-1} \\ &= O(1)\{1 + \tfrac{1}{4}(k - l)^2\}^{-1} \sum_{n=N}^{\infty} [\{1 + (n - k)^2\}^{-1} + \{1 + (n - l)^2\}^{-1}] \\ &= O(1)\{1 + (k - l)^2\}^{-1}. \end{aligned} \tag{3.15}$$

Using this result repeatedly we see that (3.14) is

$$O(1) \sum \{1 + (n_1 - n_3)^2\}^{-1} \{1 + (n_3 - n_5)^2\}^{-1} \cdots \{1 + (n_r - n_1)^2\}^{-1},$$

where r is either q or $q - 1$. Using (3.15) this reduces to

$$\begin{aligned} O(1) &\sum_{n_r=N}^{\infty} \sum_{n_1=1}^{N} \{1 + (n_1 - n_r)^2\}^{-2} \\ &= O(1) \sum_{n=1}^{\infty} [\{1 + n^2\}^{-2} + \{1 + (n + 1)^2\}^{-2} + \cdots + \{1 + (n + N - 1)^2\}^{-2}] \\ &< O(1) \sum_{1}^{\infty} n(n^2 + 1)^{-2} = O(1) \end{aligned}$$

and the result is established.

Now comparing (3.11) *with* (3.12) *we see that the cumulants of* $N^{-1}x' W_j x$ *and* $N^{-1}x' A_j x$, *neglecting a term of order* N^{-s} *differ only by replacing an integral by its approximating sum. This shows that at least the result of Theorem* 5 *gives an asymptotically valid approximation to the distribution of the r(j) when the population is noncircular.* It leads also to the two following comments.

1. The mean and variance of the statistics $N^{-1}x'A_jx$ (on the appropriate, noncircular, hypothesis) may be evaluated precisely and an adjustment made to these statistics so that they have, to a sufficient order of accuracy, the mean and variance of the distribution of the $N^{-1}x'W_jx$ on the fictitious circular assumption. These adjustments will involve quantities of order N^{-1}. Theorem 5 shows that the cumulants higher than the second for the quantities $N^{-1}x'A_jx$ differ from those obtained for the $N^{-1}x'W_jx$ from the circular model by quantities of lower order than N^{-1}. Thus after the adjustments of which we have just spoken all cumulants will differ from those obtained on the circular assumption by quantities of lower order than N^{-1} and indeed it is easy to see that this holds uniformly in the order of the cumulant. If we use the circularly defined population in deriving the joint distribution of the $N^{-1}x'W_jx$, we shall have an approximation, involving an error which is $o(N^{-1})$, to the distribution of the, modified, $N^{-1}x'A_jx$ (i.e., modified so as to have an appropriate mean and variance) on the noncircular hypothesis.

2. The result even suggests that it may be preferable to replace the function $\log \tilde{\phi}(\theta) = \tilde{\psi}(\theta)$ by an integral to which it provides an approximating sum and to derive the approximate distribution from this function instead of from $\tilde{\psi}(\theta)$. Indeed, if this is done, the cumulants higher than the second will be in error by quantities which are $O(N^{-2})$. This argument needs more careful investigation, perhaps, but there is another compelling reason for doing what we have just said and this is the fact that the characteristic function turns out in many cases to correspond to a tractable distribution which thus replaces the intractable one which is obtained from (3.6) and (3.9). The first person to make use of this idea seems to have been Koopmans (1942). We refer the reader to the footnote on p. 342 for other relevant references. We illustrate with an example.

Following Leipnik (1947), we consider the distribution of $\tilde{r}(1)$ without any form of mean correction but with a circularly correlated universe with $q = 1$. Then

$$\log \tilde{\phi}(\theta) = -\frac{1}{2}\sum_0^{N-1} \log\left\{1 - 2i\,\frac{\theta_0 + \theta_1 \cos \omega_j}{1 + \rho^2 - 2\rho\cos\omega_j}\right\},$$

where we have, as is usual, put $\beta(1) = -\rho$. This is approximately

$$-\frac{N}{4\pi}\int_0^{2\pi} \log\left\{1 - 2i\,\frac{\theta_0 + \theta_1\cos\lambda}{1 + \rho^2 - 2\rho\cos\lambda}\right\} d\lambda$$

$$= -\frac{N}{4\pi}\left\{\int_0^{2\pi} \log\{y - 2z\cos\lambda\}\, d\lambda - \int_0^{2\pi} \log(1 + \rho^2 - 2\rho\cos\lambda)\, d\lambda\right\}$$

$$y = 1 + \rho^2 - 2i\theta_0, \qquad z = (\rho + i\theta_1).$$

We evaluate this as

$$-\tfrac{1}{2}\log\left(\frac{y+\sqrt{y^2-4z^2}}{2}\right)^N$$

so that the joint characteristic function is

$$\{\tfrac{1}{2}(y+\sqrt{y^2-4z^2})\}^{-\frac{1}{2}N}.$$

For $\theta_0 = \theta_1 = 0$ this is unity as is appropriate. It is fairly evident that it is a characteristic function since the change from the sum to the integral cannot change its positive definite nature. In any case we shall prove this by inverting it.

Thus we must evaluate

$$\frac{1}{4\pi^2}\iint_{-\infty}^{\infty} e^{-i(\theta_0 s_0+\theta_1 s_1)}\left(\frac{y+\sqrt{y^2-4z^2}}{2}\right)^{-\frac{1}{2}N} d\theta_0\, d\theta_1.$$

(The integrand is evidently absolutely integrable for $N \geq 3$.) This is

$$\frac{2}{4\pi^2}\iint e^{-\frac{1}{2}ys_0-zs_1}\left(\frac{y+\sqrt{y^2-4z}}{2}\right)^{-\frac{1}{2}N} dy\, dz e^{-\frac{1}{2}\{(1+\rho^2)s_0-2\rho s_1\}},$$

which may be evaluated as

$$\frac{N s_0^{-\frac{1}{2}N-1}(s_0^{\,2}-s_1^{\,2})^{\frac{1}{2}(N-1)}}{2^{\frac{1}{2}(N+2)}\Pi^{\frac{1}{2}}(\frac{1}{2}N-\frac{1}{2})!}\, e^{-\frac{1}{2}((1+\rho^2)s_0-2\rho s_1)}.$$

Now we obtain, by integrating out s_0, the distribution of $\tilde{r}(1)$ as

$$\text{(3.16)} \qquad \frac{(\frac{1}{2}N)!}{\pi^{\frac{1}{2}}(\frac{1}{2}N-\frac{1}{2})!}(1-\tilde{r}^2)^{\frac{1}{2}(N-1)}(1+\rho^2-2\rho\tilde{r})^{-\frac{1}{2}N}.$$

When $\rho = 0$, this is the density of a correlation from $(N + 3)$ pairs of observations. If, as in the early part of this section, we evaluate the mean and variance of $\tilde{r}(1)$, with no mean correction and on the hypothesis of independence, we find that $\mathcal{E}(\tilde{r}(1)) = 0$, var $(\tilde{r}(1)) = (N + 2)^{-1}$ which agrees with this result since $(N + 2)^{-1}$ is the variance of a correlation from $N + 3$ pairs of observations. As mentioned earlier, if mean corrections are made we should add $(N + 2)^{-1}$ to $\tilde{r}(1)$ before using it as such a correlation.

The extension of this result to the consideration of higher order autocorrelations and partial autocorrelations is difficult for the integral approximation to the joint characteristic function is difficult to invert. Daniels (1956), however, obtained an approximation to the distribution of the partial circular autocorrelation $\tilde{r}(j|1, \ldots, j-1)$ on the assumption that the $x(n)$ are generated by the circularly defined process defined above (i.e., that with covariance matrix $\tilde{\Gamma}_N$). We here define the circular partial autocorrelations,

$\tilde{r}(j \mid 1, \ldots, j-1)$, as the partial correlation of $x(n)$ with $x(n-j)$, after removing $x(n-1), \ldots, x(n-j+1)$ by regression, where we define $x(n-k)$ for $n-k \leq 0$ as $x(N-n+k)$. Thus†

$$\tilde{r}(j \mid 1, \ldots, j-1) = \frac{(-)^{j+1} |\tilde{C}_{j+1}^{(1,j+1)}|}{|\tilde{C}_j|},$$

where $\tilde{C}_j$ is the $j \times j$ matrix with $\tilde{c}(u-v) = N^{-1}x'PW_jPx$ in row u column v and P is the matrix which effects the mean correction. We now have

Theorem 6.‡ *Let* $x(1), \ldots, x(n)$ *be Gaussian and circularly autocorrelated (with covariance matrix* $\tilde{\Gamma}_N$ *given by* (3.7)) *and let* $\beta_q = \delta_q = 0$. *Then* $\tilde{r}(q \mid 1, \ldots, q-1)$ *has density which, neglecting terms which are* $o(N^{-1})$, *is*

$$B(\tfrac{1}{2}, \tfrac{1}{2}N)^{-1}\{1 - \tilde{r}(q \mid 1, \ldots, q-1)\} \times \{1 - \tilde{r}^2(q \mid 1, \ldots, q-1)\}^{\frac{1}{2}N-1}, \qquad q \text{ odd},$$

$$\{B(\tfrac{1}{2}, \tfrac{1}{2}N - \tfrac{1}{2}) + B(\tfrac{3}{2}, \tfrac{1}{2}N + \tfrac{1}{2})\}^{-1}\{1 - \tilde{r}(q \mid 1, \ldots, q-1)\}^2 \times \{1 - \tilde{r}^2(q \mid 1, \ldots, q-1)\}^{\frac{1}{2}(N-3)}, \qquad q \text{ even}.$$

We observe that the distribution is independent of the $\tilde{\gamma}(n)$ and thus the distribution of $\tilde{r}(q \mid 1, \ldots, q-1)$ is that got under the null hypothesis, to the order of approximations of the theorem. We also point out that by Theorem 5 these results are approximations to those appropriate to non-circularly defined statistics in the *noncircular* case (i.e., with covariance matrix Γ_N not $\tilde{\Gamma}_N$). We discuss these questions further after the proof.

Proof. We may assume $\mathcal{E}(x(n)) = 0$. We discuss first the case where no mean corrections are made. We shall put, for simplicity of printing, during this proof only

$$c_j = x'W_jx, \qquad r_j = \frac{c_j}{c_0},$$

but shall continue to use $\tilde{r}(j \mid 1, \ldots, j-1)$ for the partial autocorrelations even when they are computed without mean correction. We shall call P_j the matrix with r_{u-v} in row u, column v for $u, v = 1, \ldots, j$. We first form the Laplace transform

$$m(s_0, s_1, \ldots, s_q) = \mathcal{E}\left\{\exp \sum_0^q s_jc_j\right\} = |\tilde{\Gamma}_N|^{-\frac{1}{2}} \left|\tilde{\Gamma}_N^{-1} - 2\sum_0^q s_jW_j\right|^{-\frac{1}{2}}.$$

It is now convenient to express $|\tilde{\Gamma}_N|^{-\frac{1}{2}}$ in terms of the zeros, $z_1, z_2, \ldots, z_q$

† We remind the reader that in this chapter we use $|\tilde{C}_k^{(i,j)}|$ for the minor of the element in row i, column j of $\tilde{C}_k$.

‡ The proof of this theorem is difficult. The theorem is not used after this section.

of the equation $\sum \beta(j)z^{q-j} = 0$, which zeros are all less than unity in modulus by assumption. Indeed $\tilde{\Gamma}_N^{-1}$ is of the form BB' where B is the circulant

$$B = \sum_0^q \beta(j)U^{q-j}.$$

This has determinant

$$\prod_{j=1}^q (1 - z_j^N).$$

This can be seen by considering

$$\left[\prod_1^q (z^N - z_j^N)\right]_{z=1} = \left\{\prod_{j=1}^q \prod_{k=0}^{N-1} (z - z_j e^{-i\omega_k})\right\}_{z=1}$$
$$= \prod_{k=0}^{N-1} \left\{\sum_0^q \beta(j)e^{i(q-j)\omega_k}\right\},$$

so that

$$|\tilde{\Gamma}_N|^{-1/2} = \prod_{j=1}^q (1 - z_j^N).$$

Daniels then introduces the transformation (see (3.7) for the δ_j)

$$\delta_0 - 2s_0 = k\sum_0^q \alpha_j^2, \qquad \alpha_0 = 1,$$

$$\delta_j - s_j = k\sum_{u=0}^{q-j} \alpha_u\alpha_{u+j}, \qquad j = 1, \ldots, q.$$

This implies that

$$BB' - S = kAA',$$

where S is the circulant $2\sum_0^q s_j W_j$, whereas A is the circulant constructed from the α_j as B is from the $\beta(j)$. Since $BB' - S$ is a circulant it is clear that it can be written in the form CC' where C is a circulant, at least for certain values of the s_j. Indeed, this may be done in many ways, since there will be one such factorization for every factorization of

$$\sum_{-q}^q (\delta_j - s_j)e^{ij\lambda} - s_0, \qquad \delta_j = \delta_{-j}, \qquad s_j = s_{-j},$$

for at $\lambda = 2\pi k/N$, $k = 0, \ldots, N-1$, these are the eigenvalues of the matrix to be factored. Since this expression is positive for s_j real and sufficiently small, the discussion in Chapter II, Section 5 shows that this expression may be factored in the form

$$\left|\sum_0^q \alpha_j e^{ij\lambda}\right|^2,$$

so that the zeros of $\sum \alpha_j z^j$ lie outside or on the unit circle. A zero will lie on the unit circle only for special values of s_j, having zero (Lebesgue) measure, in the $(2q + 2)$ dimensional space in which the real and imaginary parts of the s_j lie.† Thus

$$|BB' - S| = k^N \prod_1^N (1 - w_j^N)^2,$$

where the w_j are of less than unit modulus (except for special values of the s_j). We may also express k in terms of the $r_j = c_j/c_0$ since

$$\delta_0 + 2r_1\delta_1 + \cdots + 2r_q\delta_q - 2\sum_{j=0}^{q} r_j s_j = k\left\{\sum_0^q \alpha_j^2 + 2r_1 \sum \alpha_u \alpha_{u+1} + \cdots + 2r_q \alpha_q\right\}.$$

Introducing the vectors r with r_j in the jth place, β with $\beta(j)$ in the jth place, and α with α_j in the jth place and calling $u = \sum r_j s_j$ we find that

$$k = \frac{Q(\beta, r) - 2u}{Q(\alpha, r)},$$

$$Q(x, r) = 1 - 2x'r + x'P_q x.$$

Thus finally the Laplace transformation is

$$\frac{Q(\alpha, r)^{N/2}}{\{Q(\beta, r) - 2u\}^{N/2}} \prod_0^q \left(\frac{1 - z_j^N}{1 - w_j^N}\right). \tag{3.17}$$

Now we change variables from $s_0, s_1, \ldots, s_q$ to $u, s_1, \ldots, s_q$ or equivalently $u, \alpha_1, \ldots, \alpha_q$. The Jacobian of this transformation is

$$\frac{\partial(s_0, s_1, \ldots, s_q)}{\partial(k, \alpha_1, \ldots, \alpha_q)} \cdot \left(\frac{-2}{Q(\alpha, r)}\right).$$

The first factor is

$$(-)^{q-1}k^q \begin{vmatrix} \frac{1}{2}(1 + \alpha_1^2 + \cdots + \alpha_q^2) & \alpha_1 & \alpha_2 & \cdots & \alpha_{q-1} & \alpha_q \\ \alpha_1 & 1 + \alpha_2 & \alpha_1 + \alpha_3 & \cdots & \alpha_{q-2} + \alpha_q & \alpha_{q-1} \\ \cdot & & & & & \\ \cdot & & & & & \\ \cdot & & & & & \\ \alpha_q & 0 & & & 0 & 1 \end{vmatrix},$$

which, following Daniels, we call $\frac{1}{2}(-)^{q-1}k^q J(\alpha)$. Now we have the density

† The α_j and k are defined (by analytic continuation, for example) as analytic functions of the s_j outside of this product of intervals on the real axis and indeed for the whole product of complex planes omitting the parts of the real axis for which $\{\Sigma(\delta_j - s_j) \exp ij\lambda\} - s_0$ is not positive.

we wish by the inverse Laplace transformation

$$\prod_1^q (1 - z_j^N) \frac{1}{(2\pi i)^{q+1}} \int \cdots \int \frac{\{Q(\alpha, r)\}^{\frac{1}{2}N-q-1} J(\alpha)}{\{Q(\beta, r) - 2u\}^{\frac{1}{2}N-q} \prod (1 - w_j^N)} \times \exp -uc_0 \, du \, d\alpha,$$

where we have written $d\alpha$ for $d\alpha_1 \, d\alpha_2 \cdots d\alpha_q$. We may now invert with respect to the α_j. If $\phi(u, \alpha)$ is the Laplace transform being inverted then the joint density of the c_j satisfies

$$f(c_0, c_1, \ldots, c_q) = \frac{1}{2\pi i} \int \left\{ \frac{1}{(2\pi i)^q} \int \cdots \int \phi(u, \alpha) \, d\alpha \right\} e^{-uc_0} \, du,$$

so that

$$\int_0^\infty f(c_0, r_1 c_0, \ldots. r_q c_0) e^{uc_0} \, dc_0 = \frac{1}{(2\pi i)^q} \int \cdots \int \phi(u, \alpha) \, d\alpha.$$

By differentiating q times with respect to u and putting u equal to zero we obtain the joint density of the r_j

$$\frac{(\frac{1}{2}N - 1)!}{(\frac{1}{2}N - q - 1)!} \frac{\prod_1^q (1 - z_j^N)}{Q(\beta, r)^{\frac{1}{2}N}} \frac{1}{(2\pi i)^q} \int \cdots \int \frac{Q(\alpha, r)^{\frac{1}{2}N-q-1} J(\alpha)}{\prod (1 - w_j^N)} \, d\alpha.$$

Now we deform the contours of integration so that they pass over saddle-points of the numerator, $Q(\alpha, r)$, which will become large in modulus relative to the values of the integrands on the remainders of the contours as N increases, and thus obtains a first term in an asymptotic expansion for the joint density of $r_1, \ldots, r_q$. The saddlepoints are discovered by solving

$$\frac{\partial Q(\alpha, r)}{\partial \alpha_j} = 0, \qquad j = 1, \ldots, m,$$

that is,

$$r + P_q \alpha = 0,$$

which gives†

$$\hat{\alpha} = -P_q^{-1} r,$$

so that

$$Q(\hat{\alpha}, r) = 1 - r' P_q^{-1} r' = \frac{|P_{q+1}|}{|P_q|}.$$

The contours of integration are chosen to be the lines of steepest descent from the saddlepoints. These curves must satisfy $\mathscr{I}\{Q(\alpha, r)\} = 0$, as may be seen by putting $Q(\alpha, r) = \exp \{\log Q(\alpha, r)\}$ and observing that the level

† It is easily seen that $\hat{\alpha}$ as defined in the next line lies within the region over which α has been allowed to vary, at least for N sufficiently large.

curves of constant modulus are of a form such that $\log\{\mathcal{R}(Q(\alpha, r))\}$ is constant, i.e., such that $\mathcal{R}(Q(\alpha, r))$ is constant. Since the lines of steepest descent must cross these lines at right angles we must have $\mathcal{I}(Q(\alpha, r)) =$ constant for such a line of steepest descent.† But these curves must also pass through the point $\hat{\alpha}$ at which $Q(\alpha, r)$ is real and thus we must have $\mathcal{I}\{Q(\alpha, r)\} = 0$. If $\alpha = \hat{\alpha} + i\eta$, then this condition is satisfied for then

$$\mathcal{I}Q(\hat{\alpha} + i\eta, r) = 2\eta'(r + P_q\hat{\alpha}) = 0.$$

Thus on the lines of steepest descent $Q(\alpha, r) = |P_{q+1}|/|P_q| - \eta' P_q \eta$, which is certainly a decreasing function of η, since P_q is positive definite. Daniels then ignores the factors $\Pi(1 - z_j^N)$, $\Pi(1 - w_j^N)$. Certainly the first will approach unity quickly, but the second will do so also, for $\hat{\alpha}$ is evidently the usual least-squares estimate of β, and for $\alpha = \beta$ the zeros are of less than unit modulus, whereas, as N increases, $\hat{\alpha}$ will converge with probability 1 to β. We have proved this in the noncircular case, but for the circular case the proof is not essentially different. The neglect of these two factors will introduce an error that goes exponentially to zero with N. As far as the range over which the η_j vary is concerned we observe that, as $\mathcal{I}(s_j) \to \pm\infty$, $Q(\alpha, r) \to 0$. Thus we must have η such that it varies over the whole range defined by $\eta' P_q \eta \leq |P_{q+1}|/|P_q|$. Thus we have, again writing $d\eta$ for the product of differentials,

$$f(r_1, \ldots, r_q) \approx \frac{1}{(2\pi)^q} \frac{(\frac{1}{2}N - 1)!}{(\frac{1}{2}N - q - 1)!} \{Q(\beta, r)\}^{-\frac{1}{2}N}$$
$$\times \int \cdots \int J(\alpha) \left\{\frac{|P_{q+1}|}{|P_q|} - \eta' P_q \eta\right\}^{\frac{1}{2}N-q-1} d\eta.$$

The expression $J(\alpha)$ is a polynomial in the α_j and thus in the η_j. If we replace it by the first term in its Taylor expansion about $\hat{\alpha}$, namely $J(\hat{\alpha})$, we neglect terms of the type

$$c(k_1, \ldots, k_q) \iint_{\Sigma x_j^2 \leq 1} (1 - \sum x_j^2)^{\frac{1}{2}N-q-1} \prod x_j^{k_j}\, dx,$$

where the x_j are new variables introduced by a linear transformation that reduces $\{|P_q|/|P_{q+1}|\}P_q$ to the unit matrix. Thus the error involved in replacing $J(\alpha)$ by $J(\hat{\alpha})$ is of order N^{-1}. However, as Daniels points out, the neglected *factor* will be of the form

$$(1 + \hat{c}N^{-1} + \cdots),$$

† See De Bruijn (1958, p. 83), for example.

where $\hat{c}$ involves the r_j. It can therefore be written

$$\left\{1 + cN^{-1} + N^{-1} \sum_j a_j(r_j - \tilde{\rho}(j)) + \cdots\right\}$$

and absorbing the first term into the constant factor in the density we shall find, and recognizing that $r_j - \tilde{\rho}(j) = O(N^{-1/2})$ (in the sense that its standard deviation of this order), we are led to write

$$f(r_1, \ldots, r_q) = C \frac{J(\hat{\alpha})}{\{Q(\beta, r)\}^{1/2 N}} \int \cdots \int \left\{\frac{|P_{q+1}|}{|P_q|} - \eta' P_q \eta\right\}^{1/2 N - q - 1} d\eta(1 + o(N^{-1})).$$

There is little point in a more accurate evaluation, since in any case $f(.)$ is an approximation, to $O(N^{-1})$, to the density we want because the true system is not circular. Carrying out the integration, we obtain

$$C \frac{J(\hat{\alpha})}{\{Q(\beta, r)\}^{1/2 N}} \frac{|P_{q+1}|^{1/2(N-q)-1}}{|P_q|^{1/2(N-q-1)}} (1 + o(N^{-1})), \tag{3.18}$$

where C is a (new) constant, since the integral is

$$\left\{\frac{|P_{q+1}|}{|P_q|}\right\}^{1/2 N - q - 1} \left\{\frac{|P_q|}{|P_{q+1}|}\right\}^{1/2 q} |P_q|^{1/2} \int_{\Sigma x_j^2 \le 1} (1 - \sum x_j^2)^{1/2 N - q - 1}\, dx.$$

We shall not evaluate C, since in the form (3.18) the density is of little use. We observe first that by Theorem 6 in Chapter I,

$$1 - \hat{R}^2(q \mid q - 1) = \frac{|P_{q+1}|}{|P_q|} = \prod_1^q (1 - \hat{r}^2(j \mid 1, \ldots, j - 1)).$$

(Here we have continued to use $\hat{R}^2(q \mid q - 1)$ for the multiple correlation of $x(n)$ with $x(n - 1), \ldots, x(n - q)$ when this is defined in terms of the r_j.) It follows immediately (by induction) that

$$|P_{q+1}| = \prod_1^q (1 - \hat{r}^2(j \mid 1, \ldots, j - 1))^{q-j+1}. \tag{3.19}$$

On the other hand,

$$J(\hat{\alpha}) = \frac{|P_{q+1}|}{|P_q|} |D|, \qquad \hat{\alpha} = -D'r,$$

where D is the matrix in the bottom right-hand corner of $J(\hat{\alpha})$. (We leave this as an exercise to this chapter). Since $r = -P_q\hat{\alpha}$ we find† that the Jacobian

† Since $\hat{\alpha} + D'r = 0 = r + P_q\hat{\alpha}$ we have $P_q = D'\partial r/\partial\hat{\alpha}$.

$|\partial r/\partial\hat{\alpha}|$ is $-|P_q|/|D|$. Thus after transformation from $r_1, \ldots, r_q$ to $\hat{\alpha}_1, \ldots, \hat{\alpha}_q$ we have the density

$$C\{Q(\beta, r)\}^{-\frac{1}{2}N} \frac{|P_{q+1}|^{\frac{1}{2}(N-q)}}{|P_q|^{\frac{1}{2}(N-q-1)}}.$$

We now transform from $\hat{\alpha}_1, \ldots, \hat{\alpha}_q$ to the partial circular autocorrelations, $\tilde{r}(j \mid 1, \ldots, j-1)$, $j = 1, \ldots, q$. We again leave as an exercise the proof of the fact that the Jacobian of this transformation is

$$\prod_{j \text{ odd}} (1 - \tilde{r}^2(j \mid 1, \ldots, j-1))^{\frac{1}{2}(j-1)} \times \prod_{j \text{ even}} \{(1 - \tilde{r}(j \mid 1, \ldots, j-1))(1 - \tilde{r}_j^{\,2} \mid 1, \ldots, j-1)^{\frac{1}{2}j-1}\},$$

where the products are over all such j for $1 \le j \le q$. Then, using (3.19), we obtain for the joint density of the partial circular autocorrelations (without mean correction) the density

$$CQ(\beta, r)^{-\frac{1}{2}N} \prod_{j \text{ odd}} (1 - \tilde{r}^2(j \mid 1, \ldots, j-1))^{\frac{1}{2}(N-1)} \times \prod_{j \text{ even}} \{(1 - \tilde{r}^2(j \mid 1, \ldots, j-1))^{\frac{1}{2}N-1}(1 - \tilde{r}(j \mid 1, \ldots, j-1)\}$$

where $Q(\beta, r)$ must also be reexpressed in terms of the $\tilde{r}(j \mid 1, \ldots, j-1)$. However, when $\beta(q) = 0$, $Q(\beta, r)$ depends only on $r_1, \ldots, r_{q-1}$, as is easily seen, and only on $\tilde{r}(j \mid 1, \ldots, j-1)$, $j = 1, \ldots, q-1$. Daniels' result shows that the density for $\tilde{r}(q \mid 1, \ldots, q-1)$, when $\beta(q) = 0$, is proportional to

$$(3.20) \quad \begin{cases} \{1 - \tilde{r}^2(q \mid 1, \ldots, q-1)\}^{\frac{1}{2}(N-1)}, & q \text{ odd} \\ \{1 - \tilde{r}^2(q \mid 1, \ldots, q-1)\}^{\frac{1}{2}N-1}\{1 - \tilde{r}(q \mid 1, \ldots, q-1)\}, & q \text{ even.} \end{cases}$$

The effect of fitting a mean is now easily allowed for. With the circular definition this amounts to changing from c_j to $c_j - N\bar{x}^2$. Correspondingly we now have

$$m(s_0, s_1, \ldots, s_q) = |\tilde{\Gamma}_N|^{-\frac{1}{2}} |\tilde{\Gamma}_N^{-1} - 2\sum_0^q s_j W_j + \left(2N^{-1}\sum_0^q s_j\right)\tilde{1}\tilde{1}'|^{-\frac{1}{2}}.$$

The second term on the right is

$$\left|\tilde{\Gamma}_N^{-1} - 2\sum_0^q s_j W_j\right|^{-\frac{1}{2}} \left\{1 + 2N^{-1}\sum_0^q s_j \tilde{1}'\left(\tilde{\Gamma}_N^{-1} - 2\sum_0^q s_j W_j\right)^{-1}\tilde{1}\right\}^{-\frac{1}{2}}.$$

This last factor accounts for all of the effect of the mean correction. Moreover, $\tilde{1}'(\tilde{\Gamma}_N^{-1} - 2\sum_0^q s_j W_j)^{-1}\tilde{1}$ is just N by the sum of the elements in the first row of the matrix of the form and this sum is just the reciprocal of the same sum for the reciprocal of that matrix. The factor incorporating the effects of the mean correction is

$$\left(1 + 2N^{-1}\sum_0^q s_j\left[N\left\{\left(\sum_0^q \beta(j)\right)^2 - 2\sum_0^q s_j\right\}^{-1}\right]\right)^{-\frac{1}{2}} = \left(\frac{\{\sum_0^q \beta(j)\}^2}{[\{\sum_0^q \beta(j)\}^2 - 2\sum_0^q s_j]}\right)^{-\frac{1}{2}}$$
$$= \left(\frac{\{\sum_0^q \beta(j)\}^2}{k(\sum \alpha_j)^2}\right)^{-\frac{1}{2}}.$$

Now remembering that in the Laplace transform (3.17) we ignored the factor

$$\prod_1^q \left(\frac{1 - z_j^N}{1 - w_j^N}\right)$$

and that

$$k = \frac{Q(\beta, r) - 2u}{Q(\alpha, r)}$$

we see that the net effect of mean correction is to change N to $(N - 1)$ in (3.17) and to introduce the factor $(\sum \alpha_j)/(\sum \beta(j))$. Absorbing $(\sum \beta(j))^{-1}$ into a constant factor, which can be later determined, we have only to handle $\sum \alpha_j$. Precisely as just before (3.18) we may replace this by $\sum \hat{\alpha}_j$ and take it outside of the integral in inverting the Laplace transform. Now we have (see exercise 5 to this chapter)

$$\text{(3.21)} \quad \sum_0^q \hat{\alpha}_j = (1 - \tilde{r}(1))(1 - \tilde{r}(2 \mid 1)) \cdots (1 - \tilde{r}(q \mid 1, \ldots, q - 1)),$$

and we see from (3.20) that the densities for $\tilde{r}(q \mid 1, \ldots, q - 1)$ are proportional to

$$\{1 - \tilde{r}(q \mid 1, \ldots, q - 1)\}\{1 - \tilde{r}^2(q \mid 1, \ldots, q - 1)\}^{\frac{1}{2}N-1}, \qquad q \text{ odd}$$
$$\{1 - \tilde{r}(q \mid 1, \ldots, q - 1)\}^2\{1 - \tilde{r}^2(q \mid 1, \ldots, q - 1)\}^{\frac{1}{2}(N-3)}, \qquad q \text{ even};$$

evaluating the constants we obtain the theorem.

Of course, the adjunction of (3.21) [to the formula obtained for no mean correction with N replaced by $(N - 1)$] will give us the joint distribution for all partial autocorrelations to the same order of accuracy, but, as we have said, this is not a useful result compared with that of the theorem.

The simplest case of q odd is that in which $q = 1$, and we have then seen that in this case, using $\tilde{r}(1) + (N + 2)^{-1}$ (with mean corrections) as an ordinary correlation from $(N + 3)$ pairs of observations, it is quite adequate for $N \geq 20$ (and in fact for N smaller under most circumstances). When q is even, the simplest case is $\tilde{r}(2 \mid 1)$. The density from the theorem for q even has mean $-2/(N + 1)$ and variance $(N - 1)(N + 3)\{(N + 1)^2(N + 2)\}^{-1}$. This is $(N + 2)^{-1} + O(N^{-3})$. This suggests that $\tilde{r}(2 \mid 1) + 2/(N + 1)$ be treated as an ordinary correlation from $(N + 3)$ observations. The case q odd gives the mean as $-(N - 1)^{-1}$ and the variance as $(N - 2)/(N - 1)^2$, as is easily seen. For $q = 1$ we know the exact results $-(N - 1)^{-1}$ and $(N^2 - 3N)/\{(N + 1)(N - 1)^2\}$. The latter differs by $O(N^{-2})$ from that obtainable from (3.20) so that there is no contradiction. The following table compares the significance points obtained from (3.20) for $q = 2$ with those obtained by treating $\tilde{r}(2 \mid 1) + 2/(N + 1)$ as an ordinary correlation from $(N + 3)$ pairs of observations.

Table 2 FIVE PER CENT POINT FOR TWO-SIDED TEST WITH EQUAL TAIL AREAS

		Number of observations		
		15	20	25
Distribution (3.19)	Upper	0.36	0.32	0.31
	Lower	−0.62	−0.54	−0.46
$\tilde{r}(2 \mid 1) + 2/(N + 1)$ as correlation from $N + 3$ observations	Upper	0.35	0.32	0.30
	Lower	−0.60	−0.50	−0.46

The agreement is evidently fairly good.

This suggests the rule that $\tilde{r}(q \mid 1, \ldots, q - 1)$ have added to it $2/N$ when q is even and $1/N$ when q is odd and be treated as a correlation from $N + 3$ pairs of observations. However, one feels intuitively that as q increases the "degrees of freedom" should decrease and not remain fixed as is implied by the use of $N + 3$. A more accurate approximation would be required to validate this, however.

The basic apparatus for testing the significance of $\beta(q)$ in an autoregressive model, and thus for determining the order of the autoregression is therefore, in a sense, complete. We shall not deal in any greater detail with the multiple decision problem involved. Two further points should be mentioned, however. The first of these is that Jenkins has suggested, on the basis it seems of not very strong evidence, that the distributions (3.20) are better approximations to the distributions of noncircular statistics than of circular statistics.† The evidence is based on the mean and variance for $r(1)$, which are $-(N - 1)^{-1}$ and $(N + 1)^{-1} + O(N^{-3})$. The latter is the same as that got from (3.19) to order N^{-3} and the former differs from the value got from (3.20) by N^{-2} as for $\tilde{r}(1)$. The second point is that in many fields when dealing with the test based on $r(1)$ or $\tilde{r}(1)$ (i.e., that $\beta(1) = 0$ and the data is

† The reason why this will be so may be gauged from Theorem 5 for this compares circular statistics for a circular population with noncircular statistics for a noncircular population. See the discussion following that theorem.

generally independent) it is customary to adopt a one-sided test (as we have implied in forming Table 1). This is because a negative true first autocorrelation is rare in practice. However, there seems to be no reason for using a one-sided test with higher order autocorrelations.

Beyond this point relatively little is known. The distribution (3.12) may be used to give a confidence interval for $\rho = \beta(1)$, in a first-order autoregression and a more accurate approximation for the, relevant, noncircular case has been obtained by Daniels (1956) and Jenkins (1956). Jenkins has used these results to suggest the use of the transformation from $r(1)$, ρ to y, θ given by

$$r(1) = \sin y, \qquad \rho = \sin \theta,$$

so that $(y - \theta)$ is used as normal with zero mean and variance N^{-1}. (Again the approximation has been shown by Jenkins to be better for $r(1)$ than for $\tilde{r}(1)$.) This enables a confidence interval to be allotted to ρ.

Some additional work has also been done by Chanda (1962, 1964) on tests of goodness of fit of the type discussed in Section 2. (See (3.11) and the discussion of Quenouille's test following it.) In the later paper cited Chanda investigates an asymptotic expansion for the distribution of a test due to Bartlett and Diananda (1950), which requires a prior knowledge of the autoregressive parameters on the null hypothesis. In the earlier paper he evaluates the mean and variance of the chi-square statistic

$$\chi_s^2 = \sum_{j=q+1}^{q+s} d_j^2$$

(see Section 2). He obtains these moments to order N^{-1} and forms a modified statistic

$$\chi_M^2 = \frac{A\chi_s^2}{B},$$

where

$$A = \mathcal{E}(\chi_s^2), \qquad B = \tfrac{1}{2}\operatorname{var}(\chi_s^2),$$

which he then treats as chi-square with degrees of freedom A^2/B. (The statistic χ_M^2 has expectation A^2/B and variance $(A^2 2B)/B^2 = 2A^2/B$, as is appropriate.) Of course A and B depend upon the unknown parameters of the hypothesized autoregressive process. He then compares the significance points for χ_s^2, obtained via χ_M^2, with those obtained by using χ_s^2 as chi-square with s degrees of freedom; for example, for Gaussian data with $N = 20$, $p = 1$, $s = 6$, $\rho = \pm 0.8$ (where ρ is the autoregressive parameter) the 5% point is, from χ_M^2, 13.1, whereas the 5% point for χ_6^2 is 12.6. For $N = 60$, $p = 1$, $s = 12$, $\rho = \pm 0.8$ the two significance points are 21.3 and

21.0. In second-order autoregressive cases the oscillatory situation is studied (see Chapter I, Section 4) and using the notation, a, θ of the section mentioned for $n = 20$, $a = \pm 0.08$, $\theta = \frac{1}{4}\pi$, $s = 6$ the significance point becomes 15.2 compared to 12.6. (This is much the largest discrepancy found.) For $n = 60$, $s = 12$, $a = \pm 0.8$, $\theta = \frac{1}{4}\pi$ the two significance points are 22.5 and 21.0. Chanda also examines the effects of nonnormality.

McGregor (1962) applied methods similar to those of Daniels to study the distribution of the coefficient of correlation between two Gaussian simple Markoff processes

$$x_1(n) = \rho_1 x_1(n-1) + \epsilon_1(n), \qquad x_2(n) = \rho_2 x_2(n-1) + \epsilon_2(n)$$

when the $\epsilon_2(n)$ are independent Gaussian with zero mean and unit variance. Except for the lack of a mean correction, this is a natural statistic to use to test the independence of $x_1(n)$ and $x_2(n)$, though, of course, the power of the test is directed against alternatives near that of a first-order autoregression for the vector process $x(n)$.

We will not reproduce the very complicated density here (which is accurate neglecting terms of order N^{-1}). The mean of the distribution is zero and the variance is near to $N^{-1}(1 + \rho_1\rho_2)/(1 - \rho_1\rho_2)$, which is the variance for the limiting normal distribution. (See Chapter IV, exercise 8.) The form of the distribution suggests that a satisfactory procedure should be to use r as an ordinary correlation from $N(1 - \rho_1\rho_2)/(1 + \rho_1\rho_2)$ observations. (Of course, ρ_1 and ρ_2 would need to be estimated from the data and mean corrections would have to be made.) This procedure was suggested by Bartlett (1935). However, when ρ_1 and ρ_2 are both nonzero, the test statistic $r_{12}(0)$ gives an inefficient test of the hypothesis that $x_1(n)$ and $x_2(n)$ are independent, and indeed a statistic may be found which, under many circumstances, will be more efficient than $r_{12}(0)$ and whose distribution, on Gaussian assumption, is exactly known (see Hannan, 1955b). This is obtained by regressing $x_2(2n)$, $n = 1, \ldots, [\frac{1}{2}(N-1)]$, on $\{x_2(2n-1) + x_2(2n+1)\}$, $x_1(2n)$, and $\{x_1(2n-1) + x_1(2n+1)\}$ and using the coefficient of $x_1(2n)$ as the test statistic. [We assume here that $x_2(n)$ is generated by a Gaussian simple Markoff process but need make no such assumption about $x_1(n)$.] We discuss this test and related exact tests for serial correlation in a series of exercises in this chapter as well as in the final chapter. In any case, more efficient tests than these exact tests may be obtained. We refer the reader again to the reference just cited as well as to the last chapter.

No other investigations of the asymptotic properties of statistics appropriate to the case $p > 1$ seems to have been made, even for the case $q = 1$, and we leave the subject in this somewhat unsatisfactory state (so far as $q > 1$).

4. MOVING AVERAGES AND MIXED AUTOREGRESSIVE, MOVING AVERAGE MODELS. INTRODUCTION

As has previously been said, considerable use has recently been made of mixed autoregressive moving average models. We refer the reader to Box and Jenkins (1970) for a detailed working out of inference procedures for such models, particularly in cases where the number of parameters to be estimated is small. We shall give an account here of a somewhat different kind and shall not attempt to duplicate their work. We begin by a survey of the problems involved in fitting these models, for the case $p = 1$.

A main motivation for the use of mixed models is the fact that they are more general than autoregressive models so that with a small amount of data, when economy in the introduction of parameters is important, their use may be essential. The main problem associated with their use is due to the difficulty in fitting the model. All techniques used ultimately go back to the method of maximum likelihood. We illustrate by the simplest case

$$x(n) = \epsilon(n) + \alpha\epsilon(n-1), \qquad |\alpha| < 1, \qquad \mathcal{E}(\epsilon(m)\,\epsilon(n)) = \delta_m{}^n\sigma^2,$$

where $x(n)$ is scalar and $\epsilon(n)$ is Gaussian. The likelihood function now is

$$(2\pi)^{-\frac{1}{2}N}\{\det(\Gamma_N)\}^{-\frac{1}{2}} \exp -\tfrac{1}{2}x'\Gamma_N{}^{-1}x,$$

where Γ_N has $\sigma^2(1 + \alpha^2)$ in the main diagonal, $\alpha\sigma^2$ in the first diagonal above and below the main diagonal and zeros elsewhere. The determinant $\Delta_N = \sigma^{-2N} \det(\Gamma_N)$ satisfies

$$\Delta_N = (1 + \alpha^2)\Delta_{N-1} - \alpha^2\Delta_{N-2}, \qquad \Delta_0 = 1, \qquad \Delta_1 = 1 + \alpha^2,$$

from which we obtain

$$\Delta_N = \frac{1 - \alpha^{2N+2}}{1 - \alpha^2}.$$

The matrix $\Gamma_N{}^{-1}$ is a rather complicated expression (see Shaman, 1969) but it is near to a matrix with $\sigma^{-2}(1 - \alpha^2)^{-1}(-\alpha)^{j-k}$ in row j column k, so that the log likelihood function is *near* to

$$\log L = c - \frac{N}{2\log\sigma^2} + \tfrac{1}{2}\log(1 - \alpha^2) - \frac{1}{2\sigma^2(1 - \alpha^2)}\sum_{-N+1}^{N-1}(-\alpha)^j s_j,$$

where

$$s_j = \sum_{1}^{N-j} x(n)\,x(n+j).$$

This gives for the maximum likelihood solution the equations

$$\hat{\sigma}^2 = \frac{1}{N}\frac{\sum_{-N+1}^{N-1}(-\hat{\alpha})^j s_j}{(1-\hat{\alpha}^2)},$$

$$\left\{-\hat{\alpha}(1-\hat{\alpha}^2) + \frac{\hat{\alpha}}{\hat{\sigma}^2}\sum_{-N+1}^{N-1}(-)^j\hat{\alpha}^j s_j - \frac{(1-\hat{\alpha}^2)}{2\hat{\sigma}^2}\sum_{-N+1}^{N-1}(-)^j j\hat{\alpha}^{j-1}s_j\right\} = 0.$$

Even if the solution is further approximated by dropping the first term in the last expression (on the heuristic grounds that it is $O(1)$ whereas the other will be $O(N)$) the resulting equation is extremely difficult to solve. Once one moves beyond the simplest case of a first order moving average, and considers, for example, a mixed autoregressive, moving average model, the solution by these methods becomes very difficult to manage. Box and Jenkins (1970) have in fact proceeded by direct evaluation of the likelihood function, or at least of the quadratic form in the exponent, dropping the analogue of the term $\frac{1}{2}\log(1-\alpha^2)$, on the same grounds as those given above.

In a mixed, autoregressive, moving average, model an identification problem arises, of the same nature as has earlier arisen for autoregressions alone. We achieve a unique form of representation by considering the spectrum, which must be of the form

$$f(\lambda) = \frac{\sigma^2\,|\sum_0^s \alpha(k)e^{ik\lambda}|^2}{2\pi\,|\sum_0^q \beta(j)e^{ij\lambda}|^2}, \qquad \alpha(0) = \beta(0) = 1,$$

if $x(n)$ satisfies

$$\sum_0^q \beta(j)x(n-j) = \sum_0^s \alpha(k)\epsilon(n-k), \tag{4.1}$$

the $\epsilon(n)$ being independent and identically distributed with zero mean and variance σ^2. We require that all solutions of

$$g(z) = 0, \qquad g(z) = \sum_0^s \alpha(k)z^k,$$

lie on or outside the unit circle and all solutions of

$$h(z) = 0, \qquad h(z) = \sum_0^q \beta(j)z^j,$$

lie outside the unit circle and that $g(z)$ and $h(z)$ have no zero in common. Given that the system is of the form (1) (with the $\epsilon(n)$ as stated) and the first condition is satisfied (i.e., the condition relating to the zeros of $g(z)$ and $h(z)$), the requirement that $g(z)$ and $h(z)$ have no common zero can clearly be imposed also. Except in the Gaussian case, however, it is not necessarily true that if $x(n)$ satisfies a relation of form (4.1), with $h(z)$ having no zeros on the unit circle, then all of the conditions on the zeros of $h(z)$ and $g(z)$ can be made to be satisfied (with $\epsilon(n)$ independent and not merely orthogonal).

For such a model as (4.1) Box and Jenkins calculate an approximation to the sum of squares,

$$\sum_1^N \epsilon(n)^2,$$

at least for the case in which all zeros of $g(z)$ also lie outside of the unit circle. This sum of squares is the basic component in the logarithm of the likelihood function (on Gaussian assumptions), as can be seen by considering (4.1) as a transformation from $\epsilon(n)$, $n = -s + 1, \ldots, N$ to $x(n)$, $n = 1, \ldots, N$. Box and Jenkins calculate the sum of squares by forming

$$e(n) = -\sum_1^s \alpha(k)e(n-k) + x(n), \qquad n = -q + 1, \ldots, N,$$

commencing from s initial values $e(-q + 1 - s), \ldots, e(-q)$. Then $\epsilon(n)$ is calculated as

$$\epsilon(n) = \sum_0^q \beta(j)e(n-j).$$

If the zeros of the equation

$$\sum_0^s \alpha(k)z^{s-k} = 0$$

are all a good deal less than unity, then, when N is large, the result should depend only to a minor degree on the initial values, and finally a good approximation to

$$\sum_1^N \epsilon(n)^2,$$

and thus to the main component in the log likelihood function, will have been found. Then the minimum of this expression (which is, of course, a function of $\alpha(1), \ldots, \alpha(s)$, $\beta(1), \ldots, \beta(q)$ and the initial values) is found by a search over a reasonably fine grid and thus what should be a good approximation to the maximum likelihood estimators is discovered. The method will apparently lead to a very great computational burden if $(q + s)$ is large (when N must be large also); for example, with $q = s = 5$ and $N = 200$, the examination of a grid of 30 points for each parameter requires something greater than 10^{18} multiplications, which is not a feasible calculation. Methods based on steepest descents may reduce the labour but the task of evaluating the large number of derivatives is arduous.† Thus the method seems restricted to cases where $(q + s)$ is small (perhaps no greater than 3) or where the number of unknown parameters is small, as is the case when one or both $g(z)$ and $h(z)$ have $(z - 1)$ as a factor, perhaps repeated. Box and Jenkins have in fact used such models very successfully and have fitted models of high order in q and s by this device of taking $(z - 1)$ as a factor‡ (i.e., of representing the filter described by $g(z)$, say, as composed of two factors, of which one is a differencing operator). We refer the reader to the work of these authors for further details. We shall also discuss below some of the problems associated with this (and any other technique) in the mixed case.

We shall work with mean corrected quantities throughout this section and so may assume that $x(n)$ and $\epsilon(n)$ have zero mean vectors.

We shall begin by examining some existing techniques, other than those already discussed, for estimating the parameters in a rational spectrum. In the Section 5 we shall present a treatment based on maximum likelihood (assuming normality) but with the likelihood function expressed, approximately, in spectral terms, so as to simplify the problem. The technique will be to find an estimator which, at least asymptotically, has the same properties as the maximum likelihood estimator. These properties will not be restricted to the Gaussian case. (Thus we shall establish the asymptotic normality of the estimators assuming only that the innovation sequence, $\epsilon(n)$, is serially independently and identically distributed with finite variance.) The estimators are only approximations to the maximum likelihood estimators, on Gaussian assumptions, both because the spectral form of the likelihood which is used is only an approximation and because even then the likelihood equations are highly nonlinear and can only be iteratively solved. These are not major points for in the first place they will apply (the first to a lesser extent perhaps) even when the likelihood is not expressed in spectral form and in the second place because the data will not be Gaussian, in practice, so that the maximum likelihood estimator has no "sacred" meaning.

We shall begin our discussion by considering the scalar moving average case partly because it is so much simpler, partly so as to motivate the more elaborate considerations for the mixed case. Thus our model is

$$x(n) = \sum_{0}^{s} \alpha(k)\epsilon(n - k), \qquad \alpha(0) = 1,$$

† In any case Box and Jenkins emphasize that not merely the maximum likelihood estimator is required but also that they wish to examine the likelihood function in its entirety, so that the calculation of $\Sigma\epsilon(n)^2$ for a grid of parameter values is part of their routine. No doubt there is much to be said for this.

‡ They also consider cases where $(z - \exp i\theta)$, θ known, is a factor, for example in connection with seasonal variation.

where $g(z)$ is subjected to the restriction named above. We shall in fact assume that $g(z)$ has no zeros in *or on* the unit circle. It is not easy to imagine circumstances where a zero on the unit circle can arise without it being known a priori. Indeed on prior grounds one would imagine that this case will arise with probability zero. If it is known that there is such a zero and the location of the zero is known the methods given below may be modified (by a preliminary filtering of the data) so that the estimates of the $\alpha(k)$ will retain the properties we shall derive. We shall not discuss this further.

We commence from the natural estimator of the spectrum of $x(n)$, namely,†

$$\hat{f}(\lambda) = \frac{1}{2\pi} \sum_{-s}^{s} c(j) e^{-ij\lambda}. \tag{4.2}$$

Assuming only, as we shall, that the $\epsilon(n)$ are i.i.d. $(0, \sigma^2)$ this will converge with probability 1 to $f(\lambda)$. It is immediately apparent that $\hat{f}(\lambda)$ is not necessarily nonnegative, but this will be so if s is taken large enough for with $s = N$ we get the periodogram, $I(\lambda)$. However, this last observation is not of much value for the methods we present below depend upon s being fixed a priori. Provided $\hat{f}(\lambda)$ is positive we have an immediate procedure for estimating the $\alpha(k)$, namely the determination of the, essentially unique, factorization

$$\hat{f}(\lambda) = \frac{\hat{\sigma}^2}{2\pi} \left| \sum_{0}^{s} \hat{\alpha}(k) e^{ik\lambda} \right|^2 = \frac{\hat{\sigma}^2}{2\pi} |\hat{g}(e^{i\lambda})|^2, \qquad \hat{\alpha}(0) = 1, \tag{4.3}$$

wherein the $\hat{\alpha}(k)$ are such that $\hat{g}(z)$ has all of its zeros outside the unit circle. Provided there are no zeros of $g(z)$ on the circle, which we have assumed, then evidently for N large enough $\hat{f}(\lambda)$ *will* be positive and we have found initial estimates of the $\alpha(k)$. These may be very inefficient, however, as we shall shortly show, and unless s is small the labor of finding the estimates is not inconsiderable. Since we present an efficient method in Section 5 (efficient in a sense to be made precise in that section) we shall discuss only a special case here, to demonstrate the inefficiency of the $\hat{\alpha}(k)$. We consider the case $s = 1$ when the equation (4.3) for $\alpha(1)$, which we now call α, becomes

$$r(1) = \frac{\hat{\alpha}}{1 + \hat{\alpha}^2}$$

which has a real solution only if $|r(1)| \le \frac{1}{2}$. Then

$$\hat{\alpha} = \frac{2r(1)}{1 + (1 - 4r(1)^2)^{1/2}}.$$

† Throughout this section and the next we shall use a hat for an inefficient estimate and a tilde for a reasonably efficient one.

We may find the (asymptotic) variance of $\hat{\alpha}$ by a direct argument but here it will suffice to point out that $\sqrt{N}(\hat{\alpha} - \alpha)$ is asymptotically normal with variance

$$\frac{1 + \alpha^2 + 4\alpha^4 + \alpha^6 + \alpha}{(1 - \alpha^2)^2}. \tag{4.4}$$

To show this first put $\{1 - 4r(1)^2\}^{1/2} = \phi(r)$, $\{1 - 4\rho(1)^2\}^{1/2} = \phi(\rho)$, for simplicity of printing; then

$$\begin{aligned}
\sqrt{N}(\hat{\alpha} - \alpha) &\\
&= \sqrt{N}\left\{\frac{2r(1)}{1 + \phi(r)} - \frac{2\rho(1)}{1 + \phi(\rho)}\right\}\\
&= \sqrt{N}\big(r(1) - \rho(1)\big)\frac{2}{1 + \phi(\rho)} + \frac{2r(1)\sqrt{N}\{\phi(\rho) - \phi(r)\}}{\{1 + \phi(r)\}\{1 + \phi(\rho)\}}\\
&= \sqrt{N}\big(r(1) - \rho(1)\big)\frac{2}{1 + \phi(\rho)} + \frac{8r(1)\sqrt{N}\{r(1)^2 - \rho(1)^2\}}{\{1 + \phi(r)\}\{1 + \phi(\rho)\}\{\phi(r) + \phi(\rho)\}}\\
&= \sqrt{N}\big(r(1) - \rho(1)\big)\left\{\frac{2}{1 + \phi(\rho)} + \frac{8r(1)\big(r(1) + \rho(1)\big)}{\{1 + \phi(r)\}\{1 + \phi(\rho)\}\{\phi(r) + \phi(\rho)\}}\right\}.
\end{aligned}$$

The last factor converges with probability one to $(1 + \alpha^2)^2/(1 - \alpha^2)$ since it converges to

$$\frac{\alpha}{\rho(1)} + \frac{2\alpha^2}{\phi(\rho)} = 1 + \alpha^2 + \frac{2\alpha^2(1 + \alpha^2)}{1 - \alpha^2} = \frac{(1 + \alpha^2)^2}{1 - \alpha^2}.$$

However, we already know (see the end of Chapter V) that $\sqrt{N}\big(r(1) - \rho(1)\big)$ is asymptotically $N\big(0, 1 - 3\rho(1)^2 + 4\rho(1)^4\big)$. Simplifying this we arrive at the expression (4.4). We shall obtain, in the next Section, an estimator, $\tilde{\alpha}$, for which $\sqrt{N}(\tilde{\alpha} - \alpha)$ is asymptotically $N(0, 1 - \alpha^2)$ so that the asymptotic efficiency of $\tilde{\alpha}$ relative to $\hat{\alpha}$ is $(1 - \alpha^2)^3/(1 + \alpha^2 + 4\alpha^4 + \alpha^6 + \alpha^8)$ which is 0.76 for $\alpha = \frac{1}{4}$ and 0.26 for $\alpha = \frac{1}{2}$. Unless α is quite small the efficiency is unacceptably low.

The first person to suggest a more efficient estimator (other than maximum likelihood) was Durbin (1959) who pointed out that, under the conditions given above, one could proceed by carrying out a *long* autoregression of $x(n)$ on $x(n - 1)$, $x(n - 2), \ldots, x(n - m)$. If m is sufficiently large all further autoregressive coefficients are effectively zero and we may commence from the joint distribution of these autoregressive coefficients. Let us call b_j the jth of these (this being at variance with our previous notation but necessary because of the place of autoregression elsewhere in the present section). Proceeding heuristically (we prefer to present a precise treatment of another method, later) we see that the b_j have a joint distribution which is normal with mean β_j (the "true" autoregression) and covariance matrix $N^{-1}\sigma^2\Gamma_m^{-1}$. This matrix, Γ_m^{-1}, as we have indicated in Section 3 has elements in the jth diagonal above the main diagonal and in the kth row

$$\sum_u \beta_u \beta_{u+j},$$

where the sum is for $0 \leq u \leq \min(k-1, m-j, m-k-j)$. Also, since

$$x(n) = \sum_0^s \alpha(j)\epsilon(n-j),$$

then

$$\sum \beta_j x(n-j) = \epsilon(n),$$

so that, ignoring β_j for $j > m$,

$$\left|\sum_0^s \alpha(j)e^{ij\lambda}\right|^2 \left|\sum_0^m \beta_j e^{ij\lambda}\right|^2 = 1;$$

that is,

$$\sum_0^s \alpha(j)e^{ij\lambda} \left|\sum_0^m \beta_j e^{ij\lambda}\right|^2 = \left(\sum_0^s \alpha(j)e^{-ij\lambda}\right)^{-1},$$

so that

$$\sum_{j=0}^s \alpha(j) \sum_u \beta_u \beta_{u+k-j} = 0, \qquad k = 1, 2, \ldots, m, \tag{4.5}$$

since the right-hand side of the previous equation may be expanded in a series of nonnegative powers of $\exp -i\lambda$. Inserting our estimates b_j of the β_j we obtain estimates $a(k)$ of the $\alpha(k)$ from (4.5). Using the covariance matrix for the b_j established above, the covariance matrix of the $a(k)$ may be obtained. It is not easy to prove an asymptotic theorem concerning these estimators because of the difficulty in determining m, though no doubt such a theorem could be established. It will, of course, be necessary to allow m to increase indefinitely with N if an asymptotic result is to be established since if m is held fixed the method will not be consistent. A problem with the method is the determination of m, for which an appropriate value depends on the true $\alpha(j)$.

We discuss next a method due to Walker (1962) which has the same kind of asymptotic properties as are claimed for Durbin's technique and for which an asymptotic theory is fairly easily obtained. The defect of Walker's method is that it leads, not to estimates of the $\alpha(k)$ directly but instead to estimates of the $\rho(j)$. These, of course, are to be better estimates than the $r(j)$. Then the $\alpha(k)$ have to be obtained by factoring an estimate of $f(\lambda)$

obtained from these improved estimates of the $\rho(j)$. If s is large this requires the solution of a polynomial of high degree, which is unpleasant. There is also the problem that the estimate of $f(\lambda)$ may not factor.

Walker's method uses the first $s + k$ autocorrelations, where k is fixed in advance. The choice of k affects the efficiency of the method, however, so that in truth one needs to consider an estimation of k from the data. We shall discuss this point again later. The joint distribution of the first $(s + k)$ of the $\sqrt{N}(r(j) - \rho(j))$ is asymptotically normal with zero mean. The covariance between the lth and mth of these statistics (obtainable from exercise 6 to Chapter IV) is

$$\mu(l, m) = \sum_{t=-\infty}^{\infty} \{\rho(t)\,\rho(t + l - m) + \rho(t + l)\,\rho(t - m) - 2\rho(t)\,\rho(l)\,\rho(t - m) - 2\rho(t)\,\rho(m)\,\rho(t - l) + 2\rho(t)^2\,\rho(l)\,\rho(m)\}.$$

Since $\rho(j) = 0$, $|j| > s$, only the first s of the $r(j)$ are directly available† to estimate the $\rho(j)$. One may consider the factoring of the joint distribution of the $r(j)$ into the distribution of the $r(j), j = 1, \ldots, s$, for given values of the $r(j), j = s + 1, \ldots, s + k$ and the joint distribution of this last set. It is clearly the first factor from which the information concerning the $\rho(j)$ is to be derived. This leads directly to the use of the residuals from the regression of $r(j), j = 1, \ldots, s$, on the remaining k of these autocorrelations as the estimate of the $\rho(j)$. The regression coefficients are not known and must be estimated from the data. We shall not go into details. The final estimate is

$$\hat{r} = r - \hat{Q}\hat{M}_k^{-1} r_k, \tag{4.6}$$

where

(i) r has $r(j)$ in the jth row, $j = 1, \ldots, s$;

(ii) $\hat{Q}$ has $\hat{\mu}(j, s + m)$ in row j, column $m, j = 1, \ldots, s;\ m = 1, \ldots, k$, where $\hat{\mu}$ is got from μ by replacing $\rho(j)$ by $r(j)$;

(iii) $\hat{M}_k$ has $\hat{\mu}(s + l, s + m)$ in row l, column $m, l, m = 1, \ldots, k$;

(iv) r_k has $r(s + j)$ in row $j, j = 1, \ldots, k$.

If ρ has $\rho(j)$ in the jth row, $j = 1, \ldots, s$, then it is not difficult to show that $\sqrt{N}(\hat{r} - \rho)$ is asymptotically normal but we shall not quote the covariance matrix for what we need is not $\hat{r}$ but an estimate of α. For this it is necessary

† If the $\mu(l, m)$ were known numbers but the $\rho(j)$ were not (which is somewhat contradictory) the $r(j)$ for $j > s$ would be of the nature of ancillary statistics. In practice an iterative procedure is needed, where the $\mu(l, m)$ are estimated from a previous iteration, so that this comment *is* relevant.

to factorize the estimated spectral density, or equivalently to factorize,

$$\sum_{-s}^{s} \hat{r}(j)e^{ij\lambda}.$$

If s is large this is not a simple task, needless to say. (However, an algorithm for this factorization is given in Wilson, 1969.) Moreover, one must now convert the covariance matrix for the $\hat{r}(j)$, $j = 1, \ldots, s$, into one for the estimates of the $\alpha(j)$. This results in even more complicated expressions involving the estimate of the Jacobian of the transformation from the $\hat{r}(j)$ to the estimates of the $\alpha(j)$. We shall proceed no further but instead move to the next section where we discuss some asymptotically satisfactory estimates obtained via spectral methods.

5. THE ESTIMATION OF MOVING AVERAGE AND MIXED MOVING AVERAGE AUTOREGRESSIVE MODELS USING SPECTRAL METHODS

The techniques we shall use are based on estimates of the spectrum. This may seem strange at first sight for the tradition has been to use "time domain" techniques to estimate the parameters of processes with rational spectra, that is, techniques based upon autocovariances, and these certainly are satisfactory for autoregressions. However, we already know that the transformation from the $x(n)$ to the $w(\omega_k)$ simplifies considerations since, approximately, it diagonalizes the covariance matrix. It can be expected to simplify the equations of estimation in the present case also. In fact the equations which we derive reduce to the equations of estimation for autoregressions, previously introduced, in the autoregressive case.

We shall show that the estimation methods we introduce are asymptotically efficient (and the same applies to the methods of Section 2) in the following sense. If the data is Gaussian, we may calculate the maximum likelihood estimators and consider their limiting distribution. The estimators themselves may be used whether the data is Gaussian or not. Their asymptotic distribution will turn out to be independent of the Gaussian assumption given (only) the requirement that the $\epsilon(n)$ are i.i.d. $(0, G)$. We shall show that our estimators have asymptotically the same distribution as the maximum likelihood estimator under this same condition. It is in this sense that we speak of them as asymptotically efficient. Of course the maximum likelihood estimator is computationally too difficult to be directly used otherwise we should use it.

(a) Moving Averages. The Case $p = 1$

We proceed as follows, operating heuristically to obtain estimators and later proving that they have desirable properties. The likelihood function on

Gaussian assumptions is predominantly a function of the quadratic form $x'\Gamma_N^{-1}x$, as we have earlier suggested. This is of the form

$$w^*(P\Gamma_N P^*)^{-1}w,$$

where w has $w(\omega_j)$ in row j (see Chapter V, Section 2), whereas the unitary matrix P has $(N)^{-\frac{1}{2}} \exp in\omega_j$ in row j column n. The matrix $P\Gamma_N P^*$ will be very near to diagonal with $f(\omega_j)$ in the jth place in the main diagonal, in the sense that the omitted term, in each element, will be $O(N^{-1})$.† Indeed, the element in row u, column v, is

$$\frac{1}{N}\sum\sum \gamma(n-m)e^{i(n\omega_u - m\omega_v)} = \sum_{-s}^{s}\gamma(j)e^{ij\omega_u}\left\{N^{-1}\sum_m e^{im(\omega_u-\omega_v)}\right\}.$$

where m lies between 1 and $N-j$ for j positive or zero and $-j+1$ and N for j negative. For $\omega_u \neq \omega_v$ this is evidently of order N^{-1} while for $u = v$ it is

$$\sum_{-s}^{s}\gamma(j)\left(1-\frac{|j|}{N}\right)e^{ij\omega_u},$$

which differs from $2\pi f(\lambda)$ by a quantity of order N^{-1}. Thus to order N^{-1} we have to minimize

$$\sum_0^{N-1}\frac{I(\omega_j)}{f(\omega_j)}.$$

This leads to the equations

$$\sum_{l=0}^{s}\tilde{\alpha}(l)\sum_{j=0}^{N-1}\frac{I(\omega_j)\cos(k-l)\omega_j}{\tilde{f}(\omega_j)^2} = 0, \qquad k = 1, \ldots, s, \ \tilde{\alpha}(0) = 1,$$

where now

$$\tilde{f}(\omega_j) = \frac{\tilde{\sigma}^2}{2\pi}\left|\sum_0^s \tilde{\alpha}(u)e^{iu\omega_j}\right|^2.$$

These equations would be very difficult to solve. Since $\tilde{f}(\omega_j)^{-2}$ enters, apparently, as a weight function we are led to replace it by $\hat{f}(\omega_j)^{-2}$ so that our estimation equations become

$$(5.1)\qquad \sum_{l=0}^{s}\hat{\alpha}^{(1)}(l)\sum_{j=0}^{N-1}\frac{I(\omega_j)\cos(k-l)\omega_j}{\hat{f}(\omega_j)^2} = 0, \qquad k = 1, \ldots, s, \ \hat{\alpha}(0) = 1.$$

Here $\hat{f}(\omega_j)$ *is computed from* (4.2). (It would be preferable to replace it by an estimate obtained from Walker's estimates $\hat{r}(j)$ [see (4.6)] if these were available. However, it will be no more difficult, computationally, to obtain the *asymptotically* efficient estimator of α to which $\hat{f}$ will lead us and a second

† See Theorem 7, Chapter III, for a related result.

iteration, commencing with this, will probably be preferable to computing $\hat{r}$ and beginning from that.) The superscript (1) on $\hat{\alpha}^{(1)}(l)$ is meant to differentiate it from the first step estimator $\hat{\alpha}(l)$, got from the $c(j)$, which we *might* have called $\hat{\alpha}^{(1)}(l)$.

As we show later (5.1) does *not* give an improved estimator. Somewhat surprisingly, we obtain an asymptotically efficient estimator by forming the vector

$$\tilde{\alpha}^{(1)} = 2\hat{\alpha} - \hat{\alpha}^{(1)}. \tag{5.2}$$

[Of course, if we use Walker's estimates, $\hat{r}(j)$, to obtain $\hat{\alpha}^{(1)}$ we shall also have to replace $\hat{\alpha}$ by his estimates of α in (5.2).] An estimate of $f(\lambda)$ will have to be factored to get $\hat{\alpha}$. *As we shall later show this factorization can be avoided.*

We may write (5.1) in the form

$$\hat{\alpha}^{(1)} = -\hat{A}^{-1}\hat{a} \tag{5.3}$$

where $\hat{A}$ has $\hat{a}_{k-l}$ in row k, column l, $k, l = 1, \ldots, s$, while $\hat{a}$ has $\hat{a}_k$ in row k, $k = 1, \ldots, s$, and

$$\hat{a}_k = N^{-1} \sum_{j=0}^{N-1} \frac{I(\omega_j)}{\hat{f}(\omega_j)^2} \cos k\omega_j, \qquad k = 0, 1, \ldots, s. \tag{5.4}$$

Computationally it will be preferable to replace $\hat{a}_k$ by

$$\frac{I(0)}{\hat{f}(0)^2} + 2 \sum_{1}^{[\frac{1}{2}(N-1)]} \frac{I(\omega_j) \cos k\omega_j}{\hat{f}^2(\omega_j)} + \frac{(-)^k I(\pi)}{\hat{f}^2(\pi)}, \tag{5.5}$$

where the last term occurs only when N is even and the first will be zero if mean corrections are made (which will almost always be the case). Indeed we might also replace $I(\omega_j)$ by

$$\left(\sum_1^N x(n) \cos n\omega_j\right)^2 + \left(\sum_1^N x(n) \sin n\omega_j\right)^2.$$

The elimination of N^{-1} in (5.4) and of the factor $(2\pi N)^{-1}$ in the $I(\omega_j)$ will not affect (5.3), of course. The formula (5.3) can be viewed as the usual least squares formula for $-\alpha$ but with covariances computed from (5.4).

It is instructive to put the formula for an autoregression in this form. If the $c(n)$ are replaced by $(1 - |n|/N)c(n)$ (see Section 2 immediately below the proof of Theorem 1) then, for $p = 1$, the formula of Section 2 become

$$\hat{\beta} = -\hat{B}^{-1}\hat{b},$$

where the $\hat{b}_k$ out of which $\hat{B}$ and $\hat{b}$ are constructed are

$$\hat{b}_k = N^{-1} \sum_{j=0}^{N-1} I(\omega_j) \cos k\omega_j, \qquad k = 0, 1, \ldots, q.$$

The division by $\hat{f}(\omega_j)$ in (5.4) is designed to change the spectrum to a multiple of $|g(\omega_j)|^{-2}$, i.e., to the spectrum of an autoregression.

We introduce the matrix Φ with $\phi(k-l)$ in row k, column l, $k, l = 1, \ldots, s$, where

$$\phi(k) = \frac{1}{2\pi}\int_{-\pi}^{\pi} \frac{1}{|g(e^{i\lambda})|^2} e^{ik\lambda}\, d\lambda.$$

Then we have the following theorem, whose proof we have put in the appendix to this chapter.

Theorem 7. *Let $x(n)$, for $p = 1$, be generated by a moving average of order s for which $g(z)$ has all of its zeros outside of the unit circle and $\epsilon(n)$ is i.i.d. $(0, \sigma^2)$. If $\tilde{\alpha}^{(1)}$ is defined by (5.2) then $\tilde{\alpha}^{(1)}$ converges almost surely to α and $\sqrt{N}(\tilde{\alpha}^{(1)} - \alpha)$ is asymptotically normal with zero mean and covariance matrix Φ^{-1}. The matrix Φ^{-1} is consistently estimated by*

$$\hat{\Phi}^{-1} = \left\{\sum_0^s \hat{\alpha}(j)\hat{a}_j\right\}\hat{A}^{-1}$$

where $\hat{\alpha}$ is obtained (see (4.3)) by factoring (4.2), $\hat{A}$ is defined below (5.3) and $\hat{a}_j$ by (5.4).

A number of comments need to be made in relation to this theorem.

1. In the first place it will often be worthwhile carrying through further iterations of the process, so that $\tilde{\alpha}^{(1)}$ is used to estimate $|g|^2$ from the insertion of which estimate, $|\tilde{g}^{(1)}|^2$ let us say, in (5.1) in place of $\hat{f}$, will result an estimate $\hat{\tilde{\alpha}}^{(1)}$ which is used to give a further estimate $\tilde{\alpha}^{(2)} = 2\tilde{\alpha}^{(1)} - \hat{\tilde{\alpha}}^{(1)}$. Of course $\tilde{\alpha}^{(2)}$ will have the same asymptotic properties as $\tilde{\alpha}^{(1)}$. Experience (which we report below) suggests that quite a few iterations may be needed in some cases.

In more detail, we obtain $\hat{\tilde{\alpha}}^{(1)}$ by means of

$$\hat{\tilde{\alpha}}^{(1)} = -\tilde{A}^{(1)^{-1}}\tilde{a}^{(1)}, \tag{5.6}$$

wherein $\tilde{A}^{(1)}$ has $\tilde{a}_{k-l}^{(1)}$ in row k, column l, $k, l = 1, \ldots, s$, whereas $\tilde{a}^{(1)}$ has $\tilde{a}_k^{(1)}$ in the kth place, and where

$$\tilde{a}_k^{(1)} = N^{-1}\sum_{j=0}^{N-1} \frac{I(\omega_j)}{|\tilde{g}^{(1)}(\omega_j)|^4} \cos k\omega_j, \qquad k = 0, 1, \ldots, s; \tag{5.7}$$

$$\tilde{g}^{(1)}(\omega_j) = \sum_0^s \tilde{\alpha}^{(1)}(k)e^{ik\omega_j}.$$

(The $\tilde{a}_k^{(1)}$ and $\hat{a}_k$ do not estimate the same thing, but rather they estimate parameters differing by a factor $(2\pi)^{-2}\sigma^4$, but the risk of confusion has to be accepted in the interests of a simpler notation.)

Now we put

$$\tilde{\boldsymbol{\alpha}}^{(2)} = 2\tilde{\boldsymbol{\alpha}}^{(1)} - \hat{\tilde{\boldsymbol{\alpha}}}^{(1)}.$$

In the following we assume that r iterations are completed.

2. We might also proceed in the following fashion, which avoids the problem of factoring the spectral density.

(i) We form $\hat{\boldsymbol{\alpha}}^{(1)}$ exactly as in (5.3).

(ii) We then form $\hat{\boldsymbol{\alpha}}^{(2)}$, first using $\hat{\boldsymbol{\alpha}}^{(1)}$ to form $\hat{g}^{(1)}$ as $\tilde{\boldsymbol{\alpha}}^{(1)}$ was used to form $\tilde{g}^{(1)}$ in (5.7). Then the $\hat{a}_k^{(1)}$ are formed using $\hat{g}^{(1)}$ as $\tilde{a}^{(1)}$ was formed from $\tilde{g}^{(1)}$ in (5.7). Then we obtain $\hat{\boldsymbol{\alpha}}^{(2)}$ as

$$\hat{\boldsymbol{\alpha}}^{(2)} = -\hat{A}^{(1)^{-1}}\hat{a}^{(1)}.$$

Finally

$$\tilde{\boldsymbol{\alpha}}^{(1)} = 2\hat{\boldsymbol{\alpha}}^{(1)} - \hat{\boldsymbol{\alpha}}^{(2)},$$

where we have, somewhat ambiguously, continued to use $\tilde{\boldsymbol{\alpha}}^{(1)}$ for our first asymptotically efficient estimator. All of our statements of the theorem and in the notes below hold true for either definition of $\tilde{\boldsymbol{\alpha}}^{(1)}$.

Further iterations may now be completed, commencing from $\tilde{\boldsymbol{\alpha}}^{(1)}$, exactly as in paragraph 1 just preceding.

3. The estimate $\hat{\Phi}^{-1}$ is not asymptotically efficient since it is founded upon $\hat{\boldsymbol{\alpha}}$, which is not efficient. An asymptotically efficient estimator may be obtained by using the $\tilde{\alpha}^{(r)}(j)$ to estimate $|g|^{-2}$ and hence the $\phi(j)$ and Φ^{-1}. An alternative asymptotically efficient estimate of Φ^{-1} is

$$\tilde{\Phi}^{(r)^{-1}} = \left\{\sum_0^s \tilde{\alpha}^{(r)}(j)\tilde{a}_j^{(r)^{-1}}\right\}\hat{A}^{(r)^{-1}} \tag{5.8}$$

If a further iteration is to be completed all of the computations for $\tilde{\Phi}^{(r)-1}$ will have been done.

4. The computations are quite reasonable and lead directly to estimates of the $\alpha(j)$. In cases where a moving average is being fitted the number of observations is unlikely to be very large, for with very large samples the same type of objections to these models will arise as was discussed at the beginning of Chapter V. In such circumstances, for $N \leq 500$ let us say, the computation of the $I(\omega_j)$ is no great task. Once this is done the remaining calculations are equivalent to that for the estimation of a regression on r variables. In any case, if sample sizes too great for the direct computation of $I(\omega_j)$ are encountered, then $I(\omega_j')$, $\omega_j' = 2\pi j/N'$, $N \leq N' = 2^s < 2N$ and N' could replace $I(\omega_j)$ and N in (5.4). The asymptotic properties of

the estimator will not be affected, as is evident from our proof. In fact, the summation could be taken over $\lambda_j = \pi j/M$ and smoothed estimates of spectra of the form of the $\hat{f}(\lambda_j)$ discussed in the Chapter V used in place of the periodogram. We shall, however, continue to work in terms of $I(\omega_j)$, for simplicity of presentation of the results and because this seems the best procedure to use for smaller sample sizes.

5. We shall show that the procedure leads to asymptotically efficient estimation procedures. These results are of limited value for really small sample sizes. A more exact distribution theory for such sample sizes is needed.

6. It is fairly apparent that if a fixed finite number of terms is omitted from the inner summation in (5.1) the asymptotic properties of the estimators will not be affected. Thus, if $a(l)$ is the estimate obtained when a set of p values of ω_j is omitted, then

$$(5.9)\quad \sum_{l=1}^{s} \sqrt{N}\{\hat{\alpha}^{(1)}(l) - a(l)\}\left\{\frac{1}{N}\sum_{j=0}^{N-1}\frac{I(\omega_j)\cos(k-l)\omega_j}{\hat{f}(\omega_j)^2}\right\} = -\sum_{0}^{s} a(l)\frac{1}{\sqrt{N}}\sum{}'\frac{I(\omega_j)\cos(k-l)\omega_j}{\hat{f}(\omega_j)^2},$$

where $\sum'$ is a sum over the p omitted values. The right-hand side converges in probability to zero while the matrix, with entries

$$\frac{1}{N}\sum_{j=0}^{N-1}\frac{I(\omega_j)\cos(k-l)\omega_j}{\hat{f}(\omega_j)^2}$$

in row k, column l, $k, l = 1, \ldots, r$, will be shown, in the Appendix to this Chapter, to converge with probability 1 to a nonsingular limit. Thus $\sqrt{N}(\hat{\alpha}^{(1)}(l) - a(l))$ converges in probability to zero. (The same will undoubtedly be true if a set of values is omitted which is dependent on N if the right hand term in (5.9) converges in probability to zero, but we do not pursue this point.) In particular this observation shows that a simple mean correction has no effect on our results for this corresponds to dropping the term for $j = 0$ from the inner summation in (5.5).

7. It is easily seen that the variance of $\sqrt{N}(\tilde{\alpha}^{(r)}(s) - \alpha(s))$ in its limiting distribution is $\{1 - \alpha(s)^2\}$, for it is the variance of the sth autoregressive coefficient in an autoregression of that order when the sth autoregressive constant is $\alpha(s)$. [See Section 2, just above (5.10).] To test $\alpha(s) = 0$ we form $\sqrt{N}\,\tilde{\alpha}^{(r)}(s)/(1 - \tilde{\alpha}^{(r)}(s)^2)^{1/2}$, or perhaps

$$\frac{\sqrt{(N-s)}\,\tilde{\alpha}^{(r)}(s)}{\sqrt{1-\tilde{\alpha}^{(r)}(s)^2}},$$

and use this as t with $(N - s)$ degrees of freedom. Of course our theory tells us only that this is asymptotically $N(0, 1)$ if $\alpha(s) = 0$ but a vague conjecture is that the use of t will be better if N is not large.

8. The $\tilde{\alpha}^{(i)}(k)$ provide an estimate of $g(z)$ having all of its zeros outside of the unit circle as also does $\hat{\tilde{\alpha}}^{(1)}$, for example. Indeed any estimate got from equations of the form of (5.1) will do this. To see this we refer to exercise 7 to this chapter which shows that this will be so, for $\hat{\alpha}^{(1)}$ for example, provided the matrix, with entries $\hat{a}_{k-l}$ in row k column l, k, $l = 0, \ldots, s$, is positive definite. But if this is not so it means that

$$\sum_{j=0}^{N-1} \frac{I(\omega_j)}{\hat{f}(\omega_j)^2} \sum_0^s\sum u_k u_l e^{i(k-l)\omega_j} = 0,$$

for some u_k not all zero; but since

$$\sum_0^s u_k e^{ik\omega_j}.$$

can be zero for at most s values of ω_j (unless the u_k are all zero), the matrix can fail to be positive definite only if $(N - s)$ values of $I(\omega_j)$ are null. Excluding the trivial case the result follows.

It does not seem to be true that $\tilde{\alpha}^{(1)}$, for example, satisfies the same condition but if the process is iterated, as suggested under 1 and 2 above, eventually $\tilde{\alpha}^{(i)} = 2\tilde{\alpha}^{(i-1)} - \hat{\tilde{\alpha}}^{(i-1)}$ will be arbitrarily close to $\tilde{\alpha}^{(i-1)}$ and $\hat{\tilde{\alpha}}^{(i)}$, the latter of which does satisfy the condition. In this sense the procedure provides estimates which give an estimate of $g(z)$ with all zeros outside of the unit circle.

It should perhaps be emphasized that points at which $f(\lambda)$ is very small are associated with a large amount of information in relation to the estimation of a moving average and that it is important to allow them to enter into the estimation procedure. We discuss a test of goodness of fit for models of the type discussed here (as well as others) in the last section of this chapter. Of course, such a test could be constructed from a test of $\tilde{\alpha}^{(r)}(s + 1), \ldots,$ $\tilde{\alpha}^{(r)}(s + k)$, for k large, by analogy with the autoregressive case. The interested reader may also consult Walker (1967) for a test procedure comparing autoregressive and moving average models.

(b) Moving Averages. The Case $p > 1$

We now consider the case $p > 1$. We deal only with the extension of our own method as presented in Theorem 7. (The examination of the likelihood function would now be difficult because of the number of parameters involved.) Then we are considering

$$x(n) = \sum_0^s A(j)\epsilon(n-j), \qquad A(0) = I_p, \tag{5.10}$$

where the $\epsilon(n)$ are i.i.d. $(0, G)$. It is, of course, possible that while $x(n)$ has p components $\epsilon(n)$ has fewer. This case needs consideration but we omit it here and, equivalently, assume $\det\{f(\lambda)\} \not\equiv 0$. We then know that there is a unique factorization of $f(\lambda)$,

$$f(\lambda) = \frac{1}{2\pi}\left(\sum_0^s A(j)e^{ij\lambda}\right)G\left(\sum_0^s A(j)e^{ij\lambda}\right)^*,$$

if we require that $\det\{\sum A(j)z^j\} = \det\{g(z)\}$ has no zeros in the unit circle (and $A(0) = I_p$). We in fact require that all zeros lie outside the unit circle. We commence our estimation procedure from

$$\hat{f}(\lambda) = \frac{1}{2\pi}\sum_{-s}^{s} C(j)e^{ij\lambda},$$

which converges with probability one to $f(\lambda)$. Proceeding heuristically, we must minimize

$$\sum_j \operatorname{tr}\,\{g(e^{i\omega_j})^{-1}\, I(\omega_j)\, g^*(e^{i\omega_j})^{-1}G^{-1}\},$$

wherein $I(\omega_j)$ was defined in (3.8) in Chapter IV. This formula is got, as before, from a transformation of the $x_j(n)$ accomplished by the finite Fourier transform. The derivative with respect to $\alpha_{uv}(k)$, the typical element of $A(k)$, is, except for a constant factor,

$$\sum_j \operatorname{tr}\,\{f(\omega_j)^{-1}\, I(\omega_j) f(\omega_j)^{-1}E_{uv}Gg^*(e^{i\omega_j})e^{ik\omega_j}\},$$

where E_{uv} has zero elements save for a unit in row u column v. This leads to

$$\sum_{l=0}^{s}\left\{\sum_j \hat{f}(\omega_j)^{-1}\, I(\omega_j)\hat{f}(\omega_j)^{-1}e^{i(l-k)\omega_j}\right\}\hat{A}^{(1)}(l) = 0, \qquad k = 1, \ldots, s, \quad \hat{A}^{(1)}(0) = I_p. \tag{5.11}$$

Then our first asymptotically efficient estimate is

$$\tilde{A}^{(1)}(l) = 2\hat{A}(l) - \hat{A}^{(1)}(l), \qquad l = 0, \ldots, s. \tag{5.12}$$

To obtain $\hat{A}(l)$ we must factorize $\hat{f}(\lambda)$. Robinson (1967, pp. 190–200), describes a computer program for such a factorization. We may, however, wish to avoid doing this both because of the labor involved and, more importantly because we may not be sure that $\hat{f}(\lambda)$ will factor. (It will be difficult to check whether it is nonnegative definite for all λ.) We discuss a procedure which avoids this factorization later.

In order to state our theorem we need to introduce some special notation. Let us call α_s the vector with $\alpha_{uv}(k)$ in row $(k-1)p^2 + (u-1)p + v$ and $\hat{\alpha}_s^{(1)}$, $\tilde{\alpha}_s^{(1)} \cdots$ the corresponding estimates. We call $\hat{\Phi}_s$ the matrix having $\hat{\Phi}(k-l)$ as the block, of p rows and columns, in the (k, l)th place, $k, l = 1, \ldots, s$, where

$$2\pi\hat{\Phi}(k) = \frac{1}{N}\sum_j \hat{f}^{-1}I\hat{f}^{-1}e^{-ik\omega_j}, \qquad k = 0, 1, \ldots, s.$$

In writing this formula we have omitted the argument ω_j from $\hat{f}$ and I (though not, of course, from the exponential function). The symbol I in the formula stands for $I(\omega_j)$ and should not be confused with the symbol for the identity matrix, I_p, which will always bear a subscript.

Finally $\hat{\phi}_s$ is formed from the $\hat{\Phi}(k)$ precisely as α_s was from the elements of the $A(l)$, namely as a column vector with $\hat{\phi}_{uv}(k)$ in row

$$(k-1)p^2 + (u-1)p + v.$$

Now

$$(\hat{\Phi}_s \otimes I_p)\hat{\alpha}_s^{(1)} = -\hat{\phi}_s. \tag{5.13}$$

We put

$$\Phi(k) = \frac{1}{2\pi}\int_{-\pi}^{\pi} g^{*-1}G^{-1}g^{-1}e^{-ik\lambda}\,d\lambda.$$

Then Φ_s is formed from the $\Phi(k)$ as was $\hat{\Phi}_s$ from the $\hat{\Phi}(k)$.

Though it seems likely that this can be avoided, if the proof of Theorem 7 is to be carried over fairly directly to Theorem 7′ it seems necessary to assume the existence of fourth moments for the $\epsilon(n)$. This is partly because of the occurrence of $\hat{G}$ in the formula.

Theorem 7′. *If $x(n)$ is generated by* (5.10) *and the fourth moments of the $\epsilon(n)$ are finite then $\tilde{\alpha}_s^{(1)}$ converges almost surely to α_s and $\sqrt{N}(\tilde{\alpha}_s^{(1)} - \alpha_s)$ is asymptotically normal with mean vector zero and covariance matrix $(\Phi_s^{-1} \otimes G^{-1})$.*

The covariance matrix is consistently estimated through $\hat{\Phi}_s$ and the estimate

$$\hat{G}_s^{(1)-1} = \frac{1}{2\pi N}\sum_{l=0}^{s}\left\{\sum_{j=0}^{N-1}\hat{f}^{-1}\,I\,\hat{f}^{-1}e^{il\omega_j}\right\}\hat{A}^{(1)}(l)$$

of G^{-1}.

Notes. 1. We have inserted an s subscript to indicate the order of the moving average fitted. An alternative estimation formula for G^{-1} would be

$$\frac{2\pi}{N}\sum_j \hat{g}^{-1}I\hat{g}^{*-1}$$

but this would require a further series of matrix inversions. The $\Phi(k)$ of Theorem 7′ and the $\phi(k)$ of Theorem 7 are not fully analogous as in the latter case G is a scalar and thus the formula $\Phi_s \otimes G$ may be adjusted so that it does not appear. Indeed it is $\Phi_s \otimes G$ which is analogous to Φ of Theorem 7.

2. If the function $\hat{f}$ is not factored we proceed as follows. We obtain $\hat{\alpha}_s^{(1)}$ by (5.13) and compute

$$\hat{g}^{(1)}(\omega_j) = \sum_0^s \hat{A}^{(1)}(k) e^{ik\omega_j},$$

$$\hat{G}_s^{(1)^{-1}} = \frac{1}{2\pi} \sum_{l=0}^{s} \left\{ \frac{1}{N} \sum_{j=0}^{N-1} \hat{f}^{-1} I \hat{f}^{-1} e^{il\omega_j} \right\} \hat{A}^{(1)}(l).$$

Then we put $\hat{f}^{(1)} = (2\pi)^{-1} \hat{g}^{(1)} \hat{G}^{(1)} \hat{g}^{(1)*}$ and solve

$$\sum_{l=0}^{s} \left\{ \sum_j \hat{f}^{(1)^{-1}} I \hat{f}^{(1)^{-1}} e^{i(l-k)\omega_j} \right\} \hat{A}^{(2)}(l) = 0, \qquad k = 1, \ldots, s.$$

Finally

$$\tilde{A}^{(1)}(l) = 2\hat{A}^{(1)}(l) - \hat{A}^{(2)}(l).$$

(Again we have used the same symbols for two different, asymptotically efficient, estimators. The justification is that the theorem stated above is true for either one, only one of the two will be computed and some economy in notation is needed.) Now, of course

$$2\pi \hat{\Phi}^{(1)}(k) = N^{-1} \sum_j \hat{f}^{(1)^{-1}} I \hat{f}^{(1)^{-1}} e^{-ik\omega_j}$$

and $\hat{\Phi}_s^{(1)}$, $\hat{\phi}_s^{(1)}$ are defined in these terms, so that

$$\hat{\alpha}_s^{(2)} = -(\hat{\Phi}_s^{(1)^{-1}} \otimes I_p) \hat{\phi}_s^{(1)}.$$

3. Again further iterations will be performed commencing from the $\tilde{A}^{(1)}(l)$ (using either definition) in place of $\hat{A}^{(1)}(l)$ as in Note 2 above. In that case we shall estimate $\Phi_s^{-1} \otimes G^{-1}$ by means of $\tilde{\Phi}_s^{(r)}$ and

$$\tilde{G}_s^{(r)^{-1}} = \sum_{l=0}^{s} \left\{ \frac{1}{N} \sum_j \tilde{f}^{(r)^{-1}} I \tilde{f}^{(r)^{-1}} e^{il\omega_j} \right\} \tilde{A}^{(r)}(l).$$

4. The comments 5, 6, 7, and 8 below Theorem 7, suitably modified, extend also to the present situation.

5. Some computational details are as follows. Let $\hat{f}(\omega_j) = \frac{1}{2}\{\hat{c}(\omega_j) - i\hat{q}(\omega_j)\}$. Then $\hat{c}$ is symmetric and $\hat{q}$ skew symmetric and $\hat{c}(\omega_{N-j}) = \hat{c}(\omega_j)$, $\hat{q}(\omega_{N-j}) = -\hat{q}(\omega_j)$. Now $\hat{f}(\omega_j)^{-1} = \{U_j - iV_j\}$. Then dropping the ω_j argument for simplicity, we see, from exercise 9 of Chapter V, that

$$U_j = 2(\hat{c} + \hat{q}\hat{c}^{-1}\hat{q})^{-1}, \qquad \hat{V}_j = 2\hat{c}\hat{q}(\hat{c} + \hat{q}\hat{c}^{-1}\hat{q})^{-1}.$$

Putting $I(\omega_j) = (X_j - iY_j)$, then

$$\hat{\Phi}(k) = 2N^{-1} \sum_{j=0}^{[\frac{1}{2}N]} \delta_j\{(U_jX_jU_j - U_jY_jV_j - V_jX_jV_j - V_jY_jU_j)\cos k\omega_j + (U_jX_jV_j + U_jY_jU_j + V_jX_jU_j - V_jY_jV_j)\sin k\omega_j\},$$

where $\delta_j = \frac{1}{2}$ for $j = 0$ or $\frac{1}{2}N$ and is otherwise unity.

The inversions of $\hat{\Phi}_s$, $\hat{\Phi}_s^{(1)}$, and so on, are not quite so large a task as might be imagined, especially if the moving average is fitted by a step-by-step increase in s. Let us put

$$\hat{\Phi}_s = \begin{vmatrix} \hat{\Phi}_{s-1} & D \\ D' & \hat{\Phi}(s) \end{vmatrix},$$

where D is constituted by a column of $(s-1)$ blocks, the jth being $\hat{\Phi}(j-s)$. The inverse of $\hat{\Phi}_s$ is symmetric and of the form

$$\begin{bmatrix} \hat{\Phi}_{s-1}^{-1} + \hat{\Phi}_{s-1}^{-1} D\{\hat{\Phi}(s) - D'\hat{\Phi}_{s-1}^{-1} D\}^{-1} D'\hat{\Phi}_{s-1}^{-1} & -\hat{\Phi}_{s-1}^{-1} D\{\hat{\Phi}(s) - D'\hat{\Phi}_{s-1}^{-1} D\}^{-1} \\ & \{\hat{\Phi}(s) - D'\hat{\Phi}_{s-1}^{-1} D\}^{-1} \end{bmatrix},$$

and if $\hat{\Phi}_{s-1}^{-1}$ is available from the preceding step this requires only the inversion of a $(p \times p)$ matrix (plus some matrix multiplications).

6. Theorem 7′ enables us to construct appropriate test statistics and inference procedures for moving average models and indeed the close correspondence between Theorems 1 and 7′ shows fairly clearly what the nature of the tests is to be; for example, to test $A(s) = 0$ the obvious test statistic is the quadratic form in the elements of $\tilde{A}^{(1)}$ with the inverse of the covariance matrix of $\tilde{A}^{(1)}$ as weight matrix. This estimate of the covariance matrix is

$$\{\hat{\Phi}(s) - D'\hat{\Phi}_{s-1}^{-1} D\}^{-1} \otimes \hat{G}^{-1},$$

so that the relevant quadratic form becomes

$$N \operatorname{tr}\{\tilde{A}^{(1)}(s)'\{\hat{\Phi}(s) - D'\hat{\Phi}_{s-1}D\}\tilde{A}^{(1)}(s)\hat{G}\}. \tag{5.14}$$

It follows from Theorem 7′ that as N increases this will have a distribution converging to that of chi-square with p^2 degrees of freedom. Of course, if $\tilde{A}^{(r)}(s)$ is computed, we shall also have computed estimates $\tilde{\Phi}^{(r)}(s)$, $\tilde{G}^{(r)}$ and these will be used in place of $\tilde{A}^{(1)}$, $\hat{\Phi}$, $\hat{G}$.

7. The case $p > 1$ is one which presents considerable computing problems because of the large number of matrix inversions when N is large [see (5.11)], which will be especially troublesome if p is not quite small. For this reason one may wish to replace expressions such as (5.11) by computationally

simpler forms. Thus one might replace $2\pi\hat{\Phi}(k)$ by

$$2\pi\hat{\Phi}^{(F)}(k) = \frac{1}{M}\sum_{j=0}^{M-1}\hat{f}^{-1}(\lambda_j)\hat{f}^{(F)}(\lambda_j)\hat{f}^{-1}(\lambda_j)e^{-ik\lambda_j}, \qquad k = 0, 1, \ldots, s,$$

$$\lambda_j = \frac{2\pi j}{M}, \qquad N = mM,$$

wherein $\hat{f}^{(F)}(\lambda_j)$ is the finite Fourier transform estimate in Chapter V, Section 4, Example 1, computed from m of the $I(\omega_j)$. It is apparent that $\hat{\Phi}(k)$ and $\hat{\Phi}^{(F)}(k)$ differ only because of the variation in $\hat{f}(\omega_j)$, $\exp ik\omega_j$, over each "band," of width $2\pi m/N$, over which the $I(\omega_k)$ have been averaged to give $\hat{f}^{(F)}(\lambda_j)$. If $M = O(N^{1/2-\epsilon})$, $\epsilon > 0$, then $\sqrt{N}/m = O(N^{-\epsilon})$ so that $\sqrt{N}$ by the oscillation of any component of $\hat{f}^{-1}(\lambda)$ over any band of width $2\pi m/N$, and similarly $\sqrt{N}$ by the oscillation in $\exp(-ik\lambda)$ over such a band, is $O(N^{-\epsilon})$. Thus

$$\sqrt{N}\,\|\hat{\Phi}^{(F)}(k) - \hat{\Phi}(k)\| \le KN^{-\epsilon}\sum_{j=0}^{N-1}\|I(\omega_j)\| \le K_1N^{-\epsilon}\sum_{j=1}^{p}c_j(0) \to 0.$$

Thus we may replace $\hat{\Phi}$ by $\hat{\Phi}^{(F)}$, and similarly for $\hat{\Phi}^{(1)}$ and so on, and reduce our calculations and number of matrix inversions. In other words, in all sums over j we may replace ω_j, $j = 0, 1, \ldots, N-1$, by λ_j, $j = 0, 1, \ldots, M-1$ and $I(\omega_j)$ by $\hat{f}^{(F)}(\lambda_j)$.

(c) The Mixed Moving-Average, Autoregressive Case for $p = 1$

We now assume that (4.1) holds where $g(z)$ and $h(z)$, defined below (4.1), satisfy the conditions there stated on their zeros.† There is one problem which arises immediately. If $\beta(q) = \alpha(s) = 0$ and we attempt to fit (4.1) then we shall be in a situation where our restrictions are insufficient to identify the $\alpha(k)$ and $\beta(j)$, for evidently we may now introduce a factor, $|1 + \alpha \exp i\lambda|^2$, $0 \le |\alpha| < 1$, in both numerator and denominator of (4.1) without violating any of the conditions imposed. Thus, in the Gaussian case, the likelihood function is constant along the lines (one line for each set $\alpha(k)$, $k = 1, \ldots, s-1$, $\beta(j)$, $j = 1, \ldots, q-1$)

$$\alpha(k) + \alpha\alpha(k-1),\ k = 1, \ldots, s;\ \beta(j) + \alpha\beta(j-1),$$
$$j = 1, \ldots, q;\quad \beta(q) = \alpha(s) = 0;\quad 0 \le \alpha < 1. \tag{5.15}$$

If truly $\beta(q) = \alpha(s) = 0$ there will therefore tend to be maxima near one of

† In addition we require that $g(z)$ have no zeros on the unit circle. This case is dealt with by Durbin (1960) and Walker (1962) and Walker, in particular, establishes properties for his estimators analogous to those for the case $q = 0$.

these lines and it will be a matter of chance which local maximum predominates and the estimates may not, as $N \to \infty$, converge to the values $\alpha(k)$, $\beta(j)$ satisfying $\alpha(s) = \beta(q) = 0$. The same must be true of any method based upon second moments since $f(\lambda)$ is constant along this same line. The fact that the likelihood is near to a maximum along a line such as (5.15) might be discernible from an examination of the likelihood function as a whole. However, this would be difficult in case $q + s > 2$ (and the possibility in case $q + s \leq 2$ seems of no real interest). To be sure this circumstance may not matter very much for presumably a vector of estimates, $\hat{\beta}(j)$, $\hat{\alpha}(k)$, obtained by techniques founded on second order quantities will approach the line (5.15) as $N \to \infty$. However, the theory is considerably complicated by a consideration of this case, $\beta(q) = \alpha(s) = 0$, and we thus omit it. This omission does have one very slightly unfortunate aspect. It means that in fitting a model of the form of (4.1) we cannot in successive trials, simultaneously increase both q and s by unity. For the test we should then wish to apply to determine whether both should have been so increased will be invalid since the distribution of the test statistic on the null hypothesis (that q and s are both one less than the latest value tried) will be difficult to obtain since it will involve the effect due to the chance location of the set of estimates near to the line (5.15). However, there will be no such trouble if at each stage only q or s is increased by unity. Though the possibility that $\beta(q)$ and $\alpha(s)$ are both strictly zero does not seem of great interest from an applied point of view yet the instability this situation introduces suggests that when these constants are small estimation will again be difficult. This possibility $\big($that $\alpha(s) = \beta(q) = 0\big)$ will show up in our asymptotic considerations through a near singular estimated covariance matrix for the estimates.

We henceforth assume that either $\beta(q)$ or $\alpha(s)$ is nonnull in addition to requiring that $h(z)$ and $g(z)$ have no zeros in or on the unit circle.

It is reasonable to commence by estimating the $\alpha(k)$, $\beta(j)$ from the first $q + s$ autocovariances. We know that

$$x(n) = \sum_{k=0}^{s} \alpha(k) \sum_{u=0}^{\infty} \lambda(u)\, \epsilon(n - u - k)$$

(see Chapter I, Section 3), where $\lambda(u)$ satisfies

$$\sum_{0}^{q} \beta(j)\lambda(u - j) = 0, \qquad u = 1, 2, 3, \ldots,$$

and the initial conditions $\lambda(0) = 1$, $\lambda(u) = 0$, $u < 0$. The $\lambda(u)$ may be expressed in terms of the $\beta(j)$ exactly as before. We now have the equations

$$(5.16) \qquad \sum_{0}^{q} \beta(j)\gamma(j - k) = \sigma^2 \sum\sum_{0}^{s} \alpha(l)\, \alpha(m)\lambda(l - m - k), \qquad k = 1, 2, \ldots, q + s,$$

whose right side is null for $k > s$ (since then $l - m - k$ is necessarily negative). Thus the last q equations

$$\sum_0^q \beta(j)\gamma(j-k) = 0, \qquad k = s+1, \ldots, s+q, \qquad \beta(0) = 1,$$

serve to determine the $\beta(j)$ since, provided one of $\alpha(s)$ and $\beta(q)$ is non-null, the matrix, having $\gamma(j-k)$ in row $k-s$, column j, is nonsingular. (We leave this as an exercise.) Thus we may consistently estimate the $\beta(j)$ from

$$\sum_0^q \hat{\beta}(j)\, c(j-k) = 0, \qquad k = s+1, \ldots, s+q.$$

Evidently $\hat{\beta}(j)$ converges to $\beta(j)$ with probability 1 since, as we know, the $c(j)$ so converge. Now we could insert the $\hat{\beta}(j)$ in (5.16) to obtain estimates of the $\hat{\alpha}(k)$ but we need only proceed as follows in order to obtain the estimates necessary for our later procedure.

Step 1. We form

$$\hat{c}_y(n) = \sum_{k,l=0}^{q}\sum \hat{\beta}(k)\hat{\beta}(l)c(n+k-l) = \hat{c}_y(-n), \qquad n = 0, 1, \ldots, s,$$

and *for later use*

$$\hat{h}(e^{i\omega_j}) = \sum_0^q \hat{\beta}(k)e^{ik\omega_j}, \qquad j = 0, 1, \ldots, N-1;$$

then

$$\hat{f}_y(\lambda) = \frac{1}{2\pi}\sum_{-s}^{s} \hat{c}_y(n)e^{-in\lambda} = \frac{1}{2\pi}\left\{\hat{c}_y(0) + 2\sum_1^s \hat{c}_y(n)\cos n\lambda\right\}.$$

If this is not positive for all λ, we have an indication either that our model is incorrect or our sample size is too small. As $N \to \infty$, evidently $\hat{f}_y(\lambda)$ converges with probability 1 to

$$\frac{\sigma^2}{2\pi}\,|g(e^{i\lambda})|^2.$$

If we do factor $\hat{f}_y$, we shall have

$$\hat{f}_y = \frac{\hat{\sigma}^2}{2\pi}\,|\hat{g}|^2 = \frac{\hat{\sigma}^2}{2\pi}\left|\sum_0^s \hat{\alpha}(j)e^{ij\lambda}\right|^2, \qquad \hat{\alpha}(0) = 1.$$

If we do *not* factor $\hat{f}_y$, we proceed as follows. We form†

(5.17) $$\hat{a}(k) = \frac{1}{N}\sum_j I\,|\hat{h}|^2\hat{f}_y^{-2}\cos k\omega_j, \qquad k = 0, 1, \ldots, s,$$

† Again we have omitted the symbol ω_j in I, $\hat{h}$ and $\hat{f}_y$ and again we warn the reader that I stands for $I(\omega_j)$ and not the unit matrix whose symbol will always bear a subscript.

and then $\hat{A}$, $\hat{a}$, as for $q = 0$. Then $\hat{\alpha} = -\hat{A}^{-1}\hat{a}$. We next define

$$\hat{g}(e^{i\omega_j}) = \sum_k \hat{\alpha}(k)e^{ik\omega_j}.$$

We have now used the pair of symbols $\hat{\alpha}$, $\hat{g}$ for estimates of the same pair of things (α, g) computed in different ways. This varies from our procedure in case $q = 0$ but it is here convenient and it is hoped that it will cause no confusion. Finally, we compute

$$\hat{b}(k) = \frac{1}{N}\sum_{j=0}^{N-1} I\, |\hat{g}|^{-2} \cos k\omega_j, \qquad k = 0, 1, \ldots, q. \tag{5.18}$$

Then

$$\hat{\beta}^{(1)} = -\hat{B}^{-1}\hat{b},$$

wherein $\hat{B}$ has $\hat{b}(k - l)$ in row k, column l, whereas $\hat{b}$ has $\hat{b}(k)$ in row k.

Step 2. Now we form

$$|\hat{h}^{(1)}|^2 = \left|\sum_0^q \hat{\beta}^{(1)}(k)e^{ik\omega_j}\right|^2,$$

$$\hat{a}^{(1)}(k) = \frac{1}{N}\sum I\, |\hat{h}^{(1)}|^2\, |\hat{g}|^{-4} \cos k\omega_j, \qquad k = 0, 1, \ldots, s. \tag{5.19}$$

We then form $\hat{A}^{(1)}$, $\hat{a}^{(1)}$ from these quantities. (In fact $\hat{a}^{(1)}(k)$ estimates something differing by a constant from that which $\hat{a}(k)$ estimates but we retain this notation for simplicity.) Then we form

$$\hat{\alpha}^{(1)} = -\hat{A}^{(1)-1}\hat{a}^{(1)}.$$

To obtain $\tilde{\alpha}^{(1)}$ we must combine $\hat{\alpha}^{(1)}$ and $\hat{\alpha}$ (computed according to either procedure). For this purpose we need the matrix $\hat{\Omega}$ with entries $\hat{\omega}(k - l)$, $k = 1, \ldots, q$, $l = 1, \ldots, s$,

$$\hat{\omega}(k) = \frac{1}{N}\sum_j (\hat{g}\bar{\hat{h}})^{-1}e^{-ik\omega} = \frac{1}{2\pi}\int_{-\pi}^{\pi} (\hat{g}(e^{i\lambda})\bar{\hat{h}}(e^{i\lambda}))^{-1}e^{-ik\lambda}\, d\lambda. \tag{5.20}$$

Then

$$\tilde{\alpha}^{(1)} = [\{I_s - \hat{A}^{-1}\hat{\Omega}'\hat{B}^{-1}\hat{\Omega}\}^{-1}(\hat{\alpha} - \hat{\alpha}^{(1)})] + \hat{\alpha}.$$

Finally we form $\tilde{g}$ from $\tilde{\alpha}^{(1)}$ as we formed $\hat{g}$ and $\hat{\alpha}$ and then $\tilde{B}$, $\tilde{b}$ from $\tilde{g}$ in the same way as $\hat{B}$, $\hat{b}$ were formed using $\hat{g}$. Then

$$\tilde{\beta}^{(1)} = -\tilde{B}^{-1}\tilde{b}$$

We introduce the matrices Φ, Ψ, Ω, of which Φ is as before and Ψ, Ω have $\psi(k - l)$, k, $l = 1, \ldots, q$, $\omega(k - l)$, $k = 1, \ldots, q$; $l = 1, \ldots, s$, in row k, column l, where

$$\psi(k) = \frac{1}{2\pi}\int_{-\pi}^{\pi} |h|^{-2} e^{-ik\lambda}\, d\lambda,$$

$$\omega(k) = \frac{1}{2\pi}\int_{-\pi}^{\pi} (g\bar{h})^{-1} e^{-ik\lambda}\, d\lambda.$$

In the appendix to this Chapter we prove the following theorem.

Theorem 8. *If $x(n)$ is generated by (2.1) where $g(z)$ and $h(z)$ have no zeros in or on the unit circle and no common zero and both $\alpha(s)$ and $\beta(q)$ are not zero, then $\tilde{\alpha}^{(1)}$, $\tilde{\beta}^{(1)}$ converge almost surely to α, β and*

$$\sqrt{N}\begin{pmatrix} \tilde{\alpha}^{(1)} - \alpha \\ \tilde{\beta}^{(1)} - \beta \end{pmatrix}$$

is asymptotically normal with zero mean vector and covariance matrix

$$\begin{bmatrix} \Phi & -\Omega' \\ -\Omega & \Psi \end{bmatrix}^{-1}.$$

The matrix $\hat{\Omega}$ (see (5.20)) consistently estimates Ω while Φ and Ψ are consistently estimated by

$$\hat{\Phi} = \left\{\sum_0^s \hat{\alpha}(k)\hat{a}(k)\right\}^{-1}\hat{A}, \qquad \hat{\Psi} = \left\{\sum_0^q \hat{\beta}^{(1)}(k)\hat{b}(k)\right\}^{-1}\hat{B}.$$

We make here a number of remarks.

1. As we show in the appendix, when $\alpha(s) = \beta(q) = 0$, the covariance matrix is singular. This corresponds to the lack of identification of the vectors α, β. The same occurs when $g(z)$ and $h(z)$ have a common zero.

2. Further iterations may be performed commencing from $\tilde{\alpha}^{(1)}$, $\tilde{\beta}^{(1)}$ in place of $\hat{\alpha}$, $\hat{\beta}^{(1)}$, at the start of step 2. Thus $\tilde{h}$, computed from $\tilde{\beta}^{(1)}$, replaces $\hat{h}^{(1)}$ and $\tilde{g}$ (already computed) replaces $\hat{g}$. If this has been done we shall have computed estimators $\tilde{\Phi}$, $\tilde{\Psi}$, $\tilde{\Omega}$ in place of $\hat{\Phi}$, $\hat{\Psi}$, $\hat{\Omega}$ and now these, first, estimators will be asymptotically efficient, in the sense that they are based on asymptotically efficient estimators of α and β.

3. Comments 5, 6, and 8 relating to the case $q = 0$ remain basically true here also. In particular mean corrections, which make $I(0)$ into zero, have no effect on the asymptotic results.

4. Most of the computations are relatively straightforward and are of the same form as earlier described for $q = 0$. An exception is $\hat{\Omega}$. Here we may simplify the expression for $\hat{w}(k)$ as follows. Let us suppose that $q \geq s$ and

that the zeros of $\hat{h}(z)$ are $\hat{z}_j, j = 1, \ldots, q$. Then

$$\hat{w}(k) = \hat{\beta}(q)^{-1} \sum_{j=1}^{q} \{\hat{g}(\hat{z}_j^{-1})^{-1} \hat{z}_j^{k-1} \prod_{\substack{i=1 \\ i \neq j}}^{q} (\hat{z}_i - \hat{z}_j)^{-1}\}, \qquad -s + 1 \le k \le q - 1.$$

If $q < s$ we call $\hat{w}_j$ the zeros of $\hat{g}(z), j = 1, \ldots, s$, and obtain

$$\hat{w}(k) = \hat{\alpha}(s)^{-1} \sum_{j=1}^{s} \{\hat{h}(\hat{w}_j^{-1})^{-1} \hat{w}_j^{-k-1} \prod_{\substack{i=1 \\ i \neq j}}^{s} (\hat{w}_i - \hat{w}_j)^{-1}\}, \qquad -s + 1 \le k \le q - 1.$$

Of course in practice it may be easier to expand $\hat{g}$ and $\bar{\hat{h}}$ in powers of $\exp i\lambda$ and $\exp -i\lambda$ and collect together the coefficient of $\exp ik\lambda$ in the product of these two functions. In the case where $p = q = 1$ then $\hat{w}(0) = \{1 - \hat{\alpha}(1)\hat{\beta}(1)\}^{-1}$. When $\tilde{w}(k)$ are required we replace $\hat{z}_j$ (or $\hat{w}_j$) by $\tilde{z}_j$ (or $\tilde{w}_j$) where these are the zeros of $\tilde{h}$ (or $\tilde{g}$). If h or g is of high degree it may be easier to compute $\hat{w}(k)$ or $\tilde{w}(k)$ directly from the middle term in (5.20). For example putting $\hat{g} = \hat{u} + i\hat{v}$, $\hat{h} = \hat{y} + i\hat{z}$ (where $\hat{g}$, $\hat{h}$, $\hat{u}$, $\hat{v}$, $\hat{y}$, $\hat{z}$ are functions of w_j) then

$$\hat{w}(k) = 2 \sum_{j=0}^{[1/2N]} \delta_j \{|\hat{g}|^2 \, |\hat{h}|^2 \big((\hat{u}\hat{y} + \hat{v}\hat{z}) \cos kw_j + (\hat{u}\hat{z} - \hat{v}\hat{y}) \sin kw_j\big)\}$$

where δ_j is as in the fifth note after Theorem 7 and the term for $j = 0$ will be omitted if mean corrections are made (as will almost certainly be the case). Since $\hat{g}$, $\hat{h}$ are already computed this calculation is not a large task.

5. It does not seem possible to give such a neat form to the test of significance for $\alpha(s) = 0$ or $\beta(q) = 0$ as was possible for the cases, respectively, $q = 0$ or $s = 0$. The question of an appropriate method for fitting a model of the mixed autoregressive moving average type (i.e., the determination of q and s) needs consideration as there is no obvious analog of the type of procedure used when q or s is prescribed as zero. One procedure would be to successively increase q and s by unity, but there is evidently some need also to reconsider earlier values at each stage. Perhaps a reasonable procedure would be to test the significance of the last parameter fitted, say $\tilde{\beta}^{(1)}(q)$, and if this is significant to successively examine the significance of the parameters of the moving average part beginning from the last of these fitted, namely $\tilde{\alpha}^{(1)}(s)$. Certainly this should be done after the last increase in q or s is made. If after a certain point it is decided to increase q and s no further, since no significant improvement in the fit is discerned and, say, q was the last index it was found worthwhile to increase, then $\tilde{\alpha}^{(1)}(s)$, $\tilde{\alpha}^{(1)}(s - 1)$, and so on, should be examined to determine whether they add anything to the explanation. If some such procedure is not adopted, there must be a considerable risk of overfitting. Of course such an overfitting need not badly lower the accuracy of a prediction from such an estimated relation.

problem best by comparison with the case of an autoregression where overfitting raises the variances of the estimates of needed parameters. In the mixed model overfitting may have equally drastic consequences. Consider, for example, the case where truly $q = 1$ and $s = 0$ but we assume $q = 1, s = 1$. If we had known the truth, we should have estimated $\beta(1)$ through an autoregression and have obtained an estimate, $\tilde{\beta}_0(1)$, such that $\sqrt{N}(\tilde{\beta}_0(1) - \beta(1))$ is asymptotically normal with variance $(1 - \beta(1)^2)$. If we assume $s = 1$ then we obtain $\tilde{\beta}_1(1)$ with $\sqrt{N}(\tilde{\beta}_1(1) - \beta(1))$ asymptotically normal with variance, from Theorem 8, which is the element in the bottom right hand corner of

$$\begin{bmatrix} 1 & -1 \\ -1 & \{1 - \beta(1)^2\}^{-1} \end{bmatrix}^{-1}.$$

Thus the variance is $\beta(1)^{-2}\{1 - \beta(1)^2\}$ and the efficiency of $\tilde{\beta}_1(1)$ is measured by $\beta(1)^2 < 1$. It is to be expected that trouble will be struck when $\alpha(1) = 0$ and $\beta(1)$ is small since then we are near to an unidentified situation but the efficiency is disturbingly low for even quite moderate values of $\beta(1)$. Of course if $\alpha(1) \neq 0$ then $\tilde{\beta}_0(1)$ is not consistent. However for $\alpha(1)$ small it is conceivable that, when N is not very large, $\tilde{\beta}_0(1)$ will be a better estimate of $\beta(1)$ than will $\tilde{\beta}_1(1)$. In such circumstances, also, the use of $q = 2, s = 0$ may give even better results. These comments suggest that mixed models are likely to be most important when the zeros of $g(z)$ lie near to the unit circle. Unfortunately, this is also the case where N may need to be quite large before the asymptotic theory becomes valid. These models should probably be used with care.

(d) The Case $p > 1$

The techniques introduced above extend to the case $p > 1, q > 0$, but the computations now become extensive and we shall not enter into the details here. The problem of identification must be considered before the problem of estimation, and we refer the reader to Hannan (1969b) for a discussion of that subject.

Though there are, no doubt, circumstances in which a vector mixed moving-average autoregressive model will give a much better fit with a given number of constants than either a moving-average or autoregressive model, the computational complications are so great that it can be doubted whether the more complicated model will be used, and we do not feel that the techniques at the present stage are sufficiently near to being practically useful to be included in this book.

We conclude this section by describing the results of a sampling experiment for the case $p = 1$. We took 20 sequences of 100 observations each from the process

$$x(n) - 0.8x(n-1) = \epsilon(n) + 0.5\epsilon(n-1)$$

and used the method described above which avoids factorization of the spectral density matrix [though this would have been easy to do in the present case except for the possibility that $\hat{f}_y(\lambda)$ would not factorize]. The iterations were continued up to the formation of $\tilde{\alpha}^{(5)}$ and $\tilde{\beta}^{(4)}$. The mean of the $\tilde{\alpha}^{(5)}$ was 0.504 and the variance 0.011 (compared to a theoretical value of 0.009). The mean of the $\tilde{\beta}^{(4)}$ was -0.768 and the variance was 0.003 (compared to a theoretical value of 0.004). Both variances are quite reasonable, compared with their theoretical values, for 20 samples. However, $\tilde{\beta}^{(4)}$ seems to be biased toward zero. Of the 20 samples, 13 had "stabilized" by $\tilde{\alpha}^{(1)}$, $\tilde{\beta}^{(1)}$, in the sense that further iterations hardly changed them. By $\tilde{\alpha}^{(2)}$, $\tilde{\beta}^{(2)}$, 15 had stabilized and by $\tilde{\alpha}^{(3)}$, $\tilde{\beta}^{(3)}$ 19 had stabilized. The exception gave the sequence of values

$\hat{\alpha}^{(1)}$	$\hat{\alpha}^{(2)}$	$\tilde{\alpha}^{(1)}$	$\tilde{\alpha}^{(2)}$	$\tilde{\alpha}^{(3)}$	$\tilde{\alpha}^{(4)}$	$\tilde{\alpha}^{(5)}$
0.985	0.998	0.962	0.917	0.828	0.666	0.498
$\hat{\beta}^{(1)}$	$\hat{\beta}^{(1)}$	$\tilde{\beta}^{(1)}$	$\tilde{\beta}^{(2)}$	$\tilde{\beta}^{(3)}$	$\tilde{\beta}^{(4)}$	
−0.725	0.897	0.519	−0.088	−0.514	−0.705	

It seems probable that a further iteration should have been carried out. (The number of iterations was decided upon a priori. In practice one would be guided by the stability of the values.) Choosing a random number between 1 and 20 we choose the following sequence (the 8th) to illustrate the procedure in a more stable case. For this series the values were

$\hat{\alpha}^{(1)}$	$\hat{\alpha}^{(2)}$	$\tilde{\alpha}^{(1)}$	$\tilde{\alpha}^{(2)}$	$\tilde{\alpha}^{(3)}$	$\tilde{\alpha}^{(4)}$	$\tilde{\alpha}^{(5)}$
0.441	0.520	0.338	0.329	0.329	0.329	0.329
$\hat{\beta}$	$\hat{\beta}^{(1)}$	$\tilde{\beta}^{(1)}$	$\tilde{\beta}^{(2)}$	$\tilde{\beta}^{(3)}$	$\tilde{\beta}^{(4)}$	
−0.756	−0.712	−0.752	−0.755	−0.755	−0.755	

6. GENERAL THEORIES OF ESTIMATION FOR FINITE PARAMETER MODELS

We now discuss some very general theories of estimation for finite parameter models due to Whittle (1951), Walker (1964). This work seems to me to be important. It is technical, however, and we shall present the results here without proofs since they are not directly used elsewhere in the book.† The interested reader may refer for the proofs to Walker (1964). The book by Whittle (1951) is of great importance as an originator of ideas concerning inference with time series data. The general theory of inference for time series had already been considered by Grenander (1950) in a very influential thesis. The characteristic feature of Whittle's work is a thorough integration of classical inferential procedures with spectral theory and it is for this reason that we emphasize it here. To begin with, we consider a Gaussian process $x(n)$, scalar with a spectral density $f(\lambda)$, which is continuous in $[-\pi, \pi]$ and not zero there. Then certainly

$$x(n) = \sum_0^\infty \alpha(j)\epsilon(n-j),$$

where the $\epsilon(n)$ are N.I.D. $(0, \sigma^2)$ (this representation may be taken to be the canonical representation occurring in the Wold decomposition). Then

$$\sigma^2 = \exp \frac{1}{2\pi} \int_{-\pi}^{\pi} \log \{2\pi f(\lambda)\}\, d\lambda. \tag{6.1}$$

We assume that, in addition to σ^2, the spectrum $f(\lambda)$, depends on q parameters $\theta_1, \ldots, \theta_q$, which collectively we shall call θ. Then the likelihood function is

$$-\tfrac{1}{2} \log \det \big(V_N(\theta)\big) - \tfrac{1}{2} N \log 2\pi\sigma^2 - \frac{x' V_N^{-1}(\theta) x}{2\sigma^2},$$
$$x' = \big(x(1), \ldots, x(N)\big),$$

where $V_N(\theta) = \sigma^{-2}\Gamma_N$ which evidently depends only upon θ. Now by a result of Grenander and Szego of the same type as was proved in Theorem 7, Chapter III,

$$\frac{1}{N} \log \det \big(V_N(\theta)\big) = \frac{1}{2\pi} \int_{-\pi}^{\pi} \log \left(\frac{2\pi}{\sigma^2} f(\lambda)\right) d\lambda + O(N^{-1})$$

Since the first term is a constant independent of θ [from (6.1)], we again see that to order N^{-1} the first term may be ignored. This leads to the consideration of

$$N \log \sigma^2 + \frac{x' V_N^{-1}(\theta) x}{\sigma^2}.$$

† The results refer to the scalar case, but may no doubt be extended to the vector case, and make other assumptions (e.g., the existence of fourth moments) which have not been found necessary in our particular case. Moreover, the method of maximum likelihood applied directly (even in the modified form discussed below) leads to difficult nonlinear equations. For these reasons we have preferred to establish our results directly in each case rather than call on the general results of this section.

This gives us, by the method of maximum likelihood applied to this expression

$$\hat{\sigma}^2 = \frac{1}{N} x' V_N^{-1}(\hat{\theta}) x,$$

whereas the estimate of θ, namely $\hat{\theta}$, is found by minimizing the quadratic form $x'V_N^{-1}(\theta)x$. This expression itself is not easy to handle because of the difficulty in finding an explicit expression for the element of $V_N^{-1}(\theta)$. Then, following Whittle (1951), we adopt the device previously used, which is to consider

$$U_N(x, \theta) = N \sum_{-N+1}^{N-1} \beta(n, \theta)\left(1 - \frac{|n|}{N}\right) c(n),$$

where $\beta(n, \theta)$ is the typical coefficient in the expression of $\{g(z)\, g(z^{-1})\}^{-1}$ in a Laurent expansion (assumed valid in an annulus including the unit circle), and

$$g(z) = \sum_0^{\infty} \alpha(j) z^j.$$

We call this $g(z, \theta)$ when we need to emphasize the dependence on θ. Here we have used an argument previously adopted (see Section 5), namely, to express $V_N(\theta)$ approximately in the form

$$V_N(\theta) \approx PDP^*,$$

where D has $\{2\pi f(\omega_j)/\sigma^2\}$ in the jth place in the main diagonal, whereas P is the unitary matrix with

$$\frac{1}{\sqrt{N}} e^{in\omega_j}, \qquad j = 0, 1, \ldots, N-1,$$

in the nth row and the jth column. Then $V_N^{-1}(\theta) \approx PD^{-1}P^*$ and

$$x'PD^{-1}P^*x = \sum_j |g(e^{i\omega_j})|^{-2}\, I(\omega_j),$$

which gives the required formula. Thus we are finally led to consider the minimization of $N \log \sigma^2 + U_N(x, \theta)/\sigma^2$ with respect to θ and σ^2. This is the same procedure used in Section 4.

This technique may be applied irrespective of whether $x(n)$ is Gaussian. Walker considers the case in which the $\epsilon(n)$ are i.i.d. $(0, \sigma^2)$ with finite fourth moment and the parameters θ lie in a closed bounded set in q-dimensional Euclidean space. It is assumed that $|g(z, \theta_1)|^2$, $|g(z, \theta_2)|^2$ are not equal almost everywhere in $[-\pi, \pi]$ if $\theta_1 \neq \theta_2$ and, as before, that $f(\lambda)$ and $f^{-1}(\lambda)$ are continuous in $[-\pi, \pi]$. We now call $\hat{\sigma}_N^2$, $\hat{\theta}_N$ the modified maximum-likelihood solutions obtained from

$$(6.2)\qquad -L_N(\theta, \sigma^2) = N \log \sigma^2 + \frac{U_N(x, \theta)}{\sigma^2}.$$

Walker establishes the following results, originally obtained by Whittle (1951, 1962).

1. $(\hat{\theta}_N, \hat{\sigma}_N^2)$ converges in probability to (θ, σ^2).
2. Under assumptions of continuity for the derivatives up to order 3 with respect to the θ_i of the function $|g(z, \theta)|^{-2}$ in the neighborhood of the true value θ_0 and assuming that

$$\sum_0^\infty n\, |\alpha(n, \theta_0)| < \infty,$$

the vector $N^{-1/2}(\hat{\theta}_N - \theta_0)$ has a distribution that converges as $N \to \infty$ to the normal distribution with zero-mean vector and covariance matrix W_0^{-1}, where W_0 has entries $w_{ij}^{(0)}$:

$$(6.3)\qquad w_{ij}^{(0)} = \frac{1}{4\pi} \int_{-\pi}^{\pi} \left(\frac{\partial \log |g(e^{i\lambda}, \theta)|^2}{\partial \theta_i}\right)_0 \left(\frac{\partial \log |g(e^{i\lambda}, \theta)|^2}{\partial \theta_j}\right)_0 d\lambda,$$

assuming this to be nonsingular. (The subscript refers to evaluation of the derivatives at $\theta = \theta_0$.)

This expression (6.3) is an interesting one, to say the least. It is easy to check that it gives the covariance matrix for a mixed-moving-average autoregression in Theorem 3 and that we have verified the asymptotic equivalence of our results to those from the maximum-likelihood procedure.

7. TESTS OF GOODNESS OF FIT

We have already spoken of tests of goodness of fit for some of the finite parameter models discussed in earlier sections of this chapter, that is tests of an hypothesized model against a very general range of alternatives. One method of testing the goodness of fit of an autoregression of order q is, as we have suggested, to compare it with one of order $q + t$, for t large, by testing the significance of the matrices of coefficients for lags above q by a test of the type of a partial multiple correlation. The same procedure may also be applied to the moving-average case. The mixed case presents some difficulty because of the restriction on the true orders of the autoregressive and moving average parts but of course one could always construct a test by comparing a mixed model of orders (q, s) against one of order $(q + t, s)$ or of order $(q, s + t)$, once there is good evidence that (say in the first case) the qth autoregressive coefficient matrix is nonsingular. Since these procedures fall within the range of what was considered in earlier sections we discuss them no further here. We shall begin by discussing an extension of a procedure due to Quenouille (1947), discussed briefly above, to the case $p \geq 1$, the extension being due to Bartlett and Rajalakshman (1953). Thus we are considering a goodness of fit test for a multiple autoregressive scheme which is of the same nature as that earlier discussed in this section but computationally simpler. We hope to achieve this by means of the formation of a moving-average operation, of length $2q$, on the matrices $C(n)$ (the moving average having matrix coefficients) which produces a sequence of matrices so that after appropriate normalization there is no correlation between elements from different matrices in the sequence. Consider the possible factorizations of a spectral density matrix for an autoregression

$$f(\lambda)^{-1} = 2\pi h^{(j)}(e^{i\lambda})^* G^{(j)-1} h^{(j)}(e^{i\lambda}),$$

where $h^{(j)}(z)$ is a polynomial in z of degree q (the order of the autoregression). Of course, only one such factorization corresponds to $h^{(j)}(z)$ having a determinant with all of its zeros outside of the unit circle and $h^{(j)}(0) = I_p$. We call that factor $h(z)$ as before. Consider now the sequence

$$D^{(j)}(u) = \frac{1}{N} \sum_k h(e^{i\omega_k}) I(\omega_k) h^{(j)}(e^{i\omega_k})^* e^{-iu\omega_k}, \qquad u > 0. \tag{7.1}$$

We proceed heuristically and leave the easy precise justification of the end results until later. Then we see that by replacing $I(\omega_k)$ with its expectation, which is near $f(\omega_k)$, and approximating the sum by an integral, the mean of $D^{(j)}(u)$ will ultimately be

$$\frac{1}{2\pi} \int_{-\pi}^{\pi} h h^{-1} G h^{*-1} h^{(j)*} e^{-iu\lambda}\, d\lambda = 0, \qquad u > 0.$$

It we use the formula for the covariances of the components of $I(\omega_k)$ from Corollary 1, Chapter V, and again approximate a sum by an integral, the variance will ultimately be had from

$$N\mathcal{E}\big(d_{pq}^{(j)}(u)\, d_{rs}^{(j)}(v)\big) \to 2\pi \int_{-\pi}^{\pi} \{h_{pa} \bar{h}_{qb}^{(j)} h_{rc} \bar{h}_{sd}^{(j)} f_{ad} f_{cb} e^{-i(u+v)\lambda} + h_{pa} \bar{h}_{qb}^{(j)} h_{rc} \bar{h}_{sd}^{(j)}(e^{-i\lambda}) f_{ac} f_{db} e^{-i(u-v)\lambda}\}\, d\lambda, \tag{7.2}$$

where we have used a tensor notation (summing over repeated subscripts) and have inserted the argument of the function in the one case in which it is exp $(-i\lambda)$ *and not* exp $i\lambda$. This is easily seen to be zero for $u \neq v$ if $h^{(j)} = h$. For

$u = v$, $u, v > 0$, it is $\sigma_{pr}\sigma_{sq}$. This case is not of great interest because the effects of estimating the function h are not easily allowed for. However, there is another case in which (7.2) is null for $u \neq v$ and this is where $h^{(j)}$ corresponds to that unique factorization in which $h^{(j)}(z)$ has a determinant with all of its zeros *inside* the unit circle and $\{z^{-q}h^{(j)}(z)\}_{z=\infty} = I_p$. (Let us call this $h^{(1)}$.) Once again $h^{(1)}$ has real coefficient matrices so that $h^{(1)}(\exp -i\lambda) = \overline{h^{(1)}}$ and we obtain for (7.2)

$$\frac{1}{2\pi}\int_{-\pi}^{\pi} \{\sigma_{pe}\bar{h}^{de}\bar{h}_{sd}^{(1)}\sigma_{rf}\bar{h}^{cf}\bar{h}_{qb}^{(1)}e^{-i(u+v)\lambda} + \sigma_{pr}g_{sq}^{(1)}e^{-i(u-v)\lambda}\}\, d\lambda = 0, \qquad u \neq v,$$

where $g_{sq}^{(1)}$ is the typical element of $G^{(1)}$, $f(\lambda) = (2\pi)^{-1}h^{(1)^{-1}}G^{(1)}h^{(1)*^{-1}}$, and $\bar{h}^{de}$ is the typical element of h^{*-1}. When $u = v$, the expression is

$$\sigma_{pr}g_{sq}^{(1)}. \tag{7.3}$$

This second case is the one we consider further, since it provides the generalization of Quenouille's test. (The first case generalizes a test due to Bartlett and Diananda (1950) but is less useful, since the effects of estimating h (and $h^{(1)}$) are not easily allowed for.)

Now

$$\begin{aligned} D^{(1)}(u) &= \sum_0^q\sum B(j)C(j-k-u)B^{(1)}(k)'\left(1 - \frac{|j-k-u|}{N}\right) \\ &= \sum_0^q\sum B(j)C'(u+k-j)B^{(1)}(k)'\left(1 - \frac{|u+k-j|}{N}\right), \end{aligned} \tag{7.4}$$

where the $B^{(1)}$ are the coefficient matrices of $h^{(1)}$.

Theorem 9. *Let $x(n)$ be generated by an autoregression of order q and $\hat{D}^{(1)}(u)$ be obtained from $D^{(1)}(u)$ by the use of the estimates $\hat{B}(j)$ of the $B(j)$ (given by (2.6)) to estimate both the $B(j)$ and $B^{(1)}(k)'$ in (7.4). Then for any fixed $d > 0$ the elements of all of the matrices $N^{1/2}\hat{D}^{(1)}(u)$, $0 < u \leq d$, are asymptotically jointly normal with zero means and covariances which are zero as between two elements for differing u and otherwise is $\sigma_{pr}g_{sq}^{(1)}$ as between the elements in row p, column q, and row r, column s. Here $g_{ij}^{(1)}$ is the typical element of $G^{(1)}$ which occurs in the factorization*

$$f(\lambda) = \frac{1}{2\pi}h^{(1)^{-1}}G^{(1)}h^{(1)*^{-1}}, \tag{7.5}$$

where $h^{(1)}(z)$ is a matrix of polynomials of z of degree q with determinant having all of its zeros inside the unit circle and with $\{z^{-q}h^{(1)}(z)\}_{z=\infty} = I_p$.

Thus

$$N \sum_{1}^{d} \operatorname{tr} \{\hat{G}^{-1}\hat{D}^{(1)}(u)\hat{G}^{(1)^{-1}}\hat{D}^{(1)}(u)'\}$$

is asymptotically distributed as chi-square with dp^2 d.f. and may be used to test the goodness of fit of a qth order autoregression to the data.

Proof. We have

$$\sqrt{N} \sum_{0}^{q}\sum B(j)C(j-k-u)B^{(1)}(k)'\left(1 - \frac{|j-k-u|}{N}\right)$$
$$= \frac{1}{\sqrt{N}} \sum_{j,k=0}^{q} B(j) \sum_{1}^{N} x(n-j+u)x'(n-k)B^{(1)}(k)' + O(N^{-\frac{1}{2}}),$$

wherein we have in the first term added a fixed finite number of terms to bring that expression to a simple form, just as in the proof of Theorem 1. The first term is

$$\frac{1}{\sqrt{N}} \sum_{n=1}^{N} \epsilon(n+u)\epsilon'^{(1)}(n), \tag{7.6}$$

where $\epsilon^{(1)}(n)$ is the sequence corresponding to $B^{(1)}(k)$, that is,

$$\sum_{0}^{q} B^{(1)}(k)x(n-k) = \epsilon^{(1)}(n), \qquad \mathcal{E}\left(\epsilon^{(1)}(n)\epsilon^{(1)}(m)'\right) = \delta_m{}^n G^{(1)}.$$

Since $x(n-k)$ is independent of $\epsilon(n+u)$, $k \geq 0$, $u > 0$, it is evident that $\epsilon^{(1)}(n)$ is independent of $\epsilon(n+u)$, $u > 0$. Of course, $\epsilon^{(1)}(n)$ may also be expressed as a linear combination of $\epsilon(n-j), j \geq 0$, with coefficient matrices decreasing exponentially. Thus the expression (7.6) is of precisely the same nature as $e_{ij}(v)$ in Theorem 1 and the asymptotic normality is established in the same way as in that theorem. Moreover, the typical covariance for the elements of (7.6) is

$$N^{-1}\mathcal{E}\left\{\sum_{m=1}^{N} \epsilon_p(m+u)\epsilon_q^{(1)}(m) \sum_{n=1}^{N} \epsilon_r(n+v)\epsilon_s^{(1)}(n)\right\}, \qquad u, v > 0.$$

Since $\epsilon_s^{(1)}(n)$, for $m \geq n$, is independent of $\epsilon_p(m+u)$, $\epsilon_r(n+v)$ and orthogonal to $\epsilon_q^{(1)}(m)$ if $m > n$, we see that the only contribution to the expectation can come when $m = n$ and $u = v$ when we obtain

$$N^{-1} \sum_{m=1}^{N} \sigma_{pr} g_{qs}^{(1)} = \sigma_{pr} g_{qs}^{(1)}$$

as required.

Now consider the effects of inserting estimates of $B(j)$ and $B^{(1)}(k)$, the latter being got by factoring $\hat{f}(\lambda)$, as stated in the theorem. We have

$$(7.7)\quad \sqrt{N}\{\hat{D}^{(1)}(u) - D^{(1)}(u)\}$$

$$= \sqrt{N} \sum_0^q\sum \{\hat{B}(j) - B(j)\}C'(u+q-k-j)\hat{B}^{(1)}(q-k)'$$

$$\times \left(1 - \frac{|u+q-k-j|}{N}\right) + \sqrt{N} \sum_0^q\sum B(j)C'(u+q-k-j)$$

$$\times \{\hat{B}^{(1)}(q-k)' - B^{(1)}(q-k)'\}\left(1 - \frac{|u+q-k-j|}{N}\right), u > 0.$$

Now $\{\hat{B}(j) - B(j)\}$ converges with probability 1 to zero and so therefore must $\{\hat{B}^{(1)}(k)' - B^{(1)}(k)'\}$. On the other hand, putting $v = u + q$, $v > q$ and taking the case in which we have made no mean correction for simplicity, then

$$\sqrt{N}\sum_{j=0}^{q} B(j)\left(1 - \frac{|v-k-j|}{N}\right)C'(v-k-j)$$

$$= \frac{1}{\sqrt{N}}\sum_{j=0}^{q} B(j) \sum_n{}' x(n+v-k-j)\, x(n)'.$$

Now, as in the proof of Theorem 1, we add a finite number of terms to the inner sum and obtain

$$\frac{1}{\sqrt{N}}\sum_{n=1}^{N} \epsilon(n+v-k)\, x(n)' \qquad v - k > 0;$$

by the same argument used in proving Theorem 1 this is asymptotically normal with zero mean. The second term on the right in (7.7) converges in probability to zero and the same may be shown for the first term in a similar manner. Thus (7.7) converges in probability to zero and thus the elements of $\sqrt{N}\,\hat{D}^{(1)}(u)$, $u > 0$, have the asymptotic distributional properties stated in the theorem.

Now, by writing $\hat{D}^{(1)}(u)$ as a column $\hat{d}^{(1)}(u)$ with the (u, v)th element in row $(u-1)p + v$ we see that the covariance matrix of $\sqrt{N}\,\hat{d}^{(1)}(u)$ is $G \otimes G^{(1)}$. Thus

$$N\, d^{(1)}(u)'(\hat{G} \otimes \hat{G}^{(1)})^{-1}\, d^{(1)}(u) = N \operatorname{tr}\,(\hat{G}^{-1}\hat{D}^{(1)}(u)\hat{G}^{(1)^{-1}}\hat{D}^{(1)}(u)')$$

is asymptotically distributed as chi-square with p^2 d.f., and these quantities are asymptotically independently distributed for $u = 1, 2, \ldots, d$, where d

is fixed a priori. Thus

$$N \sum_1^d \operatorname{tr}\left(\hat{G}^{-1}\hat{D}^{(1)}(u)\hat{G}^{(1)^{-1}}\hat{D}^{(1)}(u)'\right)$$

is distributed as chi-square with dp^2 d.f. This completes the proof.

The remaining problem is that of computing the $\hat{B}^{(1)}(k)$ and the actual mechanism of the test. In connection with the first question we observe that we have, using $\hat{f}$ for the estimate of f from the $\hat{B}(j)$,

$$\begin{aligned}\int \{\textstyle\sum \hat{B}^{(1)}(k)e^{ik\lambda}\}\hat{f}(\lambda)e^{iv\lambda}\,d\lambda &= 0, \qquad v > -q,\\ &= \hat{G}^{(1)}, \qquad v = -q,\end{aligned}$$

for the integrand is

$$\hat{G}^{(1)}\left(\sum \hat{B}^{(1)}(k)'e^{-ik\lambda}\right)^{-1}e^{iv\lambda} = \hat{G}^{(1)}\left(\sum \hat{B}^{(1)}(k)'e^{i(q-k)\lambda}\right)^{-1}e^{i(v+q)\lambda}$$

and the matrix function $(\sum \hat{B}^{(1)}(k)z^{q-k})^{-1}$ is holomorphic within the unit circle.† Thus we obtain

$$\begin{aligned}(7.8)\quad \sum \hat{B}^{(1)}(k)C(k+v) &= 0, \qquad v = -q+1, \ldots, 0, \qquad \hat{B}^{(1)}(q) = I_p,\\ &= \hat{G}^{(1)}, \qquad v = -q.\end{aligned}$$

These equations may also be written as

$$\begin{aligned}\sum \hat{B}^{(1)}(q-k)C'(k-u) &= 0, \qquad u = 1, \ldots, q, \qquad \hat{B}^{(1)}(q) = I_p,\\ &= \hat{G}^{(1)}, \qquad u = 0,\end{aligned}$$

so that the $\hat{B}^{(1)}(q-k)$ corresponds to the $C'(j)$ in the same way as $\hat{B}^{(1)}(k)$ corresponds to the $C(j)$. Now we proceed as follows.

1. Compute $\hat{B}(j)$, $\hat{B}^{(1)}(k)$, $\hat{G}$, $\hat{G}^{(1)}$ according to (2.6), (2.8), and (7.8). If one is testing the goodness of fit of a qth order autoregression it is very likely that one will have previously tested the significance of the matrix $\hat{B}(q+1)$ in a $(q+1)$th-order model. In this case exercise 8 shows that already the quantities $\hat{B}^{(1)}(j)$ and $\hat{G}^{(1)}$ will have been computed in the process of inverting C_{q+1}.
2. Compute the quantities

$$\hat{D}(u) = \sqrt{(N-u+q)}\sum_0^q\sum \hat{B}(j)C(j-k-u)\hat{B}^{(1)}(k)'.$$

(We have replaced the factor $(1 - |u+k-j|/N)$ by unity and have modified

† In fact this will be, strictly, true only if the $C(n)$ are defined by using a divisor N^{-1} and not $(N-|n|)^{-1}$.

the factor $\sqrt{N}$ accordingly. Experience with the case $p = 1$ suggests that this may be preferable, though it is a "moot" point.)†

3. Compute

$$\sum_1^d \operatorname{tr} \hat{G}^{-1} D(u) \hat{G}^{(1)^{-1}} \hat{D}(u)',$$

which is to be treated as chi-square with $p^2 d$ degrees of freedom.

In case $p = 1$ the formulae are somewhat simpler and have been given in Section 2, in the paragraphs immediately preceding (7.12).

As we have pointed out in a footnote on page 338 this extension of Quenouille's test may have reduced power compared to the likelihood ratio test, its virtue being the reduction in calculations and simplicity of carrying through the test *for any d*, once $\hat{B}(j)$, $\hat{B}^{(1)}(G)$, $\hat{G}, \hat{G}^{(1)}$ have been found. The reason for the reduction in power can be seen from the final computation under 3 above, for this is based on the sum of squares of the elements in $\hat{G}^{-\frac{1}{2}}\hat{D}(u)\hat{G}^{(1)-\frac{1}{2}}$. This sum of squares is *of the nature of* a sum of squares due to regression divided by a residual sum of squares. In the likelihood-ratio test the corresponding term will use a residual sum of squares from an autoregression of order $q + u$ and if the null hypothesis is not true this will be much smaller than that for the test being discussed. This is because the residual sum of squares used for that test is based on $\hat{G}$ which comes from an autoregression of order only q and may be too large if the null hypothesis is not true.

Two further points are the following:

1. It may be shown that this test is asymptotically equivalent to that of the form of a partial multiple autocorrelation treated after (2.14) in Section 2. Its virtue is some computational simplification.

2. One may, of course, find the (normalized) eigenvectors $\hat{\phi}_j$ of $\hat{G}$ and $\hat{\psi}_k$ of $\hat{G}^{(1)}$. If $\hat{\lambda}_j$, $\hat{\mu}_k$ are the associated eigenvalues the quantities

$$e_{j,k}(u) = \frac{\hat{\phi}_j' \hat{D} \hat{\psi}_k}{\hat{\lambda}_j \hat{\mu}_k},$$

which are individually and independently distributed, asymptotically, as normal variates with zero mean and unit variance.

A number of other goodness of fit procedures have been suggested. We refer the reader to Bartlett (1954) Bartlett and Diananda (1950), Grenander and Rosenblatt (1957), Hannan (1960) (especially section IV.3), for a discussion of some of these.

† See the footnote on p. 403.

8. CONTINUOUS TIME PROCESSES AND DISCRETE APPROXIMATIONS

In Chapter I, Section 3 we discussed stochastic differential equations and in Chapter II Section 4, their relation to rational spectra. We are led here to consider the relation of such spectra to the problems of the present chapter, via the sampling of the corresponding continuous time phenomenon. We consider the scalar case. Thus we consider a spectrum of the form

$$(8.1)\quad f(\mu) = \frac{\sigma^2}{2\pi}\frac{|\sum_j \alpha(j)(i\mu)^j|^2}{|\sum_k \beta(k)(i\mu)^k|^2}, \qquad -\infty < \mu < \infty, \qquad \alpha(0) = \beta(0) = 1,$$

where the polynomial in the denominator is of a degree at least one higher than that in the numerator. (If that were not so, $f(u)$ would not be integrable.) We may decompose $f(\mu)$ into partial fractions

$$\frac{\sigma^2}{2\pi}\left\{\sum_{k=1}^{r}\sum_{j=1}^{q_k}\frac{a_{k,j}}{(i\mu - z_k)^j} + \sum_{k=1}^{r}\sum_{j=1}^{q_k}\frac{\bar{a}_{k,j}}{(-i\mu - \bar{z}_k)^j}\right\},$$

where q_k is the multiplicity of the zero z_k, having negative real part, of the polynomial $q(i\mu) = \sum \beta(k)(i\mu)^k$. (See Van der Waerden, 1949, p. 88.) Since

$$\gamma(t) = \int_{-\infty}^{\infty} e^{it\mu} f(\mu)\, d\mu,$$

we obtain the expression

$$\gamma(t) = \sigma^2 \sum_{k=1}^{r}\sum_{j=1}^{q_k} a_{k,j} e^{tz_k}\frac{t^{j-1}}{(j-1)!}, \qquad t \geq 0,$$
$$\gamma(-t) = \gamma(t).$$

Now, if we sample at points $t = 0, \pm 1, \pm 2, \ldots$, we obtain

$$\gamma(n) = \sigma^2 \sum_{k=1}^{r}\sum_{j=1}^{q_k} a_{k,j} e^{nz_k}\frac{n^{j-1}}{(j-1)!} = \gamma(-n), \qquad n \geq 0.$$

Thus

$$f^{(1)}(\lambda) = \frac{1}{2\pi}\sum_{-\infty}^{\infty}\gamma(n)e^{-in\lambda}$$
$$= \frac{\sigma^2}{2\pi}\sum_{k=1}^{r}\sum_{j=1}^{q_k}\frac{a_{k,j}}{(j-1)!}\left\{\left[\frac{\partial^{j-1}}{\partial u^{j-1}}(1 - e^u)^{-1}\right]_{u=z_k - i\lambda}\right.$$
$$\left. + \left[\frac{\partial^{j-1}}{\partial u^{j-1}}(1 - e^u)^{-1}\right]_{u=z_k + i\lambda} - \delta_1^{\,j}\right\},$$

which is evidently a rational spectrum with constants determined by the $a_{k,j}$ and z_k. In particular, for $q_k \equiv 1$

$$f^{(1)}(\lambda) = \frac{\sigma^2}{2\pi} \sum_{k=1}^{r} \frac{a_k(1 - e^{2z_k})}{1 + e^{2z_k} - 2e^{z_k} \cos \lambda}.$$

The terms for which z_k is not real group in conjugate pairs.

Thus a method for proceeding to estimate the constants in $f(\mu)$ would be to estimate the constants in $f^{(1)}(\lambda)$ by the technique of Section 4 and thence to obtain the $a_{k,j}$, z_k and so the $\alpha(j)$, $\beta(k)$ in $f(\mu)$. Some problems arise for the autoregressive, moving average constants in $f^{(1)}(\lambda)$ may be greater in number than occur in $f(\mu)$ so that some constraints on them may be needed.

Example 1

$$f(\mu) = \frac{\sigma^2}{2\pi} \frac{1}{|1 - \beta^{-1}(i\mu)|^2}, \qquad \beta < 0.$$

Then $a_1 = -\frac{1}{2}\beta$, $z_1 = \beta$,

$$f^{(1)}(\lambda) = \frac{\sigma^2}{2\pi} \frac{-\frac{1}{2}\beta(1 - e^{2\beta})}{1 + e^{2\beta} - 2e^{\beta} \cos \lambda},$$

which is the spectrum of an autoregression with autocorrelation $\exp \beta$. In this case the procedure is quite straightforward.

Example 2

$$f(\mu) = \frac{\sigma^2}{2\pi} \frac{1}{|(1 - \beta_1^{-1} i\mu)(1 - \beta_2^{-1} iu)|^2}, \qquad \beta_1 < 0, \qquad \beta_2 < 0.$$

Then $a_1 = -2\beta_1^2(\beta_1^2 - \beta_2^2)$, $a_2 = -2\beta_2^2(\beta_2^2 - \beta_1^2)$, $z_1 = \beta_1$, $z_2 = \beta_2$. Let us put $b_1 = \exp \beta_1$, $b_2 = \exp \beta_2$:

$$f^{(1)}(\lambda) = \frac{\sigma^2}{2\pi} \left\{ \frac{a_1(1 - b_1^2)}{1 + b_1^2 - 2b_1 \cos \lambda} + \frac{a_2(1 - b_2^2)}{1 + b_2^2 - 2b_2 \cos \lambda} \right\}$$

and the numerator of $f^{(1)}(\lambda)$, when written out as a rational function in its lowest terms, will not, in general, be a constant. Thus the coefficients of the discrete time, autoregressive-moving average process are functions of a smaller number of parameters and the efficient estimation problem is quite a complex one. One procedure would be to estimate $f^{(1)}(\lambda)$, as a rational spectrum, by the procedures of this section when the sampling interval has been small, relative to the time lags over which autocorrelations in the original process are substantial. (Thus each $\exp z_k$ will be near to unity in modulus.) This estimate of $f^{(1)}(\lambda)$ can then be treated as an estimate of $f(\mu)$ and the estimates of the constants σ^2 $\alpha(j)$, $\beta(k)$ could be obtained from that.

The matter needs further investigation. Alternatively one may neglect the constraints imposed by the fact that there are truly only three unknown parameters (β_1, β_2, and σ^2) and use the autoregressive constants in the estimate of $f^{(1)}(\lambda)$ to estimate β_1 and β_2.

For a discussion of the vector case along these lines (without discussion of the problem just mentioned) see Phillips (1959).

EXERCISES

1. Prove that, $\hat{\sigma}_q^{\,2}$ being the estimate of the variance of $\epsilon(n)$ is an autoregression of order q (with $p = 1$),

$$\hat{\sigma}_q^{\,2} = \frac{\det(C_{q+1})}{\det(C_q)}.$$

2. Prove that the mean and variance of von Neumann's ratio, on the null hypothesis that the $x(n)$ are N.I.D. (μ, σ^2), are 2 and $4(N-2)/(N^2-1)$, respectively.

3. Show that Γ_N^{-1}, for $x(n)$ generated by an autoregression of order q, $N > q$, has, when $\sigma^2 = 1$, in the kth row and the jth diagonal above the main diagonal, the entry

$$\begin{aligned} &\sum_u \beta(u)\,\beta(u+j), && 0 \le j \le q, \\ &0, && q < j, \end{aligned}$$

where the sum is over u from 0 to $\min\{k-1, q-j, N-j-k\}$.

4. Prove that $J(\hat{\alpha}) = \{|P_{q+1}|/|P_q|\}\,|D|$ in the notation of the first displayed formula below (3.14). (Write $J(\hat{\alpha})$ in the form

$$J(\hat{\alpha}) = \left|\begin{array}{c|c} 1 & \hat{\alpha} \\ \hline \hat{\alpha} & D \end{array}\right|$$

and use $\hat{\alpha} = -D'r$.)

5. Obtain the Jacobian of the transformation from $\hat{\alpha}_1, \ldots, \hat{\alpha}_q$ to

$$\hat{r}(j \mid 1, \ldots, j-1),$$

$j = 1, \ldots, q$, required during the proof of Theorem 6. (To do this write $\hat{\alpha}_{j,q}$ for $\hat{\alpha}_j$ where q is the order of the fitted autoregression and show that

$$\hat{\alpha}_{j,q} = \hat{\alpha}_{j,q-1} + \hat{\alpha}_{q-j,q-1}\hat{\alpha}_{q,q},$$

using the relation $r_q = -P_q\hat{\alpha}_q$ where the q subscripts refer to the order of the fitted autoregression, once more. This defines a transformation from $\hat{\alpha}_{j,q}, j = 1, \ldots, q$, to $\hat{\alpha}_{j,q-1}, j = 1, \ldots, q-1$, and $\hat{\alpha}_{qq} = \hat{r}(q \mid 1, \ldots, q-1)$. The Jacobian of this transformation is then easily obtained as

$$\begin{aligned} H_q &= \{1 - r^2(q \mid 1, \ldots, q-1)\}^{\frac{1}{2}(q-1)}, && q \text{ odd}, \\ &= \{1 - r(q \mid 1, \ldots, q-1)\}\{1 - r^2(q \mid 1, \ldots, q-1)\}^{\frac{1}{2}q-1}, && q \text{ even}. \end{aligned}$$

Now repeating the process with q replaced by $q - 1$ we obtain H_{q-1} and the Jacobian of the required transformation is $H_q H_{q-1} \cdots H_2$.)

Show also that

$$\sum_0^q \hat{\alpha}_j = \prod_1^q \{1 - r(j \mid 1, \ldots, j-1)\}.$$

6. Assume $x(n)$ to be autoregressive and Gaussian of order q and that $N = (M+1)q$. Consider the conditional distribution of $y(m) = x((q+1)m)$ given $x(m(q+1)+j)$, $x(m(q+1)-j)$, $j = 1, \ldots, q$, $m = 1, \ldots, M$. Show that the expectation of $y(m)$ in the conditional distribution is linear in

$$z_j(m) = \tfrac{1}{2}\{x(m(q+1)+j) + x(m(q+1)-j)\}$$

and that the coefficient of $z_q(m)$ is null if and only if the autoregression is of order q. Derive, thence, an exact test for $\beta(q) = 0$. Show that the asymptotic relative efficiency (ARE) of this test as compared to that based on $r(q \mid 1, \ldots, q-1)$ [i.e., $\hat{\beta}(q)$] is

$$\left\{\frac{2}{q+1} \sum_0^{q-1} \beta^2(j)\right\}^{-1}.$$

[For ARE see Fraser (1957). For a further discussion of this problem see Hannan (1969c) and references therein.]

7. Let P_{q+1} have $\rho(j-k)$ in row j column k and put $\beta = -P_q^{-1} p_q$ where p_q has $\rho(j)$ in the jth place. Show that the necessary and sufficient condition that

$$\sum_0^q \beta(j) z^j, \qquad \beta(0) = 1$$

have all of its zeros outside of the unit circle is the condition that P_{q+1} be positive definite.

Hints. Having determined the $\beta(j)$ from P_{q+1} use the relation $\sum \beta(j)\rho(k-j) = 0$ to generate a sequence $\rho(n) = \rho(-n)$ for all n. Composing P_{q+j} out of these show that it is positive definite by establishing the relation $\det(P_{q+j})/\det(P_{q+j-1}) = \det(P_{q+1})/\det(P_q) > 0$. Show thence that the $\rho(n)$ are a positive definite sequence and the Fourier coefficients of an autoregressive spectral density function. Factoring this to obtain $\tilde{\beta}(j)$ for which $\sum \tilde{\beta}(j) z^j$ has all of its zeros outside of the circle show that $\tilde{\beta}(j)$ is determined by the $\rho(n)$, $n = 0, 1, \ldots, q$ and thus equals $\beta(j)$.

8. If a positive definite matrix A_q is partitioned as

$$\left[\begin{array}{c|c} A & D \\ \hline D' & E \end{array}\right],$$

where A and E are square, show that A_q^{-1} is

$$\left[\begin{array}{c|c} A^{-1} + A^{-1}D(E - D'A^{-1}D)^{-1}D'A^{-1} & -A^{-1}D(E - D'A^{-1}D)^{-1} \\ \hline & (E - D'A^{-1}D)^{-1} \end{array}\right].$$

Hence show that Γ_q^{-1}, for an autoregression of order $q - 1$, has as entry in the top left hand corner the matrix $\Gamma_{q-1}^{-1} + \sum_1^{q-1} B^{(1)\prime}(j)G^{(1)^{-1}}B^{(1)}(j)$, in the right-hand column of blocks the matrices $B^{(1)\prime}(j)G^{(1)^{-1}}$, $j = 0, 1, \ldots, q - 2$, and in the bottom righthand corner $G^{(1)^{-1}}$ where the matrices $B^{(1)}(j)$ are $G^{(1)}$ are defined in Section 8. Discuss the relevance of this to the computing problems of this chapter.

APPENDIX

Proof of Theorem 8

We consider only Theorem 8 in this appendix. Theorem 7 is essentially contained in Theorem 8. We shall not give a separate proof of Theorem 7′. The proof follows fairly easily from that of Theorem 8, though it is simpler in some ways and more complex in others.

We shall give the proof as if $\hat{\alpha}$, based on a factorization of $\hat{f}_y$, is used. The proof for the case where $\hat{\alpha}$ is obtained from (5.17) is hardly different and we shall discuss it later. We introduce the estimate $\dot{\beta} = -B^{-1}b$ wherein B and b are defined in the same way as $\hat{B}$, $\hat{b}$ but replacing $\hat{g}$ by g. Of course $\dot{\beta}$ is not computable. We also introduce $\dot{\alpha} = -A^{-1}a$ where A and a are defined in the same way as $\hat{A}^{(1)}$, $\hat{a}^{(1)}$ but with g and h in place of $\hat{g}$, $\hat{h}^{(1)}$, in (5.19). Again $\dot{\alpha}$ is not computable. Our procedure is to express $\tilde{\alpha}^{(1)}$, $\hat{\beta}^{(1)}$ in terms of $\dot{\alpha}$, $\dot{\beta}$ whose asymptotic distribution is relatively easy to obtain. We first observe that $\hat{B}$, B converge with probability one to $(\sigma^2/2\pi)\Psi$ while $\hat{b}$, b converge in the same way to $(\sigma^2/2\pi)\psi$, where ψ is a vector of q components with $\psi(k)$ in the kth row. To show this we first observe that $x(n)$ is mixing and thus ergodic (see Chapter IV) and thus $c(n)$ converges almost surely to $\gamma(n)$. Hence the $\hat{\beta}(k)$ converge almost surely to the $\beta(k)$ and $2\pi\hat{f}_y$ converges almost surely to $\sigma^2\,|g|^2$. In turn, the zeros of $\hat{g}$ converge almost surely to those of g and thus for N sufficiently large they lie outside of the unit circle. Thus for $N > N_0$, let us say,

$$|\hat{g}|^{-2} = \sum_{-\infty}^{\infty} \hat{\delta}(u)e^{iu\lambda}, \qquad |\hat{\delta}(u)| < K\gamma^{-|u|}, \qquad \gamma > 1.$$

(Of course, N_0 will depend on the realization.) Also

$$\hat{\delta}(u) = \frac{1}{2\pi}\int_{-\pi}^{\pi} |\hat{g}|^{-2}e^{-iu\lambda}\,d\lambda,$$

and since $|\hat{g}|^{-2}$ is uniformly bounded for $N > N_0$ then $\hat{\delta}(u)$ converges almost surely to $\delta(u)$, for each u, where the $\delta(u)$ are the Fourier coefficients of $|g|^{-2}$. We shall continue to use K for a finite positive constant, not necessarily always the same one. Now $\hat{B}$ has, in row k, column l, the element

$$(1) \qquad \frac{1}{2\pi}\sum_{v=-N+1}^{N-1} c(v)\left(1 - \frac{|v|}{N}\right)\sum_{j=-\infty}^{\infty} \hat{\delta}(l - k - v + jN).$$

Since $|c(v)(1 - |v|/N) < c(0)$ and since $\delta(l - k - v + jN)$ converges almost surely to zero with N for $j \neq 0$ and to $\delta(l - k - v)$ for $j = 0$ then, by dominated convergence, (1) converges almost surely to

$$\frac{1}{2\pi} \sum_{\infty}^{-\infty} \gamma(v)\delta(l - k - v).$$

But this is $(2\pi)^{-1}$ by the $(k - l)$th Fourier coefficient of $f|g|^{-2} = (2\pi)^{-1}\sigma^2|h|^{-2}$ so that (1) converges almost surely to $(\sigma^2/2\pi)\psi\ (k - l)$. Thus $\hat{B}$, $\hat{b}$ converges almost surely, respectively, to $(\sigma^2/2\pi)\Psi$, $(\sigma^2/2\pi)\psi$. A similar, but much simpler, argument establishes the required result for B and b.

We have

$$\sqrt{N}(\hat{\beta}^{(1)} - \dot{\beta}) = -\sqrt{N}\{\hat{B}^{-1}(B - \hat{B})B^{-1}\hat{b} + B^{-1}(\hat{b} - b)\}, \tag{2}$$

where $N^{1/2}(B - \hat{B})$ and $N^{1/2}(b - \hat{b})$ have typical element

$$\text{(3)}\quad N^{-1/2}[I \cdot e^{ik\omega_j}\{|g|^{-2}\,\hat{g}^{-1}(\hat{g} - g) + |\hat{g}|^{-2}\,\bar{g}^{-1}(\hat{\bar{g}} - \bar{g})\}]$$

$$= \sum_{v=1}^{s} \sqrt{N}(\hat{\alpha}(v) - \alpha(v))\left\{N^{-1}\sum_j I|g|^{-2}\,\hat{g}^{-1}e^{i(k+v)\omega_j} + N^{-1}\sum_j I\cdot|\hat{g}|^{-2}\,\bar{g}^{-1}e^{i(k-v)\omega_j}\right\}.$$

Observing that the second term within the bracketed factor is real and thus may be conjugated then by almost the same proof as for the case of the $\hat{b}(k)$ the bracketed factor may be shown to converge almost surely to

$$\frac{\sigma^2}{4\pi^2}\int_{-\pi}^{\pi} |h|^{-2}g^{-1}e^{iv\lambda}\{e^{ik\lambda} + e^{-ik\lambda}\}d\lambda = \frac{\sigma^2}{2\pi^2}\int_{-\pi}^{\pi} (|h|^2\,g)^{-1}e^{iv\lambda}\cos k\lambda\, d\lambda.$$

On the other hand, $\sqrt{N}(\hat{\alpha} - \alpha)$ converges in distribution (indeed is asymptotically normal, see Section 4) so that the right side of (3) may be replaced by

$$\sum_{v=1}^{s} \sqrt{N}(\hat{\alpha}(v) - \alpha(v))\frac{\sigma^2}{2\pi^2}\int_{-\pi}^{\pi} (|h|^2\,g)^{-1}e^{iv\lambda}\cos k\lambda\, d\lambda,$$

the neglected term converging in probability to zero.

Now since $-B^{-1}\hat{b}$ converges almost surely to β and $\hat{B}^{-1}$, B^{-1} both converge almost surely to $(2\pi/\sigma^2)\Psi^{-1}$ we may replace (2) by $(2\pi/\sigma^2)\Psi w$ where the vector w has in the kth place

$$w(k) = \sum_{v=1}^{s} \sqrt{N}(\hat{\alpha}(v) - \alpha(v))\left\{\frac{\sigma^2}{2\pi^2}\int_{-\pi}^{\pi} (|h|^2\,g)^{-1}e^{iv\lambda}\left(\sum_{1}^{q}\beta(l)\cos(k - l)\lambda + \cos k\lambda\right)d\lambda\right\}$$

$$= \sum_{v=1}^{s} \sqrt{N}(\hat{\alpha}(v) - \alpha(v))\frac{\sigma^2}{2\pi}\,\omega(k - v).$$

Thus

$$\sqrt{N}(\hat{\beta}^{(1)} - \dot{\beta}) - \Psi^{-1}\Omega\sqrt{N}(\hat{\alpha} - \alpha)$$

converges in probability to zero.

In the same way we may show that

$$\sqrt{N}(\hat{\alpha}^{(1)} - \dot{\alpha}) - 2\sqrt{N}(\hat{\alpha} - \alpha) + \Phi^{-1}\Omega'\sqrt{N}(\hat{\beta}^{(1)} - \beta)$$

converges in probability to zero. We leave the reader to check through the routine of this proof which follows that for $\sqrt{N}(\hat{\beta}^{(1)} - (\dot{\beta})$ with no essential changes. Thus neglecting quantities which converge in probability to zero we have

$$\sqrt{N}(\hat{\alpha}^{(1)} - \alpha) = \sqrt{N}(\dot{\alpha} - \alpha) + 2\sqrt{N}(\hat{\alpha} - \alpha) - \Phi^{-1}\Omega'\sqrt{N}(\hat{\beta}^{(1)} - \beta)$$

$$\sqrt{N}(\hat{\beta}^{(1)} - \beta) = \sqrt{N}(\dot{\beta} - \beta) + \Psi^{-1}\Omega\sqrt{N}(\hat{\alpha} - \alpha).$$

Thus

(4) $$\sqrt{N}(\tilde{\alpha}^{(1)} - \alpha) = -\sqrt{N}\{I_s - \Phi^{-1}\Omega'\Psi^{-1}\Omega\}^{-1}\{(\dot{\alpha} - \alpha) - \Phi^{-1}\Omega'(\dot{\beta} - \beta)\},$$

again neglecting quantities which converge in probability to zero. (This depends upon the fact to be established later that $\sqrt{N}(\dot{\alpha} - \alpha)$ and $\sqrt{N}(\dot{\beta} - \beta)$ are asymptotically normal plus the fact that $\hat{\Phi}$, $\hat{\Psi}$, $\hat{\Omega}$ converge almost surely to Φ, Ψ, Ω.) Since $\hat{\alpha}^{(1)}$ and $\hat{\alpha}$ converge almost surely to α (the former since $\hat{A}$ and $\hat{a}$ converge almost surely to $(2\pi/\sigma^2)\Phi$, $(2\pi/\sigma^2)\phi$, respectively) then $\tilde{\alpha}^{(1)}$ converges almost surely to α. Thus we can now repeat the argument beginning from $\sqrt{N}(\tilde{\beta}^{(1)} - \dot{\beta})$ and using (4). Following the same steps we find that, neglecting similar quantities,

(5) $$\sqrt{N}(\tilde{\beta}^{(1)} - \beta) = \sqrt{N}\{I_q - \Psi^{-1}\Omega\Phi^{-1}\Omega'\}^{-1}\{(\dot{\beta} - \beta) - \Psi^{-1}\Omega(\dot{\alpha} - \alpha)\}.$$

It remains to investigate the asymptotic distribution of $\sqrt{N}(\dot{\alpha} - \alpha)$, $\sqrt{N}(\dot{\beta} - \beta)$ and use (4) and (5) to convert this into the asymptotic distribution for

$$\sqrt{N}(\tilde{\alpha}^{(1)} - \alpha), \quad \sqrt{N}(\tilde{\beta}^{(1)} - \beta).$$

Now

$$\sqrt{N}(\dot{\alpha} - \alpha) = -\sqrt{N}A^{-1}(a + A\alpha)$$

and $\sqrt{N}(a + A\alpha)$ has typical element

$$\frac{\sigma^2}{2\pi}u_k = N^{-\frac{1}{2}}\sum_j I\,|h|^2\,|g|^{-2}g^{-1}e^{ik\omega_j}, \qquad k = 1, \ldots, s,$$

since $1 + \sum \alpha(l)\exp(-il\omega_j) = \bar{g}(\omega_j)$. Since $(\sigma^2/2\pi)A$ converges almost surely to Φ we may replace $\sqrt{N}(\dot{\alpha} - \alpha)$ by $-\Phi^{-1}u$ where u has u_k as a typical element. Similarly we may replace $\sqrt{N}(\dot{\beta} - \beta)$ by $-\Psi^{-1}v$, where v has typical element

$$v_k = \frac{2\pi}{\sigma^2}N^{-\frac{1}{2}}\sum_j I\,\bar{h}|g|^{-2}\,e^{jk\omega_j}$$

$$= \frac{2\pi}{\sigma^2}N^{-\frac{1}{2}}\sum_j I|h|^2\,|g|^{-2}\,h^{-1}e^{ik\omega_j}.$$

Put

$$\left|\frac{h}{g}\right|^2 = \left|\sum_0^\infty f_u e^{iu\lambda}\right|^2, \qquad h^{-1} = \sum_j^\infty h_u e^{iu\lambda},$$

where both the f_u and h_u sequences decrease exponentially. Then

$$v_k = \frac{N^{1/2}}{\sigma^2} \sum_{u=-N+1}^{N-1} c(n)\left(1 - \frac{|n|}{N}\right) \sum_j \sum_u \sum_v f_u f_{u+v} h_{v-k+n+jN},$$

where the sum is over all j, u, v for which the summand is defined (or equivalently for all u, v, j subject to $f_a = 0, a < 0;\ h_b = 0, b < 0$). We may show that we may neglect the terms for $j \neq 0$ since $\sum_j N^r h_{v-k+n+jN}$ converges to zero for each fixed $v, k, n, r > 0$ and $j \neq 0$ (since h_u converges to zero exponentially). Thus we may consider

$$v_k' = \frac{N^{1/2}}{\sigma^2} \sum_{-N+1}^{N-1} c(n)\left(1 - \frac{n}{N}\right) \sum_u \sum_v f_u f_{u+v} h_{v-k+n}.$$

Rearranging, this is

$$\sigma^{-2} N^{-1/2} \sum_{l,m=1}^{N}\sum x(l)\, x(m) \sum_u \sum_v f_u f_{u+v} h_{v-k+l-m}$$

$$= \sigma^{-2} N^{-1/2} \sum_{l,m=1}^{N}\sum x(l)\, x(m) \sum_w \sum_t f_{w-m} f_{w+t-l} h_{t-k}$$

$$= \sigma^{-2} N^{-1/2} \sum_w \sum_t h_{t-k} \sum_j f_j x(w - t - j) \sum_l f_l x(w - l),$$

where the summations are over all j, l, w, t subject to j and l being nonnegative, $t > k$ and the requirement that both $w - l$, $w - t - j$ shall lie between 1 and N inclusive. If we restrict w to the range $1 \leq w \leq N$ we omit terms whose expected modulus (as we see from the second last displayed formula) is dominated by

$$KN^{-1/2} \sum_{w=N}^{\infty} \sum_t a_f^{-w} a_f^{-w+t} a_h^{-t+k}\, \mathcal{E}\left\{\left|\sum_{l,m}\sum x(l)\, x(m)\, a_f^{m+l}\right|\right\}$$

where a_f^{-u} is the exponential rate of decrease of f_u and a_h^{-u} of h_u. This is not greater than

$$KN^{-1/2} \sum_{w=N}^{\infty} \sum_t a_f^{-w+N} a_f^{-w+N+t} a_h^{-t+k} = O(N^{-1/2}).$$

Thus we now consider

$$v_k'' = \sigma^{-2} N^{-1/2} \sum_{w=1}^{N} \sum_t h_{t-k} \sum_j f_j x(w - t - j) \sum_l f_l x(w - l).$$

Now

$$\sum_{l=0}^{\infty} f_l x(w - l) = \epsilon(w)$$

and this differs from the corresponding sum occurring in v_k'' by

$$\sum_{l=w}^{\infty} f_l x(w - l),$$

whose standard deviation is dominated by

$$K \sum_w^{\infty} a_f^{-j} = K a_f^{-w}.$$

Replacing the sums over j and l with $\epsilon(w)$ and $\epsilon(w + t)$ adds to v_k'' a quantity whose standard deviation is dominated by

$$KN^{-\frac{1}{2}} \sum_w \sum_t h_{t-k} a_f^{-2w} = O(N^{-\frac{1}{2}}).$$

Finally we are left with

$$\sigma^{-2} N^{-\frac{1}{2}} \sum_{w=1}^{N} \sum_t h_{t-k} \epsilon(w - t) \epsilon(w) = \sigma^{-2} N^{-\frac{1}{2}} \sum_{n=1}^{N} \epsilon(n)\, y(n - k),$$

where

$$y(w - k) = \sum_0^{\infty} h_j \epsilon(w - k - j).$$

This means that $y(n)$ is generated by the autoregressive relation

$$\sum_0^{q} \beta(j)\, y(n - j) = \epsilon(n).$$

Similarly and to the same approximation we may replace u_k with

$$\sigma^{-2} N^{-\frac{1}{2}} \sum_1^{N} \epsilon(n)\, z(n - k),$$

where $z(n)$ is generated by

$$\sum_0^{s} \alpha(j)\, z(n - j) = \epsilon(n).$$

Now from the proof of Theorem 1 we see that $\Phi^{-1}u$, $\Psi^{-1}v$ are asymptotically jointly normal with zero mean and covariance matrix

$$\begin{bmatrix} \Phi^{-1} & \Phi^{-1}\Omega'\Psi^{-1} \\ \Psi^{-1}\Omega\Phi^{-1} & \Psi^{-1} \end{bmatrix},$$

since u and v are of the same forms (save for the constant σ^2) as are considered in that section.

Thus finally it remains only to compute the asymptotic covariance matrix for $\sqrt{N}(\tilde{\alpha}^{(1)} - \alpha)$, $\sqrt{N}(\tilde{\beta}^{(1)} - \beta)$ from (4) and (5). This is

$$\begin{bmatrix} \Phi & -\Omega' \\ -\Omega & \Psi \end{bmatrix}^{-1}.$$

It can be seen from the proof that if we had begun from an estimate $\hat{\alpha}$ not got by factoring $\hat{f}_y$ but rather from $-\hat{A}^{-1}\hat{a}$ [see (5.17)] then the argument would proceed almost without change for we need only to know that $\hat{\alpha}$ converges almost surely to α and $\sqrt{N}(\hat{\alpha} - \alpha)$ is asymptotically normal and the proof of the latter for $\hat{\alpha}$ is of the same form as for $\hat{\beta}^{(1)}$.

We now point out that the covariance matrix will be singular if both $\alpha(s)$ and $\beta(q) = 0$. For then, calling α the vector with $\alpha(j)$ in the $(j + 1)$st place for $j = 0$,

$,\ldots, s-1$ and β the vector with $\beta(k)$ in the $(k+1)$st place, $k = 0,\ldots, q-1$ we see that

$$\Phi\alpha = \Omega'\beta, \qquad \Psi\beta = \Omega\alpha,$$

Thus

$$\begin{bmatrix} \Phi^{-1} & \Phi^{-1}\Omega'\Psi^{-1} \\ \Psi^{-1}\Omega\Phi^{-1} & \Psi^{-1} \end{bmatrix} \begin{pmatrix} -\phi \\ \psi \end{pmatrix} = 0$$

and our result is established. To the extent that the estimate of this matrix, hence of the covariance matrix of the estimates $\hat{\alpha}$, $\hat{\beta}$, is near to singular we shall have an indication that $\alpha(s) = \beta(q) = 0$. This subject merits further investigation but we shall not go further into it here.

CHAPTER VII

Regression Methods

1. INTRODUCTION

One of the main purposes of the present chapter is to deal with the situation wherein a mean correction, possibly time dependent, needs to be made to the vector sequence before the analysis is commenced. We need to discuss how to make that correction as well as the effect it will have upon the estimates of spectra, for example, made from the mean corrected quantities. To some extent, indeed to a large extent from a practical viewpoint, these investigations can also be viewed as accounting for the possibility that the spectra are not absolutely continuous, for the appropriate method by which to account for spectral jumps is evidently regression.

In addition the use of linear models relating sets of vector time series is important for its own sake; for example, "distributed lag" models, of the form†

$$z(n) = \sum_{0}^{\infty} B(j)' \, y(n-j) + x(n), \tag{1.1}$$

wherein the matrices $B(j)$ "converge to zero" at some appropriate rate and $x(n)$ is stationary and totally independent of the $y(n)$ sequence, have become of considerable importance in economics in recent years. (We take $z(n)$ to be a vector of p components and $y(n)$ to be of q components so that the $B(j)'$ are $p \times q$ and $x(n)$ is $p \times 1$.) Throughout this chapter we will use a notation typified by (1.1). Thus $x(n)$ will be the unobserved *stationary* residual, $y(n)$ will be the "regressor" sequence and $z(n)$ will be the "regressand" sequence.‡ We shall always assume that $\mathcal{E}(y(m)x(n)') \equiv 0$ but often will assume also that the two sequences are totally independent. Of course, other specifications will also be made.

† We have found it convenient to call $B(j)'$ the matrix of coefficients of $y(n-j)$. This is basically because when $p = 1$ and there are no lags the vector of regression coefficients is customarily taken as a column vector. This notation to some extent disagrees with that used for autoregressions.

‡ Though "regrediend" may be more appropriate, and was once forced upon the author by an editor, the less pretentious sounding regressand is preferred here.

If an asymptotic theory is to be constructed we need, at least, to say something about the limiting behavior of the moments of the $y(n)$ sequence. One could, for example, require this to be stationary and ergodic. However, such a specification would eliminate too many important cases (for example, $y_j(n) = n^j$). The specification we shall adopt is that given in Section 3 of Chapter IV, namely conditions (a), (b), (c) or (a), (b)′, (c) (See also Chapter II, Section 6.) We shall speak of the first of these sets as "Grenander's conditions" as they appeared first in a paper by Grenander (1954).

2. THE EFFICIENCY OF LEAST SQUARES. FIXED SAMPLE SIZE

We begin by considering the regression relation

$$z(n) = B'y(n) + x(n), \qquad n = 1, \ldots, N, \tag{2.1}$$

which we rewrite in the tensor notation

$$z = (I_p \otimes Y)\beta + x,$$

wherein $z_i(n)$ is in row $(i - 1)p + n$ of z, $x_i(n)$ is in row $(i - 1)p + n$ of x, Y has $y_j(n)$ in row n column j, β has β_{ij} in row $(j - 1)q + i$ (The columns of B are placed successively down β as in Chapter IV, Section 3). We shall continue to use Γ_p for the covariance matrix of the vector x since we shall be emphasizing the case which arises when $x(n)$ comes from a stationary process. However, now we take Γ_p to consist of p^2 blocks, of which a typical one contains the N^2 entries $\gamma_{ij}(u - v)$, $u, v = 1, \ldots, N$. This model (2.1) evidently incorporates (1.1) insofar as all $B(j)$ in that relation are null from some point on and there are no restrictions on the $B(j)$. In fact, the situation in which there are linear restrictions is not unimportant and we indicate how it may be introduced into our considerations. Thus we consider r linear restrictions of the form

$$\operatorname{tr}(BA_j) = 0, \qquad j = 1, \ldots, r,$$

A_j being a $(q \times p)$ real matrix. *We agree once and for all to call trivial, such linear restrictions as amount to a mere presumption that a certain column of B' is null, or that this is so after a known change of variables* (whereby $y(n)$ is replaced by $Cy(n)$ and B' by $B'C^{-1}$ so that it is a column of $B'C^{-1}$ which is null). This is evidently a reasonable terminology. Let α_j be the column got from A_j as β was from B and let E project onto the orthogonal complement of the space spanned by the α_j in pq dimensional real vector space. Then $E\beta = \beta$ is the relation embodying the linear restrictions. We now have

$$z = (I_p \otimes Y)E\beta + x = U\beta + x, \qquad U = (I_p \otimes Y)E.$$

Let us first consider the general model $z = U\beta + x$ and later specialize to the case where $U = (I_p \otimes Y)E$. We shall continue to use Γ_p for the covariance matrix of the vector x *though we do not at first restrict it to be of the special form introduced below* (2.1). We shall always take Γ_p to be nonsingular. We shall use A^{-1} for the generalized inverse of a symmetric matrix A (see Section 2 of the Mathematical Appendix). We shall assume that β belongs to the space spanned by the rows of U for if E projects onto that space $U\beta = UE\beta$ and we may replace β by $E\beta$. Of course if $Y'Y$ is nonsingular and $U = (I_p \otimes Y)E$ then our notation does not conflict and the E previously introduced does project onto the space spanned by the rows of U.

We call $\tilde{\beta}$ the best linear unbiased estimate (BLUE) of β, namely that estimate which is linear in z, has expectation β and whose covariance matrix is as small as possible in the usual ordering of symmetric matrices. It is well known† that

(2.2) $$\tilde{\beta} = (U'\Gamma_p^{-1}U)^{-1}U'\Gamma_p^{-1}z$$

with covariance matrix

(2.3) $$\mathcal{E}\{(\tilde{\beta} - \beta)(\tilde{\beta} - \beta)'\} = (U'\Gamma_p^{-1}U)^{-1}.$$

If Γ_p had been scalar (i.e., a scalar multiple of the identity matrix) then we should have computed

(2.4) $$\hat{\beta} = (U'U)^{-1}U'z.$$

In general (i.e., for any nonsingular Γ_p) this has covariance matrix

$$\mathcal{E}\{(\hat{\beta} - \beta)(\hat{\beta} - \beta)'\} = (U'U)^{-1}U'\Gamma_p^{-1}U(U'U)^{-1}.$$

Of course, when $U = (I_p \otimes Y)$ the least squares estimate $\hat{\beta}$, when arranged as a matrix $\hat{B}$, having the same relation to $\hat{\beta}$ as B does to β (see Section 1) is

$$\hat{B} = (Y'Y)^{-1}Y'Z,$$

where Z has $z_j(n)$ in row n, column j.

We let u_{ij} and x_k be typical elements of the matrix U and the vector x, respectively. We now have the following theorem, whose proof, in the main, follows Watson (1967)

Theorem 1. *Let* $z = U\beta + x$ *where* $\mathcal{E}(x) = 0$, $\mathcal{E}(xx') = \Gamma_p$, *where* Γ_p *is nonsingular, and* $\mathcal{E}(u_{ij}x_k) \equiv 0$. *Define* $\tilde{\beta}$, $\hat{\beta}$ *by* (2.2) *and* (2.4), *respectively, where* $(U'\Gamma_p^{-1}U)^{-1}$ *and* $(U'U)^{-1}$ *are the generalized inverses of the indicated matrices. Let* $\mathcal{M}(U)$ *be the linear space spanned by the columns of* U *and let* $\mathcal{U}$ *be the largest subspace of* $\mathcal{M}(U)$ *for which* $\Gamma_p\mathcal{U} \subset \mathcal{M}(U)$. *Then the*

† See, for example, Rao (1965, section 4a).

necessary and sufficient condition that $\alpha'\hat{\beta} = \alpha'\tilde{\beta}$, *almost surely, is the condition that* $U(U'U)^{-1}\alpha \in \mathcal{U}$. *In particular the necessary and sufficient condition that* $\tilde{\beta} = \hat{\beta}$, *almost surely, is the condition that* $\Gamma_p\mathcal{M}(U) = \mathcal{M}(U)$. *This is equivalent to saying that* $\mathcal{M}(U)$ *is spanned by eigenvectors of* Γ_p.

Note. It may be observed that saying that $\alpha'\hat{\beta} = \alpha'\tilde{\beta}$ almost surely is the same as saying that they have the same variance for the former clearly implies the latter while the proof of the theorem which shows that $\alpha'\tilde{\beta}$ is the BLUE for $\alpha'\beta$ shows also that it is essentially unique. The same goes for $\hat{\beta}$ and $\tilde{\beta}$, i.e., $\hat{\beta} = \tilde{\beta}$ a.s. is equivalent to

$$(U'U)^{-1}U'\Gamma_p U(U'U)^{-1} = (U'\Gamma_p^{-1}U)^{-1}.$$

Proof. If $\Gamma_p\mathcal{U} \subset \mathcal{M}(U)$ then, for some vector a,

$$\Gamma_p U(U'U)^{-1}\alpha = Ua.$$

We may assume that $(U'U)^{-1}U'Ua = a$ and $(U'U)^{-1}U'U\alpha = \alpha$. Thus we obtain

$$\alpha'(U'U)^{-1}U'\Gamma_p U(U'U)^{-1}\alpha + \alpha'(U'U)^{-1}U'Ua = \alpha'a.$$

Also

$$(U'U)(U'U)^{-1}\alpha = U'\Gamma_p^{-1}Ua,$$

that is,

$$(U'\Gamma_p^{-1}U)^{-1}\alpha = a,$$

since Γ_p is nonsingular and thus the range of $U'\Gamma_p^{-1}U$ is the same as the range of U'. Thus

$$\alpha'(U'\Gamma_p^{-1}U)^{-1}\alpha = \alpha'a.$$

Thus $\alpha'\hat{\beta}$ and $\alpha'\tilde{\beta}$ have the same variance and are almost surely equal.

To establish the necessity we observe that if $\Gamma_p U(U'U)^{-1}\alpha$ is not orthogonal to $\mathcal{M}(U)^{\perp}$ then choosing $c \in \mathcal{M}(U)^{\perp}$ we may consider $z'\{U(U'U)^{-1}\alpha + \epsilon c\}$ which has expectation $\beta'\alpha + \epsilon\beta'U'c = \beta'\alpha$, and variance

$$\alpha'(U'U)^{-1}(U'\Gamma_p U)(U'U)^{-1}\alpha + 2\epsilon c'\Gamma_p U(U'U)^{-1}\alpha + \epsilon^2 c'\Gamma_p c$$

and choosing ϵ small enough and negative this may be made less than the first term, which is impossible since then $z'\{U(U'U)^{-1}\alpha + c\}$ is an unbiased estimator of $\alpha'\beta$ with smaller variance than the BLUE. Thus $\Gamma_p U(U'U)^{-1}\alpha$ is orthogonal to $\mathcal{M}(U)^{\perp}$ and thus $\Gamma_p U(U'U)^{-1}\alpha \in \mathcal{M}(U)$, that is, $U(U'U)^{-1}\alpha \in \mathcal{U}$ as required.

The statement that $\tilde{\beta} = \hat{\beta}$, almost surely, is the same as the statement that $U(U'U)^{-1}\alpha \in \mathcal{U}$ for all α so that $\mathcal{U}$ is of the same dimension as $\mathcal{M}(U)$ and since $\mathcal{U} \subset \mathcal{M}(U)$ then $\mathcal{U} = \mathcal{M}(U)$ and the necessary and sufficient condition that $\tilde{\beta} = \hat{\beta}$ almost surely is $\Gamma_p\mathcal{M}(U) = \mathcal{M}(U)$. Since Γ_p is symmetric it follows that $\mathcal{M}(U)$ is spanned by eigenvectors of Γ_p, for this is true of any invariant subspace for a symmetric matrix. This completes the proof.

Now consider the case where $\Gamma_p = (G \otimes I_N)$, i.e., where $x(n)$ is generated by a process of uncorrelated random vectors with constant covariance matrix, G, and $Y'Y$ is nonsingular. Now Γ_p commutes with $(I_p \otimes YY')$ and since it is symmetric it must leave $\mathcal{R}(I_p \otimes YY')$ invariant and thus also $\mathcal{R}(I_p \otimes Y)$ invariant.† But $\Gamma_p(I_p \otimes Y)E = (G \otimes Y)E = (I_p \otimes Y)(G \otimes I_q)E$ so that Γ_p can fail to leave $\mathcal{R}((I_p \otimes Y)E)$ invariant only if $(G \otimes I_q)$ fails to leave $\mathcal{R}(E)$ invariant. On the other hand let Γ_p leave $\mathcal{R}((I_p \otimes Y)E)$ invariant. Then for all x there is a y so that

$$\Gamma(I_p \otimes Y)\, Ex = (I_p \otimes Y)\, Ey.$$

But the left-hand side is $(I_p \otimes Y)(G \otimes I_q)Ex$. If $(G \otimes I_q)$ does not leave $\mathcal{R}(E)$ invariant then we may choose x so that $z = (G \otimes I_q)Ex \notin \mathcal{R}(E)$ and $(I_p \otimes Y)z \in \mathcal{R}((I_p \otimes Y)E)$, i.e., $(I_p \otimes Y)z = (I_p \otimes Y)Ey_1$. Putting $z = Ey_1 + z_1$ we see that $z_1 \neq 0$ (since $z \notin \mathcal{R}(E)$) and $(I_p \otimes Y)z_1 = 0$, and this is impossible since $I_p \otimes Y'Y$ is nonsingular. Thus the condition of the theorem now becomes the condition that $(G \otimes I_q)$ leaves $\mathcal{R}(E)$ invariant and since E is a projection and $G \otimes I_q$ is symmetric this is equivalent to the requirement that $(G \otimes I_q)$ commutes with E, i.e., $(G \otimes I_q)E\beta = E(G \otimes I_q)\beta$, i.e., GB' satisfies the restrictions if B does. If this is so for all nonnegative G it must be so for all symmetric G since any such matrix is a difference of nonnegative matrices.

We leave it as an exercise for the reader to check that the only matrix E which is idempotent and which commutes with $G \otimes I_q$, for all symmetric G, is of the form $I_p \otimes F$ where F is idempotent. Thus the restrictions amount to $(I_p \otimes F)\beta = 0$, i.e., $B'F = 0$. This is equivalent to saying that the linear restrictions are trivial.

Theorem 2. *If* $\Gamma_p = (G \otimes I_N)$ *and* $U = (I_p \otimes Y)E$ *with* $Y'Y$ *nonsingular and* E *idempotent then in order that* $\tilde{\beta} = \hat{\beta}$, *a.s., it is necessary and sufficient that* $G \otimes I_q$ *commutes with* E, *that is* GB' *satisfies the linear restrictions imposed by* E *if* B' *does so. If this is to hold for all nonnegative definite* G *it is necessary and sufficient that the restrictions be trivial.*

To illustrate the nature of this theorem, consider the case where $z_1(n) = \mu + x_1(n)$, $z_2(n) = x_2(n)$ where $x_1(n)$ and $x_2(n)$ have unit variance and the only nonvanishing correlation is $\mathcal{E}(x_1(n)x_2(n)) = \rho \neq 0$. Now $p = 2$, $q = 1$ with $y(n) \equiv 1$. The linear restriction puts $\beta_{21} = 0$ ($\beta_{11} = \mu$ in our present notation). The least-squares estimate of μ is $\bar{z}_1$ with variance N^{-1}.

† By $\mathcal{R}(A)$ we mean the range of the operator A.

The BLUE is $\bar{z}_1 - \rho\bar{z}_2$ with variance $(1 - \rho^2)N^{-1}$. It may be observed that the theorem is trivial in case $p = 1$ for then the Γ_p of the theorem is a scalar matrix and correspondingly the linear restrictions are always trivial. The theorem shows that, in considering the possible efficiency of least squares, we may restrict ourselves to the case where no linear restrictions are made since only with trivial restrictions can efficiency hold and these can always be eliminated by redefinition of $y(n)$.

We consider first the case $p = 1, q = 1$. What we wish to do is to see how inefficient least squares can be. We introduce the efficiency

$$e = \frac{(y'y)^2}{(y'\Gamma y)(y'\Gamma^{-1}y)}, \tag{2.5}$$

which is the ratio of the variance of $\tilde{\beta}$ to that of $\hat{\beta}$. *We have there written y in place of the single column matrix Y.* We seek to minimize e by varying the vector y. In other words, we shall measure how, with a given stationary process generating the residuals, the vector on which we regress may be chosen so as to make the efficiency of least squares as small as possible. It may appear, at first sight, more interesting to keep y fixed and vary Γ and to minimize the efficiency for variation in Γ. However, we may then always choose Γ so that $y = a\phi + b\psi$ where ϕ and ψ are normalized eigenvectors of Γ for eigenvalues μ and ν and $a \neq 0$, $b \neq 0$. Then the efficiency of $\hat{\beta}$ is

$$(a^2 + b^2)^2\{(a^2\mu + b^2\nu)(a^2\mu^{-1} + b^2\nu^{-1})\}^{-1}$$
$$= (a^2 + b^2)^2\left\{a^4 + b^4 + a^2b^2\left(\frac{\mu}{\nu} + \frac{\nu}{\mu}\right)\right\}^{-1}$$

which may be made as small as is desired by choosing μ/ν large. Thus no useful conclusion is reached. As we shall see keeping Γ fixed and varying the regressors does, on the other hand, lead to useful results. We follow Watson (1955) and shall see that e is minimized when $y = 2^{-1/2}\{\phi_1 \pm \phi_N\}$ where the ϕ_j are the eigenvectors of Γ and are arranged in order so that the corresponding eigenvalues, μ_j, are in increasing order; $\mu_1 \leq \mu_2 \leq \cdots \leq \mu_N$. The corresponding minimum value of e is

$$\left[\frac{1}{2}\left\{\left(\frac{\mu_1}{\mu_N}\right)^{1/2} + \left(\frac{\mu_N}{\mu_1}\right)^{1/2}\right\}\right]^{2} = \frac{4\mu_1\mu_N}{(\mu_1 + \mu_N)^2}. \tag{2.6}$$

Let, in general, $y = \sum x_j\phi_j$. Then

$$e = (\sum x_j^2)^2\{\sum \mu_j x_j^2 \sum \mu_j^{-1}x_j^2\}^{-1}.$$

If some μ_j are equal, we amalgamate those terms in each sum and write the resulting expression as

$$(\sum' w_j)^2\{\sum' \mu_j w_j \sum' \mu_j^{-1} w_j\}^{-1},$$

where $\sum'$ sums over the set of unequal eigenvalues of Γ and w_j is the sum of the x_k^2 for the same eigenvalue μ_j. If there is only one term in $\sum'$, then evidently $\Gamma = y(0)I_N$ and $e = 1$. If there are precisely two terms it is a trivial matter to obtain the minimizing values of w_j and to show that these give the value of e given by (2.6). If there are more than two terms we now show that the maximum cannot be taken with more than two w_j nonnull. This will then complete the proof. Let M be the maximized value of e^{-1} and this be taken at a point where $w_j = w_{j,0}$ where $w_{j,0} \neq 0, j = 1, 2, 3$ (let us say). Then, taking all w_j, $j \neq 1, 2, 3$, at their optimal values, we must have the following relation (since the maximum value of the function now being considered is at an interior point of the region of variation of the w_j, $j = 1, 2, 3$):

$$\left[\frac{d}{dw_k}\{M^{-1}(\textstyle\sum' w_j)^2 - \sum' \mu_j w_j \sum' \mu_j^{-1} w_j\}\right]_0 = 0, \qquad k = 1, 2, 3.$$

This gives us

$$2M^{-1}\textstyle\sum' w_{j,0} - \mu_k \sum' \mu_j^{-1} w_{j,0} - \mu_k^{-1} \sum' \mu_j w_{j,0} = 0, \qquad k = 1, 2, 3.$$

The determinant of this system, as a set of equations in $\sum' w_{j,0}$, $\sum' \mu_j^{-1} w_{j,0}$, $\sum' \mu_j w_{j,0}$, is

$$-\frac{2M^{-1}}{\mu_1\mu_2\mu_3}(\mu_1 - \mu_2)(\mu_1 - \mu_3)(\mu_2 - \mu_3) \neq 0,$$

so that no solution is possible unless $\sum' w_{j,0} = \sum' \mu_j w_{j,0} = \sum' \mu_j^{-1} w_{j,0} = 0$ and since the rank of the matrix of this system of equation is three then $w_{j,0} \equiv 0$, which is a contradiction. Thus the result (2.6) holds.

If $p = 1$, $q > 1$ we are led to consider an arbitrary linear combination, $\alpha'\beta$ of the elements of the vector to be estimated and to measure the efficiency of least squares by minimizing this efficiency by varying both Y and α. We may assume that $Y'Y = P$, the projection onto the space spanned by the columns of Y', and the ratio to be minimized is thus

$$e(\alpha, Y) = \alpha'(Y'\Gamma^{-1}Y)^{-1}\alpha/\alpha'Y'\Gamma Y\alpha, \qquad P\alpha = \alpha.$$

But if $\alpha'Y'Y\alpha = 1$, as we may assume,† then

$$\alpha'(Y'\Gamma^{-1}Y)^{-1}\alpha \geq (\alpha'Y'\Gamma^{-1}Y\alpha)^{-1}$$

by the inequality between the arithmetic and harmonic means. Indeed if $Y'\Gamma^{-1}Y$ has eigenvalues ω_i then

$$\alpha'(Y'\Gamma^{-1}Y)^{-1}\alpha = \textstyle\sum u_i^2\omega_i^{-1} \geq (\sum u_i^2\omega_i)^{-1}, \qquad \sum u_i^2 = 1,$$

† Since $\alpha'\beta = \alpha'(Y'Y)^{-1}Y'Y\beta = \alpha'P\beta$ so that we may take $P\alpha = \alpha$ and thus assuming $\alpha'P\alpha = 1$ merely involves a change of scale.

where the u_i are the components of α when that vector is expressed linearly in terms of the eigenvectors of $Y'\Gamma^{-1}Y$. Thus

$$e(\alpha, Y) \geq \{\alpha' Y'\Gamma^{-1} Y\alpha \cdot \alpha' Y'\Gamma Y\alpha\}^{-1}.$$

Since $\alpha' Y' Y\alpha = 1$ we know that the right side is not less than the expression (2.6). However, it is easy to see that this lower bound is attained for we may take Y to be composed of q columns of which the first is $2^{-\frac{1}{2}}\{\phi_1 + \phi_N\}$, the second is $2^{-\frac{1}{2}}\{\phi_2 + \phi_{N-1}\}$ and so on while α has unity in the first place and zeros elsewhere.

It is not easy simply to characterize the efficiency of least squares when $p > 1$.

Theorem 3. *When $p = 1$ an, attainable, lower bound to the efficiency (relative to the BLUE) with which any linear combination, $\alpha'\beta$, may be estimated by least squares is given by* (2.6).

This lower bound to the efficiency will usually be much too strict and in most circumstances the actual efficiency will be much higher. We close this section with an example designed to bring home the meaning of the result.

Let us assume that the true nature of $x(n)$ (in case $p = q = 1$) is a second-order moving average

$$x(n) = \epsilon(n) + \alpha(1)\epsilon(n-1) + \alpha(2)\epsilon(n-2),$$

and we assume it to be a first-order moving average but with the same first autocorrelation. Consider the case where $\alpha(1) = 0.2$, $\alpha(2) = -0.2$, when $\rho(1) = 0.15$, $\rho(2) = -0.19$. In this case e is 0.76. (This efficiency is the limiting value attained as N increases but would be very near to correct even for quite small values of N.) If $\alpha(1) = 0.8$, $\alpha(2) = -0.2$ then $\rho(1) = 0.38$, $\rho(2) = -0.12$ and $e = 0.00$. (Again as a limit as N increases.) The apparently strange thing is that here the two correlograms agree slightly better than in the first case! The explanation is, of course, to be found in the nature of the spectrum, for the μ_j are, as we know from Theorem 7, Chapter III, approximately the values of $2\pi f(\lambda)$ (the spectrum of $x(n)$) at the points $2\pi k/N$, $k = 0, 1, \ldots, [\frac{1}{2}N]$. In the second case $f(\lambda)$ is zero (at $\lambda = \pi$) so that necessarily e becomes zero as N increases. This is a striking exemplification of the explanatory power of Fourier methods, a virtue of these methods which is most apparent in the problems of the present chapter.

When $p > 1$ it is not easy to give a characterization of the efficiency of least squares of the type given for $p = 1$ in Theorem 3. In general we might consider the efficiency of $\phi'\hat{B}\psi$ as an estimator of $\phi' B\psi$. This is not the general linear form in the elements of B. Moreover, the expression for the variance of $\phi'\tilde{B}\psi$ is not easy to handle. We can obtain an upper bound for the efficiency of $\hat{B}$ by comparing the variance of $\phi'\hat{B}\psi$ with that of the estimate obtained by estimating $B\psi$ by the best linear unbiased procedure appropriate to the observables $\zeta(n) = \psi' z(n)$, the residuals now being $\xi(n) = \psi' x(n)$. If Γ_ψ is the matrix with $\mathcal{E}(\xi(m)\xi(m+n))$ in the nth diagonal above and below the main diagonal and $\mu_j(\psi)$ are the corresponding eigenvalues, $\mu_1(\psi) \le \mu_2(\psi) \le \cdots \le \mu_N(\psi)$, then

$$\inf_{\phi,\psi,Y} \{\text{eff}\,(\phi'\hat{B}\psi)\} \le \inf_{\psi} \frac{4\mu_1(\psi)\mu_N(\psi)}{(\mu_1(\psi) + \mu_N(\psi))^2}, \qquad \phi'\phi = \psi'\psi = 1.$$

Thus a measure of how bad least squares may be has been obtained. Since the spectrum of the $\xi(n)$ process is $\psi' f(\lambda)\psi$, we see that an approximate evaluation of the efficiency is got by replacing $\mu_1(\psi)$ and $\mu_N(\psi)$ by the least and greatest values of $\psi' f(\lambda)\psi$ as ψ and λ vary.

3. THE EFFICIENCY OF LEAST SQUARES. ASYMPTOTIC THEORY

We now consider the situation where the $y(n)$, $n = 1, \ldots, N$, constitute parts of sequences satisfying the conditions (a), (b), and (c) of Chapter IV, Section 3. Thus we now write $y_j^{(N)}(n)$, $n = 1, \ldots, N$; $j = 1, \ldots, q$, for the elements of the regressor vectors. Often the (N) superscript can be dropped, of course, since the $y_j^{(N)}(n)$ will not depend upon it. We will usually not insert that superscript unless dependence upon it is to be emphasized. Some such restrictions as those due to Grenander are essential if asymptotic results are to be obtained and if these results are to be expressed in Fourier terms, as seems desirable, then Grenander's conditions seem near to minimal. They allow us to consider an extremely wide range of cases, as we have already emphasized. Indeed any ergodic stationary process clearly satisfies the conditions as also do the variables corresponding to a polynomial trend. One example of some importance is that which occurs when the

$$\mu_k(n) = \sum_j \beta_{kj} y_j(n), \qquad k = 1, \ldots, p,$$

are solutions of a system of homogeneous difference equations

$$\sum_0^q B(j) y(n-j) = 0$$

for which all zeros of

$$\det\left\{\sum_0^q B(j) z^j\right\}$$

lie *on* the unit circle. The reason why this case arises is that for zeros outside the unit circle the contribution eventually fades to zero while for zeros inside of the circle the contribution eventually explodes, in such a way that neither case can be fitted into the scheme of Grenander's conditions. Such situations will always need somewhat special consideration. In any case, if the zeros are at the points $\exp i\theta_j$, $j = 0, 1, \ldots, s$ (we shall usually take $\theta_0 = 0$) and if the jth of these has multiplicity q_j, then the mean $\mu_k(n)$ is a linear combination of terms of the form

$$\begin{aligned} &n^u \cos \theta_j n, \qquad u = 0, 1, \ldots, q_j - 1, \\ &n^u \sin \theta_j n, \qquad u = 0, 1, \ldots, q_j - 1, \end{aligned} \qquad j = 0, 1, \ldots, s,$$

$$q_0 + 2\sum_{j=1}^{s} q_j = q.$$

Then these q sequences become the $y_j(n)$ and their coefficients, in the expression for $\mu_k(n)$, become the β_{kj}. Of course, if $\theta_j = 0, \pm\pi$ then no sine terms occur. In order to discover the nature of $M(\lambda)$ (see Chapter IV, Section 3) we need to evaluate

$$\lim_{N\to\infty} \frac{\sum_{m=1}^{N} [m^u(m+n)^v \cos \theta_j m \cos \{\theta_k(m+n)\}]}{|\sum_{m=1}^{N} \{m^{2u} \cos^2 \theta_j m\} \sum_{m=1}^{N} \{m^{2v} \cos^2 \theta_k m\}|^{1/2}}$$

and similar expressions with $\cos \theta_j m$ replaced by $\sin \theta_j m$ or $\cos \{\theta_k(m + n)\}$ and $\cos \theta_k m$ replaced by the corresponding sine terms. We leave it as an exercise for the reader to show that this is zero for $\theta_j \neq \theta_k$ and that when $\theta_j = \theta_k$ we obtain, for fixed u, v, the following results, where each combination of cosine and sine is shown.

	$\sin \theta_j(m + n)$	$\cos \theta_j(m + n)$
$\sin \theta_j m$	$a(u, v) \cos \theta_j n$	$-a(u, v) \sin \theta_j n$
$\cos \theta_j m$	$a(u, v) \sin \theta_j n$	$a(u, v) \cos \theta_j n$

where

$$a(u, v) = (u + v + 1)^{-1}\{(2u + 1)(2v + 1)\}^{1/2}.$$

Thus, if we arrange the regressor variables (i.e., those of the form $n^u \cos \theta_j n$ and $n^u \sin \theta_j n$) down $y(n)$ so that all those for the same frequency, θ_j, occur together (taking them by increasing frequency from top to bottom) and so that for each u the cosine term follows immediately after the sine term and the u values increase, then $M(\lambda)$ will increase only at the points $\pm\theta_j$ and at these points the jump in M will consist of a matrix with zeros outside of a block, of dimension $2q_j$, for $\theta_j \neq 0$, π, or q_j, for $\theta_j = 0, \pi$ in the jth place down the diagonal. For $+\theta_j$ ($\theta_j \neq 0, \pi$) this block will be composed of q_j^2 blocks of 2×2 matrices of which a typical one, in the u, vth place,

$u, v = 0, 1, \ldots, q_j - 1$, will be

$$\frac{a(u, v)}{2}\begin{bmatrix} 1 & i \\ -i & 1 \end{bmatrix},$$

with that at $-\theta_j$ being of the same form but conjugated. For $\theta_j = 0, \pi$ the matrix has $a(u, v)$ in row u, column v, $u, v = 0, \ldots, q_j - 1$. Calling M_j the jump at θ_j, we see that M_j has as its non-null blocks

$$A \otimes \tfrac{1}{2}\begin{bmatrix} 1 & i \\ -i & 1 \end{bmatrix},$$

where A has $a(u, v)$ as its typical entry. We have dealt with this case in detail for it is central to our treatment given below.

Though it would not be difficult to give an analysis in general terms so that we could allow β to be subjected to the linear constraints summarized in $E\beta = \beta$, nevertheless we shall refrain from doing so in order to make the exposition of an already complicated subject somewhat simpler. However, in relation to other theorems proved below it is useful first to extend Theorem 2. We wish now to take $Y^{(N)}$ in place of Y and to make the same assertion (still taking $\Gamma_p = I_N \otimes G$) about the limiting behavior of $\hat{\beta}$ and $\tilde{\beta}$ as N increases. We wish to assert that the covariance matrices of these two estimates, after appropriate normalization, converge to the same limit as $N \to \infty$ if and only if E leaves the restrictions invariant. The remainder of the theorem will then follow as before. The only problem which now arises comes from the normalization, for now we must consider

$$(I_p \otimes D_N)\hat{\beta}, \qquad (I_p \otimes D_N)\tilde{\beta},$$

where D_N was defined in Chapter IV, Section 3. We now require that $I_p \otimes D_N$ commutes with E. Unless this condition is imposed it seems difficult to get a neat asymptotic result. The condition itself seems a natural one and does not seem to restrict the applicability of the results in any substantial way. Indeed saying that $I_p \otimes D_N$ commutes with E is the same as saying that $D_N B$ satisfies the restrictions if B does. Insofar as the restrictions affect only a certain row of B clearly this will be so (since D_N is diagonal). If a restriction, for example, requires the equality of two elements in different rows of B then the same condition will hold for $D_N B$ (if it holds for B) if the corresponding diagonal elements of D_N are equal. This can always be achieved provided these two elements, $d_j(N)$, $d_k(N)$ (let us say) have a ratio converging to some finite, nonzero limit. It hardly seems likely that such a condition of equality of two elements will be required unless the two corresponding $y_j^{(N)}(n)$, $y_k^{(N)}(n)$ do behave in this fashion. Once this condition is imposed we easily evaluate

$$(3.1)\quad \lim_{N\to\infty}(I_p \otimes D_N)\operatorname{cov}(\hat{\beta})(I_p \otimes D_N) = \{E(I_p \otimes R(0))E\}^{-1}(G \otimes R(0))\{E(I_p \otimes R(0))E\}^{-1}$$

$$(3.2)\quad \lim_{N\to\infty}(I_p \otimes D_N)\operatorname{cov}(\tilde{\beta})(I_p \otimes D_N) = \{E(G^{-1} \otimes R(0))E\}^{-1};$$

for example, the latter is

$$\lim_{N\to\infty}(I_p \otimes D_N)\big(E(I_p \otimes Y')(G^{-1} \otimes I_N)(I_p \otimes Y)E\big)^{-1}(I_p \otimes D_N) = \lim_{N\to\infty}\big(E(G^{-1} \otimes D_N^{-1}Y'YD_N^{-1})E\big)^{-1} = \big(E(G^{-1} \otimes R(0))E\big)^{-1}.$$

Of course, all of the inverses are generalized inverses. Now to show the equality of the right-hand terms of (3.1) and (3.2) is in no way different from showing the equality, a.e., of $\tilde{\beta}$ and $\hat{\beta}$ under the circumstances of Section 2, when $Y'Y$ is replaced by $R(0)$, and evidently the necessary and sufficient condition is once more that $(G \otimes I_q)$ commute with E.

Theorem 4. *If $\Gamma_p = (G \otimes I_N)$, $R(0)$ is nonsingular and $I_p \otimes D_N$ commutes with E then in order that $\hat{\beta}$ and $\tilde{\beta}$ have asymptotically the same covariance matrix it is necessary and sufficient that $G \otimes I_q$ commutes with E (i.e., GB' satisfies the linear restrictions if B does so). If this is to be so for all G it is necessary and sufficient that the linear restrictions be trivial.*

In particular the theorem shows that for least squares to be efficient when $x(n)$ can have an arbitrary spectrum then it is necessary that the linear restrictions be trivial. The same will be true if $x(n)$ is required to have, say, an arbitrary continuous spectrum or say an arbitrary continuous spectrum with $f(\lambda) > 0$, $\lambda \in (-\pi, \pi]$, for in the latter case, though we now include only the Γ_p of the form $(G \otimes I_N)$ with $G > 0$, the last part of the argument in Theorem 2 will still hold, for if $G \otimes I_q$ commutes with E for any $G > 0$ it will evidently do so for all $G \geq 0$. Thus, in seeking for a condition that will ensure that least squares is asymptotically efficient for all $x(n)$ with sufficiently general spectral density matrix, we are entitled to restrict ourselves to the case in which there are no nontrivial linear constraints. (Of course, there is the intermediate problem of finding what the situation is for a *particular* $f_x(\lambda)$ but though we consider this in the unconstrained case it seems too complicated to be worth separate consideration in the other case.)

We have, in Theorem 8 in Chapter IV, established that when $f(\lambda)$ is piecewise continuous with no jumps at discontinuities of $f(\lambda)$ then

$$(3.3)\quad \lim_{N\to\infty}(I_p \otimes D_N)\operatorname{cov}(\hat{\beta})(I_p \otimes D_N) = \big(I_p \otimes R(0)\big)^{-1}\int_{-\pi}^{\pi} 2\pi f(\lambda) \otimes M'(d\lambda)\big(I_p \otimes R(0)\big)^{-1}.$$

The proof given there was for the case in which $z(n)$ is replaced by $x(n)$ in the computation of $\hat{\beta}$. However,

$$(\hat{B} - B) = (Y'Y)^{-1}Y'Z - (Y'Y)^{-1}Y'YB$$
$$= (Y'Y)^{-1}Y'X,$$

where X has $x_j(n)$ in row n, column j and thus Theorem 8 in Chapter IV covers our present considerations. We now prove the complementary result for $\tilde{\beta}$.

Theorem 5. *If* $R(0)$ *is nonsingular and* $x(n)$ *is stationary with a.c. spectrum,* $f(\lambda) > 0$, $\lambda \in (-\pi, \pi]$, *which is piecewise continuous with no discontinuities at the points of jump of* $M(\lambda)$, *then*

$$\lim_{N\to\infty} (I_p \otimes D_N) \operatorname{cov}(\tilde{\beta})(I_p \otimes D_N) = \left\{\int_{-\pi}^{\pi} \{2\pi f(\lambda)\}^{-1} \otimes M'(d\lambda)\right\}^{-1}. \tag{3.4}$$

Proof. The proof we give is, substantially, a modification of that in Grenander and Rosenblatt (1957), their treatment being for the case $p = 1$. We observe that if $f(\lambda)^{-1}$ is continuous we can, when $f(\lambda) > 0$, $\lambda \in [-\pi, \pi]$, find spectral densities $f^{(i)}(\lambda)^{-1}$, $i = 1, 2$, of moving averages so that

$$0 < f^{(1)}(\lambda)^{-1} \le f(\lambda)^{-1} \le f^{(2)}(\lambda)^{-1}$$

and

$$f^{(2)}(\lambda)^{-1} - f^{(1)}(\lambda)^{-1} < \epsilon I_p.$$

Here these inequalities are to be interpreted in the usual way as between Hermitian matrices; for example, we may approximate $f(\lambda)^{-1} + \frac{1}{4}\epsilon I_p$ uniformly (in λ), and $f(\lambda)^{-1} - \frac{1}{4}\epsilon I_p$ similarly, using the Cesaro means of the Fourier series for the elements of $f(\lambda)^{-1} \pm \frac{1}{4}\epsilon I_p$. Taking the uniform approximations sufficiently close the result follows. Thus we have approximated $f(\lambda)$ above and below by autoregressive spectra so that the above inequalities hold. We first show that

$$\lim_{N\to\infty} (I_p \otimes D_N)\{(I_p \otimes Y')\Gamma_p^{(i)^{-1}}(I_p \otimes Y)\}^{-1}(I_p \otimes D_N) \tag{3.5}$$
$$= \left[\int_{-\pi}^{\pi} \{2\pi f^{(i)}(\lambda)\}^{-1} \otimes M'(d\lambda)\right]^{-1}, \qquad i = 1, 2,$$

where $\Gamma_p^{(i)}$ corresponds to $f^{(i)}(\lambda)$ as Γ_p does to $f(\lambda)$.

The matrix being inverted on the right of (3.5) is nonsingular for $u \neq 0$ being any nonnull vector in pq-dimensional space,

$$u^* \int f^{(i)}(\lambda)^{-1} \otimes M'(d\lambda) u \ge a \int u^*(I \otimes M'(d\lambda))\, u$$
$$\ge a u^* u \int \mu(d\lambda),$$

where aI_p is a lower bound to $f^{(i)}(\lambda)^{-1}$, $a > 0$, and $\mu(\lambda)$ is used for the least eigenvalue of $M(\lambda)$, which clearly increases to the least eigenvalue of $R(0)$, which is positive. Thus we replace (3.5) with a corresponding expression in which the matrix on either side is inverted. Let $x(n)$ be stationary and satisfy the equation

$$\sum_{j=0}^{r} B(j)x(n-j) = \eta(n),$$

where the $B(j)$ are the autoregressive matrices corresponding, say, to $f^{(1)}(\lambda)$ and the $\eta(n)$ are independent and identically distributed random vectors with mean zero and covariance matrix unity (we do *not* take $B(0) = I_p$ for convenience). Let $\tilde{x}$ have $x_k(n)$ in the $(n-1)p + k$th place. *Thus we have now reordered the elements of x taking them in dictionary order according to n first and then k.* Let

$$\Delta x = \left[\begin{array}{c:c} \Delta_1 & 0 \\ \hdashline \multicolumn{2}{c}{B} \end{array}\right] x = \xi,$$

where B consists of $(N - r)p$ rows and Np columns and consists of blocks of $p \times p$ matrices, of which a typical one in the kth row of blocks of B and the lth column of blocks of B is $B(r + k - l)$, with this replaced by the null matrix if $(r + k - l)$ does not lie between 0 and r (inclusive). The matrix Δ_1 is chosen by the Gram Schmidt method† so that the first rp rows of ξ consist of orthonormal random variables, orthogonal also to all of those in the last $(N - r)p$ rows, these latter being got from $\eta(r + 1)$, $\eta(r + 2), \ldots,$ $\eta(n)$, successively placed down the column. Hence

$$\Gamma^{(1)-1} = \Delta'\Delta,$$

where $\Gamma^{(1)}$ is *not* $\Gamma_p^{(1)}$ but instead has N^2 blocks, the typical one being $\Gamma^{(1)}(m - n)$. Thus $\Gamma_p^{(1)}$ is got from $\Gamma^{(1)}$ by reordering the elements $\gamma_{ij}^{(1)}(m - n)$, in placing them by rows for example, taking m first and then i instead of the reverse and by columns in taking n first and then j. In evaluating (3.5) (or rather its inverse), it is thus best to perform the same permutation of the elements of rows and columns for each component matrix. Having evaluated it, we shall reverse the permutations. Thus we consider

$$\begin{aligned}(D_N^{-1} \otimes I_p)&(Y' \otimes I_p)\Delta'\Delta(Y \otimes I_p)(D_N^{-1} \otimes I_p) \\ &= \left\{\sum_{m,n=1}^{N}\sum (D_N^{-1}y(m)\, y(n)' D_N^{-1}) \otimes \sum_{t=-\infty}^{\infty} B'(t)\, B(t + m - n)\right\} + R_N,\end{aligned}$$

where R_N is a remainder involving the terms in Δ and Δ' from Δ_1. There are only $4r^2$ blocks in $\Delta'\Delta$ which involve elements of Δ_1 (those lying in the first

† See Section 2 of the Mathematical Appendix.

$2r$ rows and columns of blocks). It follows immediately that R_N converges with probability 1 to zero. The sum over t in the first part of the last displayed expression is of course a finite sum and we have written it in that fashion only so as to avoid writing down the limits of the summation for the nonnull terms. Moreover,

$$4\pi^2 \sum B'(t)\, B(t+u) = \int_{-\pi}^{\pi} {f^{(1)}}^{-1}(\lambda) e^{-iu\lambda}\, d\lambda.$$

Omitting R_N the remaining term in the sum to be evaluated is

$$\sum_{u=-r}^{r} \left\{ \sum{}' D_N^{-1} y(m) y'(m-u) D_N^{-1} \otimes \sum_{u=-\infty}^{\infty} B'(t)\, B(t+u) \right\},$$

where the sum $\sum'$ is over m such that m and $m-u$ lie between 1 and N. This converges to

$$\sum_u \int_{-\pi}^{\pi} e^{-iu\lambda}\, M(d\lambda) \otimes \sum B'(t)\, B(t+u) = \int_{-\pi}^{\pi} M(d\lambda) \otimes \{2\pi f^{(1)}(-\lambda)\}^{-1}$$
$$= \int_{-\pi}^{\pi} M'(d\lambda) \otimes \{2\pi f^{(1)}(\lambda)\}^{-1}.$$

Reversing the order of the indices of the tensors once more (as in our discussion of the relation between $\Gamma^{(1)}$ and $\Gamma_p^{(1)}$) we obtain the result we need. We obtain a corresponding result for $f^{(2)}(\lambda)$. The remainder of the proof for $f(\lambda) > 0$, $\lambda \in [-\pi, \pi]$ and the extension to the case where $f(\lambda)$ may be only piecewise continuous is almost precisely the same as in Theorem 8, Chapter IV. This completes the proof.

If $f(\lambda) \not> 0$ one may replace $f(\lambda)$ by $f(\lambda) + \epsilon I_p$, and, from a consideration of (3.5) (with Γ_p replacing $\Gamma_p^{(i)}$), see that as ϵ decreases to zero we obtain the covariance matrix for $(I_p \otimes D_N)(\tilde{\beta} - \beta)$ as a limit from (3.4). However, this limit will certainly not in general be got by inserting generalized inverses in (3.4). Indeed take $p = 1$ and $q = 1$ when the right-hand terms of (3.3) and (3.4) become, respectively,

$$\int_{-\pi}^{\pi} 2\pi f(\lambda)\, m(d\lambda), \qquad \left\{ \int_{-\pi}^{\pi} (2\pi f(\lambda))^{-1}\, m(d\lambda) \right\}^{-1}.$$

If $f(\lambda)$ is null on a set on which our $m(\lambda)$ increases then the second of these expressions, when $f(\lambda)$ is replaced by $f(\lambda) + \epsilon$, will approach zero as ϵ tends to zero.

Let us relate these results to those of the last section, again beginning with $p = q = 1$. From the argument leading to (3.6) of the last section as a lower bound to the efficiency of least squares we see that, when $p = 1$, a lower bound to the asymptotic efficiency in the stationary case is got by confining the increase in $m(\lambda)$ to two sets, those, respectively, where $f(\lambda)$ is greatest (say f_u) and where it is least (say f_l), the total increase on the two sets to be equal. This may be seen by replacing the two integrals by approximating Riemann–Stieltjes sums and applying the result, just referred to, of the last section. (It is, of course, possible to construct an $m(\lambda)$ increasing, say, at only two points in $(0, \pi)$. For example, $y(n) = \cos n\lambda_1 + \cos n\lambda_2$ will provide such an $m(\lambda)$.) Thus if the Riemann–Stieltjes sum for the first integral is

$$\sum 2\pi f(\omega_j) m_j, \qquad \omega_j = \frac{2\pi j}{N}$$

then m_j replaces $(x_j^2/\sum x_j^2)$ in (2.6) while $2\pi f(\omega_j)$ takes the place of μ_j. Since we may approximate to the two integrals arbitrarily closely by such Riemann–Stieltjes sums and $f(\lambda)$ is assumed continuous (or has no discontinuities at jumps in $m(\lambda)$) the truth of the statement is apparent. For $p = 1, q > 1$ we at first are led to consider $b'\hat{\beta}$ and $b'\tilde{\beta}$. However, these will have variances converging to zero. Thus we are led to take $\alpha' D_N \hat{\beta}$ and $\alpha' D_N \tilde{\beta}$, that is, to consider a sequence of linear combinations of the estimates with coefficients, given by $D_N \alpha$, which increase in such a way as to ensure that the variance of the linear form does not tend to zero. The two variances to be compared are now

$$\alpha' R(0)^{-1} \int_{-\pi}^{\pi} 2\pi f(\lambda)\, M'(d\lambda) R(0)^{-1} \alpha, \qquad \alpha' \left\{ \int \{2\pi f(\lambda)\}^{-1}\, M'(d\lambda) \right\}^{-1} \alpha.$$

If we put $\alpha = R(0)^{1/2} c$ then we must compare

$$c' \int_{-\pi}^{\pi} 2\pi f(\lambda)\, N'(d\lambda) c, \qquad c' \left\{ \int \{2\pi f(\lambda)\}^{-1}\, N(d\lambda) \right\}^{-1} c,$$

where $N(\lambda) = R(0)^{-1/2} M(\lambda) R(0)^{-1/2}$. Now once more the comparison, if we replace the integrals by approximating Riemann–Stieltjes sums, is of the same form as in Section 2 and thus we obtain the following theorem:

Theorem 6. *When $p = 1$ and $f(\lambda)$ is continuous or has no discontinuities at jumps in $M(\lambda)$, then the asymptotic efficiency (relative to BLUE) with which any linear combination $\alpha' D_N \beta$ may be estimated by least squares is bounded below by*

$$\frac{4 f_l f_u}{(f_l + f_u)^2}. \tag{3.6}$$

As we have indicated in Section 2 we have been unable to find an analogous result when $p > 1$. Almost exactly as at the end of that section we may consider $(\phi \otimes D_N \psi)\beta$ and the corresponding estimates, $(\phi \otimes D_N \psi)\hat{\beta}$, $(\phi \otimes D_N \psi)\tilde{\beta}$. Again varying ϕ and ψ as well as $M(\lambda)$ we obtain an *upper* bound to the minimum possible asymptotic efficiency of least squares as

$$\inf_{\psi} \frac{4f_l(\psi)\,f_u(\psi)}{(f_l(\psi)+f_u(\psi))^2},$$

where $f_l(\psi)$ is the minimum value of $\psi' f(\lambda)\psi$ and $f_u(\psi)$ is the maximum value.

We now begin the study of the converse problem, namely that of discovering when least squares is efficient. We once more introduce the matrix $N(\lambda) = R(0)^{-\frac{1}{2}}M(\lambda)R(0)^{-\frac{1}{2}}$ so that $N(\lambda)$ is Hermitian with nonnegative increments and satisfies

$$\int_{-\pi}^{\pi} N(d\lambda) = I_q.$$

Now we have the following theorem [Grenander and Rosenblatt, 1957]. (By the support of $N(\lambda)$ we mean the set in which $N(\lambda)$ is increasing.)

Theorem 7. *The support of $N(\lambda)$ may be decomposed into not more than q disjoint sets S_j so that putting*

$$N_j = \int_{S_j} N(d\lambda)$$

the N_j form an essentially unique maximal set of mutually annihilating idempotents.

Note. By saying that the N_j are mutually annihilating idempotents we mean that they satisfy

$$N_j^* = N_j, \qquad N_j N_k = \delta_j^k N_j, \qquad \sum_j N_j = I_q. \tag{3.7}$$

The set is maximal in the sense that no further decomposition of the S_j can lead to a larger set with the same property. The decomposition is essentially unique in the sense that any alternative decomposition leading to a set with the same properties would be based upon sets S_j' so that to the symmetric difference between S_j and S_j' (i.e., the set of points belonging to one of these two sets and not the other) the measure function $\operatorname{tr}\{N(\lambda)\}$ allots zero measure.

Proof. Put $n(\lambda) = q^{-1}\operatorname{tr}(N(\lambda))$. It may be possible to decompose $(-\pi, \pi]$ into two sets S, S' so that,

$$S \cup S' = (-\pi, \pi],$$

$$N(S) = \int_S dN(\lambda), \qquad N(S') = \int_{S'} dN(\lambda), \qquad N(S) + N(S') = I_q,$$

$$N(S)^2 = N(S), \qquad N(S')^2 = N(S').$$

Since

$$N(S) = N(S)I_q = N(S)^2 + N(S)\,N(S') = N(S) + N(S)\,N(S'),$$

then $N(S)$, $N(S')$ are mutually annihilating Hermitian idempotents. We may now perhaps decompose S and or S' further in the same way. The process must stop after a finite number of steps since there can be at most q nonnull mutually annihilating Hermitian matrices in q-dimensional space. It is conceivable that if a different decomposition had been chosen at some step in the process a different final result would have been obtained but we proceed to show that in fact that cannot be so. Let S_j, $j = 1, \ldots, r \leq q$ be the sets of a first maximal decomposition (one which can be decomposed no further) and S_j', $j = 1, \ldots, r' \leq q$ be the sets of a second such decomposition. Now we may evidently reduce all of the $N(S_j)$ to diagonal form by one and the same unitary transformation, $N(S_j) \to UN(S_j)U^*$. We may also choose U so that $N(S_1)$ has units in the first r_1 places and zeros elsewhere, $N(S_2)$ has units in diagonal places numbered $r_1 + 1$ to $r_1 + r_2$ and zeros elsewhere and so on. Let us rename the transformed matrices as $N(S_j)$ and rename $UN(S_j')U^*$ as $N(S_j')$ also. Now $N(\lambda)$, $\lambda \in S_j$, cannot have nonnull elements outside of the rows and columns numbered $r_1 + r_2 + \cdots + r_{j-1} + 1$ to $r_1 + r_2 + \cdots + r_j$, or at least this is true almost everywhere with respect to $n(\lambda)$ (or rather the measure it induces on $(-\pi, \pi]$). Now let S_k' overlap with two S_j, say S_l, S_m, in the sense that the overlaps constitute sets of nonnull measure with respect to $n(\lambda)$. If

$$N_{k,l} = \int_{S_k' \cap S_l} N d(\lambda)$$

is idempotent, so must be the integral of $N(d\lambda)$ over the complement of S_k' in S_l as may be seen by reducing $N_{k,l}$ to diagonal form by a unitary transformation. But $N_{k,l}$ is not null since $N(S_k' \cap S_l)$ has not got zero measure with respect to $n(\lambda)$. This is a contradiction since $N(S_k')$ is certainly the direct sum of the $N(S_k' \cap S_j)$. Then the decomposition is essentially unique and the theorem is proved.

The sets S_j are called the elements of the spectrum of the regressor set but we shall speak of the elements of the spectrum, for short. It may be asked what these S_j mean. This can be illustrated by the case $q = 2$. Let us assume that the number of S_j is 2 (the maximum possible). If $S \subset S_1$ then $N(S) = \int_s N(d\lambda) \leq N_1$. Thus $N(S)N_2 = 0$ and we may decompose the two-dimensional complex vector space in which each $N(S)$ operates into two orthogonal, one-dimensional subspaces so that if $S \subset S_1$, $N(S)$ is the null operator on the second and leaves both invariant. If $S \subset S_2$ then $N(S)$ leaves the second invariant and is the null operator on the first. Thus $N(\lambda)$ is, for each λ, diagonal with the support of the first element in the main diagonal in S_1 and of the second in S_2. Thus we have effectively replaced our two initial sequences $y_1(n)$, $y_2(n)$, by two new *complex* combinations (the same ones for each n) so that these new sequences are not merely incoherent at each frequency but, even more specially have their spectral masses concentrated on disjoint subsets. (These statements about spectra and coherence relate to the generalized spectra as defined in relation to Grenander's conditions.) They correspond to two signals sent over channels so that there is no interference.

We now prove a simple central theorem which is an extension to $p > 1$ of a theorem due to Grenander and Rosenblatt (1957, p. 244).

Theorem 8. *Let $f(\lambda)$ be continuous with* $\det\{f(\lambda)\} > 0$, $\lambda \in (-\pi, \pi]$. *Then $\tilde{\beta}$ and $\hat{\beta}$ have asymptotically the same covariance matrix*† *if and only if $f(\lambda)$ is constant on the elements of the spectrum.*

Equality of the two covariance matrices is equivalent to

$$\{I_p \otimes R(0)\}^{-1}\int_{-\pi}^{\pi} 2\pi f(\lambda) \otimes M'(d\lambda)\{I_p \otimes R(0)\}^{-1}\int_{-\pi}^{\pi} \{2\pi f(\lambda)\}^{-1} \otimes M'(d\lambda) = I_{pq}$$

which, upon premultiplication by $\{I_p \otimes R(0)\}^{1/2}$ and postmultiplication by its inverse becomes

$$\int_{-\pi}^{\pi} f(\lambda) \otimes N'(d\lambda)\int_{-\pi}^{\pi} f(\lambda)^{-1} \otimes N'(d\lambda) = I_{pq}. \tag{3.8}$$

If the condition of the theorem holds then we see that the left side of the last expression is

$$(\textstyle\sum f_j \otimes N_j')(\sum f_j^{-1} \otimes N_j'),$$

where f_j is the constant value of $f(\lambda)$ on S_j. Because of (3.7), this is

$$\textstyle\sum f_j f_j^{-1} \otimes N_j' = I_p \otimes I_q = I_{pq},$$

as required.

To establish necessity we again introduce $n(\lambda) = q^{-1} \operatorname{tr} N(\lambda)$. Then $n(\lambda)$ defines a probability distribution on $(-\pi, \pi]$. The diagonal elements, $n_{jj}(\lambda)$, of $N(\lambda)$ are clearly absolutely continuous with respect to $n(\lambda)$ and the same is true of the off-diagonal elements, $n_{jk}(\lambda)$. Thus we may put $N(d\lambda) = p(\lambda)n(d\lambda)$ and rewrite (3.8) as

$$\int_{-\pi}^{\pi} \{f(\lambda) \otimes p(\lambda)\}\, n(d\lambda)\int_{-\pi}^{\pi} \{f(\lambda)^{-1} \otimes p(\lambda)\}\, n(d\lambda) = I_{pq}.$$

If we take the trace of both sides we obtain

$$\iint_{-\pi}^{\pi} \operatorname{tr}\, \{f(\lambda)\, f(\mu)^{-1}\} \operatorname{tr}\, \{p(\lambda)\, p(\mu)\}\, n(d\lambda)\, n(d\mu) = pq. \tag{3.9}$$

† This is somewhat loosely phrased. We mean, of course, that after normalization, through D_N the two matrices are asymptotically equal.

Now

$$\operatorname{tr}\{f(\lambda)f(\mu)^{-1}+f(\mu)f(\lambda)^{-1}\}\geq 2p$$

with equality holding if and only if $f(\lambda)=f(\mu)$. Indeed the left side is

$$\operatorname{tr}\{f(\mu)^{-\frac{1}{2}}f(\lambda)f(\mu)^{-\frac{1}{2}}+f(\mu)^{\frac{1}{2}}f(\lambda)^{-1}f(\mu)^{\frac{1}{2}}\}=\sum(\nu_j+\nu_j^{-1})$$

where the ν_j are the eigenvalues of $f(\mu)^{-\frac{1}{2}}f(\lambda)f(\mu)^{-\frac{1}{2}}$. Since $x+x^{-1}\geq 2$ and equality holds only when $x=1$ the result follows. Thus the left side of (3.9) is greater than or equal to

$$p\iint \operatorname{tr}\{p(\lambda)\,p(\mu)\}\,n(d\lambda)\,n(d\mu)=p\operatorname{tr}\left\{\iint p(\lambda)\,p(\mu)\,n(d\lambda)\,n(d\mu)\right\}$$

$$=p\operatorname{tr}(I_q)=pq,$$

with equality holding if and only if $\operatorname{tr}\{p(\lambda)\,p(\mu)\}$ is zero wherever $f(\lambda)\neq f(\mu)$, possibly neglecting a set of zero measure with respect to $n(d\lambda)\,n(d\mu)$. The $p(\lambda)$ are elements of a finite (q^2)-dimensional vector space, the inner product being $\operatorname{tr}\big(p(\lambda)\,p(\mu)^*\big)=\operatorname{tr}\big(p(\lambda)\,p(\mu)\big)$. Thus there can be only a finite number of different $p(\lambda)$ which annihilate each other. Thus, neglecting a set of zero measure with respect to $n(d\lambda)\,n(d\mu)$, $f(\lambda)$ must take on only finitely many different values. Call S_j the jth set of constancy for $f(\lambda)$ (neglecting a set of measure zero) and

$$N_j=\int_{S_j}p(\lambda)\,n(d\lambda).$$

Then the N_j are Hermitian nonnegative, annihilate each other and sum to I_q. Thus they are also idempotent since $N_j=N_jI_q=N_j\sum_k N_k=N_j^2$. Thus the theorem is established.

Now we deduce the following theorem, which is again an extension of a theorem due to Grenander and Rosenblatt (1957):

Theorem 9. *The necessary and sufficient condition that $\hat\beta$ and $\tilde\beta$ have asymptotically the same covariance matrix for all continuous*† *$f(\lambda)$ with* $\det\{f(\lambda)\}>0$, *$\lambda\in(-\pi,\pi]$ is, for $p=1$, that the elements of the spectrum be points, or pairs of points (in $(-\pi,\pi]$) symmetrically placed with respect to the origin and in case $p>1$ that they be individual points in* $(-\pi,\pi]$.

The proof of this is obvious, the difference between the cases $p=1$ and $p>1$ arising because, for $p=1$, $f(\lambda)=f(-\lambda)$, whereas this is not necessarily true for $p>1$.

† The theorem remains true if the phrase "all continuous $f(\lambda)$" is replaced by "a family of $f(\lambda)$ which separates the points of $(-\pi,\pi]$, for $p>1$ or of $[0,\pi]$, for $p=1$." For example for $p=1$ this family may consist of the single spectrum for a first-order autoregression with $\beta(1)\neq 0$.

We consider further the situation where the result of Theorem 9 holds. If $\theta_j \geq 0$ is a point of the spectrum of the regressor set then so must be $-\theta_j$ $(\theta_j \neq \pi)$ and necessarily the jumps in $N(\lambda)$ at $\pm\theta_j$ are conjugates of each other. If $\theta_j = 0, \pi$ then N_j is real, symmetric, and idempotent. If $\theta_j \neq 0, \pi$ then the jump in $N(\lambda)$ at that point is of the form $U_j + iV_j$ where U_j is symmetric and V_j is skew symmetric. If $p = 1$ so that the idempotent associated with the set $\{\theta_j, -\theta_j\}$ is $2U_j$ then $2U_j$ is idempotent. If $U_j + iV_j$ is idempotent, as must be the case when $p > 1$ and *may* be so for $p = 1$, we call it N_j and observe that $\bar{N}_j$ is also an Hermitian idempotent and $N_j\bar{N}_j = 0$. Thus $U_j^2 = -V_j^2$, $U_jV_j = V_jU_j$. Also $N_j^2 = N_j$ so that $2U_j^2 = U_j$, $2U_jV_j = V_j$. Thus $2U_j$ is again idempotent. The matrices $2U_j$ (corresponding to $\theta_j \in (0, \pi)$) are mutually annihilating. Indeed, $N_jN_k = 0 = \bar{N}_jN_k$ $(j \neq k)$. (Of course, $N_j = \bar{N}_j$ if the element of the spectrum is $\{\theta_j, -\theta_j\}$ but if it is $\{\theta_j\}$ then this is not so.) This shows that $U_jU_k = 0$, $(j \neq k)$. Together with the U_j for $\theta_j = 0$ or π (if these occur) the $2U_j$ thus form a set of mutually annihilating idempotents summing to I_q. We may thus decompose the vector space, $\mathfrak{X}$ let us say, in which they operate into mutually perpendicular subspaces, $\mathfrak{X}_j$, say, in the jth of which $2U_j$ becomes E_j, the identity operator for that subspace (and null on the other subspaces) and the same is true for U_j for $j = 0, \pi$. Since $U_j + iV_j$ is Hermitian nonnegative, the V_j are also null outside of the $\mathfrak{X}_j$ (and skew symmetric there). When $U_j + iV_j = N_j$ (i.e., $p > 1$ or $\{\theta_j\}$ is an element of the spectrum if $p = 1$) more can be said about V_j, as the relations deduced above show. Indeed then $(2V_j)^2 = -(2U_j)^2$ so that $(2V_j)^2$ is $-E_j$. Thus $2V_j$ satisfies $2V_j(2V_j)' = -(2V_j)^2 = E_j$ and $2V_j$ is also orthogonal on $\mathfrak{X}_j$. Thus as an operator on $\mathfrak{X}_j$ it is a direct sum of rotations in two-dimensional perpendicular subspaces plus, possibly, the identity operator in a one-dimensional subspace. [See, for example, Boerner (1963, p. 13, Theorem 5.2e).] Since $(2V_j)^2 = -E_j$, there can be no such one-dimensional subspace and the rotation angles must all be $\frac{1}{2}\pi$. Thus $\mathfrak{X}_j$ is even-dimensional and, in a suitable basis for $\mathfrak{X}_j$,

$$(3.10)\qquad 2V_j = \begin{bmatrix} 0 & 1 & 0 & 0 & & & \\ -1 & 0 & 0 & 0 & & & \\ 0 & 0 & 0 & 1 & & & \\ 0 & 0 & -1 & 0 & & & \\ & & & & \cdot & & \\ & & & & & \cdot & \\ & & & & & & \cdot \\ & & & & & 0 & 1 \\ & & & & & -1 & 0 \end{bmatrix},$$

We have thus proved the following.

Corollary 1. *If the result of Theorem* 9 *holds we may decompose the q dimensional (real) vector space $\mathfrak{X}$ in which the U_j, V_j, operate into mutually perpendicular subspaces $\mathfrak{X}_j$, with identity operators E_j, one for each $\theta_j \geq 0$ in the spectrum of the regressor set, so that U_j, V_j, operate in $\mathfrak{X}_j$ (and are null in $\mathfrak{X}_k$, $k \neq j$) and there $U_j = E_j$, $\theta_j = 0$, π; $2U_j = E_j$, $\theta_j \neq 0$, π while V_j is skew symmetric. If $\{\theta_j\}$ is itself an element of the spectrum (as must be the case for $p > 1$ but may also be so for $p = 1$) then $\mathfrak{X}_j$ is even-dimensional and V_j is of the form of* (3.10).

We leave it to the reader to check that an example of a set of regressors satisfying the conditions of the corollary (for $p > 1$) is provided by the set described at the beginning of this section, the dimensions of the $\mathfrak{X}_j$ being $2p_j$, where the p_j are the multiplicities of the zeros $\exp i\theta_j$ of the characteristic polynomial. Thus, as far as the case $p > 1$ is concerned, the asymptotic spectral behavior of any regressor set which satisfies Theorem 9 is that of a set of solutions to a difference equation with zeros, for its characteristic polynomial, lying upon the unit circle. Of course, other sequences will also give rise to the same $M(\lambda)$. It is interesting to observe that, when $p > 1$, $\mathfrak{X}_j$ must be even-dimensional if Theorem 9 is to hold. In particular, if $\cos n\theta_j$ occurs amongst the $y_j(n)$, so must $\sin n\theta_j$. If $p = 1$, this does not have to be so, as pointed out by Rosenblatt (1959).

The conditions under which least squares is asymptotically efficient are clearly highly special. Moreover we must emphasize the asymptotic nature of the result. For the case in which only $\cos n\theta_j$ or $\sin n\theta_j$ occur (and not $n^u \cos n\theta_j$, for example, with $u > 0$) it seems that the asymptotic theory is a reasonable approximation even for N not very large. In connection with the estimation of a seasonal oscillation Terrell and Tuckwell (1969) investigated the efficiency of least squares when the $y_k(n)$ were $\cos 2\pi jn/12$, $\sin 2\pi jn/12$, $j = 1, \ldots, 6$ and $p = 1$, while $x(n)$ is generated by

$$x(n) - 1.1x(n-1) + 0.3x(n-2) = \epsilon(n),$$

the roots of the characteristic equation

$$z^2 - 1.1z + 0.3$$

being -0.6 and -0.5. The data was transformed by the subtraction from $z(n)$ of a centered 12-term moving average. This does not affect the regressor variables $y_j(n)$, of course, and results in new residuals with a spectrum having a fairly sharp peak near $\pi/6$. The actual efficiencies relative to BLUE for the least squares estimates of α_j, $j = 1, \ldots, 6$ and β_j, $j = 1, \ldots, 5$ are shown below. Here α_j is the coefficient of $\cos 2\pi j/12$ while β_j is that for $\sin 2\pi j/12$. Four values of N are tabulated. (Here N is the number of observations actually used in the regression, i.e., 12 less than the number available before filtering.)

Table 1 RELATIVE EFFICIENCIES FOR AUTOREGRESSIVE RESIDUALS

N	α_1	β_1	α_2	β_2	α_3	β_3	α_4	β_4	α_5	β_5	α_6
36	0.9453	0.9236	0.9323	0.9038	0.9366	0.9366	0.8934	0.9556	0.8632	0.9865	0.9166
48	0.9523	0.9318	0.9458	0.9223	0.9495	0.9495	0.9137	0.9648	0.8882	0.9895	0.9331
60	0.9584	0.9395	0.9549	0.9351	0.9582	0.9582	0.9279	0.9710	0.9060	0.9914	0.9444
72	0.9634	0.9460	0.9614	0.9443	0.9644	0.9644	0.9382	0.9753	0.9191	0.9927	0.9525

It is of some interest to compare the actual variances of the BLUE with that got from the asymptotic formula. These are, for α_1, for example, 0.86, 0.89, 0.91, 0.92 for the four values of N used in Table 1, so that even for $N = 72$ the actual variance was a good deal lower than that given by the asymptotic formula.

Rosenblatt (1956b) investigated the case where $p = 1, q = 2$ with $y_1(n) \equiv 1$, $y_2(n) = n$ and $x(n) = \rho x(n-1) + \epsilon(n)$. Once more least squares is asymptotically efficient. Rosenblatt's tabulations are quite extensive and cover $N = 10, 15, 20, 50$ and, for each N, $\rho = -0.8(0.2)0.8$. We present three extracts. In each case (a) indicates the covariance matrix for least squares, (b) for BLUE, and (c) is that got from the asymptotic formula. The three columns correspond to the three matrix elements namely (1, 1) for the constant term, (2, 2) for the coefficient of n and (1, 2) for the covariance between the two estimates.

Table 2

	(1, 1)	(1, 2)	(2, 2)
		$N = 10, \rho = 0.8$	
(a)	3.48	−0.36	0.07
(b)	3.17	−0.32	0.06
(c)	10.00	−1.50	0.30
		$N = 15, \rho = -0.6$	
(a)	0.15	−0.15	0.002
(b)	0.13	−0.13	0.002
(c)	0.10	−0.10	0.001
		$N = 50, \rho = 0.4$	
(a)	0.22	−0.006	0.0003
(b)	0.21	−0.006	0.0002
(c)	0.22	−0.007	0.0003

Needless to say the results show a better agreement between the LSE and BLUE and of both with the asymptotic results as N increases and $|\rho|$ gets smaller, as the above table shows. For $N = 50$ all results are fairly good. By far the greatest discrepancy is that shown for $N = 10$, $\rho = 0.8$. As might be expected the agreement with the asymptotic theory tends to be relatively worst in the third column.

A third investigation is that by Chipman and others (1968). They consider the case $p = q = 1$ and estimators of the mean μ when $x(n) = \rho x(n-1) + \epsilon(n)$. We may take $x(n)$ to have unit variance. The least squares estimate, $\bar{z}$, has variance

$$\frac{1}{N}\sum_{-N+1}^{N-1} \gamma(n)\left(1 - \frac{|n|}{N}\right) = \frac{1}{N}\sum_{-N+1}^{N-1} \rho^{(n)}\left(1 - \frac{|n|}{N}\right)$$
$$= \frac{1}{N^2}\frac{N(1-\rho^2) - 2\rho(1-\rho^N)}{(1-\rho)^2}.$$

Since the new variates $z(1)$, $(1-\rho)^{-1}\{z(n) - \rho z(n-1)\}$ are uncorrelated with mean μ and variances unity (for $z(1)$) and $(1+\rho)/(1-\rho)$ (for the remainder) we obtain for the variance of $\tilde{\mu}$

$$\frac{1+\rho}{2\rho + N(1-\rho)}.$$

Thus the efficiency of $\bar{z}$ is

$$e = \frac{N^2(1+\rho)(1-\rho)^2}{\{2\rho + N(1-\rho)\}\{N(1-\rho^2) - 2\rho(1-\rho^N)\}}.$$

For $0 \le \rho \le 1$ it is shown in the reference just given that e is always greater than 0.87769. Of course, for any ρ, $-1 < \rho < 1$, e approaches unity as N increases.

This completes our discussion of the efficiency of least squares and we move on, in the next section, to the more important problem of practicable procedures for the computation of efficient estimators in the circumstances which are likely to obtain, where least squares will *not* be efficient, but we do not know Γ_p. At the same time we discuss the asymptotic normality of our estimates. We have already given such a discussion for least squares in Chapter IV, Section 4 (Theorems 10 and 10′) and we refer the reader to that section. We shall use some of those results again in the next section.

4. THE EFFICIENT ESTIMATION OF REGRESSIONS

One procedure by which one might proceed to estimate a regression more efficiently is the following. One might begin by assuming a simple model for the process generating $x(n)$, such as an autoregressive model. One then uses the LSE to estimate B and obtains estimate, $\hat{x}(n)$, for the residuals. Then the structure of the process generating the $x(n)$ is estimated from the $\hat{x}(n)$. Then Γ_p is estimated from this and an approximation to BLUE is obtained. If, in fact, $x(n)$ is generated by a finite parameter model of the kind assumed it is fairly evident that this will lead to an asymptotically efficient procedure. In the case where $p = 1$ the use of an autoregressive model in this way is equivalent to the use of the transformation

$$z(n) \to \sum_0^a \hat{\beta}(j)\, z(n-j), \qquad y_k(n) \to \sum_0^a \hat{\beta}(j)\, y_k(n-j);$$
$$k = 1, \ldots, q; \qquad n = a+1, \ldots, N,$$

where the $\hat{\beta}(j)$ are the estimates of the autoregressive parameters got from the residuals from the least squares regression. (We have used the symbol a for the order of the autoregression since q is used for another purpose in this chapter.) Often, of course, one will commence by filtering the data, for example, by a first difference filter (this being applied to all of the sequences involved) in order to reduce the residuals to something which, it is hoped, will be nearer to a realization of a process of uncorrelated random variables.

When $p > 1$ it is evident that in general one will not be able to proceed in the manner of the last paragraph and that one essentially will have to invert the matrix $\hat{\Gamma}_p$. There is room for ingenuity in doing this and integrating it with the estimation procedure for B, but we proceed no further along that line here.†

Of course, we can always proceed by direct consideration of the likelihood function on Gaussian assumptions. This will tend to reduce to the methods of the last two paragraphs unless the extensive calculations are performed to actually discover the maximum of the likelihood function. If N is really small then possibly such a procedure should be adopted, with a simple but plausible model chosen for the residual $x(n)$. In such a situation the total number of parameters involved, over and above regression parameters, will be small and actual maximization of the likelihood function may be feasible.

We shall devote the remainder of this section to what are essentially large sample techniques (though the scope for the use of the methods is fairly wide). There are a number of problems that arise in connection with regression, especially with time series perhaps, which need to be watched. In the first place it is possible that the regression relation holds good only over

† We shall, in Section 5 (subsection b) consider another technique which, in case $x(n) = B(1)x(n-1) + \epsilon(n)$, amounts to the use of the least squares regression of $z(n)$ on $z(n-1)$, $y(n)$ and $y(n-1)$, the coefficient matrices for the first two regressor vectors estimating $B(1)$ and B.

part of the frequency range. Thus the $z(n)$ may in fact be composed of two constituents, one mainly consisting, let us say, of intermediate frequencies and one mainly of low and high frequencies. It may only be the first part that is related to the $y_j(n)$. Related to this is the possibility of noise effects, especially in the $y_j(n)$. It is well known that we cannot estimate the matrix B consistently from second-order quantities alone if there is noise superimposed on the $y_j(n)$ before observation of them and additional information is needed. We discuss this further below but mention here that it is not improbable that these noise effects will be dominant at certain frequencies so that again a technique which may be flexibly used to eliminate certain frequencies is called for. The natural technique to use seems to be one based on estimates of spectra, and it is this technique that we shall describe here. It may be emphasized that what we shall be doing is basically the same as what has been suggested at the beginning of this section but working with the Fourier transformed data and replacing autoregressive procedures for estimating the structure of the $x(n)$ process by those of spectral estimation. Insofar as the computation of spectra is a first step in any examination of time series data the later calculations will turn out to be computationally rather simple and in common with standard regression procedures will yield at once both the estimate of B and the estimate of covariance matrix of its estimator. We shall show, under suitable conditions, that the estimates are asymptotically normal so as to complete, in an asymptotic fashion, the inference procedure.

We shall use $\hat{f}_x$ for the estimate of the matrix of spectra and cross spectra of the residuals obtained from the residuals, $\hat{x}(n)$, from the least squares regression. This may, of course, be calculated directly from the estimate $\hat{B}$ as

$$\hat{f}_x = \hat{f}_z + \hat{B}'\hat{f}_y\hat{B} - \hat{f}_{zy}\hat{B} - \hat{B}'\hat{f}_{yz}, \tag{4.1}$$

where we, now and henceforth, assume that f_z (*the matrix of spectra and cross spectra for* $z(n)$), f_y (*the matrix of spectra and cross spectra for* $y(n)$) *and* f_{zy} (*the matrix of cross spectra having that between* $z_j(n)$ *and* $y_k(n)$ *in row* j, *column* k) *are all estimated by one and the same procedure among those discussed in Chapter V.*† Here a comment is called for since the $y_j(n)$ are to be assumed only to satisfy Grenander's conditions and not necessarily to be generated by stationary processes so that f_y and f_{zy} may not be well defined. Even if f_y, for example, is interpreted in terms of the $M(\lambda)$ introduced in Section 3, we will not now require that $M(\lambda)$ be absolutely continuous (since a main purpose of regression may be to remove from the data the cause of jumps in the spectrum). The quantities $\hat{f}_{zy}$, $\hat{f}_y$ can, however, be computed

† This condition could be relaxed but it seems unimportant and simplifies the proof.

in any case and we shall later show that the quantity $\hat{f}_x$ is, within the present context, a satisfactory estimate of f_x.

There is another way in which f_x might be estimated. This is by forming

$$\check{f}_x = \hat{f}_z - \hat{f}_{zy}\hat{f}_y^{-1}\hat{f}_{yz}. \tag{4.2}$$

In the case $p = q = 1$ this reduces to the standard estimate of $f_z(1 - \sigma_{zy}^2)$. The essential point about it is that the estimate $\check{f}_x$ is, subject to certain qualifications, a valid estimate of the residual spectrum after the effects of $y(n)$ have been eliminated whether the dependence of $z(n)$ on $y(n)$ is instantaneous (as our regression model implies) or not. On the other hand, if the model is valid, it is obviously not so good an estimate as (4.1). Moreover this estimate seems to have meaning only when $y(n)$ is itself generated by a stationary process. We confine our considerations to $\hat{f}_x$.

The estimate of β which we propose (and whose meaning we discuss a little later) is†

$$\tilde{\beta} = \left[\frac{1}{2M}\sum_{-M+1}^{M}\left\{\hat{f}_x^{-1}\left(\frac{\pi k}{M}\right) \otimes \hat{f}_y\left(\frac{-\pi k}{M}\right)\right\}\right]^{-1} \cdot \left[\frac{1}{2M}\sum_{-M+1}^{M}\left\{\hat{f}_x\left(\frac{\pi k}{M}\right) \otimes I_q\right\}^{-1} \hat{e}_{zy}\left(\frac{\pi k}{M}\right)\right], \tag{4.3}$$

where by $\hat{e}_{zy}$ we mean the matrix $\hat{f}_{zy}$ rearranged as a column vector, with the rows of $\hat{f}_{zy}$ arranged down it as a column, successively. (The reason for the need to put $-\pi k/M$ in place of $\pi k/M$ in the second tensor factor in the first matrix corresponds to that which makes us put $M'(\lambda)$ in place of $M(\lambda)$ in (3.3) and (3.4).) As we have said we shall discuss computational details later.

We assume that

$$\det\left(f_x(\lambda)\right) \geq a > 0, \qquad \lambda \in (-\pi, \pi] \tag{4.4}$$

and that $f_x(\lambda)$ is continuous. These conditions could be relaxed under suitable conditions on $M(\lambda)$ but we leave the details of such a relaxation to the reader. It follows that

$$\int f_x(\lambda)^{-1} \otimes M'(d\lambda)$$

is nonsingular.

We assume that $y(n)$ satisfies (a), (b), and (c) of Chapter IV, Section 3 (Grenander's conditions) and that

$$x(n) = \sum_{-\infty}^{\infty} A(j)\,\epsilon(n-j), \qquad \sum_{-\infty}^{\infty} \|A(j)\| < \infty, \tag{4.5}$$

† Henceforth we shall use $\tilde{\beta}$ to mean the estimate defined by (4.3), though we have previously used this for the BLUE, which is not usually computable.

where the $\epsilon(n)$ are independent and identically distributed with zero mean vector and covariance matrix G. It would no doubt be possible to extend the result of Theorem 10, below, to the alternative situation of Chapter IV, Section 3 (of Theorem 10′) but we have not done that.

We also assume that

$$\text{(4.6)} \quad \overline{\lim_{N\to\infty}} \left(\frac{N}{M}\right)^{1/2} \|\mathcal{E}\{\tilde{f}_x(\lambda) - f_x(\lambda)\}\| \leq a < \infty, \qquad M \to \infty, \qquad \frac{M}{N^{1/2}} \to 0,$$

where we have used the symbol $\tilde{f}(\lambda)$ for the, noncomputable, estimate of $f(\lambda)$ (using the same estimator procedure as for the computed spectra) obtained from the unobservable $x(n)$. Thus we are assuming, in particular, that the bias in $\tilde{f}_x$ is of at most of the same order of magnitude as the standard deviation. This of course involves the assumption that M does not increase too slowly. On the other hand the condition $M/\sqrt{N} \to 0$ implies that it does not increase too fast. Thus these conditions imply some restriction upon the smoothness of $f(\lambda)$ over and above the continuity of that function which is implied by (4.5).

Some such restrictions are necessary to avoid the dominance of the effects on the limiting distribution of the estimation procedure for $f_x(\lambda)$. The conditions seem quite mild ones, though they are subject to the usual objections to asymptotic results. The conditions can be weakened further if the $y(n)$ process is further restricted, for example by assuming that it also is generated by a relation of the form of (4.5).

Theorem 10. *If $x(n)$ is generated by* (4.5) *and $f_x(\lambda)$ satisfies* (4.4) *and the condition* (4.6) *is satisfied then the distribution of $(I_p \otimes D_N)(\tilde{\beta} - \beta)$, where $\tilde{\beta}$ is defined by* (4.3), *converges as $N \to \infty$ to a multivariate normal distribution with zero mean vector and covariance matrix given by the right side of* (3.4). *This covariance matrix is consistently estimated by*

$$\text{(4.7)} \quad N^{-1}(I_p \otimes D_N)\left\{\frac{1}{2M}\sum_{-M+1}^{M} \hat{f}_x^{-1}\left(\frac{\pi k}{M}\right) \otimes \hat{f}_y\left(\frac{-\pi k}{M}\right)\right\}^{-1} (I_p \otimes D_N).$$

We have placed the proof of this theorem in the appendix to this chapter.

It may be observed that the calculation for the determination of (4.7) will have already been done in the process of computing $\tilde{\beta}$ so that no further calculations are needed. It may also be pointed out that the calculations for the determination of the covariance matrix of $\hat{\beta}$ are of the same order of magnitude so that in fact little additional computing is needed above that for $\hat{\beta}$ if one is to compute the estimate of the covariance matrix in both cases. Of course parametric models might lead to reduced calculations but in any case the computations for (4.3) (and (4.7)) are of a minor nature by present day standards.

It may be worthwhile discussing the meaning of (4.3). This is most easily seen when $p = q = 1$. Then

$$\tilde{\beta} = \frac{1/2M \sum_{-M+1}^{M} \hat{f}_{zy}/\hat{f}_x}{1/2M \sum_{-M+1}^{M} \hat{f}_y/\hat{f}_x},$$

where we have omitted the argument $\pi k/M$ for simplicity of presentation. This is a weighted average of the $\hat{\beta}(\pi k/M)$ (i.e., the complex regression coefficients by frequency band) with weight function which is $(\hat{f}_y/\hat{f}_x)$. The signal to noise ratio for z is, at frequency λ, $\beta^2 f_y/f_x$ (assuming $y(n)$ to be stationary so that f_y has a meaning) so that basically we are estimating β as a weighted average of estimates by frequency band, weighting these estimates by estimates of the signal to noise ratio. The general formula (4.3) is an elaboration of this simple case.

Before going on to computational details let us first discuss "mean corrections." Let us introduce the model

$$z(n) = B'y(n) + \Delta'\big(w(n) - D'y(n)\big) + x(n), \tag{4.8}$$

wherein $u(n) = w(n) - D'y(n)$ is also generated by a relation of the form of (4.5) (with different $A(j)$ and $\epsilon(n)$ of course) for which $u(m)$ is independent of $x(n)$ for all m, n. We assume $y(n)$ to satisfy the conditions (a), (b)$'$, and (c) of Chapter IV, Section 3 and, in relation to both the spectra of $x(n)$ and of $u(n)$, as $f(\lambda)$ in Theorem 8, that $y(n)$ satisfies the conditions of that theorem. We have in mind, of course, the situation where $y(n)$ satisfies Theorem 9 and most especially is of the form described at the beginning of Section 3 (see Corollary 1) so that a trend and periodic component (and/or trending periodic component) is being removed. The case $y(n)$ scalar and identically unity is, at least, always relevant. Then almost precisely as in the proof of Theorem 9 of Chapter IV, Section 3,

$$\lim_{N\to\infty} \frac{\sum_{m=1}^{N} u(m)\, y(m+n)}{\left(\sum_1^N u(m)^2 \sum_1^N y(m)^2\right)^{1/2}} = 0$$

almost surely, so that the vector

$$\begin{pmatrix} y(n) \\ \cdots \\ u(n) \end{pmatrix}$$

also satisfies Grenander's conditions, with

$$M(\lambda) = \begin{bmatrix} M_y(\lambda) & 0 \\ 0 & M_u(\lambda) \end{bmatrix}$$

and $M_u(d\lambda) = \gamma_u(0)^{-1} f_u(\lambda)\, d\lambda$.

We now adopt the following regression procedure†

(i) We carry out the least squares regressions of $w(n)$ on $y(n)$ to estimate D, as $\hat{D}$ and of $z(n)$ on $y(n)$ to estimate B as $\hat{B}$.

(ii) We then form $\hat{z}(n) = z(n) - \hat{B}'y(n)$, $\hat{u}(n) = w(n) - \hat{D}'y(n)$ and we estimate Δ by means of formula (4.3) with $z(n)$ replaced by $\hat{z}(n)$ and $y(n)$ replaced by $\hat{u}(n)$ ($\hat{f}_x$ being got from (4.1) with these same replacements). Call the resulting estimate $\tilde{\Delta}$.

Theorem 11. *The estimates of B, Δ, D in* (4.8) *by the above method, under the conditions stated below* (4.8), *are asymptotically efficient in the sense that they have the same asymptotic distributional properties as for the BLUE procedures applied (separately) to* $z(n) = B'y(n) + \Delta'u(n) + x(n)$ *and* $w(n) = D'y(n) + u(n)$.

We omit the proof of this theorem.‡ The first of the two regressions stated at the end of the Theorem cannot be carried out, since $u(n)$ is not observed. The point of the theorem is that no efficiency is lost, asymptotically, if trend and periodic components (or trending periodic components) are first removed by least-squares regression and B' is then estimated from the residuals. This will greatly reduce the calculations in some cases. It would be better to have Theorem 11 stated for more general $u(n)$, but some restriction near to that of the theorem seems to be needed for a general result. For particular cases, such as $y(n)$ scalar and $y(n) \equiv 1$, the restriction on $u(n)$ is not needed. However, we leave the result as it stands and go no further in this direction but proceed to computational details for (4.3). For the case in which $p = 1$ the computational details are straightforward. *Here and often below we shall put*

$$\delta_k = 1, \qquad k \neq 0, M; \qquad \delta_k = \tfrac{1}{2}, \qquad k = 0, M.$$

Then we form the matrix and vector

$$W = \frac{1}{2M} \sum_0^M \delta_k \hat{f}_x^{-1} \hat{c}_{yy}, \qquad w = \frac{1}{2M} \sum_0^M \delta_k \hat{f}_x^{-1} \hat{c}_{zy},$$

where $\hat{c}_{yy}$ is the matrix of cospectral estimates for the $y(n)$ and $\hat{c}_{zy}$ is the vector of cospectra for z with the $y(n)$. Both are computed after mean correction, at least. Then

$$\tilde{\beta} = W^{-1}w,$$

† Theorem 10 applies to the estimation of (4.8). We choose to perform the regression in two steps, and by least squares at the first step, so as to reduce the amount of calculation.

‡ See Section 5 for closely related considerations.

and this may be treated (asymptotically) as if it were normal with mean β and covariance matrix

$$N^{-1}W^{-1}.$$

For $p > 1$ the method is not fully analogous with the classical least squares procedure since the matrix to be inverted is $pq \times pq$ and not $q \times q$. Also quadrature spectra now enter into the calculations. We must first form (again making mean corrections)

$$\frac{1}{2M}\sum_{-M+1}^{M} \hat{f}_x^{-1}\hat{f}_{zy} = \frac{1}{M}\sum_{0}^{M} \delta_k\, \Re(\hat{f}_x^{-1}\hat{f}_{zy}).$$

Writing $\hat{f}_x$ as $\frac{1}{2}(\hat{c} - i\hat{q})$ $(k \neq 0, M)$ we know that $\hat{f}_x^{-1}$ corresponds to $2(u - iv)$, where

$$\begin{bmatrix} \hat{c} & \hat{q} \\ -\hat{q} & \hat{c} \end{bmatrix}^{-1} = \begin{bmatrix} \hat{u} & \hat{v} \\ -\hat{v} & \hat{u} \end{bmatrix}$$

and

$$\hat{u} = (\hat{c} + \hat{q}\hat{c}^{-1}\hat{q})^{-1}, \qquad \hat{v} = -\hat{c}^{-1}\hat{q}(\hat{c} + \hat{q}\hat{c}^{-1}\hat{q})^{-1}.$$

Thus

$$(4.9)\qquad \frac{1}{M}\sum_{0}^{M} \delta_k\, \Re(\hat{f}_x^{-1}\hat{f}_{zy}) = \frac{1}{M}\sum_{0}^{M} \delta_k(\hat{u}\hat{c}_{zy} - \hat{v}\hat{q}_{zy}),$$

where $\hat{c}_{zy}$ and $\hat{q}_{zy}$ are again co- and quadrature components.

Call w the column vector got by writing the rows of the matrix (4.9) successively down a column.

We next form

$$(4.10)\qquad \frac{1}{2M}\sum_{-M+1}^{M} \left\{\hat{f}_x^{-1}\left(\frac{\pi k}{M}\right) \otimes \hat{f}_y\left(\frac{-\pi k}{M}\right)\right\} = \frac{1}{M}\sum_{0}^{M} \delta_k\{\hat{u} \otimes \hat{c}_y - \hat{v} \otimes \hat{q}_y\} = W.$$

When this matrix has been inverted it must be applied to the vector w to get

$$\tilde{\beta} = W^{-1}w$$

and $\tilde{\beta}$ may be treated as being normal with mean β and covariance matrix $N^{-1}W^{-1}$.

Some points which may be made in relation to this estimation procedure are as follows.

1. The method is appropriate, as is evident from the statement of the theorem, in cases in which $y(n)$ is a deterministic sequence. In case $p = 1$ and all $y_j(n)$ are of the form $\cos n\lambda$ or $\sin n\lambda$ there is little or no point in carrying through the calculations, since we can hardly expect the results to differ from those obtained from direct least squares regression. This is in accordance with the results of Theorem 5 in Section 3. In cases where N is

small there is sometimes efficiency to be gained, as our discussion in the last section shows. The techniques of this section are intrinsically incapable of regaining that since to do so would require an alteration, of the shape of the spectrum of $x(n)$, *inside* a band of width $2\pi/N$. However, there are other cases in which least squares is asymptotically efficient and $p = 1$ and in which the techniques of this section would be useful. This would be the case, for example, for a polynomial regression when N is not too large (relative to the degree of the polynomial). In this case $\hat{f}_y(\lambda)$ may tend to be appreciable quite a distance away from $\lambda = 0$, though asymptotically it is only $\lambda = 0$ which contributes to the variance. In such a situation it might be worthwhile confining the summations to a range of values of $\pi k/M$ for which $\hat{f}_y(\lambda)$ is appreciable, omitting the remainder of the range. This can be judged from the expression

$$\frac{\left|\sum_1^N n^u e^{in\lambda}\right|^2}{\left(\sum_1^N n^u\right)^2} \sim \frac{(u+1)^2}{N^2\lambda^2}, \qquad \lambda > 0.$$

The use of the complete formula (omitting no part of the range) will in principle cause no loss of efficiency and in practice should not matter either, but involves needless computations.

2. A complication that may develop in practice is that the linear relation (2.1) may hold only over part of the range, perhaps because of substantial errors of observation [in the $y_j(n)$] over the remaining part. We are natuıally led to confine the summations in (4.3) to those values of k for which the relation (2.1) does hold. Again under reasonable circumstances the result of Theorem 10 will continue to hold, with the integration in (3.4) correspondingly over the subrange included in the estimation procedure. The inference procedure stays the same, of course, since the matrix W^{-1} is still obtained from the matrix used to estimate β, the summation going over the subrange.†

3. Tests of significance of the elements of $\tilde{\beta}$ and confidence intervals procedures follow immediately from the result of Theorem 10. Of course, we face the usual multiple comparisons problem if pq is large in that a large number of, nonindependent, tests may be carried out. We do not discuss this in general here since it then has no new features as compared to the classical situation. (See Scheffé 1959, Chapter 3.) However there is one situation which is related to the present one and which does call for comment and that is the one where the regression is over $\cos 2\pi kn/N$ and $\sin 2\pi kn/N$, *but k is chosen after reference to the data.* This is a situation which does arise in practice, since someone may examine data and believe he sees a

† For a more detailed discussion of errors of observation we refer the reader to Hannan (1963).

periodic component in it (i.e., a jump in the spectrum). He then examines the periodogram and perceives a very large value at or near the frequency suggested by the original study of the data. Now, of course the largest out of a large number of $I(\omega_j)$ has been chosen and care must be taken with the inference procedure. We discuss this problem in some detail in Section 6. However, we first consider (in Section 5) the effects of regression procedures on later treatments of the data (as discussed in earlier chapters), since they are needed for the discussion in Section 6.

4. We may also use these methods for the efficient estimation of (2.1) under the linear restrictions introduced in Section 2, namely

$$\alpha_j'\beta = \operatorname{tr}(BA_j) = 0, \qquad j = 1, \ldots, r. \tag{4.11}$$

Once more we introduce the matrix, E, which projects onto the orthogonal complement of the part of pq dimensional real vector space spanned by the α_j. We now consider the BLUE for (2.1) subject to (4.11). To this end we introduce

$$\tilde{\beta} = (EWE)^{-1}w \tag{4.12}$$

where W and w are precisely as defined in (4.10) and through (4.9). The matrix inverse is the generalized inverse and we discuss its computation below. As a fairly immediate corrollary to Theorem 10, we have the following.

Corollary 2. *Let the conditions of Theorem 10 be satisfied but also B be constrained by (4.11) where E commutes with $I_p \otimes D_N$. If $\tilde{\beta}$ is given by (4.12) then the distribution of $(I_p \otimes D_N)(\tilde{\beta} - \beta)$ converges as $N \to \infty$ to a multivariate normal distribution with zero mean vector and covariance matrix*

$$\left\{E\left[\int_{-\pi}^{\pi}(2\pi f(\lambda))^{-1} \otimes M'(d\lambda)\right]E\right\}^{-1}$$

which is consistently estimated by

$$N^{-1}(I_p \otimes D_N)(EWE)^{-1}(I_p \otimes D_N).$$

The reasonability of the requirement that E commute with $I_p \otimes D_N$ was discussed just above (3.1). The theorem asserts that $\tilde{\beta}$, given by (4.12), may, for N large, be treated as normal with mean vector β and covariance matrix $N^{-1}(EWE)^{-1}$.

The computation of W and w was discussed above. Let us orthonormalize the α_j to reach a new set φ_j satisfying $\varphi_j\varphi_k = \delta_j^k$. This may be relatively easy to do in the type of circumstance covered by the corollary as the α_j are likely to be vectors composed mainly of zeros. Then

$$F = I_{pq} - E = \sum_{j=1}^{r} \varphi_j\varphi_j'$$

and

$$(EWE)^{-1} = (EWE + F)^{-1} - F.$$

Here $(EWE + F)$ will be non-singular.

The corollary well exemplifies the usefulness of Fourier methods as can be seen by considering the solution of the problem through a direct application of the BLUE procedure.

One further comment may be made. Let the φ_j be divided into two sets, for $j = 1, \ldots, r_1$ and $j = r_1 + 1, \ldots, r = r_1 + r_2$. We have in mind the circumstances where we maintain the validity of r_1 constraints and wish to test the remainder so that the $\varphi_j, j = 1, \ldots, r_1$, have been chosen by orthc-normalizing $\alpha_j, j = 1, \ldots, r_1$. Let $\tilde{\beta}_0$ be computed via (4.12) but using

$$E_1 = I_{pq} - F_1, \qquad F_1 = \sum_{j=1}^{r_1} \varphi_j \varphi_j'$$

in place of E. To test the validity of the remaining constraints we use

$$N^{-1}\tilde{\beta}_0'(F_2(E_1WE_1)^{-1}F_2)^{-1}\tilde{\beta}_0, \qquad F_2 = F - F_1,$$

as chi-square with r_2 degrees of freedom.

Of course

$$(F_2(E_1WE_1)^{-1}F_2)^{-1} = (F_2(E_1WE_1)^{-1}F_2 + (I_{pq} - F_2))^{-1} - (I_{pq} - F_2).$$

5. THE EFFECTS OF REGRESSION PROCEDURES ON ANALYSIS OF RESIDUALS

The title of this section is not fully appropriate since we cover the case of a mixed regression of the form typified in the simplest case by

$$z(n) = \rho z(n-1) + \beta y(n) + x(n)$$

as well as models such as

$$\big(z(n) - \beta y(n)\big) = \rho\{z(n-1) - \beta y(n-1)\} + x(n).$$

In the first of these two cases $z(n-1)$ and $y(n)$ stand on more or less the same footing whereas in the second it is the residuals from the linear relation of $z(n)$ with $y(n)$ which are related by a simple autoregression. We shall come back to this kind of problem later and first discuss spectral methods. Throughout the section we assume $f_x(\lambda)$ continuous, though this condition could be considerably relaxed.

(a) We consider, needless to say, the circumstance in which Grenander's conditions hold for the $y_j(n)$. What we wish to discuss first is the effect of using $\hat{x}(n)$ in place of $x(n)$ in the analysis of residuals, where $\hat{x}(n)$ is the observed vector of residuals from the regression. Consider

$$(N-j)\,b_{uv}(j) = \sum_1^{N-j} \hat{x}_u(n)\hat{x}_v(n+j) - \sum_1^{N-j} x_u(n)\,x_v(n+j)$$

$$= -\sum_1^{N-j} \{\hat{\mu}_u(n) - \mu_u(n)\}\,x_v(n+j)$$

$$-\sum_1^{N-j} x_u(n)\{\hat{\mu}_v(n+j) - \mu_v(n+j)\}$$

$$+\sum_1^{N-j} \{\hat{\mu}_u(n) - \mu_u(n)\}\{\hat{\mu}_v(n+j) - \mu_v(n+j)\},$$

where $\hat{\mu}_u(n)$ is the estimate, by regression, of $\mu_u(n)$, the mean of $z_u(n)$. The first term, for example, using the least-squares estimates of the regression coefficients, is

$$-\sum_{k=1}^{q} \{\hat{\beta}_{uk} - \beta_{uk}\} \sum_{n=1}^{N-j} y_k(n)\,x_v(n+j) = -\sum_{k=1}^{q} d_k(N)\{\hat{\beta}_{uk} - \beta_{uk}\} \times \left[d_k^{-1}(N) \sum_{n=1}^{N-j} y_k(n)\,x_v(n+j)\right].$$

(Though we appear to deal here only with the least squares estimate the result will be seen to be general and we have dealt with this case only for notational convenience.) The variance of the last expression (in square brackets) is easily seen to be dominated by the maximum value of $f_v(\lambda)$ in $(-\pi, \pi]$. On the other hand $d_k(N)\,\{\hat{\beta}_{uk} - \beta_{uk}\}$ has variance which is, as we know, $O(1)$. Thus the first term is $O(1)$, uniformly in j, in the sense that the mean-square error is bounded by a constant independent of j for all $|j| < N$. (In fact, the mean square error is bounded by

$$c\left\{\frac{\sum_1^{N-j} y_k(n)^2}{d_k^2(N)}\right\},$$

where c is independent of j.) The same is evidently true of the second term. The third is of the form

$$\sum\sum_{k,l=1}^{q} (\hat{\beta}_{uk} - \beta_{uk})(\hat{\beta}_{vl} - \beta_{vl})\left\{\sum_1^{N-j} y_k(n)\,y_l(n+j)\right\},$$

which is easily seen to be of the same nature and indeed to have mean square bounded by

$$c\left\{\frac{\sum_1^{N-j} y_k(n)^2}{d_k^2(N)} \cdot \frac{\sum_{j+1}^{N} y_l(n)^2}{d_l^2(N)}\right\}^{1/2}.$$

Now the error in an estimate of $f_{uv}(\lambda)$ of the general form of (4.2) in Chapter

V, due to using $\hat{x}(n)$, will be

$$\hat{e}(\lambda) = \frac{1}{2\pi} \sum_{-N+1}^{N-1} k_n e^{-in\lambda} \frac{1}{N} b_{uv}(n). \tag{5.1}$$

In the case in which $k_n = k(n/M)$ with $|k(x)|$ integrable (i.e., cases (4.3), (4.4), (4.5), (4.6), (4.7), and (4.9) of Chapter V, Section 4) it is immediate that the root mean square error of (5.1) is $O(M/N)$. In the case of formulas (4.1), (4.2), and (4.8) of Chapter V, Section 4 we see that the root mean-square error of (5.1) is dominated by

$$\frac{1}{2\pi} \frac{M}{N} \left\{ \frac{1}{M} \sum_{-N+1}^{N-1} |k_n| \right\} \leq c \frac{M}{N} \log\left(\frac{N}{M}\right) \to 0.$$

We have not established an asymptotic variance formula for $\tilde{\beta}$ (computed according to (4.3)) but we know that $d_k(N)(\tilde{\beta} - \beta)$ is asymptotically normal so that $\max_j \{(M/N)^\epsilon (N-j)\, |b_{uv}(j)|\}$ converges in probability to zero for all $\epsilon > 0$.

Theorem 12. *Let $f_x(\lambda)$ be continuous and the $y_j(n)$ satisfy Grenander's conditions. Let $\hat{f}_x(\lambda)$ be obtained by any of the methods of Chapter V, Section 4 using the calculated residuals $\hat{x}(n)$ obtained by inserting an estimate for B' in* (2.1). *If the LSE or BLUE procedures are used to obtain that estimate then if $\hat{e}(\lambda)$, given by* (5.1), *is the error due to use of $\hat{x}(n)$ in place of $x(n)$* var $(\max_j \hat{e}(\lambda)) = O(M/N)$ *for estimation procedures other than* (4.1), (4.2), *and* (4.8) *of Chapter V, Section* 4 *and is $O((M/N)\log(N/M))$ for those three. For B estimated by $\tilde{B}$ got from formula* (4.3) *then* $\max_\lambda \{\hat{e}(\lambda)\}/\{M/N\}^{1-\epsilon}$ *converges in probability to zero for all* $\epsilon > 0$.

This result should be interpreted in the light of the root mean-square error for the estimate computed from $x(n)$. We know the variance to be $O(M/N)$ so that if the $(\text{bias})^2$ is of no higher order than this root mean-square error then the bias is $O(\sqrt{M/N})$ and the effect from the regression is asymptotically negligible. This means in turn that the effects upon latter statistical procedures, using the $\hat{f}(\lambda)$, from the regression, will also be negligible for small M/N.

However, there are cases in which the effects of the regression are concentrated at particular frequencies when also these effects will be concentrated in a particular band and may be worth accounting for, and easily accounted for. An example is the mean correction, which makes $I(0) = 0$ but does not affect any $I(\omega_j)$, $j \neq 0$. It seems plausible now that we should replace $\hat{f}(0)$, computed from $\hat{x}(n)$, with $(1 - (1/m))^{-1}\hat{f}(0)$, since if m bands of width $2\pi/N$ were used to form $\hat{f}(0)$ by the FFT method, of which one contributes zero, we should certainly divide by $(m-1)$ not m. Since $m^{-1} = 2M/N$ our

Theorem 12 is not contradicted, but the correction is so easily made that it might as well be used. Since we here concentrate on circumstances wherein the generalized spectrum of the $y_j(n)$ is a jump spectrum, where, as we have seen in Theorem 9, we are likely to use direct least squares, we deal only with the case where the regression is computed using $\hat{\beta}$. We call the periodogram, computed from $\hat{x}(n)$, $\hat{I}(\lambda)$ and consider

$$\begin{aligned}(5.2)\quad \hat{I}_{uv}(\lambda) &= I_{uv}(\lambda) - l^*(\lambda)\, Y(Y'Y)^{-1}Y'x_u x_v'\, l(\lambda)\\ &\quad - l^*(\lambda)x_v x_u'\, Y(Y'Y)^{-1}Y'\, l(\lambda) + l^*(\lambda)\, Y(Y'Y)^{-1}Y'x_u x_v'\, Y'(Y'Y)^{-1}Y'\, l(\lambda)\end{aligned}$$

wherein the vector $l(\lambda)$ *has* $(2\pi N)^{-1/2} \exp -i\lambda n$ *in the* nth *place and the vector* x_u *has* $x_u(n)$ *in the* nth *place.* Taking expectations the second term becomes

$$-2\pi N \int_{-\pi}^{\pi} l^*(\lambda)\, Y(Y'Y)^{-1}Y'\, l(\theta)\, l^*(\theta)\, l(\lambda)\, f_{uv}(\theta)\, d\theta.$$

Putting $w(\lambda, y)$ for the vector with

$$(5.3)\qquad \frac{1}{d_k(N)} \sum_1^N y_k(n) e^{in\lambda}$$

in the kth place, this expectation is seen to converge to

$$-w(\lambda, y)^*\, R(0)^{-1}\, w(\lambda, y) f_{uv}(\lambda).$$

The third term has expectation converging to the same value. The last term has expectation converging to

$$w(\lambda, y)^*\, R(0)^{-1} \int_{-\pi}^{\pi} f_{uv}(\theta)\, M(d\theta)\, R(0)^{-1}\, w(\lambda, y).$$

We consider only the case in which $N(\theta) = R(0)^{-1/2}\, M(\theta)\, R(0)^{-1/2}$ jumps at points λ_p by amounts N_p and otherwise does not increase. (The mixed case could easily be treated but does not seem sufficiently important.) Thus the last expression is

$$w(\lambda, y)^*\, R(0)^{-1/2} \left(\sum f_{uv}(\lambda_p) N_p\right) R(0)^{-1/2}\, w(\lambda, y).$$

Let us assume that $w(\lambda, y)^*\, w(\lambda, y)$ converges to zero at all points save those at which $N(\theta)$ jumps, which will be true for polynomial or trigonometric regression. Then the last displayed expression converges to

$$f_{uv}(\lambda)\, w(\lambda, y)^*\, R(0)^{-1}\, w(\lambda, y),$$

so that the overall expectation of the last three terms in (5.2) is

$$-w(\lambda, y)^*\, R(0)^{-1}\, w(\lambda, y) f_{uv}(\lambda).$$

We now consider an estimate

$$\hat{\hat{f}}(\lambda) = \int K_N(\lambda - \theta)\, \hat{I}_{uv}(\theta)\, d\theta. \tag{5.4}$$

This has expectation converging to

$$\int K_N(\lambda - \theta)\, f_{uv}(\theta)\, d\theta - \int K_N(\lambda - \theta)\, w(\theta, y)^*\, R(0)^{-1}\, w(\theta, y)\, f_{uv}(\theta)\, d\theta. \tag{5.5}$$

We are interested in the circumstances in which $f_{uv}(\theta)$ is smooth enough in relation to $K_N(\lambda - \theta)$ for us to be able to write (5.5) in the form

$$f_{uv}(\lambda)\left\{1 - \int K_N(\lambda - \theta)\, w(\theta, y)^*\, R(0)^{-1}\, w(\theta, y)\, d\theta\right\}. \tag{5.6}$$

Of course, as we already know, the correction that (5.6) implies is of smaller magnitude than the root mean-square error in $\hat{\hat{f}}_{uv}$. However, it may not be uncommon, in practice, for the effect from (5.6) to be sizable near some points because of the nature of the regression used (e.g., the fitting of a high order polynomial trend) and the fact that M is rather large compared with N. Since the second factor in (5.6) is a known number, judgment can be used in determining whether to apply it. Thus our final conclusion is as follows (we prefer not to state it in the form of a theorem).

If an estimate of the spectrum of the form (5.4) *is constructed, in which* $\hat{I}(\lambda)$ *is computed from* $\hat{x}(n)$ *and the* $\hat{x}(n)$ *are the residuals from a least-squares regression on regressor vectors* $y(n)$, *whose spectrum* $M(\lambda)$ *jumps at a finite number of points but otherwise does not increase, and* $f_{uv}(\lambda)$ *is relatively smooth* [*see* (5.5) *and* (5.6)], *whereas* $w(\theta, \lambda)$ *is a vector with* kth *element given by* (5.3), *then* (5.6) *to a good approximation will describe the bias effect of the use of* $\hat{x}(n)$ *in place of* $x(n)$.

Example

An important case is that of a regression in the form of

$$\sum_0^{q_1} \mu_j n^j + \sum_{q_1+1}^{q_1+q_2} (\alpha_j \cos n\theta_j + \beta_j \sin n\theta_j).$$

We leave it to the reader to show that the second factor in (5.6) then becomes, to a high order of approximation,

$$1 - \frac{2\pi}{N}\left[(q_1 + 1)\, K_N(\lambda) + \sum_1^{q_2} \{K_N(\lambda - \theta_j) + K_N(\lambda + \theta_j)\}\right]; \tag{5.7}$$

for example, for the FFT estimate $K_N(\lambda) = M/\pi$, $|\lambda| \le \pi/M$ and otherwise zero so that $2\pi K_N(\lambda)/N = 2M/N = m^{-1}$, $|\lambda| < \pi/M$, and otherwise zero, which shows that for a mean correction our factor is $(1 - m^{-1})$ as our earlier argument suggested. This factor (5.7) will be very near to unity away from $\lambda = 0$ or θ_j. At $\lambda = 0$ it differs from unity by a quantity of order $(q_1 + 1)/m$ which may be considerable in practice. *The expression* (5.7) *is used by dividing* $\hat{\hat{f}}(\lambda)$ *by it, of course.*

(b) In the course of our proof of Theorem 12 we showed that the serial covariances $\hat{c}_{uv}(j)$, when computed from the residuals $\hat{x}_u(n)$ from a regression on vectors $y(n)$ satisfying Grenander's conditions, differed from the $c_{uv}(j)$, computed from the unobservable $x(n)$, by quantities that are of the order $(N-j)^{-1}$. Thus certainly $\sqrt{N}(\hat{c}_{uv}(j) - c_{uv}(j))$ converges in probability to zero. We may now consider the model

$$\sum_{j=0}^{s} B(j)\{z(n-j) - By(n-j)\} = \epsilon(n), \qquad B(0) = I_p, \tag{5.8}$$

where again the $y(n)$ satisfy Grenander's conditions. The natural estimation procedure is to regress $z(n)$ on the $y(n)$, either by least squares (or BLUE) or using the asymptotically efficient method of Theorem 10 and, calling $\hat{x}(n)$ the calculated residuals from that regression, to then compute the estimates of the $B(j)$ by means of (2.6) in Chapter VI, but with $\hat{x}(n)$ replacing $(x(n) - \bar{x})$ in the computation of the $c_{uv}(j)$. The result follows immediately.†

Theorem 13. *If the coefficient matrices* $B(j)$ *in* (8) *are estimated through formula* (2.6) *of Chapter VI but using the calculated residuals,* $\hat{x}(n)$, *from the LSE or BLUE regression procedures, or the procedure of Theorem* 10, *in place of* $x(n) - \bar{x}$ *then* $\sqrt{N}(\hat{\beta}_q - \beta_q)$ *again has the asymptotic distribution of Theorem* 1 *of Chapter VI.*

We now go on to discuss an alternative model, widely used in economics. This is of the form

$$\sum_{j=0}^{s} B(j)\, z(n-j) + \Delta y(n) = \epsilon(n), \qquad B(0) = I_p. \tag{5.9}$$

We have already indicated one reason for considering this type of model, in the footnote to page 439. Thus we might be initially concerned with the model

$$z(n) = \Delta_1 \tilde{y}(n) + x(n),$$

† No doubt a more careful analysis would show that, at least when least-squares regression is used and the conditions (a), (b)′, and (c) (of Chapter IV, Section 3) hold, then $\hat{\beta}_q$ converges almost surely to β_q.

where we postulate that

$$\sum_0^s B(j)\, x(n-j) = \epsilon(n), \qquad B(0) = I_p.$$

Now we are led to form

$$\sum_0^s B(j)\, z(n-j) = \sum_0^s B(j)\Delta_1\, \tilde{y}(n-j) = \epsilon(n),$$

which reduces to (5.9) if $y(n)$ is a vector of $q(s+1)$ components with $\tilde{y}_u(n-j)$ in row $jq+u$ and

$$\Delta = [\Delta_1 \vdots B(1)\Delta_1 \vdots \cdots \vdots B(s)\Delta_1].$$

If we ignore the restrictions imposed upon the elements of Δ it is apparent that we may suffer an asymptotic loss of efficiency. To avoid this we may use the method of Theorem 10. In any case (5.9) provides a model from which an estimation procedure for Δ_1 and the $B(j)$ may be obtained.

Before going on to consider it we link (5.8) and (5.9). The following theorem is easily proved (and we therefore omit that proof).†

Theorem 14. *If $y(n)$ satisfies Grenander's conditions then in order that* (5.9) *should, for any $B(j)$, Δ, be representable in the form* (5.8) *it is necessary and sufficient that the $y_j(n)$ be of the form $n^v \cos \theta_j n$, $n^v \sin \theta_j n$, $v = 0, 1, \ldots, q_j - 1, j = 1, \ldots, r$ where both sine and cosine terms occur if one does so.*

This does not mean that (5.8) is applicable only when the $y_j(n)$ are of the kind stated in the theorem but merely that when that is so (5.8) is always relevant. In any case, because of Theorem 14 we are led to modify (5.9) in a way analogous to that introduced in Section 4 [see (4.8)] to form

$$(5.10) \quad \sum_0^s B(j)\{z(n-j) - B'y(n-j)\} = \Delta'\{w(n) - D'y(n)\} + \epsilon(n),$$
$$B(0) = I_p.$$

Now $u(n) = w(n) - D'y(n)$ is assumed to be generated by a linear process, that is, of the form

$$\sum_{-\infty}^{\infty} A(j)\, \eta(n-j), \qquad \sum_{-\infty}^{\infty} \|A(j)\| < \infty,$$

where the $\eta(n)$ are i.i.d. $(0, H)$ and are independent of $\epsilon(m)$ for all m, n. We have in mind the situation in which trends and periodic components (e.g., due to seasonal influences) have been removed by the regression of $w(n)$ on $y(n)$, but we do not restrict $y(n)$ to be of this special form, since it is

† See Exercise 4.

not necessary. It would be possible to be more general and to assume less concerning $x(n)$ (allowing its elements to have sums of squares increasing faster than N), but, especially in the vector case, this seems to lead to a rather complicated problem. Of course, q may be unity and the only component of $y(n)$ identically unity. We shall use r for the number of components in $x(n)$. *We also must impose a mild condition on the spectrum of* $x(n)$, *namely that the intersection of the null spaces of* $f_x(\lambda)$ *(with* λ *varying) shall be null;* for example, if the zero lag covariance matrix of $x(n)$ is nonsingular, this will be so.

The method of estimation we discuss is that based upon regression of $z(n)$ and $w(n)$ on $y(n)$ to determine estimates $\hat{B}$, $\hat{D}$ of B, D followed by the use of

$$\hat{z}(n) = z(n) - \hat{B}y(n), \qquad \hat{x}(n) = w(n) - \hat{D}y(n). \tag{5.11}$$

This first regression procedure may be direct least squares (and we have this principally in mind) but it may also be, for example, the asymptotically efficient procedure of the last section. We next determine the $B(j)$ and Δ by direct least squares regression of $\hat{z}(n)$ on the $\hat{z}(n-j)$, $j = 1, \ldots, s$ and $\hat{x}(n)$. In doing this we shall, of course, replace expressions such as

$$\frac{1}{N-s}\sum_{n=s+1}^{N} \hat{z}_j(n-u)\,\hat{x}_k(n-v),$$

with expressions of the form

$$\frac{1}{N-u+v}\sum_{1}^{N-u+v} \hat{z}_j(n)\,\hat{x}_k(n+u-v), \qquad 0 \leq u - v. \tag{5.12}$$

In the vector case, in particular, the calculations are fairly extensive. The matrix to be inverted is of the form

$$\begin{bmatrix} \hat{C}_s & \hat{C}_{zx} \\ \hat{C}_{xz} & \hat{C}_x(0) \end{bmatrix}, \tag{5.13}$$

where $\hat{C}_s$ is $ps \times ps$ and composed of s^2 blocks $\hat{C}_z(u-v)$, where this contains all serial covariances for lag $(u-v)$ for the $\hat{z}(n)$ series, while $\hat{C}_{zx}$ is $ps \times r$ and contains s blocks of $(p \times r)$ matrices of which a typical one is $\hat{C}_{zx}(k)$, this being composed of the elements (5.12) for $(u-v) = k$. Finally $\hat{C}_x$ is $r \times r$ and has the zero lag covariances of the $\hat{x}(n)$ series as its elements. The inversion of (5.13) may be reduced to the inversion of $\hat{C}_s$, $\hat{C}_x(0)$ and of $[\hat{C}_x(0) - \hat{C}_{xz}\hat{C}_s^{-1}\hat{C}_{zx}]$. The major task will thus be the inversion of $\hat{C}_s$. We have described how this may be done iteratively, as would be needed in any case if the order s is to be determined by successively fitting terms of higher order.

So far as the properties of the estimates $\hat{B}$, $\hat{D}$ are concerned these have already been discussed in earlier sections of this chapter. We confine ourselves therefore to the $B(j)$ and Δ. Rather than give an elaborate statement we can confine ourselves to

Theorem 15. *The joint distribution of the elements of* $\sqrt{N}\{\hat{B}(j) - B(j)\}$ $j = 1, \ldots, s$ *and* $\sqrt{N}\{\hat{\Delta} - \Delta\}$ *obtained by the methods described above converges to a multivariate normal distribution. As a result the asymptotic treatment of the estimates* $\hat{B}(j)$, $\hat{\Delta}$ *will be the same as that for a system of regressions carried out for the classical linear model, as if the vectors* $\hat{x}(n)$, $\hat{z}(n-j), j > 0$, *had been vectors of fixed numbers.*

Proof. The proof is almost the same as that of Theorem 1, Chapter VI. First taking B and D in (5.10) as null, we introduce the equations of estimation

$$(5.14)\qquad \begin{cases} \sum_0^s \hat{B}(j)\, C_z(j-k) + \hat{\Delta}\, C_{wz}(-k) = 0, \qquad k = 1, \ldots, s, \\ \sum_0^s \hat{B}(j)\, C_{zw}(j) + \hat{\Delta}\, C_w = 0, \end{cases}$$

where $C_z(n)$ has the sample serial covariances, $c_{uv}(n)$, for z in the (u, v)th place and $C_{wz}(-k)$ has the sample serial covariance between $w_u(n)$ and $z_v(n-k)$ in the same place, while C_w (which might be called $C_{ww}(0)$) has the zero-lag covariances of the $w_u(n)$ as its elements. We estimate G by

$$(5.15)\qquad \sum_0^s \hat{B}(j)\, C_z(j) + \hat{\Delta}\, C_{wz}(0) = \hat{G}.$$

We may write the equations (5.14) in the form

$$\left(I_p \otimes \begin{bmatrix} C_z & C_{zw} \\ C_{wz} & C_w \end{bmatrix}\right)\begin{pmatrix} \hat{\beta} \\ \hat{\delta} \end{pmatrix} = -\begin{pmatrix} c \\ c_{wz} \end{pmatrix}$$

or

$$(I_p \otimes C)\begin{pmatrix} \hat{\beta} \\ \hat{\delta} \end{pmatrix} = -c.$$

Here C_z has s rows and columns of blocks, with $C_z(u-v)$ in the (u, v)th place. (It corresponds to C_q of Chapter VI, Section 2.) C_{zw} has s rows and 1 column of blocks with $C'_{wz}(-k)$ in the kth row. (We may put $C'_{zw}(-k) = C_{zw}(k)$). Of course, $C_{zw} = C'_{wz}$. The column c_z has $c_{uv}(n)$ in row $(u-1)ps + (n-1)p + v$, whereas c_{wz} has the covariance between $z_u(n)$ and $w_v(n)$ in row $(u-1)r + v$; $\hat{\beta}$ is as in Chapter VI, Section 2 and $\hat{\delta}$ has δ_{uv} in row $(u-1)r + v$. Now, exactly as in the proof of Theorem 1, in Chapter VI $(I_p \otimes C)$ converges with probability 1 to a nonsingular matrix, $I_p \otimes \Gamma$, and

c converges with probability 1 to γ where Γ and γ are defined in terms of true covariances and

$$-(I_p \otimes \Gamma)^{-1}\gamma = \begin{pmatrix} \beta \\ \delta \end{pmatrix}.$$

Correspondingly $\hat{G}$ converges with probability 1 to G. Again, as in the proof of Theorem 1 in Chapter VI, we may introduce end terms so that, for example, the serial covariance $c_{uv}(m-n)$ in the block (m, n) of C_z is replaced by

$$\frac{1}{N}\sum_1^N z_u(l-m)\, z_v(l-n).$$

We place "hats" on all matrices to indicate this change.

Correspondingly we may replace $\hat{\beta}$, $\hat{\delta}$ by b, d so that

$$(I_p \otimes \hat{C})\begin{pmatrix} b \\ d \end{pmatrix} = -\hat{c}$$

and

$$\sqrt{N}\left\{\begin{pmatrix} b \\ d \end{pmatrix} - \begin{pmatrix} \hat{\beta} \\ \hat{\delta} \end{pmatrix}\right\}$$

converges to zero in probability. We then form

$$\sqrt{N}(I_p \otimes \hat{C})\begin{pmatrix} b-\beta \\ d-\delta \end{pmatrix} = -e$$

and, observing that

$$(I_p \otimes \hat{C})\begin{pmatrix} \beta \\ \delta \end{pmatrix}$$

is the tensor form of

$$\frac{1}{N}\left\{\sum_j B(j)\left(\sum_1^N z(n-j)\, z'(n-k)\right) + \Delta \sum_1^N w(n)\, z'(n-k)\right\} = \frac{1}{N}\sum_1^N \epsilon(n)\, z'(n-k),$$

$$\frac{1}{N}\left\{\sum_j B(j)\left(\sum_1^N z(n-j)\, w'(n)\right) + \Delta \sum_1^N w(n)\, w'(n)\right\} = \frac{1}{N}\sum_1^N \epsilon(n)\, w'(n),$$

we see that e is constructed, as a tensor, from the quantities in the matrices

$$\frac{1}{\sqrt{N}}\sum_1^N \epsilon(n)\, z'(n-k), \qquad \frac{1}{\sqrt{N}}\sum_1^N \epsilon(n)\, w'(n).$$

It is easy to see that $\mathcal{E}(e) = 0$ and $\mathcal{E}(ee') = (G \otimes \Gamma)$ which are consistently estimated by $\hat{G} \otimes C$. The proof of the asymptotic normality of e is as in the proof of Theorem 1 in Chapter VI. This establishes Theorem 12, in case B and D are null. If they are not null, they may be estimated as described before the theorem. The proof that this estimation does not affect the statement in Theorem 12 is along the lines of the part (a) of this section. (See Theorem 13.) We omit the details.

(c) We next consider the problem of testing the residuals from a regression for independence, restricting ourselves to the case $p = 1$. In a sense our earlier developments allow us to do this for (a) of this section shows that asymptotically the effects of regression do not affect tests of independence of the residuals $x(n)$ suitable against the alternative that this is generated by a first-order autoregression. Similarly, one might test the independence of the $\epsilon(n)$ in the model (5.10) by including a lag, $s + 1$, in the model and testing the significance of $\hat{B}(s + 1)$. Our purpose here is to obtain more precise results than these and it is for this reason that we have restricted ourselves to $p = 1$. We consider first the model (2.1) for $p = 1$, the null hypothesis being that the $x(n)$ are N.I.D. $(0, \sigma^2)$. A natural statistic to use against the alternative that $x(n) = \rho x(n-1) + \epsilon(n)$ is

$$r_A = \frac{\hat{x}'A\hat{x}}{\hat{x}'\hat{x}}, \tag{5.16}$$

where $\hat{x}$ is the vector of residuals from the regression of $z(n)$ on the $y_j(n)$ and A is a matrix of the kind introduced in Chapter VI, Section 3, which makes r_A into the first autocorrelation of the $\hat{x}(n)$ or something closely related to it.

Theorem 16. *Let r_A be defined by* (5.16) *where $\hat{x} = Qx$, $Q = I_N - Y(Y'Y)^{-1}Y'$, and x is a vector of N components of N.I.D.* (0, 1) *variates. Let the non-null eigenvalues of QAQ be $\nu_1 \geq \nu_2 \geq \cdots \geq \nu_{N-q}$*

$$r_A = \frac{x'QAQx}{x'Qx} = \frac{\sum_1^{N-q} \nu_j \xi_j^2}{\sum_1^{N-q} \xi_j^2}$$

where the ξ_j are N.I.D. (0, 1). *The distribution of r_A may be discovered from* (3.6) *in Chapter VI and has density*

$$\frac{N-q-2}{2} \sum_{p=1}^{k} (\nu_p - r_A)^{\frac{1}{2}(N-q-4)} \prod_{j=1}^{\frac{1}{2}(N-q-1)} (\nu_p - \nu_j); \quad \nu_{k+1} \leq r_A \leq \nu_k. \tag{5.17}$$

Proof. Evidently QAQ and Q commute since Q is idempotent. Thus they may be brought to diagonal form by the same orthogonal transformation. Let P be the orthogonal matrix which does this. The diagonal elements of

PQP' are all unity or zero and we agree to have the unit ones in the first $N - q$ places. Then if $\xi = Px$ the formula for r_A follows immediately. We omit the details of the derivation of (5.17).

Thus in principle the problem is solved. *In practice there seems no reason why the eigenvalues of QAQ, hence the probability of getting a value of r_A greater than that observed, should not be computed.* So far no program seems to be available. Of course, some intermediate procedure requiring fewer computations might be preferable; for example, the mean and variance of r_A are easily found by the methods in Chapter VI, Section 3. We discuss these for the case in which

$$A = A_d = \begin{bmatrix} 1 & -1 & 0 & \cdots & 0 & 0 \\ -1 & 2 & -1 & \cdots & 0 & 0 \\ \cdot & \cdot & \cdot & \cdots & \cdot & \cdot \\ \cdot & \cdot & \cdot & \cdot & \cdot & \cdot \\ 0 & 0 & 0 & \cdots & 2 & -1 \\ 0 & 0 & 0 & \cdots & -1 & 1 \end{bmatrix},$$

as suggested by Durbin and Watson (1950, 1951). (This makes r_A into a multiple of von Neumann's ratio.) We then use, for the statistic r_A, the symbol d introduced by Durbin and Watson. By the methods of Theorem 2 in Chapter VI, if

$$R = 2(N-1) - \operatorname{tr}\{Y'A_d\, Y(Y'Y)^{-1}\},$$
$$S = 2(N-4) - 2\operatorname{tr}\{Y'A_d^2\, Y(Y'Y)^{-1}\} + \operatorname{tr}[\{Y'A_d\, Y(Y'Y)^{-1}\}^2],$$

then

$$\mathcal{E}(d) = \frac{R}{N-q}, \qquad \operatorname{var}(d) = \frac{2\{S - R\mathcal{E}(d)\}}{\{(N-q)(N-q+2)\}}.$$

(In calculating R and S the mean correction may be neglected, since A_d is chosen so that $\tilde{1}$ is an eigenvector for eigenvalue zero.) Durbin and Watson suggest that $\frac{1}{4}d$ be used as a beta variate, that is, one with density

$$\text{(5.18)} \qquad \frac{\Gamma(p+q)}{\Gamma(p)\,\Gamma(q)}\, x^{p-1}(1-x)^{q-1}, \qquad 0 < x < 1, \qquad p > 0, \qquad q > 0.$$

(Of course, $0 < d < 4$, hence the use of $\frac{1}{4}d$.) Since this density has mean $p/(p+q)$ and variance $pq/\{(p+q)^2(p+q+1)\}$ (Cramer, 1946, p. 244) p and q may be found by adjusting these to give the density (5.18) the mean and variance for d. An alternative suggested by Henshaw (1966) [see also Theil and Nagar (1961)] is as follows. Consider

$$x = \frac{d - \nu_{N-q}}{\nu_1 - \nu_{N-q}}. \tag{5.19}$$

This will be in [0, 1], precisely, since the maximum and minimum values for d are ν_1, ν_{N-q}. If this were a beta variate (it is not), since

$$\mathcal{E}(x) = \frac{\mathcal{E}(d) - \nu_{N-q}}{\nu_1 - \nu_{N-q}}, \qquad \text{var}\,(x) = \frac{\text{var}\,(d)}{(\nu_1 - \nu_{N-q})^2},$$

and if p and q, the beta parameters, can be obtained in some other way, we may solve for ν_1 and ν_{N-q}. We may get estimates of suitable p and q from the coefficients of skewness and kurtosis for d, which will be the same as for x since these coefficients are scale and location free. The beta distribution fitted in this way evidently has the correct mean and variance and the correct skewness and kurtosis, but not the correct range. It is evidently quite an adequate procedure but we leave the reader to discover the details from Henshaw's paper. An alternative would be; find ν_1 and ν_{N-q} and use x as defined by (5.19) as a beta variate with the correct mean and variance. This will give d the correct range and mean and variance. The data will not be truly normal in fact and the simple procedure of Durbin and Watson first described is probably adequate. The test of significance based on d has been widely used (and misused) by economists mainly on the basis of an ingenious suggestion put forward by Durbin and Watson, which we now discuss. Let $\theta_1 > \theta_2 \cdots > \theta_{N-1}$ be the eigenvalues of A_d, other than the zero eigenvalue. Then

$$\theta_{j+q-1} \le v_j \le \theta_j \tag{5.20}$$

(there being q regressor vectors y_j of which one is $\tilde{1}$). This follows from the fact† that the jth eigenvalue λ_j of a quadratic form $x'Bx$ is the minimum value that the maximum possible value of the form can assume (subject to $x'x = 1$) when x is subjected to $(j - 1)$ linear constraints (the minimum being over all such sets of constraints). Now, if k additional fixed constraints are imposed, certainly λ_j is not increased, but it cannot be decreased below λ_{j+k}, since k of the $j + k - 1$ constraints are fixed. Taking $B = A_d$, $k = (q - 1)$ and the linear constraints to be those imposed by the requirement that $x = Qy$, we obtain the result (5.20). Thus

$$d_l = \frac{\sum_1^{N-q} \theta_{j+q-1}\xi_j^2}{\sum_1^{N-q} \xi_j^2} \le d \le \frac{\sum_1^{N-q} \theta_j\xi_j^2}{\sum_1^{N-q} \xi_j^2} = d_u.$$

† See Kato (1966, p. 60).

Now the bounding statistic d_l and d_u have distributions depending only upon q and N and not on the actual $y_j(n)$ and their significance points can be tabulated once and for all. This was done for the 1%, 2.5% and 5% points for a one sided test with $N = 15(1)40(5)100$ and $q - 1 = 1(1)5$. (Durbin and Watson use k' for our $q - 1$ and n for our N.) Small values of d correspond to positive autocorrelation (since $d \approx 2(1 - r(1))$ and thus a one-sided test with the critical region consisting of small values of d is customarily used. If d is below the significance point for d_l it is certainly significant at the chosen level and if above that for d_u it is certainly not significant at that level. If it lies between the two the issue is in doubt. In this case, one of the procedures previously suggested may be used to locate a good approximation to the significance point. Thus only when the test is indecisive need further computations be carried out.

It has been pointed out that if the $y_j(n)$ have their (generalized) spectrum concentrated at the origin of frequencies then the significance point got from d_u must be near to the true significance point. (See Hannan and Terrell, 1968 and references therein.) In particular, this will be true for $y_j(n)$ of the form n^{j-1}. The reason for this is, roughly, that the regression procedure, in Fourier terms, resembles a regression on eigenvectors of A_d for the lowest eigenvalues. The point is worth noting since when the $y_j(n)$, as is often the case in economics, have spectra very concentrated at the origin of frequencies and a value of d is observed near to the significance point for d_l but just above it then d is almost certainly significant. However, we do not emphasize the point or go into details since there is no great problem in locating the true significance point fairly accurately and one will prefer to calculate the actual level at which the observed value is just significant rather than only to know that the level is above (say) the 0.05 or 0.01 levels.

We have previously said that this test has been misused by economists and this is true. In particular, it has been used when the residuals are obtained from a regression in which have been included, as regressors, lagged values of the $z(n)$, i.e., a regression of the form of (5.9), but with $p = 1$. It is clear that now d does not have the distribution upon which the above considerations are based. More importantly, the statistic is not even appropriate. Let us consider the model (5.8) which is more directly handled. It will be clear that if the statistic is inappropriate for (5.8) it will be inappropriate for (5.9) also. We replace d by $\hat{r}(1)$, the first autocorrelation of the residuals, to which it is equivalent, neglecting "end effects." Now

$$(5.21) \qquad \frac{1}{N}\sum_{1}^{N}\hat{\epsilon}(n)\,\hat{\epsilon}(n-1) = \hat{c}(1) - \hat{c}_s'\hat{C}_s^{-1}\hat{c}_s(-1) - \hat{c}_s'\hat{C}_s^{-1}\hat{c}_s(+1) + \hat{c}_s'\hat{C}_s^{-1}K_s\hat{C}_s^{-1}\hat{c}_s,$$

where $\hat{c}(1)$, $\hat{C}_s$, $\hat{c}_s$ are as defined in Chapter VI, Section 2 without the circumflexes, but are now computed from the observed residuals $\hat{x}(n)$ (instead of from the actual $x(n)$) from the regression of $z(n)$ on $y(n)$. The vector $\hat{c}_s(-1)$ has $\hat{c}(j-1)$ in the jth place and similarly $\hat{c}_s(+1)$ has $\hat{c}(j+1)$ in that position. The matrix K_s has $\hat{c}(i+1-j)$ in row i, column j. It is easily checked that the second term is $-\hat{c}(1)$ and thus cancels with the first. The last is

$$(\hat{c}(2), \hat{c}(3), \ldots, \hat{c}(s) \,:\, \hat{c}_s'\hat{C}_s^{-1} d_s)\hat{C}_s^{-1}\hat{c}_s,$$

where d_s is $\hat{c}_s$ with its elements in the reverse order. Since the third term is

$$(\hat{c}(2), \hat{c}(3), \ldots, \hat{c}(s) : \hat{c}(s+1))\ \hat{C}_s^{-1}\hat{c}_s,$$

we get for (5.21) the expression

$$(0\ 0, \ldots, 0 \mid -\hat{c}(s+1) + \hat{c}_s'\hat{C}_s^{-1} d_s)\hat{C}_s^{-1}\hat{c}_s.$$

We leave it to the reader to check that

$$-\hat{c}(s+1) + \hat{c}_s'\hat{C}_s^{-1} d_s = \frac{\det\{\hat{C}_{s+2}^{1,s+2}\}}{\det\{\hat{C}_s\}}$$

and that the element in the sth row of $\hat{C}_s^{-1}\hat{c}_s$ is

$$(-)^{s+1}\frac{\det\{\hat{C}_{s+1}^{1,s+1}\}}{\det\{\hat{C}_s\}}.$$

Since (see (2.9) in Chapter VI)

$$\frac{1}{N}\sum_1^N \hat{\epsilon}(n)^2 = \frac{\det\{\hat{C}_{s+1}\}}{\det\{\hat{C}_s\}},$$

we see that $\hat{r}(1)$ is

$$(5.22)\quad (-)^s\left\{(-)^{s+2}\frac{\det(\hat{C}_{s+2}^{1,s+2})}{\det(\hat{C}_{s+1})}\right\}\left\{(-)^{s+1}\frac{\det(\hat{C}_{s+1}^{1,s+1})}{\det(\hat{C}_s)}\right\}$$
$$= (-)^s\,\hat{r}(s+1 \mid 1, \ldots, s)\,\hat{r}(s \mid 1, \ldots, s-1).$$

Here $\hat{r}(s \mid 1, \ldots, s-1)$, for example, is the sth partial autocorrelation between $z(n)$ and $z(n-s)$ eliminating $z(n-1), \ldots, z(n-s+1)$, after removing the effects of regression. From what was said under (a) the effects of the regression will be asymptotically negligible so that we are effectively considering $z(n) - \beta' y(n)$. In any case, (5.22) shows that the statistic is inappropriate since a nonsignificant result will tend to be got if

$$\hat{r}(s \mid 1, \ldots, s-1)$$

is small even if $\hat{r}(s+1 \mid 1, \ldots, s)$ is large; for example, let $s = 1$ and $\beta = 0$ and

$$z(n) + \beta(2)z(n-2) = \epsilon(n), \qquad \beta(2) \neq 0.$$

Now $\rho(1) = \rho(1 \mid 0) = 0$ but $\rho(2 \mid 1) = \beta(2) \neq 0$. Thus having fitted a first-order autoregression we shall test the serial association of the residuals and have a good chance of finding them serially independent contrary to the facts. The appropriate statistic is $\hat{r}(s + 1 \mid 1, \ldots, s)$, as we have suggested.

There is some need to determine more accurately the distribution of statistics such as $\hat{r}(s + 1 \mid 1, \ldots, s)$, computed on the basis of the model (5.8), at least with $p = 1$. The problem is difficult enough with a simple mean correction, as our discussions in Chapter VI showed, even when circular definitions are adopted so that $\tilde{1}$ is an eigenvector of the matrices involved in the definition of the quadratic forms. Exact results are obtainable only when a circular definition is adopted. An alternative to the extension of techniques, such as those of Daniels, for the discovery of asymptotic expansions for the distribution is to find such an asymptotic expansion for the lower moments. In principle this is easily done but in practice the labor is formidable. Some attempt in this direction is made in Hannan and Terrell (1968).

6. TESTS FOR PERIODICITIES

As we mentioned earlier there is a particular hypothesis testing problem which calls for special attention. This arises when a time series $z(n)$ is observed and it is then thought that it contains one or more strictly periodic components. This is a sufficiently common phenomenon to merit special study and, at the same time, it is a problem calling for an hypothesis testing approach, for failing the production of any convincing prior reason for the existence of such a periodicity one is very much inclined to doubt its existence.

The problem was first considered by Fisher (1929) who discussed the case where the model is

$$z(n) = \mu + \rho \cos(n\theta + \phi) + \epsilon(n), \tag{6.1}$$

where the $\epsilon(n)$ are N.I.D. $(0, \sigma^2)$. The null hypothesis is that $\rho = 0$. The alternative hypothesis is that $\rho > 0$ but $\mu, \rho, \theta, \phi, \sigma^2$ are unknown. Fixing θ the maximum of the likelihood is obtained for $\hat{u}, \hat{\rho}, \hat{\phi}$, got by regressing $z(n)$ on the unit constant and $\cos n\theta$, $\sin n\theta$. The absolute maximum is then got by varying θ. Correspondingly the likelihood ratio becomes a function of the ratio of the maximum of the regression sum of squares for $\cos n\theta$, $\sin n\theta$, after mean correction, (maximizing over θ) to the corrected sum of squares for $z(n)$. This is very near to

$$\max\left\{\frac{2\pi I(\pi)}{\sum (z(n) - \bar{z})^2}, \frac{\max_{0<\lambda<\pi} 4\pi I(\lambda)}{\sum (z(n) - \bar{z})^2}\right\}, \quad I(\lambda) = \frac{1}{2\pi N} \left|\sum z(n) e^{in\lambda}\right|^2, \tag{6.2}$$

the difference being entirely due to the fact that $\sum \cos n\lambda$, $\sum \sin n\lambda$, $\sum (\cos n\lambda \sin n\lambda)$ are not quite zero while $\sum \cos^2 n\lambda$, $\sum \sin^2 n\lambda$ are not quite $N/2$. The errors due to the neglect of these differences are of order N^{-1} by the maximized expression, (6.2). We shall see later that this maximized expression (6.2) is of order $\log N$ so that the neglect of the terms such $\sum \cos n\lambda$ will not matter asymptotically. Moreover, for $\lambda = 2\pi j/N$, $j = 1, \ldots, [\frac{1}{2}N]$ there is no neglect. In any case the distribution of (6.2) seems extraordinarily hard to obtain. (See Walker (1965) for some details.) What has been done instead is to consider the quantity

$$\max_{0<j\le[\frac{1}{2}(N-1)]} \left\{\frac{4\pi I(2\pi j/N)}{\sum (z(n) - \bar{z})^2}\right\}, \tag{6.3}$$

where we have now eliminated the frequency π. (The elimination of the frequency 0 is of no concern, since a mean correction will certainly have to be made.) The elimination of π is made to avoid the slight additional complexity associated with this other frequency for which phase is meaningless and with which is associated only one instead of two degrees of freedom. We are concerned mainly with some asymptotic results in this section and the limiting distribution, on the null hypothesis, will not be affected by the inclusion or exclusion of π. Correspondingly we could take our test statistic as the greater of (6.3) and $2\pi I(\pi)/\sum \{z(n) - \bar{z}\}^2$ in case N is even without affecting the validity of the asymptotic result. To simplify the derivations of the exact results we first obtain we shall, however, assume $N = 2P + 1$ so that the maximum in (6.3) is

$$\max_{1\le j\le P} \left\{\frac{4\pi I(\omega_j)}{\sum (z(n) - \bar{z})^2}\right\}.$$

We commence by obtaining the distribution of this on the assumption that $\rho = 0$ (see (6.1)). We follow Whittle (1951). For other derivations the reader may consult Fisher (1929) and Feller (1966). We put

$$x_j = \frac{I(\omega_j)}{\sum_1^P I(\omega_j)} = \frac{y_j}{\sum y_j},$$

where the y_j are independently and identically distributed with the exponential distribution (i.e., having density $\exp -x$). Thus $y_j = \{2\pi I_N(\omega_j)/\sigma^2\}$. Let the random variables g_j be the x_j arranged in order of magnitude so that

$$g_1 \ge g_2 \ge \cdots \ge g_r \ge \cdots \ge g_P.$$

We shall obtain the distribution of g_r on the assumption that $z(n) = \mu + \epsilon(n)$. We could commence with the model

$$z(n) = \mu + \sum_{j=1}^{r} \rho_j \cos (n\theta_j + \phi_j) + \epsilon(n)$$

and restrict ourselves to θ_j of the form $2\pi j/N$. Then the likelihood ratio test for the hypothesis that at most $(r-1)$ of the ρ_j are nonnull will be founded on the rth greatest $I(\omega_j)$, but it will involve only the $M-r+1$ smallest of these quantities, the other $(r-1)$ being sufficient for the $(r-1)$ unspecified ρ_j and ϕ_j. Thus this is not a reason for considering g_r. However, if, say, both g_1 and g_2 were large and were nearly equal, it might be of interest to examine g_2 for significance, as the second greatest; for example, if g_1 and g_2 correspond to adjacent ω_j this might give a better test against a θ midway between these two frequencies. (This is, of course, a test containing a subjective element and really a consideration of the absolute maximum of $I(\theta)$ is called for.)

We obtain the characteristic function of g_r^{-1} as

$$\mathcal{E}\left\{\exp\left(\frac{i\theta}{g_r}\right)\right\} = P\binom{P-1}{r-1}\int_{y_r=0}^{\infty}\int_{y_{r-1}=y_r}^{\infty}\cdots\int_{y_1=y_r}^{\infty}\int_{y_{r+1}=0}^{y_r}\cdots\int_{y_P=0}^{y_r}\exp\left\{i\theta\frac{\sum y_j}{y_r}-\sum y_j\right\}\prod dy_j.$$

The first factor arises from the fact that any one out of the P of the y_j may give the greatest x_j together with the fact that there are $\binom{P-1}{r-1}$ ways of choosing the $(r-1)$ of the $(P-1)$ remaining y_j which are to be greater than the rth greatest one. This expression is easily evaluated as

$$\phi_r(\theta) = \frac{P!}{(P-r)!\,(r-1)!}\int_0^{\infty}\frac{e^{r(i\theta-y)}\{1-e^{i\theta-y}\}^{P-r}}{(1-i\theta/y)^{P-1}}\,dy.$$

For $P > 2$ this is evidently absolutely integrable as a function of θ so that the frequency function of g_r^{-1} is

$$\frac{1}{2\pi}\frac{P!}{(P-r)!\,(r-1)!}\int_{-\infty}^{\infty}\exp(-i\theta g_r^{-1})\phi_r(\theta)\,d\theta$$
$$= \frac{r}{2\pi}\binom{P}{r}\sum_0^{P-r}(-)^j\binom{P-r}{j}\int_{-\infty}^{\infty}\int_0^{\infty}\frac{\exp\{-i\theta/g_r+(i\theta-y)(j+r)\}}{(1-i\theta/y)^{P-1}}\,dy\,d\theta.$$

Consider

$$\int_{-\infty}^{\infty}\frac{\exp i\theta\{j+r-g_r^{-1}\}}{(1-i\theta/y)^{P-1}}\,d\theta.$$

This vanishes for $j+r > g_r^{-1}$ since the pole at $\theta = -iy$ lies in the bottom half plane. For $j+r < g_r^{-1}$ the last integral is

$$2\pi y^{P-1}\frac{\{g_r^{-1}-j-r\}^{P-2}\exp\{y(j+r-g_r^{-1})\}}{(P-2)!}$$

so that we obtain the density function of g_r^{-1} as

$$r\binom{P}{r}\sum_{k=r}^{[g_r^{-1}]}\frac{(-)^{k-r}\binom{P-r}{k-r}(g_r^{-1}-k)^{P-2}}{(P-2)!}\int_0^\infty\left\{\exp\left(\frac{-y}{g_r}\right)\right\}y^{P-1}\,dy$$
$$= r\binom{P}{r}(P-1)\sum_{k=r}^{[g_r^{-1}]}(-)^{k-r}\binom{P-r}{k-r}(1-kg_r)^{P-2}g_r^{\,2}.$$

Adjoining the Jacobian g_r^{-2} and integrating from x to r^{-1}, we obtain

$$P(g_r > x) = \frac{P!}{(r-1)!}\sum_{k=r}^{[1/x]}\frac{(-)^{k-r}(1-kx)^{P-1}}{k(P-k)!\,(k-r)!}. \tag{6.4}$$

(To obtain this result we integrate $(1-kg_r)^{P-2}$ over the range from x to k^{-1} since this term comes into the sum defining the density for g_r only when $g_r < k^{-1}$.) For $r = 1$ the 5 and 1% significance points were tabulated in Fisher (1950, p. 16.59a) for $P = 5(1)50$. For $r = 2$ the 5% points were tabulated for $P = 3(1)10(5)50$ in Fisher (1940).

We may, asymptotically, consider

$$Pg_r - \log P = \frac{y_{(r)}}{s} - \log P,$$

where $s = P^{-1}\sum y_j$ and we have now used $y_{(r)}$ for the rth largest of the y_j. Since $(\log P)(s^{-1}-1)$ converges in probability to zero we are led to consider $(y_{(r)} - \log P)/s$ and in the same way may finally consider $(y_{(r)} - \log P)$. Since $(y_j - \log P)$ has the density

$$P(y_j - \log P \le x) = \left(1 - \frac{e^{-x}}{P}\right),$$

then

$$\lim_{N\to\infty} P\{g_r > P^{-1}(x+\log P)\} = \lim_{N\to\infty} P(y_{(r)} - \log P > x)$$
$$= \lim_{N\to\infty}\left\{1 - \frac{\sum_{j=0}^{r-1}\binom{P}{j}(1-e^{-x}/P)^{P-j}e^{-xj}}{P^j}\right\}$$
$$= 1 - \exp\{-e^{-x}\} \ne 0, \qquad x > 0.$$

For $r = 1$ and $P = 50$ the 5 and 1% points from the exact distribution are 0.160 and 0.131. The corresponding values of x are 2.9702, 4.6002, and thus the significance points for g_r from the asymptotic formula are 0.170, 0.137, which are only moderate approximations.

We summarize these results in† the following theorem:

Theorem 17. *If* $\epsilon(n)$ *is N.I.D.* $(0, \sigma^2)$ *and*

$$I(\epsilon, \omega_j) = \frac{1}{2\pi N}\left|\sum_1^N \epsilon(n)e^{in\omega_j}\right|^2, \qquad N = 2P + 1$$

then putting

$$g_r = r\text{th greatest of } \frac{I(\epsilon, \omega_j)}{\sum_{j=1}^M I(\epsilon, \omega_j)}, \qquad j = 1, \ldots, P$$

we have $P(g_r > x)$ *given by* (6.4). *As* $N \to \infty$ $P(g_r > P^{-1}(x + \log P))$ *converges to* $1 - \exp(-e^{-x})$ *and in this case we may take* $0 \le j \le [\frac{1}{2}N]$.

The fact that the values $j = 0$ and $\frac{1}{2}N$ (N even) may be included in the asymptotic result follows immediately from the fact that $I(\epsilon, 0)$, $I(\epsilon, \pi)$ are distributed as multiples of a chi-square variate with one degree of freedom so that with arbitrarily high probability they will be less than some suitable constant.

The test as it stands is of limited value because it is assumed that $\epsilon(n)$ has uniform spectrum and is Gaussian. The first defect was, substantially, removed by Whittle (1951). We follow Walker (1965), who also pointed out how unimportant is the assumption of normality.

We assume that

$$z(n) = \mu + \rho\cos(n\theta + \phi) + x(n),$$

where‡

$$(6.5) \qquad x(n) = \sum_{-\infty}^{\infty} \alpha(j)\,\epsilon(n-j), \qquad \sum_{-\infty}^{\infty} |\alpha(j)|\,|j|^\delta < \infty, \qquad \delta > 0.$$

Then since the $\epsilon(n)$ are normal we may apply the results of Section 2 in Chapter V, which show that the kth absolute moment of

$$r_N(\lambda) = I(\lambda) - \frac{2\pi}{\sigma^2} f(\lambda)\, I(\epsilon, \lambda)$$

is of order $N^{-\delta k}$, $\delta \le \frac{1}{2}$; $N^{-\frac{1}{2}k}$, $\delta \ge \frac{1}{2}$, uniformly in λ. Now

$$P\{\max_{1 \le j \le M} |r_N(\omega_j)| \ge a\} \le \tfrac{1}{2}NP\{|r_N(\omega_1)| \ge a\} \le CNa^{-k}N^{-\delta k}, \qquad \delta \le \tfrac{1}{2}$$

† We change our notation from $I(\lambda)$ to $I(\epsilon, \lambda)$ because of the need to distinguish the statement of this theorem from that of later ones.

‡ Since we are here taking $x(n)$ to be Gaussian, the assumption that $x(n)$ comes from a linear process is not troublesome. The condition on the $\alpha(j)$ seems mild but could no doubt be weakened further.

with the same formula, but δ is replaced by $\frac{1}{2}$, for $\delta \geq \frac{1}{2}$. (Here we have used Markov's inequality.) Thus for all $a > 0$, $\delta > 0$, by taking k large enough, we see that

$$\lim_{N\to\infty} P\{\max_{1\leq j\leq M} |r_N(\omega_j)| \geq a\} = 0$$

Let $H(\lambda) = I(\lambda)/f(\lambda)$. Now assume further that $f(\lambda) \geq a > 0$, $\lambda \in (-\pi, \pi]$, and consider

$$I(\epsilon, \omega_j) - |H(\omega_j) - I(\epsilon, \omega_j)| \leq H(\omega_j) = \frac{I(\omega_j)}{f(\omega_j)}$$
$$\leq I(\epsilon, \omega_j) + |H(\omega_j) - I(\epsilon, \omega_j)|$$

and observe that

$$\max_j |H(\omega_j) - I(\epsilon, \omega_j)|$$

converges in probability to zero by what has just been shown. Thus we have proved the following theorem:

Theorem 18. *If $x(n)$ is generated by* (6.5) *with the $\epsilon(n)$ as in Theorem* 17 *and $f(\lambda) \geq a > 0$, $\lambda \in (-\pi, \pi]$, then, for each $x > 0$,*

$$\lim_{N\to\infty} P\left\{\max_{0\leq j\leq[\frac{1}{2}N]} H(\omega_j) - \log P > x\right\} = 1 - \exp(-e^{-x})$$

$$\lim_{N\to\infty} P\left\{\max_{0\leq j\leq[\frac{1}{2}N]} \left\{\frac{H(\omega_j)}{P^{-1}\sum_0^{[\frac{1}{2}N]} H(\omega_j)}\right\} - \log P > x\right\} = 1 - \exp(-e^{-x}).$$

The second statement of the theorem follows from the fact that,

$$P^{-1}\sum_0^{[\frac{1}{2}N]} H(\omega_j) = \left\{\frac{2\pi}{\sigma^2 P}\right\}\sum I(\epsilon, \omega_j) + P^{-1}\sum\left(\frac{r_N(\omega_j)}{f(\omega_j)}\right).$$

Since the last term evidently converges in probability to zero and the first converges to unity with probability 1, then $1 - P^{-1}\sum H(\omega_j)$ converges in probability to zero, which establishes the result.

The second form of test statistic in the theorem is introduced because it seems the most natural analog of that of Fisher's theorem (Theorem 17). Thus we replace $I(\omega_j)$ with $I(\omega_j)/f(\omega_j)$, which is approximately $I(\epsilon, \omega_j)$, and proceed to use it as if it had been $I(\epsilon, \omega_j)$. This idea is due to Whittle (1951), as previously mentioned.

There are two further steps which have to be taken. The first of these is to replace $f(\omega_j)$ by some estimate and the second is to relax the requirement of normality upon the $\epsilon(n)$. The first step is easily taken. Consider, for example, the case in which $x(n)$ is autoregressive of order p. Then

$$(6.6)\quad \mathcal{E}\{\max_{\lambda} |\hat{f}(\lambda) - f(\lambda)|\} = \mathcal{E}\left\{\max_{\lambda} \frac{1}{2\pi}\left| \sum_0^p\sum \{\hat{\beta}(j)\,\hat{\beta}(k) - \beta(j)\,\beta(k)\}\, e^{i(j-k)\lambda}\right|\right\}$$
$$\leq \frac{1}{2\pi} \sum_0^p\sum \mathcal{E}\{|\hat{\beta}(j)\,\hat{\beta}(k) - \beta(j)\,\beta(k)|\}$$
$$= O(p^2 N^{-1}).$$

Thus by the same argument we may replace $f(\omega_j)$ by $\hat{f}(\omega_j)$ when $x(n)$ is generated by an autoregression. We call the resulting statistic $\hat{H}(\omega_j)$. The theorems discussed in Chapter VI, Section 5 allows this result to be extended to other finite parameter models. We shall not discuss that further here for the following kind of reason. If there is a jump in the spectrum of $x(n)$ at θ then it is very likely that this will lead to an estimate of the autoregressive spectrum which is large in the neighborhood of θ. (This is discussed in an exercise to this chapter.) This will, of course, lead to a reduction in the power of the test. For this reason it will be necessary to remove the effects of the jump from the series $x(n)$ by regression (on $\cos n\hat{\theta}$, $\sin n\hat{\theta}$, $\hat{\theta}$ being the estimated frequency) before estimating $f(\lambda)$. This, of course, involves us in an iterative procedure. Presumably one would search the spectrum for the maximum $I(\omega_j)$, compute $\hat{f}(\omega_j)$ (as an autoregressive estimate) *after* removing the effects of $\cos n\omega_j$, $\sin n\omega_j$ by regression and then compute $\hat{H}(\omega_j)$. It is conceivable that one will not now have located the maximum of $\hat{H}(\omega_j)$. If one wishes to be sure of doing that one will have to form $\hat{H}(\omega_j)$, as just described, at every point ω_j. This requires a separate autoregressive estimation procedure at every point.† This is not by any means an impossible computational procedure, particularly when it is remembered that an autoregressive model (or any other finite parameter model) is only likely to be used when N is not very large.

An alternative procedure is to use a smoothed estimate of $f(\lambda)$ and the natural procedure is to use the FFT or Cooley–Tukey procedures (see Chapter V). If N is within the range which makes this computationally feasible the FFT will almost certainly be used as, in principle, $I(\omega_j)$ is needed for the test in any case. We discuss, first, a procedure based upon the FFT and later how other procedures may be used in case N is too large to make the computation of all $I(\omega_j)$ economical. We assume that with each N we have associated an m (or $M = [N/2m]$) which is determined not only by N but also

† The fact needs checking that (6.6) still holds when the $\hat{\beta}(j)$, and hence $\hat{f}(\lambda)$, are got after removing $\cos n\lambda$, $\sin n\lambda$, by regression, from $x(n)$. It is easy to see that this is so.

by the assumed "bandwidth" of the spectrum, $f(\lambda)$, of the residuals, $x(n)$, in the relation $z(n) = \mu + \rho \cos(n\theta + \phi) + x(n)$. The integer m has, of course, the same meaning as in Chapter V. It is clear that it is pointless to consider testing $\rho = 0$ without some prior concept of the bandwidth of $f(\lambda)$ for it will be impossible to distinguish a jump in the spectrum from an arbitrarily high, thin peak in its absolutely continuous component on the basis of a finite amount of data. Our assumption concerning this bandwidth is embodied in the choice of m. We assume that as $N \to \infty$, $m \to \infty$ and $m/N \to 0$. We consider the estimate

$$\tilde{f}(\omega_j) = \frac{1}{m} \sum_k{}' I(\omega_{j-k}), \tag{6.7}$$

where $\sum'$ is over m values of k such that the ω_{j-k} are symmetrically located around† ω_j but so that ω_j is not included in the sum.

We now point out that

$$\max_{0 \le j \le [\frac{1}{2}N]} |\tilde{f}(\omega_j) - f(\omega_j)|$$

converges in probability to zero. Certainly

$$\lim_{N\to\infty} \max_j |\mathcal{E}\{\tilde{f}(\omega_j) - f(\omega_j)\}| = \lim_{N\to\infty} \max_j \left|\mathcal{E}\left\{\frac{1}{m} \sum_k I(\omega_{j-k}) - f(\omega_j)\right\}\right|,$$

where now we have included $I(\omega_j)$ in the sum. By Theorem 10 in Chapter V the last expression is $O(M^{-q})$, where $2mM = N$ and $q = \min(\delta, 1)$. Moreover, in place of $I(\omega_{j-k}) - \mathcal{E}(I(\omega_{j-k})$ we may consider

$$|\sum \alpha(p) \exp(ip\omega_{j-k})|^2 \left\{I(\epsilon, \omega_{j-k}) - \frac{\sigma^2}{2\pi}\right\},$$

since we know that

$$\max_j \{|\sum \alpha(p) \exp(ip\omega_{j-k})|^2 I(\epsilon, \omega_{j-k}) - I(\omega_{j-k})\}$$

converges in probability to zero. It is easy to show that

$$\max_j \left\{I(\epsilon, \omega_j) - \frac{\sigma^2}{2\pi}\right\}$$

converges in probability to zero since the $(2\pi/\sigma^2)I(\epsilon, \omega_j)$ are independently distributed as exponential variates. Since $|\sum \alpha(p) \exp(ip\lambda)|^2$ is a continuous function of λ it follows that

$$\max_{0 \le j \le [\frac{1}{2}N]} |\tilde{f}(\omega_j) - f(\omega_j)|$$

† For m odd, so far as our theoretical development is concerned, we may adopt an arbitrary convention such as that there is one more ω_{j-k} to the left of ω_j than to the right.

converges in probability to zero and we may use this $\tilde{f}(\omega_j)$ in place of $f(\omega_j)$ in Theorem 18, that is, by replacing $H(\omega_j)$ by $\tilde{H}(\omega_j) = I(\omega_j)/\tilde{f}(\omega_j)$.

In fact, though we shall not go into details here, the same applies to any estimate of $f(\omega_j)$ made by any of the methods of Chapter V. This is evidently true of the Cooley–Tukey estimator, for which the proof is almost the same as in the present section, but estimators of the remaining types of Chapter V may be handled in a similar way through formula (4.3) of Chapter V vis.,

$$\hat{f}_K(\lambda) = \int K_N(\lambda - \theta) I(\theta)\, d\theta.$$

In this circumstance we would be led to use

$$\tilde{f}_K(\omega_j) = \frac{\hat{f}_K(\omega_j) - (2\pi/N)K_N(0)\, I(\omega_j)}{1 - (2\pi/N)K_N(0)} \tag{6.8}$$

and to use it to form $\tilde{H}_K(\omega_j) = I(\omega_j)/\tilde{f}_K(\omega_j)$. So far as the asymptotic distribution on the null hypothesis is concerned there will be no effect from this adjustment but the power of the test will be improved.† We state the result as

Theorem 19. *If $x(n)$ is generated by* (6.5) *with $\epsilon(n)$ as in Theorem* 17 *and $\tilde{f}(\omega_j)$ is computed according to* (6.7), *or more generally* (6.8), *and $M \to \infty$, $M/N \to 0$ as $N \to \infty$ then the conclusions of theorem* 18 *hold true if H is replaced by $\tilde{H} = I/\tilde{f}$ (or more generally $\tilde{H}_K = I/\tilde{f}_K$).*

If the periodogram is too expensive to compute then, of course, we shall have to proceed in other ways. We might act as follows. Compute the quantities $I(\omega_j')$, $\omega_j' = 2\pi j/N'$, $N \le N' = 2^s < 2N$, $j = 0, 1, \ldots, 2^{s-1}$, and the Cooley–Tukey estimator

$$\tilde{f}^{(c)}(\omega_j') = \frac{1}{m'} \sum_k{}' I(\omega_{j-k}'),$$

where again $\sum'$ omits the term $k = 0$. Then find the maximum of $I(\omega_j')/\tilde{f}^{(c)}(\omega_j')$. Let ω' be the maximizing value of ω_j' and let ω be the value of $2\pi j/N$ nearest to ω'. We might then form $I(\omega)/\tilde{f}^{(c)}(\omega')$ and use this in place of the maximum of $\tilde{H} = I/\tilde{f}$. This test can be invalid only insofar as ω is not the value which maximizes $\tilde{H}$ but since this means that we err on the side of caution in allocating a significance level to our result we may not be too concerned about it.

† The degree to which the adjustments embodied in changing from $\hat{f}$ to $\tilde{f}$ will be effective in increasing power will depend upon the location of a jump θ. If it lies halfway between two points ω_j, ω_{j+1} the reduction in power may still be considerable even after the use of $\tilde{f}$. For a further discussion see Nicholls (1967).

The conclusion of Theorem 19 is remarkably general as our assumptions concerning $x(n)$ have been quite minor, *with one exception*, this being that $x(n)$ is Gaussian. Following Walker (1965) we now point out that this assumption is itself of minor importance so that, to a satisfactory approximation, the results extend to the case where the $\epsilon(n)$ are independent and identically distributed but not normal. What we shall do is to show that under the circumstances discussed in Theorem 17, provided moments of $\epsilon(n)$ of sufficiently high order exist, then the probability got under the (wrong) assumption of normality will, under the circumstances relevant to a test of significance, be very near to the true probability. Of course, Theorems 18 and 19 are based upon assumptions of normality also, but only insofar as moments of any order exist for that distribution. We shall not restate Theorems 18 and 19 but they certainly continue to hold in the same approximate fashion (in the sense that $I(\epsilon, \omega_j)$ may be replaced by $I(\omega_j)/\tilde{f}(\omega_j)$ without asymptotic distributions being affected) if either moments of a high enough order exist or some stronger condition, such as

$$\sum_{-\infty}^{\infty} |\alpha(n)|\,|n|^{1/2} < \infty$$

is imposed.

We consider

$$P\left\{\max_j \sigma^{-2} 2\pi I(\omega_j, \epsilon) > \log \frac{P}{p}\right\}.$$

Let A_j be the event that $\sigma^{-2}2\pi I(\omega_j, \epsilon) > \log (P/p)$. Then (Feller, 1957, p. 100)

$$\sum_{j=1}^{P} P(A_j) - \sum_{j>k=1}^{P} P(A_j \cap A_k) \le P\left(\bigcup_j A_j\right) \le \sum_{j=1}^{P} P(A_j).$$

However, if an absolute moment of order ≥ 6 of the $\epsilon(n)$ is finite then it follows from a slight generalization of a theorem in Cramér (1937, p. 81) that uniformly in j

$$\lim_{N\to\infty} \frac{P(A_j)}{\exp(-\log(P/p))} = 1$$

and similarly that

$$\lim_{N\to\infty} \frac{P(A_j \cap A_k)}{\exp(-2\log(P/p))} = 1.$$

Thus we have, for N sufficiently large,

$$(6.9)\quad p - \tfrac{1}{2}p^2 < P \exp\left(-\log\frac{P}{p}\right) - \tfrac{1}{2}P(P-1)\exp\left(-2\log\frac{P}{p}\right)$$
$$\le P\left\{\max_j \sigma^{-2}2\pi I(\omega_j, \epsilon) > \log\frac{P}{p}\right\} \le P\exp\left(-\log\frac{P}{p}\right) = p,$$

since, for example,

$$\overline{\lim_{N\to\infty}} P\left\{\max_j \sigma^{-2} 2\pi I(\omega_j, \epsilon) > \log \frac{P}{p}\right\} \leq p.$$

The relation (6.9) is the approximation we sought for.

This result shows that the effects of nonnormality must be rather small (provided the appropriate moments exist) since p will be a small number (e.g., 0.05 or 0.01) so that $\frac{1}{2}p^2$ will be very small indeed. Thus if in using Theorem 19, say, we find that the significance level of our observed $\max_j H(\omega_j)$ is p (in the sense that the asymptotic theory gives us this chance of exceeding the observed value) then we shall be entitled to assume that the true level of significance lies between $p - \frac{1}{2}p^2$ and p (subject to the usual qualifications concerning the use of asymptotic results). For practical purposes the knowledge that the true significance level lies within this range should be sufficient.

Other procedures for testing for periodicities have been suggested, in particular by Priestley (1962). Priestley's technique, which we do not describe in detail, is founded upon the use of the difference between two smoothed spectral estimators, one being based on a much greater degree of smoothing (i.e., smaller M and larger m) than the other. Let M_1 and M_2, $M_1 \gg M_2$, be the two relevant constants and $P(\lambda)$ the corresponding difference between the two spectral estimators (the one for the smaller M_2 being subtracted). Then the test statistic is

$$\max_k C(M_1, M_2, N) \sum_{j=1}^{k} P\left(\frac{2\pi j}{M_1}\right),$$

where $C(M_1, M_2, N)$ is a suitable sequence of constants. Some power comparisons are made in Priestley (1962) between his test and the test earlier discussed, *upon the assumption that the jump occurs at a point* $2\pi j/M_1$. Taking $M_1 = [\frac{1}{2}N]$, $M_2 = [\log N]$ his test then becomes most powerful (but only when this is done). However, even then the comparison is invalid for if it were known that the jump occurred at such a point then $I(\lambda)$ would be computed only at the point $2\pi j/M_1$ and the test based upon his $I(\lambda)$ values would again be the more powerful. Priestly suggests that in practice his test be used by first locating the point, λ_0, at which $P(\lambda)$ is at its greatest value and then using

$$\max_k C(M_1, M_2, N) \sum_{j=1}^{k} P\left(\lambda_0 + \frac{2\pi j}{M_1}\right).$$

However, it seems as difficult here to determine the effects of the use of this most suitable grid of points, $\lambda_0 + 2\pi j/M_1$, as it would be to determine the distribution of $\max I(\lambda)$ (the maximum being over all $\lambda \in (-\pi, \pi]$). Moreover in cases when M_1 is large (i.e., near to $\frac{1}{2}N$), which are the only cases in which Priestley's test appears to have (or can be expected to have) power comparable to that of the test based on $I(\lambda)$, there seems no reason to believe that the effects of so locating the grid will be of a different order of magnitude from those arising from using max $I(\lambda)$ in place of max $I(\omega_j)$ in the test we have presented.

We conclude this section with a short discussion of the multivariate situation. This is considerably more complex for two reasons. The first is that the exact distribution of the likelihood ratio statistic is not easy to find (in the circumstances corresponding to Theorem 17) and the second is that a wider range of situations presents itself so that a wider range of test statistics might be used. This latter circumstance arises from the fact that the apparent periodicity might have been observed in a variety of ways. It might, for example, have been observed in the two (let us say) components of the vector series, or in the coherence between them (perhaps in a circumstance where normally it might be thought that there would be zero coherence). The phase relationships between the two series are also relevant since if a force induces a periodicity in (say) two series it is not unreasonable, in some circumstances, to expect that these two oscillations should be in phase.

The first situation we are led to investigate is that in which it is believed that the relevant model is

$$z_j(n) = \mu_j + \rho_j \cos(n\theta + \phi_j) + \epsilon_j(n), \qquad j = 1, \ldots, p,$$

wherein the ρ_j, ϕ_j, μ_j but also θ are unknown. The $\epsilon_j(n)$ are assumed to be N.I.D. $(0, G)$, where G is nonsingular. If θ is required to be of the form $2\pi k/N$, $N = 2P + 1$, then the likelihood ratio test for the hypothesis $\rho_j \equiv 0$ is based upon the statistic

$$\min_k \{\Lambda_k(2P, p, 2)\} = \min_k \frac{\det\{\sum_j I(\omega_j) - I(\omega_k) - I(\omega_{-k})\}}{\det\{\sum_j I(\omega_j)\}},$$

wherein the sum is over values of j from 1 to $P = \frac{1}{2}(N - 1)$. The distribution of

$$\frac{1 - \sqrt{\Lambda_k(2P, p, 2)}}{\sqrt{\Lambda_k(2P, p, 2)}} \frac{N - p - 2}{p}$$

is known to be the F distribution with $2p$ and $2(N - p - 2)$ degrees of freedom. However, the distribution of the minimum value of this statistic is not easy to obtain. As N increases, the distribution of

$$x_k = -(N - 3 - \tfrac{1}{2}p) \ln \{\Lambda_k(2P, p, 2)\}$$

approaches that of chi square with $2p$ degrees of freedom. An asymptotic treatment of this hypothesis testing problem could undoubtedly be founded upon this result. One would, of course, need a theorem analogous to

Theorem 19 above before a useful result is arrived at. Again this could be founded upon Theorem 2 in Chapter V. If G is assumed to be diagonal (which corresponds to the series being incoherent on the null hypothesis, in the situation analogous to Theorems 18 and 19) the relevant test statistic becomes a product of those for each component of the vector taken individually In the case $p = 2$ some investigations of this situation have been made in Nicholls (1969). In this type of situation the phase angle is also of some interest since phase agreement at the frequency at which the periodograms maximize is added evidence for the presence of a jump. This aspect of the problem is also discussed by Nicholls.

7. DISTRIBUTED LAG RELATIONSHIPS

In economics, in particular, in recent years there has been considerable interest in models of the type

$$z(n) = \alpha + \sum_{j=-s}^{t} B(j)'y(n-j) + x(n), \tag{7.1}$$

wherein $z(n)$, $x(n)$ are vectors of p components and $y(n)$ is a vector of q components. We say more about their stochastic nature later. In fact the models tend to be more special than (7.1), which indeed falls within the framework erected in Sections 1 to 5 of this chapter. A typical scalaı case is that in which†

$$z(n) = \alpha_0' + \alpha_1 \sum_{0}^{\infty} \beta^j y(n-j) + x(n),$$

so that the coefficients are prescribed in terms of three constants, α_0', α_1, β. Such models, where the $B(j)$, perhaps infinite in number, are functions of only a relatively few parameters have an obvious appeal in situations where the sample size is not large so that the investment (with attendant risks) in a special model may lead to a substantial reward (if the model is appropriate). The literature surrounding such models is now quite large [see Griliches (1967) for a survey and Amemiya and Fuller (1967) for some additional work].‡ We shall not, therefore, attempt to be definitive in our treatment of the subject in the space available here. The methods we discuss have a general applicability and are not restricted in their application to an econometric context. Indeed they arose, in the author's experience, through an oceanographic statistical problem.

We begin by discussing (7.1). Throughout this section we confine ourselves to the use of a simple mean correction for $z(n)$ and $y(n)$, though the methods

† We use α_0' as a notation so as to reserve α_0 for another use at the end of the section.
‡ See also Dhrymes (1969), Fishman (1969).

and, no doubt, the theorem extend to more elaborate corrections. We assume, to begin with, that $x(n)$ and $y(m)$ are independent for all m, n and that $x(n)$ is generated by a linear process

$$x(n) = \sum_{-\infty}^{\infty} A(k)\,\epsilon(n-k), \qquad \sum_{-\infty}^{\infty} \|A(k)\| < \infty, \tag{7.2}$$

where the $\epsilon(n)$ are vectors of p components which are i.i.d. $(0, G)$. As we have already said, the model (7.1) can now be regarded as a special case of (2.1) since we have only to take the $y(n)$ as there introduced to be made up of a partitioned vector of $s + t + 1$ partitioned parts, the typical part being the vector $y(n-j)$ occurring in (7.1). However this procedure could be computationally very troublesome if s or t were large and in addition leads to a rather complicated and cumbersome procedure when s and t have to be obtained by a process of successively adding new terms to the regression. Of course, least squares procedure could be used but if the accuracy of the estimates is to be measured, as we have already said, the computational burden is as great as for the asymptotically efficient method discussed in Section 4. For these reasons another procedure has been suggested (Hannan, 1963, 1967c). We discuss the case $p = q = 1$ first so as to show more clearly the meaning of the method. Having estimated the $\beta(j)$ we estimate α by

$$\hat{\alpha} = \bar{z} - \sum \hat{\beta}(j)\bar{y}.$$

We have the relation between spectra

$$f_{zy}(\lambda) = \left\{\sum_{-s}^{t} \beta(j)e^{ij\lambda}\right\} f_y(\lambda)$$

if we assume, as we shall, that $y(n)$ is stationary with absolutely continuous spectrum. This suggests estimating the $\beta(j)$ from the approximate relation

$$\hat{f}_{zy}(\lambda_k) = \sum_{-s}^{t} \hat{\beta}(j)e^{ij\lambda_k}\hat{f}_y(\lambda_k), \qquad \lambda_k = \frac{2\pi k}{2M},$$

where, as we have said earlier, $\hat{f}_{zy}$, $\hat{f}_y$ are computed from mean corrected data. *We discuss important conditions which need to be imposed on the spectral estimation procedure and on $y(n)$ in Theorem 20 below.* Thus

$$\hat{\beta}(j) = \frac{1}{2M} \sum_{k=-M+1}^{M} \frac{\hat{f}_{zy}(\lambda_k)}{\hat{f}_y(\lambda_k)} e^{-ij\lambda_k}. \tag{7.3}$$

Since $\hat{f}_{zy} = \frac{1}{2}(\hat{c}_{zy} - i\hat{q}_{zy})$, this formula reduces to

$$\hat{\beta}(j) = \frac{1}{2M} \sum_{0}^{M} \delta_k \frac{(\hat{c}_{zy}(\lambda_k)\cos j\lambda_k - \hat{q}_{zy}(\lambda_k)\sin j\lambda_k)}{\hat{f}_y(\lambda_k)}. \tag{7.4}$$

The advantage of the procedure is obvious. Each $\hat{\beta}(j)$ is individually obtained and s or t may be increased without recomputing any $\hat{\beta}(j)$ obtained earlier. Moreover, as we shall see under suitable conditions, the covariance matrix of the $\sqrt{N}(\hat{\beta}(j) - \beta(j))$ is

$$\left[\frac{1}{2\pi}\int_{-\pi}^{\pi}\frac{f_x(\lambda)}{f_y(\lambda)}e^{i(j-k)\lambda}\,d\lambda\right], \tag{7.5}$$

the element in row j, column k being shown. (We shall indicate, in Theorem 20, conditions under which the $\hat{\beta}(j)$ are asymptotically normal with this covariance matrix.) This is to be compared to

$$\left[\frac{1}{2\pi}\int_{-\pi}^{\pi}\frac{f_y(\lambda)}{f_x(\lambda)}e^{i(j-k)\lambda}\,d\lambda\right]^{-1}, \tag{7.6}$$

which is the asymptotic covariance matrix for $\sqrt{N}(\tilde{\beta}(j) - \beta(j))$, where $\tilde{\beta}(j)$ is the best linear unbiased estimate. (This is easily got from (3.4).) The two are equal for arbitrary s, t when and only when $f_y(\lambda)/f_x(\lambda)$ is constant almost everywhere. One can thus expect the method to be fairly efficient if $f_y(\lambda)/f_x(\lambda)$ does not vary too greatly. This may be contrasted with the least squares procedure which will be efficient if $f_x(\lambda)$ is constant. The three methods will therefore be equal when both spectra are constant. The matrix (7.5) (or (7.6) for that matter) may be easily estimated in the present instance before the estimation procedure for the $\beta(j)$ is commenced, for it is, again showing the entry in row j column k,

$$\left[\frac{1}{2\pi}\int_{-\pi}^{\pi}\frac{f_z(\lambda)}{f_y(\lambda)}\{1 - \sigma^2(\lambda)\}e^{i(j-k)\lambda}\,d\lambda\right].$$

(The situation is special because the variables upon which we are regressing are lagged values of a single variable.) This may be estimated by

$$\left[\frac{1}{2M}\sum_{u=-M+1}^{M}\frac{\hat{f}_z(\lambda_u)}{\hat{f}_y(\lambda_u)}\{1 - \hat{\sigma}^2(\lambda_u)\}e^{i(j-k)\lambda_u}\right], \tag{7.7}$$

$$= \left[\frac{1}{M}\sum_{u=0}^{M}\delta_u\left\{\frac{\hat{f}_z(\lambda_u)}{\hat{f}_y(\lambda_u)}\left(1 - \hat{\sigma}^2(\lambda_u)\right)\cos(j-k)\lambda_u\right\}\right],$$

the elements in any diagonal being independent of the row in which they lie. There is another point which merits further investigation than we give here. We have

$$\sum_{-M+1}^{M}\hat{\beta}(j)^2 = \frac{1}{2M}\sum_{-M+1}^{M}\frac{\hat{\sigma}^2(\lambda_u)\hat{f}_z(\lambda_u)}{\hat{f}_y(\lambda_u)}. \tag{7.8}$$

Of course, no one will suggest that the method being discussed may be used for arbitrary s and t. Clearly a valid asymptotic theory will require s

and t fixed (as we shall require) or at least $(s + t)$ increasing only slowly compared with M. However, (7.8) does give some indication how large remaining $\hat{\beta}(j)$ can be. Thus, if s and t are small and the sum of squares of the computed $\hat{\beta}(j)$ is a while (7.8) is b, we know at least that the next few $\hat{\beta}(j)$ must have sum of squares not greater than $(b - a)$. Moreover,†

$$\frac{1}{2M} \sum_{-M+1}^{M} \frac{\hat{f}_z(\lambda_u)}{\hat{f}_y(\lambda_u)} - \{\text{var}\,(\hat{\beta}(j))\}^{\wedge} = b,$$

$$\frac{(b - a)}{\text{var}\,(\hat{\beta}(j))^{\wedge}} = \frac{(b - a)}{(c - b)},$$

$$c = \frac{1}{2M} \sum_{-M+1}^{M} \frac{\hat{f}_z(\lambda_u)}{\hat{f}_y(\lambda_u)},$$

so that knowing $a, b, c, a \leq b \leq c$, we may gauge whether it will be necessary to include more terms in the estimation procedure. Clearly a "goodness of fit" test is called for (and available) here but we shall pursue the matter no further.

A numerical example with $p = q - 1$ wherein $z(n)$ is sea level and $y(n)$ is atmospheric pressure (at Sydney, Australia) gave the following results. Here $N = 548$, $M = 30$, $s = t = 3$

$j =$	-3	-2	-1	0	1	2	3
$\hat{\beta}(j) =$	0.068	0.068	0.111*	−0.646*	0.069	−0.304*	0.021

The value of (7.7), for $j = k$, is $(0.041)^2$ so that the quantities with asterisks are significantly different from zero. The right side of (7.8) is 0.624. The sum of squares of the 7 of the $\hat{\beta}(j)$ computed is 0.537 so that most of the variation in $z(n)$ which can be explained by the $y(n - j)$ is accounted for. In this case the ratio f_x/f_y appears near to constant but f_x is very far from that. If f_x/f_y were precisely constant then the $\hat{\beta}(j)$ are asymptotically independent (under the conditions we discuss below). Thus we could test the next few $\hat{\beta}(j)$ by

$$\frac{\sum' \hat{\beta}(j)^2}{\text{var}\,(\hat{\beta}(j))^{\wedge}},$$

where $\sum'$ sums over the additional j values (say ν in number). This is approximately distributed as chi-square with ν d.f. We know that the numerator is not greater than 0.087. The denominator is 0.0017 so that the ratio is not greater than (approximately) 50. We are tempted to use this as chi-square with 53 d.f., since there are 53 remaining coefficients and one is inclined to

† We put the circumflex in the following expression outside of the bracket, to indicate that it is the estimate of var $(\hat{\beta}(j))$ which we are considering.

believe that the remaining coefficients represent random variations from zero values. The validity of this procedure needs investigation, of course. Of course no great labor would be involved in discovering a much greater number of $\hat{\beta}(j)$.

Unfortunately there is a not insignificant bias problem with the method. This problem arises from the fact that filtering a process and then computing a spectral estimate does *not* give the same result as computing the spectrum of the unfiltered process and multiplying this by the square of the gain of the filter.

We consider only the case in which $y(n)$ is stationary to the fourth order with absolutely continuous spectrum and

$$(7.9)\quad f_y(\lambda) = \frac{1}{2\pi} \sum_{-\infty}^{\infty} \Gamma_y(n) e^{-in\lambda}, \quad \sum_{-\infty}^{\infty} |n|^r \, \|\Gamma_y(n)\| < \infty, \quad \det\big(f_y(\lambda)\big) \geq a > 0;$$

$y(n)$ also has fourth cumulant function that satisfies

$$(7.10)\qquad \sum\sum\sum_{n_1, n_2, n_3 = -\infty}^{\infty} |k_{ijkl}(0, n_1, n_2, n_3)| < \infty.$$

However, we also need a condition on the spectral estimation procedure to control the bias effect mentioned above, namely,

$$(7.11)\qquad \overline{\lim_{\delta \to 0}} \frac{k(x + \delta) - k(x)}{\delta x^u} \leq a < \infty.$$

We introduce the estimation equations

$$(7.3)'\quad \hat{B}'(j) = \frac{1}{2M} \sum_{k=-M+1}^{M} \hat{f}_{zy}(\lambda_k) \hat{f}_y(\lambda_k)^{-1} e^{-ij\lambda}, \qquad \hat{\alpha} = \bar{z} - \sum_{-s}^{t} \hat{B}'(j)\bar{y}.$$

Thus

$$(7.4)'\quad \hat{B}'(j) = \frac{1}{M} \sum_{0}^{M} \delta_k \{ \big(\hat{c}_{zy}(\lambda_k)\, \hat{u}_y(\lambda_k) - \hat{q}_{zy}(\lambda_k)\, \hat{v}_y(\lambda_k)\big) \cos j\lambda_k - \big(\hat{q}_{zy}(\lambda_k)\, \hat{u}_y(\lambda_k) + \hat{c}_{zy}(\lambda_k)\, \hat{v}_y(\lambda_k)\big) \sin j\lambda_k \},$$

where

$$\hat{u}_y = (\hat{c}_y + \hat{q}_y \hat{c}_y^{-1} \hat{q}_y)^{-1}, \qquad \hat{v}_y = -\hat{c}_y \hat{q}_y (\hat{c}_y + \hat{q}_y \hat{c}_y^{-1} \hat{q}_y)^{-1},$$

as in Section 4. If $p = 1$, $\hat{B}'(j)$, $\hat{c}_{zy}$, $\hat{q}_y$, are row vectors. In general these are matrices of p rows and q columns.

We introduce the column vectors $\beta(j)$, $\hat{\beta}(j)$ which are related to $B(j)$, $\hat{B}(j)$ as are β, $\hat{\beta}$ to B, $\hat{B}$ in Section 2, namely such that $\beta(j)$ has $\beta_{uv}(j)$ in row $(v - 1)q + u$. In case $p = 1$, $\beta(j) = B(j)$ of course. Now we shall show that the covariances between the elements of $\sqrt{N}(\hat{\beta}(j) - \beta(j))$ and $\sqrt{N}\hat{\beta}((k) - \beta(k))$ are given, asymptotically, by the matrices

$$(7.5)' \qquad \frac{1}{2\pi}\int_{-\pi}^{\pi} \{f_x(\lambda) \otimes f_y'(\lambda)^{-1}\} e^{-i(j-k)\lambda}\, d\lambda,$$

which may be estimated by

$$(7.7)' \quad \frac{1}{M}\sum_{0}^{M} \delta_k\{(a \otimes u_y + b \otimes v_y)\cos(j-k)\lambda_u + (a \otimes v_y - b \otimes u_y)\sin(j-k)\lambda_u\},$$

wherein we have suppressed the argument symbol, λ_u, in a, b, u_y, v_y, and

$$a = c_{zy}u_y c_{yz} - q_{zy}v_y c_{yz} - q_{zy}u_y q_{yz} + c_{zy}v_y q_{yz},$$
$$b = q_{zy}u_y c_{yz} + c_{zy}v_y c_{yz} - c_{zy}u_y q_{yz} + q_{zy}v_y q_{yz}.$$

The relation analogous to (7.8) is

$$(7.8)' \qquad \sum_{-M+1}^{M} \operatorname{tr}(\hat{B}(j)'\, \hat{B}(j)) = \frac{1}{2M}\sum_{-M+1}^{M} \operatorname{tr}(\hat{f}_{zy}\hat{f}_y^{-2}\hat{f}_{yz}),$$

whose detailed expression, in terms of the real and imaginary part of $\hat{f}_y$ and $\hat{f}_{zy}$, we omit.

Now we have†

Theorem 20. *Let $z(n)$, $y(n)$ and $x(n)$ be related by* (7.1) *where $x(n)$ is generated by* (7.2) *and $y(n)$ is fourth order stationary with spectrum satisfying* (7.9) *and fourth cumulant function satisfying* (7.10). *Let* $\lim_{N\to\infty} M^2/N = 0$ *and let spectra and cross spectra be estimated by a process satisfying one of the following conditions;*

(i) *The $\hat{f}$ are of the type* (4.2) *in Chapter V with k_n satisfying* (4.4) *in Chapter V and with $k(x) = 0$, $|x| > 1$ and satisfying* (7.11) *for $u \le r + [r]$ and*

$$(7.12) \qquad \lim_{N\to\infty} \frac{N}{M^q} = 0, \qquad q = 2u + 2.$$

(ii) *The $\hat{f}$ are obtained by the truncated formula with*

$$(7.13) \qquad \overline{\lim_{N\to\infty}} \frac{N}{M^q} < \infty, \qquad q = 2\{r + [r]\}.$$

(iii) *The $\hat{f}$ are obtained from the FFT, Cooley-Tukey or Daniell procedures (see examples* 1, 2, *and* 8 *of Chapter V, Section* 4) *and $r \ge 1$ while*

$$(7.14) \qquad \lim_{N\to\infty} \frac{N}{M^4} = 0.$$

† For the Gaussian case a much stronger theorem is proved in Wahba (1969).

Let the $B(j)$ and α be estimated by (7.3)′ *and let* $\beta(j)$, $\hat{\beta}(j)$ *be defined in terms of* $B(j)$, $\hat{B}(j)$ *as above* (7.5)′. *Then as* $N \to \infty$ *the joint distribution of the elements of* $\sqrt{N}(\hat{\beta}(j) - \beta(j))$ *converges to the multivariate normal distribution with mean vector and covariance matrix having the* (j, k)th *block of entries corresponding to covariances between* $\hat{\beta}(j)$ *and* $\hat{\beta}(k)$ *which is* (7.5)′. *This may be consistently estimated by* (7.7)′.

We have placed the proof of this theorem in the appendix to this chapter for it is fairly tedious and is also closely related to the proof of Theorem 10 which is also in the appendix.

The most satisfactory result from an asymptotic point of view is probably (ii) for here (7.11) does not arise and thus provided $M^2/N \to 0$ and r in (7.9) is sufficiently large then the theorem holds. The only procedures satisfying the conditions (i) of the theorem among those listed as examples in Chapter V, Section 4 are the Tukey-Hanning and Parzen procedures for which u cannot be greater than unity so that we must have (7.9) with $r = 1$ and $N = o(M^4)$. The same is true for the Cooley-Tukey and FFT procedures. Thus in all of these cases M must tread the fairly narrow line between $M^2/N \to 0$ and $M^4/N \to \infty$. The Bartlett procedure (example 4 of Chapter V, Section 4) does not satisfy the conditions of the theorem. Of course, once more these are asymptotic results and must be treated with caution appropriate to such results. The nett conclusion is that the method should be satisfactory for estimation procedures 1, 2, 3, 5, 6, 8 (of Chapter V, Section 4) provided $f_y(\lambda)$ is sufficiently smooth and M is large enough to ensure that bias effects are negligible and small enough to ensure that the variances decrease appropriately.

Since the case $p = 1$, $q > 1$ is probably of special importance we give some computing details for this below. The formulae are

$$(7.4)'' \qquad \hat{\beta}'(j) = \frac{1}{M}\sum_0^M \delta_k\{(\hat{c}_{zy}(\lambda_k)\,\hat{u}_y(\lambda_k) - \hat{q}_{zy}(\lambda_k)\,\hat{v}_y(\lambda_k))\cos j\lambda_k - (\hat{q}_{zy}(\lambda_k)\,\hat{u}_y(\lambda_k) + \hat{c}_{zy}(\lambda_k)\,\hat{v}_y(\lambda_k))\sin j\lambda_k\},$$

where

$$\hat{u}_y = (\hat{c}_y + \hat{q}_y\hat{c}_y^{-1}\hat{q}_y)^{-1}, \qquad \hat{v}_y = -\hat{c}_y\hat{q}_y(\hat{c}_y + \hat{q}_y\hat{c}_y^{-1}\hat{q}_y)^{-1}.$$

We estimate the covariance matrix between $\sqrt{N}(\hat{\beta}(j) - \beta(j))$ and $\sqrt{N}(\hat{\beta}(k) - \beta(k))$ by means of

$$(7.7)'' \qquad \frac{1}{M}\sum_0^M \delta_u\hat{w}(\lambda_u)\{\hat{u}_y(\lambda_u)\cos(j-k)\lambda_u + \hat{v}_y(\lambda_u)\sin(j-k)\lambda_u\},$$

where

$$\hat{w}(\lambda) = \hat{f}_z(\lambda) - \hat{f}_{zy}(\lambda)\hat{f}_y(\lambda)^{-1}\hat{f}_{yz}(\lambda)$$
$$= \hat{f}_z(\lambda) - \tfrac{1}{2}\{\hat{c}_{zy}(\lambda)\,\hat{u}_y(\lambda) - \hat{q}_{zy}\,\hat{v}_y(\lambda)\}\hat{c}_{yz}(\lambda)$$
$$- \tfrac{1}{2}\{\hat{c}_{zy}(\lambda)\,\hat{v}_y(\lambda) + \hat{q}_{zy}(\lambda)\,\hat{u}_y(\lambda)\}\,\hat{q}_{yz}(\lambda).$$

We also have

$$\sum_{-M+1}^{M} \hat{\beta}(j)'\,\hat{\beta}(j) = \frac{1}{2M}\sum_{-M+1}^{M}\{\hat{f}_z(\lambda_u) - \hat{w}(\lambda_u)\} \tag{7.8$''$}$$
$$= \frac{1}{M}\sum_{0}^{M}\delta_u\{\hat{f}_z(\lambda_u) - \hat{w}(\lambda_u)\},$$

so that the $\hat{f}_z(\lambda_u)$ and $\hat{w}(\lambda_u)$ may be used to assess how large the remaining $\hat{\beta}(j)$ can be.

We close this section by returning to the relation (7.2). This can be reduced to the simple form

$$\begin{aligned} z(n) &= \alpha_0 + \alpha_1 y(n) + \beta z(n-1) + u(n), \qquad \alpha_0 = \alpha_0'(1-\beta), \\ u(n) &= x(n) - \beta x(n-1). \end{aligned} \tag{7.15}$$

If the $u(n)$ are independent then we already know that we may satisfactorily estimate (7.15) by means of least squares regression (see Theorem 15). This case is therefore hardly relevant. There does not seem to be a very good reason for commencing from (7.2) rather than (7.15) itself. (If the spectrum of $x(n)$ is known then the neglect of that information in (7.15) would lead to a loss of information, but this does not seem to be a very likely circumstance.) If the $u(n)$ are thought not to be independent it seems natural to include further lagged variables (of $z(n)$ and $y(n)$) on the right of (7.15) in the hope of once more bringing (7.15) to the form covered by Theorem 15. In any case a simple estimation procedure for (7.15) has been provided by Liviatan (1963). We put $\alpha_0 = 0$ (and drop the subscript on α_1) for notational simplicity *but the use of mean corrected sums of squares and cross products below would make this unnecessary*. Then we introduce the equations

$$\begin{aligned} \sum_2^N z(n)\,y(n) &= \hat{\alpha}\sum_2^N y(n)^2 + \hat{\beta}\sum_2^N z(n-1)\,y(n), \\ \sum_2^N z(n)\,y(n-1) &= \hat{\alpha}\sum_2^N y(n)\,y(n-1) + \sum_1^{N-1} z(n)\,y(n). \end{aligned} \tag{7.16}$$

We have chosen the ranges of summation so that (7.13) when combined with (7.12) leads to

$$\begin{aligned} (\hat{\alpha}-\alpha)\sum y(n)^2 + (\hat{\beta}-\beta)\sum z(n-1)\,y(n) &= \sum u(n)\,y(n), \\ (\hat{\alpha}-\alpha)\sum y(n)\,y(n-1) + (\hat{\beta}-\beta)\sum z(n)\,y(n) &= \sum u(n)\,y(n-1). \end{aligned} \tag{7.17}$$

However, no difference will be made in the final conclusions we state if the sums of squares and cross products are adjusted to contain the maximum number of terms; therefore we replace the first row of (7.16), for example with

$$N^{-1}\sum_1^N y(n)\,z(n) = \hat{\alpha}_1 N^{-1}\sum_1^N y(n)^2 + \hat{\beta}(N-1)^{-1}\sum_2^N y(n)\,z(n-1).$$

Indeed such a replacement should probably be made.

We assume that

$$x(n) = \sum_{-\infty}^{\infty}\alpha(k)\,\epsilon(n-k), \qquad \sum_{-\infty}^{\infty}|\alpha(k)| < \infty,$$

with the $\epsilon(n)$ serially independent and identically distributed as usual, and further that $y(n)$ satisfies Grenander's conditions. Then these conditions are satisfied by the sequence

$$\sum_0^{\infty}\beta^j\,y(n-j),$$

it being assumed that $y(n)$ is zero for $n < 0$.

Putting

$$d(N) = \sum_1^N y(n)^2,$$

we may arrange (7.17) in the form

$$d(N)\begin{pmatrix}(\hat{\alpha}-\alpha)\\(\hat{\beta}-\beta)\end{pmatrix} = \hat{A}^{-1}\hat{B}\,d(N)b,$$

where

$$\hat{A} = d(N)^{-2}\begin{bmatrix}\sum y(n)^2 & \sum y(n-1)\,z(n)\\ \sum y(n)\,y(n-1) & \sum y(n)\,z(n)\end{bmatrix},$$

$$\hat{B} = d(N)^{-2}\begin{bmatrix}\sum y(n)^2 & \sum y(n)\,y(n-1)\\ \sum y(n)\,y(n-1) & \sum y^2(n)\end{bmatrix},$$

whereas b is just the vector of regression coefficients of $u(n)$ on $y(n)$ and $y(n-1)$. Now $\hat{A}^{-1}\hat{B}$ converges in probability to

$$\begin{bmatrix}1 & \alpha\sum\beta^j\rho_y(j-1)\\ \rho_y(1) & \alpha\sum\beta^j\rho_y(j)\end{bmatrix}^{-1}\begin{bmatrix}1 & \rho_y(1)\\ \rho_y(1) & 1\end{bmatrix} = A^{-1}B,$$

provided A is nonsingular (which we assume). On the other hand, the distribution of $d(N)b$ is known to converge to a normal distribution with zero mean and covariance matrix

$$B^{-1}2\pi\int_{-\pi}^{\pi} f_u(\lambda)\,M(d\lambda)B^{-1}. \tag{7.18}$$

Indeed this follows immediately from Theorem 8, Chapter IV, if we remember that $B = R(0)$ in this case. Here $M(\lambda)$ is the spectral distribution function for the correlation sequence of the vector with components $y(n)$, $y(n-1)$. This is

$$M(d\lambda) = \begin{bmatrix} 1 & e^{i\lambda} \\ e^{-i\lambda} & 1 \end{bmatrix} m(d\lambda),$$

where

$$\rho_y(n) = \int_{-\pi}^{\pi} e^{-in\lambda}\, m(d\lambda),$$

so that the joint distribution of $d(N)(\hat{\alpha}_1 - \alpha_1)$, $d(N)(\hat{\beta} - \beta)$ is asymptotically normal with zero mean vector and covariance matrix

$$A^{-1} 2\pi \int_{-\pi}^{\pi} f_u(\lambda) \begin{bmatrix} 1 & e^{i\lambda} \\ e^{-i\lambda} & 1 \end{bmatrix} m(d\lambda) A^{-1}.$$

If $f_u(\lambda) = \gamma_u/2\pi$ then this becomes $\gamma_u A^{-1}$. We shall not discuss this method further but leave it to the reader to see how (7.18) may be estimated. (See Hannan, 1965.) Evidently Liviatan's treatment will extend to more elaborate models. Indeed, if we consider that $\beta(j)$ satisfy

$$\sum_0^{\infty} \beta(j) z^j = \frac{p(z)}{q(z)}, \qquad p(z) = \sum_0^{p} p_j z^j, \qquad q(z) = \sum_0^{q} q_j z^j, \qquad q(0) = 1,$$

where $q(z)$ has no zeros in or on the unit circle, we obtain

$$\sum q_j\, z(n-j) = \sum p_j\, y(n-j) + u(n),$$
$$u(n) = \sum q_j\, x(n-j).$$

Again we leave it to the reader to elaborate Liviatan's estimation procedure and the asymptotic treatment of its statistical properties which we have given.

Liviatan's procedure is not asymptotically efficient. In simple cases we can approach the problem via the method of maximum likelihood (on Gaussian assumptions), for example, when $x(n)$ is assumed serially independent. Even in these simple cases the equations of maximum likelihood are nonlinear and are not easy to solve. In the simple model (7.2), when $x(n)$ is not narrowly prescribed, an application of maximum likelihood to the Fourier transformed data after the (approximate) fashion of Section 4, Chapter VI, leads to a reasonable computational procedure (see Hannan, 1965) which gives asymptotically more efficient estimators than Liviatan's. However, the case is too special and the details too tedious to be worth discussion here. It is evident that if the $x(n)$ are assumed to be generated by a moving average scheme (or a mixed moving average-autoregressive scheme with generating function $r(z)/q(z)$) then, assuming the generating function for $\sum \beta(j)z^j$ to be $p(z)/q(z)$, the problem reduces to a generalization of the mixed moving-average autoregressive scheme studied in Chapter VI, with an additional term of the form $\sum r_j y(n-j)$. (Putting $r(z) = \sum r_j z^j$.) In this form the problem seems solvable but we shall not attempt its solution here.

We conclude by mentioning that the models of this section may be generalized to allow for time-dependent means. Thus in place of (7.1) we may consider

$$\{z(n) - \Delta_1 y(n)\} = \sum_{-s}^{t} B(j)\{w(n-j) - \Delta_2 y(n-j)\} + x(n),$$

where $y(n)$ satisfies what we have called Grenander's conditions. In this case we initially regress $z(n)$ on $y(n)$ and $w(n)$ on $y(n)$ and carry out the remaining procedures with the residuals. The regression may be accomplished either by least squares or, say, by the efficient method of Section 5. In the same way as in Section 5 it may be shown that the effects of the regression will not asymptotically affect the procedure for the estimation of the $B(j)$.

EXERCISES

1. Show that the following statements are equivalent.
   (a) $\mathcal{A}$ is a linear subspace of the finite dimensional space $\mathfrak{X}$ and $A\mathcal{A} \subset \mathcal{A}$ where A is symmetric.
   (b) $\mathcal{A}$ is spanned by eigenvectors of A.
2. Show by a direct argument that

$$\int f(\lambda)\, m(d\lambda) \int f(\lambda)^{-1}\, m(d\lambda),$$

wherein $m(\lambda)$, as in Section 3, is minimized by taking all of the weight distributed by $m(\lambda)$ at the two sets of points in $(-\pi, \pi]$ at which $f(\lambda)$ is least and is greatest (half of the weight going to each set).

3. Let $\hat{\beta}$ and $\tilde{\beta}$ be as in Theorem 9 and $f(\lambda)$ the spectrum of the stationary process generated by

$$x(n) = \rho x(n-1) + \epsilon(n)$$

where $\epsilon(n)$ is as usual and $|\rho| < 1$ is a nonzero scalar. Show that $\hat{\beta}$ and $\tilde{\beta}$ have asymptotically the same covariance matrix if and only if the elements of the spectrum are points or pairs of points symmetrically placed with respect to the origin.

4. Prove Theorem 14.

5. Find the coefficients of skewness and kurtosis of the statistic x of (5.19) and hence describe in detail the procedure outlined below that formula for finding an approximation to the significance point for d.

6. Let

$$z(n) = \mu + \rho \cos (n\theta + \phi) + \epsilon(n).$$

Find the contribution to $\mathcal{E}\{I(\omega_j)\}$ from the term $\rho \cos (n\theta + \phi)$ and to $\mathcal{E}\{I(\theta)\}$. Hence discuss the effects of using the maximum over the grid ω_j in place of the actual maximum of $I(\lambda)$ in Section 6.

7. Hannan (1961b). Let

$$z(n) = y(n) + x(n),$$

$$\sum_{0}^{q} \beta(j)\, x(n-j) = \epsilon(n), \quad \beta(0) = 1; \quad y(n) = \sum_{-p}^{p} \alpha_u e^{in\theta_u}; \quad \alpha_u = \bar{\alpha}_{-u}, \theta_u = -\theta_u,$$

$$\mathcal{E}(\epsilon(n)) = 0, \mathcal{E}(\epsilon(n)^2) = \sigma^2, \qquad \mathcal{E}(x(m)\, x(m+n)) = \gamma(n),$$

where the $\epsilon(n)$ are as usual. Let $\hat{\beta}$ be the estimate of the vector β, of q components, having $\beta(j)$ in the jth place, got from the usual autoregressive equations with no mean correction. Show that $\hat{\beta}$ converges in probability to

$$\beta - \sum_u |\alpha_u|^2 h(\theta_u) \left\{\Gamma_q + \sum_u |\alpha_u|^2 A_u\right\}^{-1} a_u,$$

where

$$h(\theta_u) = \sum \beta(j) e^{-ij\theta_u},$$

Γ_q has $\gamma(m-n)$ in row m, column n, $m, n = 1, \ldots, q$, while A_u has $\exp i(m-n)\theta_u$ in that place and a_u is a vector of q components with $\exp im\theta_u$ in the mth place. To show this show first that

$$\sum_{j=0}^{q} \left\{\beta_j N^{-1/2} \sum_{m=1}^{N} z(m-j)\, z(m-k)\right\}$$

converges in probability to a random variable (with finite mean and variance) plus

$$N^{1/2} \sum_u |\alpha_u|^2 e^{ik\theta_u} h(\theta_u).$$

Hence show that $\hat{h}(\lambda)$, the estimate of $h(\lambda)$ got from the $\hat{\beta}(j)$, converges in probability to

$$h(\lambda) - \sum_u |\alpha_u|^2 h(\theta_u) a_\lambda^* \left\{\Gamma_q + \sum_u |\alpha_u|^2 A_u\right\}^{-1} a_u,$$

where a_λ has $\exp in\lambda$ in the nth place.

Now let $\beta(j) = 0, j > q_0$ and take $q \gg q_0$. Use the approximation

$$\gamma(n) = \int_{-\pi}^{\pi} e^{in\lambda} f(\lambda)\, d\lambda \approx \frac{2\pi}{q} \sum e^{is\lambda_j} f(\lambda_j),$$

$$-\pi < \lambda_j = \frac{2\pi j}{q} \leq \pi.$$

Assume that θ_u is among the λ_j. Using these approximations evaluate the probability limit of $\hat{h}(\lambda_j)$ as

$$\frac{h(\theta_u)}{\{1 + q\,|h(\theta_u)|^2\,|\alpha_u|^2/\gamma(0)\}}, \qquad \lambda_j = \theta_u,$$

$$h(\lambda_j), \qquad \lambda_j \neq \theta_u, \qquad u = -p, \ldots, p.$$

Remembering that the spectral density is $(\sigma^2/2\pi)\,|h(\lambda)|^{-2}$, and that if $y(n)$ is present the order of the autoregression generating $x(n)$ is very likely to be grossly overstated, discuss the relevance of this result to the use of an autoregressive procedure for the estimation of the spectral density of $x(n)$ in connection with a test for the presence of $y(n)$.

APPENDIX

1 *Proof of Theorem* 10

We shall prove this theorem assuming $x(n)$ to be generated by a linear process with $\sum \|A(j)\| < \infty$.

We consider first

$$\frac{1}{2M}\sum_{-M+1}^{M}\left\{\hat{f}_x\left(\frac{\pi k}{M}\right) \otimes I_q\right\}^{-1} \hat{e}_{zy}\left(\frac{\pi k}{M}\right), \tag{1}$$

which we write as

$$\frac{1}{2M}\sum_k \{\hat{f}_x \otimes I_q\}^{-1}\hat{e}_{zy}$$

for simplicity of printing. We shall do the same with other sums of the same form. We observe that $\hat{e}_{zy} = (I_p \otimes \hat{f}_y')\beta + \hat{e}_{xy}$, where $\hat{e}_{xy}$ is got from $\hat{f}_{xy}$ as $\hat{e}_{zy}$ was from $\hat{f}_{zy}$. Thus we see that

$$\tilde{\beta} = \beta + \left\{\frac{1}{2M}\sum_k (\hat{f}_x^{-1} \otimes \hat{f}_y')\right\}^{-1}\left\{\frac{1}{2M}\sum_k (\hat{f}_x^{-1} \otimes I_q)\hat{e}_{xy}\right\},$$

and we need consider only the second term. First consider

$$N^{-1}(I_p \otimes D_N)\left\{\frac{1}{2M}\sum \hat{f}_x^{-1} \otimes \hat{f}_y'\right\}^{-1}(I_p \otimes D_N). \tag{2}$$

To save repetitions of the proof we consider only the case in which estimation procedures for spectra are of the form

$$\hat{f}_{uv} = \sum_{-M+1}^{M} k\left(\frac{n}{M}\right) c_{uv}(n)\left(1 - \frac{|n|}{N}\right)e^{in\lambda}.$$

The Cooley-Tukey procedure, for example, may be handled similarly. We have made the lower limit $-M+1$ to simplify formulae below. The effect is negligible.

We first observe that

$$\max \|\hat{f}_x(\lambda) - f_x(\lambda)\|$$

converges in probability to zero, where $\|A\|$ means the norm of A. Indeed, using $\tilde{f}_x(\lambda)$ for the estimate of $f_x(\lambda)$ constructed from the actual $x(n)$ and not estimates of them, we have, from (4.1),

$$\{\hat{f}_x - f_x\} = \{\tilde{f}_x - f_x\} - (\hat{B} - B)'\hat{f}_{yx} - \hat{f}_{xy}(\hat{B} - B) + (\hat{B} - B)'\hat{f}_y(\hat{B} - B).$$

However, the last three terms converge in probability to zero uniformly in λ; for example, we write the third term in the form

$$(I_p \otimes \hat{f}_{xy})(\hat{\beta} - \beta) = (I_p \otimes \hat{f}_{xy} D_N^{-1})(I_p \otimes D_N)(\hat{\beta} - \beta).$$

The typical entry in $\hat{f}_{xy} D_N^{-1}$ is

$$\frac{1}{N}\frac{1}{2\pi}\sum_{-M+1}^{M} k\left(\frac{n}{M}\right)\left\{\frac{\sum_m x_j(m)\, y_k(m+n)}{d_k(N)}\right\} e^{-in\lambda}.$$

Now

$$\mathcal{E}\left\{\max_\lambda \left|\frac{1}{2\pi N}\sum_{-M+1}^{M} k\left(\frac{n}{M}\right)\left\{\frac{\sum_m x_j(m)\, y_k(m+n)}{d_k(N)}\right\} e^{-in\lambda}\right|\right\}$$

$$\le \frac{1}{2\pi N}\sum_{-M+1}^{M}\left|k\left(\frac{n}{M}\right)\right| \mathcal{E}\left\{\left|\frac{\sum_m x_j(m) y_k(m+n)}{d_k(N)}\right|\right\} = O\left(\frac{M}{N}\right),$$

since $\sum x_j(m)y_k(m+n)/d_k(N)$ has mean zero and standard deviation not greater than the supremum of the spectrum for $x_j(n)$. Since $(I_p \otimes D_N)(\hat{\beta} - \beta)$ converges in distribution then

$$\max_\lambda \|\hat{f}_{xy}(\hat{B} - B)\|$$

converges in probability to zero as required. The other two terms may be treated similarly and we see that max $\|\hat{f}_x(\lambda) - f_x(\lambda)\|$ converges in probability to zero if

$$\max_\lambda |\tilde{f}_{ij}(\lambda) - f_{ij}(\lambda)|, \qquad i, j = 1, \ldots, p,$$

does so, where the f_{ij}, $\tilde{f}_{ij}$ are for the $x(n)$ sequence, of course. Consider, for example, the case where, dropping the (i, j) subscript for convenience,

$$\tilde{f}(\lambda) = \frac{1}{2\pi}\sum_{-M+1}^{M} k\left(\frac{n}{M}\right) c(n)\left(1 - \frac{|n|}{N}\right)e^{-in\lambda}.$$

Then

$$\max_\lambda |\tilde{f}(\lambda) - f(\lambda)| \le \max_\lambda |\tilde{f}(\lambda) - \mathcal{E}(\tilde{f}(\lambda))| + \max_\lambda |\mathcal{E}(\tilde{f}(\lambda)) - f(\lambda)|.$$

The second term on the right converges to zero by Theorem 10, Chapter V. The first is dominated by

$$\frac{1}{2\pi}\sum_{-M+1}^{M}\left|k\left(\frac{n}{M}\right)\right| |c(n) - \gamma(n)| \left(1 - \frac{|n|}{N}\right),$$

whose expectation is dominated by

$$\left(\frac{K}{\sqrt{N}}\right)\sum_{-M+1}^{M}\left|k\left(\frac{n}{M}\right)\right|,$$

since the standard deviation of $c(n)$ is $O((N-n)^{-1/2})$. Since $M/\sqrt{N} \to 0$ the result is established.

Thus outside of a set S in the sample space of all histories of $x(n)$, whose probability measure may be taken arbitrarily small for N large,

$$\|\hat{f}_x(\lambda)\| \geq a > 0.$$

Now

$$\begin{aligned}
&\left\|\frac{N}{2M}\sum_k \{(\hat{f}_x^{-1} - f_x^{-1}) \otimes D_N^{-1} f_y' D_N^{-1}\}\right\|^2 \\
&\qquad \leq \left\{\frac{1}{2M}\sum_k \|\hat{f}_x^{-1} - f_x^{-1}\|^2\right\}\left\{\frac{N}{2M}\sum_k \|D_N^{-1} f_y' D_N^{-1}\|^2\right\} \\
&\qquad \leq \frac{1}{2M}\sum_k \{\|\hat{f}_x^{-1}\|^2 \|\hat{f}_x - f\|^2 \|f_x^{-1}\|^2\} \\
&\qquad\quad \times \frac{N}{2M}\sum_k \operatorname{tr}\{(D_N^{-1} f_y' D_N^{-1})^2\}.
\end{aligned}$$

Outside of S this is dominated by

$$K\sum_k \|\hat{f}_x - f_x\|^2, \tag{3}$$

since

$$\frac{N}{(2M)^2}\sum_k \operatorname{tr}\{(D_N^{-1} f_y D_N^{-1})^2\} \leq K,$$

where again we have used K as a general finite positive constant, not always the same one. But

$$K\sum_k \|\hat{f}_x - f_x\|^2 \leq K\sum_k \|\hat{f}_x - \tilde{f}_x\|^2 + K\sum_k \|\tilde{f}_x - f_x\|^2.$$

Since $\|\hat{f}_x - \tilde{f}_x\|^2 \leq \operatorname{tr}\{(\hat{f}_x - \tilde{f}_x)^2\}$, which is easily seen to be $O(N^{-1})$, whereas $\operatorname{tr}\{(\tilde{f}_x - f_x)^2\}$ has expectation $O(M/N)$, and since $M^2/N \to 0$, we see that the expectation of (3) taken over the whole sample space is $o(1)$. Thus we have reduced (2) to the form

$$N^{-1}(I_p \otimes D_N)\left\{\frac{1}{2M}\sum_k (f_x^{-1} \otimes f_y')\right\}^{-1}(I_p \otimes D_N). \tag{4}$$

However

$$f_x^{-1}(\lambda) = \frac{1}{2\pi}\sum_{-\infty}^{\infty}\Delta(j)e^{ij\lambda}, \qquad \sum_{-\infty}^{\infty}\|\Delta(j)\| < \infty.$$

This follows from the fact that f_x has an absolutely convergent Fourier series and has determinant which is never null. Thus each element of f_x^{-1} is a rational function of functions which have absolutely convergent Fourier series and its denominator is never zero. It follows from a famous theorem due to Wiener (see, for example, Naimark, 1960, p. 205) that $f_x^{-1}(\lambda)$ also has an absolutely convergent Fourier series. The expression (4) thus is

$$\left\{\frac{1}{2\pi}\sum_{-\infty}^{\infty}\Delta(j)\otimes\frac{1}{2\pi}\hat{R}'(j)k\left(\frac{j}{M}\right)\right\}^{-1},$$

where by $\hat{R}(j)$ we mean the matrix with entries

$$\hat{\rho}_{uv}(j) = \frac{\sum_1^{N-j} y_u(m)y_v(m+j)}{\{\sum_1^N y_u(m)^2\sum_1^N y_v(m)^2\}^{1/2}} = \hat{\rho}_{vu}(-j) = \hat{P}_{uv}(j+2lm), \qquad j \geq 0.$$

By dominated convergence this converges to

$$\left\{\left(\frac{1}{2\pi}\right)^2\sum_{-\infty}^{\infty}\Delta(j)\otimes R(-j)\right\}^{-1}$$

$$= \left\{\left(\frac{1}{2\pi}\right)^2\sum_{-\infty}^{\infty}\Delta(j)\otimes\int_{-\pi}^{\pi}e^{ij\lambda}\,M'(d\lambda)\right\}^{-1} = \left\{\frac{1}{2\pi}\int_{-\pi}^{\pi}f_x^{-1}(\lambda)\otimes M'(d\lambda)\right\}^{-1}.$$

Thus we have established that (2) converges in probability to

$$\left\{\int\left\{2\pi f_x(\lambda)\right\}^{-1}\otimes M'(d\lambda)\right\}^{-1}.$$

We need to consider therefore

$$\frac{N}{2M}\sum_k\{\hat{f}_x\otimes D_N\}^{-1}\hat{e}_{xy}.$$

Now we again seek to replace $\hat{f}_x$ with f_x and thus consider

$$\frac{N}{2M}\sum\{f_x^{-1}\otimes I_q\}\{(\hat{f}_x - f_x)\otimes I_q\}\{\hat{f}_x^{-1}\otimes I_q\}(I_p\otimes D_N^{-1})\hat{e}_{xy}, \tag{5}$$

and again outside of S its norm is dominated by

$$\left(\left[K\frac{N}{2M}\sum_k\|\hat{f}_x - f_x\|^2\right]\left[\frac{1}{2M}\sum\|(I_p\otimes N^{-1/2}D_N)^{-1}e_{xy}\|^2\right]\right)^{1/2}. \tag{6}$$

The second factor under the square root sign is

$$N^{-1}\sum_{v=1}^{q}\sum_{u=1}^{p}\sum_{j=-M+1}^{M}k^2\left(\frac{j}{M}\right)\left\{\frac{N^{-1/2}\sum' x_u(n)\,y_v(n+j)}{\sqrt{N^{-1}\sum y_v(n)^2}}\right\}^2,$$

where $\sum'$ is over $1 \le n,\ n + j \le N$. This has expectation dominated by

$$KN^{-1} \sum_v \sum_u \sum_j k^2\left(\frac{j}{M}\right) = O\left(\frac{M}{N}\right).$$

Thus (6) converges in probability to zero since

$$K \sum_k \|\hat{f}_x - f_x\|^2$$

does so, as we have already seen.

We are left with the problem of establishing a central limit theorem for

$$N(I_p \otimes D_N)^{-1}\left\{\frac{1}{2M}\sum_k (f_x^{-1} \otimes I_q)\hat{e}_{xy}\right\}. \tag{7}$$

This is

$$N(I_p \otimes D_N)^{-1} \frac{1}{4\pi^2} \sum_{-\infty}^{\infty} (\Delta(j) \otimes I_q)\, k\left(\frac{j}{M}\right) \hat{c}_{xy}(j),$$

where $c_{xy}(j)$ is constructed from the matrix $C_{xy}(j)$, whose typical entry is the sample covariance between $x_u(n)$ and $y_v(n + j)$ in the same way as $\hat{e}_{xy}$ was constructed from $\hat{f}_{xy}$ and we put $\hat{c}_{xy}(j + 2lm) = \hat{c}_{xy}(j)$ for all integers l. This is

$$\frac{N}{(2\pi)^2}(I_p \otimes D_N)^{-1} \sum_{-\infty}^{\infty} (\Delta(j) \otimes I_q)\, k\left(\frac{j}{M}\right) \frac{1}{N} \sum{}' x(n) \otimes y(n + j), \tag{8}$$

the sum being over n such that n and $n + j$ lie between 1 and N, inclusive. Now

$$\mathcal{E}[N^{-1}\{\sum{}' x(n) \otimes (N^{-1/2}D_N)^{-1}y(n + j)\}\{\sum{}' x(n) \otimes (N^{-1/2}D_N)^{-1}y(n + k)\}']$$

converges to

$$\sum_{-\infty}^{\infty} \Gamma_x(u) \otimes R(u - j + k)$$

boundedly for each fixed j, k. Thus the variance of (7) converges to

$$\frac{1}{(2\pi)^4} \sum_{-\infty}^{\infty}\sum (\Delta(j) \otimes I_q) \sum_{-\infty}^{\infty} \Gamma(u) \otimes R(u - j + k)\,(\Delta(k) \otimes I_q)'$$

$$= \frac{1}{(2\pi)^4} \sum_{u=-\infty}^{\infty} \left[\sum_{k,j=-\infty}^{\infty}\sum \Delta(j)\Gamma(u + j - k)\Delta(k)'\right] \otimes R(u)$$

$$= \frac{1}{(2\pi)^2} \sum_{-\infty}^{\infty} \Delta(u)' \otimes R(u) = \frac{1}{(2\pi)^2} \int_{-\pi}^{\pi} \sum \Delta(u)' e^{-iu\lambda} \otimes M'(d\lambda).$$

This is

$$\int_{-\pi}^{\pi} (2\pi f(\lambda))^{-1} \otimes M'(d\lambda)$$

since

$$\sum_j \Delta(j)\Gamma(u + j - k) = \delta_u^k 4\pi^2.$$

It follows that we may replace (8) with

$$\frac{1}{(2\pi)^2}\sum_{-P}^{P}(\Delta(j)\otimes I_q)\,k\left(\frac{j}{M}\right)\sum_n x(n)\otimes D_N^{-1}y(n+j), \tag{9}$$

and the remaining neglected terms have a norm whose expectation may be made arbitrarily small by making P sufficiently large. It follows from Bernstein's lemma (see the proof of Theorem 14 in the Appendix to Chapter IV) that we need only establish the central limit theorem for the fixed finite number of terms in (9), and this has already been done in the proof of Theorem 10, Chapter IV. Indeed, we may make the inner sum run over $1 \le n \le N$ by adding an asymptotically negligible term and reduce ourselves to the consideration of the joint normality of the elements of

$$\sum_{n=1}^{N} x(n)\otimes \tilde{D}_N^{-1}\tilde{y}(n),$$

where $\tilde{D}_N = I_{2P+1}\otimes D_N$ and

$$\tilde{y}(n) = \begin{pmatrix} y(n-P) \\ y(n-P+1) \\ \cdot \\ \cdot \\ \cdot \\ y(n+P) \end{pmatrix}$$

This may be rearranged in the form

$$\tilde{D}_N^{-1}\sum_{N=1}^{N}\tilde{y}(n)\,x(n)',$$

when the proof of its asymptotic normality is reduced to that for a vector of least squares regression coefficient under the conditions of Theorem 10 in Chapter IV.

2 *Proof of Theorem* 20

We consider, for simplicity of notation, the case where $y(n)$ and $z(n)$ are known to have zero means. A check through the proof will show that the use of mean corrected data does not affect the results. Consider

$$\sqrt{N}\{\hat{\beta}(j)-\beta(j)\} = \frac{\sqrt{N}}{2M}\sum_k (I_p\otimes \hat{f}_y^{-1})\left\{\hat{e}_{zy} - (I_p\otimes \hat{f}_y)\sum_l \beta(l)e^{-il\lambda_k}\right\}e^{ij\lambda_k}. \tag{10}$$

Here we have omitted the argument variable, $\lambda_k = \pi k/M$, except in the exponentials and also the limits of summation for k and l, namely, $-M+1 \le k \le M$ and $s \le l \le t$. The column vector $\hat{e}_{zy}$ is obtained from $\hat{f}_{zy}$ as $\beta(j)$ is from $B(j)$. We wish first to show that we may replace the expression within braces in (10) by $\hat{e}_{xy}$ in the sense that the error introduced by doing that converges in probability to zero. This expression is obtained, as was $\hat{e}_{zy}$ from $\hat{f}_{zy}$, from

$$\hat{f}_{zy} - \sum_l B(l)'\hat{f}_y e^{il\lambda_k},$$

and we are thus led to consider

$$(11)\quad \sqrt{N}\Big\{\hat{f}_{yz} - \hat{f}_{yx} - \sum_l \hat{f}_y B(l)e^{-il\lambda_k}\Big\}$$
$$= \sqrt{N}\frac{1}{2\pi}\Big[\sum_l \Big\{\sum_{n=-\infty}^{\infty} (k_n - k_{n-l})e^{-i(n-l)\lambda_k}N^{-1} \times \sum_m{}' y(m)y(m-l+n)'\Big\} B(l)e^{-il\lambda_k}\Big] + R,$$

wherein $\sum'$ is over $1 \le m, m + n \le N$. We have taken the sum over n over the full range to avoid writing down the limits separately for the two terms although $k_n = 0$ for $|n| > N$. The residual, R, is

$$\sqrt{N}\frac{1}{2\pi}\Big[\sum_l \Big\{\sum_{-\infty}^{\infty} k_n e^{-in\lambda_k}N^{-1}(\textstyle\sum'' - \sum')y(m)\,y(m+n)'\Big\} B(l)e^{-il\lambda_k},$$

where $\sum''$ is for $1 \le m, m + n - l \le N$. Thus there are at most l terms involved in the difference of the two sums and the standard deviation of R is evidently $O(M/\sqrt{N})$ and its contribution to (11) is thus negligible. Since we know from the proof of Theorem 10 that $\max_\lambda \|\hat{f}_y(\lambda) - f_y(\lambda)\|$ converges in probability to zero, then evidently we may replace (10) with

$$(12)\qquad \frac{\sqrt{N}}{2M}\sum_k (I_p \otimes \hat{f}_y^{-1})\,\hat{d}e^{ij\lambda_k} + \frac{\sqrt{N}}{2M}\sum_k (I_p \otimes \hat{f}_y^{-1})\,\hat{e}_{xy}e^{ij\lambda_k},$$

where $\hat{d}$ corresponds to $\hat{D}$ as $\beta(j)$ does to $B(j)$ and $\hat{D}'$ is given by the right side of (11) with R omitted. We now wish to show that the first term is negligible. Let $D' = \mathcal{E}(\hat{D}')$, so that

$$D' = \frac{\sqrt{N}}{2\pi}\sum_l \Big\{\sum_{-\infty}^{\infty} (k_n - k_{n-l})e^{-i(n-l)\lambda_k}\Big(1 - \frac{|n|}{N}\Big)\Gamma_y(n-l)\Big\} B(l)e^{-il\lambda_k},$$

and let d correspond to D as $\hat{d}$ does to $\hat{D}$. First consider

$$(13)\qquad \frac{\sqrt{N}}{2M}\sum_k (I_p \otimes \hat{f}_y^{-1})\,(\hat{d} - d)e^{ij\lambda_k}.$$

Now

$$|k_n - k_{n-l}| < cM^{-1}, \qquad 0 < c < \infty,$$

and consequently, in much the same way as the variance of a spectral estimator is found, it may be shown that $\sqrt{(N/M)}\{M(\hat{d} - d)\}$ has standard deviation bounded by a constant independent of N. Using the observation about $\hat{f}_y$ previously made it follows that (13) converges in probability to zero and we must now consider

$$(14)\qquad \frac{\sqrt{N}}{2M}\sum_k (I_p \otimes \hat{f}_y^{-1})\,de^{ij\lambda_k} = \frac{\sqrt{N}}{2M}\sum_k (I_p \otimes f_y^{-1})\,de^{ij\lambda_k} + \frac{\sqrt{N}}{2M}\sum_k (I_p \otimes (\hat{f}_y^{-1} - f_y^{-1}))\,de^{ij\lambda_k}.$$

The second term converges in probability to zero. Indeed it is

$$\frac{1}{2M}\sum_k (I_p \otimes \hat{f}_y^{-1}\left(\frac{N}{M}\right)^{1/2}(f_y - \hat{f}_y)f_y^{-1})\sqrt{M}\, de^{ij\lambda_k}.$$

Now outside of a set S in the sample space of all histories of the process, with arbitiarily small probability for N large, $\|\hat{f}_y\| \geq b > 0$, $\|f_y\| \geq b > 0$. On the other hand, $(N/M)^{1/2}(\hat{f}_y - f_y)$ has elements with standard deviations which are bounded by a constant. Also $\sqrt{M}d$ has elements which converge to zero since $\sqrt{M}(k_n - k_{n-l})$ does so and

$$\sum_{-\infty}^{\infty} \|\Gamma_y(n)\|$$

certainly converges. Thus we have only to prove that the first term on the right in (14) converges to zero in order to be left only with the second term in (12). This is, rewritten as a matrix,

$$\sum_l \frac{\sqrt{N}}{2M}\sum_k \left\{\sum_{-\infty}^{\infty}(k_n - k_{n-l})e^{-in\lambda_k}\left(1 - \frac{|n|}{N}\right)\Gamma_y(n-l)f_y^{-1}B(l)e^{ij\lambda_k}\right\}. \tag{15}$$

Since det $(f_y) \geq c > 0$ and f_y has an absolutely convergent Fourier series then so does f_y^{-1} (by the same argument as was used in the proof of Theorem 10). Moreover, since

$$\sum_{-\infty}^{\infty} |n|^r \|\Gamma_y(n)\| < \infty,$$

then f_y is $[r]$ times continuously differentiable and its $[r]$th derivative has an absolutely convergent Fourier series. Then the same must be true of f_y^{-1}. Let us put

$$f_y^{-1} = \sum_{-\infty}^{\infty} L(u)e^{iu\lambda}.$$

Then

$$\sum_{-\infty} |u|^{[r]}\|L(u)\| < \infty.$$

Now, using the orthogonality properties of the exp $(-in\lambda_k)$, then (15) is

$$\sum_l \left[\sqrt{N}\sum_{-\infty}^{\infty}\left\{(k_n - k_{n-l})\left(1 - \frac{|n|}{N}\right)\Gamma_y(n-l)\sum_{v=-\infty}^{\infty} L(n - j + 2vM)\right\}\right]B(l),$$

and we must consider the typical term

$$\sqrt{N}\sum_{-\infty}^{\infty}\left\{(k_n - k_{n-l})\left(1 - \frac{|n|}{N}\right)\Gamma_y(n-l)\sum_{v=-\infty}^{\infty} L(n - j + 2vM)\right\}. \tag{16}$$

If $k_n = 0$, $|n| > M$, this is

$$\sqrt{N}\sum_{-M}^{M}(k_n - k_{n-l})\left(1 - \frac{n}{N}\right)\Gamma_y(n-l)\{L(n-j) + o(M^{-[r]})\} \tag{17}$$

dominated by

$$\frac{\sqrt{N}}{M^{q+1}} \sum_{-M}^{M} \left| \frac{(k_n - k_{n-l})}{(l/M)(n/M)^q} \right| |n|^q \, \|\Gamma_y(n-l)\| \, \|L(n-j)\|, \qquad q = r + [r],$$

which converges to zero by assumptions (7.9), (7.11), and (7.12).

For the truncated estimator (16) contains in the outer summation only $2l$ terms, a typical one being

$$\sqrt{N}\left(1 - \frac{M+1}{N}\right)\Gamma_y(M+1-l)\,\{L(n-j) + o(M^{-[r]})\}.$$

This is $o(\sqrt{N}\,M^{-(r+[r])})$ and thus converges to zero by (7.13).

For the F.F.T., Cooley-Tukey and Daniell procedures we decompose (16) breaking up the outer sum into two ranges of summation, the first for $2jM \leq n \leq (2j+1)M$, $j = 0, \pm 1$, and the second for $(2j+1)M \leq n \leq (2j+2)M$, $j = 0, \pm 1, \ldots$. The first of these may be put in the form (17) for $q = 1$ (but with n varying over all values of n in the first range of summation) so that the factor

$$\overline{\lim_{N\to\infty}}\, |(k_n - k_{n-l})/(nl/M^2)| < a$$

and thus is asymptotically negligible by condition (7.14) if (7.9) is satisfied for $r = 1$. The remaining component is dominated by

$$C\sqrt{N}\sum{}'' \, |k_n - k_{n-l}| \, \|\Gamma_y(n-l)\|, \tag{18}$$

where C is a finite constant. (We cannot now assert that the inner sum is asymptotically small as it will always contain terms which are $O(1)$.) However, now $|k_n - k_{n-l}| = O(M^{-3})$, since, for example, for $M \leq n \leq 2M$, putting $n = 2M - u$, we have for the Daniell estimator

$$k_n - k_{n-l} = \frac{\sin \frac{1}{2}\pi(u/M)}{\pi(1 - \frac{1}{2}(u/M))} - \frac{(\sin \frac{1}{2}\pi(u-l)/M)}{\pi(1 - \frac{1}{2}(u-l)/M)} = O(M^{-3}).$$

Thus (18) is $O(M^{-1})$ if (7.14) is satisfied.

Thus in all cases we are left with the expression

$$\frac{\sqrt{N}}{2M}\sum_k (I_p \otimes \hat{f}_y^{-1})\hat{e}_{xy}e^{ij\lambda_k}.$$

The proof of the asymptotic normality of this is precisely the same as for Theorem 10.

Mathematical Appendix

1. INTRODUCTION

The content of this book could be as well described if, instead of the title we have given it, we had called it "The Second-Order Properties of Random Functions." Thus it is substantially concerned with families, $\{x(t)\}$, where $t \in R$ (the real line), of random variables having finite mean square. From this it follows that it is concerned with families of functions, defined upon a measure space, which are all square integrable with respect to the measure. Thus the apparatus of Hilbert space is an essential piece of our mathematical equipment. In the next section we therefore discuss Hilbert spaces. We do not attempt precisely to develop and prove all of the propositions which we there enunciate but restrict our proofs to those properties which are needed and which cannot be found, easily, in standard texts.

A central place in the theory with which the book is concerned is played by the stationary random functions. For the study of these, Fourier methods play a premier part and in the third section of this Appendix we introduce the notions, notations and definitions needed, without again attempting to give complete proofs of properties whose proofs are readily available in the literature. In connection with the prediction theory of random functions the theories of Hilbert space (and of Banach space) come together with Fourier methods through the study of functions square integrable on the boundary of the unit circle and holomorphic within its interior and we also discuss these in Section 3.

In the fourth section we survey, again substantially without proof, the notion of generalized function. No extensive use of generalized functions is made in this book and the sections involving the use of this concept could be omitted without affecting the accessibility to the reader of the remainder of the book. However, the concept of generalized function leads to the concept of generalized random process which seems sure to play an extremely important part in the future development of the theory of random processes. At a later time a more complete integration of this concept into the theory would be both more desirable and more possible but in this work we shall restrict our discussion of these topics.

Finally, in Section 5 we survey some, basically very elementary, properties of tensor products of vector spaces. Many of the sets of parameters occurring in multiple time series (such as matrices of regression coefficients) are basically of the nature of a tensor and this is most apparent when their covariance and distributional properties are considered.

2. HILBERT SPACE AND BANACH SPACE

The classic example of a Hilbert space arises in the following fashion. We consider a space Ω upon a Borel field, $\mathcal{A}$, of sets in which is defined a measure† μ and the class of all complex-valued, measurable functions $x(\omega)$ on Ω for which

$$\|x\|^2 = \int_\Omega |x(\omega)|^2 \, \mu(d\omega) < \infty. \tag{2.1}$$

This class of functions we call $L_2(\mu)$ (or L_2 when we do not need to emphasize which measure is involved). We introduce the notation

$$(x, y) = \int_\Omega x(\omega) \, \overline{y(\omega)} \, \mu(d\omega),$$

so that $\|x\|^2 = (x, x)$. We do not distinguish between two elements x_1, x_2, of L_2 which are equivalent in the sense that their difference satisfies $\|x_1 - x_2\|^2 = 0$. Thus the elements of L_2 may henceforth be considered as classes of equivalent functions. Then it is a classic fact (see, for example, Halmos (1950)) that L_2 has the following properties:

(A) L_2 is a (linear) vector space, that is, along with $x_1, x_2 \in L_2$, this space also contains any linear combination $\alpha_1 x_1 + \alpha_2 x_2$, where the α_j are complex numbers.

(B) The "inner product" (x, y) satisfies (i) $(x, y) = \overline{(y, x)}$; (ii) $(\alpha x, y) = \alpha(x, y)$, $(x + y, z) = (x, z) + (y, z)$ and is such that the quantity $\|x\|$ defined by it is a "norm"; that is, (i) $\|x\| \geq 0$, $\|x\| = 0$ iff $x = 0$, (ii) $\|x + y\| \leq \|x\| + \|y\|$ (triangle inequality), (iii) $\|\alpha x\| = |\alpha| \, \|x\|$.

(C) L_2 is complete in the norm, $\|x\|$. By this it is meant that if $x_n \in L_2$, $n = 1, 2, \ldots$ and $\lim_{n,m\to\infty} \|x_n - x_m\| = 0$ (i.e., x_n forms a Cauchy sequence) then there is an $x \in L_2$ so that $\lim_{n\to\infty} \|x - x_n\| = 0$.

In general we call a Hilbert space any space, $\mathcal{H}$, of elements x which is a linear space over the complex field (i.e., which satisfies (A)), in which an inner product is defined satisfying the conditions in (B), and which is complete in the sense of (C) above. If α is restricted to lie in the real field and

† We assume a knowledge of the theory of measure and integration equivalent to that contained in most modern courses in pure probability theory. All that we require is contained in Halmos (1950).

$(x, y) = (y, x)$ we call $\mathcal{H}$ a real Hilbert space. The space of real valued, measurable, functions satisfying (2.1) is an example of a real Hilbert space. It is evident that the inner product is a continuous function of its arguments for if $x_n \to x$, $y_n \to y$ then $|(x, y) - (x_n, y_n)| = |(x - x_n, y) + (x_n, y - y_n)|$ and by Schwartz's inequality, $|(x, y)| \leq \|x\| \, \|y\|$, (which follows easily from the properties of the inner product) then $|(x, y) - (x_n, y_n)|$ converges to zero as n increases.

By a closed subspace, $\mathcal{M}$, of the Hilbert space, $\mathcal{H}$, we mean a subset of elements of $\mathcal{H}$ which is closed under the formation of linear combinations but also is topologically closed in the sense that the limit of any Cauchy sequence of elements of $\mathcal{M}$ converges to an element of $\mathcal{M}$. Thus $\mathcal{M}$ is a Hilbert space in its own right. By the orthogonal complement of $\mathcal{M}$ in $\mathcal{H}$, which we call $\mathcal{M}^\perp$, we mean the set of all y such that $(y, x) = 0$, $x \in \mathcal{M}$. This is also a subspace and any $z \in \mathcal{H}$ can be uniquely decomposed as $z = x + y$, $x \in \mathcal{M}$, $y \in \mathcal{M}^\perp$. We thus say that $\mathcal{H}$ is the orthogonal direct sum of $\mathcal{M}$ and $\mathcal{M}^\perp$ and write $\mathcal{H} = \mathcal{M} \oplus \mathcal{M}^\perp$. Evidently $\|z\|^2 = \|x\|^2 + \|y\|^2$. This definition extends to any sequence of pairwise mutually orthogonal subspaces of $\mathcal{H}$. Thus if $\mathcal{M}_j$, $j = 1, 2, \ldots$, is a sequence of subspaces of $\mathcal{H}$ for which $(x_j, x_k) \equiv 0$, $x_j \in \mathcal{M}_j$, $x_k \in \mathcal{M}_k$, for all pairs (j, k), then we form the subspace $\mathcal{M}$ which is the orthogonal direct sum of the subspaces, $\mathcal{M}_j$, $\mathcal{M} = \sum_\oplus \mathcal{M}_j$, in the sense that $\mathcal{M}$ consists of all x of the form $\sum x_j$, $x_j \in \mathcal{M}_j$ for which $\sum \|x_j\|^2 < \infty$. In particular $\mathcal{M}$ may be $\mathcal{H}$, itself. More generally we may say that $\mathcal{H}$ is the (not necessarily orthogonal) direct sum of the subspaces $\mathcal{M}_1, \mathcal{M}_2$ if each element of $\mathcal{H}$ can be *uniquely* represented as a sum of an element of $\mathcal{M}_1$ and an element of $\mathcal{M}_2$, and this definition extends immediately to any finite number of summands. Finally, we say that a subspace $\mathcal{M}$ is "spanned" by the set, S, of vectors in it (or that the set S is dense in $\mathcal{H}$) if for every element x of $\mathcal{M}$ and every $\epsilon > 0$ there is an x_n of the form

$$x_n = \sum_1^n \alpha_j y_j, \qquad y_j \in S, \tag{2.2}$$

so that $\|x - x_n\| < \epsilon$. The α_j will, in general, depend on the subset of n elements of S chosen, of course.

Throughout this book we are concerned only with separable Hilbert spaces, that is, spaces $\mathcal{H}$ in which there exists a denumerable set of elements $y_j, j = 1, 2, \ldots$, which is dense in $\mathcal{H}$; for example, if Ω is an N-dimensional Euclidean space R^N and μ is a Lebesgue-Stieltjes measure, so that μ may be defined through a distribution function F on Ω (not necessarily of finite total variation), then $L_2(\mu)$ [which we shall also call $L_2(F)$] is necessarily separable. In a separable space we may always introduce an orthonormal basis (complete orthonormal sequence), that is, a denumerable dense set of elements

ϕ_j satisfying $(\phi_j, \phi_k) = \delta_j^k$, where δ_j^k is zero for $j \neq k$ and is unity otherwise. This may be accomplished via the Gram-Schmidt orthonormalization process which replaces a sequence y_j, $j = 1, 2, \ldots$, with the orthonormal sequence

$$\phi_1 = \frac{y_1}{\|y_1\|}; \qquad \phi_n = \frac{\{y_n - \sum_1^{n-1} (y_n, \phi_j)\phi_j\}}{\|y_n - \sum_1^{n-1} (y_n, \phi_j)\phi_j\|}, \qquad n > 1.$$

If we commence from a dense set y_j, then, since y_n may be expressed linearly in terms of the ϕ_j, $j = 1, \ldots, n$, it is evident that the ϕ_j are also dense and our assertion concerning the existence of an orthonormal basis in a separable space is seen to be true. The representation (2.2) may be simplified, and for every $x \in \mathcal{H}$ and every $\epsilon > 0$ there is an n_0 so that

$$\|x - \sum_1^n (x, \phi_j)\phi_j\| < \epsilon, \qquad n > n_0; \tag{2.3}$$

the coefficient α_j, which may now be taken to be (x, ϕ_j), does not depend on n. The validity of (2.3) depends on the almost obvious fact, which we record here for later use, that $(x, y_j) \equiv 0$, for the y_j forming a sequence dense in $\mathcal{H}$, implies $x = 0$. We see from (2.3) that we may write

$$x = \sum_1^\infty (x, \phi_j)\phi_j, \qquad \|x\|^2 = \sum_1^\infty |(x, \phi_j)|^2. \tag{2.4}$$

The nature of the representation (2.4) must be understood; for example, when the ϕ_j are functions on the real line, square integrable with respect to μ, it is *not* implied that the right hand side of the first part of (2.4) converges pointwise. The convergence considered is that described by (2.3), which in the case of functions squaie integrable respect to μ on Ω would be called mean-square convergence with weighting $\mu(d\lambda)$ or mean square convergence (μ). This being understood, the representation (2.4) gives us an intuitively simple picture of a separable Hilbert space. Evidently, the orthonormal basis, ϕ_j, being chosen, every $x \in \mathcal{H}$ may be equally well represented as a sequence $\{(x, \phi_j)\}$ satisfying the second relation in (2.4) while given any sequence $\{\alpha_j\}$ satisfying $\sum |\alpha_j|^2 < \infty$ then $\sum \alpha_j \phi_j$ is an element of $\mathcal{H}$. This correspondence between elements of $\mathcal{H}$ and sequences $\{\alpha_j\}$ is 1:1 and, with the algebraic operations on sequences being defined in the obvious way,

$$\alpha\{\alpha_j\} + \beta\{\beta_j\} = \{\alpha\alpha_j + \beta\beta_j\}$$

and the inner product for sequences by means of

$$(\{\alpha_j\}, \{\beta\}) = \sum \alpha_j \bar{\beta}_j,$$

this 1:1 correspondence becomes an isomorphism of $\mathcal{H}$ with the Hilbert space, usually called l_2, of all such sequences. This space, l_2, is, of course, an obvious generalization of the concept of a finite-dimensional complex vector space, with the usual inner product, and makes it apparent that the concept of a separable Hilbert space is by no means esoteric. An example of a denumerable dense set, in the case of the space L_2 for μ, a Lebesgue-Stieltjes measure in R^n, is provided by the "simple" or "jump" functions that take in a finite set of different values on rectangle sets whose corner points have rational coordinates.

We now discuss an example which further illustrates these concepts and which will be used elsewhere in the book. We consider a $(p \times p)$ matrix, $F(\lambda)$, with the entry in row j, column k being $F_{jk}(\lambda)$. We take $F_{jk}(\lambda)$ to be a complex function of bounded variation† on R. By this we mean that we have $F_{jk} = U_{jk} + iV_{jk}$ where U_{jk} and V_{jk} are real-valued functions of bounded variation. Thus, for example, we may put $U_{jk} = U_{jk}^+ - U_{jk}^-$, where U_{jk}^+ and U_{jk}^- are the upper and lower variations of U_{jk} and are both nondecreasing functions of finite total increase on R. Moreover, the supports of these two parts of U_{jk} (i.e., the sets where these functions are increasing) are nonintersecting. Moreover, we require that, for each $\lambda \in R$, the matrix $F(\lambda)$ is to be Hermitian and have nonnegative increments, that is, $F(\lambda_1) - F(\lambda_2) \geq 0$, $\lambda_1 \geq \lambda_2$ (we shall consistently use the notation $A \geq 0$ to mean that an Hermitian matrix is nonnegative definite). We put $m(\lambda) = \operatorname{tr} F(\lambda)$ where tr A is the trace of the matrix A. Now each element of $F(\lambda)$ is absolutely continuous (a.c.) with respect to $m(\lambda)$ so that each of U_{jk}^+, U_{jk}^-, V_{jk}^+, V_{jk}^- (for all j, k) allots zero measure to any measurable‡ set to which $m(\lambda)$ allots zero measure. This is obvious for the diagonal elements of $F(\lambda)$ since these are necessarily real valued nondecreasing functions which add to $m(\lambda)$. We also have

$$|F_{jk}(\lambda_1) - F_{jk}(\lambda_2)|^2 \leq \{F_{jj}(\lambda_1) - F_{jj}(\lambda_2)\}\{F_{kk}(\lambda_1) - F_{kk}(\lambda_2)\} \leq \{m(\lambda_1) - m(\lambda_2)\}^2$$

because of the Hermitian and nonnegative nature of $F(\lambda)$. Taking account of the form of $F_{jk}(\lambda)$ and of the nonintersecting nature of the supports of the upper and lower variations of its components, we have

$$U_{jk}^+(\lambda_1) - U_{jk}^+(\lambda_2) \leq m(\lambda_1) - m(\lambda_2),$$

with the same relation holding for all of the other components. This implies the result we require. Thus by the theorem of Radon-Nikodym we may put

$$F(d\lambda) = f(\lambda)\, m(d\lambda),$$

† It is not necessary in what follows that the total variation of F_{jk} be finite, but this is the only case we need for this book.

‡ That is, measurable with respect to the usual Borel field on R.

where $f(\lambda)$ is a matrix of elements $f_{jk}(\lambda) = u_{jk}(\lambda) + iv_{jk}(\lambda)$ and its real and imaginary parts are the Radon-Nikodym derivatives of F_{jk} with respect to m. The matrix, $f(\lambda)$, is clearly Hermitian and nonnegative, almost everywhere, with respect to the measure $m(\lambda)$. (We write a.e. (m) to abbreviate this type of phrase.) We may thus take it to be Hermitian nonnegative everywhere.

We now define the space $L_2(F)$ as the space of all vectors (of p components) of measurable, complex-valued functions, $x(\lambda)$, for which

$$\int_{-\infty}^{\infty} x^*(\lambda)\, f(\lambda)\, x(\lambda)\, m(d\lambda) < \infty.$$

(By A^* we mean of course the transposed conjugate of the matrix A, and we shall often use this notation even when A is real.) We shall also write integrals of this type, when convenient, in the form

$$\int_{-\infty}^{\infty} x^*(\lambda)\, F(d\lambda)\, x(\lambda).$$

It is clear that the set of all such vectors x forms a linear space, where now αx means of course the vector obtained by multiplying each element of $x(\lambda)$ by the complex number α. We define, x, y belonging to $L_2(F)$, the inner product by

$$(x, y) = \int_{-\infty}^{\infty} x(\lambda)^*\, F(d\lambda)\, y(\lambda).$$

To prove that $L_2(F)$ is a Hilbert space only the completeness property (c) needs to be checked. This may be accomplished by factoring the Hermitian nonnegative matrix $f(\lambda)$ as $\sqrt{f(\lambda)} \cdot \sqrt{f(\lambda)}$. Here and elsewhere in the book we mean by $\sqrt{A}$ the square root of the Hermitian nonnegative matrix A obtained by transforming that matrix to diagonal form D by means of a unitary operator Q

$$QAQ^* = D$$

and choosing $\sqrt{A} = Q^*\sqrt{D}Q$ where $\sqrt{D}$ is obtained by taking the nonnegative square root of all of the (necessarily nonnegative) elements of D. The fact that $\sqrt{f(\lambda)}$ is a measurable function (with respect to the usual Borel field on R) follows from elementary properties of measurable functions and the proof will be left to the reader. Now, if $x_n(\lambda)$ is a sequence that satisfies

$$\lim_{m,n\to\infty} \|x_m - x_n\|^2 = 0,$$

$\sqrt{f(\lambda)}\, x_n(\lambda)$ is a sequence of vectors whose components form, individually, Cauchy sequences with respect to $m(\lambda)$, and thus there is a vector $y(\lambda)$ with components square integrable with respect to $m(\lambda)$ so that the components of $\sqrt{f(\lambda)}\, x_n(\lambda)$ converge in mean square (weighting $m(d\lambda)$) to $y(\lambda)$. If $f(\lambda)$ is nonsingular almost everywhere, then by putting $x(\lambda) = \{\sqrt{f(\lambda)}\}^{-1}\, y(\lambda)$ we see that this vector satisfies $\|x\|^2 < \infty$ and

$$\lim_{n\to\infty} \int_{-\infty}^{\infty} \{x_n(\lambda) - x(\lambda)\}^* F(d\lambda)\{x_n(\lambda) - x(\lambda)\} = 0,$$

so that $L_2(F)$ is seen to be complete. If $f(\lambda)$ is singular on a set of measure not zero, we may proceed as follows. In the first place two elements $x_1(\lambda)$, $x_2(\lambda)$ of $L_2(F)$ are, of course, to be regarded as equivalent if $\|x_1 - x_2\| = 0$, so that in particular we do not alter the element x of $L_2(F)$ if we add to the function $x(\lambda)$, which defines it, a vector function, $y(\lambda)$, that is zero except on the set where $f(\lambda)$ is singular and there is an eigenvector of $f(\lambda)$ for zero eigenvalue. We call the generalized† inverse A^{-1} of the Hermitian matrix A the matrix obtained by forming D as above, defining D^{-1} by replacing all nonzero elements with their reciprocals and putting $A^{-1} = Q^*D^{-1}Q$. Then, provided $y(\lambda)$ belongs to the range of $\sqrt{f(\lambda)}$ [a.e. (m)] the vector $x(\lambda) = \{\sqrt{f(\lambda)}^{-1}\, y(\lambda)$ satisfies $\{\sqrt{f(\lambda)}\}\, x(\lambda) = y(\lambda)$ [a.e. (m)], so that we see that $L_2(F)$ is complete as in the nonsingular case. That $y(\lambda)$ belongs to the range of $\sqrt{f(\lambda)}$, a.e. (m), follows from the evident fact that $\sqrt{f(\lambda)}\, x_n(\lambda)$ does so together with the well known fact‡ that from any mean-square convergent sequence we can extract a subsequence converging to the same limit almost everywhere, for then, the range of $\sqrt{f(\lambda)}$ being a vector subspace of p-dimensional complex space and therefore closed, the limit $y(\lambda)$ belongs to that range.

The scope of the result just proved may be extended slightly in a rather trivial, but useful, fashion by considering the space of all $(p \times q)$ matrices $X(\lambda)$ for which

$$\operatorname{tr}\left\{\int_{-\infty}^{\infty} X^*(\lambda)\, F(d\lambda)\, X(\lambda)\right\} < \infty.$$

With the usual definition of linear operations and the inner product

$$(X, Y) = \operatorname{tr}\left\{\int_{-\infty}^{\infty} X^*(\lambda)\, F(d\lambda)\, Y(\lambda)\right\}$$

this is easily seen from the foregoing results to be a Hilbert space. Indeed, it is evidently no more than the direct sum of q copies of the Hilbert space $L_2(F)$ just defined.

† We use the generalized inverse only for Hermitian matrices when its properties are near enough to those of a true inverse for a special notation not to be needed.

‡ See, for example, Munroe (1953, p. 226).

We shall not make extensive use of the spectral theory of operators in a Hilbert space in this book, though the spectral theory for stationary processes, introduced in Chapter II, could be wholly based upon such a foundation. However, we do need the notion of a projection operator in a Hilbert space and this we now proceed to define.

By a bounded linear operator A in a Hilbert space $\mathcal{H}$ we mean a law that associates with each vector $x \in \mathcal{H}$ a vector (in $\mathcal{H}$), to be denoted Ax, so that

$$A(\alpha x + \beta y) = \alpha Ax + \beta Ay$$

and for which

$$\|Ax\| \leq M\|x\|, \qquad x \in \mathcal{H},$$

where M is a positive real number. The smallest M that satisfies this last inequality we call $\|A\|$. In a finite-dimensional vector space $\|A\|$ is max $\sqrt{\lambda_j}$ where the λ_j are the eigenvalues of A^*A. (The definition of A^* is given just below.) It is easily seen that $\|A\|$ satisfies the requirements for a norm, given under B in the definition of a Hilbert space. (The set $\mathcal{B}(\mathcal{H})$ of all such bounded operators in $\mathcal{H}$ is not, however, a Hilbert space itself though it is, of course a linear space and indeed a Banach space.) We shall use the concept of an unbounded opreator in this book only in certain sections and we mention an example here mainly to bring home the nature of the restriction of boundedness. An example is that of all square integrable functions† on the interval $[0, 1]$ and the operation of differentiation. This is a linear operation which takes $\lambda^n(2n + 1)^{1/2}$ (which has unit norm) into a vector with norm $n\{(2n + 1)/(2n - 1)\}^{1/2}$ and thus is evidently not bounded. It is also not defined for all elements of the Hilbert space though it is defined in a set of functions dense in that space. We now return to the consideration of bounded operators. Along with such an operator we consider its adjoint A^* defined uniquely by the relation

$$(Ax, y) = (x, A^*y).$$

Then A^* is also a bounded operator. In the case of a finite-dimensional vector space, when A is described in coordinates by a matrix, the adjoint matrix A^* is the transposed conjugate of A. If $A = A^*$, A is said to be Hermitian. If the only solution of the equation $Ax = 0$ is $x = 0$, the inverse operator A^{-1} may be defined. In general A^{-1} is not bounded, as is shown by the example of the operator A on the square integrable functions on $[0, 1]$ defined by $Af(\lambda) = \lambda f(\lambda)$. The necessary and sufficient condition that

† If we do not qualify the phrase "square integrable function" we mean this to refer to functions square integrable with respect to Lebesgue measure. The same goes for "almost everywhere" and "absolutely continuous" and for all expressions, for example $\|x\|_\infty$, defined in terms of them.

A^{-1} be bounded is that $\|Ax\| \geq m\|x\|$ for some $m > 0$. However, if $A^{-1} = A^*$, then A^{-1} is necessarily bounded and is then said to be unitary.

As we have already said we shall not be concerned with unbounded operators outside of certain starred sections which may be omitted. If a linear operator A is not bounded its domain, $\mathfrak{D}(A)$, namely the set of $x \in \mathcal{H}$ for which Ax is defined, cannot constitute the whole of $\mathcal{H}$. Among the unbounded linear operators the closed operators are of especial importance. An operator is closed if $x_n \in \mathfrak{D}(A)$ and $x_n \to x$, $Ax_n \to y$ implies that $x \in \mathfrak{D}(A)$ and $Ax = y$. Of course any bounded linear operator is closed.

Let $\mathcal{M}$ be a subspace of $\mathcal{H}$ and decompose each $x \in \mathcal{H}$ as $x = x_1 + x_2$, $x_1 \in \mathcal{M}$, $x_2 \in \mathcal{M}^\perp$. Then the operator E defined by $Ex = x_1$ is a linear operator with $\|E\| = 1$ which is called the perpendicular projection onto $\mathcal{M}$. (We shall deal only with perpendicular projections in this book.) It is easy to see that $E^2 = E = E^*$ and little more difficult to show that any operator having these last two properties is a projection. If $\mathcal{M}$ is spanned by a single vector x then $Ey = \{(y, x)/\|x\|^2\}x$ and in general the projection onto a finite dimensional subspace is accomplished by "regressing" upon a basis for that subspace. The following properties of projections are easily established. (In all cases E_i projects on $\mathcal{M}_i$):

1. If E projects on $\mathcal{M}$, then $(I - E)$ projects on $\mathcal{M}^\perp$.
2. $E_1 + E_2$ is a projection (on $\mathcal{M}_1 \oplus \mathcal{M}_2$) iff $E_1E_2 = E_2E_1$; that is, iff $\mathcal{M}_1$ and $\mathcal{M}_2$ are orthogonal.
3. If $\mathcal{M}_1 \subset \mathcal{M}_2$, then $E_1E_2 = E_2E_1 = E_1$ and $E_2 - E_1$ projects on the orthogonal complement of $\mathcal{M}_1$ in $\mathcal{M}_2$.

To conclude this section we mention that there are vector spaces which are complete with respect to norms but whose norms are not defined via an inner product (so that the concept of orthogonality is not meaningful). Such spaces are called Banach spaces. An important example is the space $L_p(\mu)$, $p > 1$ of functions $x(\omega)$ on a measure space Ω for which

$$\|x\|_p = \left\{\int |x(\omega)|^p \mu(d\omega)\right\}^{1/p} < \infty.$$

For spaces L_p, $p \geqslant 1$, the important inequalities of Holder and Minkowski obtain. These are, respectively,

$$\int_\Omega |x(\omega)y(\omega)|\mu(d\omega) \leq \|x\|_p \, \|y\|_q, \qquad x \in L_p, \qquad y \in L_q, \qquad \frac{1}{p} + \frac{1}{q} = 1,$$

$$\|x + y\|_p \leq \|x\|_p + \|y\|_p.$$

The latter, of course, shows $\|\cdot\|_p$ is to be a norm, since it evidently has the other necessary properties. The spaces L_p, $p \geq 1$, are in fact complete in these norms and are thus Banach spaces. Finally, we define the norm

$\|x\|_\infty$ as the essential supremium of x, that is, the smallest number greater than $|x(\omega)|$ for almost all $\omega(\mu)$. The space L_∞ of all functions for which this norm is finite also constitutes a Banach space.

One of the central theorems of the theory of Banach spaces is the Hahn-Banach theorem which concerns bounded linear mappings of the Banach space into the complex numbers, that is, laws $x \in \mathcal{B} \to F(x)$, $F(x)$ being a complex number, so that $F(\alpha x + \beta y) = \alpha F(x) + \beta F(y)$ for which $F(x) \leq \|F\|\ \|x\|$, $\|F\|$ being the smallest positive real number for which this inequality holds (the boundedness of F implying that $\|F\|$ is finite). $\|F\|$ is called the norm of F. The Hahn–Banach theorem asserts that if $\mathcal{X}$ is a subspace of $\mathcal{B}$ and F is a bounded linear functional on $\mathcal{X}$ then F can be extended to constitute a linear functional on all of $\mathcal{B}$ of the same norm.

Finally, we mention that the space L_p^* (the adjoint space) of all continuous linear functionals on $L_p(p \geq 1)$ can be identified with L_q, $1/p + 1/q = 1$,via

$$F(x) = \int x(\omega)y(\omega)d\omega, \qquad x \in L_p, \qquad y \in L_q, \qquad \frac{1}{p} + \frac{1}{q} = 1,$$

in the sense that $\|F\|$ (as defined for linear functionals in L_p) is just $\|y\|_q$ for the $y \in L_q$ which defines it. Thus L_p^* is also a Banach space. In particular, since $\frac{1}{2} + \frac{1}{2} = 1$, the space $\mathcal{H}^*$ of all linear functionals on $\mathcal{H}$ can be identified with $\mathcal{H}$ itself.†

3. FOURIER METHODS

By the Fourier series of an integrable function $f(\lambda)$, $\lambda \in [-\pi, \pi]$ we mean the formal series

$$f(\lambda) \sim \sum_0^\infty \{\alpha_n \cos n\lambda + \beta_n \sin n\lambda\},$$

where

$$\alpha_n = \frac{1}{\pi}\int_{-\pi}^{\pi} f(\lambda)\cos n\lambda\, d\lambda, \qquad n \neq 0; \qquad = \frac{1}{2\pi}\int_{-\pi}^{\pi} f(\lambda)\, d\lambda, \qquad n = 0,$$

$$\beta_n = \frac{1}{\pi}\int_{-\pi}^{\pi} f(\lambda)\sin n\lambda\, d\lambda.$$

Of course, β_0 may be taken to be zero. The series can be written in the form

$$(3.1)\quad f(\lambda) \sim \sum_{-\infty}^{\infty} \gamma_n e^{-in\lambda}, \quad \gamma_n = \tfrac{1}{2}(\alpha_n + i\beta_n), \quad n \neq 0, \quad \gamma_0 = \alpha_0, \quad \gamma_{-n} = \bar{\gamma}_n,$$

† This follows only for separable Hilbert spaces by our arguments (by recognizing that any such $\mathcal{H}$ is isomorphic to some L_2; for example, the square integrable functions on [0, 1]). However, the result is true for all Hilbert spaces separable or otherwise.

where now

$$\gamma_n = \frac{1}{2\pi}\int_{-\pi}^{\pi} e^{in\lambda} f(\lambda)\,d\lambda.$$

If, for example, $f(\lambda)$ is continuous and of bounded variation in $[-\pi, \pi]$, then (3.1) will converge pointwise to $f(\lambda)$ (except possibly at $\lambda = \pi$). However, we are not very concerned with the complex question of the pointwise convergence of the series. Instead, we shall consider other ways in which, given the coefficients γ_n, we may determine $f(\lambda)$, and in particular we mention that these coefficients *do* determine that function (a.e.), whether the series converges pointwise or not. We begin by using some of the theory of the last section and consider the space L_2 of square integrable functions on $[-\pi, \pi]$, among which are the functions $\exp(-in\lambda)$, $n = 0, \pm 1, \dots$, and these are orthonormal with respect to the inner product

$$(f, g) = \frac{1}{2\pi}\int_{-\pi}^{\pi} f(\lambda)\,\overline{g(\lambda)}\,d\lambda.$$

They constitute, in fact, a complete orthonormal sequence. Assuming this for the moment, we then know that

$$f(\lambda) = \sum_{-\infty}^{\infty} \gamma_n e^{-in\lambda}, \qquad \sum |\gamma_n|^2 = \frac{1}{2\pi}\int_{-\pi}^{\pi} |f(\lambda)|^2\,d\lambda, \tag{3.2}$$

if we interpret the first relation in the sense of mean-square convergence.

There are two other ways which we shall consider of obtaining $f(\lambda)$ from the γ_n. The first of these is obtained by replacing the partial sum

$$s_N = \sum_{-N}^{N} \gamma_n e^{-in\lambda}$$

by the Cesaro sum $\sigma_N = N^{-1}(s_0 + s_1 + \cdots + s_{N-1})$ and investigating the convergence of σ_N to f.

We have

$$\sigma_N(\lambda) = \sum_{-N}^{N}\left(1 - \frac{|n|}{N}\right)\gamma_n e^{-in\lambda} = \int_{-\pi}^{\pi} f(\theta)\frac{1}{2\pi}\sum_{-N}^{N}\left(1 - \frac{|n|}{N}\right)e^{-in(\lambda-\theta)}\,d\theta$$

$$= \frac{1}{2\pi}\int_{-\pi}^{\pi} f(\theta) L_N(\lambda - \theta)\,d\theta$$

where

$$L_N(\lambda) = \sum_{-N}^{N}\left(1 - \frac{|n|}{N}\right)e^{-in\lambda}$$

$$= \frac{1}{N}\sum_{j,k=0}^{N-1}\sum e^{-i(j-k)\lambda} = \frac{1}{N}\left|\sum_{0}^{N-1} e^{ij\lambda}\right|^2$$

$$= \frac{1}{N}\left(\frac{\sin \frac{1}{2}N\lambda}{\sin \frac{1}{2}\lambda}\right)^2.$$

This function, $L_N(\lambda)$, is called Fejér's kernel and is (i) evidently positive, (ii) integrates to 2π (from the first expression for it), and (iii) converges uniformly to zero outside of any nondegenerate interval containing the origin, as $N \to \infty$.

The last property follows from the fact that

$$L_N(\lambda) \leq \frac{1}{N(\sin \frac{1}{2}\delta)^2}, \qquad \delta \leq |\lambda| \leq \pi.$$

Evidently $\sigma_N(\lambda)$ is uniformly bounded if $\|f\|_\infty < \infty$.

Consider the expression

$$\begin{aligned}\sigma_N(\lambda) - f(\lambda) &= \frac{1}{2\pi}\int_{-\pi}^{\pi} f(\theta)\, L_N(\lambda - \theta) - f(\lambda)\, d\theta \\ &= \frac{1}{2\pi}\int_{-\pi}^{\pi} \{f(\lambda - \theta) - f(\lambda)\}\, L_N(\theta)\, d\theta,\end{aligned}$$

where we continue $f(\lambda)$ on periodically outside $[-\pi, \pi]$. Then we have

$$\begin{aligned}\|\sigma_N - f\|^2 &= \frac{1}{2\pi}\int_{-\pi}^{\pi} \{\sigma_N(\lambda) - f(\lambda)\}^2\, d\lambda \\ &= \left(\frac{1}{2\pi}\right)^2 \int_{-\pi}^{\pi}\left[\int_{-\pi}^{\pi} \{f(\lambda - \theta) - f(\lambda)\}\, L_N(\theta)\, d\theta\right]\{\sigma_N(\lambda) - f(\lambda)\}\, d\lambda \\ &= \frac{1}{2\pi}\int_{-\pi}^{\pi} L_N(\theta) \frac{1}{2\pi}\int_{-\pi}^{\pi} \{f(\lambda - \theta) - f(\lambda)\}\{\sigma_N(\lambda) - f(\lambda)\}\, d\lambda\, d\theta \\ &\leq \frac{1}{2\pi}\int_{-\pi}^{\pi} L_N(\theta)\, \|f_\theta - f\|\, d\theta \left[\frac{1}{2\pi}\int_{-\pi}^{\pi} \{\sigma_N(\lambda) - f(\lambda)\}^2\, d\lambda\right]^{1/2},\end{aligned}$$

where we have written f_θ for the function $f(\lambda - \theta)$ considered as a function of λ. Thus

$$\|\sigma_N - f\| \leq \frac{1}{2\pi}\int_{-\pi}^{\pi} L_N(\theta)\, \|f_\theta - f\|\, d\theta \leq \sup_{|\theta| < \delta} \|f_\theta - f\| + 2\, \|f\|_\infty \sup_{|\theta| \geq \delta} L_N(\theta).$$

If f is essentially bounded and is continuous in a closed interval containing λ, $a \leq \lambda \leq a + b$, then evidently the last displayed expression converges uniformly to zero for λ in that interval. Thus $\sigma_N(\lambda)$ converges uniformly to $f(\lambda)$ in any interval of continuity of that function, and in particular if $f(\lambda)$ is continuous, and, $f(-\pi) = f(\pi)$, then σ_N converges uniformly to $f(\lambda)$ in $[-\pi, \pi]$. Thus we have constructed a sequence which accomplishes what Weierstrass' famous theorem asserts can be accomplished. If $f \in L_p$, we have for $g \in L_q$, $p^{-1} + q^{-1} = 1$,

$$\left|\frac{1}{2\pi}\int_{-\pi}^{\pi}\{\sigma_N(\lambda) - f(\lambda)\}\, g(\lambda)\, d\lambda\right|$$
$$= \left|\frac{1}{(2\pi)^2}\int_{-\pi}^{\pi}\int_{-\pi}^{\pi}\{f(\lambda - \theta) - f(\lambda)\}\, g(\lambda)\, L_N(\theta)\, d\theta\, d\lambda\right|$$
$$\leq \frac{1}{2\pi}\int_{-\pi}^{\pi} L_N(\theta)\left|\frac{1}{2\pi}\int_{-\pi}^{\pi}\{f(\lambda - \theta) - f(\lambda)\}\, g(\lambda)\, d\lambda\right| d\theta$$
$$\leq \frac{\|g\|_q}{2\pi}\int_{-\pi}^{\pi} \|f_\theta - f\|_p\, L_N(\theta)\, d\theta$$

by Holder's inequality, where we have again written f_θ for $f(\lambda - \theta)$, considered as a function of λ.

We now call on the Hahn–Banach theorem. We take the subspace of L_p in that theorem to be that spanned by $\{\sigma_N(\lambda) - f(\lambda)\}$ and the linear functional on this subspace to be that one which allots $\|\sigma_N(\lambda) - f(\lambda)\|_p$ to $\sigma_N(\lambda) - f(\lambda)$. Then this functional is certainly of norm unity on this (one-dimensional) subspace and the theorem asserts that this functional can be extended to all of L_p to have the same norm. Then, taking g above to be the function in L_q defining the extended functional, we have

$$\|\sigma_N - f\|_p \leq \frac{1}{2\pi}\int_{-\pi}^{\pi} \|f_\theta - f\|_p\, L_N(\theta)\, d\theta \tag{3.3}$$
$$\leq \sup_{|\theta|<\delta} \|f_\theta - f\|_p + 2\|f\|_p \sup_{|\theta|\geq\delta} L_N(\theta).$$

However, $\|f_\theta - f\|_p \leq \|f_\theta - \hat{f}_\theta\|_p + \|\hat{f}_\theta - f\|_p$, where we choose $\hat{f}_\theta(\lambda)$ to be $f_\theta(\lambda)$ if $|f_\theta(\lambda)| < A$ and to be A otherwise. By choosing A sufficiently large the first norm may be made less than $\frac{1}{2}\epsilon$ (uniformly in θ) by Lebesgue's dominated convergence theorem. By the same theorem, as $\theta \to 0$, the second norm converges to $\|\hat{f} - f\|_p$, where $\hat{f}$ is defined in terms of f in the same way as for $\hat{f}_\theta$ in terms of f_θ. Thus

$$\lim_{\theta\to 0} \|f_\theta - f\|_p \leq \tfrac{1}{2}\epsilon + \tfrac{1}{2}\epsilon = \epsilon.$$

We have shown that $\|\sigma_N - f\|_p$ converges to zero with N; that is, that the Cesaro sum of the Fourier series of a function in L_p converges in the L_p norm to that function. This, incidentally, establishes the completeness of the $\exp(in\theta)$, $n = 0, \pm 1, \ldots$, in L_2.

An alternative method of "summing" the Fourier series of $f(\lambda)$ is obtained by considering the function

$$\sum_0^\infty \gamma_n z^n, \qquad |z| < 1.$$

Since the γ_n are certainly bounded this converges and defines a function holomorphic within the unit circle. The same is true also of

$$\sum_0^\infty \gamma_{-n} z^n.$$

This leads us to hope that, as $r \to 1$, in some appropriate sense

$$\sum_0^\infty \gamma_n r^n e^{-i\lambda n} + \sum_0^\infty \gamma_{-n} r^n e^{i\lambda n} - \gamma_0$$

will converge to $f(\lambda)$. We write this last expression (the Abel sum of the Fourier series) as

$$\begin{aligned} u_r(e^{i\lambda}) &= \sum_{-\infty}^{\infty} \gamma_n r^{|n|} e^{-in\lambda} = \frac{1}{2\pi}\int_{-\pi}^{\pi} f(\theta) \sum_{-\infty}^{\infty} r^{|n|} e^{in(\theta-\lambda)}\, d\theta \\ &= \frac{1}{2\pi}\int_{-\pi}^{\pi} f(\theta) P_r(\theta - \lambda)\, d\theta, \end{aligned}$$

where, writing $\mathcal{R}(z)$ for the real part of z,

$$\begin{aligned} P_r(\lambda) &= \sum_{-\infty}^{\infty} r^{|n|} e^{in\lambda} \\ &= \left\{ 2\mathcal{R}\left(\frac{1}{1 - re^{i\lambda}}\right) - 1 \right\} = \frac{1 - r^2}{1 + r^2 - 2r\cos\lambda}. \end{aligned}$$

This function is called Poisson's kernel. It is not difficult to see that it has precisely the same three basic properties as Fejér's kernel has, with $r \to 1$ replacing $N \to \infty$. Thus the proofs we have given for Cesaro summability of Fourier series carry straight over to Abel summability. Thus we know that Cesaro (Abel) sums of the Fourier series of a continuous function (with $f(\pi) = f(-\pi)$) converges uniformly to that function as $N \to \infty$ $(r \to 1)$, that these sums converge to $f(\lambda)$ in the L_p norm if $f \in L_p$, $p \geq 1$ and that both sets of sums are uniformly bounded if $f \in L_\infty$.

It is considerably more difficult to prove the (important) theorem that both the Cesaro and Abel sums converge almost everywhere to $f(\lambda)$ if $f(\lambda)$ is integrable. We shall not have the occasion to use this theorem but we shall use the fact that we can find a subsequence of the sequence of Cesaro (or Abel) means which converges boundedly, a.e., if $f(\lambda)$ is a bounded measureable function. This is much easier to prove for under these conditions the Cesaro (or Abel) means converge in measure and boundedly and it is easy to prove that an a.e. convergent subsequence can be obtained from a sequence which converges in measure (see Halmos, 1950, p. 93).

Finally, in this discussion of Fourier series we mention the Hardy classes of functions H_p (we are concerned only with H_1 and H_2); H_p consists of the functions holomorphic within the unit circle and whose L_p norms are bounded as $r \to 1$. This definition is equivalent to the following for $p \geq 1$ (though the proof is difficult for $p = 1$)†, which is the only one that we shall use. $H_p, p \geq 1$, is the class of functions that belongs to L_p on the unit circle and whose Fourier coefficients γ_n vanish for $n < 0$. It is evident that H_p is a Banach space, relative to the L_p norm on the circle, and a closed subspace of L_p. A narrower class of functions, A, is that of the functions continuous on the unit circle and for which $\gamma_n = 0, n < 0$. Since the uniform limit of a sequence of continuous functions is a continuous function A is a Banach space relative to the "uniform" norm $\|f\|_\infty$, where

$$\|f\|_\infty = \sup_{\pi \leq \lambda \leq \pi} |f(\lambda)| = \sup_{|z|<1} |g(z)|, \qquad f(\lambda) = g(e^{i\lambda}).$$

(The second equality follows from the maximum modulus principle.) The space A has the additional important property that it is closed under the formation of products of functions as well as under the formation of linear combinations.

We turn now to the consideration of Fourier integrals. Thus we replace the interval $[-\pi, \pi]$ with $(-\infty, \infty)$ and all unqualified terms a.e., L_p, etc., refer to functions in the latter space with Lebesgue measure as the measure. For $f \in L_1$ then the Fourier transform is of course defined as

$$\int_{-\infty}^{\infty} f(t) e^{i\lambda t}\, dt,$$

where we now use t as the "running variable" in $(-\infty, \infty)$. This transform is well known in probability theory in which, $f(t)$ being a probability density function, the Fourier transform of $f(t)$ is called the characteristic function. Of course, this definition is then expanded to encompass the Fourier transform of a nondecreasing function F of finite total increase (e.g., a probability distribution) for which the Fourier transform

$$\int_{-\infty}^{\infty} e^{i\lambda t}\, F(dt)$$

certainly exists. We assume as known the classic theorem, called the continuity theorem for characteristic functions (or probability distributions), which asserts that a sequence of distributions converges at all points of continuity to a (proper) limit distribution iff the sequence of characteristic functions converges pointwise to a limit function that is continuous at the origin. The limiting distribution function is then uniquely defined if it is required to be continuous on the right and has as characteristic function the limit of the sequence of these functions. This theorem is stated and proved in

† The result is proved in Chapter III.

texts on the theory of probability for probability distribution, but of course it remains true for any sequence of nondecreasing functions all having the same finite total increase, whether or not it be unity.

We wish to consider functions $f(t) \in L_2$. For these the integral

$$\int_{-\infty}^{\infty} f(t)e^{i\lambda t}\,dt$$

may not converge, but

$$\int_{-T}^{T} f(t)e^{i\lambda t}\,dt \tag{3.4}$$

certainly does since f is integrable over any finite interval. We put

$$Uf(t) = \frac{1}{\sqrt{2\pi}}\int_{-\infty}^{\infty} f(t)e^{i\lambda t}\,dt = \frac{1}{\sqrt{2\pi}}\hat{f}(\lambda),$$

where the right-hand side is defined as the limit, of the sequence of integrals over $[-T, T]$, *in mean square as* $T \to \infty$. We need to establish the validity of this definition. At the same time we consider

$$Vf(t) = \frac{1}{\sqrt{2\pi}}\int_{-\infty}^{\infty} f(t)e^{-i\lambda t}\,dt,$$

defined in the same way. We show that U and V are unitary transformations of L_2 and moreover that $U = V^* = V^{-1}$. Thus we shall have proved that the Fourier transformation, defined as above, maps L_2 onto itself, so that

$$(f, g) = (\hat{f}, \hat{g});$$

that is,

$$\int_{-\infty}^{\infty} f(t)\,\overline{g(t)}\,dt = \frac{1}{2\pi}\int_{-\infty}^{\infty} \hat{f}(\lambda)\,\overline{\hat{g}(\lambda)}\,d\lambda,$$

which is called Plancherel's formula, and, moreover, that

$$f(t) = \frac{1}{2\pi}\int_{-\infty}^{\infty} \hat{f}(\lambda)e^{-i\lambda t}\,d\lambda$$

(the Fourier inversion formula for functions in L_2); that is, $V^{-1} = U$. These are clearly the analogs of the relations between γ_n and $f(\lambda_n)$ when $f(\lambda)$ is square integrable on $[-\pi, \pi]$ (see (3.2)).

We call $e_\lambda(t)$ the function which is sign λ for $t \in (0, \lambda)$ and is zero otherwise and put

$$Ue_\lambda(t) = \frac{1}{\sqrt{2\pi}}\frac{e^{i\lambda t} - 1}{it}, \qquad Ve_\lambda(t) = \frac{1}{\sqrt{2\pi}}\frac{e^{-i\lambda t} - 1}{-it}.$$

It is easily checked that

$$(Ue_\lambda, Ue_\mu) = (Ve_\lambda, Ve_\mu) = (e_\lambda, e_\mu),$$
$$(Ve_\lambda, e_\mu) = (e_\lambda, Ue_\mu)$$

since

$$\begin{aligned}
\frac{1}{2\pi}\int_{-\infty}^{\infty}\frac{(e^{i\lambda t}-1)(e^{-i\mu t}-1)}{t^2}\,dt
&= \frac{1}{2\pi}\int_{-\infty}^{\infty}\frac{\cos(\lambda-\mu)t - \cos\lambda t - \cos\mu t + 1}{t^2}\,dt \\
&= \frac{1}{2\pi}\{|\lambda| + |\mu| - |\lambda-\mu|\}\int_{-\infty}^{\infty}\frac{\sin^2 t}{t^2}\,dt \\
&= \tfrac{1}{2}\{|\lambda| + |\mu| - |\lambda-\mu|\} \\
&= 0 \text{ if } \lambda\mu \le 0, \quad = \min\{|\lambda|, |\mu|\} \text{ if } \lambda\mu \ge 0.
\end{aligned}$$

The transformations extend by linearity to linear combinations of functions of the form e_λ, that is, to step functions. On this linear space they are isometric (preserve the inner product) and are adjoint to each other. The step functions are dense in L_2 as we mentioned in the first section. These properties of U and V are retained, therefore, when they are extended by continuity to all of L_2. Thus, if $f_n \to f$ (in L_2), evidently Uf_n is a Cauchy sequence (since $\|Uf_n - Uf_m\| = \|f_n - f_m\|$) and converges to an element we call Uf. Then, if $g_n \to g$, we have $(Uf, Ug) = \lim_n (Uf_n, Ug_n) = \lim_n (f_n, g_n) = (f, g)$ and U is isometric on L_2. The other properties are similarly proved to hold: $U^*U = V^*V = I$ and $U = V^*$ so that U has a bounded inverse V and is unitary. We have from the very definition of U

$$\begin{aligned}
(Uf, e_\lambda) &= \int_0^\lambda Uf(t)\,dt \\
&= \frac{1}{\sqrt{2\pi}}\int_{-\infty}^{\infty}\frac{e^{i\lambda t}-1}{it}f(t)\,dt.
\end{aligned}$$

Thus, differentiating,

$$Uf = \frac{d}{d\lambda}\frac{1}{\sqrt{2\pi}}\int_{-\infty}^{\infty}\frac{e^{i\lambda t}-1}{it}f(t)\,dt,$$

but

$$\frac{1}{\sqrt{2\pi}}\lim_{h\to 0}\frac{1}{h}\int_{-\theta}^{\theta}\frac{e^{i(\lambda+h)t}-e^{i\lambda t}}{it}f(t)\,dt = \frac{1}{\sqrt{2\pi}}\lim_{h\to 0}\int_{-\theta}^{\theta}\frac{\sin\frac{1}{2}ht}{\frac{1}{2}ht}\,e^{\frac{1}{2}iht}e^{it\lambda}f(t)\,dt,$$

and the integrand is bounded in modulus by $|f(t)|$. Thus we may take the limit under the integral sign, by dominated convergence, to get

$$\frac{1}{\sqrt{2\pi}}\int_{-\theta}^{\theta} e^{it\lambda} f(t)\,dt. \tag{3.5}$$

The function $f_\theta(t)$, which is $f(t)$ for $|t| < \theta$ and otherwise zero, is obviously convergent in L_2 to $f(t)$. Since the last expression is $Uf_\theta(t)$ and U is unitary, we see that

$$Uf(t) = \frac{1}{\sqrt{2\pi}}\int_{-\infty}^{\infty} e^{it\lambda} f(t)\,dt,$$

where the right-hand side is the limit of the sequence of expressions (3.5) as $\theta \to \infty$ (in L_2). Thus we have established what we require.

We may, of course, wish to replace the convergence in mean square (with respect to Lebesgue measure) of the expression (3.4) by convergence in mean square with respect to some other measure on R and to consider only those functions $f(t)$ for which this convergence takes place. This class may be wider or narrower than the space L_2 but, of course, no analogue of Plancherel's theorem will now hold for this theorem is tied to the use of Lebesgue measure in the definition of the convergence, in mean square, of the expressions (3.4).

Once or twice we shall have occasion to use certain facts concerning the relation between the smoothness of a periodic function $f(\lambda)$, $\lambda \in [-\pi, \pi]$, and the rate at which its Fourier coefficients decrease. The former is describable in terms of the modulus of continuity of f, namely,

$$w(\delta, f) = \sup |f(\lambda_1) - f(\lambda_2)|, \qquad |\lambda_1 - \lambda_2| < \delta \qquad (\text{mod } 2\pi),$$

If $w(\delta, f) \leq c\delta^\alpha$ for some c and $\alpha > 0$ then f is said to satisfy a Lipshitz condition of that order (written $f \in \text{Lip}\ \alpha$). Of course, $w(\delta, f)$ must converge to zero with δ if f is continuous. On the other hand, if $\alpha > 1$, the derivative of $f(\lambda)$ exists and is zero and f is a constant. Only $0 \leq \alpha < 1$ is of interest (and for $\alpha = 0$ the condition merely means that $f(\lambda)$ is bounded). It may be shown (see Zygmund, 1959) that $f \in \text{Lip}\ \alpha$ implies $|\gamma(n)| = O(n^{-\alpha})$, $\alpha < 1$; $|\gamma(n)| = o(n^{-1})$, $\alpha = 1$. Moreover, $f \in \text{Lip}\ \alpha$ implies that $|f - \sigma_n| = O(n^{-\alpha})$, $\alpha < 1$; $= O(\log n/n)$, $\alpha = 1$, both uniformly in λ.

To conclude this section we mention finally that the concept of Cesaro summability extends to Fourier transforms. We restrict our attention to continuous, integrable, functions, $f(t)$, since this will suffice for our purposes. Thus, putting

$$\hat{f}(\lambda) = \int_{-\infty}^{\infty} f(t)e^{i\lambda t}\,dt,$$

we attempt to recover $f(t)$ as the limit

$$\lim_{\Lambda\to\infty} \frac{1}{2\pi} \int_{-\Lambda}^{\Lambda} \left(1 - \frac{|\lambda|}{\Lambda}\right) \hat{f}(\lambda) e^{-it\lambda}\, d\lambda.$$

The proof that in fact this limit is $f(t)$ for all t and that the convergence is uniform and bounded in t does not differ in any essential respect from that given for the case of functions on a finite interval for we may put

$$\frac{1}{2\pi} \int_{-\Lambda}^{\Lambda} \left(1 - \frac{|\lambda|}{\Lambda}\right) \hat{f}(\lambda) e^{-it\lambda}\, d\lambda = \int_{-\infty}^{\infty} f(s) \left\{ \frac{1}{2\pi} \int_{-\Lambda}^{\Lambda} \left(1 - \frac{|\lambda|}{\Lambda}\right) e^{i(s-t)\lambda}\, d\lambda \right\} ds,$$

where the inner factor is easily seen to be

$$\frac{\sin^2 \frac{1}{2}(s - t)\Lambda}{2\pi\{\frac{1}{2}(s - t)\}^2 \Lambda},$$

which has the basic three properties required for the proof, that it be positive, integrate to unity and converge uniformly to zero outside of any open interval containing t.

4. GENERALIZED FUNCTIONS

The concept of generalized functions is usually introduced through an explanation of the need to make precise the notion of the Dirac delta function, $\delta(t)$, which is "zero for all $t \neq 0$ but infinite at $t = 0$ so that

$$\int \delta(t)\, dt = 1."$$

This definition is, of course, meaningless as it stands, but the end terms of the formula it implies,

$$\phi(t) \to \int_{-\infty}^{\infty} \phi(t)\, \delta(t)\, dt = \phi(0), \tag{4.1}$$

are not, and it is via this formula that Schwartz made the idea of the delta function, and other generalized functions, precise. Thus a generalized function is defined to be a continuous linear functional, $\phi \to \Phi(\phi)$, on a suitable space of base (or testing) functions. Of course, we now need to say what this space of functions is to be and what we mean by continuous. In this we follow Gelfand and Shilov (1964). The three spaces with which we are concerned are K, Z, and S.

K: This is the (linear) space of all infinitely differentiable functions on R with compact support, that is, which are zero off a bounded set. A sequence of functions $\phi_n \in K$ is said to converge to a function ϕ iff the supports of the sequence are contained in one and the same compact set and the sequence, and all sequences of its derivatives, converge uniformly. Then the limit function ϕ also belongs to K, as is well known. By the continuity of the linear functional Φ we mean that $\Phi(\phi_n) \to \Phi(\phi)$ if $\phi_n \to \phi$ in the sense of the convergence in K. This space of generalized functions we call K'. Clearly (4.1) defines an element of K' and thus (4.1) serves to define the Dirac delta function. Any locally integrable function, that is, any function $f(t)$ which is integrable over any finite range, serves to define a generalized function through the formula

$$\phi \to \Phi_f(\phi) = \int f(t)\, \phi(t)\, dt. \tag{4.2}$$

Thus the locally integrable functions can be embedded in K' in this way. At the same time it is clear that some functions are "left out in the cold" (e.g., $f(t) = t^{-1}$) by this embedding. For a discussion of this type of problem we refer the reader to the reference cited. The first main use of the theory of generalized functions arises in connection with their differentiations. We define Φ', the derivative of Φ, by means of $\Phi'(\phi) = -\Phi(\phi')$ whose meaning can be perceived by integrating, by parts, (4.2), with f replaced by f', f being assumed differentiable. Then Φ' is also seen to belong to K' so that each $\Phi \in K'$ is infinitely differentiable.

Z: The linear space Z consists of all entire analytic functions $\phi(z)$ such that for any r

$$|z^r \phi(z)| \leq c e^{a|y|}, \qquad z = x + iy,$$

where a depends on ϕ and c on ϕ and r. This is also described by saying that ϕ is a function of exponential type ($|\phi| \leq c \exp a|z|$) which is rapidly decreasing on the real axis. The importance of Z arises from the fact that the image of K under Laplace† transformation is Z and the correspondence thus established is 1:1. For Z convergence of a sequence ϕ_n to zero is defined by requiring that

$$|z^r \phi_n(z)| \leq c e^{a|y|},$$

where now a does *not* depend upon n, and for every q and r

$$\lim_{n\to\infty} \max_x |(1 + |x|^2)^r\, \phi_n^{(q)}(x)| = 0.$$

This definition of convergence is motivated by the fact that it corresponds precisely to saying that a sequence of elements of Z converges when and only when the sequence of preimages in K under Fourier transformation does so. We call $\hat{\phi} \in Z$ the Fourier transform of $\phi \in K$. Now Z' is defined just as for

† For elements of K the Fourier transform defines the Laplace transform so that we may speak of $\hat{\phi}$, as a function of z, as the Fourier transform.

K' as the space of continuous linear functionals on Z. We now define the Fourier transform of $\Phi \in K'$ by means of

$$\hat{\Phi}(\hat{\phi}) = 2\pi\, \Phi(\phi).$$

The motivation for this definition is clearly the Plancherel theorem, as can be seen by taking Φ as defined by, for example, another element $\phi \in K$ through (4.2). We may observe that the image of ϕ' under Fourier transformation is $it\hat{\phi}$ so that

$$\hat{\Phi}(it\hat{\phi}) = 2\pi\, \Phi(\phi') = -2\pi\, \Phi'(\phi).$$

S: The space S is slightly wider than the space K and consists of all functions on R which are infinitely differentiable and which, together with all of their derivatives, are rapidly decreasing; that is, for each $r \geq 0$ and q

$$\lim_{|x| \to \infty} |(1 + |x|^2)^r\, \phi^{(q)}(x)| = 0, \qquad \phi \in S.$$

Convergence for S is defined by saying that the sequence $\phi_n \to 0$ iff, for each $r \geq 0$ and q,

$$\lim_{n \to \infty} \max_x |(1 + |x|^2)^r\, \phi_n^{(q)}(x)| = 0.$$

The importance of S, a linear space, arises from the fact that S is mapped in a 1:1 and bicontinuous fashion onto itself by means of the Fourier transform. Once more we define S' as the space of continuous, linear functionals on S and the Fourier transform and derivative of elements of S' by the same formula as before.

5. TENSOR PRODUCTS

Let $\mathfrak{X}$ and $\mathfrak{Y}$ be finite-dimensional vector spaces over the complex field of dimensions p and q. We proceed to define a tensor product $\mathfrak{U} = \mathfrak{X} \otimes \mathfrak{Y}$ of these vector spaces. This is to be a vector space over the complex field and a mapping ϕ of the product space $\mathfrak{X} \times \mathfrak{Y}$ onto $\mathfrak{U}$ with the following properties†:

(i) ϕ is bilinear, that is, if $x_i \in \mathfrak{X}$, $y_i \in \mathfrak{Y}$, and a_i, b_i are complex numbers:

$$\phi(a_1x_1 + a_2x_2, y) = a_1\phi(x_1, y) + a_2\phi(x_2, y),$$
$$\phi(x, b_1y_1 + b_2y_2) = b_1\phi(x, y_1) + b_2\phi(x, y_2).$$

(ii) If $x_1, \ldots, x_p$ form a basis for $\mathfrak{X}$ and $y_1, \ldots, y_q$ form a basis for $\mathfrak{Y}$ then $\phi(x_i, y_j)$ form a basis for $\mathfrak{U}$.

We shall write $x \otimes y$ for the image of (x, y) under ϕ.

† For further details see, for example, Greub (1967).

It is not difficult to show that if $\mathcal{X}$ and $\mathcal{Y}$ are realized as spaces of all columns of p and q complex numbers, respectively, then $\mathcal{X} \otimes \mathcal{Y}$ may be realized as a vector space of all pq-dimensional columns of complex numbers and if x has components x_i and y has components y_j then the image $x \otimes y$ may be chosen to have, in this realization, the number $x_i y_j$ in row $(i-1)q+j$. An alternative realization for $\mathcal{X} \otimes \mathcal{Y}$ is got by mapping (x, y) onto xy' so that $\mathcal{X} \otimes \mathcal{Y}$ is realized as the space of all matrices of complex numbers of p rows and q columns.

If A and B are linear operators in $\mathcal{X}$ and $\mathcal{Y}$, we define the tensor product $A \otimes B$ as the operator in $\mathcal{X} \otimes \mathcal{Y}$ defined by

$$A \otimes B\,(x \otimes y) = Ax \otimes By.$$

Then

$$(A_1 \otimes B_1)(A_2 \otimes B_2)x \otimes y = A_1 \otimes B_1(A_2 x \otimes B_2 y) = A_1 A_2 x \otimes B_1 B_2 y,$$

so that

$$(A_1 \otimes B_1)(A_2 \otimes B_2) = A_1 A_2 \otimes B_1 B_2.$$

Also

$$(a_1 A_1 + a_2 A_2) \otimes B = a_1 A_1 \otimes B + a_2 A_2 \otimes B$$

and similarly for $b_1 B_1 + b_2 B_2$.

If $\mathcal{X}$ and $\mathcal{Y}$ are realized as vector spaces of p-tuples and q-tuples, respectively, of complex numbers then A and B are realized as matrices of p and q, respectively, rows and columns. Then when $\mathcal{X} \otimes \mathcal{Y}$ is realized as a column of pq-tuples as described in the previous paragraph $A \otimes B$ is easily seen to be realized as the partitioned matrix

$$\begin{bmatrix} a_{11}B & a_{12}B & \cdots & a_{1p}B \\ a_{21}B & a_{22}B & \cdots & a_{2p}B \\ \cdots & \cdots & \cdots & \cdots \\ a_{p1}B & a_{p2}B & \cdots & a_{pp}B \end{bmatrix},$$

that is, the matrix with $a_{ij}b_{kl}$ in row $(i-1)q+k$, column $(j-1)q+l$. This matrix is often called the Kronecker product of A and B. On the other hand, if $\mathcal{X} \otimes \mathcal{Y}$ is realized as the space of matrices of p rows and q columns then $A \otimes B(x \otimes y)$ is evidently realized as

$$Axy'B'$$

and the linear mapping $A \otimes B$ of $x \otimes y$ is got by the transformation $C \rightarrow ACB'$, $C \in \mathcal{X} \otimes \mathcal{Y}$.

If ϕ_i is an eigenvector of the operator A for eigenvalue λ_i and ψ_j is an eigenvector of B for eigenvalue μ_j then from the second realization, for example, it is evident that $\phi_i \otimes \psi_j$ is an eigenvector of $A \otimes B$ with eigenvalue

$\lambda_i\mu_j$. If λ_i is of multiplicity m_i and μ_j of multiplicity n_j then $\lambda_i\mu_j$ has multiplicity m_in_j. This may be seen by taking A, B as realized by matrices and reducing A, B to Jordan canonical form by the transformation $A \to PAP^{-1}$, $B \to QBQ^{-1}$. In this canonical form PAP^{-1} and QBQ^{-1} have null elements below the main diagonal and their eigenvalues in appropriate multiplicity, in the main diagonal. Thus if $A \otimes B$ is taken in the first matrix realization then $A \otimes B \to (PAP^{-1}) \otimes (QBQ^{-1}) = (P \otimes Q)(A \otimes B)(P \otimes Q)^{-1}$ again reduces $A \otimes B$ to upper triangular form and evidently $A \otimes B$ has $\lambda_i\mu_j$ with multiplicity m_in_j.

It follows, in particular, that $\det(A \otimes B) = \{\det(A)\}^q\{\det(B)\}^p$ and $\operatorname{tr}(A \otimes B) = \operatorname{tr}(A)\operatorname{tr}(B)$.

These definitions may be extended to an arbitrary number of tensor factors so that we may form $\mathcal{X}_1 \otimes \mathcal{X}_2 \otimes \cdots \otimes \mathcal{X}_r$. These may be defined inductively. Similarly we define $A_1 \otimes A_2 \otimes \cdots \otimes A_r$ if A_j acts in $\mathcal{X}_j$ as a linear transformation. We shall not list the extensions to $r > 2$ of the properties listed above since these are all fairly obvious. For example, if the eigenvalues of A_j are $\lambda_{j,i}$, with multiplicity $n_{j,i}$ then the eigenvalues of $A_1 \otimes A_2 \otimes \cdots \otimes A_r$ and their multiplicities are

$$\prod_{j=1}^{r} \lambda_{j,i(j)}, \qquad \prod_{j=1}^{r} n_{j,i(j)},$$

where all possible choices $i(1), \ldots, i(r)$ are allowed.

It may be observed from the abstract definition first given that $\mathcal{X} \otimes \mathcal{Y}$ and $\mathcal{Y} \otimes \mathcal{X}$ are isomorphic as vector spaces and similarly there is an isomorphism between $\mathcal{X}_1 \otimes \mathcal{X}_2 \otimes \cdots \otimes \mathcal{X}_r$ and the tensor product of the $\mathcal{X}_j$ taken in any other order. If $\mathcal{X} \otimes \mathcal{Y}$ (and $\mathcal{Y} \otimes \mathcal{X}$) are realized as spaces of $p \times q$ (and $q \times p$) matrices then the isomorphism of $\mathcal{X} \otimes \mathcal{Y}$ with $\mathcal{Y} \otimes \mathcal{X}$ is clearly attained through $C \in \mathcal{X} \otimes \mathcal{Y} \leftrightarrow C' \in \mathcal{Y} \otimes \mathcal{X}$. This isomorphism defines a corresponding isomorphism of the algebra of linear operators in $\mathcal{X} \otimes \mathcal{Y}$ into the algebra of linear operators in $\mathcal{Y} \otimes \mathcal{X}$ and clearly under this correspondence $A \otimes B \leftrightarrow B \otimes A$. The generalization to more than two factors is evident.

Bibliography

Abramowitz, M., and I. A. Stegun (1964), *Handbook of Mathematical Functions*, Nat. Bur. Standards, App. Math. Ser. No. 55, Washington, U.S. Govt. Printing Office.

Achieser, N. I. (1956), *Theory of Approximation*, New York, Frederick Ungar.

Amemiya, T., and W. A. Fuller (1967), "A comparative study of alternative estimators of a distributed lag model," *Econometrica*, **35,** 509–529.

Amos, D. E., and L. H. Koopmans (1963), *Tables of the Distribution of the Coefficient of Coherence for Stationary Bivariate Gaussian Processes*, Washington, Office of Technical Services, Dept. of Commerce.

Anderson, R. L. (1942), "Distribution of the serial correlation coefficient," *Ann. Math. Statist.*, **13,** 1–13.

——— and T. W. Anderson (1950), "Distribution of the circular serial correlation coefficient for residuals from a fitted Fourier series," *Ann. Math. Statist.*, **21,** 59–81.

Anderson, T. W. (1948), "On the theory of testing serial correlation," *Skand. Aktuarietidskr.*, **31,** 81–115.

——— (1958), *Introduction to Multivariate Statistical Analysis*, New York, John Wiley.

——— (1970), *Time Series Analysis*, New York, John Wiley.

——— and A. M. Walker (1964), "On the asymptotic distribution of the autocorrelations of a sample from a linear stochastic process," *Ann. Math. Statist.*, **35,** 1296–1303.

Bartlett, M. S. (1935), "Some aspects of the time correlation problem in regard to tests of significance," *J. R. Statist. Soc.* **98,** 536–543.

——— (1954), "Problemes de l'analyse spectrale des series temporelles stationnaires," *Publications de L'Institut de Statistique de l'Univ. de Paris*, **III-3,** 119–134.

——— (1955), *An Introduction to Stochastic Processes*, Cambridge University Press.

——— (1963), "The spectral analysis of point processes," *J. R. Statist. Soc.*, **25,** 264–296.

Bartlett, M. S., and P. H. Diananda (1950), "Extensions of Quenouille's test for autoregressive schemes," *J. R. Statist. Soc.*, Ser. B, **12,** 108–115.

——— and J. Medhi (1955), "On the efficiency of procedures for smoothing periodograms from time series with continuous spectra," *Biometrika*, **42,** 143–150.

——— and D. V. Rajalakshman (1953), "Goodness of fit tests for simultaneous autoregressive series," *J. R. Statist. Soc.*, Ser. B, **15,** 107–124.

Bernstein, S. (1926), "Sur l'extension du théorème limite du calcul des probabilités aux sommes de quantités dépendantes," *Math. Ann.*, **97,** 1–59.

Billingsley, P. (1965), *Ergodic Theory and Information*, New York, John Wiley.

——— (1968), *Convergence of Probability Measures*, New York, John Wiley.

Blackman, R. B. (1965), *Linear Data Smoothing and Prediction*, Reading, Mass., Addison-Wesley.

——— and J. W. Tukey (1959), *The Measurement of Power Spectra*, New York, Dover.

Bôcher, M. (1907), *An Introduction to Higher Algebra*, New York, Macmillan.

Boerner, H. (1963), *Representations of Groups*, Amsterdam, North Holland.

Bonnet, G. (1965), "Theorie de l'information—sur l'interpolation optimale d'une fonction aléatoire échantillonnée," *C.R. Acad. Sci. Paris*, **260,** 784–787.

Box, G. E. P., and G. M. Jenkins (1962), "Some statistical aspects of adaptive optimization and control," *J. R. Statist Soc.*, **24,** 297–343.

——— (1970), *Time Series, Forecasting and Control*, San Francisco, Holden-Day.

Brillinger, D. R. (1965), "An introduction to polyspectra," *Ann. Math. Statist.*, **36,** 1351–1374.

——— and M. Rosenblatt (1967a), "Computation and interpretation of k-th order spectra," pp. 189–232 in *Advanced Seminar on Spectral Analysis of Time Series* (ed. B. Harris), New York, John Wiley.

——— (1967b), "Asymptotic theory of estimates of k-th order spectra," pp. 153–188 in *Advanced Seminar on Spectral Analysis of Time Series* (ed. B. Harris), New York, John Wiley.

Chanda, K. C. (1961), "Comparative efficiencies of methods of estimating parameters in linear autoregressive schemes," *Biometrika*, **48,** 427–432.

——— (1962), "On bounds of serial correlations," *Ann. Math. Statist.*, **33,** 1457–1460.

——— (1964), "Asymptotic expansions for tests of goodness of fit for linear autoregressive schemes," *Biometrika*, **51,** 459–465.

Chipman, J. S., K. R. Kadiyala, A. Madansky, and J. W. Pratt (1968), "Efficiency of the sample mean when residuals follow a first-order stationary Markoff process," *J. Amer. Statist. Assoc.*, **63,** 1237–1246.

Cooley, J. W., and J. W. Tukey (1965), "An algorithm for the machine calculation of complex Fourier series," *Mathematics of Computation*, **19,** 297–301.

Cox, D. R., and P. A. W. Lewis (1966), *The Statistical Analysis of Series of Events*, London, Methuen.

——— and H. D. Miller (1965), *The Theory of Stochastic Processes*, London, Methuen.

Cramér, H. (1937), *Random Vatiables and Probability Distributions*, Cambridge University Press.

——— (1946), *Mathematical Methods of Statistics*, Princeton University Press.

——— and M. R. Leadbetter (1967), *Stationary and Related Stochastic Processes*, New York, John Wiley.

Daniels, H. E. (1956), "The approximate distributions of serial correlation coefficients," *Biometrika*, **43,** 169–185.

De Bruijn, N. G. (1958), *Asymptotic Methods in Analysis*, Amsterdam, North-Holland.

Deutsch, R. (1962), *Nonlinear Transformations of Random Processes*, Englewood Cliffs, N. J., Prentice-Hall.

Dhrymes, P. J. (1969), "Efficient estimation of distributed lags with autocorrelated errors," *Int. Economic. Rev.* **10,** 47–67.

Diananda, P. H. (1954), "The central limit theorem for m-dependent variables asymptotically stationary to second order," *Proc. Camb. Phil. Soc.*, **50,** 287–292.

Dixon, W. J. (1944), "Further contributions to the problem of serial correlation," *Ann. Math. Statist.*, **15,** 119–144.

Doob, J. L. (1953), *Stochastic Processes*, New York, John Wiley.

Durbin, J. (1959), "Efficient estimators of parameters in moving average models," *Biometrika*, **46,** 306–316.

——— (1960), "The fitting of time series models," *Rev. Int. Statist. Inst.*, **28,** 233–244.

——— and G. S. Watson (1950), "Testing for serial correlation in least squares regression." I., *Biometrika*, **37,** 409–428.

——— (1951), "Testing for serial correlation in least squares regression." II., *Biometrika*, **38,** 159–178.

Eicker, F. (1967), "Limit theorems for regressions with unequal and dependent errors," *Proc. Fifth Berkeley Sympos. Math. Statist. and Probability* (Berkeley Calif. 1965/66), Vol. 1: Statistics, pp. 59–82, Berkeley, Univ. California Press.

Feller, W. (1957), *An Introduction to Probability Theory and its Applications*, Vol. I (2nd Ed.), New York, John Wiley.

——— (1966), *An Introduction to Probability Theory and its Applications*, Vol. II, New York, John Wiley.

Fisher, R. A. (1929), "Tests of significance in harmonic analysis," *Proc. Roy. Soc., Ser.* A, **125,** 54–59.

——— (1940), "On the similarity of the distributions found for the test of significance in harmonic analysis, and in Steven's problem in geometrical probability," *Annals of Eugenics*, **10,** 14–17.

——— (1950), *Contributions to Mathematical Statistics*, New York, John Wiley.

Fishman, G. S. (1969), *Spectral Methods in Econometrics*, Santa Monica, Rand Corporation.

Fisz, M. (1963), *Probability Theory and Mathematical Statistics*, New York, John Wiley.

Fraser, D. A. S. (1957), *Non Parametric Methods in Statistics*, New York, John Wiley.

Freeman, H. (1965), *Discrete Time Systems*, New York, John Wiley.

Gelfand, I. M., and G. E. Shilov (1964), *Generalized Functions*, Vol. I, New York, Academic Press.

——— and N. Ya. Vilenkin (1964), *Generalised Functions, Vol. 4, Applications to Harmonic Analysis*, Translated by A. Feinstein, New York, Academic Press.

Gnedenko, B. V., and A. N. Kolmogoroff (1954), *Limit Distributions for Sums of Independent Random Variables*, Reading, Mass., Addison-Wesley.

Goodman, N. R. (1963), "Statistical analysis based on a certain multivariate complex Gaussian distribution (an introduction)," *Ann. Math. Statist.*, **33,** 152–177.

Goursat, E., and E. R. Hedrick (1904), *A Course in Mathematical Analysis*, Boston, Ginn.

Granger, C. W. J., and M. Hatanaka (1964), *Spectral Analysis of Economic Time Series*, Princeton University Press.

Grenander, U. (1950), "Stochastic processes and statistical inference," *Ark. Mat.*, **1,** 195–277.

——— (1951), "On empirical spectral analysis of stochastic processes," *Ark. Mat.*, **1,** 503–531.

——— (1954), "On the estimation of regression coefficients in the case of an autocorrelated disturbance," *Ann. Math. Statist.*, **25,** 252–272.

——— and M. Rosenblatt (1957), *Statistical Analysis of Stationary Time Series*, New York, John Wiley.

——— and G. Szego (1958), *Toeplitz Forms and Their Applications*, Berkeley, Univ. California Press.

Greub, W. (1967), *Multilinear Algebra*, Berlin, Springer-Verlag.

Griliches, Z. (1967), "Distributed lags: a survey," *Econometrica*, **35,** 16–49.

Groves, G. W., and E. J. Hannan (1968), "Time series regression of sea level on weather," *Rev. Geophysics*, **6,** 129–174.

Halmos, P. R. (1947), *Finite Dimensional Vector Spaces* (2nd Ed.), Princeton University Press.

——— (1950), *Measure Theory*, New York, Van Nostrand.

Hannan, E. J. (1960), *Time Series Analysis*, London, Methuen.

——— (1961a), "A central limit theorem for systems of regressions," *Proc. Camb. Phil. Soc.*, **57,** 583–588.

——— (1961b), "Testing for a jump in the spectral function," *J. R. Statist. Soc.*, **23,** 394–404.

——— (1963) "Regression for time series," (Proc. Symp. on) *Time Series Analysis* (ed. M. Rosenblatt), New York, John Wiley.

——— (1963), "Regression for time series with errors of measurement," *Biometrika*, **50,** 293–302.

——— (1965a), "Group representations and applied probability," *J. App. Prob.*, **2,** 1–68. (Also issued as Vol. 3 of Methuen's Review Series in Applied Probability, London, Methuen).

——— (1965b), "The estimation of relationships involving distributed lags," *Econometrica*, **33,** 206–224.

——— (1967a), "The concept of a filter," *Proc. Camb. Phil. Soc.*, **63,** 221–227.

——— (1967b), "Measurement of a wandering signal amid noise," *J. App. Prob.*, **4,** 90–102.

——— (1967c), "The estimation of a lagged regression relation," *Biometrika*, **54,** 409–418.

——— (1969a), "Fourier methods and random processes," *Bull. Int. Statist. Inst.*, **42,** 475–496.

——— (1969b), "The identification of vector mixed autoregressive-moving average systems," *Biometrika*, **56,** 223–225.

——— (1969c), "A note on an exact test for trend and serial correlation," *Econometrica*, **37,** 485–489.

——— and R. D. Terrell (1968), "Testing for serial correlation after least squares regression," *Econometrica*, **36,** 133–150.

——— (1970), "The seasonal adjustment of economic time series," *Int. Ec. Rev.*, **11,** 1–29.

Hasselmann, K. F., W. H. Munk and G. J. F. MacDonald (1963), "Bispectra of ocean waves," (Proc. Symp. on) *Time Series Analysis* (ed. M. Rosenblatt), New York, John Wiley.

Heble, M. P. (1961), "A regression problem concerning stationary processes," *Trans. Amer. Math. Soc.*, **99,** 350–371.

Helgason, S. (1962), *Differential Geometry and Symmetric Spaces*, New York, Academic Press.

Helson, H. (1964), *Lectures on Invariant Subspaces*, New York, Academic Press.

——— and D. Lowdenslager (1958), "Prediction theory and Fourier series in several variables," *Acta Math.*, **99,** 165–202.

Henshaw, R. C. (1966), "Testing single-equation least squares regression models for autocorrelated disturbances," *Econometrica*, **34,** 646–660.

Hewitt, E., and K. Stromberg (1965), *Real and Abstract Analysis*, New York, Springer-Verlag.

Hille, E., and R. S. Phillips (1957), *Functional Analysis and Semi-Groups*, Providence, American Mathematical Society.

Hoffman, K. (1962), *Banach Spaces of Analytic Functions*, Englewood Cliffs, N.J., Prentice-Hall.

Hsu, P. L. (1941), "On the limiting distribution of the roots of a determinantal equation," *J. London Math. Soc.*, **16,** 183–194.

James, A. T. (1964), "Distributions of matrix variates and latent roots derived from normal samples," *Ann. Math. Statist.*, **35,** 475–501.

Jenkins, G. M. (1954a), "Tests of hypotheses in the linear autoregressive model. Null hypothesis distributions in the Yule scheme," *Biometrika*, **41,** 405–419.

——— (1954b), "An angular transformation for the serial correlation coefficient," *Biometrika*, **41,** 261–265.

——— (1956), "Tests of hypotheses in the linear autoregressive model. Null distributions for higher order schemes; non-null distributions," *Biometrika*, **43,** 186–199.

——— and D. G. Watts (1968), *Spectral Analysis*, San Francisco, Holden-Day.

Kalman, R. E. (1960), "A new approach to linear filtering and prediction problems," *Trans. Amer. Soc. Mech. Eng., J. Basic Engineering*, **82,** 35–45.

——— (1963), "New methods of Wiener filtering theory," pp. 270–388 in *Proc. First Symp. on Eng. Applications of Random Function Th. and Prob.* (Eds. J. L. Boganoff and F. Kozin), New York, John Wiley.

——— and R. S. Bucy (1961), "New results in linear filtering and prediction theory," *Trans. Amer. Soc. Mech. Eng., J. Basic Engineering* **83,** 95–108.

Kato, T. (1966), *Perturbation Theory for Linear Operators*, Berlin, Springer-Verlag.

Kendall, M. G., and A. S. Stuart (1966), *The Advanced Theory of Statistics Vol.* **3**, London, Griffin.

Khatri, C. G. (1966), "A note on a large sample distribution of a transformed multiple correlation," *Ann. Inst. Statist. Math.*, **18,** 375–380.

Koopmans, T. C. (1942), "Serial correlation and quadratic forms in normal variables," *Ann. Math. Statist.*, **13,** 14–33.

Leipnik, R. B. (1947), "Distribution of the serial correlation coefficient in a circularly correlated universe," *Ann. Math. Statist.*, **18,** 80–87.

Liviatan, N. (1963), "Consistent estimation of distributed lags," *International Economic Review*, **4,** 44–52.

Loève, M. (1960), *Probability Theory* (2nd Ed.), Princeton, Van Nostrand.

Lorentz, G. G. (1966), *Approximation of Functions*, New York, Holt, Rinehart and Winston.

Macduffee, C. C. (1956), *The Theory of Matrices*, New York, Chelsea.

McGregor, J. R. (1962), "The approximate distribution of the correlation between two stationary linear Markov series," *Biometrika*, **49,** 379–388.

Mackey, G. W. (1968), *Induced Representations of Groups and Quantum Mechanics*, New York, Benjamin.

Madow, W. G. (1945), "Note on the distribution of the serial correlation coefficient," *Ann. Math. Statist.* **16,** 308–310.

Mann, H. B. (1953), *Introduction to the Theory of Stochastic Processes Depending on a Continuous Parameter*, Nat. Bur. Standards, App. Math. Ser. No. 24, Washington, U.S. Govt. Printing Office.

——— and A. Wald (1943), "On the statistical treatment of linear stochastic difference equations," *Econometrica*, **11,** 173–220.

Masani, P. (1959a), "Cramer's theorem on monotone matrix valued functions and the Wold decomposition," in *Probability and Statistics*, the Harald Cramer volume (Ed. U. Grenander), Stockholm, Almqvist and Wiksell.

——— (1959b), "I. Sur la fonction génératrice d'un processus stochastique vectoriel," "II. Isomorphie entre les domaines temporel et spectral d'un processus vectoriel, régulier," "III. Sur les fonctions matricielles de la classe de Hardy H_2," "IV. Sur les fonctions matricielles de la classe de Hardy H_2," *C.R. Acad. Sci.*, Paris, **249,** 360–362, 496–498, 873–875, 906–907.

——— (1960), "Une généralisation pour les fonctions matricielles de la classe de Hardy H_2 d'un theoreme de Nevanlinna," *C.R. Acad. Sci.*, Paris, **251**, 318–320.

Moran, P. A. P. (1947), "Some theorems on time series I," *Biometrika*, **34,** 281–291.

Munk, W. H., and D. E. Cartwright (1966), "Tidal spectroscopy and prediction," *Phil. Trans.*, **259,** 533–581.

Naimark, M. A. (1960), *Normed Rings*, Groningen, Noordhoff.

——— (1964), *Linear Representations of the Lorentz Group*, New York, Macmillan.

von Neumann, J. (1941), "Distribution of the ratio of the mean square successive difference to the variance," *Ann. Math. Statist.*, **12,** 367–395.

Nerlove, M. (1964) "Spectral analysis of seasonal adjustment procedures," *Econometrica*, **32,** 241–286.

Neyman, J. (1954), "Discussion on Symposium on Interval Estimation," *J. R. Statist. Soc.*, **16,** 216–218.

Nicholls, D. F. (1967), "Estimation of the spectral density function when testing for a jump in the spectrum," *Aust. J. Statist.*, **9,** 103–108.

——— (1969), "Testing for a jump in co-spectra," *Aust. J. Statist.*, **11,** 7–13.

Olshen, R. A. (1967), "Asymptotic properties of the periodogram of a discrete stationary process," *J. App. Prob.*, **4,** 508–528.

Orcutt, G. H. (1948), "A study of the autogressive nature of the time series used for Tinbergen's model of the economic system of the United States, 1919–32," *J. R. Statist. Soc.*, Ser. B, **10,** 1–45.

Owen, D. B. (1962), *Handbook of Statistical Tables*, Reading, Mass., Addison-Wesley.

Paley, R. E. A. C., and N. Wiener (1934), *Fourier Transforms in the Complex Domain*, Providence, Amer. Math. Soc.

Parzen, E. (1957), "On consistent estimates of the spectrum of a stationary time series," *Ann. Math. Statist.*, **28,** 329–348.

——— (1961), "An approach to time series analysis," *Ann. Math. Statist.*, **32,** 951–989.

Phillips, A. W. (1959), "The estimation of parameters in a system of stochastic differential equations," *Biometrika*, **46,** 67–76.

Potapov, V. P. (1958), "The multiplicative structure of J-contractive matrix functions," (in Translations *Amer. Math. Soc.*, Ser. 2, **15,** 131–244).

Priestley, M. B. (1962), "Analysis of stationary processes with mixed spectra—II," *J. R. Statist. Soc.*, Ser. B, **24,** 511–529.

Quenouille, M. H. (1947), "A large-sample test for the goodness of fit of autoregressive schemes," *J. R. Statist. Soc.*, Ser. A, **110,** 123–129.

Rao, C. R. (1965), *Linear Statistical Inference and its Applications*, New York, John Wiley.

Riesz, F., and B. Sz-Nagy (1956), *Functional Analysis*, London, Blackie.

Robertson, J. B. (1968), "Orthogonal decompositions of multivariate weakly stationary stochastic processes," Canad. J. Math., **20,** 368–383.

Robinson, E. A. (1962), *Random Wavelets and Cybernetic Systems*, London, Griffin.

——— (1967), *Multi-Channel Time Series Analysis*, San Francisco, Holden-Day.

Rosenblatt, M. (1956a), "A central limit theorem and a strong mixing condition," *Proc. Nat. Acad. Sci.*, U.S.A., **42,** 43–47.

——— (1956b), "Some regression problems in time series analysis," *Proc. of Third Berkeley Symp. on Math Statist. and Prob.* 1954, **1,** 165–186, Berkeley, Univ. of Calif. Press.

——— (1959), "Statistical analysis of stochastic processes with stationary residual," in *Probability and Statistics*, the Harald Cramér volume (Ed. U. Grenander), pp. 246–275, Uppsala, Almqvist and Wiksell.

——— (1961), "Some comments on narrow band-pass filters," *Quart. Appl. Math.*, **18,** 387–393.

Rozanov, Yu. A. (1967), *Stationary Random Processes*, San Francisco, Holden-Day.

Rubin, H. (1945), "On the distribution of the serial correlation coefficient," *Ann. Math. Statist.*, **16,** 211–215.

Scheffé, H. (1959), *The Analysis of Variance*, New York, John Wiley.

Shaman, P. (1969), "On the inverse of the covariance matrix of a first order moving average," *Biometrika*, **56,** 595–600.

Shapiro, H. S., and R. A. Silverman (1960), "Alias-free sampling of random noise," J. Soc. Indust. Appl. Math., **8,** 225–248.

Tate, R. F., and G. W. Klett (1959), "Optimal confidence intervals for the variance of a normal distribution," *J. Amer. Statist. Assoc.*, **54,** 674–682.

Terrell, R. D., and N. E. Tuckwell (1969), "The efficiency of least squares in estimating a stable seasonal pattern," Research report available from Stats. Dept. SGS, ANU, Canberra, Australia.

Theil, H., and A. L. Nagar (1961), "Testing the independence of regression disturbances," *J. Amer. Statist. Assoc.*, **56,** 793–806.

Tintner, G. (1940), *The Variate Difference Method*, Cowles Comm. Monograph No. 5, Evanston, Ill., The Principia Press.

Vilenkin, N. J. (1968), *Special Functions and the Theory of Group Representations*, Province, Amer. Math. Soc.

van der Waerden, B. L. (1949), *Modern Algebra*, New York, Frederick Ungar.

Wahba, Grace (1968) "On the distribution of some statistics useful in the analysis of jointly stationary time series," *Ann. Math. Statist.*, **39,** 1849–1862.

——— (1969), "Estimation of the coefficients in a multidimensional distributed lag model," *Econometrica*, **31,** 398–407.

Wainstein, L. A., and V. D. Zubakov (1962), *Extraction of Signals from Noise*, Englewood Cliffs, N.J., Prentice-Hall.

Walker, A. M. (1962), "Large sample estimation of parameters for autoregressive processes with moving average residuals," *Biometrika*, **49,** 117–132.

——— (1964), "Asymptotic properties of least-squares estimates of parameters of the spectrum of a stationary non-deterministic time-series," *J. Aust. Math. Soc.*, **4,** 363–384.

——— (1965), "Some asymptotic results for the periodogram of a stationary time series," *J. Aust. Math. Soc.*, **5,** 107–128.

——— (1967), "Some tests of separate families of hypotheses in time series analysis," *Biometrika*, **54,** 39–68.

Watson, G. S. (1955), "Serial correlation in regression analysis I," *Biometrika*, **42,** 327–341.

——— (1956), "On the joint distribution of the circular serial correlation coefficients," *Biometrika*, **43,** 161–168.

——— (1967), "Linear least squares regression," *Ann. Math. Statist.*, **38,** 1679–1699.

Whittaker, E. T., and G. N. Watson (1946), *A Course of Modern Analysis*, Cambridge University Press.

Whittle, P. (1951), *Hypothesis Testing in Time Series Analysis*, Thesis, Uppsala University, Almqvist and Wiksell, Uppsala, Hafner, New York.

——— (1962), "On the convergence to normality of quadratic forms in independent variables," *Theor. Verojaotnst. i Primenen.*, **9,** 113–118.
Wiener, N. (1923), "Differential space," *J. Math. Phys.*, **2,** 131–174.
——— (1933), *The Fourier Integral*, New York, Dover.
——— (1958), Nonlinear Problems in Random Theory, Cambridge, Mass., *M.I.T.* Press.
——— and P. Masani (1957), "The prediction theory of multivariate stochastic processes, I," *Acta. Math.*, **98,** 111–150.
——— (1958), "The prediction theory of multivariate stochastic processes, II," *Acta. Math.*, **99,** 93–137.
Wilson, G. (1969), Factorization of a covariance generating function. *SIAM J. Numerical Analysis*, **6,** 1–7.
Wold, H. (1938), *A Study in the Analysis of Stationary Time Series*, Uppsala, Almqvist and Wiksell.
——— (1965), *Bibliography on Time Series and Stochastic Processes*, London, Oliver and Boyd.
Woodroofe, M. B., and J. W. Van Ness (1967), "The maximum deviation of sample spectral densities," *Ann. Math. Statist.*, **38,** 1559–1569.
Yaglom, A. M. (1961), "Second order homogeneous random fields," *Fourth Berkeley Symposium on Mathematical Statistics and Probability*, **2,** 593–622, Berkeley, Univ. California Press.
——— (1962), *An Introduction to the Theory of Stationary Random Functions*, Englewood Cliffs, N.J., Prentice-Hall.
Yosida, K. (1965), *Functional Analysis*, Berlin, Springer-Verlag.
Zymund, A. (1959), *Trigonometric Series, Vol. I,* Cambridge University Press.

Table of Notations

Index